INTERMEDIATE ALGEBRA

A TEXT/WORKBOOK

Charles P. McKeague

CUESTA COLLEGE

INTERMEDIATE ALGEBRA

A TEXT/WORKBOOK

NINTH EDITION

Charles P. McKeague

CUESTA COLLEGE

BROOKS/COLE
CENGAGE Learning™

Australia • Brazil • Japan • Korea • Mexico • Singapore • Spain • United Kingdom • United States

BROOKS/COLE
CENGAGE Learning™

Intermediate Algebra: A Text/Workbook,
Ninth Edition
Charles P. McKeague

Acquisitions Editor: Marc Bove

Developmental Editor: Shaun Williams

Assistant Editor: Carrie Jones

Editorial Assistant: Zachary Crockett

Media Editor: Bryon Spencer

Marketing Manager: Laura McGinn

Marketing Assistant: Shannon Maier

Marketing Communications Manager:
  Darlene Macanan

Content Project Manager: Jennifer Risden

Design Director: Rob Hugel

Art Director: Vernon Boes

Print Buyer: Becky Cross

Rights Acquisitions Specialist: Dean Dauphinais

Production Service: XYZ Textbooks

Text Designer: Diane Beasley

Photo Researcher: Bill Smith Group

Copy Editor: Katherine Shields, XYZ Textbooks

Illustrator: Kristina Chung, XYZ Textbooks

Cover Designer: Irene Morris

Cover Image: Pete McArthur

Compositor: Donna Looper, XYZ Textbooks

For product information and technology assistance, contact us at
Cengage Learning Customer & Sales Support, 1-800-354-9706
For permission to use material from this text or product,
submit all requests online at **www.cengage.com/permissions**
Further permissions questions can be emailed to
permissionrequest@cengage.com

Library of Congress Control Number: 2011930156

ISBN-13: 978-1-133-10364-6
ISBN-10: 1-133-10364-2

Brooks/Cole
20 Davis Drive
Belmont, CA 94002-3098
USA

Cengage Learning is a leading provider of customized learning solutions with office locations around the globe, including Singapore, the United Kingdom, Australia, Mexico, Brazil, and Japan. Locate your local office at:
www.cengage.com/global

Cengage Learning products are represented in Canada by Nelson Education, Ltd.

To learn more about Brooks/Cole, visit **www.cengage.com/brookscole**
Purchase any of our products at your local college store or at our preferred online store **www.cengagebrain.com**

Printed in the United States of America
1 2 3 4 5 6 7 15 14 13 12 11

To my Grandchildren,

Marissa Rochelle McKeague
Kendra Lauren McKeague
Justin Patrick McKeague
Ethan Matthew McKeague
Brooke Abby Jacobs
Ava Lauren Jacobs
Charles Patterson Jacobs

Brief Contents

Contents

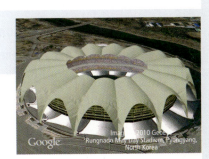

3 Equations and Inequalities in Two Variables

4 Systems of Linear Equations and Inequalities

5 Exponents and Polynomials

6 Rational Expressions and Rational Functions

9 Exponential and Logarithmic Functions

10 Conic Sections

Preface to the Instructor

The Intermediate Algebra Course as a Bridge to Further Success

Intermediate algebra is a bridge course. This course and its syllabus bring the student to the level of ability required of college students, while getting them ready to make a successful start in college algebra or precalculus. After eight successful editions, we have developed several interlocking, proven features that will improve students' chances of success in the course.

New to This Edition

CHAPTER INTRODUCTIONS

Each chapter opens with a revised introduction in which an exciting real-world application is used to stimulate interest in the chapter. We expand on these opening applications later in the chapter.

KEY WORDS

At the beginning of each chapter, we have provided a concise list of key words and definitions the student will encounter throughout the chapter.

CHAPTER OUTLINES

An outline of objectives arranged by section appears at the beginning of each chapter. This outline helps the student organize expectations for the chapter into short-term goals and prepare for the work ahead.

OBJECTIVES

In order to help organize topics within each section, we have greatly enhanced our use of objectives in both the sections and in the accompanying problem sets. Objectives for each chapter are shown immediately preceding the first section in each chapter. Section objectives are shown at the start of each section.

TICKET TO SUCCESS

Previously included as Getting Ready for Class, these reviewed and revised questions require the student to provide written responses. They reinforce the idea of reading the section before coming to class, as their answers can truly be their ticket to this course. The questions now appear at the beginning of each section to provide the student a chance to become more involved learners as they read through the section, keeping in mind important concepts.

SECTION OPENERS

Similar to the chapter introductions, each section in the book opens with a revised introduction that includes an interesting real-world application. The presence of these applications helps the student engage in and relate to the mathematics they are learning, and feel confident about moving forward. We expand on these opening applications later in the chapter as well.

MOVING TOWARD SUCCESS

Each problem set now starts with this new feature that includes a motivational quote followed by a few questions to focus the student on success. Because good study habits are essential to the success of math students, this feature is prominently displayed at the start of every problem set.

CHALLENGE PROJECTS

This new feature appears at the end of select chapters in addition to the existing group and research projects. These projects revolve around the use of the Google Earth® online program, and help students apply concepts they have learned in the chapter to real-life locations around the world.

Organization of Problem Sets

The problem sets begin with drill problems that are linked to the section objectives, and are then followed by the categories of problems discussed below.

Applying the Concepts Students are always curious about how the mathematics they are learning can be applied, so we have included inviting applications, some with illustrations, in most of the problem sets in the book and have labeled them to show students the array of uses of mathematics.

Getting Ready for the Next Section Many students think of mathematics as a collection of discrete, unrelated topics. As instructors, we know that this is not the case. The Getting Ready for the Next Section problems reinforce the cumulative, connected nature of this course by showing how the concepts and techniques flow one from another. These problems review all of the material that students will need in order to be successful in the next section, and gently prepare students to move forward.

Maintaining Your Skills One of the major themes of our book is continuous review. We strive to continuously hone techniques learned earlier by keeping the important concepts in the forefront of the course. The Maintaining Your Skills problems review material from the previous chapter, or they review problems that form the foundation of the course.

End-of-Chapter Summary, Review, and Assessment

We have learned that students are more comfortable with a chapter that sums up what they have learned thoroughly and accessibly, and reinforces concepts and techniques well. To help students grasp concepts and get more practice, each chapter ends with the following features that together give a comprehensive reexamination of the chapter.

Chapter Summary The chapter summary recaps all main points from the chapter in a visually appealing grid. In the margin, next to each topic where appropriate, is an example that illustrates the type of problem associated with the topic being reviewed. When students prepare for a test, they can use the chapter summary as a guide to the main concepts of the chapter.

Chapter Review Following the chapter summary in each chapter is the chapter review. It contains an extensive set of problems that review all the main topics in the chapter. This feature can be used flexibly, as assigned review, as a recommended self-test for students as they prepare for examinations, or as an in-class quiz or test.

Cumulative Review Starting in Chapter 2, following the chapter review is a set of problems that reviews material from preceding chapters. This keeps students current with past topics and helps them retain the information they study.

Chapter Test This set of problems is representative of all the main points of the chapter. These don't contain as many problems as the chapter review, and should be completed in 50 minutes.

Chapter Projects Each chapter closes with a pair of projects. One is a group project, suitable for students to work on in class. The second project is a research project for students to do outside of class and tends to be open ended.

Additional Features of the Book

Blueprint for Problem Solving Found in the main text, this feature is a detailed outline of steps required to successfully attempt application problems. Intended as a guide to problem solving in general, the blueprint takes the student through the solution process of various kinds of applications.

Early Coverage of Functions Functions are introduced in Chapter 3 and then integrated in the rest of the text. This feature forms a bridge to college algebra by requiring students to work with functions and function notation throughout the course.

Facts from Geometry Many of the important facts from geometry are listed under this heading. In most cases, an example or two accompanies each of the facts to give students a chance to see how topics from geometry are related to the algebra they are learning.

Supplements for the Instructor

If you are interested in any of the supplements below, please contact your sales representative.

Annotated Instructor's Edition
ISBN-10: 1-133-11025-8 | ISBN-13: 978-1-133-11025-5
This special instructor's version of the text contains answers next to exercises and instructor notes at the appropriate location.

Complete Solutions Manual   ISBN-10: 1-133-49086-7 | ISBN-13: 978-1-133-49086-9
This manual contains complete solutions for all problems in the text.

Cengage Instructor's Resource Binder for Algebra Activities
ISBN-10: 0-538-73675-5 | ISBN-13: 978-0-538-73675-6
NEW! Each section of the main text is discussed in uniquely designed Teaching Guides containing instruction tips, examples, activities, worksheets, overheads, assessments, and solutions to all worksheets and activities.

Enhanced WebAssign   ISBN-10: 0-538-73810-3 | ISBN-13: 978-0-538-73810-1
Exclusively from Cengage Learning, Enhanced WebAssign® combines the exceptional Mathematics content that you know and love with the most powerful online homework solution, WebAssign. Enhanced WebAssign engages students with immediate feedback and rich tutorial content helping students to develop a deeper conceptual understanding of their subject matter. Online assignments can be built by selecting from thousands of text-specific problems or supplemented with problems from any Cengage Learning textbook.

PowerLecture with ExamView® Algorithmic Equations
ISBN-10: 1-133-49120-0 | ISBN-13: 978-1-133-49120-0
This CD-ROM (or DVD) provides the instructor with dynamic media tools for teaching. Create, deliver, and customize tests (both print and online) in minutes with ExamView® Computerized Testing Featuring Algorithmic Equations. Easily build solution sets for homework or exams using Solution Builder's online solutions manual. Microsoft® PowerPoint® lecture slides and figures from the book are also included on this CD-ROM (or DVD).

Solution Builder This online instructor database offers complete worked solutions to all exercises in the text, allowing you to create customized, secure solutions printouts (in PDF format) matched exactly to the problems you assign in class. Visit http://www.cengage.com/solutionbuilder.

Text-Specific DVDs ISBN-10: 1-133-49083-2 | ISBN-13: 978-1-133-49083-8
This set of text-specific DVDs features segments taught by the author and worked-out solutions to many examples in the book. Available to instructors only.

For the Student

Enhanced WebAssign   ISBN-10: 0-538-73810-3 | ISBN-13: 978-0-538-73810-1
Exclusively from Cengage Learning, Enhanced WebAssign® combines the exceptional Mathematics content that you know and love with the most powerful online homework solution, WebAssign. Enhanced WebAssign engages students with immediate feedback and rich tutorial content helping students to develop a deeper conceptual understanding of their subject matter. Online assignments can be built by selecting from thousands of text-specific problems or supplemented with problems from any Cengage Learning textbook.

Student Solutions Manual ISBN-10: 1-133-49119-7 | ISBN-13: 978-1-133-49119-4
This Manual contains complete annotated solutions to all odd problems in the problem sets and all chapter review and chapter test exercises.

Student Workbook ISBN-10: 1-133-52533-4 | ISBN-13: 978-1-133-52533-2
Get a head start with this hands-on resource! The Student Workbook is packed with assessments, activities, and worksheets to help you maximize your study efforts.

Acknowledgments

I would like to thank my editor at Cenage Learning, Marc Bove for his help and encouragement and ensuring a good working relationship with the editorial and marketing group at Cengage. Jennifer Risden continues to keep us on track with production and, as always, is the consummate professional. Donna Looper, head of production in my office, along with Staci Truelson did a fantastic job of keeping us organized and efficient. They are both a pleasure to work with. Special thanks to the other members of our team Mary Skutley, Kaela SooHoo, Mike Landrum, Katherine Shields, Kendra Nomoto, and Christina Machado; all of whom played an important roll in the production of this book.

Pat McKeague
September 2011

Preface to the Student

I often find my students asking themselves the question "Why can't I understand this stuff the first time?" The answer is "You're not expected to." Learning a topic in mathematics isn't always accomplished the first time around. There are many instances when you will find yourself reading over new material a number of times before you can begin to work problems. That's just the way things are in mathematics. If you don't understand a topic the first time you see it, that doesn't mean you won't succeed in this course. Understanding mathematics takes time. The process of understanding requires reading the book, studying the examples, working problems, and getting your questions answered.

How to Be Successful in Mathematics

1. If you are in a lecture class, be sure to attend all class sessions on time. You cannot know exactly what goes on in class unless you are there. Missing class and then expecting to find out what went on from someone else is not the same as being there yourself.

2. Read the book. It is best to read the section that will be covered in class beforehand. Reading in advance, even if you do not understand everything you read, is still better than going to class with no idea of what will be discussed.

3. Work problems every day and check your answers. The key to success in mathematics is working problems. The more problems you work, the better you will become at working them. The answers to the odd-numbered problems are given in the back of the book. When you have finished an assignment, be sure to compare your answers with those in the book. If you have made a mistake, find out what it is, and correct it.

4. Do it on your own. Don't be misled into thinking someone else's work is your own. Having someone else show you how to work a problem is not the same as working the same problem yourself. It is okay to get help when you are stuck. As a matter of fact, it is a good idea. Just be sure you do the work yourself.

5. Review every day. After you have finished the problems your instructor has assigned, take another 15 minutes and review a section you have already completed. The more you review, the longer you will retain the material you have learned.

6. Don't expect to understand every new topic the first time you see it. Sometimes you will understand everything you are doing, and sometimes you won't. Expecting to understand each new topic the first time you see it can lead to disappointment and frustration. The process of understanding takes time. You will need to read the book, work problems, and get your questions answered.

7. Spend as much time as it takes for you to master the material. No set formula exists for the exact amount of time you need to spend on mathematics to master it. You will find out as you go along what is or isn't enough time for you. If you end up spending 2 or more hours on each section in order to master the material there, then that's how much time it takes; trying to get by with less will not work.

8. Relax. Take a deep breath and work each problem one step at a time. It's probably not as difficult as you think.

Basic Properties and Definitions

The image above shows current gasoline prices, in U.S. dollars, in different parts of Europe. Considering that the average price per gallon in the United States is $3.97, the retail price for gasoline in European countries is much higher than in the United States.

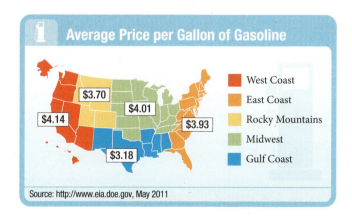

   Throughout the U.S., the average price per gallon also varies. The chart here shows the average retail price per gallon of gasoline in different parts of the country. In this chapter we will begin our work with real numbers and their properties. When you finish this chapter you will be able to take numbers, like the ones listed above, and construct tables and graphs for comparison.

Preview

Key Words	Definition
Set	A collection of objects or things
Real Number	Any number that can be represented on the real number line
Factor	A number that divides another evenly
Prime Number	A positive number whose only positive factors are 1 and itself.
Opposites	Any two real numbers that are the same distance from 0 on the number line but in opposite directions
Reciprocals	Any two real numbers whose product is 1
Absolute Value	The distance of a number from 0 on the number line
Least Common Denominator	The smallest number that is divisible by all the denominators

Chapter Outline

1.1 Fundamental Definitions and Notation
- A Translate expressions written in English to algebraic expressions.
- B Expand and multiply numbers raised to positive integer exponents.
- C Simplify expressions using the rule for order of operations.
- D Find the value of an expression.

1.2 Sets and Venn Diagrams
- A Identify elements of a set.
- B Find the union and intersection of two sets.
- C Use Venn diagrams.

1.3 The Real Numbers
- A List the elements in subsets of the real numbers.
- B Factor positive integers into the product of primes.
- C Reduce fractions to lowest terms.

1.4 Simple and Compound Inequalities
- A Graph simple and compound inequalities.
- B Translate sentences and phrases written in English into inequalities.

1.5 Arithmetic with Real Numbers
- A Find the opposite, the reciprocal, and the absolute value of a real number.
- B Add and subtract real numbers.
- C Multiply and divide real numbers.
- D Solve arithmetic problems that contain fractions.

1.6 Properties of Real Numbers
- A Recognize and apply the properties of real numbers.
- B Apply the rule for order of operations.
- C Simplify algebraic expressions.
- D Find the value of an algebraic expression.

1.7 Recognizing Patterns
- A Recognize a pattern in a sequence of numbers.
- B Extend an arithmetic sequence.
- C Extend a geometric sequence.
- D Recognize and extend the Fibonacci sequence.

Fundamental Definitions and Notation

A Translate expressions written in English into algebraic expressions.

B Expand and multiply numbers raised to positive integer exponents.

C Simplify expressions using the rule for order of operations.

D Find the value of an expression.

TICKET TO SUCCESS

Each section of the book will begin with some problems and questions like the ones below. Think about them while you read through the following section. Before you go to class, answer each problem or question with a written response using complete sentences. Writing about mathematics is a valuable exercise. As with all problems in this course, approach these writing exercises with a positive point of view. You will get better at giving written responses to questions as the course progresses. Even if you never feel comfortable writing about mathematics, just attempting the process will increase your understanding and ability in this course.

Keep these questions in mind as you read through the section. Then respond in your own words and in complete sentences.

1. Why is it important to translate expressions written in symbols into the English language?
2. What is the exponent in the expression 5^4 and what does it mean?
3. Write the four steps in the rule for order of operations.
4. How would you find the value of an algebraic expression?

iTunes released The Beatles music for sale in November 2010. By analyzing the following bar chart, we can see a dramatic increase in digital music sales during that same month. The increase was measured at a 12.4% surge. But even if we didn't know the exact percent increase, the bar chart still clearly shows that November sales were greater than those of the previous three months; in other words, the taller the bar on the chart, the greater the sales.

From the information in the chart, we can find many relationships between the sales percentages. For example, the sales percentage in January 2011 is less than the sales percentage in March 2011. In mathematics, we use symbols to represent relationships between quantities. If we let P represent the sales percentage per month, then the relationship between the January and March sales numbers can be written this way:

$$P(\text{January}) < P(\text{March})$$

In this section, we will review the fundamental definitions and symbols used in mathematics, such as letting letters represent numbers or quantities.

A Translating Expressions

We have compiled the following list of many of the basic symbols and definitions we will be using throughout the book.

Comparison Symbols

	In Symbols	In Words
Equal	$a = b$	a is equal to b
Not equal	$a \neq b$	a is not equal to b
Less than	$a < b$	a is less than b
Less than or equal	$a \leq b$	a is less than or equal to b
Not less than	$a \nless b$	a is not less than b
Greater than	$a > b$	a is greater than b
Greater than or equal	$a \geq b$	a is greater than or equal to b
Not greater than	$a \ngtr b$	a is not greater than b
Equivalent	$a \Leftrightarrow b$	a is equivalent to b

Operation Symbols

	In Symbols	In Words
Addition	$a + b$	The sum of a and b
Subtraction	$a - b$	The difference of a and b
Multiplication	$ab, a \cdot b, a(b), (a)(b),$ or $(a)b$	The product of a and b
Division	$a \div b, a/b,$ or $\dfrac{a}{b}$	The quotient of a and b

Some key words from the above list are *sum, difference, product,* and *quotient.* They are used frequently in mathematics. For instance, we may say the product of 3 and 4 is 12. We mean both the expressions $3 \cdot 4$ and 12 are called the products of 3 and 4. The important idea here is that the word *product* implies multiplication, regardless of whether it is written $3 \cdot 4$, 12, 3(4), (3)4, or (3)(4).

When we let a letter, such as x, stand for a number or group of numbers, then we say x is a *variable* because the value it takes on can vary. In the lists of notation and symbols just presented, the letters a and b are variables that represent the numbers we will work with in this book. In the next example, we show some translations between sentences written in English and their equivalent expressions in algebra. Note that all the sentences contain at least one variable.

EXAMPLE 1

In English	In Symbols
The sum of x and 5 is less than 2.	$x + 5 < 2$
The product of 3 and x is 21.	$3x = 21$
The quotient of y and 6 is 4.	$\frac{y}{6} = 4$
Twice the difference of b and 7 is greater than 5.	$2(b - 7) > 5$
The difference of twice b and 7 is greater than 5.	$2b - 7 > 5$ ■

B Exponents

Consider the expression 3^4. The 3 is called the *base* and the 4 is called the *exponent*. The exponent 4 tells us the number of times the base appears in the product. That is:

$$3^4 = 3 \cdot 3 \cdot 3 \cdot 3 = 81$$

The expression 3^4 is said to be in *exponential form*, whereas $3 \cdot 3 \cdot 3 \cdot 3$ is said to be in *expanded form*.

EXAMPLE 2　Expand and multiply 5^2.

SOLUTION　$5^2 = 5 \cdot 5 = 25$　　　　Base 5, exponent 2　　　■

EXAMPLE 3　Expand and multiply 2^5.

SOLUTION　$2^5 = 2 \cdot 2 \cdot 2 \cdot 2 \cdot 2 = 32$　　　Base 2, exponent 5　　　■

EXAMPLE 4　Expand and multiply 4^3.

SOLUTION　$4^3 = 4 \cdot 4 \cdot 4 = 64$　　　Base 4, exponent 3　　　■

C Order of Operations

It is important when evaluating arithmetic expressions in mathematics that each expression have only one answer in reduced form. Consider the expression

$$3 \cdot 7 + 2$$

If we find the product of 3 and 7 first, then add 2, the answer is 23. On the other hand, if we first combine the 7 and 2, then multiply by 3, we have 27. The problem seems to have two distinct answers depending on whether we multiply first or add first. To avoid this situation, we will decide that multiplication in a situation like this will always be done before addition. In this case, only the first answer, 23, is correct.

PRACTICE PROBLEMS

1. Write a statement, using symbols, that is equivalent to each English statement.
   a. The sum of x and 2 is greater than 5.
   b. The product of 2 and y is 18.
   c. The quotient of a and 3 is 7.
   d. Twice the product of y and z is less than 2.
   e. The product of twice y and z is less than 2.

2. Expand and multiply 4^2.

3. Expand and multiply 2^4.

4. Expand and multiply 3^3.

Answers

1. a. $x + 2 > 5$　b. $2y = 18$
   c. $\frac{a}{3} = 7$　d. $2(yz) < 2$
   e. $(2y)z < 2$
2. 16
3. 16
4. 27

The complete set of rules for evaluating expressions follows.

NOTE
This rule is very important. We will use it many times throughout the book. It is a simple rule to follow. First, we evaluate any numbers with exponents; then we multiply and divide; and finally we add and subtract, always working from left to right when more than one of the same operation symbol occurs in a problem.

> **Rule** Order of Operations
> When evaluating a mathematical expression, we will perform the operations in the following order:
> 1. Begin with the expression in the innermost parentheses or brackets and work our way out.
> 2. Evaluate, or simplify, all numbers with exponents, working from left to right if more than one of these expressions is present.
> 3. Work all multiplications and divisions left to right.
> 4. Perform all additions and subtractions left to right.

Simplify each expression.
5. $6 + 2(3 + 4)$

EXAMPLE 5 Simplify the expression $5 + 3(2 + 4)$.

SOLUTION
$$5 + 3(2 + 4) = 5 + 3(6) \quad \text{Simplify inside parentheses.}$$
$$= 5 + 18 \quad \text{Then multiply.}$$
$$= 23 \quad \text{Add.} \quad ■$$

6. $5 \cdot 3^2 - 2 \cdot 4^2$

EXAMPLE 6 Simplify the expression $5 \cdot 2^3 - 4 \cdot 3^2$.

SOLUTION
$$5 \cdot 2^3 - 4 \cdot 3^2 = 5 \cdot 8 - 4 \cdot 9 \quad \text{Simplify exponents left to right.}$$
$$= 40 - 36 \quad \text{Multiply left to right.}$$
$$= 4 \quad \text{Subtract.} \quad ■$$

7. $30 - (2 \cdot 3^2 - 8)$

EXAMPLE 7 Simplify the expression $20 - (2 \cdot 5^2 - 30)$.

SOLUTION $20 - (2 \cdot 5^2 - 30) = 20 - (2 \cdot 25 - 30)$ } Simplify inside parentheses, evaluating exponents first, then multiplying.
$$= 20 - (50 - 30)$$
$$= 20 - (20) \quad \text{Subtract.}$$
$$= 0 \quad ■$$

8. $60 + 20 \div 2 - 40$

EXAMPLE 8 Simplify the expression $40 - 20 \div 5 + 8$.

SOLUTION $40 - 20 \div 5 + 8 = 40 - 4 + 8 \quad$ Divide first.
$$= 36 + 8$$ } Then, add and subtract left to right.
$$= 44 \quad ■$$

9. $3 + 5[2 + (7 \cdot 2 - 10)]$.

EXAMPLE 9 Simplify the expression $2 + 4[5 + (3 \cdot 2 - 2)]$.

SOLUTION $2 + 4[5 + (3 \cdot 2 - 2)] = 2 + 4[5 + (6 - 2)]$ } Simplify inside innermost parentheses.
$$= 2 + 4(5 + 4)$$
$$= 2 + 4(9)$$
$$= 2 + 36 \quad \text{Then multiply.}$$
$$= 38 \quad \text{Add.} \quad ■$$

Answers
5. 20
6. 13
7. 20
8. 30
9. 33

D Finding the Value of an Algebraic Expression

An *algebraic expression* is a combination of numbers, variables, and operation symbols. For example, each of the following is an algebraic expression

$$5x \qquad x^2 + 5 \qquad 4a^2b^3 \qquad 5t^2 - 6t + 3$$

An expression such as $x^2 + 5$ will take on different values depending on what number we substitute for x. For example,

When →	$x = 3$	When →	$x = 7$
the expression →	$x^2 + 5$	the expression →	$x^2 + 5$
becomes →	$3^2 + 5 = 9 + 5$	becomes →	$7^2 + 5 = 49 + 5$
	$= 14$		$= 54$

As you can see, our expression is 14 when x is 3, and when x is 7, our expression is 54. In each case, the value of the expression is found by replacing the variable with a number.

EXAMPLE 10   Evaluate the expression $a^2 + 8a + 16$ when a is 0, 1, 2, 3, and 4.

SOLUTION   We can organize our work efficiently by using a table.

When a is	The expression $a^2 + 8a + 16$ becomes
0	$0^2 + 8 \cdot 0 + 16 = 0 + 0 + 16 = 16$
1	$1^2 + 8 \cdot 1 + 16 = 1 + 8 + 16 = 25$
2	$2^2 + 8 \cdot 2 + 16 = 4 + 16 + 16 = 36$
3	$3^2 + 8 \cdot 3 + 16 = 9 + 24 + 16 = 49$
4	$4^2 + 8 \cdot 4 + 16 = 16 + 32 + 16 = 64$

∎

10. Evaluate $(a + 4)^2$ when $a = 0$, 1, 2, 3, 4.

EXAMPLE 11   Find the value of the expression $3x + 4y + 5$ when x is 6 and y is 7.

SOLUTION   We substitute the given values of x and y into the expression and simplify the result.

When →	$x = 6$ and $y = 7$
the expression →	$3x + 4y + 5$
becomes →	$3 \cdot 6 + 4 \cdot 7 + 5 = 18 + 28 + 5$
	$= 51$

∎

11. Evaluate $2x + 7y - 5$ when x is 4 and y is 1.

Answers
10. 16, 25, 36, 49, 64
11. 10

Problem Set 1.1

Moving Toward Success

"Success is a science; if you have the conditions, you get the result."
—Oscar Wilde, 1854–1900, Irish dramatist and poet

1. How will setting short-term goals help you reach your long-term goals for this class?
2. How do you plan to achieve your short-term goals?

A Translate each of the following statements into symbols. [Example 1]

1. The sum of x and 5 is 2.

2. The sum of y and -3 is 9.

3. The difference of 6 and x is y.

4. The difference of x and 6 is $-y$.

5. The product of t and 2 is less than y.

6. The product of $5x$ and y is equal to z.

7. The sum of x and y is less than the difference of x and y.

8. Twice the sum of a and b is 15.

9. Three times the difference of x and 5 is more than y.

10. The product of x and y is greater than or equal to the quotient of x and y.

B Expand and multiply. [Examples 2–4]

11. 6^2

12. 8^2

13. 10^2

14. 10^3

15. 2^3

16. 5^3

17. 2^4

18. 1^4

19. 10^4

20. 4^3

21. 11^2

22. 10^5

C The problems that follow are intended to give you practice using the rule for order of operations. Simplify each expression. [Examples 5–9]

23. **a.** $3 \cdot 5 + 4$

b. $3(5 + 4)$

c. $3 \cdot 5 + 3 \cdot 4$

24. **a.** $3 \cdot 7 - 6$

b. $3(7 - 6)$

c. $3 \cdot 7 - 3 \cdot 6$

25. **a.** $6 + 3 \cdot 4 - 2$

b. $6 + 3(4 - 2)$

c. $(6 + 3)(4 - 2)$

26. a. $8 + 2 \cdot 7 - 3$

  b. $8 + 2(7 - 3)$

  c. $(8 + 2)(7 - 3)$

27. a. $(7 - 4)(7 + 4)$

  b. $7^2 - 4^2$

28. a. $(8 - 5)(8 + 5)$

  b. $8^2 - 5^2$

29. a. $(5 + 7)^2$

  b. $5^2 + 7^2$

  c. $5^2 + 2 \cdot 5 \cdot 7 + 7^2$

30. a. $(8 - 3)^2$

  b. $8^2 - 3^2$

  c. $8^2 - 2 \cdot 8 \cdot 3 + 3^2$

31. a. $2 + 3 \cdot 2^2 + 3^2$

  b. $2 + 3(2^2 + 3^2)$

  c. $(2 + 3)(2^2 + 3^2)$

32. a. $3 + 4 \cdot 4^2 + 5^2$

  b. $3 + 4(4^2 + 5^2)$

  c. $(3 + 4)(4^2 + 5^2)$

33. a. $40 - 10 \div 5 + 1$

  b. $(40 - 10) \div 5 + 1$

  c. $(40 - 10) \div (5 + 1)$

34. a. $20 - 10 \div 2 + 3$

  b. $(20 - 10) \div 2 + 3$

  c. $(20 - 10) \div (2 + 3)$

35. a. $40 + [10 - (4 - 2)]$

  b. $40 - 10 - 4 - 2$

36. a. $50 - [17 - (8 - 3)]$

  b. $50 - 17 - 8 - 3$

37. a. $3 + 2(2 \cdot 3^2 + 1)$

  b. $(3 + 2)(2 \cdot 3^2 + 1)$

38. a. $4 + 5(3 \cdot 2^2 + 5)$

  b. $(4 + 5)(3 \cdot 2^2 + 5)$

C The problems below will make certain you are using the rule for order of operations correctly. [Examples 5–9]

39. $5 \cdot 10^3 + 4 \cdot 10^2 + 3 \cdot 10 + 1$

40. $6 \cdot 10^3 + 5 \cdot 10^2 + 4 \cdot 10 + 3$

41. $3[2 + 4(5 + 2 \cdot 3)]$

42. $2[4 + 2(6 + 3 \cdot 5)]$

43. $6[3 + 2(5 \cdot 3 - 10)]$

44. $8[7 + 2(6 \cdot 9 - 14)]$

45. $5(7 \cdot 4 - 3 \cdot 4) + 8(5 \cdot 9 - 4 \cdot 9)$

46. $4(3 \cdot 9 - 2 \cdot 9) + 5(6 \cdot 8 - 5 \cdot 8)$

47. $25 - 17 + 3$

48. $38 - 19 + 1$

49. $109 - 36 + 14$

50. $200 - 150 + 20$

51. $20 - 13 - 3$

52. $57 - 18 - 8$

53. $63 - 37 - 4$

54. $71 - 11 - 1$

55. $36 \div 9 \cdot 4$

56. $28 \div 7 \cdot 2$

57. $75 \div 3 \cdot 25$

58. $48 \div 3 \cdot 2$

59. $64 \div 16 \div 4$

60. $125 \div 25 \div 5$

61. $75 \div 25 \div 5$

62. $36 \div 12 \div 4$

The problems below are problems you will see later in the book. Simplify each expression without using a calculator.

63. $18{,}000 - 9{,}300$

64. $18{,}000 - 4{,}500$

65. $3.45 + 2.6 - 1.004$

66. $24.3 + 6(8.1)$

67. $275 \div 55$

68. $4.8 \div 2.4$

69. $4(2)(4)^2$

70. $230(5) - 20(5)^2$

71. $250(5) - 25(5)^2$

72. $3(3)^2 + 2(3) - 1$

73. $5 \cdot 2^3 - 3 \cdot 2^2 + 4 \cdot 2 - 5$

74. $125 \cdot 2^2$

75. $125 \cdot 2^3$

76. $7.5(10)$

77. $500(1.5)$

78. $39.3 \cdot 60$

79. $5(0.10)$

80. $2(0.25)$

81. $0.20(8)$

82. $0.30(12)$

83. $0.08(4,000)$

84. $0.09(6,000)$

We are assuming that you know how to use a calculator to do simple arithmetic problems. Use a calculator to simplify each expression. If rounding is necessary, round your answers to the nearest ten thousandth (4 places past the decimal point). You will see these problems again later in the book.

Simplify.

85. $0.6931 + 1.0986$

86. $1.6094 + 1.9459$

87. $3(0.6931)$

88. $2(1.9459)$

89. $250(3.14)$

90. $165(3.14)$

91. $4,628 \div 25$

92. $7,546 \div 35$

93. $65,000 \div 5,280$

94. $2,358 \div 5,280$

95. $1 - 0.8413$

96. $1.2052 - 1$

97. $16(3.5)^2$

98. $4(3.14)3^2$

99. $11.5(130) - 0.05(130)^2$

100. $10(130) - 0.04(130)^2$

D [Examples 10–11]

101. Find the value of each expression when x is 5.

   a. $x + 2$

   b. $2x$

   c. x^2

   d. 2^x

102. Find the value of each expression when x is 3.

   a. $x + 5$

   b. $5x$

   c. x^5

   d. 5^x

103. Find the value of each expression when x is 10.

   a. $x^2 + 2x + 1$

   b. $(x + 1)^2$

   c. $x^2 + 1$

   d. $(x - 1)^2$

104. Find the value of each expression when x is 8.

   a. $x^2 - 6x + 9$

   b. $(x - 3)^2$

   c. $x^2 - 3$

   d. $x^2 - 9$

105. Find the value of $b^2 - 4ac$ if

   a. $a = 2, b = 5,$ and $c = 3$

   b. $a = 10, b = 60, c = 30$

   c. $a = 0.4, b = 1,$ and $c = 0.3$

106. Find the value of $6x + 5y + 4$ if

   a. $x = 3$ and $y = 2$

   b. $x = 2$ and $y = 3$

   c. $x = 0$ and $y = 0$

107. Find the value of $2x^2 - 3y + 2$ if

   a. $x = -1$ and $y = 3$

   b. $x = 3$ and $y = -1$

   c. $x = 4$ and $y = 9$

108. Find the value of $x^2 + 4y - 6$ if

   a. $x = 3$ and $y = -7$

   b. $x = -5$ and $y = -4$

   c. $x = 4$ and $y = 3$

109. Find the value of $x^2 + 3x + 4$ if

   a. $x = -4$

   b. $x = 2$

   c. $x = -5$

110. Find the value of $2x - 4y^2 + 9$ if

   a. $x = -7$ and $y = -2$

   b. $x = 3$ and $y = 4$

   c. $x = -4$ and $y = 3$

Extending the Concepts

Many of the problem sets in this book end with a few problems like the ones that follow. These problems challenge you to extend your knowledge of the material in the problem set. In most cases, there are no examples in the text similar to these problems. You should approach them with a positive point of view because even though you may not complete them correctly, just the process of attempting them will increase your knowledge and ability in algebra.

111. Fermat's Last Theorem The year 2001 was the 400th anniversary of the birth of the French mathematician Pierre de Fermat. He has become famous for what has come to be called Fermat's last theorem. This theorem states that if n is an integer greater than 2, then there are no positive integers $x, y,$ and z that will make the formula $x^n + y^n = z^n$ true.

However, there are many ways to make the formula $x^n + y^n = z^n$ true when n is 1 or 2. Show that this formula is true for each case below.

a. $n = 1, x = 5, y = 8,$ and $z = 13$

b. $n = 1, x = 2, y = 3,$ and $z = 5$

c. $n = 2, x = 3, y = 4,$ and $z = 5$

d. $n = 2, x = 7, y = 24,$ and $z = 25$

112. Internet The chart shows the number of internet users in different regions of the world. Use it to fill in the blank with < or > to make the statement true about the number of users.

a. North America _____ Latin America

b. Europe _____ North America + Latin America

c. Asia − Europe _____ North America

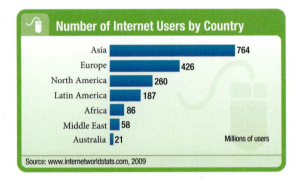

Number of Internet Users by Country

Region	Millions of users
Asia	764
Europe	426
North America	260
Latin America	187
Africa	86
Middle East	58
Australia	21

Source: www.internetworldstats.com, 2009

Sets and Venn Diagrams

TICKET TO SUCCESS

Keep these questions in mind as you read through the section. Then respond in your own words and in complete sentences.

1. What is the difference between a set and a subset?

2. What does the symbol $\in$ mean?

3. What is the difference between a union and an intersection?

4. What is a Venn diagram?

Robert Kneschke/Shutterstock.com

A survey of seventy-three Internet users asked them to name their favorite social networking sites. The following were the four answer choices:

a. Facebook **b.** Twitter **c.** Both **d.** Neither

The results of this survey are displayed in an illustration called a Venn diagram, shown here:

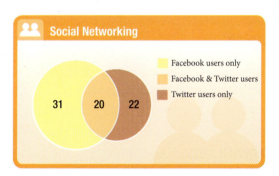

From the illustration we see that there are more Facebook users than Twitter users and that 20 people chose "Both," as represented by the light orange area. None of the users surveyed selected "Neither." Everyday information like this can be represented mathematically with collections of data called sets. In this section we will work with sets and use Venn diagrams to give visual representations of the relationships between different sets.

A Sets and Subsets

The concept of a set can be considered the starting point for all the branches of mathematics.

> **Definition**
>
> A **set** is a collection of objects or things. The objects in the set are called **elements**, or **members**, of the set.

Sets are usually denoted by capital letters, and elements of sets are denoted by lowercase letters. We use braces, { }, to enclose the elements of a set.

To show that an element is contained in a set, we use the symbol $\in$. That is,

$$x \in A \text{ is read "}x \text{ is an element (member) of set } A\text{"}$$

For example, if A is the set {1, 2, 3}, then $2 \in A$. However, $5 \notin A$ means 5 is not an element of set A.

> **Definition**
>
> Set A is a **subset** of set B, written $A \subset B$, if every element in A is also an element of B. That is:
>
> $$A \subset B \quad \text{if and only if} \quad A \text{ is contained in } B$$

PRACTICE PROBLEMS

Identify each statement as either true or false.

1. The set {2, 4, 6, . . . } is an example of an infinite set.

2. The set {2, 4, 8} is a subset of {2, 4, 6, . . . }.

EXAMPLE 1

The set of numbers used to count things is {1, 2, 3, . . . }. The dots mean the set continues indefinitely in the same manner. This is an example of an *infinite set*. ∎

EXAMPLE 2

The set of all numbers represented by the dots on the faces of a regular die is {1, 2, 3, 4, 5, 6}. This set is a subset of the set in Example 1. It is an example of a *finite set* because it has a limited number of elements. ∎

> **Definition**
>
> The set with no members is called the **empty**, or **null**, **set.** It is denoted by the symbol ∅. The empty set is considered a subset of every set.

B Operations with Sets

Two basic operations are used to combine sets: *union* and *intersection*.

> **Definition**
>
> The **union** of two sets A and B, written $A \cup B$, is the set of all elements that are either in A or in B, or in both A and B. The key word here is *or*. For an element to be in $A \cup B$, it must be in A or B. In symbols, the definition looks like this:
>
> $$x \in A \cup B \quad \text{if and only if} \quad x \in A \text{ or } x \in B$$

Answers

1. True

2. True

Definition

The **intersection** of two sets A and B, written $A \cap B$, is the set of elements in both A and B. The key word in this definition is the word *and*. For an element to be in $A \cap B$, it must be in both A and B. In symbols:

$$x \in A \cap B \qquad \text{if and only if} \qquad x \in A \text{ and } x \in B$$

EXAMPLES   Let $A = \{1, 3, 5\}$, $B = \{0, 2, 4\}$, and $C = \{1, 2, 3, \ldots\}$.

3. $A \cup B = \{0, 1, 2, 3, 4, 5\}$

4. $A \cap B = \varnothing$ (A and B have no elements in common.)

5. $A \cap C = \{1, 3, 5\} = A$

6. $B \cup C = \{0, 1, 2, 3\ldots\}$ ■

For the following problems, let $A = \{1, 2, 3\}$, $B = \{0, 3, 7\}$, and $C = \{0, 1, 2, 3, \ldots\}$.

3. $A \cup B$

4. $A \cap B$

5. $A \cap C$

6. $B \cap C$

Another notation we can use to describe sets is called *set-builder* notation. Here is how we write our definition for the union of two sets A and B using set-builder notation:

$$A \cup B = \{x \mid x \in A \text{ or } x \in B\}$$

The right side of this statement is read "the set of all x such that x is a member of A or x is a member of B." As you can see, the vertical line after the first x is read "such that."

EXAMPLE 7   If $A = \{1, 2, 3, 4, 5, 6\}$, find $C = \{x \mid x \in A \text{ and } x \geq 4\}$.

SOLUTION   We are looking for all the elements of A that are also greater than or equal to 4. They are 4, 5, and 6. Using set notation, we have

$$C = \{4, 5, 6\}$$ ■

7. Let $A = \{2, 4, 6, 8, 10\}$ and find $B = \{x \mid x \in A \text{ and } x < 8\}$.

C Venn Diagrams

Venn diagrams are diagrams, or pictures, that represent the union and intersection of sets. Each of the following diagrams is a Venn diagram. The shaded region in the first diagram shows the union of two sets A and B, while the shaded part of the second diagram shows only their intersection.

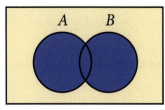

$A \cup B$

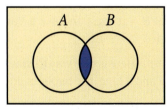

$A \cap B$

In each diagram, we think of the elements of set A as being the points inside the circle labeled A. The elements of set B are the points inside the circle labeled B. We enclose the two sets within a rectangle to indicate that there are still other elements that are neither in A nor in B. The rectangle is sometimes referred to as the *universal set*.

Answers

3. $\{0, 1, 2, 3, 7\}$

4. $\{3\}$

5. $\{1, 2, 3\}$

6. $\{0, 3, 7\}$

7. $\{2, 4, 6\}$

As you might expect, not all sets intersect. Here is the definition we use for non-intersecting sets, along with a Venn diagram that illustrates their relationship.

> **Definition**
>
> Two sets with no elements in common are said to be **disjoint** or **mutually exclusive**. Two sets are disjoint if their intersection is the empty set.

A and B are disjoint if and only if
$A \cap B = \varnothing$

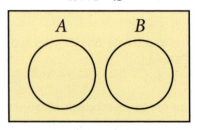

A and B are disjoint

The next example involves a deck of playing cards. A regular deck of cards contains 52 cards, in four suits: hearts (♥), clubs (♣), diamonds (♦), and spades (♠). The numbered cards are numbered 2 through 10. The face cards are jacks, queens, and kings. There are four aces.

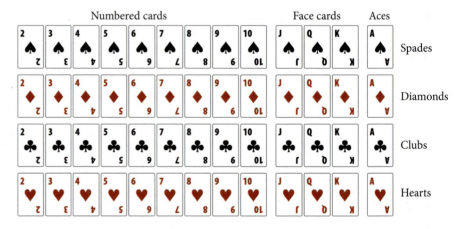

EXAMPLE 8 Suppose you have a deck of 52 playing cards. Let set $A = \{Aces\}$ and set $B = \{Kings\}$ and use a Venn diagram to show that A and B are mutually exclusive.

SOLUTION Here is one way to draw the Venn diagram. Since a card cannot be both an ace and a king, the two sets do not intersect.

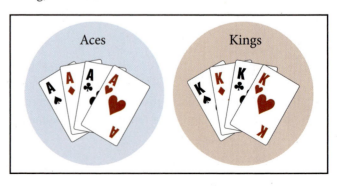

8. Suppose a sample space is a deck of cards. Let $A = \{Spades\}$ and set $B = \{Hearts\}$. Use a Venn diagram to show A and B are mutually exclusive.

Answers
8–11. For answers to practice problems that include graphs, see the section titled Solutions to Selected Practice Problems at the back of the book.

EXAMPLE 9 Use a Venn diagram to show the intersection of the sets $A = \{Aces\}$ and $B = \{Spades\}$ from a deck of playing cards.

SOLUTION There is one card that is in both sets. It is the ace of spades.

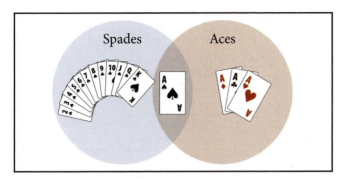

Another relationship between sets that can be represented with Venn diagrams is the subset relationship. The following diagram shows that A is a subset of B.

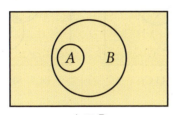

$$A \subset B$$

Along with giving a graphical representation of union and intersection, Venn diagrams can be used to test the validity of statements involving combinations of sets and operations on sets.

EXAMPLE 10 Use Venn diagrams to check the following equality. (Assume no two sets are disjoint.):

$$A \cap (B \cup C) = (A \cap B) \cup (A \cap C)$$

SOLUTION We begin by making a Venn diagram of the left side with $B \cup C$ shaded in with vertical lines.

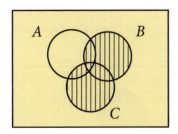

Using the same diagram, we now shade in set A with horizontal lines.

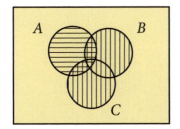

9. Use a Venn diagram to show the intersection of the sets $A = \{Kings\}$ and $B = \{Diamonds\}$ from the sample space of a deck of playing cards.

10. Use a Venn diagram to check the following equality. (Assume no two sets are disjoint).

$$A \cup (B \cap C) = (A \cup B) \cap (A \cup C)$$

The region containing both vertical and horizontal lines is the intersection of A with $B \cup C$, or $A \cap (B \cup C)$.

Next, we diagram the right side and shade in $A \cap B$ with horizontal lines and $A \cap C$ with vertical lines.

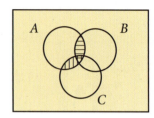

Any region containing vertical or horizontal lines is part of the union of $A \cap B$ and $A \cap C$, or $(A \cap B) \cup (A \cap C)$.

The original statement appears to be true.

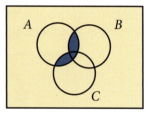

$$A \cap (B \cup C) \qquad (A \cap B) \cup (A \cap C)$$

■

11. Let A and B be two intersecting sets neither of which is a subset of the other. Use a Venn diagram to illustrate the set

$$\{x \mid x \notin A \text{ or } x \in B\}$$

EXAMPLE 11 Let A and B be two intersecting sets, neither of which is a subset of the other. Use a Venn diagram to illustrate the set

$$\{x \mid x \in A \text{ and } x \notin B\}$$

SOLUTION Using vertical lines to indicate all the elements in A and horizontal lines to show everything that is not in B we have

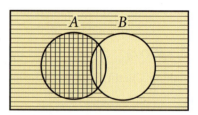

Since the connecting word is "and" we want the region that contains both vertical and horizontal lines.

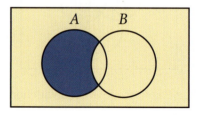

■

Problem Set 1.2

Moving Toward Success

"Knowledge has to be improved, challenged, and increased constantly, or it vanishes."

—Peter F. Drucker, 1909–2005, Austrian-born management consultant and author

1. Before reading a section from the first to the last page, do you first scan it briefly, reviewing?

2. Why should you pay attention to the italicized or bold words, and the information in colored boxes?

B For the following problems, let $A = \{0, 2, 4, 6\}$, $B = \{1, 2, 3, 4, 5\}$, and $C = \{1, 3, 5, 7\}$. [Examples 3–7]

1. $A \cup B$

2. $A \cup C$

3. $A \cap B$

4. $A \cap C$

5. $B \cap C$

6. $B \cup C$

7. $A \cup (B \cap C)$

8. $C \cup (A \cap B)$

9. $\{x \mid x \in A \text{ and } x < 4\}$

10. $\{x \mid x \in B \text{ and } x > 3\}$

11. $\{x \mid x \in A \text{ and } x \notin B\}$

12. $\{x \mid x \in B \text{ and } x \notin C\}$

13. $\{x \mid x \in A \text{ or } x \in C\}$

14. $\{x \mid x \in A \text{ or } x \in B\}$

15. $\{x \mid x \in B \text{ and } x \neq 3\}$

16. $\{x \mid x \in C \text{ and } x \neq 5\}$

C Consider a regular deck of cards to construct the following diagrams. [Examples 8, 9]

17. Construct a Venn diagram showing set A, which is the set of queens, and set B, which is the set of Kings.

18. Construct a Venn diagram showing set A, which is a set of queens, and set B, which is a set of hearts.

19. Construct a Venn diagram showing the following sets: $A = \{\text{Jacks}\}$, $B = \{\text{Diamonds}\}$

20. Construct a Venn diagram showing the following sets: $A = \{\text{Jacks}\}$, $B = \{\text{Tens}\}$

21. Construct a Venn diagram showing the following sets: $A = \{\text{Jacks}\}$, $B = \{\text{Queens}\}$, $C = \{\text{Kings}\}$

22. Construct a Venn diagram showing the following sets: $A = \{\text{Jacks}\}$, $B = \{\text{Queens}\}$, $C = \{\text{Hearts}\}$

C Use Venn diagrams to show each of the following regions. Assume sets A, B, and C all intersect one another. [Example 10]

23. $(A \cup B) \cup C$

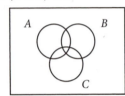

24. $(A \cap B) \cap C$

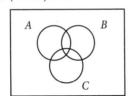

25. $A \cap (B \cap C)$

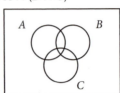

26. $A \cup (B \cup C)$

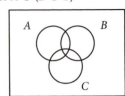

27. $A \cap (B \cup C)$

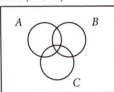

28. $(A \cap B) \cup C$

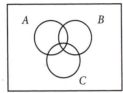

29. $A \cup (B \cap C)$

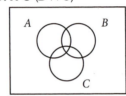

30. $(A \cup B) \cap C$

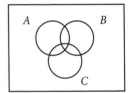

31. Use a Venn diagram to show that if $A \subset B$, then $A \cap B = A$.

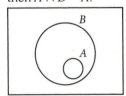

32. Use a Venn diagram to show that if $A \subset B$, then $A \cup B = B$.

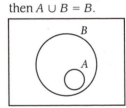

C Let A and B be two intersecting sets neither of which is a subset of the other. Use Venn diagrams to illustrate each of the following sets. [Example 11]

33. $\{x \mid x \in A \text{ and } x \in B\}$

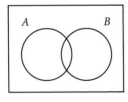

34. $\{x \mid x \in A \text{ or } x \in B\}$

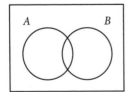

35. $\{x \mid x \notin A \text{ and } x \in B\}$

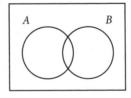

36. $\{x \mid x \notin A \text{ and } x \notin B\}$

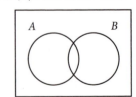

37. $\{x \mid x \in A \text{ or } x \notin B\}$

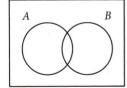

38. $\{x \mid x \notin A \text{ or } x \notin B\}$

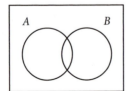

39. $\{x \mid x \in A \cap B\}$

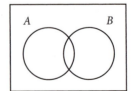

40. $\{x \mid x \in A \cup B\}$

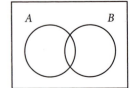

41. **Social Networking** The Venn Diagram shows a survey of people who use Facebook, Twitter, and both. If the set of Facebook users is A and the set of Twitter users is B and 100 people were surveyed, how many people fall into the following sets?

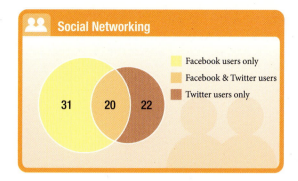

a. $\{x \mid x \notin A \cup B\}$

b. $\{x \mid x \in A \text{ and } x \notin A \cap B\}$

c. $\{x \mid x \notin A \cap B\}$

d. $\{x \mid x \in B \text{ or } x \notin A \cup B\}$

Maintaining your Skills

Perform the indicated operations.

42. $42 \div 3$

43. $189 \div 9$

44. $246 \div 2$

45. $156 \div 4$

46. $336 \div 7$

47. $3 \cdot 5 \cdot 7$

48. $2^2 \cdot 7 \cdot 11$

49. $3^2 \cdot 7 \cdot 17$

The Real Numbers

OBJECTIVES

A List the elements in subsets of the real numbers.

B Factor positive integers into the product of primes.

C Reduce fractions to lowest terms.

TICKET TO SUCCESS

Keep these questions in mind as you read through the section. Then respond in your own words and in complete sentences.

1. What is a real number?
2. What are rational numbers and what is the notation used to write them?
3. How are multiplication and factoring related?
4. How would you reduce the fraction $\frac{180}{210}$ to lowest terms?

Mariusz S./Shutterstock.com

Hang gliding is a sport that continues to be popular in many parts of the world. One type of gliding requires a pilot to use a foot-launch. The pilot kicks off from a spot on a hill, and the craft either looses or gains altitude, based on the effects of gravity and rising air masses. If we let the craft's takeoff point equal zero, we can use positive and negative numbers to represent the altitude of the craft at any point during the flight. We can then plot these numbers on a real number line. In this section, we explore real numbers and different ways to represent them, including the real number line.

The Real Number Line

The *real number line* is constructed by drawing a straight line and labeling a convenient point with the number 0. Positive numbers are in increasing order to the right of 0; negative numbers are in decreasing order to the left of 0. The point on the line corresponding to 0 is called the *origin*.

The numbers associated with the points on the line are called *coordinates* of those points. Every point on the line has a number associated with it. The set of all these numbers makes up the set of real numbers.

> **NOTE**
> The numbers on the number line increase in size as we move to the right. When we compare the size of two numbers on the number line, the one on the left is always the smaller number.

> **Definition**
>
> A **real number** is any number that is the coordinate of a point on the real number line.

PRACTICE PROBLEMS

1. Locate the numbers -3, -1.5, $-\frac{1}{4}$, 1, and 2.75 on the number line.

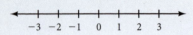

EXAMPLE 1   Locate the numbers -4.5, -0.75, $\frac{1}{2}$, $\sqrt{2}$, π, and 4.1 on the real number line.

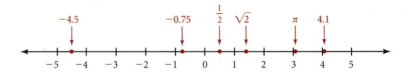

A   Subsets of the Real Numbers

Next, we consider some of the more important subsets of the real numbers. Each set listed here is a subset of the real numbers.

Counting (or natural) numbers = $\{1, 2, 3, \ldots\}$

Whole numbers = $\{0, 1, 2, 3, \ldots\}$

Integers = $\{\ldots, -3, -2, -1, 0, 1, 2, 3, \ldots\}$

Rational numbers = $\left\{ \dfrac{a}{b} \,\middle|\, a \text{ and } b \text{ are integers}, b \neq 0 \right\}$

> **NOTE**
> As you can see, the whole numbers are the counting numbers and 0 together, whereas the integers are the whole numbers along with the opposites of all the counting numbers. We can say that the counting numbers are a subset of the whole numbers, and the whole numbers are a subset of the integers.

Remember, the notation used to write the rational numbers is read "the set of numbers $\frac{a}{b}$, such that a and b are integers and b is not equal to 0." Any number that can be written in the form

$$\frac{\text{Integer}}{\text{Integer}}$$

is a rational number. Rational numbers are numbers that can be written as the ratio of two integers. Each of the following is a rational number:

$\dfrac{3}{4}$ Because it is the ratio of the integers 3 and 4

-8 Because it can be written as the ratio of -8 to 1

0.75 Because it is the ratio of 75 to 100

$0.333\ldots$ Because it can be written as the ratio of 1 to 3

Still other numbers on the number line are not members of the subsets we have listed so far. They are real numbers, but they cannot be written as the ratio of two integers; that is, they are not rational numbers. For that reason, we call them *irrational numbers*.

> **NOTE**
> We can find decimal approximations to some irrational numbers by using a calculator or a table. For example, on an eight-digit calculator
>
> $\sqrt{2} = 1.4142136$
>
> This is not exactly $\sqrt{2}$ but simply an approximation to it. There is no decimal that gives $\sqrt{2}$ exactly.

Irrational numbers = $\{x \mid x \text{ is real, but not rational}\}$

The following are irrational numbers:

$$\sqrt{2}, \quad -\sqrt{3}, \quad 4 + 2\sqrt{3}, \quad \pi, \quad \pi + 5\sqrt{6}$$

Answer

1. See Solutions Section.

EXAMPLE 2 For the set $\{-5, -3.5, 0, \frac{3}{4}, \sqrt{3}, \sqrt{5}, 9\}$, list the numbers that are (a) whole numbers, (b) integers, (c) rational numbers, (d) irrational numbers, and (e) real numbers.

SOLUTION

a. Whole numbers: 0, 9

b. Integers: −5, 0, 9

c. Rational numbers: $-5, -3.5, 0, \frac{3}{4}, 9$

d. Irrational numbers: $\sqrt{3}, \sqrt{5}$

e. They are all real numbers. ∎

The following diagram gives a visual representation of the relationships among subsets of the real numbers.

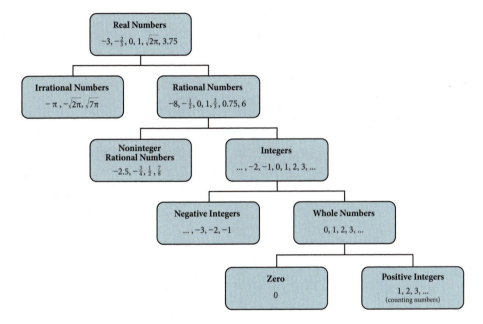

B Prime Numbers and Factoring

The following diagram shows the relationship between multiplication and factoring:

Multiplication

Factors: $3 \cdot 4 = 12;$ Product

Factoring

When we read the problem from left to right, we say the *product* of 3 and 4 is 12, or we multiply 3 and 4 to get 12. When we read the problem in the other direction, from right to left, we say we have *factored* 12 into 3 times 4, or 3 and 4 are *factors* of 12. The number 12 can be factored still further.

$$12 = 4 \cdot 3$$

Definition

If a and b represent integers, then a is said to be a **factor** (or divisor) of b if a divides b evenly—that is, if a divides b with no remainder.

A **prime number** is any positive integer larger than 1 whose only positive factors (divisors) are 1 and itself. An integer greater than 1 that is not prime is said to be **composite.**

2. For the set $\left\{-8, -2.1, -\sqrt{2}, 0, \frac{5}{8}, 3\right\}$ choose the following:

a. Whole numbers

b. Integers

c. Rational numbers

d. Irrational numbers

e. Real numbers

3. Factor 420 into the product of primes.

4. Reduce $\frac{231}{770}$ to lowest terms.

Here is a list of the first few prime numbers:

$$\text{Prime numbers} = \{2, 3, 5, 7, 11, 13, 17, 19, 23, 29, 31, 37, 41, \dots\}$$

When a number is not prime, we can factor it into the product of prime numbers. To factor a number into the product of primes, we simply factor it until it cannot be factored further.

EXAMPLE 3   Factor 525 into the product of primes.

SOLUTION   Because 525 ends in 5, it is divisible by 5.

$$525 = 5 \cdot 105$$
$$= 5 \cdot 5 \cdot 21$$
$$= 5 \cdot 5 \cdot 3 \cdot 7$$
$$= 3 \cdot 5^2 \cdot 7$$
∎

C Reducing to Lowest Terms

EXAMPLE 4   Reduce $\frac{210}{231}$ to lowest terms.

SOLUTION   First we factor 210 and 231 into the product of prime factors. Then we reduce to lowest terms by dividing the numerator and denominator by any factors they have in common.

$$\frac{210}{231} = \frac{2 \cdot 3 \cdot 5 \cdot 7}{3 \cdot 7 \cdot 11} \quad \text{Factor the numerator and denominator completely.}$$

$$= \frac{2 \cdot \cancel{3} \cdot 5 \cdot \cancel{7}}{\cancel{3} \cdot \cancel{7} \cdot 11} \quad \text{Divide the numerator and denominator by } 3 \cdot 7.$$

$$= \frac{2 \cdot 5}{11}$$

$$= \frac{10}{11}$$

The small lines we have drawn through the factors that are common to the numerator and denominator are used to indicate that we have divided the numerator and denominator by those factors. ∎

Answers
3. $2^2 \cdot 3 \cdot 5 \cdot 7$
4. $\frac{3}{10}$

Problem Set 1.3

Moving Toward Success

"All growth depends upon activity. There is no development physically or intellectually without effort, and effort means work."

—Calvin Coolidge, 1872–1933, 30th President of the United States

1. Why are the Ticket to Success questions helpful to read before reading the section?
2. Why should you write out your answers to the Ticket to Success questions before class?

A For $\{-6, -5.2, -\sqrt{7}, -\pi, 0, 1, 2, 2.3, \frac{9}{2}, \sqrt{17}\}$, list all the elements of the set that are named in each of the following problems. [Example 2]

1. Counting numbers
2. Whole numbers
3. Rational numbers
4. Integers

5. Irrational numbers
6. Real numbers
7. Nonnegative integers
8. Positive integers

B Factor each number into the product of prime factors. [Example 3]

9. 60
10. 154
11. 266
12. 385

13. 111
14. 735
15. 369
16. 1,155

C Reduce each fraction to lowest terms. [Example 4]

17. $\dfrac{165}{385}$
18. $\dfrac{550}{735}$
19. $\dfrac{385}{735}$
20. $\dfrac{266}{285}$

21. $\dfrac{111}{185}$
22. $\dfrac{279}{310}$
23. $\dfrac{525}{630}$
24. $\dfrac{205}{369}$

25. $\dfrac{75}{135}$
26. $\dfrac{38}{30}$
27. $\dfrac{6}{8}$
28. $\dfrac{10}{25}$

29. $\dfrac{200}{5}$
30. $\dfrac{240}{6}$
31. $\dfrac{10}{22}$
32. $\dfrac{39}{13}$

Applying the Concepts

33. **Goldbach's Conjecture** The letter shown here was written by Christian Goldbach in 1742. In the letter Goldbach indicates that every even number greater than two can be written as the sum of two prime numbers. For example $4 = 2 + 2$, $6 = 3 + 3$, and $8 = 3 + 5$. His assertion has never been proven, although most mathematicians believe it is true. Because of this, it has come to be known as Goldbach's Conjecture. In March 2000, the publishing firm of Faber and Faber offered a $1 million prize to anyone who could prove Goldbach's Conjecture by March 20, 2002, but no one did. Show that Goldbach's Conjecture holds for each of the following even numbers.

    a. 10

    b. 16

    c. 24

    d. 36

34. **Perfect Numbers** More than 2,200 years ago, the Greek mathematician Euclid wrote: "A perfect number is a whole number that is equal to the sum of all its divisors, except itself." The first perfect number is 6 because $6 = 1 + 2 + 3$. The next perfect number is between 20 and 30. Find it.

35. **Commission** Suppose you have a job that pays commission, and that each week you make more than $100 in commission, but no more than $200. Write two equations, using x, one that gives the maximum amount of commission you make in a week, and one showing the least amount you make in a week.

36. **Hang Gliding** A hang glider launches from a hill. He slowly circles downward, decreasing his altitude by 47 feet. Use a positive or negative number to describe his new position relative to his starting position.

37. **Hang Gliding** A hang glider launches from a hill. She immediately hits an updraft and rises 36 feet. Use a positive or negative number to describe her new position relative to her starting position.

38. **Hours of Work** Suppose you have a job that requires that you work at least 20 hours but less than 40 hours per week. Write two equations, using t, one that gives that maximum amount of hours you work in a week, and one showing the least amount of hours you work in a week.

Simple and Compound Inequalities

1.4

31

OBJECTIVES

A Graph simple and compound inequalities.

B Translate sentences and phrases written in English into inequalities.

TICKET TO SUCCESS

Keep these questions in mind as you read through the section. Then respond in your own words and in complete sentences.

1. When would you use a bracket instead of a parenthesis to graph an inequality on a number line?
2. How would you graph $\{x \mid x < 4\}$?
3. What is a continued inequality?
4. What is a compound inequality?

Ambient Ideas/Shutterstock.com

The average temperature in Denver, Colorado ranges from a low of 16°F in January to a high of 88°F in July. This range is visually represented on the number line below.

In this section, we will look at inequalities and use them to represent ranges of numbers, including temperatures.

A Graphing Inequalities

We can use the real number line to give a visual representation to inequality statements.

EXAMPLE 1 Graph $\{x \mid x \le 3\}$.

SOLUTION We want to graph all the real numbers less than or equal to 3—that is, all the real numbers below 3 and including 3. We label 0 on the number line for reference, as well as 3 since the latter is what we call the endpoint. The graph is shown here:

We use a right bracket at 3 to show that 3 is part of the solution set.

NOTE
In this book we will refer to real numbers as being on the real number line. Actually, real numbers are *not* on the line; only the points representing them are on the line. We can save some writing, however, if we simply refer to real numbers as being on the number line.

PRACTICE PROBLEMS

1. Graph $\{x \mid x \le 2\}$.

Answer
1. See Solutions Section.

Note You may have come from an algebra class in which open and closed circles were used at the endpoints of number line graphs. If so, you would show the graph for Example 1 this way, using a closed circle at 3 to show that 3 is part of the graph.

The two number line graphs are equivalent; they both show all the real numbers that are less than or equal to 3. ∎

2. Graph $\{x \mid x < 2\}$.

EXAMPLE 2   Graph $\{x \mid x < 3\}$.

SOLUTION   The graph is identical to the graph in Example 1 except at the endpoint 3. In this case we use a parenthesis that opens to the left to show that 3 is not part of the graph.

Note For those of you who are used to open/closed circles at the endpoints of number line graphs, here is the equivalent graph in that format.

∎

The following table further clarifies the relationship between number line graphs that use parentheses and brackets at the endpoints and those that use open and closed circles at the endpoints.

Inequality notation	Graph using parentheses/brackets.	Graph using open and closed circles.
$x < 2$		
$x \leq 2$		
$x \geq -3$		
$x > -3$		

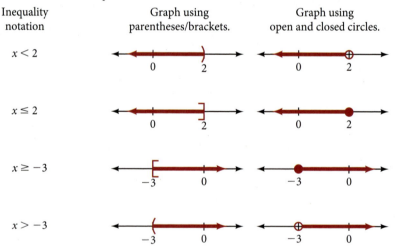

In this book, we will use the parentheses/brackets method of graphing inequalities because that method is better suited to the type of problems we will work in intermediate algebra.

Previously, we defined the *union* of two sets A and B to be the set of all elements that are in either A or B. Remember, the word *or* is the key word in the definition. The *intersection* of two sets A and B is the set of all elements contained in both A and B, the key word here being *and*. We can put the words *and* and *or* together with our methods of graphing inequalities to graph some *compound inequalities*.

Answer
2. See Solutions Section.

EXAMPLE 3   Graph {x | x ≤ −2 or x > 3}.

SOLUTION   The two inequalities connected by the word *or* are referred to as a *compound inequality.* We begin by graphing each inequality separately.

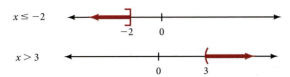

Because the two inequalities are connected by the word *or,* we graph their union; that is, we graph all points on either graph.

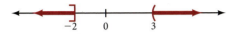

EXAMPLE 4   Graph {x | x > −1 and x < 2}.

SOLUTION   We first graph each inequality separately.

Because the two inequalities are connected by the word *and,* we graph their intersection—the part they have in common.

Note Sometimes compound inequalities that use the word *and* as the connecting word can be written in a shorter form. For example, the compound inequality −3 ≤ x and x ≤ 4 can be written −3 ≤ x ≤ 4. The word *and* does not appear when an inequality is written in this form. It is implied. Inequalities of the form −3 ≤ x ≤ 4 are called *continued inequalities.* This new notation is useful because it takes fewer symbols to write it. The graph of −3 ≤ x ≤ 4 is

EXAMPLE 5   Graph {x | 1 ≤ x < 2}.

SOLUTION   The word *and* is implied in the continued inequality 1 ≤ x < 2; that is, the continued inequality 1 ≤ x < 2 is equivalent to 1 ≤ x and x < 2. Therefore, we graph all the numbers between 1 and 2 on the number line, including 1 but not including 2.

The table that follows shows the connection between number line graphs for a variety of continued inequalities. Again, we have included the graphs with open and closed circles for those of you who have used this type of graph previously. Remember, however, that in this book we will be using the parentheses/brackets method of graphing.

3. Graph {x | x ≤ −1 or x > 2}.

NOTE
It is not absolutely necessary to show these first two graphs. It is simply helpful to do so. As you get more practice at this type of graphing, you can easily omit them.

4. Graph {x | x > −2 and x < 3}.

5. Graph {x | −2 < x ≤ 1}.

Answers
3–5. See Solutions Section.

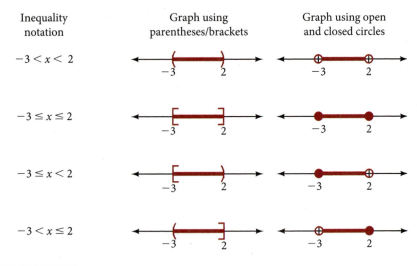

Inequality notation	Graph using parentheses/brackets	Graph using open and closed circles
$-3 < x < 2$		
$-3 \leq x \leq 2$		
$-3 \leq x < 2$		
$-3 < x \leq 2$		

6. Graph $\{x \mid x \leq -1 \text{ or } 2 \leq x \leq 4\}$.

EXAMPLE 6 Graph $\{x \mid x < -2 \text{ or } 2 < x < 6\}$.

SOLUTION Here we have a combination of compound and continued inequalities. We want to graph all real numbers that are either less than -2 or between 2 and 6.

B Translating Inequalities

In addition to the phrases that translate directly into inequality statements, we have the following translations:

In Words	In Symbols
x is at least 40.	$x \geq 40$
x is at most 30.	$x \leq 30$
x is no more than 20.	$x \leq 20$
x is no less than 10.	$x \geq 10$
x is between 4 and 5.	$4 < x < 5$

In the last case, we can include the endpoints 4 and 5 by using "x is between 4 and 5, inclusive," which translates to $4 \leq x \leq 5$.

7. Repeat Example 7 if you work more than 10 hours but less than 20 hours each week.

EXAMPLE 7 Suppose you have a part-time job that requires that you work at least 10 hours, but no more than 20 hours, each week. Use the letter t to write an inequality that shows the number of hours you work per week.

SOLUTION If t is at least 10 but no more than 20, then $10 \leq t$ and $t \leq 20$, or equivalently, $10 \leq t \leq 20$. Note that the word *but,* as used here, has the same meaning as the word *and.*

8. Suppose your weight ranges from a low of 125 pounds to a high of 131 pounds over a 6-month period. Write an inequality using W that gives this range of weight.

EXAMPLE 8 If the highest temperature on Tuesday was 76°F and the lowest temperature was 55°F, write an inequality using the letter x that gives the range of temperatures on Tuesday.

SOLUTION Since the smallest value of x is 55 and the largest value of x is 76, then $55 \leq x \leq 76$. We could say that the temperature on Tuesday was between 55°F and 76°F, inclusive.

Answers
6. See Solutions Section.
7. $10 < t < 20$
8. $125 \leq W \leq 131$

Problem Set 1.4

Moving Toward Success

"No person was ever honored for what he received. Honor has been the reward for what he gave."

—Calvin Coolidge, 1872–1933, 30th President of the United States

1. How can utilizing a study partner or group benefit you in this class?

2. How would a study group provide emotional support during this class?

A Graph the following on a real number line. [Examples 1–4]

1. $\{x \mid x < 1\}$

2. $\{x \mid x > -2\}$

3. $\{x \mid x \leq 1\}$

4. $\{x \mid x \geq -2\}$

5. $\{x \mid x \geq 4\}$

6. $\{x \mid x \leq -3\}$

7. $\{x \mid x > 4\}$

8. $\{x \mid x < -3\}$

9. $\{x \mid x > 0\}$

10. $\{x \mid x < 0\}$

11. $\{x \mid 3 \geq x\}$

12. $\{x \mid 4 \leq x\}$

13. $\{x \mid x \leq -3 \text{ or } x \geq 1\}$

14. $\{x \mid x < 1 \text{ or } x > 4\}$

15. $\{x \mid -3 \leq x \text{ and } x \leq 1\}$

16. $\{x \mid 1 < x \text{ and } x < 4\}$

17. $\{x \mid -3 < x \text{ and } x < 1\}$

18. $\{x \mid 1 \leq x \text{ and } x \leq 4\}$

19. $\{x \mid x < -1 \text{ or } x \geq 3\}$

20. $\{x \mid x < 0 \text{ or } x \geq 3\}$

21. $\{x \mid x \leq -1 \text{ and } x \geq 3\}$

22. $\{x \mid x \leq 0 \text{ and } x \geq 3\}$

23. $\{x \mid x > -4 \text{ and } x < 2\}$

24. $\{x \mid x > -3 \text{ and } x < 0\}$

A Graph the following on a real number line. [Examples 4–5]

25. $\{x \mid -1 \leq x \leq 2\}$

26. $\{x \mid -2 \leq x \leq 1\}$

27. $\{x \mid -1 < x < 2\}$

28. $\{x \mid -2 < x < 1\}$

29. $\{x \mid -4 < x \leq 1\}$

30. $\{x \mid -1 < x \leq 5\}$

A Graph each of the following. [Example 6]

31. $\{x \mid x < -3 \text{ or } 2 < x < 4\}$

32. $\{x \mid -4 \leq x \leq -2 \text{ or } x \geq 3\}$

33. $\{x \mid x \leq -5 \text{ or } 0 \leq x \leq 3\}$

34. $\{x \mid -3 < x < 0 \text{ or } x > 5\}$ **35.** $\{x \mid -5 < x < -2 \text{ or } 2 < x < 5\}$ **36.** $\{x \mid -3 \leq x \leq -1 \text{ or } 1 \leq x \leq 3\}$

B Translate each of the following phrases into an equivalent inequality. [Examples 7–8]

37. x is at least 5.   **38.** x is at least -2.   **39.** x is no more than -3.   **40.** x is no more than 8.

41. x is at most 4.   **42.** x is at most -5.   **43.** x is between -4 and 4.   **44.** x is between -3 and 3.

45. x is between -4 and 4, inclusive.   **46.** x is between -3 and 3, inclusive.

47. Name two numbers that are 5 units from 2 on the number line.   **48.** Name two numbers that are 6 units from -3 on the number line.

49. Write an inequality that gives all numbers that are less than 5 units from 8 on the number line.   **50.** Write an inequality that gives all numbers that are less than 3 units from 5 on the number line.

51. Write an inequality that gives all numbers that are more than 5 units from 8 on the number line.   **52.** Write an inequality that gives all numbers that are more than 3 units from 5 on the number line.

53. Engine Temperature The temperature gauge on a car keeps track of the temperature of the water that cools the engine. The temperature gauge shown here registers temperatures from 50°F to 270°F. If F is the temperature of the water in the car engine, write an inequality that shows all the temperatures that can register on the temperature gauge below.

54. Triangle Inequality The *triangle inequality* is a property that is true for all triangles. It states that in any triangle, the sum of the lengths of any two sides is always greater than the length of the third side. Below is a triangle with sides of length $x, y,$ and z. Use the triangle inequality to write three inequalities using $x, y,$ and z.

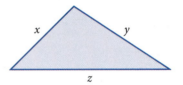

55. Temperature The record temperatures in the state of Alaska are a low of $-80°$F and a high of $100°$F. Graph an inequality that describes this information.

56. Temperature The record temperatures in the state of Colorado are a low of $-61°$F and a high of $114°$F. Graph an inequality that describes this information.

Arithmetic with Real Numbers

OBJECTIVES

A Find the opposite, the reciprocal, and the absolute value of a real number.

B Add and subtract real numbers.

C Multiply and divide real numbers.

D Solve arithmetic problems that contain fractions.

TICKET TO SUCCESS

Keep these questions in mind as you read through the section. Then respond in your own words and in complete sentences.

1. How do you add two real numbers with opposite signs?
2. What is a reciprocal?
3. Explain in words how to divide real numbers.
4. What is a least common denominator, or LCD?

Ilja Masik/Shutterstock.com

The temperature at the airport is 70°F. A plane takes off and reaches its cruising altitude of 28,000 feet, where the temperature is −40°F. Find the difference in the temperatures at takeoff and at cruising altitude.

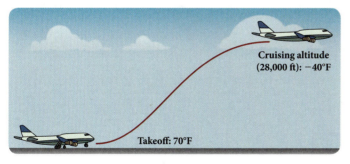

Cruising altitude
(28,000 ft): −40°F

Takeoff: 70°F

We know intuitively that the difference in temperature is 110°F. If we write this problem using symbols, we have

$$70 - (-40) = 110$$

In this section, we review the rules for arithmetic with real numbers, which will include problems such as this one. Before we review these rules, let's review some basic definitions and vocabulary.

A Opposites, Reciprocals, and Absolute Value

The negative sign in front of a number can be read in a variety of different ways. It can be read as "negative" or "the opposite of." We say −4 is the opposite of 4, or negative 4. The one we use will depend on the situation. For instance, the expression −(−3) is best read "the opposite of negative 3." Because the opposite of −3 is 3, we have −(−3) = 3. In general, if a is any positive real number, then

$$-(-a) = a \qquad \text{The opposite of a negative is a positive.}$$

> **Definition**
>
> Any two real numbers the same distance from 0, but in opposite directions from 0 on the number line, are called **opposites,** or additive inverses.

PRACTICE PROBLEMS

1. Give the opposite of each of the following numbers.
   a. 5
   b. $\frac{1}{4}$
   c. −3
   d. $-\sqrt{5}$

EXAMPLE 1   The numbers −3 and 3 are opposites. So are π and $-\pi$, $\frac{3}{4}$ and $-\frac{3}{4}$, and $\sqrt{2}$ and $-\sqrt{2}$.   ■

In a previous math class, you worked problems involving the multiplication of fractions. This idea is useful in understanding another basic definition: the *reciprocal* of a number.

> **Definition**
>
> Any two real numbers whose product is 1 are called **reciprocals,** or multiplicative inverses.

2. Give the reciprocal of 5.

EXAMPLE 2   Give the reciprocal of 3.

SOLUTION

Number	Reciprocal	
3	$\frac{1}{3}$	Because $3 \cdot \frac{1}{3} = \frac{3}{1} \cdot \frac{1}{3} = \frac{3}{3} = 1$ ■

3. Give the reciprocal of $\frac{1}{4}$.

EXAMPLE 3   Give the reciprocal of $\frac{1}{6}$.

SOLUTION

Number	Reciprocal	
$\frac{1}{6}$	6	Because $\frac{1}{6} \cdot 6 = \frac{1}{6} \cdot \frac{6}{1} = \frac{6}{6} = 1$ ■

4. Give the reciprocal of $\frac{3}{4}$.

EXAMPLE 4   Give the reciprocal of $\frac{4}{5}$.

SOLUTION

Number	Reciprocal	
$\frac{4}{5}$	$\frac{5}{4}$	Because $\frac{4}{5} \cdot \frac{5}{4} = \frac{20}{20} = 1$ ■

5. Give the reciprocal of x if $x \neq 0$.

EXAMPLE 5   Give the reciprocal of a.

SOLUTION

Number	Reciprocal	
a	$\frac{1}{a}$	Because $a \cdot \frac{1}{a} = \frac{a}{1} \cdot \frac{1}{a} = \frac{a}{a} = 1$ ($a \neq 0$) ■

Answers
1. a. −5 b. $-\frac{1}{4}$ c. 3 d. $\sqrt{5}$
2. $\frac{1}{5}$
3. 4
4. $\frac{4}{3}$
5. $\frac{1}{x}$

Although we will not develop multiplication with negative numbers until later in this section, you should know that the reciprocal of a negative number is also a negative number. For example, the reciprocal of −5 is $-\frac{1}{5}$.

The third basic definition we will cover is that of *absolute value*.

> **Definition**
>
> The **absolute value** of a number (also called its magnitude) is the distance the number is from 0 on the number line. If x represents a real number, then the absolute value of x is written $|x|$.

This definition of absolute value is geometric in form since it defines absolute value in terms of the number line. Here is an alternative definition of absolute value that is algebraic in form since it involves only symbols.

> **Alternate Definition**
>
> If x represents a real number, then the **absolute value** of x is written $|x|$, and is given by
>
> $$|x| = \begin{cases} x & \text{if } x \geq 0 \\ -x & \text{if } x < 0 \end{cases}$$

If the original number is positive or 0, then its absolute value is the number itself. If the number is negative, its absolute value is its opposite (which must be positive).

EXAMPLE 6   Write $|5|$ without absolute value symbols.

SOLUTION $|5| = 5$ ■

EXAMPLE 7   Write $|-2|$ without absolute value symbols.

SOLUTION $|-2| = 2$ ■

EXAMPLE 8   Write $\left|-\frac{1}{2}\right|$ without absolute value symbols.

SOLUTION $\left|-\frac{1}{2}\right| = \frac{1}{2}$ ■

EXAMPLE 9   Write $-|-3|$ without absolute value symbols.

SOLUTION $-|-3| = -3$ ■

EXAMPLE 10   Write $-|5|$ without absolute value symbols.

SOLUTION $-|5| = -5$ ■

EXAMPLE 11   Write $-|-\sqrt{2}|$ without absolute value symbols.

SOLUTION $-|-\sqrt{2}| = -\sqrt{2}$ ■

Write the following without absolute value symbols.

6. $|7|$

7. $|-4|$

8. $\left|-\frac{2}{3}\right|$

9. $-|-6|$

10. $-|7|$

11. $-|-\sqrt{3}|$

Answers

6. 7

7. 4

8. $\frac{2}{3}$

9. -6

10. -7

11. $-\sqrt{3}$

B Adding and Subtracting Real Numbers

Now let's review the rules for arithmetic with real numbers and the justification for those rules. Reviewing these rules will only strengthen your mathematics foundation and help you tackle more complicated concepts later in the book. We can justify the rules for addition of real numbers geometrically by use of the real number line. Consider the sum of -5 and 3:

$$-5 + 3$$

We can interpret this expression as meaning "start at the origin and move 5 units in the negative direction and then 3 units in the positive direction." With the aid of a number line we can visualize the process.

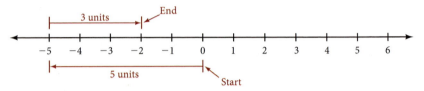

Because the process ends at -2, we say the sum of -5 and 3 is -2:

$$-5 + 3 = -2$$

We can use the real number line in this way to add any combination of positive and negative numbers.

The sum of -4 and -2, $-4 + (-2)$, can be interpreted as starting at the origin, moving 4 units in the negative direction, and then 2 more units in the negative direction:

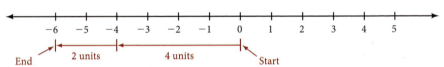

Because the process ends at -6, we say the sum of -4 and -2 is -6:

$$-4 + (-2) = -6$$

We can eliminate actually drawing a number line by simply visualizing it mentally. The following example gives the results of all possible sums of positive and negative 5 and 7.

EXAMPLE 12 Add all combinations of positive and negative 5 and 7.

SOLUTION $5 + 7 = 12$

$-5 + 7 = 2$

$5 + (-7) = -2$

$-5 + (-7) = -12$ ∎

Looking closely at the relationships in Example 12 (and trying other similar examples if necessary), we can arrive at the following rule for adding two real numbers.

> **Rule Adding Two Real Numbers**
>
> With the *same* sign:
>
> **Step 1** Add their absolute values.
>
> **Step 2** Attach their common sign. If both numbers are positive, their sum is positive; if both numbers are negative, their sum is negative.
>
> With *opposite* signs:
>
> **Step 1** Subtract the smaller absolute value from the larger.
>
> **Step 2** Attach the sign of the number whose absolute value is larger.

NOTE
We are showing addition of real numbers on the number line to justify the rule we will write for addition of positive and negative numbers. You may want to skip ahead and read the rule on the next page first, and then come back and read through this discussion again. The discussion here is the "why" behind the rule.

12. Add.

$3 + 5$

$-3 + 5$

$3 + (-5)$

$-3 + (-5)$

NOTE
This rule is the most important rule we have had so far. It is very important that you use it exactly the way it is written. Your goal is to become fast and accurate at adding positive and negative numbers. When you have finished reading this section and working the problems in the problem set, you should have attained that goal.

Answers
12. 8, 2, −2, −8

In order to have as few rules as possible, we will not attempt to list new rules for the *difference* of two real numbers. We will instead define it in terms of addition and apply the rule for addition.

> **Definition**
>
> If a and b are any two real numbers, then the **difference** of a and b, written $a - b$, is given by
>
> $$a - b = a + (-b)$$
>
> To subtract b, add the opposite of b.

We define the process of subtracting b from a as being equivalent to adding the opposite of b to a. In short, we say, "subtraction is addition of the opposite."

EXAMPLE 13 Subtract $5 - 3$.

SOLUTION $5 - 3 = 5 + (-3)$ Subtracting 3 is equivalent to adding −3.
$\qquad\qquad = 2$ ∎

EXAMPLE 14 Subtract $-7 - 6$.

SOLUTION $-7 - 6 = -7 + (-6)$ Subtracting 6 is equivalent to adding −6.
$\qquad\qquad = -13$ ∎

EXAMPLE 15 Subtract $9 - (-2)$.

SOLUTION $9 - (-2) = 9 + 2$ Subtracting −2 is equivalent to adding 2.
$\qquad\qquad = 11$ ∎

EXAMPLE 16 Subtract $-6 - (-5)$.

SOLUTION $-6 - (-5) = -6 + 5$ Subtracting −5 is equivalent to adding 5.
$\qquad\qquad = -1$ ∎

EXAMPLE 17 Subtract -3 from -9.

SOLUTION   We must be sure to write the numbers in the correct order. Because we are subtracting -3, the problem looks like this when translated into symbols:

$\qquad -9 - (-3) = -9 + 3$ Change to addition of the opposite.
$\qquad\qquad\quad = -6$ Add. ∎

EXAMPLE 18 Add -4 to the difference of -2 and 5.

SOLUTION   The difference of -2 and 5 is written $-2 - 5$. Adding -4 to that difference gives us

$\qquad (-2 - 5) + (-4) = -7 + (-4)$ Simplify inside parentheses.
$\qquad\qquad\qquad\quad = -11$ Add. ∎

13. Subtract $7 - 4$.

14. Subtract $-6 - 3$.

15. Subtract $8 - (-2)$.

16. Subtract $-4 - (-6)$.

17. Subtract -5 from -3.

18. Add -8 to the difference of -7 and 4.

Answers
13. 3
14. −9
15. 10
16. 2
17. 2
18. −19

C Multiplying and Dividing Real Numbers

Multiplication with whole numbers is simply a shorthand way of writing repeated addition.

For example, $3(-2)$ can be evaluated as follows:

$$3(-2) = -2 + (-2) + (-2) = -6$$

We can evaluate the product $-3(2)$ in a similar manner if we first apply the commutative property of multiplication, covered later in this chapter:

$$-3(2) = 2(-3) = -3 + (-3) = -6$$

NOTE

This discussion is to show why the rule for multiplication of real numbers is written the way it is. Even if you already know how to multiply positive and negative numbers, it is a good idea to review the "why" that is behind it all.

From these results it seems reasonable to say that the product of a positive and a negative is a negative number.

The last case we must consider is the product of two negative numbers, such as $-3(-2)$. To evaluate this product we will look at the expression $-3[2 + (-2)]$ in two different ways. First, since $2 + (-2) = 0$, we know the expression $-3[2 + (-2)]$ is equal to 0. On the other hand, we can apply the distributive property, which we will also work with later in the chapter, to get

$$-3[2 + (-2)] = -3(2) + (-3)(-2) = -6 + ?$$

Because we know the expression is equal to 0, it must be true that our ? is 6 because 6 is the only number we can add to -6 to get 0. Therefore, we have

$$-3(-2) = 6$$

Here is a summary of what we have so far:

If the original numbers have		the answer is
the same sign	$3(2) = 6$	positive
different signs	$3(-2) = -6$	negative
different signs	$-3(2) = -6$	negative
the same sign	$-3(-2) = 6$	positive

Rule Multiplying Two Real Numbers

Step 1 Multiply their absolute values.

Step 2 If the two numbers have the *same* sign, the product is positive. If the two numbers have *opposite* signs, the product is negative.

19. Multiply.
   a. $4(5)$
   b. $4(-5)$
   c. $-4(5)$
   d. $-4(-5)$

EXAMPLE 19   Multiply all combinations of positive and negative 7 and 3.

SOLUTION $7(3) = 21$

$$7(-3) = -21$$

$$-7(3) = -21$$

$$-7(-3) = 21$$

∎

Now in the case of division, dividing a by b is equivalent to multiplying a by the reciprocal of b. In short, we say, "division is multiplication by the reciprocal."

Because division is defined in terms of multiplication, the same rules hold for assigning the correct sign to a quotient as for assigning the correct sign to a product; that is, *the quotient of two numbers with like signs is positive, while the quotient of two numbers with unlike signs is negative.*

Answers

19. a. 20   **b.** -20 **c.** -20 **d.** 20

Definition

If a and b are any two real numbers, where $b \neq 0$, then the **quotient** of a and b, written $\frac{a}{b}$ is given by

$$\frac{a}{b} = a \cdot \left(\frac{1}{b}\right)$$

EXAMPLE 20   Divide each of the following:

a. $\frac{6}{3}$ **b.** $\frac{6}{-3}$ **c.** $\frac{-6}{3}$ **d.** $\frac{-6}{-3}$

SOLUTION

a. $\quad \frac{6}{3} = 6 \cdot \left(\frac{1}{3}\right) = 2$

b. $\quad \frac{6}{-3} = 6 \cdot \left(-\frac{1}{3}\right) = -2$

c. $\quad \frac{-6}{3} = -6 \cdot \left(\frac{1}{3}\right) = -2$

d. $\quad \frac{-6}{-3} = -6 \cdot \left(-\frac{1}{3}\right) = 2$ ∎

The second step in the preceding examples is written only to show that each quotient can be written as a product. It is not actually necessary to show this step when working problems.

Division with the Number 0

For every division problem an associated multiplication problem involving the same numbers exists. For example, the following two problems say the same thing about the numbers 2, 3, and 6:

Division	*Multiplication*
$\frac{6}{3} = 2$	$6 = 2(3)$

We can use this relationship between division and multiplication to clarify division involving the number 0.

First of all, dividing 0 by a number other than 0 is allowed and always results in 0. To see this, consider dividing 0 by 5. We know the answer is 0 because of the relationship between multiplication and division. This is how we write it:

$$\frac{0}{5} = 0 \qquad \text{because} \qquad 0 = 0(5)$$

On the other hand, dividing a nonzero number by 0 is not allowed in the real numbers. Suppose we were attempting to divide 5 by 0. We don't know whether there is an answer to this problem, but if there is, let's say the answer is a number that we can represent with the letter n. If 5 divided by 0 is a number n, then

$$\frac{5}{0} = n \qquad \text{and} \qquad 5 = n(0)$$

But this is impossible because no matter what number n is, when we multiply it by 0 the answer must be 0. It can never be 5. In algebra, we say expressions like $\frac{5}{0}$ are undefined because there is no answer to them; that is, division by 0 is not allowed in the real numbers.

20. Divide.

a. $\frac{12}{4}$ **b.** $\frac{12}{-4}$ **c.** $\frac{-12}{4}$ **d.** $\frac{-12}{-4}$

NOTE
These examples indicate that if a and b are positive real numbers then

$$\frac{-a}{b} = \frac{a}{-b} = -\frac{a}{b}$$

and

$$\frac{-a}{-b} = \frac{a}{b}$$

Answers
20. a. 3 **b.** −3 **c.** −3 **d.** 3

D   Review of Fractions

Before we go further with our study of real numbers, we need to review arithmetic with fractions. Recall that for the fraction $\frac{a}{b}$, a is called the *numerator* and b is called the *denominator*. To multiply two fractions, we simply multiply numerators and multiply denominators.

21. Multiply $\frac{3}{7} \cdot \frac{2}{5}$.

EXAMPLE 21   Multiply $\frac{3}{5} \cdot \frac{7}{8}$.

SOLUTION $\frac{3}{5} \cdot \frac{7}{8} = \frac{3 \cdot 7}{5 \cdot 8} = \frac{21}{40}$ ■

Next we review division with fractions.

Divide and reduce to lowest terms.

22. $\frac{3}{5} \div \frac{6}{7}$

EXAMPLE 22   Divide and reduce to lowest terms: $\frac{3}{4} \div \frac{6}{11}$.

SOLUTION $\frac{3}{4} \div \frac{6}{11} = \frac{3}{4} \cdot \frac{11}{6}$ Definition of division

$= \frac{33}{24}$ Multiply numerators, and multiply denominators.

$= \frac{11}{8}$ Divide numerator and denominator by 3.   ■

To add fractions, each fraction must have the same denominator. Recall the definition for the least common denominator shown here:

> **Definition**
>
> The **least common denominator** (LCD) for a set of denominators is the smallest number divisible by *all* the denominators.

The first step in adding fractions is to find a common denominator for all the denominators. We then rewrite each fraction (if necessary) as an equivalent fraction with the common denominator. Finally, we add the numerators and reduce to lowest terms if necessary.

23. Add $\frac{3}{14} + \frac{7}{30}$.

EXAMPLE 23   Add $\frac{5}{12} + \frac{7}{18}$.

SOLUTION   The least common denominator for the denominators 12 and 18 must be the smallest number divisible by both 12 and 18.

$$\left. \begin{array}{l} 12 = 2 \cdot 2 \cdot 3 \\ 18 = 2 \cdot 3 \cdot 3 \end{array} \right\} \text{LCD} = 2 \cdot 2 \cdot 3 \cdot 3 = 36$$

12 divides the LCD

18 divides the LCD

Next, we rewrite our original fractions as equivalent fractions with denominators of 36. To do so, we multiply each original fraction by an appropriate form of the number 1:

$$\frac{5}{12} + \frac{7}{18} = \frac{5}{12} \cdot \frac{\mathbf{3}}{\mathbf{3}} + \frac{7}{18} \cdot \frac{\mathbf{2}}{\mathbf{2}} = \frac{15}{36} + \frac{14}{36}$$

Finally, we add numerators and place the result over the common denominator, 36.

$$\frac{15}{36} + \frac{14}{36} = \frac{15 + 14}{36} = \frac{29}{36}$$ ■

Answers

21. $\frac{6}{35}$

22. $\frac{7}{10}$

23. $\frac{47}{105}$

Problem Set 1.5

Moving Toward Success

"Do you want to know who you are? Don't ask. Act! Action will delineate and define you."

—Thomas Jefferson, 1743–1826, 3rd President of the United States

1. Name two resources, other than the book and your instructor, that will help you get the grade you want in this course.

2. When working problems, is it important to look at the answers at the back of the book? Why?

A Complete the following table. [Examples 1–5]

	Number	Opposite	Reciprocal
1.	4		
2.	−3		
3.	$-\frac{1}{2}$		
4.	$\frac{5}{6}$		
5.		−5	
6.		7	
7.		$-\frac{3}{8}$	
8.		$\frac{1}{2}$	
9.			−6
10.			−3
11.			$\frac{1}{3}$
12.			$-\frac{1}{4}$

13. Name two numbers that are their own reciprocals.

14. Give the number that has no reciprocal.

15. Name the number that is its own opposite.

16. The reciprocal of a negative number is negative—true or false?

A Write each of the following without absolute value symbols. [Examples 6–11]

17. $|-2|$

18. $|-7|$

19. $\left|-\frac{3}{4}\right|$

20. $\left|\frac{5}{6}\right|$

21. $|\pi|$

22. $|-\sqrt{2}|$

23. $-|4|$

24. $-|5|$

25. $-|-2|$

26. $-|-10|$

27. $-\left|-\frac{3}{4}\right|$

28. $-\left|\frac{7}{8}\right|$

B Find each of the following sums. [Example 12]

29. $6 + (-2)$

30. $11 + (-5)$

31. $-6 + 2$

32. $-11 + 5$

33. $19 + (-7)$

34. $15 + (-3)$

35. $-17 + 4$

36. $-25 + 6$

B **D** Find each of the following differences. [Examples 13–18]

37. $-7 - 3$

38. $-6 - 9$

39. $-7 - (-3)$

40. $-6 - (-9)$

41. $\dfrac{3}{4} - \left(-\dfrac{5}{6}\right)$

42. $\dfrac{2}{3} - \left(-\dfrac{7}{5}\right)$

43. $\dfrac{11}{42} - \dfrac{17}{30}$

44. $\dfrac{13}{70} - \dfrac{19}{42}$

45. Subtract 5 from -3.

46. Subtract -3 from 5.

47. Find the difference of -4 and 8.

48. Find the difference of 8 and -4.

49. Subtract $4x$ from $-3x$.

50. Subtract $-5x$ from $7x$.

51. What number do you subtract from 5 to get -8?

52. What number do you subtract from -3 to get 9?

53. Add -7 to the difference of 2 and 9.

54. Add -3 to the difference of 9 and 2.

55. Subtract $3a$ from the sum of $8a$ and a.

56. Subtract $-3a$ from the sum of $3a$ and $5a$.

B **C** Complete each of the following tables.

57.

a	b	Sum $a + b$	Difference $a - b$	Product ab	Quotient $\dfrac{a}{b}$
3	12				
-3	12				
3	-12				
-3	-12				

58.

a	b	Sum $a + b$	Difference $a - b$	Product ab	Quotient $\dfrac{a}{b}$
8	2				
-8	2				
8	-2				
-8	-2				

C Find the following products. [Example 19]

59. $3(-5)$

60. $-3(5)$

61. $-3(-5)$

62. $4(-6)$

63. $2(-3)(4)$

64. $-2(3)(-4)$

65. $-2(5x)$

66. $-5(4x)$

67. $-\dfrac{1}{3}(-3x)$

68. $-\dfrac{1}{6}(-6x)$

69. $-\dfrac{2}{3}\left(-\dfrac{3}{2}y\right)$

70. $-\dfrac{2}{5}\left(-\dfrac{5}{2}y\right)$

C **D** Multiply the following. [Examples 19, 21]

71. $\dfrac{3}{5} \cdot \dfrac{7}{8}$

72. $\dfrac{6}{7} \cdot \dfrac{9}{5}$

73. $\dfrac{1}{3} \cdot 6$

74. $\dfrac{1}{4} \cdot 8$

75. $\left(\dfrac{2}{3}\right)^3$

76. $\left(\dfrac{4}{5}\right)^2$

77. $\left(\dfrac{1}{10}\right)^4$

78. $\left(\dfrac{1}{2}\right)^5$

79. $\dfrac{3}{5} \cdot \dfrac{4}{7} \cdot \dfrac{6}{11}$

80. $\dfrac{4}{5} \cdot \dfrac{6}{7} \cdot \dfrac{3}{11}$

81. $\dfrac{4}{3} \cdot \dfrac{3}{4}$

82. $\dfrac{5}{8} \cdot \dfrac{8}{5}$

C **D** Use the definition of division to write each division problem as a multiplication problem, then simplify. [Examples 20, 23]

83. $\dfrac{8}{-4}$

84. $\dfrac{-8}{4}$

85. $\dfrac{4}{0}$

86. $\dfrac{-7}{0}$

87. $\dfrac{0}{-3}$

88. $\dfrac{0}{5}$

89. $-\dfrac{3}{4} \div \dfrac{9}{8}$

90. $-\dfrac{2}{3} \div \dfrac{4}{9}$

91. $-8 \div \left(-\dfrac{1}{4}\right)$

92. $-12 \div \left(-\dfrac{2}{3}\right)$

93. $-40 \div \left(-\dfrac{5}{8}\right)$

94. $-30 \div \left(-\dfrac{5}{6}\right)$

95. $\dfrac{4}{9} \div (-8)$

96. $\dfrac{3}{7} \div (-6)$

D Add the following fractions. [Example 24]

97. $\dfrac{2}{5} + \dfrac{1}{15}$

98. $\dfrac{5}{8} + \dfrac{1}{4}$

99. $\dfrac{17}{30} + \dfrac{11}{42}$

100. $\dfrac{19}{42} + \dfrac{13}{70}$

101. $\dfrac{9}{48} + \dfrac{3}{54}$

102. $\dfrac{6}{28} + \dfrac{5}{42}$

103. $\dfrac{25}{84} + \dfrac{41}{90}$

104. $\dfrac{23}{70} + \dfrac{29}{84}$

105. $\dfrac{5}{72} + \dfrac{11}{84}$

106. $\dfrac{4}{90} + \dfrac{15}{180}$

107. $\dfrac{4}{39} + \dfrac{7}{65}$

108. $\dfrac{4}{75} + \dfrac{6}{105}$

109. $\dfrac{7}{54} + \dfrac{7}{60}$

110. $\dfrac{3}{112} + \dfrac{5}{56}$

The problems below are problems you will see later in the book. Simplify each expression.

111. $\dfrac{3}{14} + \dfrac{7}{30}$

112. $\dfrac{3}{10} + \dfrac{11}{42}$

113. $32\left(\dfrac{3}{4}\right) - 16\left(\dfrac{3}{4}\right)^2$

114. $32\left(\dfrac{3}{2}\right) - 16\left(\dfrac{3}{2}\right)^2$

115. $-2(-14) + 3(-4) - 1(-10)$

116. $-4(0)(-2) - (-1)(1)(1) - 1(2)(3)$

117. $1(0)(1) + 3(1)(4) + (-2)(2)(-1)$

118. $1[0 - (-1)] - 3(2 - 4) + (-2)(-2 - 0)$

119. $-3(-1 - 1) + 4(-2 + 2) - 5[2 - (-2)]$

120. $3(-2)^2 + 2(-2) - 1$

121. $4(-1)^2 + 3(-1) - 2$

122. $2(-2)^3 - 3(-2)^2 + 4(-2) - 8$

123. $5 \cdot 2^3 - 3 \cdot 2^2 + 4 \cdot 2 - 5$

124. $\dfrac{0 - 4}{0 - 2}$

125. $\dfrac{0 + 6}{0 - 3}$

126. $\dfrac{-4 - 4}{-4 - 2}$

127. $\dfrac{6 + 6}{6 - 3}$

128. $\dfrac{-6 + 6}{-6 - 3}$

129. $\dfrac{4 - 4}{4 - 2}$

130. $\dfrac{2 - 4}{2 - 2}$

131. $\dfrac{3 + 6}{3 - 3}$

132. $\dfrac{3 - (-1)}{-3 - 3}$

133. $\dfrac{-1 - 3}{3 - (-3)}$

Properties of Real Numbers

1.6

OBJECTIVES

A Recognize and apply the properties of real numbers.

B Apply the rule for order of operations.

C Simplify algebraic expressions.

D Find the value of an algebraic expression.

TICKET TO SUCCESS

Keep these questions in mind as you read through the section. Then respond in your own words and in complete sentences.

1. Explain the associative properties of addition and multiplication.
2. How would you apply the distributive property to the expression $a(b + c)$?
3. Give an example of combining similar terms.
4. What does it mean to simplify an expression?

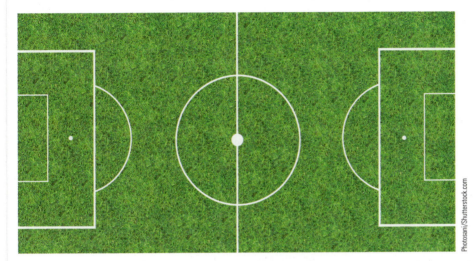

Photosani/Shutterstock.com

The area of the field shown above can be found in two ways: We can multiply its length a by its width $b + c$, or we can find the area of each half and add those to find the total area.

Area of large rectangle: $a(b + c)$

Sum of the areas of two smaller rectangles: $ab + ac$

Because the area of the large rectangle is the sum of the areas of the two smaller rectangles, we can write:

$$a(b + c) = ab + ac$$

This equation is called the *distributive property*. It is one of the properties we will be discussing in this section.

A Properties of Real Numbers

We know that adding 3 and 7 gives the same answer as adding 7 and 3. The order of two numbers in an addition problem can be changed without changing the result. This fact about numbers and addition is called the *commutative property of addition*.

For all the properties listed in this section, a, b, and c represent real numbers.

> ### Commutative Property of Addition
> *In symbols:* $a + b = b + a$
> *In words:* The *order* of the numbers in a sum does not affect the result.
>
> ### Commutative Property of Multiplication
> *In symbols:* $a \cdot b = b \cdot a$
> *In words:* The *order* of the numbers in a product does not affect the result.

Complete each statement so it is an example of the commutative property.

1. $5 + x = x + $ ____

EXAMPLE 1 The statement $3 + 7 = 7 + 3$ is an example of the commutative property of addition. ∎

2. $7 \cdot 3 = 3 \cdot$ ____

EXAMPLE 2 The statement $3 \cdot x = x \cdot 3$ is an example of the commutative property of multiplication. ∎

Another property of numbers you have used many times has to do with grouping. When adding $3 + 5 + 7$, we can add the 3 and 5 first and then the 7, or we can add the 5 and 7 first and then the 3. Mathematically, it looks like this: $(3 + 5) + 7 = 3 + (5 + 7)$. Operations that behave in this manner are called *associative* operations.

> **NOTE**
> The other two basic operations (subtraction and division) are not commutative. If we change the order in which we are subtracting or dividing two numbers, we change the result.

> ### Associative Property of Addition
> *In symbols:* $a + (b + c) = (a + b) + c$
> *In words:* The *grouping* of the numbers in a sum does not affect the result.
>
> ### Associative Property of Multiplication
> *In symbols:* $a(bc) = (ab)c$
> *In words:* The *grouping* of the numbers in a product does not affect the result

The following examples illustrate how the associative properties can be used to simplify expressions that involve both numbers and variables.

3. Simplify $5 + (7 + y)$.

EXAMPLE 3 Simplify $2 + (3 + y)$ by using the associative property.

SOLUTION
$$2 + (3 + y) = (2 + 3) + y \qquad \text{Associative property}$$
$$= 5 + y \qquad \text{Add.}$$

4. Simplify $3(2x)$.

EXAMPLE 4 Simplify $5(4x)$ by using the associative property.

SOLUTION
$$5(4x) = (5 \cdot 4)x \qquad \text{Associative property}$$
$$= 20x \qquad \text{Multiply.}$$

5. Simplify $\frac{1}{3}(3a)$.

EXAMPLE 5 Simplify $\frac{1}{4}(4a)$ by using the associative property.

SOLUTION
$$\frac{1}{4}(4a) = \left(\frac{1}{4} \cdot 4\right)a \qquad \text{Associative property}$$
$$= 1a \qquad \text{Multiply.}$$
$$= a$$
∎

Answers
1. $5 + x = x + 5$
2. $7 \cdot 3 = 3 \cdot 7$
3. $12 + y$
4. $6x$
5. a

EXAMPLE 6 Simplify $2\left(\frac{1}{2}x\right)$ by using the associative property.

SOLUTION

$$2\left(\frac{1}{2}x\right) = \left(2 \cdot \frac{1}{2}\right)x \qquad \text{Associative property}$$

$$= 1x \qquad \text{Multiply.}$$

$$= x$$

$\blacksquare$

EXAMPLE 7 Simplify $6\left(\frac{1}{3}x\right)$ by using the associative property.

SOLUTION

$$6\left(\frac{1}{3}x\right) = \left(6 \cdot \frac{1}{3}\right)x \qquad \text{Associative property}$$

$$= 2x \qquad \text{Multiply.}$$

Our next property involves both addition and multiplication. It is called the *distributive property* and is stated as follows.

> **Distributive Property**
>
> In symbols: $a(b + c) = ab + ac$
> In words: Multiplication *distributes* over addition.

You will see as we progress through the book that the distributive property is used very frequently in algebra. To see that the distributive property works, compare the following:

$$3(4 + 5) = 3(9) = 27$$

$$3(4) + 3(5) = 12 + 15 = 27$$

In both cases the result is 27. Because the results are the same, the original two expressions must be equal, or $3(4 + 5) = 3(4) + 3(5)$.

EXAMPLE 8 Apply the distributive property to $5(4x + 3)$ and then simplify the result.

SOLUTION

$$5(4x + 3) = 5(4x) + 5(3) \qquad \text{Distributive property}$$

$$= 20x + 15 \qquad \text{Multiply.} \qquad \blacksquare$$

EXAMPLE 9 Apply the distributive property to $6(3x + 2y)$ and then simplify the result.

SOLUTION

$$6(3x + 2y) = 6(3x) + 6(2y) \qquad \text{Distributive property}$$

$$= 18x + 12y \qquad \text{Multiply.}$$

EXAMPLE 10 Apply the distributive property to $\frac{1}{2}(3x + 6)$ and then simplify the result.

SOLUTION

$$\frac{1}{2}(3x + 6) = \frac{1}{2}(3x) + \frac{1}{2}(6) \qquad \text{Distributive property}$$

$$= \frac{3}{2}x + 3 \qquad \text{Multiply.} \qquad \blacksquare$$

6. Simplify $5\left(\frac{1}{5}x\right)$.

7. Simplify $9\left(\frac{2}{3}x\right)$.

Apply the distributive property and then simplify.

8. $7(6x + 8)$

9. $9(7x + 11y)$

10. $\frac{1}{3}(3x + 6)$

Answers
6. x
7. $6x$
8. $42x + 56$
9. $63x + 99y$
10. $x + 2$

Apply the distributive property, then simplify if possible.

11. $4(5y + 2) + 8$

EXAMPLE 11 Apply the distributive property to $2(3y + 4) + 2$ and then simplify the result.

SOLUTION
$$2(3y + 4) + 2 = 2(3y) + 2(4) + 2 \quad \text{Distributive property}$$
$$= 6y + 8 + 2 \quad \text{Multiply.}$$
$$= 6y + 10 \quad \text{Add.} \quad \blacksquare$$

We can combine our knowledge of the distributive property with multiplication of fractions to manipulate expressions involving fractions. Here are some examples that show how we do this.

12. $a\left(2 - \dfrac{1}{a}\right)$

EXAMPLE 12 Apply the distributive property to $a\left(1 + \dfrac{1}{a}\right)$, and then simplify if possible.

SOLUTION $\quad a\left(1 + \dfrac{1}{a}\right) = a \cdot 1 + a \cdot \dfrac{1}{a} = a + 1 \quad \blacksquare$

13. $5\left(\dfrac{1}{5}x + 8\right)$

EXAMPLE 13 Apply the distributive property to $3\left(\dfrac{1}{3}x + 5\right)$, and then simplify if possible.

SOLUTION $\quad 3\left(\dfrac{1}{3}x + 5\right) = 3 \cdot \dfrac{1}{3}x + 3 \cdot 5 = x + 15 \quad \blacksquare$

14. $12\left(\dfrac{2}{3}x + \dfrac{3}{4}y\right)$

EXAMPLE 14 Apply the distributive property to $6\dfrac{1}{3}x + \dfrac{1}{2}y$, and then simplify if possible.

SOLUTION $\quad 6\left(\dfrac{1}{3}x + \dfrac{1}{2}y\right) = 6 \cdot \dfrac{1}{3}x + 6 \cdot \dfrac{1}{2}y = 2x + 3y \quad \blacksquare$

Combining Similar Terms

The distributive property can also be used to combine similar terms. (For now, a *term* is a number, or the product of a number with one or more variables.) Similar terms are terms with the same variable part. The terms $3x$ and $5x$ are similar, as are $2y$, $7y$, and $-3y$, because the variable parts are the same.

Combine similar terms.

15. $4x + 9x$

EXAMPLE 15 Use the distributive property to combine similar terms.

SOLUTION
$$3x + 5x = (3 + 5)x \quad \text{Distributive property}$$
$$= 8x \quad \text{Add.} \quad \blacksquare$$

16. $y + 5y$

EXAMPLE 16 Use the distributive property to combine similar terms.

SOLUTION
$$3y + y = (3 + 1)y \quad \text{Distributive property}$$
$$= 4y \quad \text{Add.} \quad \blacksquare$$

Answers
11. $20y + 16$
12. $2a - 1$
13. $x + 40$
14. $8x + 9y$
15. $13x$
16. $6y$

Additive Identity Property

There exists a unique number 0 such that
$$a + 0 = a \quad \text{and} \quad 0 + a = a$$

Multiplicative Identity Property

There exists a unique number 1 such that
$$a(1) = a \quad \text{and} \quad 1(a) = a$$

Additive Inverse Property

In symbols: $a + (-a) = 0$

In words: Opposites add to 0.

Multiplicative Inverse Property

For every real number a, except 0, there exists a unique real number $\dfrac{1}{a}$ such that

In symbols: $a\left(\dfrac{1}{a}\right) = 1$

In words: Reciprocals multiply to 1.

EXAMPLE 17   State the property of real numbers that justifies each statement.

   a. $7(1) = 7$ Multiplicative identity property

   b. $4 + (-4) = 0$ Additive inverse property

   c. $6\left(\dfrac{1}{6}\right) = 1$ Multiplicative inverse property

   d. $(5 + 0) + 2 = 5 + 2$ Additive identity property   ∎

B   Order of Operations

In the examples that follow, we find a combination of operations. In each case, we use the rule for order of operations.

EXAMPLE 18   Simplify $(-2 - 3)(5 - 9)$.

SOLUTION $(-2 - 3)(5 - 9) = (-5)(-4)$ Simplify inside parentheses.

              $= 20$ Multiply.   ∎

EXAMPLE 19   Simplify $2 - 5(7 - 4) - 6$.

SOLUTION $2 - 5(7 - 4) - 6 = 2 - 5(3) - 6$ Simplify inside parentheses.

                 $= 2 - 15 - 6$ Then, multiply.

                 $= -19$ Finally, subtract, left to right.   ∎

EXAMPLE 20   Simplify $2(4 - 7)^3 + 3(-2 - 3)^2$.

SOLUTION $2(4 - 7)^3 + 3(-2 - 3)^2 = 2(-3)^3 + 3(-5)^2$ Simplify inside parentheses.

                 $= 2(-27) + 3(25)$ Evaluate numbers with exponents.

                 $= -54 + 75$ Multiply.

                 $= 21$ Add.   ∎

---

17. State the property of real numbers that justifies each statement.

   a. $2 + 0 = 2$

   b. $3(1) = 3$

   c. $-5 + 5 = 0$

   d. $\dfrac{3}{4} \cdot \dfrac{4}{3} = 1$

Simplify.

18. $(-6 - 4)(8 - 12)$

19. $3 - 7(9 - 5) - 2$

20. $3(5 - 8)^3 - 2(-1 - 1)^2$

Answers

17. a. Additive identity

    b. Multiplicative identity

    c. Additive inverse

    d. Multiplicative inverse

18. 40

19. −27

20. −89

21. Simplify $10x + 3 + 7x + 12$.

C Simplifying Expressions

We can use the commutative, associative, and distributive properties together to simplify expressions.

EXAMPLE 21 Simplify $7x + 4 + 6x + 3$.

SOLUTION We begin by applying the commutative and associative properties to group similar terms.

$$7x + 4 + 6x + 3 = (7x + 6x) + (4 + 3) \quad \text{Commutative and associative properties}$$

$$= (7 + 6)x + (4 + 3) \quad \text{Distributive property}$$

$$= 13x + 7 \quad \text{Add.} \quad \blacksquare$$

22. Simplify $9 + 5(4y + 8) + 10y$.

EXAMPLE 22 Simplify $4 + 3(2y + 5) + 8y$.

SOLUTION Because our rule for order of operations indicates that we are to multiply before adding, we must distribute the 3 across $2y + 5$ first.

$$4 + 3(2y + 5) + 8y = 4 + 6y + 15 + 8y \quad \text{Distributive property}$$

$$= (6y + 8y) + (4 + 15) \quad \text{Commutative and associative properties}$$

$$= (6 + 8)y + (4 + 15) \quad \text{Distributive property}$$

$$= 14y + 19 \quad \text{Add.} \quad \blacksquare$$

We can combine our knowledge of the properties of multiplication with our definition of division to simplify more expressions involving fractions. Here are two examples:

23. Simplify $2\left(\dfrac{x}{2}\right)$.

EXAMPLE 23 Simplify $6\left(\dfrac{t}{3}\right)$.

SOLUTION
$$6\left(\dfrac{t}{3}\right) = 6\left(\dfrac{1}{3}t\right) \quad \text{Dividing by 3 is the same as multiplying by } \dfrac{1}{3}$$

$$= \left(6 \cdot \dfrac{1}{3}\right)t \quad \text{Associative property}$$

$$= 2t \quad \text{Multiply.} \quad \blacksquare$$

24. Simplify $2\left(\dfrac{x}{2} - 3\right)$.

EXAMPLE 24 Simplify $3\left(\dfrac{t}{3} - 2\right)$.

SOLUTION
$$3\left(\dfrac{t}{3} - 2\right) = 3 \cdot \dfrac{t}{3} - 3 \cdot 2 \quad \text{Distributive property}$$

$$= t - 6 \quad \text{Multiply.} \quad \blacksquare$$

Our next examples involve more complicated fractions. The fraction bar works like parentheses to separate the numerator from the denominator. Although we don't write expressions this way, here is one way to think of the fraction bar:

$$\dfrac{-8 - 8}{-5 - 3} = (-8 - 8) \div (-5 - 3)$$

As you can see, if we apply the rule for order of operations to the expression on the right, we would work inside each set of parentheses first, then divide. Applying this to the expression on the left, we work on the numerator and denominator separately, then we divide or reduce the resulting fraction to lowest terms.

Answers
21. $17x + 15$
22. $30y + 49$
23. x
24. $x - 6$

EXAMPLE 25 Simplify $\dfrac{-8-8}{-5-3}$ as much as possible.

SOLUTION
$$\dfrac{-8-8}{-5-3} = \dfrac{-16}{-8} \qquad \text{Simplify numerator and denominator separately.}$$
$$= 2 \qquad \text{Divide.} \qquad \blacksquare$$

EXAMPLE 26 Simplify $\dfrac{-5(-4)+2(-3)}{2(-1)-5}$ as much as possible.

SOLUTION
$$\dfrac{-5(-4)+2(-3)}{2(-1)-5} = \dfrac{20-6}{-2-5}$$
$$= \dfrac{14}{-7}$$
$$= -2 \qquad \blacksquare$$

EXAMPLE 27 Simplify $\dfrac{2^3+3^3}{2^2-3^2}$ as much as possible.

SOLUTION
$$\dfrac{2^3+3^3}{2^2-3^2} = \dfrac{8+27}{4-9}$$
$$= \dfrac{35}{-5}$$
$$= -7 \qquad \blacksquare$$

Remember, since subtraction is defined in terms of addition, we can restate the distributive property in terms of subtraction; that is, if a, b, and c are real numbers, then $a(b-c) = ab - ac$.

EXAMPLE 28 Simplify $3(2y-1)+y$.

SOLUTION We begin by multiplying the 3 and $2y-1$. Then, we combine similar terms.
$$3(2y-1)+y = 6y-3+y \qquad \text{Distributive property}$$
$$= 7y-3 \qquad \text{Combine similar terms.} \qquad \blacksquare$$

EXAMPLE 29 Simplify $8-3(4x-2)+5x$.

SOLUTION First we distribute the -3 across the $4x-2$.
$$8-3(4x-2)+5x = 8-12x+6+5x$$
$$= -7x+14 \qquad \blacksquare$$

EXAMPLE 30 Simplify $5(2a+3)-(6a-4)$.

SOLUTION We begin by applying the distributive property to remove the parentheses. The expression $-(6a-4)$ can be thought of as $-1(6a-4)$. Thinking of it in this way allows us to apply the distributive property.
$$-1(6a-4) = -1(6a)-(-1)(4) = -6a+4$$

Here is the complete problem:
$$5(2a+3)-(6a-4) = 10a+15-6a+4 \qquad \text{Distributive property}$$
$$= 4a+19 \qquad \text{Combine similar terms.} \qquad \blacksquare$$

25. Simplify $\dfrac{-6-6}{-5-3}$.

26. Simplify $\dfrac{5(-6)+3(-2)}{4(-3)+3}$.

27. Simplify $\dfrac{3^3-4^3}{3^2+4^2}$.

28. Simplify $2(5y-1)-y$.

29. Simplify $6-2(5x+1)+4x$.

30. Simplify $4(3a+1)-(7a-6)$.

Answers

25. $\dfrac{3}{2}$

26. 4

27. $-\dfrac{37}{25}$

28. $9y-2$

29. $-6x+4$

30. $5a+10$

D Finding the Value of an Algebraic Expression

As we mentioned earlier in this chapter, an algebraic expression is a combination of numbers, variables, and operation symbols. Each of the following is an algebraic expression:

$$7a \qquad x^2 - y^2 \qquad 2(3t - 4) \qquad \frac{2x - 5}{6}$$

An expression such as $2(3t - 4)$ will take on different values depending on what number we substitute for t. For example, if we substitute -8 for t, then the expression $2(3t - 4)$ becomes $2[3(-8) - 4)]$, which simplifies to -56. If we apply the distributive property to $2(3t - 4)$, we have

$$2(3t - 4) = 6t - 8$$

Substituting -8 for t in the simplified expression gives us $6(-8) - 8 = -56$, which is the same result we obtained previously. As you would expect, substituting the same number into an expression, and any simplified form of that expression, will yield the same result.

31. Evaluate each expression when x is -1.
a. $7(4x - 6)$
b. $28x - 42$
c. $28x - 6$

EXAMPLE 31 Evaluate the expressions $(a + 4)^2$, $a^2 + 16$, and $a^2 + 8a + 16$ when a is -2, 0, and 3.

SOLUTION Organizing our work with a table, we have

a	$(a + 4)^2$	$a^2 + 16$	$a^2 + 8a + 16$
-2	$(-2 + 4)^2 = 4$	$(-2)^2 + 16 = 20$	$(-2)^2 + 8(-2) + 16 = 4$
0	$(0 + 4)^2 = 16$	$0^2 + 16 = 16$	$0^2 + 8(0) + 16 = 16$
3	$(3 + 4)^2 = 49$	$3^2 + 16 = 25$	$3^2 + 8(3) + 16 = 49$

When we study polynomials later in the book, you will see that the expressions $(a + 4)^2$ and $a^2 + 8a + 16$ are equivalent, and that neither one is equivalent to $a^2 + 16$. ∎

Problem Set 1.6

1. When should you review your notes?
   a. Immediately after class
   b. Each day
   c. When studying for a quiz or exam
   d. All of the above
2. Why would taking practice tests in an accurate test environment be a helpful study tool?

A Use the associative property to rewrite each of the following expressions and then simplify the result. [Examples 3–7]

1. $4 + (2 + x)$

2. $6 + (5 + 3x)$

3. $(a + 3) + 5$

4. $(4a + 5) + 7$

5. $5(3y)$

6. $7(4y)$

7. $\frac{1}{3}(3x)$

8. $\frac{1}{5}(5x)$

9. $4\left(\frac{1}{4}a\right)$

10. $7\left(\frac{1}{7}a\right)$

11. $\frac{2}{3}\left(\frac{3}{2}x\right)$

12. $\frac{4}{3}\left(\frac{3}{4}x\right)$

A Apply the distributive property to each expression. Simplify when possible. [Examples 8–14]

13. $3(x + 6)$

14. $5(x + 9)$

15. $2(6x + 4)$

16. $3(7x + 8)$

17. $5(3a + 2b)$

18. $7(2a + 3b)$

19. $\frac{1}{3}(4x + 6)$

20. $\frac{1}{2}(3x + 8)$

21. $\frac{1}{5}(10 + 5y)$

22. $\frac{1}{6}(12 + 6y)$

23. $(5t + 1)8$

24. $(3t + 2)5$

A The problems below are problems you will see later in the book. Apply the distributive property, then simplify if possible. [Examples 8–14]

25. $3(3x + y - 2z)$

26. $2(2x - y + z)$

27. $10(0.3x + 0.7y)$

28. $10(0.2x + 0.5y)$

29. $100(0.06x + 0.07y)$ **30.** $100(0.09x + 0.08y)$ **31.** $3\left(x + \dfrac{1}{3}\right)$ **32.** $5\left(x - \dfrac{1}{5}\right)$

33. $2\left(x - \dfrac{1}{2}\right)$ **34.** $7\left(x + \dfrac{1}{7}\right)$ **35.** $x\left(1 + \dfrac{2}{x}\right)$ **36.** $x\left(1 - \dfrac{1}{x}\right)$

37. $a\left(1 - \dfrac{3}{a}\right)$ **38.** $a\left(1 + \dfrac{1}{a}\right)$ **39.** $8\left(\dfrac{1}{8}x + 3\right)$ **40.** $4\left(\dfrac{1}{4}x - 9\right)$

41. $6\left(\dfrac{1}{2}x - \dfrac{1}{3}y\right)$ **42.** $12\left(\dfrac{1}{4}x - \dfrac{1}{6}y\right)$ **43.** $12\left(\dfrac{1}{4}x + \dfrac{2}{3}y\right)$ **44.** $12\left(\dfrac{2}{3}x - \dfrac{1}{4}y\right)$

45. $20\left(\dfrac{2}{5}x + \dfrac{1}{4}y\right)$ **46.** $15\left(\dfrac{2}{3}x + \dfrac{2}{5}y\right)$

A Apply the distributive property to each expression. Simplify when possible. [Examples 8–14]

47. $3(5x + 2) + 4$ **48.** $4(3x + 2) + 5$ **49.** $4(2y + 6) + 8$ **50.** $6(2y + 3) + 2$

51. $5(1 + 3t) + 4$ **52.** $2(1 + 5t) + 6$ **53.** $3 + (2 + 7x)4$ **54.** $4 + (1 + 3x)5$

A Use the commutative, associative, and distributive properties to simplify the following. [Example 15, 16, 21, 22]

55. $5a + 7 + 8a + a$ **56.** $6a + 4 + a + 4a$ **57.** $3y + y + 5 + 2y + 1$ **58.** $4y + 2y + 3 + y + 7$

59. $2(5x + 1) + 2x$ **60.** $3(4x + 1) + 9x$ **61.** $7 + 2(4y + 2)$ **62.** $6 + 3(5y + 2)$

63. $3 + 4(5a + 3) + 4a$ **64.** $8 + 2(4a + 2) + 5a$ **65.** $5x + 2(3x + 8) + 4$ **66.** $7x + 3(4x + 1) + 7$

A Identify the property or properties of real numbers that justifies each of the following. [Examples 17]

67. $3 + 2 = 2 + 3$

68. $3(ab) = (3a)b$

69. $5 \cdot x = x \cdot 5$

70. $2 + 0 = 2$

71. $4 + (-4) = 0$

72. $1(6) = 6$

73. $x + (y + 2) = (y + 2) + x$

74. $(a + 3) + 4 = a + (3 + 4)$

75. $4(5 \cdot 7) = 5(4 \cdot 7)$

76. $6(xy) = (xy)6$

77. $4 + (x + y) = (4 + y) + x$

78. $(r + 7) + s = (r + s) + 7$

79. $3(4x + 2) = 12x + 6$

80. $5\left(\dfrac{1}{5}\right) = 1$

B Using the order of operations, simplify as much as possible. [Examples 18–20]

81. $3(-4) - 2$

82. $-3(-4) - 2$

83. $4(-3) - 6(-5)$

84. $-6(-3) - 5(-7)$

85. $2 - 5(-4) - 6$

86. $3 - 8(-1) - 7$

87. $4 - 3(7 - 1) - 5$

88. $8 - 5(6 - 3) - 7$

89. $2(-3)^2 - 4(-2)$

90. $5(-2)^2 - 2(-3)^3$

91. $(2 - 8)^2 - (3 - 7)^2$

92. $(5 - 8)^2 - (4 - 8)^2$

93. $7(3 - 5)^3 - 2(4 - 7)^3$

94. $3(-7 + 9)^3 - 5(-2 + 4)^3$

95. $-3(2 - 9) - 4(6 - 1)$

96. $-5(5 - 6) - 7(2 - 8)$

97. $2 - 4[3 - 5(-1)]$

98. $6 - 5[2 - 4(-8)]$

99. $(8 - 7)[4 - 7(-2)]$

100. $(6 - 9)[15 - 3(-4)]$

101. $-3 + 4[6 - 8(-3 - 5)]$

102. $-2 + 7[2 - 6(-3 - 4)]$

103. $5 - 6[-3(2 - 9) - 4(8 - 6)]$

104. $9 - 4[-2(4 - 8) - 5(3 - 1)]$

105. $1(-2) - 2(-16) + 1(9)$

106. $6(1) - 1(-5) + 1(2)$

107. $1(1) - 3(-2) + (-2)(-2)$

A C The problems below are problems you will see later in the book. Apply the distributive property, then simplify if possible. [Examples 23, 24]

108. $-1(5 - x)$

109. $-1(a - b)$

110. $-1(7 - x)$

111. $-1(6 - y)$

112. $-3(2x - 3y)$

113. $-1(x - 2z)$

114. $6\left(\dfrac{x}{2} - 3\right)$

115. $6\left(\dfrac{x}{3} + 1\right)$

116. $12\left(\dfrac{a}{4} + \dfrac{1}{2}\right)$

117. $15\left(\dfrac{a}{3} + 2\right)$

118. $15\left(\dfrac{x}{5} + 4\right)$

119. $10\left(\dfrac{x}{2} - 9\right)$

120. $12\left(\dfrac{y}{2} + \dfrac{y}{4} + \dfrac{y}{6}\right)$

121. $12\left(\dfrac{y}{3} - \dfrac{y}{6} + \dfrac{y}{2}\right)$

C Simplify each expression. [Examples 21–24, 28–30]

122. $3(5x + 4) - x$

123. $4(7x + 3) - x$

124. $6 - 7(3 - m)$

125. $3 - 5(5 - m)$

126. $7 - 2(3x - 1) + 4x$

127. $8 - 5(2x - 3) + 4x$

128. $5(3y + 1) - (8y - 5)$

129. $4(6y + 3) - (6y - 6)$

130. $4(2 - 6x) - (3 - 4x)$

131. $7(1 - 2x) - (4 - 10x)$

132. $10 - 4(2x + 1) - (3x - 4)$

133. $7 - 2(3x + 5) - (2x - 3)$

C Simplify as much as possible. [Examples 25–27]

134. $\dfrac{3(-1) - 4(-2)}{8 - 5}$

135. $\dfrac{6(-4) - 5(-2)}{7 - 6}$

136. $\dfrac{4(-3) - 5(-2)}{8 - 6}$

137. $-9 - 5\left[\dfrac{11(-1) - 9}{4(-3) + 2(5)}\right]$

138. $6 - (-3)\left[\dfrac{2 - 4(3 - 8)}{1 - 5(1 - 3)}\right]$

139. $8 - (-7)\left[\dfrac{6 - 1(6 - 10)}{4 - 3(5 - 7)}\right]$

D Complete each of the following tables. [Example 31]

140.

x	$3(5x - 2)$	$15x - 6$	$15x - 2$
-2			
-1			
0			
1			
2			

141.

x	$(x + 1)^2$	$x^2 + 1$	$x^2 + 2x + 1$
-2			
-1			
0			
1			
2			

142. Find the value of $-\dfrac{b}{2a}$ when

   a. $a = 3, b = -6$

   b. $a = -2, b = 6$

   c. $a = -1, b = -2$

   d. $a = -0.1, b = 27$

143. Find the value of $b^2 - 4ac$ when

   a. $a = 3, b = -2$, and $c = 4$

   b. $a = 1, b = -3$, and $c = -28$

   c. $a = 1, b = -6$, and $c = 9$

   d. $a = 0.1, b = -27$, and $c = 1,700$

Use a calculator to simplify each expression. If rounding is necessary, round your answers to the nearest ten thousandth (4 places past the decimal point). You will see these problems later in the book.

144. $\dfrac{1.3802}{0.9031}$

145. $\dfrac{1.0792}{0.6990}$

146. $\dfrac{1}{2}(-0.1587)$

147. $\dfrac{1}{2}(-0.7948)$

148. $\dfrac{1}{2}\left(\dfrac{1.2}{1.4} - 1\right)$

149. $\dfrac{1}{2}\left(\dfrac{1.3}{1.1} - 1\right)$

150. $\dfrac{(6.8)(3.9)}{7.8}$

151. $\dfrac{(2.4)(1.8)}{1.2}$

152. $\dfrac{0.0005(200)}{(0.25)^2}$

153. $\dfrac{0.0006(400)}{(0.25)^2}$

154. $-500 + 27(100) - 0.1(100)^2$

155. $-500 + 27(170) - 0.1(170)^2$

156. $-0.05(130)^2 + 9.5(130) - 200$

157. $-0.04(130)^2 + 8.5(130) - 210$

Applying the Concepts

Area and the Distributive Property Find the area of each of the following rectangles in two ways: first by multiplying length and width, and then by adding the areas of the two smaller rectangles together.

158.
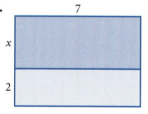

159.
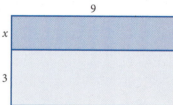

160.
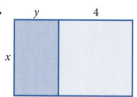

161.
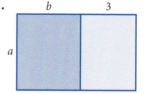

162. Soccer The chart shows the dimensions for some of the soccer fields that will be used during the 2014 World Cup. Use the chart to write an expression for the area of the Estádio Do Maracanã. Do the same for the Mineirão. Then simplify both.

World Cup Fields

	To Half-Field	Width
Estadío Do Maracanã	55 m	75 m
Estadío Mangueirão	52.5 m	68 m
Mineirão	60 m	80 m
Morenão	55 m	70 m

Source: World Cup 2014

Recognizing Patterns

TICKET TO SUCCESS

Keep these questions in mind as you read through the section. Then respond in your own words and in complete sentences.

1. What is inductive reasoning?
2. What is an arithmetic sequence?
3. Create a geometric sequence and explain how it was formed.
4. How would you use paired data to connect two sequences?

Melissa Dockstader/Shutterstock.com

A pattern is a set of recurring events, objects, or numbers. These elements repeat in a predictable manner. The quilt shown above is made by repeating the same geometric pattern a number of times. In this section, we will look at a few ways numbers can appear in a pattern.

A Patterns

Much of what we do in mathematics is concerned with recognizing patterns and classifying groups of numbers that share a common characteristic. For instance, suppose you were asked to give the next number in this sequence:

$$3, 5, 7, \ldots$$

Looking for a pattern, you may observe that each number is 2 more than the number preceding it. That being the case, the next number in the sequence will be 9 because 9 is 2 more than 7. Reasoning in this manner is called *inductive reasoning*. In mathematics, we use inductive reasoning when we notice a pattern to a sequence of numbers and then extend the sequence using the pattern.

EXAMPLE 1 Use inductive reasoning to find the next term in each sequence.
  a. 5, 8, 11, 14, . . .
  b. △, ▷, ▽, ◁, . . .
  c. 1, 4, 9, 16, . . .

PRACTICE PROBLEMS

1. Use inductive reasoning to find the next term in each sequence below.
   a. 5, 15, 45, 135, . . .
   b. ▷, ▽, ◁, △, . . .
   c. 1, 8, 27, 64 . . .

Answers
1. a. 405 **b.** ▷ **c.** 125

SOLUTION   In each case, we use the pattern we observe in the first few terms to write the next term.

    a. Each term comes from the previous term by adding 3. Therefore, the next term would be 17.

    b. The triangles rotate a quarter turn to the right each time. The next term would be a triangle that points up, △.

    c. This looks like the sequence of squares, $1^2, 2^2, 3^2, 4^2, \ldots$. The next term is $5^2 = 25$. ∎

Now that we have an intuitive idea of inductive reasoning, here is a formal definition.

> **Definition**
>
> **Inductive reasoning** is reasoning in which a conclusion is drawn based on evidence and observations that support that conclusion. In mathematics, this usually involves noticing that a few items in a group have a trait or characteristic in common and then concluding that all items in the group have that same trait.

B   Arithmetic Sequences

We can extend our work with sequences by classifying sequences that share a common characteristic. Our first classification is for sequences that are constructed by adding the same number each time.

> **Definition**
>
> An **arithmetic sequence** is a sequence of numbers in which each number (after the first number) comes from adding the same amount to the number before it.

The sequence

$$4, 7, 10, 13, \ldots$$

is an example of an arithmetic sequence, because each term is obtained from the preceding term by adding 3 each time. The number we add each time—in this case, 3—is the *common difference* because it can be obtained by subtraction.

EXAMPLE 2   Each sequence shown here is an arithmetic sequence. Find the next two numbers in each sequence.

    a. $10, 16, 22, \ldots$

    b. $\dfrac{1}{2}, 1, \dfrac{3}{2}, \ldots$

    c. $5, 0, -5, \ldots$

SOLUTION   Because we know that each sequence is arithmetic, we can look for the number that is added to each term to produce the next consecutive term.

    a. Each term is found by adding 6 to the term before it. Therefore, the next two terms will be 28 and 34.

---

2. Each sequence below is an arithmetic sequence. Find the next two numbers in each sequence.

a. $6, 16, 26, \ldots$

b. $-\dfrac{1}{2}, 0, \dfrac{1}{2}, \ldots$

c. $-5, 0, 5, \ldots$

Answers

2. a. $36, 46$ **b.** $1, \dfrac{3}{2}$ **c.** $10, 15$

b. Each term comes from the term before it by adding $\frac{1}{2}$. The fourth term will be $\frac{3}{2} + \frac{1}{2} = 2$, while the fifth term will be $2 + \frac{1}{2} = \frac{5}{2}$.

c. Each term comes from adding -5 to the term before it. Therefore, the next two terms will be $-5 + (-5) = -10$, and $-10 + (-5) = -15$. ∎

C Geometric Sequences

Our second classification of sequences with a common characteristic involves sequences that are constructed using multiplication.

Definition

A **geometric sequence** is a sequence of numbers in which each number (after the first number) comes from the number before it by multiplying by the same amount each time.

The sequence

$$4, 12, 36, 108, \ldots$$

is a geometric sequence. Each term is obtained from the previous term by multiplying by 3. The amount by which we multiply each term to obtain the next term—in this case, 3—is called the *common ratio.*

EXAMPLE 3   Each sequence shown here is a geometric sequence. Find the next number in each sequence.

a. $2, 10, 50, \ldots$

b. $3, -15, 75, \ldots$

c. $\dfrac{1}{8}, \dfrac{1}{4}, \dfrac{1}{2}, \ldots$

SOLUTION   Because each sequence is a geometric sequence, we know that each term is obtained from the previous term by multiplying by the same number each time.

a. Starting with 2, each number is obtained from the previous number by multiplying by 5 each time. The next number will be $50 \cdot 5 = 250$.

b. The sequence starts with 3. After that, each number is obtained by multiplying by -5 each time. The next number will be $75(-5) = -375$.

c. This sequence starts with $\frac{1}{8}$. Multiplying each number in the sequence by 2 produces the next number in the sequence. To extend the sequence, we multiply $\frac{1}{2}$ by 2.

$$\frac{1}{2} \cdot 2 = 1$$

The next number in the sequence is 1. ∎

3. Each sequence below is a geometric sequence. Find the next number in each sequence.

a. $10, 20, 40, \ldots$

b. $5, -15, 45, \ldots$

c. $2, 1, \dfrac{1}{2}, \ldots$

Answers

3. a. 80   **b.** -135 **c.** $\frac{1}{4}$

D The Fibonacci Sequence

There is a special sequence in mathematics named for the mathematician Fibonacci.

Fibonacci sequence: 1, 1, 2, 3, 5, 8, . . .

To construct the Fibonacci sequence, we start with two 1's. The rest of the numbers in the sequence are found by adding the two previous terms. Adding the first two terms, 1 and 1, we have 2. Then, adding 1 and 2 we have 3. In general, adding any two consecutive terms of the Fibonacci sequence gives us the next term.

A Mathematical Model

One of the reasons we study number sequences is because they can be used to model some of the patterns and events we see in the world around us. The discussion that follows shows how the Fibonacci sequence can be used to predict the number of bees in each generation of the family tree of a male honeybee. It is based on an example from Chapter 2 of the book *Mathematics: A Human Endeavor* by Harold Jacobs. If you find that you enjoy discovering patterns in mathematics, Mr. Jacobs' book has many interesting examples and problems involving patterns in mathematics.

A male honeybee has one parent, its mother, whereas a female honeybee has two parents, a mother and a father. (A male honeybee comes from an unfertilized egg; a female honeybee comes from a fertilized egg.) Using these facts, we construct the family tree of a male honeybee using ⬤ to represent a male honeybee and ⬤ to represent a female honeybee.

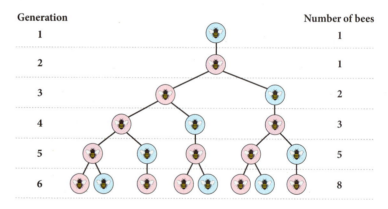

Looking at the numbers in the right column in our diagram, the sequence that gives us the number of bees in each generation of the family tree of a male honey bee is

1   1   2   3   5   8

As you can see, this is the Fibonacci sequence. We have taken our original diagram (the family tree of the male honeybee) and reduced it to a mathematical model (the Fibonacci sequence). The model can be used in place of the diagram to find the number of bees in any generation back from our first bee.

Problem Set 1.7

Moving Toward Success

"Confidence is contagious. So is lack of confidence."
—Vince Lombardi, 1913–1970, Hall of Fame American football coach

1. What is it going to take for you to be successful in mathematics?
2. Why is confidence important in mathematics?

A Here are some sequences that we will be referring to throughout the book. Find the next number in each sequence. [Example 1]

1. 1, 2, 3, 4, . . . (The sequence of counting numbers)

2. 0, 1, 2, 3, . . . (The sequence of whole numbers)

3. 2, 4, 6, 8, . . . (The sequence of even numbers)

4. 1, 3, 5, 7, . . . (The sequence of odd numbers)

5. 1, 4, 9, 16, . . . (The sequence of squares)

6. 1, 8, 27, 64, . . . (The sequence of cubes)

Find the next number in each sequence.

7. 1, 8, 15, 22, . .

8. 1, 8, 64, 512, . . .

9. 1, 8, 14, 19, . . .

10. 1, 8, 16, 25, . . .

Give one possibility for the next term in each sequence.

11. $\triangle, \triangleleft, \triangledown, \triangleright, \ldots$

12. $\Rightarrow, \Downarrow, \Leftarrow, \Uparrow, \ldots$

13. $\triangle, \square, \bigcirc, \triangle, \boxdot, \ldots$

14. $\square, \square\square, \text{⊞}, \text{⊞}, \square\square\square, \ldots$

B Each sequence shown here is an arithmetic sequence. In each case, find the next two numbers in the sequence. [Example 2]

15. 1, 5, 9, 13, . . .

16. 10, 16, 22, 28, . . .

17. 1, 0, −1, . . .

18. 6, 0, −6, . . .

19. 5, 2, −1, . . .

20. 8, 4, 0, . . .

21. $\dfrac{1}{4}, 0, -\dfrac{1}{4}, \ldots$

22. $\dfrac{2}{5}, 0, -\dfrac{2}{5}, \ldots$

23. $1, \dfrac{3}{2}, 2, \ldots$

24. $\dfrac{1}{3}, 1, \dfrac{5}{3}, \ldots$

C Each sequence shown here is a geometric sequence. In each case, find the next number in the sequence. [Example 3]

25. 1, 3, 9, . . .

26. 1, 7, 49, . . .

27. 10, −30, 90, . . .

28. 10, −20, 40, . . .

29. 1, $\dfrac{1}{2}$, $\dfrac{1}{4}$, . . .

30. 1, $\dfrac{1}{3}$, $\dfrac{1}{9}$, . . .

31. 20, 10, 5, . . .

32. 8, 4, 2, . . .

33. 5, −25, 125, . . .

34. −4, 16, −64, . . .

35. 1, $-\dfrac{1}{5}$, $\dfrac{1}{25}$, . . .

36. 1, $-\dfrac{1}{2}$, $\dfrac{1}{4}$, . . .

37. Find the next number in the sequence 4, 8, . . . if the sequence is

   a. an arithmetic sequence

   b. a geometric sequence

38. Find the next number in the sequence 1, −4, . . . if the sequence is

   a. an arithmetic sequence

   b. a geometric sequence

39. Find the 12th term of the Fibonacci sequence.

40. Find the 13th term of the Fibonacci sequence.

D Any number in the Fibonacci sequence is a *Fibonacci number*.

41. Name three Fibonacci numbers that are prime numbers.

42. Name three Fibonacci numbers that are composite numbers.

43. In the first ten terms of the Fibonacci sequence, which ones are even numbers?

44. In the first ten terms of the Fibonacci sequence, which ones are odd numbers?

The patterns in the tables below will become important when we do factoring of trinomials later in the book. Complete each table.

45.

Two Numbers a and b	Their Product ab	Their Sum $a + b$
1, −24		
−1, 24		
2, −12		
−2, 12		
3, −8		
−3, 8		
4, −6		
−4, 6		

46.

Two Numbers a and b	Their Product ab	Their Sum $a + b$
1, −54		
−1, 54		
2, −27		
−2, 27		
3, −18		
−3, 18		
6, −9		
−6, 9		

Applying the Concepts

47. Temperature and Altitude A pilot checks the weather conditions before flying and finds that the air temperature drops 3.5°F every 1,000 feet above the surface of the Earth. (The higher he flies, the colder the air.) If the air temperature is 41°F when the plane reaches 10,000 feet, write a sequence of numbers that gives the air temperature every 1,000 feet as the plane climbs from 10,000 feet to 15,000 feet. Is this sequence an arithmetic sequence?

48. Temperature and Altitude For the plane mentioned in Problem 47, at what altitude will the air temperature be 20°F?

mobil11/Shutterstock.com

49. Temperature and Altitude The weather conditions on a certain day are such that the air temperature drops 4.5°F every 1,000 feet above the surface of the Earth. If the air temperature is 41°F at 10,000 feet, write a sequence of numbers that gives the air temperature every 1,000 feet starting at 10,000 feet and ending at 5,000 feet. Is this sequence an arithmetic sequence?

50. Value of a Painting Suppose you own a painting that doubles in value every 5 years. If you bought the painting for $125 in 1995, write a sequence of numbers that gives the value of the painting every 5 years from the time you purchased it until the year 2015. Is this sequence a geometric sequence?

51. Half-Life The half-lives of two antidepressants are given below. A patient taking Antidepressant 1 tells his doctor that he begins to feel sick if he misses his morning dose, and a patient taking Antidepressant 2 tells his doctor that he doesn't notice a difference if he misses a day of taking the medication. Explain these situations in terms of the half-life of the medication.

HALF-LIFE	
Antidepressant 1	**Antidepressant 2**
11 hours	5 days

52. Half-Life Two patients are taking the antidepressants mentioned in Problem 51. Both decide to stop their current medications and take another medication. The physician tells the patient on Antidepressant 2 to simply stop taking it but instructs the patient on Antidepressant 1 to take half the normal dose for 3 days, then one-fourth the normal dose for another 3 days before stopping the medication altogether. Use half-life to explain why the doctor uses two different methods of taking the patients off their medications.

53. Half-Life The half-life of a drug is 4 hours. A patient has been taking the drug on a regular basis for a few months and then discontinues taking it. The concentration of the drug in a patient's system is 60 ng/mL when the patient stops taking the medication. Complete the table, and then use the results to construct a line graph of that data.

Hours Since Discontinuing	Concentration (ng/mL)
0	60
4	
8	
12	
16	

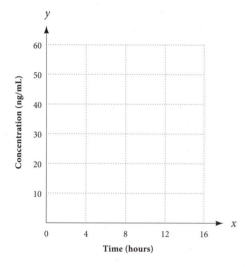

54. Half-Life The half-life of a drug is 8 hours. A patient has been taking the drug on a regular basis for a few months and then discontinues taking it. The concentration of the drug in a patient's system is 120 ng/mL when the patient stops taking the medication. Complete the table, and then use the results to construct a line graph of that data.

Hours Since Discontinuing	Concentration (ng/mL)
0	120
8	
16	
24	
32	

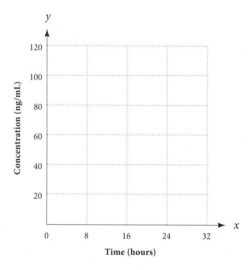

55. Boiling Point The boiling point of water at sea level is 212°F. The boiling point of water drops 1.8°F every 1,000 feet above sea level, and it rises 1.8°F every 1,000 feet below sea level. Complete the following table to write the sequence that gives the boiling points of water from 2,000 feet below sea level to 3,000 feet above sea level.

Elevation (ft)	Boiling point (°F)
−2,000	
−1,000	
0	
1,000	
2,000	
3,000	

Source: engineeringtoolbox.com

56. Chlorine Level A swimming pool needs 1 ppm (part per million) of chlorine to keep bacteria growth to a minimum. A pool contains 3 ppm chlorine and loses 20% of that amount each day if no more chlorine is added. Fill in the following table, assuming that no additional chlorine is added to a pool containing 3 ppm chlorine. (Round to the nearest hundredth.)

Day	Chlorine Level
1	
2	
3	
4	
5	
6	

Chapter 1 Summary

The margins of the chapter summaries will be used for brief examples of the topics being reviewed, whenever it is convenient.

The numbers in brackets refer to the section in which the topic can be found.

■ Exponents [1.1]

EXAMPLES

1. $2^5 = 2 \cdot 2 \cdot 2 \cdot 2 \cdot 2 = 32$
$5^2 = 5 \cdot 5 = 25$
$10^3 = 10 \cdot 10 \cdot 10 = 1{,}000$
$1^4 = 1 \cdot 1 \cdot 1 \cdot 1 = 1$

Exponents represent notation used to indicate repeated multiplication. In the expression 3^4, 3 is the *base* and 4 is the *exponent*.

$$3^4 = 3 \cdot 3 \cdot 3 \cdot 3 = 81$$

The expression 3^4 is said to be in *exponential form,* whereas the expression $3 \cdot 3 \cdot 3 \cdot 3$ is in *expanded form.*

■ Order of Operations [1.1]

2. $10 + (2 \cdot 3^2 - 4 \cdot 2)$
$= 10 + (2 \cdot 9 - 4 \cdot 2)$
$= 10 + (18 - 8)$
$= 10 + 10$
$= 20$

When evaluating a mathematical expression, we will perform the operations in the following order:

1. Begin with the expression in the innermost parentheses or brackets and work our way out.

2. Evaluate or simplify, all numbers with exponents, working from left to right if more than one of these expressions is present.

3. Work all multiplications and divisions left to right.

4. Perform all additions and subtractions left to right.

■ Sets [1.2]

3. If $A = \{0, 1, 2\}$ and $B = \{2, 3\}$, then $A \cup B = \{0, 1, 2, 3\}$ and $A \cap B = \{2\}$.

A *set* is a collection of objects or things.

The *union* of two sets A and B, written $A \cup B$, is all the elements that are in A *or* B.

The *intersection* of two sets A and B, written $A \cap B$, is the set consisting of all elements common to both A *and* B.

Set A is a *subset* of set B, written $A \subset B$, if all elements in set A are also in set B.

■ Special Sets [1.3]

4. 5 is a counting number, a whole number, an integer, a rational number, and a real number.
$\frac{3}{4}$ is a rational number and a real number.
$\sqrt{2}$ is an irrational number and a real number.

Counting numbers $= \{1, 2, 3, \dots\}$

Whole numbers $= \{0, 1, 2, 3, \dots\}$

Integers $= \{\dots, -3, -2, -1, 0, 1, 2, 3, \dots\}$

Rational numbers $= \left\{\frac{a}{b} \mid a \text{ and } b \text{ are integers}, b \neq 0\right\}$

Irrational numbers $= \{x \mid x \text{ is real, but not rational}\}$

Real numbers $= \{x \mid x \text{ is rational or } x \text{ is irrational}\}$

Prime numbers $= \{2, 3, 5, 7, 11, \dots\} = \{x \mid x \text{ is a positive integer greater than 1 whose only positive divisors are itself and 1}\}$

■ Inequalities [1.4]

5. Graph each inequality mentioned at the right.

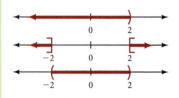

The set $\{x \mid x < 2\}$ is the set of all real numbers that are less than 2. To graph this set, we place a right parenthesis at 2 on the real number line and then draw an arrow that starts at 2 and points to the left.

The set $\{x \mid x \leq -2 \text{ or } x \geq 2\}$ is the set of all real numbers that are either less than or equal to −2 or greater than or equal to 2.

The set $\{x \mid -2 < x < 2\}$ is the set of all real numbers that are between −2 and 2; that is, the real numbers that are greater than −2 and less than 2.

■ Opposites [1.5]

6. The numbers 5 and −5 are opposites; their sum is 0.

$$5 + (-5) = 0$$

Any two real numbers the same distance from 0 on the number line, but in opposite directions from 0, are called *opposites*, or *additive inverses*. Opposites always add to 0.

■ Reciprocals [1.5]

7. The numbers 3 and $\frac{1}{3}$ are reciprocals; their product is 1.

$$3\left(\frac{1}{3}\right) = 1$$

Any two real numbers whose product is 1 are called *reciprocals*. Every real number has a reciprocal except 0.

■ Absolute Value [1.5]

8. $|5| = 5$
$|-5| = 5$

The *absolute value* of a real number is its distance from 0 on the number line. If $|x|$ represents the absolute value of x, then

$$|x| = \begin{cases} x & \text{if} & x \geq 0 \\ -x & \text{if} & x < 0 \end{cases}$$

The absolute value of a real number is never negative.

■ Addition [1.5]

9. $5 + 3 = 8$
$5 + (-3) = 2$
$-5 + 3 = -2$
$-5 + (-3) = -8$

To add two real numbers with
1. *The same sign:* Simply add absolute values and use the common sign.
2. *Different signs:* Subtract the smaller absolute value from the larger absolute value. The answer has the same sign as the number with the larger absolute value.

■ Subtraction [1.5]

10. $6 - 2 = 6 + (-2) = 4$
$6 - (-2) = 6 + 2 = 8$

If a and b are real numbers,
$$a - b = a + (-b)$$
To subtract b, add the opposite of b.

Multiplication [1.5]

11. $5(4) = 20$
$5(-4) = -20$
$-5(4) = -20$
$-5(-4) = 20$

To multiply two real numbers, simply multiply their absolute values. Like signs give a positive answer. Unlike signs give a negative answer.

Division [1.5]

12. $\frac{12}{-3} = -4$

$\frac{-12}{-3} = 4$

If a and b are real numbers and $b \neq 0$, then

$$\frac{a}{b} = a \cdot \left(\frac{1}{b}\right)$$

To divide by b, multiply by the reciprocal of b.

Properties of Real Numbers [1.6]

	For Addition	For Multiplication
Commutative	$a + b = b + a$	$ab = ba$
Associative	$a + (b + c) = (a + b) + c$	$a(bc) = (ab)c$
Identity	$a + 0 = a$	$a \cdot 1 = a$
Inverse	$a + (-a) = 0$	$a\left(\frac{1}{a}\right) = 1$
Distributive	$a(b + c) = ab + ac$	

Inductive Reasoning [1.7]

13. We use inductive reasoning when we conclude that the next number in the sequence below is 25.

$1, 4, 9, 16, \ldots$

Inductive reasoning is reasoning in which a conclusion is drawn based on evidence and observations that support that conclusion. In mathematics this usually involves noticing that a few items in a group have a trait or characteristic in common and then concluding that all items in the group have that same trait.

Arithmetic Sequence [1.7]

14. The following sequence is an arithmetic sequence because each term is obtained from the preceding term by adding 3 each time.

$4, 7, 10, 13, \ldots$

An *arithmetic sequence* is a sequence of numbers in which each number (after the first number) comes from adding the same amount to the number before it. The number we add to each term to obtain the next term is called the *common difference*.

■ **Geometric Sequence [1.7]**

15. The following sequence is a geometric sequence because each term is obtained from the previous term by multiplying by 3 each time.

4, 12, 36, 108, . . .

A *geometric sequence* is a sequence of numbers in which each number (after the first number) comes from the number before it by multiplying by the same amount each time. The amount by which we multiply each term to obtain the next term is called the *common ratio*.

🚫 **COMMON MISTAKES**

1. Interpreting absolute value as changing the sign of the number inside the absolute value symbols; that is, $|-5| = 5$, $|5| = -5$. To avoid this mistake, remember absolute value is defined as a distance and distance is always measured in positive units.

2. Confusing $-(-5)$ with $-|-5|$. The first answer is 5, but the second answer is -5.

Translate each expression into symbols. [1.1]

1. The sum of x and 2.

2. The sum of twice x and y.

Expand and multiply. [1.1]

3. 3^3 **4.** 5^3

5. 2^5 **6.** 3^4

Simplify each expression. [1.1]

7. $2 + 3 \cdot 5$ **8.** $10 - 2 \cdot 3$

9. $3 \cdot 4^2 - 2 \cdot 3^2$ **10.** $3 + 5(2 \cdot 3^2 - 10)$

Let $A = \{1, 3, 5\}$, $B = \{2, 4, 6\}$, and $C = \{0, 1, 2, 3, 4\}$, and find each of the following. [1.2]

11. $A \cup B$ **12.** $A \cap C$

13. $\{x \mid x \in A \text{ and } x \notin C\}$ **14.** $\{x \mid x \in B \text{ and } x > 4\}$

15. Locate the numbers $-4, -2.5, -1, 0, 1.5, 3.1, 4.75$ on the number line. [1.2]

For the set $\{-7, -4.2, -\sqrt{3}, 0, \frac{3}{4}, \pi, 5\}$, list all the elements that are in the following sets: [1.3]

16. Integers   **17.** Rational numbers

18. Irrational numbers

19. Factor 4,356 into the product of prime factors. [1.3]

20. Reduce $\dfrac{4,356}{5,148}$ to lowest terms. [1.3]

Graph each inequality. [1.4]

21. $\{x \mid x < -2 \text{ or } x > 3\}$ **22.** $\{x \mid x > 2 \text{ and } x < 5\}$

23. $\{x \mid 0 \le x \le 5 \text{ or } x > 10\}$

Translate each statement into an equivalent inequality. [1.4]

24. x is between 0 and 8   **25.** x is between 0 and 8, inclusive

Multiply. [1.5]

26. $\dfrac{3}{4} \cdot \dfrac{8}{5} \cdot \dfrac{5}{6}$ **27.** $\left(\dfrac{3}{4}\right)^3$

Give the opposite and reciprocal of each number. [1.5]

28. 2   **29.** $-\dfrac{2}{5}$

Simplify. [1.5]

30. $|-4|$ **31.** $|10 - 16|$

Find the following products. [1.5]

32. $6(-7)$ **33.** $-3(5)(-2)$

34. $7(3x)$ **35.** $-3(2x)$

Combine similar terms. [1.6]

36. $-2y + 4y$ **37.** $-3x - x + 7x$

Match each expression on the left with the letter of the appropriate property (or properties) on the right. [1.6]

38. $x + 3 = 3 + x$

39. $(x + 2) + 3 = x + (2 + 3)$

40. $3(x + 4) = 3(4 + x)$

41. $(5x)y = x(5y)$

42. $(x + 2) + y = (x + y) + 2$

43. $3(1) = 3$

44. $5 + 0 = 5$

45. $5 + (-5) = 0$

  a. Commutative property of addition
  b. Commutative property of multiplication
  c. Associative property of addition
  d. Associative property of multiplication
  e. Additive identity
  f. Multiplicative identity
  g. Additive inverse
  h. Multiplicative inverse

Find the following sums and differences. [1.1, 1.5, 1.6]

46. $5 - 3$ **47.** $-5 - (-3)$

48. $|-4| - |-3| + |-2|$

Apply the distributive property. [1.6]

49. $-\dfrac{1}{2}(2x - 6)$ **50.** $-3(5x - 1)$

Divide. [1.5]

51. $-\dfrac{5}{8} \div \dfrac{3}{4}$ **52.** $\dfrac{4}{7} \div (-2)$

Simplify each expression as much as possible. [1.6]

53. $2(-5) - 3$ **54.** $3(-4) - 5$

55. $\dfrac{3(-4) - 8}{-5 - 5}$ **56.** $\dfrac{9(-1)^3 - 3(-6)^2}{6 - 9}$

Find the next number in each sequence. Identify any sequences that are arithmetic and any that are geometric. [1.7]

57. $11, 8, 5, 2, \ldots$ **58.** $1, 1, 2, 3, 5, \ldots$

The numbers in brackets indicate the section(s) to which the problems correspond.

Write each of the following in symbols. [1.1]

1. Twice the sum of $3x$ and $4y$.

2. The difference of $2a$ and $3b$ is less than their sum.

Simplify each expression using the rule for order of operations. [1.1]

3. $3 \cdot 2^2 + 5 \cdot 3^2$

4. $6 + 2(4 \cdot 3 - 10)$

5. $12 - 8 \div 4 + 2 \cdot 3$

6. $20 - 4[3^2 - 2(2^3 - 6)]$

If $A = \{1, 2, 3, 4\}$, $B = \{2, 4, 6\}$, and $C = \{1, 3, 5\}$, find each of the following. [1.2]

7. $A \cap B$

8. $\{x \mid x \in B \text{ and } x \in C\}$

For the set $\{-5, -4.1, -\frac{5}{6}, -\sqrt{2}, 0, \sqrt{3}, 1, 1.8, 4\}$, list all the elements belonging to the following sets. [1.3]

9. Integers

10. Rational numbers

11. Irrational numbers

12. Whole numbers

Factor into the product of prime factors. [1.3]

13. 585

14. 620

15. Reduce $\dfrac{585}{620}$ to lowest terms. [1.3]

Write an inequality whose solution is the given graph. Write your answer using set builder notation. [1.4]

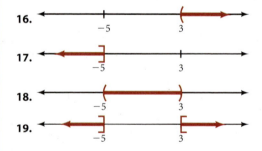

16.
$-5 \qquad 3$

17.
$-5 \qquad 3$

18.
$-5 \qquad 3$

19.
$-5 \qquad 3$

Graph each of the following. [1.4]

20. $\{x \mid x \le -1 \text{ or } x > 5\}$

21. $\{x \mid -2 \le x \le 4\}$

22. Simplify each of the following. [1.5]

a. $-(-3)$

b. $-|-2|$

State the property or properties that justify each of the following. [1.6]

23. $4 + x = x + 4$

24. $5(1) = 5$

25. $3(xy) = (3x) \cdot y$

26. $(a + 1) + b = a + (1 + b)$

Simplify each of the following as much as possible. [1.6]

27. $\dfrac{-4(-1) - (-10)}{5 - (-2)}$

28. $3 - 2\left[\dfrac{8(-1) - 7}{-3(2) - 4}\right]$

29. $-\dfrac{3}{8} + \dfrac{5}{12} - \left(-\dfrac{7}{9}\right)$

30. $-\dfrac{1}{2}(8x)$

31. $-4(3x + 2) + 7x$

32. $5(2y - 3) - (6y - 5)$

33. $3 + 4(2x - 5) - 5x$

34. $2 + 5a + 3(2a - 4)$

35. Add $-\dfrac{2}{3}$ to the product of -2 and $\dfrac{5}{6}$.

36. Subtract $\dfrac{3}{4}$ from the product of -4 and $\dfrac{7}{16}$.

Find the next number in each sequence. Identify any sequences that are arithmetic and any that are geometric. [1.7]

37. $5, -20, 80, \ldots$

38. $1, 0, -1, \ldots$

39. $7, 8, 10, 13, \ldots$

40. $1, \dfrac{1}{5}, \dfrac{1}{25}, \ldots$

The results of a survey of 120 internet users are displayed here as a Venn diagram. Use this information to answer Questions 41 and 42. [1.2]

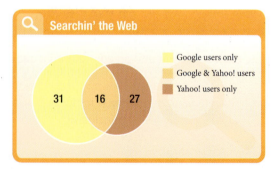

Searchin' the Web

Google users only
Google & Yahoo! users
Yahoo! users only

31   16   27

41. How many of the internet users surveyed use Yahoo! or Google or both?

42. How many of the internet users surveyed use neither Yahoo! nor Google?

Linear Equations in One Variable

OBJECTIVES

A Solve a linear equation in one variable.

TICKET TO SUCCESS

Keep these questions in mind as you read through the section. Then respond in your own words and in complete sentences.

1. What are equivalent equations?
2. What is the addition property of equality?
3. What is the multiplication property of equality?
4. Briefly explain the strategy for solving linear equations in one variable.

Adisa/Shutterstock.com

The population density of a state is calculated by using the number of people in the state compared to the area of the state. If the population of Massachusetts is 6,497,967 and the population density is 828.82 people per square mile, how many square miles does Massachusetts cover? In this section, we will begin our work with linear equations to solve problems like this one.

A Solving Linear Equations

A *linear equation in one variable* is any equation that can be put in the form

$$ax + b = c$$

where a, b, and c are constants and $a \neq 0$. For example, each of the equations

$$5x + 3 = 2 \qquad 2x = 7 \qquad 2x + 5 = 0$$

is linear because they can be put in the form $ax + b = c$. In the first equation, $5x$, 3, and 2 are called *terms* of the equation: $5x$ is a variable term; 3 and 2 are constant terms.

Furthermore, we can find a solution for the equation by substituting a number for the variable.

> **Definition**
>
> The **solution set** for an equation is the set of all numbers that when used in place of the variable make the equation a true statement.

PRACTICE PROBLEMS

1. Show that $x = 3$ is a solution to the equation $4x - 2 = 10$.

> **EXAMPLE 1**   The solution set for $2x - 3 = 9$ is {6} since replacing x with 6 makes the equation a true statement.

$$\text{When} \rightarrow \qquad x = 6$$
$$\text{the equation} \rightarrow \quad 2x - 3 = 9$$
$$\text{becomes} \rightarrow \quad 2(6) - 3 = 9$$
$$12 - 3 = 9$$
$$9 = 9 \qquad \text{A true statement} \qquad ■$$

> **Definition**
>
> Two or more equations with the same solution set are called **equivalent equations.**

2. Show that the equations $3x + 1 = 16$ and $4x - 6 = 14$ both have solution set {5}.

> **EXAMPLE 2**   The equations $2x - 5 = 9$, $x - 1 = 6$, and $x = 7$ are all equivalent equations because the solution set for each is {7}.   ■

Properties of Equality

The first property of equality states that adding the same quantity to both sides of an equation preserves equality. Or, more importantly, adding the same amount to both sides of an equation never changes the solution set. This property is called the *addition property of equality* and is stated in symbols as follows.

> **Addition Property of Equality**
>
> For any three algebraic expressions, A, B, and C,
>
> $$\text{if} \qquad A = B$$
> $$\text{then} \qquad A + C = B + C$$
>
> *In words:* Adding the same quantity to both sides of an equation will not change the solution set.

Our second new property is called the *multiplication property of equality* and is stated as follows.

> **Multiplication Property of Equality**
>
> For any three algebraic expressions A, B, and C, where $C \neq 0$,
>
> $$\text{if} \qquad A = B$$
> $$\text{then} \qquad AC = BC$$
>
> *In words:* Multiplying both sides of an equation by the same nonzero quantity will not change the solution set.

3. Solve $\frac{2}{3}x + 4 = -8$.

> **EXAMPLE 3**   Solve $\frac{3}{4}x + 5 = -4$.

SOLUTION   We begin by adding -5 to both sides of the equation. Once this has been done, we multiply both sides by the reciprocal of $\frac{3}{4}$, which is $\frac{4}{3}$.

Answers

1. See Solutions Section.

2. See Solutions Section.

3. -18

$$\frac{3}{4}x + 5 = -4$$

$$\frac{3}{4}x + 5 + (-\mathbf{5}) = -4 + (-\mathbf{5}) \qquad \text{Add } -5 \text{ to both sides.}$$

$$\frac{3}{4}x = -9$$

$$\frac{\mathbf{4}}{\mathbf{3}}\left(\frac{3}{4}x\right) = \frac{\mathbf{4}}{\mathbf{3}}(-9) \qquad \text{Multiply both sides by } \frac{4}{3}.$$

$$x = -12 \qquad\qquad \frac{4}{3}(-9) = \frac{4}{3}\left(-\frac{9}{1}\right) = -\frac{36}{3} = -12 \qquad \blacksquare$$

EXAMPLE 4 Find the solution set for $3a - 5 = -6a + 1$.

4. Find the solution set for $3a - 3 = -5a + 9$.

SOLUTION To solve for a we must isolate it on one side of the equation. Let's decide to isolate a on the left side by adding $6a$ to both sides of the equation.

$$3a - 5 = -6a + 1$$

$$3a + \mathbf{6a} - 5 = -6a + \mathbf{6a} + 1 \qquad \text{Add } 6a \text{ to both sides.}$$

$$9a - 5 = 1$$

$$9a - 5 + \mathbf{5} = 1 + \mathbf{5} \qquad \text{Add } 5 \text{ to both sides.}$$

$$9a = 6$$

$$\frac{\mathbf{1}}{\mathbf{9}}(9a) = \frac{\mathbf{1}}{\mathbf{9}}(6) \qquad \text{Multiply both sides by } \frac{1}{9}.$$

$$a = \frac{2}{3} \qquad\qquad \frac{1}{9}(6) = \frac{6}{9} = \frac{2}{3}$$

The solution set is $\left\{\frac{2}{3}\right\}$. $\blacksquare$

We can check our solution in Example 4 by replacing a in the original equation with $\frac{2}{3}$.

$$\text{When} \rightarrow \qquad a = \frac{2}{3}$$

$$\text{the equation} \rightarrow \quad 3a - 5 = -6a + 1$$

$$\text{becomes} \rightarrow \quad 3\left(\frac{2}{3}\right) - 5 = -6\left(\frac{2}{3}\right) + 1$$

$$2 - 5 = -4 + 1$$

$$-3 = -3 \qquad \text{A true statement}$$

There will be times when we solve equations and end up with a negative sign in front of the variable. The next example shows how to handle this situation.

EXAMPLE 5 Solve each equation.

a. $-x = 4$ **b.** $-y = -8$

5. Solve each equation.
a. $-x = \frac{2}{3}$
b. $-y = -4$

SOLUTION Neither equation can be considered solved because of the negative sign in front of the variable. To eliminate the negative signs, we simply multiply both sides of each equation by -1.

a. $\qquad -x = 4 \qquad$ **b.** $\qquad -y = -8$

$-\mathbf{1}(-x) = -\mathbf{1}(4) \qquad -\mathbf{1}(-y) = -\mathbf{1}(-8) \qquad$ Multiply each side by -1.

$x = -4 \qquad\qquad\qquad y = 8 \qquad\qquad \blacksquare$

NOTE
From the previous chapter, we know that multiplication by a number and division by its reciprocal always produce the same result. Because of this fact, instead of multiplying each side of our equation by $\frac{1}{9}$, we could just as easily divide each side by 9. If we did so, the last two lines in our solution would look like this:

$$\frac{9a}{9} = \frac{6}{9}$$

$$a = \frac{2}{3}$$

Answers
4. $\frac{3}{2}$
5. a. $-\frac{2}{3}$ **b.** 4

6. Solve $\frac{3}{5}x + \frac{1}{3} = -\frac{5}{6}$.

EXAMPLE 6 Solve $\frac{2}{3}x + \frac{1}{2} = -\frac{3}{8}$.

SOLUTION We can solve this equation by applying our properties and working with fractions, or we can begin by eliminating the fractions. Let's use both methods.

METHOD 1 Working with the fractions.

$$\frac{2}{3}x + \frac{1}{2} + \left(-\frac{1}{2}\right) = -\frac{3}{8} + \left(-\frac{1}{2}\right) \qquad \text{Add } -\frac{1}{2} \text{ to each side.}$$

$$\frac{2}{3}x = -\frac{7}{8} \qquad -\frac{3}{8} + \left(-\frac{1}{2}\right) = -\frac{3}{8} + \left(-\frac{4}{8}\right)$$

$$\frac{3}{2}\left(\frac{2}{3}x\right) = \frac{3}{2}\left(-\frac{7}{8}\right) \qquad \text{Multiply each side by } \frac{3}{2}.$$

$$x = -\frac{21}{16}$$

METHOD 2 Eliminating the fractions in the beginning.

Our original equation has denominators of 3, 2, and 8. The least common denominator, abbreviated LCD, for these three denominators is 24, and it has the property that all three denominators will divide it evenly. If we multiply both sides of our equation by 24, each denominator will divide into 24, and we will be left with an equation that does not contain any denominators other than 1.

$$24\left(\frac{2}{3}x + \frac{1}{2}\right) = 24\left(-\frac{3}{8}\right) \qquad \text{Multiply each side by the LCD 24.}$$

$$24\left(\frac{2}{3}x\right) + 24\left(\frac{1}{2}\right) = 24\left(-\frac{3}{8}\right) \qquad \text{Distributive property on the left side}$$

$$16x + 12 = -9 \qquad \text{Multiply.}$$

$$16x = -21 \qquad \text{Add } -12 \text{ to each side.}$$

$$x = -\frac{21}{16} \qquad \text{Multiply each side by } \frac{1}{16}.$$

CHECK To check our solution, we substitute $x = -\frac{21}{16}$ back into our original equation to obtain

NOTE
We are placing a question mark over the equal sign because we don't know yet if the expression on the left will be equal to the expression on the right.

$$\frac{2}{3}\left(-\frac{21}{16}\right) + \frac{1}{2} \stackrel{?}{=} -\frac{3}{8}$$

$$-\frac{7}{8} + \frac{1}{2} \stackrel{?}{=} -\frac{3}{8}$$

$$-\frac{7}{8} + \frac{4}{8} \stackrel{?}{=} -\frac{3}{8}$$

$$-\frac{3}{8} = -\frac{3}{8} \qquad \text{A true statement} \qquad ■$$

7. Solve.

$0.08x + 0.10(8,000 - x) = 680$

EXAMPLE 7 Solve the equation $0.06x + 0.05(10,000 - x) = 560$.

SOLUTION We can solve the equation in its original form by working with the decimals, or we can eliminate the decimals first by using the multiplication property of equality and solve the resulting equation. Here are both methods.

METHOD 1 Working with the decimals:

$$0.06x + 0.05(10,000 - x) = 560 \qquad \text{Original equation}$$

$$0.06x + 0.05(10,000) - 0.05x = 560 \qquad \text{Distributive property}$$

$$0.01x + 500 = 560 \qquad \text{Simplify the left side.}$$

Answer

6. $-\frac{35}{18}$

7. 6,000

$$0.01x + 500 + (-\mathbf{500}) = 560 + (-\mathbf{500}) \quad \text{Add } -500 \text{ to each side.}$$

$$0.01x = 60$$

$$\frac{0.01x}{\mathbf{0.01}} = \frac{60}{\mathbf{0.01}} \qquad\qquad \text{Divide each side by } 0.01.$$

$$x = 6{,}000$$

METHOD 2 Eliminating the decimals in the beginning: To move the decimal point two places to the right in $0.06x$ and 0.05, we multiply each side of the equation by 100.

$0.06x + 0.05(10{,}000 - x) = 560$	Original equation
$0.06x + 500 - 0.05x = 560$	Distributive property
$\mathbf{100}(0.06x) + \mathbf{100}(500) - \mathbf{100}(0.05x) = \mathbf{100}(560)$	Multiply each side by 100.
$6x + 50{,}000 - 5x = 56{,}000$	
$x + 50{,}000 = 56{,}000$	Simplify the left side.
$x = 6{,}000$	Add $-50{,}000$ to each side.

Using either method, the solution to our equation is 6,000.

CHECK We check our work (to be sure we have not made a mistake in applying the properties or in arithmetic) by substituting 6,000 into our original equation and simplifying each side of the result separately, as the following shows.

$$0.06(\mathbf{6{,}000}) + 0.05(10{,}000 - \mathbf{6{,}000}) \stackrel{?}{=} 560$$

$$0.06(6{,}000) + 0.05(4{,}000) \stackrel{?}{=} 560$$

$$360 + 200 \stackrel{?}{=} 560$$

$$560 = 560 \qquad \text{A true statement} \quad \blacksquare$$

Here is a list of steps to use as a guideline for solving linear equations in one variable.

Strategy   Solving Linear Equations in One Variable

Step 1   a. Use the distributive property to separate terms, if necessary.

  b. If fractions are present, consider multiplying both sides by the LCD to eliminate the fractions. If decimals are present, consider multiplying both sides by a power of 10 to clear the equation of decimals.

  c. Combine similar terms on each side of the equation.

Step 2   Use the addition property of equality to get all variable terms on one side of the equation and all constant terms on the other side. A **variable term** is a term that contains the variable. A **constant term** is a term that does not contain the variable (the number 3, for example).

Step 3   Use the multiplication property of equality to get the variable by itself on one side of the equation.

Step 4   Check your solution in the original equation to be sure that you have not made a mistake in the solution process.

As you will see as you work through the problems in the problem set, it is not always necessary to use all four steps when solving equations. The number of steps used depends on the equation. In Example 8, there are no fractions or decimals in the original equation, so Step 1b will not be used.

8. Solve for x.
$6 - 2(5x - 1) + 4x = 20$

NOTE
It would be a mistake to subtract 3 from 8 first because the rule for order of operations indicates we are to do multiplication before subtraction.

EXAMPLE 8   Solve the equation $8 - 3(4x - 2) + 5x = 35$.

SOLUTION   We must begin by distributing the -3 across the quantity $4x - 2$.

Step 1:	**a.** $8 - 3(4x - 2) + 5x = 35$	Original equation
	$8 - 12x + 6 + 5x = 35$	Distributive property
	c. $-7x + 14 = 35$	Simplify.
Step 2:	$-7x = 21$	Add -14 to each side.
Step 3:	$x = -3$	Multiply by $-\frac{1}{7}$.

Step 4:   When x is replaced by -3 in the original equation, a true statement results. Therefore, -3 is the solution to our equation.   ∎

Identities and Equations with No Solution

Two special cases are associated with solving linear equations in one variable, each of which is illustrated in the following examples.

9. Solve: $3(5x + 1) = 10 + 15x$.

EXAMPLE 9   Solve $2(3x - 4) = 3 + 6x$ for x.

SOLUTION   Applying the distributive property to the left side gives us

$6x - 8 = 3 + 6x$ Distributive property

Now, if we add $-6x$ to each side, we are left with the following

$-8 = 3$

which is a false statement. This means that there is no solution to our equation. Any number we substitute for x in the original equation will lead to a similar false statement.   ∎

10. Solve: $-4 + 8x = 2(4x - 2)$.

EXAMPLE 10   Solve $-15 + 3x = 3(x - 5)$ for x.

SOLUTION   We start by applying the distributive property to the right side.

$-15 + 3x = 3x - 15$ Distributive property

If we add $-3x$ to each side, we are left with the true statement

$-15 = -15$

In this case, our result tells us that any number we use in place of x in the original equation will lead to a true statement. Therefore, all real numbers are solutions to our equation. We say the original equation is an *identity* because the left side is always identically equal to the right side.   ∎

Answers
8. -2
9. No solution
10. All real numbers

Problem Set 2.1

Moving Toward Success

"One of the greatest discoveries a man makes, one of his great surprises, is to find he can do what he was afraid he couldn't do."

—Henry Ford, 1863–1947, American industrialist and founder of Ford Motor Company

1. What was the most important study skill you used while working through Chapter 1?

2. Why should you continue to place an importance on study skills as you work through Chapter 2?

A Solve each of the following equations. [Examples 1–8]

1. $x - 5 = 3$

2. $x + 2 = 7$

3. $2x - 4 = 6$

4. $3x - 5 = 4$

5. $7 = 4a - 1$

6. $10 = 3a - 5$

7. $3 - y = 10$

8. $5 - 2y = 11$

9. $-3 - 4x = 15$

10. $-8 - 5x = -6$

11. $-3 = 5 + 2x$

12. $-12 = 6 + 9x$

13. $-300y + 100 = 500$

14. $-20y + 80 = 30$

15. $160 = -50x - 40$

16. $110 = -60x - 50$

17. $-x = 2$

18. $-x = \dfrac{1}{2}$

19. $-a = -\dfrac{3}{4}$

20. $-a = -5$

21. $\dfrac{2}{3}x = 8$

22. $\dfrac{3}{2}x = 9$

23. $-\dfrac{3}{5}a + 2 = 8$

24. $-\dfrac{5}{3}a + 3 = 23$

25. $8 = 6 + \dfrac{2}{7}y$

26. $1 = 4 + \dfrac{3}{7}y$

27. $2x - 5 = 3x + 2$

28. $5x - 1 = 4x + 3$

29. $-3a + 2 = -2a - 1$

30. $-4a - 8 = -3a + 7$

31. $5 - 2x = 3x + 1$

32. $7 - 3x = 8x - 4$

33. $11x - 5 + 4x - 2 = 8x$

34. $2x + 7 - 3x + 4 = -2x$

35. $6 - 7(m - 3) = -1$

36. $3 - 5(2m - 5) = -2$

37. $7 + 3(x + 2) = 4(x - 1)$

38. $5 + 2(4x - 4) = 3(2x - 1)$

39. $5 = 7 - 2(3x - 1) + 4x$

40. $20 = 8 - 5(2x - 3) + 4x$

41. $\dfrac{1}{2}x + \dfrac{1}{4} = \dfrac{1}{3}x + \dfrac{5}{4}$

42. $\dfrac{2}{3}x - \dfrac{3}{4} = \dfrac{1}{6}x + \dfrac{21}{4}$

43. $-\dfrac{2}{5}x + \dfrac{2}{15} = \dfrac{2}{3}$

44. $-\dfrac{1}{6}x + \dfrac{2}{3} = \dfrac{1}{4}$

45. $\dfrac{3}{4}(8x - 4) = \dfrac{2}{3}(6x - 9)$

46. $\dfrac{3}{5}(5x + 10) = \dfrac{5}{6}(12x - 18)$

47. $\dfrac{1}{4}(12a + 1) - \dfrac{1}{4} = 5$

48. $\dfrac{2}{3}(6x - 1) + \dfrac{2}{3} = 4$

49. $0.35x - 0.2 = 0.15x + 0.1$

50. $0.25x - 0.05 = 0.2x + 0.15$

51. $0.42 - 0.18x = 0.48x - 0.24$

52. $0.3 - 0.12x = 0.18x + 0.06$

A Solve each equation, if possible. [Examples 9, 10]

53. $3x - 6 = 3(x + 4)$

54. $7x - 14 = 7(x - 2)$

55. $4y + 2 - 3y + 5 = 3 + y + 4$

56. $7y + 5 - 2y - 3 = 6 + 5y - 4$

57. $2(4t - 1) + 3 = 5t + 4 + 3t$

58. $5(2t - 1) + 1 = 2t - 4 + 8t$

Now that you have practiced solving a variety of equations, we can turn our attention to the types of equations you will see as you progress through the book. Each equation appears later in the book exactly as you see it below.

Solve each equation.

59. $3x + 2 = 0$

60. $5x - 4 = 0$

61. $0 = 6{,}400a + 70$

62. $0 = 6{,}400a + 60$

63. $x + 2 = 2x$

64. $x + 2 = 7x$

65. $0.07x = 1.4$

66. $0.02x = 0.3$

67. $5(2x + 1) = 12$

68. $4(3x - 2) = 21$

69. $50 = \dfrac{K}{48}$

70. $50 = \dfrac{K}{24}$

71. $100P = 2{,}400$

72. $3.5d = 16(3.5)^2$

73. $x + (3x + 2) = 26$

74. $2(1) + y = 4$

75. $2x - 3(3x - 5) = -6$

76. $2(2y + 6) + 3y = 5$

77. $2(2x - 3) + 2x = 45$

78. $2(4x - 10) + 2x = 12.5$

79. $3x + (x - 2) \cdot 2 = 6$

80. $2x - (x + 1) = -1$

81. $15 - 3(x - 1) = x - 2$

82. $4x - 4(x - 3) = x + 3$

83. $2(x + 3) + x = 4(x - 3)$

84. $5(y + 2) - 4(y + 1) = 3$

85. $6(y - 3) - 5(y + 2) = 8$

86. $2(x + 3) + 3(x + 5) = 2x$

87. $2(20 + x) = 3(20 - x)$

88. $6(7 + x) = 5(9 + x)$

89. $2x + 1.5(75 - x) = 127.5$

90. $x + 0.06x = 954$

91. $0.08x + 0.09(9,000 - x) = 750$

92. $0.08x + 0.09(9,000 - x) = 500$

93. $0.12x + 0.10(15,000 - x) = 1,600$

94. $0.09x + 0.11(11,000 - x) = 1,150$

95. $5\left(\dfrac{19}{15}\right) + 5y = 9$

96. $4\left(\dfrac{19}{15}\right) - 2y = 4$

97. $2\left(\dfrac{29}{22}\right) - 3y = 4$

98. $2x - 3\left(-\dfrac{5}{11}\right) = 4$

99. Work each problem according to the instructions.
 a. Solve: $8x - 5 = 0$.

 b. Solve: $8x - 5 = -5$.

 c. Add: $(8x - 5) + (2x - 5)$.

 d. Solve: $8x - 5 = 2x - 5$.

 e. Multiply: $8(x - 5)$.

 f. Solve: $8(x - 5) = 2(x - 5)$.

100. Work each problem according to the instructions.
 a. Solve: $3x + 6 = 0$.

 b. Solve: $3x + 6 = 4$.

 c. Add: $(3x + 6) + (7x + 4)$.

 d. Solve: $3x + 6 = 7x + 4$.

 e. Multiply: $3(x + 5)$.

 f. Solve: $3(x + 6) = 7(x + 2)$.

Applying the Concepts

101. Cost of a Taxi Ride The taximeter was invented in 1891 by Wilhelm Bruhn. Suppose a taxi company charges $3.85 plus $1.80 per mile for a taxi ride.

a. A woman paid a fare of $25.45. Write an equation that connects the fare the woman paid, the miles she traveled, *n*, and the charges the taximeter computes.

b. Solve the equation from part a to determine how many miles the woman traveled.

102. Coughs and Earaches In 2007, twice as many people visited their doctor because of a cough than an earache. The total number of doctor's visits for these two ailments was reported to be 45 million.

a. Let *x* represent the number of earaches reported in 2007, then write an expression using *x* for the number of coughs reported in 2007.

b. Write an equation that relates 45 million to the variable *x*.

c. Solve the equation from part b to determine the number of people who visited their doctor in 2007 to report an earache.

103. Population Density In July 2001 the population of Puerto Rico was estimated to be 3,937,000 people, with a population density of 1,125 people per square mile.

a. Let *A* represent the area of Puerto Rico in square miles, and write an equation that shows that the population is equal to the product of the area and the population density.

b. Solve the equation from part a, rounding your solution to the nearest square mile.

104. Solving Equations by Trial and Error Sometimes equations can be solved the most easily by trial and error. Solve the following equations by trial and error.

a. Find *x* and *y* if $x \cdot y + 1 = 36$, and both *x* and *y* are prime.

b. Find *w*, *t*, and *z* if $w + t + z + 10 = 52$, and *w*, *t*, and *z* are consecutive terms of a Fibonacci sequence.

c. Find *x* and *y* if $x \neq y$ and $x^y = y^x$.

105. Cars   The chart shows the fastest cars in the world. The sum of five times the speed of a car and 23 is the maximum speed of the SSC Ultimate Aero. What is the speed of the car?

106. Population Density   The chart shows the population densities of some states. Population density is given by the equation $d = \frac{x}{A}$, where d is the population density, x is the population, and A is the area of the state in square miles. If Rhode Island had a population of about 1,052,000 people, what is the area of the state? Round to the nearest square mile.

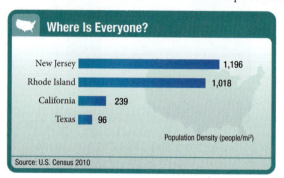

Maintaining Your Skills

From this point on, each problem set will contain a number of problems under the heading Maintaining Your Skills. These problems cover the most important skills you have learned in previous sections and chapters. By working these problems regularly, you will keep yourself current on all the topics we have covered and, possibly, need less time to study for tests and quizzes.

Identify the property (or properties) that justifies each of the following statements.

107. $ax = xa$

108. $5\left(\frac{1}{5}\right) = 1$

109. $3 + (x + y) = (3 + x) + y$

110. $3 + (x + y) = (x + y) + 3$

111. $3 + (x + y) = (3 + y) + x$

112. $7(3x - 5) = 21x - 35$

113. $5(1) = 5$

114. $5 + 0 = 5$

115. $4(xy) = 4(yx)$

116. $4(xy) = (4y)x$

117. $2 + 0 = 2$

118. $2 + (-2) = 0$

Simplify the following expressions.

119. $2x - 3y + 5 + 5x - 7$

120. $5x - 7 + 5y + 3 - 9y$

121. $3(2y - 4) + 7 - 5y$

122. $2(6x - 3) - 4x + 4$

123. $7 - 4(x - 2) + 9$

124. $3x - 5(x + 3) + 11$

Getting Ready for the Next Section

Problems under the heading, *Getting Ready for the Next Section*, are problems that you must be able to work in order to understand the material in the next section. In this case, the problems below are variations on the types of problems you have already worked in this problem set. They are exactly the types of problems you will see in explanations and examples in the next section.

Solve each equation.

125. $x \cdot 42 = 21$

126. $x \cdot 84 = 21$

127. $25 = 0.4x$

128. $35 = 0.4x$

129. $12 - 4y = 12$

130. $-6 - 3y = 6$

131. $525 = 900 - 300p$

132. $375 = 900 - 300p$

133. $486.7 = 78.5 + 31.4h$

134. $486.7 = 113.0 + 37.7h$

135. Find the value of $2x - 1$ when x is:

    a. 2   **b.** 3   **c.** 5

136. Find the value of $\dfrac{1}{x + 1}$ when x is:

    a. 1   **b.** 2   **c.** 3

Extending the Concepts

Solve each equation.

137. $\dfrac{x + 4}{5} - \dfrac{x + 3}{3} = -\dfrac{7}{15}$

138. $\dfrac{x + 1}{7} - \dfrac{x - 2}{2} = \dfrac{1}{14}$

139. $\dfrac{1}{x} - \dfrac{2}{3} = \dfrac{2}{x}$

140. $\dfrac{1}{x} - \dfrac{3}{5} = \dfrac{2}{x}$

141. $\dfrac{x + 3}{2} - \dfrac{x - 4}{4} = -\dfrac{1}{8}$

142. $\dfrac{x - 3}{5} - \dfrac{x + 1}{10} = -\dfrac{1}{10}$

143. $\dfrac{x - 1}{2} - \dfrac{x + 2}{3} = \dfrac{x + 3}{6}$

144. $\dfrac{x + 2}{4} - \dfrac{x - 1}{3} = \dfrac{x + 2}{6}$

Formulas

OBJECTIVES

A Solve a formula with numerical replacements for all but one of its variables.

B Solve formulas for the indicated variable.

C Solve basic percent problems by translating them into equations.

TICKET TO SUCCESS

Keep these questions in mind as you read through the section. Then respond in your own words and in complete sentences.

1. What is a formula in mathematics?
2. How would you solve the formula $2x + 4y = 10$ if $x = 3$?
3. Give two equivalent forms of the rate equation $d = rt$.
4. How would you find what percent of 36 is 27?

SOMATUSCAN/Shutterstock.com

Suppose you are driving home for spring break. If your average speed for the trip is 57 miles per hour and your hometown is 350 miles away, how long will it take you to arrive at your destination? In this section, we will use formulas like the rate equation to answer real-life questions, such as "Will you be home in time for dinner?"

A Solving Formulas with Given Values

> **Definition**
>
> A **formula** in mathematics is an equation that contains more than one variable.

Some formulas are probably already familiar to you—for example, the formula for the area A of a rectangle with length l and width w is $A = lw$.

To begin our work with formulas, we will consider some examples in which we are given numerical replacements for all but one of the variables.

PRACTICE PROBLEMS

1. Find y when x is -3 in $2x - 3y = 6$.

EXAMPLE 1 Find y when x is 4 in the formula $3x - 4y = 2$.

SOLUTION We substitute 4 for x in the formula and then solve for y.

$$\text{When} \rightarrow \qquad x = 4$$

$$\text{the formula} \rightarrow \quad 3x - 4y = 2$$

$$\text{becomes} \rightarrow \quad 3(4) - 4y = 2$$

$$12 - 4y = 2 \qquad \text{Multiply 3 and 4.}$$

$$-4y = -10 \qquad \text{Add } -12 \text{ to each side.}$$

$$y = \frac{5}{2} \qquad \text{Divide each side by } -4. \qquad \blacksquare$$

Note that in the last line of Example 1 we divided each side of the equation by -4. Remember that this is equivalent to multiplying each side of the equation by $-\frac{1}{4}$.

2. Repeat Example 2 if they want to sell 375 pads each week.

EXAMPLE 2 A store selling art supplies finds that they can sell x sketch pads each week at a price of p dollars each, according to the formula $x = 900 - 300p$. What price should they charge for each sketch pad if they want to sell 525 pads each week?

SOLUTION Here we are given a formula, $x = 900 - 300p$, and asked to find the value of p if x is 525. To do so, we simply substitute 525 for x and solve for p.

$$\text{When} \rightarrow \qquad x = 525$$

$$\text{the formula} \rightarrow \quad x = 900 - 300p$$

$$\text{becomes} \rightarrow \quad 525 = 900 - 300p$$

$$-375 = -300p \qquad \text{Add } -900 \text{ to each side.}$$

$$1.25 = p \qquad \text{Divide each side by } -300.$$

To sell 525 sketch pads, the store should charge \$1.25 for each pad. $\blacksquare$

3. Repeat Example 3, except: $d = 60$ miles, $r = 18$ miles per hour, and $t = 5$ hours.

EXAMPLE 3 A boat is traveling upstream against a current. If the speed of the boat in still water is r and the speed of the current is c, then the formula for the distance traveled by the boat is $d = (r - c) \cdot t$, where t is the length of time. Find c if $d = 52$ miles, $r = 16$ miles per hour, and $t = 4$ hours.

SOLUTION Substituting 52 for d, 16 for r, and 4 for t into the formula, we have

$$52 = (16 - c) \cdot 4$$

$$13 = 16 - c \qquad \text{Divide each side by 4.}$$

$$-3 = -c \qquad \text{Add } -16 \text{ to each side.}$$

$$3 = c \qquad \text{Divide each side by } -1.$$

The speed of the current is 3 miles per hour. $\blacksquare$

Answers
1. -4
2. \$1.75
3. 6 mph

FACTS FROM GEOMETRY Formulas for Area and Perimeter

To review, here are the formulas for the area and perimeter of some common geometric objects.

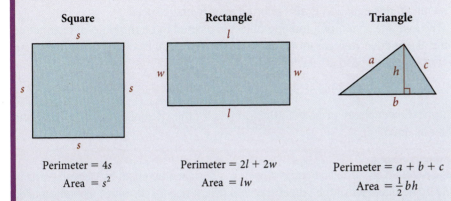

Square

Perimeter = 4s

Area = s^2

Rectangle

Perimeter = $2l + 2w$

Area = lw

Triangle

Perimeter = $a + b + c$

Area = $\frac{1}{2}bh$

The formula for perimeter gives us the distance around the outside of the object along its sides, whereas the formula for area gives us a measure of the amount of surface the object covers.

B Solving Formulas for an Indicated Variable

EXAMPLE 4   Given the formula $P = 2w + 2l$, solve for w.

SOLUTION   To solve for w, we must isolate it on one side of the equation. We can accomplish this if we delete the $2l$ term and the coefficient 2 from the right side of the equation.

To begin, we add $-2l$ to both sides.

$$P + (-2l) = 2w + 2l + (-2l)$$

$$P - 2l = 2w$$

To delete the 2 from the right side, we can multiply both sides by $\frac{1}{2}$.

$$\frac{1}{2}(P - 2l) = \frac{1}{2}(2w)$$

$$\frac{P - 2l}{2} = w$$

The two formulas

$$P = 2w + 2l \qquad \text{and} \qquad w = \frac{P - 2l}{2}$$

give the relationship between P, l, and w. They look different, but they both say the same thing about P, l, and w. The first formula gives P in terms of l and w, and the second formula gives w in terms of P and l. ■

Rate Equation and Average Speed

Now we will look at a problem that uses what is called the rate equation. You use this equation on an intuitive level when you are estimating how long it will take you to drive long distances. For example, if you drive at 50 miles per hour for 2 hours, you will travel 100 miles. Here is the rate equation:

$$\text{Distance} = \text{rate} \cdot \text{time, or } d = r \cdot t$$

4. Solve $P = 2w + 2l$ for l.

Answer

4. $l = \frac{P - 2w}{2}$

The rate equation has two equivalent forms, one of which is obtained by solving for r, while the other is obtained by solving for t. Here they are:

$$r = \frac{d}{t} \quad \text{and} \quad t = \frac{d}{r}$$

The rate in this equation is also referred to as average speed.

The average speed of a moving object is defined to be the ratio of distance to time. If you drive your car for 5 hours and travel a distance of 200 miles, then your average rate of speed is

$$\text{Average speed} = \frac{200 \text{ miles}}{5 \text{ hours}} = 40 \text{ miles per hour}$$

Our next example involves both the formula for the circumference of a circle and the rate equation.

5. Another Ferris wheel has a diameter of 200 ft. If one trip around the wheel takes 18 minutes, find the average speed of a rider. (Use 3.14 as an approximation for π.)

EXAMPLE 5   The first Ferris wheel was designed and built by George Ferris in 1893. The diameter of the wheel was 250 feet. It had 36 carriages, equally spaced around the wheel, each of which held a maximum of 40 people. One trip around the wheel took 20 minutes. Find the average speed of a rider on the first Ferris wheel.

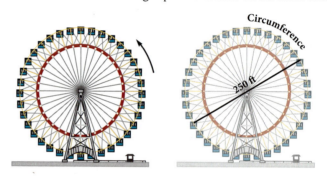

SOLUTION   We can use 3.14 as an approximation for π. The distance traveled is the circumference of the wheel, which is

$$C = 250\pi \approx 250(3.14) = 785 \text{ feet}$$

To find the average speed, we divide the distance traveled by the amount of time it took to go once around the wheel.

$$r = \frac{d}{t} = \frac{785 \text{ feet}}{20 \text{ minutes}} = 39.3 \text{ feet per minute (to the nearest tenth)}$$

Later in the book, we will convert this speed into an equivalent speed in miles per hour. ∎

> **NOTE**
> Recall that the circumference of a circle is given by the formula
> $$C = d\pi$$

6. Solve for x: $ax + 5 = cx + 3$.

> **NOTE**
> We are applying the distributive property in the same way we applied it when we first learned how to simplify $7x - 4x$. Recall that $7x - 4x = 3x$ because $7x - 4x = (7 - 4)x = 3x$ We are using the same type of reasoning when we write $ax - bx = (a - b)x$.

EXAMPLE 6   Solve $ax - 3 = bx + 5$ for x.

SOLUTION   In this example, we must begin by collecting all the variable terms on the left side of the equation and all the constant terms on the other side (just like we did when we were solving linear equations):

$$ax - 3 = bx + 5$$

$$ax - bx - 3 = 5 \qquad \text{Add } -bx \text{ to each side.}$$

$$ax - bx = 8 \qquad \text{Add 3 to each side.}$$

At this point, we need to apply the distributive property to write the left side as $(a - b)x$. After that, we divide each side by $a - b$.

$$(a - b)x = 8 \qquad \text{Distributive property}$$

$$x = \frac{8}{a - b} \qquad \text{Divide each side by } a - b. \qquad ∎$$

Answers

5. 34.9 feet per minute

6. $x = \dfrac{-2}{a - c}$

EXAMPLE 7   Solve $\dfrac{y - b}{x - 0} = m$ for y.

SOLUTION   Although we will do more extensive work with formulas like this later in the book, we need to know how to solve this particular formula for y in order to understand some things in the next chapter. We begin by simplifying the denominator on the left side and then multiplying each side of the formula by x. Doing so makes the rest of the solution process simple.

$$\dfrac{y - b}{x - 0} = m \qquad \text{Original formula}$$

$$\dfrac{y - b}{x} = m \qquad x - 0 = x$$

$$x \cdot \dfrac{y - b}{x} = m \cdot x \qquad \text{Multiply each side by } x.$$

$$y - b = mx \qquad \text{Simplify each side.}$$

$$y = mx + b \qquad \text{Add } b \text{ to each side.}$$

This is our solution. If we look back to the first step, we can justify our result on the left side of the equation this way: Dividing by x is equivalent to multiplying by its reciprocal $\frac{1}{x}$. Here is what it looks like when written out completely:

$$x \cdot \dfrac{y - b}{x} = x \cdot \dfrac{1}{x} \cdot (y - b) = 1(y - b) = y - b \qquad \blacksquare$$

EXAMPLE 8   Solve $\dfrac{y - 4}{x - 5} = 3$ for y.

SOLUTION   We proceed as we did in the previous example, but this time we clear the formula of fractions by multiplying each side of the formula by $x - 5$.

$$\dfrac{y - 4}{x - 5} = 3 \qquad \text{Original formula}$$

$$(x - 5) \cdot \dfrac{y - 4}{x - 5} = 3 \cdot (x - 5) \qquad \text{Multiply each side by } (x - 5).$$

$$y - 4 = 3x - 15 \qquad \text{Simplify each side.}$$

$$y = 3x - 11 \qquad \text{Add 4 to each side.}$$

We have solved for y. We can justify our result on the left side of the equation this way: Dividing by $x - 5$ is equivalent to multiplying by its reciprocal $\frac{1}{x - 5}$. Here are the details:

$$(x - 5) \cdot \dfrac{y - 4}{x - 5} = (x - 5) \cdot \dfrac{1}{x - 5} \cdot (y - 4) = 1(y - 4) = y - 4 \qquad \blacksquare$$

C Basic Percent Problems

The next examples in this section review how basic percent problems can be translated directly into equations. To understand these examples, we must recall that percent means "per hundred." That is, 75% is the same as $\frac{75}{100}$, 0.75, and, $\frac{3}{4}$ in reduced fraction form. Likewise, the decimal 0.25 is equivalent to 25%. To change a decimal to a percent, we move the decimal point two places to the right and write the % symbol. To change from a percent to a decimal, we drop the % symbol and move the decimal point two places to the left. The table that follows gives some of the most commonly used fractions and decimals and their equivalent percents.

7. Solve $\dfrac{y - 2}{x - 0} = \dfrac{4}{3}$ for y.

8. Solve $\dfrac{y + 8}{x - 7} = 6$ for y.

Fraction	Decimal	Percent
$\frac{1}{2}$	0.5	50%
$\frac{1}{4}$	0.25	25%
$\frac{3}{4}$	0.75	75%
$\frac{1}{3}$	$0.\overline{3}$	$33\frac{1}{3}\%$
$\frac{2}{3}$	$0.\overline{6}$	$66\frac{2}{3}\%$
$\frac{1}{5}$	0.2	20%
$\frac{2}{5}$	0.4	40%
$\frac{3}{5}$	0.6	60%
$\frac{4}{5}$	0.8	80%

9. What number is 25% of 74?

EXAMPLE 9 What number is 15% of 63?

SOLUTION To solve a problem like this, we let x equal the number in question and then translate the sentence directly into an equation. Here is how it is done:

What number is 15% of 63?

$$x = 0.15 \cdot 63$$
$$= 9.45$$

The number 9.45 is 15% of 63. ∎

10. What percent of 84 is 21?

EXAMPLE 10 What percent of 42 is 21?

SOLUTION We translate the sentence as follows:

What percent of 42 is 21?

$$x \cdot 42 = 21$$

Next, we divide each side by 42.

$$x = \frac{21}{42}$$
$$= 0.50 \text{ or } 50\%$$
∎

NOTE
We write 0.15 instead of 15% when we translate the sentence into an equation because we cannot do calculations with the % symbol. Because percent means per hundred, we think of 15% as 15 hundredths, or 0.15.

11. 35 is 40% of what number?

EXAMPLE 11 25 is 40% of what number?

SOLUTION Again, we translate the sentence directly.

25 is 40% of what number?

$$25 = 0.40 \cdot x$$

We solve the equation by dividing both sides by 0.40.

$$\frac{25}{0.40} = \frac{0.40 \cdot x}{0.40}$$
$$62.5 = x$$

25 is 40% of 62.5. ∎

Answers
9. 18.5
10. 25%
11. 87.5

Problem Set 2.2

A Use the formula $3x - 4y = 12$ to find y. [Example 1]

1. x is 0

2. x is -2

3. x is 4

4. x is -4

Use the formula $y = 2x - 3$ to find x when

5. y is 0

6. y is -3

7. y is 5

8. y is -5

A Problems 9-26 are all problems that you will see later in the text. [Example 1]

9. If $x - 2y = 4$ and $y = -\dfrac{6}{5}$, find x.

10. If $x - 2y = 4$ and $x = \dfrac{8}{5}$, find y.

11. If $2x + 3y = 6$, find y when x is 0.

12. If $2x + 3y = 6$, find x when y is 0.

13. Let $x = 160$ and $y = 0$ in $y = a(x - 80)^2 + 70$ and solve for a.

14. Let $x = 0$ and $y = 0$ in $y = a(x - 80)^2 + 70$ and solve for a.

15. Find R if $p = 1.5$ and $R = (900 - 300p)p$.

16. Find R if $p = 2.5$ and $R = (900 - 300p)p$.

17. Find P if $P = -0.1x^2 + 27x - 1,700$ and
   a. $x = 100$ **b.** $x = 170$

18. Find P if $P = -0.1x^2 + 27x - 1,820$ and
   a. $x = 130$ **b.** $x = 140$

19. Find h if $h = 16 + 32t - 16t^2$ and
   a. $t = \dfrac{1}{4}$ **b.** $t = \dfrac{7}{4}$

20. Find h if $h = 64t - 16t^2$ and
   a. $t = 1$ **b.** $t = 3$

21. Find y if $x = \dfrac{3}{2}$ and $y = -2x^2 + 6x - 5$.

22. Find y if $x = \dfrac{1}{2}$ and $y = -2x^2 + 6x - 5$.

23. If $y = Kx$, find K if $x = 5$ and $y = 15$.

24. If $d = Kt^2$, find K if $t = 2$ and $d = 64$.

25. If $V = \dfrac{K}{P}$, find K if $P = 48$ and $V = 50$.

26. If $y = Kxz^2$, find K if $x = 5$, $z = 3$, and $y = 180$.

B Solve each of the following formulas for the indicated variable. [Examples 4–8]

27. $A = lw$ for l

28. $A = \dfrac{1}{2}bh$ for b

29. $I = prt$ for t

30. $I = prt$ for r

31. $PV = nRT$ for T

32. $PV = nRT$ for R

33. $y = mx + b$ for x

34. $A = P + Prt$ for t

35. $C = \dfrac{5}{9}(F - 32)$ for F

36. $F = \dfrac{9}{5}C + 32$ for C

37. $h = vt + 16t^2$ for v

38. $h = vt - 16t^2$ for v

39. $A = a + (n - 1)d$ for d

40. $A = a + (n - 1)d$ for n

41. $2x + 3y = 6$ for y

42. $2x - 3y = 6$ for y

43. $-3x + 5y = 15$ for y

44. $-2x - 7y = 14$ for y

45. $2x - 6y + 12 = 0$ for y

46. $7x - 2y - 6 = 0$ for y

47. $ax + 4 = bx + 9$ for x

48. $ax - 5 = cx - 2$ for x

49. $A = P + Prt$ for P

50. $ax + b = cx + d$ for x

Solve for y.

51. $\dfrac{x}{8} + \dfrac{y}{2} = 1$

52. $\dfrac{x}{7} + \dfrac{y}{9} = 1$

53. $\dfrac{x}{5} + \dfrac{y}{-3} = 1$

54. $\dfrac{x}{16} + \dfrac{y}{-2} = 1$

Problems 55 through 62 are all problems that you will see later in the text. Solve each formula for y.

55. $x = 2y - 3$ **56.** $x = 4y + 1$ **57.** $y - 3 = -2(x + 4)$ **58.** $y - 2 = -3(x + 1)$

59. $y - 3 = -\frac{2}{3}(x + 3)$ **60.** $y + 1 = -\frac{2}{3}(x - 3)$ **61.** $y - 4 = -\frac{1}{2}(x + 1)$ **62.** $y - 2 = -\frac{1}{3}(x + 1)$

63. Solve for y.

a. $\dfrac{y + 1}{x - 0} = 4$

b. $\dfrac{y + 2}{x - 4} = -\dfrac{1}{2}$

c. $\dfrac{y + 3}{x - 7} = 0$

64. Solve for y.

a. $\dfrac{y - 1}{x - 0} = -3$

b. $\dfrac{y - 2}{x - 6} = \dfrac{2}{3}$

c. $\dfrac{y - 3}{x - 1} = 0$

C Translate each of the following into a linear equation and then solve the equation. [Examples 9–11]

65. What number is 54% of 38? **66.** What number is 11% of 67? **67.** What percent of 36 is 9?

68. What percent of 50 is 5? **69.** 37 is 4% of what number? **70.** 8 is 2% of what number?

The next two problems are intended to give you practice reading, and paying attention to the instructions that accompany the problems you are working. As we have mentioned previously, working these problems is an excellent way to get ready for a test or a quiz.

71. Work each problem according to the instructions.

a. Solve: $-4x + 5 = 20$.

b. Find the value of $-4x + 5$ when x is 3.

c. Solve for y: $-4x + 5y = 20$.

d. Solve for x: $-4x + 5y = 20$.

72. Work each problem according to the instructions.

a. Solve: $2x + 1 = -4$.

b. Find the value of $2x + 1$ when x is 8.

c. Solve for y: $2x + y = 20$.

d. Solve for x: $2x + y = 20$.

Applying the Concepts

73. Devin left a $4 tip for a $25 lunch with his girlfriend. What percent of the cost of lunch was the tip?

74. Janai left a $3 tip for a $15 breakfast with her boyfriend. What percent of the cost of breakfast was the tip?

75. If the sales tax is 6.5% of the purchase price, what is the sales tax on a $50 purchase?

76. If the sales tax is 7.25% of the purchase price, what is the sales tax on a $120 purchase?

77. During the annual sale at the Boot Factory, Fred purchases a pair of $94 boots for only $56.40. What is the discount rate during the sale?

78. Whole tri-tip is priced at $5.65 per pound at the local supermarket, but costs only $4.52 per pound for members of their frequent shoppers club. What is the discount rate for club members?

79. Google Earth The Google Earth map shows Crater Lake National Park in Oregon. The park covers 286.3 square miles. If the lake covers 7.2% of the park, what is the area of the lake? Round to the nearest tenth.

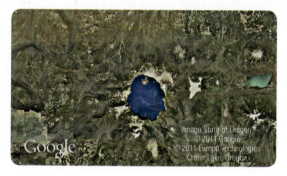

80. U.S. Energy The bar chart shows where Americans get their energy. In 2008, Americans used 99.2 quadrillion Btu of energy. How many Btu's did Americans use from natural gas? Round to the nearest tenth.

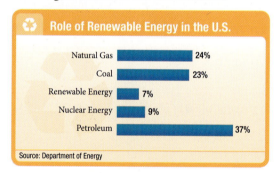

Driving The chart shows the driving distances between some of the largest East Coast cities. Use the rate equation to answer the following questions. Round to the nearest tenth.

81. If a student was traveling home for spring break driving from Washington DC to Boston and their average speed was 60 miles per hour, how long will it take for them to get home?

82. If some students were taking a road trip from Philadelphia to New York City and they made the trip in 1.75 hours, what was their average speed?

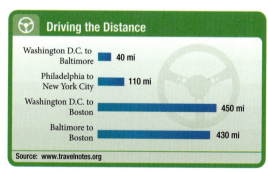

Pricing A company that manufactures ink cartridges finds that they can sell x cartridges each week at a price of p dollars each, according to the formula $x = 3{,}800 - 100p$. What price should they charge for each cartridge if they want to sell

83. 800 cartridges each week?

84. 300 cartridges each week?

85. 400 cartridges each week?

86. 900 cartridges each week?

87. Current It takes a boat 2 hours to travel 18 miles upstream against the current. If the speed of the boat in still water is 15 miles per hour, what is the speed of the current?

88. Current It takes a boat 6.5 hours to travel 117 miles upstream against the current. If the speed of the current is 5 miles per hour, what is the speed of the boat in still water?

89. Wind An airplane takes 4 hours to travel 864 miles while flying against the wind. If the speed of the airplane on a windless day is 258 miles per hour, what is the speed of the wind?

90. Wind A cyclist takes 3 hours to travel 39 miles while pedaling against the wind. If the speed of the wind is 4 miles per hour, how fast would the cyclist be able to travel on a windless day?

91. Miles/Hour A car travels 220 miles in 4 hours. What is the rate of the car in miles per hour?

92. Miles/Hour A train travels 360 miles in 5 hours. What is the rate of the train in miles per hour?

93. Kilometers/Hour It takes a car 3 hours to travel 252 kilometers. What is the rate in kilometers per hour?

94. Kilometers/Hour In 6 hours an airplane travels 4,200 kilometers. What is the rate of the airplane in kilometers per hour?

For problems 95 and 96, use 3.14 as an approximation for π. Round answers to the nearest tenth.

95. Average Speed A person riding a Ferris wheel with a diameter of 65 feet travels once around the wheel in 30 seconds. What is the average speed of the rider in feet per second?

96. Average Speed A person riding a Ferris wheel with a diameter of 102 feet travels once around the wheel in 3.5 minutes. What is the average speed of the rider in feet per minute?

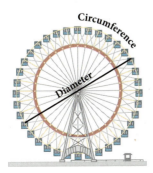

Maintaining Your Skills

Simplify using the rule for order of operations.

97. $38 - 19 + 1$

98. $200 - 150 + 20$

99. $57 - 18 - 8$

100. $71 - 11 - 1$

101. $28 \div 7 \cdot 2$

102. $48 \div 3 \cdot 2$

103. $125 \div 25 \div 5$

104. $36 \div 12 \div 4$

Solve each equation.

105. $3(2x - 9) + 15 - x = -27$

106. $17 - 2(4x + 8) + 4x = -11$

107. $8x - 7 + 5x - 12 = 7$

108. $12 - 5x + 13 + 7x = 15$

109. $\dfrac{x - 4}{3} - 5 + 2x = 24$

110. $7 - \dfrac{x + 7}{2} - 11 = -6$

Getting Ready for the Next Section

To understand all of the explanations and examples in the next section, you must be able to work the problems below.
Translate into symbols.

111. Three less than twice a number

112. Ten less than four times a number

113. The sum of x and y is 180.

114. The sum of a and b is 90.

Solve each equation.

115. $x + 2x = 90$

116. $x + 5x = 180$

117. $2(2x - 3) + 2x = 45$

118. $2(4x - 10) + 2x = 12.5$

119. $6x + 5(10,000 - x) = 56,000$

120. $x + 0.0725x = 17,481.75$

Extending the Concepts

121. Solve $\dfrac{x}{a} + \dfrac{y}{b} = 1$ for x.

122. Solve $\dfrac{x}{a} + \dfrac{y}{b} = 1$ for y.

123. Solve $\dfrac{1}{a} + \dfrac{1}{b} = \dfrac{1}{c}$ for a.

124. Solve $\dfrac{1}{a} + \dfrac{1}{b} = \dfrac{1}{c}$ for b.

125. Solve $\dfrac{2}{a} + \dfrac{2}{b} = \dfrac{2}{c}$ for a.

126. Solve $\dfrac{5}{a} + \dfrac{5}{b} = \dfrac{5}{c}$ for b.

127. Solve $\dfrac{2}{a} + \dfrac{4}{b} = \dfrac{2}{c}$ for a.

128. Solve $\dfrac{6}{a} - \dfrac{3}{b} = \dfrac{3}{c}$ for a.

Applications

2.3

OBJECTIVES

A Apply the Blueprint for Problem Solving to a variety of application problems.

B Use a formula to construct a table of paired data.

TICKET TO SUCCESS

Keep these questions in mind as you read through the section. Then respond in your own words and in complete sentences.

1. What is the first step in the Blueprint for Problem Solving?
2. Why is step 6 in the Blueprint for Problem Solving important?
3. What are complementary angles?
4. How would you use your knowledge of formulas to build a table of paired data?

kentoh/Shutterstock.com

Imagine you have been working at a local pizza restaurant. Your starting wage was $8.00 per hour. After working there for 1 year, you get a 5.5% raise. What is your new hourly wage? We frequently see applications like this in our everyday lives. In this section, we will expand our work with linear equations to look at several different kinds of applications.

A Blueprint for Problem Solving

In this section, we use the skills we have developed for solving equations to solve problems written in words. You may find that some of the examples and problems are more realistic than others. Since we are just beginning our work with application problems, even the ones that seem unrealistic are good practice. What is important in this section is the *method* we use to solve application problems, not the applications themselves. The method, or strategy, that we use to solve application problems is called the *Blueprint for Problem Solving*. It is an outline that will overlay the solution process we use on all application problems.

Strategy **Blueprint for Problem Solving**

Step 1 **Read** the problem, and then mentally **list** the items that are known and the items that are unknown.

Step 2 **Assign a variable** to one of the unknown items. (In most cases this will amount to letting x = the item that is asked for in the problem.) Then **translate** the other **information** in the problem to expressions involving the variable.

Step 3 **Reread** the problem, and then **write an equation,** using the items and variable listed in steps 1 and 2, that describes the situation.

Step 4 **Solve the equation** found in step 3.

Step 5 **Write your answer** using a complete sentence.

Step 6 **Reread** the problem, and **check** your solution with the original words in the problem.

A number of substeps occur within each of the steps in our blueprint. For instance, with steps 1 and 2 it is always a good idea to draw a diagram or picture if it helps you visualize the relationship between the items in the problem.

PRACTICE PROBLEMS

1. The width of a rectangle is 10 feet less than 4 times the length. If the perimeter is 12.5 feet, find the length and width.

EXAMPLE 1   The length of a rectangle is 3 inches less than twice the width. The perimeter is 45 inches. Find the length and width.

SOLUTION   When working problems that involve geometric figures, a sketch of the figure helps organize and visualize the problem (see Figure 1).

Step 1:   **Read and list.**
  Known items: The figure is a rectangle. The length is 3 inches less than twice the width. The perimeter is 45 inches.
  Unknown items: The length and the width

$2x - 3$

FIGURE 1

Step 2:   **Assign a variable and translate information.**
  Since the length is given in terms of the width (the length is 3 less than twice the width), we let x = the width of the rectangle. The length is 3 less than twice the width, so it must be $2x - 3$. The diagram in Figure 1 is a visual description of the relationships we have listed so far.

Step 3:   **Reread and write an equation.**
  The equation that describes the situation is

Twice the length + twice the width = perimeter
$$2(2x - 3) \quad + \quad 2x \quad = \quad 45$$

Step 4:   **Solve the equation.**
$$2(2x - 3) + 2x = 45$$
$$4x - 6 + 2x = 45$$
$$6x - 6 = 45$$
$$6x = 51$$
$$x = 8.5$$

Answers
1. length = 3.25 feet, width = 3 feet

Step 5: **Write the answer.**
The width is 8.5 inches. The length is $2x - 3 = 2(8.5) - 3 = 14$ inches.

Step 6: **Reread and check.**
If the length is 14 inches and the width is 8.5 inches, then the perimeter must be $2(14) + 2(8.5) = 28 + 17 = 45$ inches. Also, the length, 14, is 3 less than twice the width. ∎

Remember that step 1 is done mentally. Read the problem and then *mentally* list the items that you know and the items that you don't know. The purpose of step 1 is to give you direction as you begin to work application problems. Finding the solution to an application problem is a process; it doesn't happen all at once. The first step is to read the problem with a purpose in mind. That purpose is to mentally note the items that are known and the items that are unknown.

EXAMPLE 2   Pat bought a new Ford Mustang with a 5.0-liter engine. The total price, which includes the price of the car plus sales tax, was $17,481.75. If the sales tax rate is 7.25%, what was the price of the car?

SOLUTION

Step 1: **Read and list.**
Known items: The total price is $17,481.75. The sales tax rate is 7.25%, which is 0.0725 in decimal form.
Unknown item: The price of the car

Step 2: **Assign a variable and translate information.**
If we let $x =$ the price of the car, then to calculate the sales tax, we multiply the price of the car x by the sales tax rate:

$$\text{Sales tax} = (\text{sales tax rate})(\text{price of the car})$$

$$= 0.0725x$$

Step 3: **Reread and write an equation.**

$$\text{Car price} + \text{sales tax} = \text{total price}$$

$$x + 0.0725x = 17{,}481.75$$

Step 4: **Solve the equation.**

$$x + 0.0725x = 17{,}481.75$$

$$1.0725x = 17{,}481.75$$

$$x = \frac{17{,}481.75}{1.0725}$$

$$= 16{,}300.00$$

Step 5: **Write the answer.**
The price of the car is $16,300.00.

Step 6: **Reread and check.**
The price of the car is $16,300.00. The tax is $0.0725(16{,}300) = \$1{,}181.75$. Adding the retail price and the sales tax we have a total bill of $17,481.75. ∎

2. If Pat had purchased the convertible version of the Ford Mustang discussed in Example 2, the total price (the price of the car plus sales tax) would have been $23,466.30. Find the price of the convertible if the sales tax rate is 7.25%.

Answer
2. $21,880.00

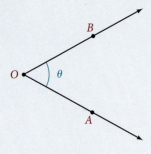

FIGURE 2

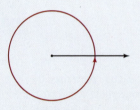

One complete revolution = 360°

FIGURE 3

FACTS FROM GEOMETRY   Angles in General

An angle is formed by two rays with the same endpoint. The common endpoint is called the *vertex* of the angle, and the rays are called the *sides* of the angle.

In Figure 2, angle θ (theta) is formed by the two rays OA and OB. The vertex of θ is O. Angle θ is also denoted as angle AOB, where the letter associated with the vertex is always the middle letter in the three letters used to denote the angle.

Degree Measure

The angle formed by rotating a ray through one complete revolution about its endpoint (Figure 3) has a measure of 360 degrees, which we write as 360°.

One degree of angle measure, written 1°, is $\frac{1}{360}$ of a complete rotation of a ray about its endpoint; there are 360° in one full rotation. (The number 360 was decided on by early civilizations because it was believed that the Earth was at the center of the universe and the sun would rotate once around the Earth every 360 days.) Similarly, 180° is half of a complete rotation, and 90° is a quarter of a full rotation. Angles that measure 90° are called *right angles,* and angles that measure 180° are called *straight angles.* If an angle measures between 0° and 90° it is called an *acute angle,* and an angle that measures between 90° and 180° is an *obtuse angle.* Figure 4 illustrates further.

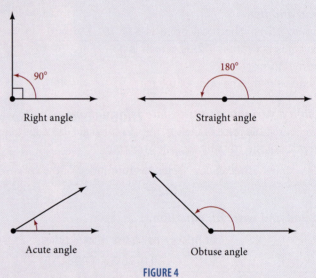

Right angle

Straight angle

Acute angle

Obtuse angle

FIGURE 4

Complementary and Supplementary Angles

If two angles add up to 90°, we call them *complementary angles,* and each is called the *complement* of the other. If two angles have a sum of 180°, we call them *supplementary angles,* and each is called the *supplement* of the other. Figure 5 illustrates the relationship between angles that are complementary and angles that are supplementary.

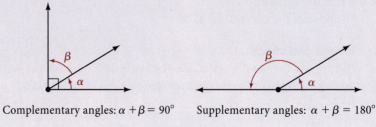

Complementary angles: $\alpha + \beta = 90°$ Supplementary angles: $\alpha + \beta = 180°$

FIGURE 5

FACTS FROM GEOMETRY Special Triangles

It is not unusual to have the terms we use in mathematics show up in the descriptions of things we find in the world around us. The flag of Puerto Rico shown here is described on the government website as "Five equal horizontal bands of red (top and bottom) alternating with white; a blue isosceles triangle based on the hoist side bears a large white five-pointed star in the center." An *isosceles triangle* as shown here and in Figure 6, is a triangle with two sides of equal length.

Angles *A* and *B* in the isosceles triangle in Figure 6 are called the *base angles:* they are the angles opposite the two equal sides. In every isosceles triangle, the base angles are equal.

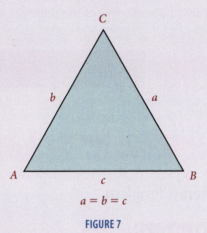

Isosceles Triangle

$a = b$

FIGURE 6

Equilateral Triangle

$a = b = c$

FIGURE 7

> **NOTE**
> As you can see from Figures 6 and 7, one way to label the important parts of a triangle is to label the vertices with capital letters and the sides with small letters: side *a* is opposite vertex *A*, side *b* is opposite vertex *B*, and side *c* is opposite vertex *C*.
> Also, because each vertex is the vertex of one of the angles of the triangle, we refer to the three interior angles as *A*, *B*, and *C*.
> Finally, in any triangle, the sum of the interior angles is 180°. For the triangles shown in Figures 6 and 7, the relationship is written
> $A + B + C = 180°$

An *equilateral triangle* (Figure 7) is a triangle with three sides of equal length. If all three sides in a triangle have the same length, then the three interior angles in the triangle also must be equal. Because the sum of the interior angles in a triangle is always 180°, the three interior angles in any equilateral triangle must be 60°.

EXAMPLE 3   Two complementary angles are such that one is twice as large as the other. Find the two angles.

SOLUTION   Applying the Blueprint for Problem Solving, we have

Step 1:   **Read and list.**
   Known items: Two complementary angles. One is twice as large as the other.
   Unknown items: The size of the angles

Step 2:   **Assign a variable and translate information.**
   Let *x* = the smaller angle. The larger angle is twice the smaller, so we represent the larger angle with 2*x*.

3. Two supplementary angles are such that one is five times as large as the other. Find the two angles.

Step 3:   Reread and write an equation.
Because the two angles are complementary, their sum is 90. Therefore,
$$x + 2x = 90$$

Step 4:   Solve the equation.
$$x + 2x = 90$$
$$3x = 90$$
$$x = 30$$

Step 5:   Write the answer.
The smaller angle is 30°, and the larger angle is $2 \cdot 30 = 60°$.

Step 6:   Reread and check.
The larger angle is twice the smaller angle, and their sum is 90°. ■

Suppose we know that the sum of two numbers is 50. If we let x represent one of the two numbers, how can we represent the other? Let's suppose for a moment that x turns out to be 30. Then the other number will be 20 because their sum is 50; that is, if two numbers add up to 50, and one of them is 30, then the other must be $50 - 30 = 20$. Generalizing this to any number x, we see that if two numbers have a sum of 50 and one of the numbers is x, then the other must be $50 - x$. The following table shows some additional examples:

If two numbers have a sum of	and one of them is	then the other must be
50	x	$50 - x$
10	y	$10 - y$
12	n	$12 - n$

Interest Problem

EXAMPLE 4   Suppose a person invests a total of $10,000 in two accounts. One account earns 5% annually, and the other earns 6% annually. If the total interest earned from both accounts in a year is $560, how much is invested in each account?

SOLUTION

Step 1:   Read and list.
Known items: Two accounts. One pays interest of 5%, and the other pays 6%. The total invested is $10,000.
Unknown items: The number of dollars invested in each individual account

Step 2:   Assign a variable and translate information.
If we let x equal the amount invested at 6%, then $10,000 - x$ is the amount invested at 5%. The total interest earned from both accounts is $560. The amount of interest earned on x dollars at 6% is $0.06x$, whereas the amount of interest earned on $10,000 - x$ dollars at 5% is $0.05(10,000 - x)$.

	Dollars at 6%	Dollars at 5%	Total
Dollars	x	$10,000 - x$	10,000
Interest	$0.06x$	$0.05(10,000 - x)$	560

4. Howard invests a total of $8,000 in two accounts. One account earns 8% annually, and the other earns 10% annually. If the total interest earned from both accounts in a year is $680, how much is invested in each account?

Answers
4. $6,000 at 8%, $2,000 at 10%

Step 3: Reread and write an equation.
The last line gives us the equation we are after.

$$0.06x + 0.05(10,000 - x) = 560$$

Step 4: Solve the equation.
To make this equation a little easier to solve, we begin by multiplying both sides by 100 to move the decimal point two places to the right

$$6x + 5(10,000 - x) = 56,000$$

$$6x + 50,000 - 5x = 56,000$$

$$x + 50,000 = 56,000$$

$$x = 6,000$$

Step 5: Write the answer.
The amount of money invested at 6% is $6,000. The amount of money invested at 5% is $10,000 - $6,000 = $4,000.

Step 6: Reread and check.
To check our results, we find the total interest from the two accounts:

The interest on $6,000 at 6% is 0.06(6,000) = 360

The interest on $4,000 at 5% is 0.05(4,000) = 200

The total interest = $560 ∎

B Table Building

We can use our knowledge of formulas to build tables of paired data. As you will see, equations or formulas that contain exactly two variables produce pairs of numbers that can be used to construct tables.

EXAMPLE 5   A piece of string 12 inches long is to be formed into a rectangle. Build a table that gives the length of the rectangle if the width is 1, 2, 3, 4, or 5 inches. Then find the area of each of the rectangles formed.

SOLUTION   Because the formula for the perimeter of a rectangle is $P = 2l + 2w$ and our piece of string is 12 inches long, the formula we will use to find the lengths for the given widths is $12 = 2l + 2w$. To solve this formula for l, we divide each side by 2 and then subtract w. The result is $l = 6 - w$. Table 1 organizes our work so that the formula we use to find l for a given value of w is shown, and we have added a last column to give us the areas of the rectangles formed. The units for the first three columns are inches, and the units for the numbers in the last column are square inches.

TABLE 1

LENGTH, WIDTH, AND AREA

Width (in.)	Length (in.)		Area (in.²)
w	$l = 6 - w$	l	$A = lw$
1	$l = 6 - 1$	5	5
2	$l = 6 - 2$	4	8
3	$l = 6 - 3$	3	9
4	$l = 6 - 4$	2	8
5	$l = 6 - 5$	1	5

∎

5. Rework Example 5 if the string is 14 inches long and the rectangle has a width of 1, 2, 3, 4, 5, or 6 inches.

14 inches

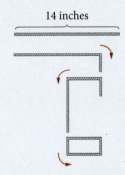

Answer
5. See Solutions Section.

Figures 8 and 9 show two bar charts constructed from the information in Table 1.

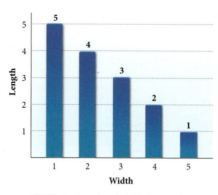

FIGURE 8 Length and width of rectangles with perimeters fixed at 12 inches

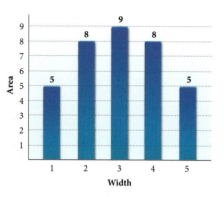

FIGURE 9 Area and width of rectangles with perimeters fixed at 12 inches

USING TECHNOLOGY

More About Example 5

A number of graphing calculators have table-building capabilities. We can let the calculator variable X represent the widths of the rectangles in Example 5. To find the lengths, we set variable Y_1 equal to $6 - X$. The area of each rectangle can be found by setting variable Y_2 equal to $X * Y_1$. To have the calculator produce the table automatically, we use a table minimum of 0 and a table increment of 1. Here is a summary of how the graphing calculator is set up:

Table Setup

Table minimum = 0
Table increment = 1
Independent variable: Auto
Dependent variable: Auto

Y Variables Setup

$Y_1 = 6 - X$
$Y_2 = X * Y_1$

The table will look like this:

X	Y_1	Y_2
0	6	0
1	5	5
2	4	8
3	3	9
4	2	8
5	1	5
6	0	0

Problem Set 2.3

Moving Toward Success

"You learn to speak by speaking, to study by studying, to run by running, to work by working; in just the same way, you learn to love by loving."

—Anatole France, 1844–1924, French writer and journalist

1. What are study skills? Give examples.
2. Why are study skills necessary for your success in this course?

Solve each application problem. Be sure to follow the steps in the Blueprint for Problem Solving. [Example 1–4]

Geometry Problems

1. **Rectangle** A rectangle is twice as long as it is wide. The perimeter is 60 feet. Find the dimensions.

2. **Rectangle** The length of a rectangle is 5 times the width. The perimeter is 48 inches. Find the dimensions.

3. **Square** A square has a perimeter of 28 feet. Find the length of each side.

4. **Square** A square has a perimeter of 36 centimeters. Find the length of each side.

5. **Triangle** A triangle has a perimeter of 23 inches. The medium side is 3 inches more than the shortest side, and the longest side is twice the shortest side. Find the shortest side.

6. **Triangle** The longest side of a triangle is two times the shortest side, whereas the medium side is 3 meters more than the shortest side. The perimeter is 27 meters. Find the dimensions.

7. **Rectangle** The length of a rectangle is 3 meters less than twice the width. The perimeter is 18 meters. Find the width.

8. **Rectangle** The length of a rectangle is 1 foot more than twice the width. The perimeter is 20 feet. Find the dimensions.

9. **Livestock Pen** A livestock pen is built in the shape of a rectangle that is twice as long as it is wide. The perimeter is 48 feet. If the material used to build the pen is $1.75 per foot for the longer sides and $2.25 per foot for the shorter sides (the shorter sides have gates, which increase the cost per foot), find the cost to build the pen.

10. **Garden** A garden is in the shape of a square with a perimeter of 42 feet. The garden is surrounded by two fences. One fence is around the perimeter of the garden, whereas the second fence is 3 feet from the first fence, as Figure 10 indicates. If the material used to build the two fences is $1.28 per foot, what is the total cost of the fences?

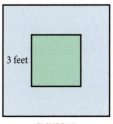

3 feet

FIGURE 10

$P = 48$ ft

Percent Problems

11. Money Shane returned from a trip to Las Vegas with $300.00, which was 50% more money than he had at the beginning of the trip. How much money did Shane have at the beginning of his trip?

12. Items Sold Every item in the Just a Dollar store is priced at $1.00. When Mary Jo opens the store, there is $125.50 in the cash register. When she counts the money in the cash register at the end of the day, the total is $1,058.60. If the sales tax rate is 8.5%, how many items were sold that day?

13. Textbook Price Suppose a college bookstore buys a textbook from a publishing company and then marks up the price they paid for the book 33% and sells it to a student at the marked-up price. If the student pays $75.00 for the textbook, what did the bookstore pay for it? Round your answer to the nearest cent.

14. Movies *The Dark Knight* grossed $15.8 million dollars on its opening weekend and had one of the most successful movie launches in history. If opening weekend accounted for 30% of the movie's gross revenue, what was the total gross revenue? Round your answer to the nearest tenth of a million dollars.

15. Hourly Wage A sheet metal worker earns $26.80 per hour after receiving a 4.5% raise. What was the sheet metal worker's hourly pay before the raise? Round your answer to the nearest cent.

16. Earnings Your annual salary for your first job out of college is $48,000. If you earn a raise of 2.5% every two years, how much will you be making in six years?

More Geometry Problems

17. Angles Two supplementary angles are such that one is eight times larger than the other. Find the two angles.

18. Angles Two complementary angles are such that one is five times larger than the other. Find the two angles.

19. Angles One angle is 12° less than four times another. Find the measure of each angle if

  a. they are complements of each other.

  b. they are supplements of each other.

20. Angles One angle is 4° more than three times another. Find the measure of each angle if

  a. they are complements of each other.

  b. they are supplements of each other.

21. Triangle A triangle is such that the largest angle is three times the smallest angle. The third angle is 9° less than the largest angle. Find the measure of each angle.

22. Triangle The smallest angle in a triangle is half of the largest angle. The third angle is 15° less than the largest angle. Find the measure of all three angles.

23. Triangle The smallest angle in a triangle is one-third of the largest angle. The third angle is 10° more than the smallest angle. Find the measure of all three angles.

24. Triangle The third angle in an isosceles triangle is half as large as each of the two base angles. Find the measure of each angle.

25. Isosceles Triangle The third angle in an isosceles triangle is 8° more than twice as large as each of the two base angles. Find the measure of each angle.

26. Isosceles Triangle The third angle in an isosceles triangle is 4° more than one fifth of each of the two base angles. Find the measure of each angle.

Interest Problems

27. Investing A woman has a total of $9,000 to invest. She invests part of the money in an account that pays 8% per year and the rest in an account that pays 9% per year. If the interest earned in the first year is $750, how much did she invest in each account?

28. Investing A man invests $12,000 in two accounts. If one account pays 10% per year and the other pays 7% per year, how much was invested in each account if the total interest earned in the first year was $960?

	Dollars at 8%	Dollars at 9%	Total
Dollars			
Interest			

	Dollars at 10%	Dollars at 7%	Total
Dollars			
Interest			

29. Investing A total of $15,000 is invested in two accounts. One of the accounts earns 12% per year, and the other earns 10% per year. If the total interest earned in the first year is $1,600, how much was invested in each account?

30. Investing A total of $11,000 is invested in two accounts. One of the two accounts pays 9% per year, and the other account pays 11% per year. If the total interest paid in the first year is $1,150, how much was invested in each account?

31. Investing Stacy has a total of $6,000 in two accounts. The total amount of interest she earns from both accounts in the first year is $500. If one of the accounts earns 8% interest per year and the other earns 9% interest per year, how much did she invest in each account?

32. Investing Travis has a total of $6,000 invested in two accounts. The total amount of interest he earns from the accounts in the first year is $410. If one account pays 6% per year and the other pays 8% per year, how much did he invest in each account?

Miscellaneous Problems

33. Ticket Prices Miguel is selling tickets to a barbecue. Adult tickets cost $6.00 and children's tickets cost $4.00. He sells six more children's tickets than adult tickets. The total amount of money he collects is $184. How many adult tickets and how many children's tickets did he sell?

	Adult	Child
Number		
Income		

34. Working Two Jobs Maggie has a job working in an office for $10 an hour and another job driving a tractor for $12 an hour. One week she works in the office twice as long as she drives the tractor. Her total income for that week is $416. How many hours did she spend at each job?

Job	Office	Tractor
Hours Worked		
Wages Earned		

35. Sales Tax A woman owns a small, cash-only business in a state that requires her to charge 6% sales tax on each item she sells. At the beginning of the day, she has $250 in the cash register. At the end of the day, she has $1,204 in the register. How much money should she send to the state government for the sales tax she collected?

36. Sales Tax A store is located in a state that requires 6% tax on all items sold. If the store brings in a total of $3,392 in one day, how much of that total was sales tax?

Table Building

37. Use $h = 32t - 16t^2$ to complete the table.

t	0	$\frac{1}{4}$	1	$\frac{7}{4}$	2
h					

38. Use $s = \dfrac{60}{t}$ to complete the table.

t	4	6	8	10
s				

39. Horse Racing The graph shows the total amount of money wagered on the Kentucky Derby. Use the information to fill in the table

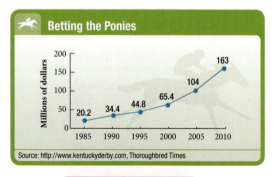

Betting the Ponies

Source: http://www.kentuckyderby.com, Thoroughbred Times

Year	Bets (millions of dollars)
1985	
1990	
1995	
2000	
2005	
2010	

Source: Kentuckyderby.com

40. BMI Sometimes body mass index is used to determine if a person is underweight, overweight, or normal. The equation used to find BMI is found by dividing a persons weight in kilograms by their height squared in meters. So if w is weight and h is height then your BMI is BMI $= \left(\dfrac{w}{h^2}\right)$. Use the formula to fill in the table. Round to the nearest tenth.

Weight (kg)	Height (m)	BMI (kg/m²)
38	1.25	
49	1.5	
53	1.75	
75	2	

Coffee Sales The chart appeared in *USA Today*. Use the information to complete the tables below.

41.

	HOT COFFEE SALES
Year	**Sales (billions of dollars)**
2005	
2006	
2007	
2008	
2009	

42.

	TOTAL COFFEE SALES
Year	**Sales (billions of dollars)**
2005	
2006	
2007	
2008	
2009	

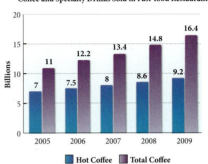

Java Sales
Coffee and Specialty Drinks Sold in Fast-food Restaurants

Hot Coffee Total Coffee

43. Distance A search is being conducted for someone guilty of a hit-and-run felony. In order to set up roadblocks at appropriate points, the police must determine how far the guilty party might have traveled during the past half-hour. Use the formula $d = rt$ with $t = 0.5$ hour to complete the following table.

Speed (miles per hour)	Distance (miles)
20	
30	
40	
50	
60	
70	

44. Speed To determine the average speed of a bullet when fired from a rifle, the time is measured from when the gun is fired until the bullet hits a target that is 1,000 feet away. Use the formula $d = rt$ with $d = 1,000$ feet to complete the following table.

Time (seconds)	Rate (feet per second)
1.00	
0.80	
0.64	
0.50	
0.40	
0.32	

45. Current A boat that can travel 10 miles per hour in still water is traveling along a stream with a current of 4 miles per hour. The distance the boat will travel upstream is given by the formula $d = (r - c) \cdot t$, and the distance it will travel downstream is given by the formula $d = (r + c) \cdot t$. Use these formulas with $r = 10$ and $c = 4$ to complete the following table.

Time (hours)	Distance Upstream (miles)	Distance Downstream (miles)
1		
2		
3		
4		
5		
6		

46. Wind A plane that can travel 300 miles per hour in still air is traveling in a wind stream with a speed of 20 miles per hour. The distance the plane will travel against the wind is given by the formula $d = (r - w) \cdot t$, and the distance it will travel with the wind is given by the formula $d = (r + w) \cdot t$. Use these formulas with $r = 300$ and $w = 20$ to complete the following table.

Time (hours)	Distance against Wind (miles)	Distance with Wind (miles)
.5		
1		
1.5		
2		
2.5		
3		

Maintaining Your Skills

Graph each of the following.

47. $\{x \mid x > -5\}$

48. $\{x \mid x \le 4\}$

49. $\{x \mid x \le -2 \text{ or } x > 5\}$

50. $\{x \mid x < 3 \text{ or } x \ge 5\}$

51. $\{x \mid x > -4 \text{ and } x < 0\}$

52. $\{x \mid x \ge 0 \text{ and } x \le 2\}$

53. $\{x \mid 1 \le x \le 4\}$

54. $\{x \mid -4 < x < -2\}$

Getting Ready for the Next Section

To understand all of the explanations and examples in the next section, you must be able to work the problems below.
Graph each inequality.

55. $x < 2$

56. $x \le 2$

57. $x \ge -3$

58. $x > -3$

Solve.

59. $-2x - 3 = 7$

60. $3x + 3 = 2x - 1$

61. $3(2x - 4) - 7x = -3x$

62. $3(2x + 5) = -3x$

Linear Inequalities in One Variable

2.4

OBJECTIVES

A Solve a linear inequality in one variable and graph the solution set.

B Write solutions to inequalities using interval notation.

C Solve a compound inequality and graph the solution set.

D Solve application problems using inequalities.

TICKET TO SUCCESS

Keep these questions in mind as you read through the section. Then respond in your own words and in complete sentences.

1. What is a linear inequality in one variable?
2. What is the addition property for inequalities?
3. What is the multiplication property for inequalities?
4. What is interval notation?

Evytar Dayan/Shutterstock.com

For Florida, the National Weather Service predicted 3 or more hours of temperature readings between 27 and 32 degrees Fahrenheit during January. Temperatures in this range would cause significant damage to Florida's $9 billion citrus industry. A citrus expert from Spain was consulted about how to manage the damage, but she needed the predictions converted to Celsius in order to render her opinion. In this section, we will begin our work with linear inequalities and their applications like this one.

A *linear inequality in one variable* is any inequality that can be put in the form

$$ax + b < c \qquad (a, b, \text{ and } c \text{ are constants}, a \neq 0)$$

where the inequality symbol $<$ can be replaced with any of the other three inequality symbols ($\leq$, $>$, or $\geq$).

Some examples of linear inequalities are

$$3x - 2 \geq 7 \qquad -5y < 25 \qquad 3(x - 4) > 2x$$

> **NOTE**
> Since subtraction is defined as addition of the opposite, our new property holds for subtraction as well as addition; that is, we can subtract the same quantity from each side of an inequality and always be sure that we have not changed the solution.

A Solving Linear Inequalities

Our first property for inequalities is similar to the addition property we used when solving equations.

Addition Property for Inequalities

For any algebraic expressions, A, B, and C,

$$\text{if} \qquad\qquad A < B$$
$$\text{then} \qquad A + C < B + C$$

In words: Adding the same quantity to both sides of an inequality will not change the solution set.

PRACTICE PROBLEMS

1. Solve $4x - 2 > 3x + 4$ and graph the solution.

NOTE
The English mathematician John Wallis (1616–1703) was the first person to use the ∞ symbol to represent infinity. When we encounter the interval (3, ∞), we read it as "the interval from 3 to infinity," and we mean the set of real numbers that are greater than three. Likewise, the interval $(-\infty, -4)$ is read "the interval from negative infinity to -4," which is all real numbers less than -4.

EXAMPLE 1   Solve $3x + 3 < 2x - 1$, and graph the solution.

SOLUTION   We use the addition property for inequalities to write all the variable terms on one side and all constant terms on the other side.

$$3x + 3 < 2x - 1$$
$$3x + (-\mathbf{2x}) + 3 < 2x + (-\mathbf{2x}) - 1 \qquad \text{Add } -2x \text{ to each side.}$$
$$x + 3 < -1$$
$$x + 3 + (-\mathbf{3}) < -1 + (-\mathbf{3}) \qquad \text{Add } -3 \text{ to each side.}$$
$$x < -4$$

The solution set is all real numbers that are less than -4. To show this, we can use set notation and write

$$\{x \mid x < -4\}$$

Or we can graph the solution set on the number line using a left-opening parenthesis at -4 to show that -4 is not part of the solution set.

This graph gives rise to the following notation, called *interval notation,* that is an alternative way to write the solution set.

$$(-\infty, -4)$$

The preceding expression indicates that the solution set is all real numbers from negative infinity up to, but not including, -4.

We have three equivalent representations for the solution set to our original inequality. Here are all three together.

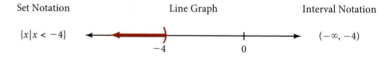

Set Notation	Line Graph	Interval Notation
$\{x \mid x < -4\}$		$(-\infty, -4)$

NOTE
The purpose of this discussion is to justify the multiplication property for inequalities. If you are having trouble understanding it, you may want to read the property itself first and then read through the discussion here. What is important is that you understand and can use the multiplication property for inequalities.

Before we state the multiplication property for inequalities, we will take a look at what happens to an inequality statement when we multiply both sides by a positive number and what happens when we multiply by a negative number.

We begin by writing three true inequality statements:

$$3 < 5 \qquad -3 < 5 \qquad -5 < -3$$

Answers
1–5. See Solutions Section.

We multiply both sides of each inequality by a positive number—say, 4:

$$4(3) < 4(5) \qquad 4(-3) < 4(5) \qquad 4(-5) < 4(-3)$$
$$12 < 20 \qquad\quad -12 < 20 \qquad\quad -20 < -12$$

Notice in each case that the resulting inequality symbol points in the same direction as the original inequality symbol. Multiplying both sides of an inequality by a positive number preserves the *sense* of the inequality.

Let's take the same three original inequalities and multiply both sides by -4.

$$3 < 5 \qquad\qquad -3 < 5 \qquad\qquad -5 < -3$$

$$-4(3) > -4(5) \qquad -4(-3) > -4(5) \qquad -4(-5) > -4(-3)$$
$$-12 > -20 \qquad\qquad 12 > -20 \qquad\qquad 20 > 12$$

Notice in this case that the resulting inequality symbol always points in the opposite direction from the original one. Multiplying both sides of an inequality by a negative number *reverses* the sense of the inequality. Keeping this in mind, we will now state the multiplication property for inequalities.

> ### Multiplication Property for Inequalities
>
> Let A, B, and C represent algebraic expressions.
>
> $$\text{If} \qquad A < B$$
> $$\text{then} \quad AC < BC \quad \text{if} \quad C \text{ is positive } (C > 0)$$
> $$\text{or} \quad\; AC > BC \quad \text{if} \quad C \text{ is negative } (C < 0)$$
>
> *In words:* Multiplying both sides of an inequality by a positive number always produces an equivalent inequality. Multiplying both sides of an inequality by a negative number reverses the sense of the inequality.

The multiplication property for inequalities does not limit what we can do with inequalities. We are still free to multiply both sides of an inequality by any nonzero number we choose. If the number we multiply by happens to be *negative*, then we *must also reverse* the direction of the inequality.

EXAMPLE 2 Find the solution set for $-2y - 3 \le 7$.

SOLUTION We begin by adding 3 to each side of the inequality.

$$-2y - 3 \le 7$$
$$-2y \le 10 \qquad \text{Add 3 to both sides.}$$

$$-\frac{1}{2}(-2y) \ge -\frac{1}{2}(10) \qquad \text{Multiply by } -\tfrac{1}{2} \text{ and reverse the}$$
$$\qquad\qquad\qquad\qquad\qquad \text{direction of the inequality symbol.}$$
$$y \ge -5$$

The solution set is all real numbers that are greater than or equal to -5. Below are three equivalent ways to represent this solution set.

Set Notation	Line Graph	Interval Notation
$\{y \mid y \ge -5\}$	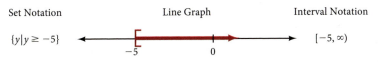	$[-5, \infty)$

Notice how a bracket is used with interval notation to show that -5 is part of the solution set. ∎

NOTE
Because division is defined as multiplication by the reciprocal, we can apply our new property to division as well as to multiplication. We can divide both sides of an inequality by any nonzero number as long as we reverse the direction of the inequality when the number we are dividing by is negative.

2. Find the solution set:
$$-3y - 2 < 7$$

When our inequalities become more complicated, we use the same basic steps we used in Section 2.1 to solve equations; that is, we simplify each side of the inequality before we apply the addition property or multiplication property. When we have solved the inequality, we graph the solution on a number line.

B Interval Notation and Graphing

The following figure shows the connection between inequalities, interval notation, and number line graphs. We have included the graphs with open and closed circles for those of you who have used this type of graph previously. In this book, we will continue to show our graphs using the parentheses/brackets method.

Inequality notation	Interval notation	Graph using parentheses/brackets	Graph using open and closed circles
$x < 2$	$(-\infty, 2)$	0 2	0 2
$x \le 2$	$(-\infty, 2]$	0 2	0 2
$x \ge -3$	$[-3, \infty)$	-3 0	-3 0
$x > -3$	$(-3, \infty)$	-3 0	-3 0

3. Solve $3(2x + 5) \le -3x$.

EXAMPLE 3   Solve $3(2x - 4) - 7x \le -3x$.

SOLUTION   We begin by using the distributive property to separate terms. Next, simplify both sides.

$$3(2x - 4) - 7x \le -3x \qquad \text{Original inequality}$$

$$6x - 12 - 7x \le -3x \qquad \text{Distributive property}$$

$$-x - 12 \le -3x \qquad 6x - 7x = (6 - 7)x = -x$$

$$-12 \le -2x \qquad \text{Add } x \text{ to both sides.}$$

$$-\frac{1}{2}(-12) \ge -\frac{1}{2}(-2x) \qquad \text{Multiply both sides by } -\tfrac{1}{2} \text{ and reverse the direction of the inequality symbol.}$$

$$6 \ge x$$

This last line is equivalent to $x \le 6$. The solution set can be represented in any of the three following ways.

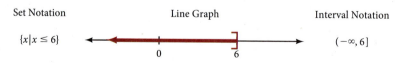

Set Notation	Line Graph	Interval Notation
$\{x \mid x \le 6\}$	0 6	$(-\infty, 6]$

NOTE
In Examples 2 and 3, notice that each time we multiplied both sides of the inequality by a negative number we also reversed the direction of the inequality symbol. Failure to do so would cause our graph to lie on the wrong side of the endpoint.

EXAMPLE 4   Solve and graph $-3 \le 2x - 5 \le 3$.

SOLUTION   We can extend our properties for addition and multiplication to cover this situation. If we add a number to the middle expression, we must add the same number to the outside expressions. If we multiply the center expression by a number, we must do the same to the outside expressions, remembering to reverse the direction of the inequality symbols if we multiply by a negative number. We begin by adding 5 to all three parts of the inequality.

$$-3 \le 2x - 5 \le 3$$
$$2 \le \quad 2x \quad \le 8 \qquad \text{Add 5 to all three members.}$$
$$1 \le \quad x \quad \le 4 \qquad \text{Multiply through by } \tfrac{1}{2}.$$

Here are three ways to write this solution set:

Set Notation   Line Graph   Interval Notation

$\{x \mid 1 \le x \le 4\}$  $[1, 4]$

C Compound Inequalities

EXAMPLE 5   Solve the compound inequality.

$$3t + 7 \le -4 \quad \text{or} \quad 3t + 7 \ge 4$$

SOLUTION   We solve each half of the compound inequality separately, and then we graph the solution set.

$$3t + 7 \le -4 \qquad \text{or} \qquad 3t + 7 \ge 4$$
$$3t \le -11 \qquad \text{or} \qquad 3t \ge -3 \qquad \text{Add } -7.$$
$$t \le -\frac{11}{3} \qquad \text{or} \qquad t \ge -1 \qquad \text{Multiply by } \tfrac{1}{3}.$$

The solution set can be written in any of the following ways:

Set Notation   Line Graph   Interval Notation

$\{t \mid t \le -\frac{11}{3} \text{ or } t \ge -1\}$  $(-\infty, -\frac{11}{3}] \cup [-1, \infty)$

This next figure shows the connection between interval notation and number line graphs for a variety of continued inequalities. Again, we have included the graphs with open and closed circles for those of you who have used this type of graph previously.

Inequality notation	Interval notation	Graph using parentheses/brackets	Graph using open and closed circles
$-3 < x < 2$	$(-3, 2)$		
$-3 \le x \le 2$	$[-3, 2]$		
$-3 \le x < 2$	$[-3, 2)$		
$-3 < x \le 2$	$(-3, 2]$		

4. Solve and graph the solution.
$-7 \le 2x + 1 \le 7$

5. Graph the solution for the compound inequality.
$3t - 6 \le -3 \quad \text{or} \quad 3t - 6 \ge 3$

D Applications

6. Use the information in Example 6 to find the price they must charge to sell at least 800 cartridges a week.

EXAMPLE 6   A company that manufactures ink cartridges for printers finds that they can sell x cartridges each week at a price of p dollars each, according to the formula $x = 1{,}300 - 100p$. What price should they charge for each cartridge if they want to sell at least 300 cartridges a week?

SOLUTION   Because x is the number of cartridges they sell each week, an inequality that corresponds to selling at least 300 cartridges a week is

$$x \geq 300$$

Substituting $1{,}300 - 100p$ for x gives us an inequality in the variable p.

$$1{,}300 - 100p \geq 300$$
$$-100p \geq -1{,}000 \qquad \text{Add } -1{,}300 \text{ to each side.}$$
$$p \leq 10 \qquad \text{Divide each side by } -100, \text{ and reverse the direction of the inequality symbol.}$$

To sell at least 300 cartridges each week, the price per cartridge should be no more than \$10; that is, selling the cartridges for \$10 or less will produce weekly sales of 300 or more cartridges. ∎

7. During one 24-hour period, the temperature in Provo, Utah, ranged from a low of 18° below zero to a high of 22° above zero Fahrenheit. Find the corresponding temperature range in degrees Celsius. In your answer, round to the nearest tenth of a degree.

EXAMPLE 7   The formula $F = \frac{9}{5}C + 32$ gives the relationship between the Celsius and Fahrenheit temperature scales. If the temperature range on a certain day is 86° to 104° Fahrenheit, what is the temperature range in degrees Celsius?

SOLUTION   From the given information we can write $86 \leq F \leq 104$. But, because F is equal to $\frac{9}{5}C + 32$, we can also write

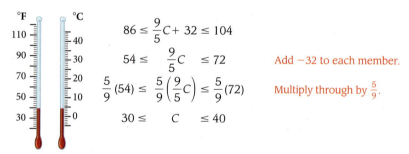

$$86 \leq \frac{9}{5}C + 32 \leq 104$$
$$54 \leq \frac{9}{5}C \leq 72 \qquad \text{Add } -32 \text{ to each member.}$$
$$\frac{5}{9}(54) \leq \frac{5}{9}\left(\frac{9}{5}C\right) \leq \frac{5}{9}(72) \qquad \text{Multiply through by } \frac{5}{9}.$$
$$30 \leq C \leq 40$$

A temperature range of 86° to 104° Fahrenheit corresponds to a temperature range of 30° to 40° Celsius. ∎

Answers

6. \$5 or less
7. $-27.8° \leq C \leq -5.6°$

Problem Set 2.4

Moving Toward Success

"Don't find fault, find a remedy."

—Henry Ford, 1863–1947, American industrialist
and founder of Ford Motor Company

1. How do you plan to use your resources list if you have difficulty with a word problem?
2. Should you do all of the assigned homework problems or just a few? Explain.

A   Solve each of the following inequalities and graph each solution. [Examples 1, 2]

1. $2x \leq 3$

2. $5x \geq -115$

3. $\frac{1}{2}x > 2$

4. $\frac{1}{3}x > 4$

5. $-5x \leq 25$

6. $-7x \geq 35$

7. $-\frac{3}{2}x > -6$

8. $-\frac{2}{3}x < -8$

9. $-12 \leq 2x$

10. $-20 \geq 4x$

11. $-1 \geq -\frac{1}{4}x$

12. $-1 \leq -\frac{1}{5}x$

13. $-3x + 1 > 10$

14. $-2x - 5 \leq 15$

15. $\frac{1}{2} - \frac{m}{12} \leq \frac{7}{12}$

16. $\frac{1}{2} - \frac{m}{10} > -\frac{1}{5}$

17. $\frac{1}{2} \geq -\frac{1}{6} - \frac{2}{9}x$

18. $\frac{9}{5} > -\frac{1}{5} - \frac{1}{2}x$

19. $-40 \leq 30 - 20y$

20. $-20 > 50 - 30y$

21. $\frac{2}{3}x - 3 < 1$

22. $\frac{3}{4}x - 2 > 7$

23. $10 - \frac{1}{2}y \leq 36$

24. $8 - \frac{1}{3}y \geq 20$

B Simplify each side first, then solve the following inequalities. Write your answers with interval notation. [Example 3]

25. $2(3y + 1) \leq -10$

26. $3(2y - 4) > 0$

27. $-(a + 1) - 4a \leq 2a - 8$

28. $-(a - 2) - 5a \leq 3a + 7$

29. $\frac{1}{3}t - \frac{1}{2}(5 - t) < 0$

30. $\frac{1}{4}t - \frac{1}{3}(2t - 5) < 0$

31. $-2 \leq 5 - 7(2a + 3)$

32. $1 < 3 - 4(3a - 1)$

33. $-\frac{1}{3}(x + 5) \leq -\frac{2}{9}(x - 1)$

34. $-\frac{1}{2}(2x + 1) \leq -\frac{3}{8}(x + 2)$

Solve each inequality. Write your answer using inequality notation.

35. $20x + 9{,}300 > 18{,}000$

36. $20x + 4{,}800 > 18{,}000$

B Solve the following continued inequalities. Use both a line graph and interval notation to write each solution set. [Example 4]

37. $-2 \leq m - 5 \leq 2$

38. $-3 \leq m + 1 \leq 3$

39. $-60 < 20a + 20 < 60$

40. $-60 < 50a - 40 < 60$

41. $0.5 \leq 0.3a - 0.7 \leq 1.1$

42. $0.1 \leq 0.4a + 0.1 \leq 0.3$

43. $3 < \frac{1}{2}x + 5 < 6$

44. $5 < \frac{1}{4}x + 1 < 9$

45. $4 < 6 + \frac{2}{3}x < 8$

46. $3 < 7 + \frac{4}{5}x < 15$

C Graph the solution sets for the following compound inequalities. Then write each solution set using interval notation.
[Example 5]

47. $x + 5 \leq -2$ or $x + 5 \geq 2$

48. $3x + 2 < -3$ or $3x + 2 > 3$

49. $5y + 1 \leq -4$ or $5y + 1 \geq 4$

50. $7y - 5 \leq -2$ or $7y - 5 \geq 2$

51. $2x + 5 < 3x - 1$ or $x - 4 > 2x + 6$

52. $3x - 1 > 2x + 4$ or $5x - 2 < 3x + 4$

Translate each of the following phrases into an equivalent inequality statement.

53. x is greater than -2 and at most 4

54. x is less than 9 and at least -3

55. x is less than -4 or at least 1

56. x is at most 1 or more than 6

57. Write each statement using inequality notation.

   a. x is always positive.

   b. x is never negative.

   c. x is greater than or equal to 0.

58. Match each expression on the left with a phrase on the right.

   a. $x^2 \geq 0$ **e.** Never true

   b. $x^2 < 0$ **f.** Sometimes true

   c. $x^2 \leq 0$ **g.** Always true

Solve each inequality by inspection, without showing any work.

59. $x^2 < 0$ **60.** $x^2 \leq 0$ **61.** $x^2 \geq 0$

62. $\dfrac{1}{x^2} \geq 0$ **63.** $\dfrac{1}{x^2} < 0$ **64.** $\dfrac{1}{x^2} = 0$

65. Work each problem according to the instructions.

   a. Evaluate when $x = 0$: $-\dfrac{1}{2}x + 1$.

   b. Solve: $-\dfrac{1}{2}x + 1 = -7$.

   c. Is 0 a solution to $-\dfrac{1}{2}x + 1 < -7$?

   d. Solve: $-\dfrac{1}{2}x + 1 < -7$.

66. Work each problem according to the instructions.

   a. Evaluate when $x = 0$: $-\dfrac{2}{3}x - 5$.

   b. Solve: $-\dfrac{2}{3}x - 5 = 1$.

   c. Is 0 a solution to $-\dfrac{2}{3}x - 5 > 1$?

   d. Solve: $-\dfrac{2}{3}x - 5 > 1$.

Applying the Concepts

Organic Groceries The map shows a range of how many organic grocery stores are found in each state. Use the chart to answer Problems 67 and 68.

67. Find the difference, d, in the number of organic stores between California and the state of Washington.

68. If Oregon had 24 organic grocery stores, what is the difference, d, between the number of stores in Oregon and the number of stores found in Nevada.

Organic Stores in the U.S.

■ 0-10
■ 11-20
■ 21-30
■ 31-40   Number of organic
■ 41-50   grocery stores
■ 51-60   per state
■ 61-70
■ 71-80
■ 81-90

Source: organic.com 2008

Maintaining Your Skills

The problems that follow review some of the more important skills you have learned in previous sections and chapters. You can consider the time you spend working these problems as time spent studying for exams.

Simplify.

69. $|-3|$

70. $|3|$

71. $-|-3|$

72. $-(-3)$

73. Give a definition for the absolute value of x that involves the number line. (This is the geometric definition.)

74. Give a definition of the absolute value of x that does not involve the number line. (This is the algebraic definition.)

Getting Ready for the Next Section

To understand all of the explanations and examples in the next section, you must be able to work the problems below.

Solve each equation.

75. $2a - 1 = -7$

76. $3x - 6 = 9$

77. $\frac{2}{3}x - 3 = 7$

78. $\frac{2}{3}x - 3 = -7$

79. $x - 5 = x - 7$

80. $x + 3 = x + 8$

81. $x - 5 = -x - 7$

82. $x + 3 = -x + 8$

Extending the Concepts

Assume a, b, and c are positive, and solve each formula for x.

83. $ax + b < c$

84. $\frac{x}{a} + \frac{y}{b} < 1$

85. $-c < ax + b < c$

86. $-1 < \frac{ax + b}{c} < 1$

Equations with Absolute Value

TICKET TO SUCCESS

Keep these questions in mind as you read through the section. Then respond in your own words and in complete sentences.

1. Why would some equations that involve absolute value have two solutions instead of one?

2. Translate $|x| = 6$ into words using the definition of absolute value.

3. Explain in words what the equation $|x - 3| = 4$ means with respect to distance on the number line.

4. When is the statement $|x| = x$ true?

Gary Blakeley/Shutterstock.com

Suppose you are going to a new friend's house for a barbecue. You begin by driving 4 blocks north and 10 blocks east, when you realize you are lost. You backtrack 5 blocks south and 8 blocks west before you stop for directions. How far are you away from your starting point? Since you are traveling in both positive and negative directions (backtracking) your answer may end up negative. Intuitively we know that distance cannot be negative, so what do you do with this information? After working through this section you will be able to use absolute value to determine the absolute value of a number and how it relates to the distance you have traveled.

A Solve Equations with Absolute Values

Previously, we defined the absolute value of x, $|x|$, to be the distance between x and 0 on the number line. The absolute value of a number measures its distance from 0.

EXAMPLE 1   Solve $|x| = 5$ for x.

SOLUTION   Using the definition of absolute value, we can read the equation as, "The distance between x and 0 on the number line is 5." If x is 5 units from 0, then x can be 5 or −5.

$$\text{If } |x| = 5 \quad \text{then } x = 5 \quad \text{or} \quad x = -5 \qquad ■$$

PRACTICE PROBLEMS

1. Solve for x: $|x| = 3$.

Answers
1. −3, 3

In general, we can see that any equation of the form $|a| = b$ is equivalent to the equations $a = b$ or $a = -b$, as long as $b > 0$.

2. Solve $|3x - 6| = 9$.

EXAMPLE 2   Solve $|2a - 1| = 7$.

SOLUTION   We can read this question as "$2a - 1$ is 7 units from 0 on the number line." The quantity $2a - 1$ must be equal to 7 or -7:

$$|2a - 1| = 7$$
$$2a - 1 = 7 \quad \text{or} \quad 2a - 1 = -7$$

We have transformed our absolute value equation into two equations that do not involve absolute value. We can solve each equation separately.

$$\begin{array}{lll}
2a - 1 = 7 & \text{or} & 2a - 1 = -7 \\
2a = 8 & \text{or} & 2a = -6 \qquad \text{Add 1 to both sides.} \\
a = 4 & \text{or} & a = -3 \qquad \text{Multiply by } \frac{1}{2}.
\end{array}$$

Our solution set is $\{4, -3\}$.

To check our solutions, we put them into the original absolute value equation:

When →	$a = 4$	When →	$a = -3$
the equation →	$|2a - 1| = 7$	the equation →	$|2a - 1| = 7$
becomes →	$|2(4) - 1| = 7$	becomes →	$|2(-3) - 1| = 7$
	$|7| = 7$		$|-7| = 7$
	$7 = 7$		$7 = 7$ ∎

3. Solve $|4x - 3| + 2 = 3$.

EXAMPLE 3   Solve $\left|\frac{2}{3}x - 3\right| + 5 = 12$.

SOLUTION   To use the definition of absolute value to solve this equation, we must isolate the absolute value on the left side of the equal sign. To do so, we add -5 to both sides of the equation to obtain

$$\left|\frac{2}{3}x - 3\right| = 7$$

Now that the equation is in the correct form, we can write

$$\begin{array}{lll}
\frac{2}{3}x - 3 = 7 & \text{or} & \frac{2}{3}x - 3 = -7 \\
\frac{2}{3}x = 10 & \text{or} & \frac{2}{3}x = -4 \qquad \text{Add 3 to both sides.} \\
x = 15 & \text{or} & x = -6 \qquad \text{Multiply by } \frac{3}{2}.
\end{array}$$

The solution set is $\{15, -6\}$. ∎

NOTE
Recall that ∅ is the symbol we use to denote the empty set. When we use it to indicate the solutions to an equation, then we are saying the equation has no solution.

4. Solve $|7a - 1| = -2$.

EXAMPLE 4   Solve $|3a - 6| = -4$.

SOLUTION   The solution set is ∅ because the left side cannot be negative and the right side is negative. No matter what we try to substitute for the variable a, the quantity $|3a - 6|$ will always be positive or zero. It can never be -4. ∎

Answers
2. $-1, 5$
3. $1, \frac{1}{2}$
4. No solution

Consider the statement $|a| = |b|$. What can we say about a and b? We know they are equal in absolute value. By the definition of absolute value, they are the same distance from 0 on the number line. They must be equal to each other or opposites of each other. In symbols, we write

$$|a| = |b| \quad \Leftrightarrow \quad a = b \quad \text{or} \quad a = -b$$

$$\uparrow \qquad\qquad \uparrow \qquad\quad \uparrow$$

Equal in   Equals   or   Opposites
absolute value

EXAMPLE 5   Solve $|x - 5| = |x - 7|$.

5. Solve $|x + 3| = |x + 8|$.

SOLUTION   The quantities $x - 5$ and $x - 7$ must be equal or they must be opposites, because their absolute values are equal:

Equals		*Opposites*
$x - 5 = x - 7$	or	$x - 5 = -(x - 7)$
$-5 = -7$		$x - 5 = -x + 7$
No solution here		$2x - 5 = 7$
		$2x = 12$
		$x = 6$

Because the first equation leads to a false statement, it will not give us a solution. (If either of the two equations were to reduce to a true statement, it would mean all real numbers would satisfy the original equation.) In this case, our only solution is $x = 6$.   ∎

Answer

5. $-\dfrac{11}{2}$

Problem Set 2.5

Moving Toward Success

"Forget the times of your distress, but never forget what they taught you."

—Herbert Gasser, 1888–1963, American physiologist

1. Do you think you will make mistakes on tests and quizzes, even when you understand the material?

2. If you make a mistake on a test or a quiz, how do you plan on learning from it?

A Use the definition of absolute value to solve each of the following equations. [Examples 1, 2]

1. $|x| = 4$

2. $|x| = 7$

3. $2 = |a|$

4. $5 = |a|$

5. $|x| = -3$

6. $|x| = -4$

7. $|a| + 2 = 3$

8. $|a| - 5 = 2$

9. $|y| + 4 = 3$

10. $|y| + 3 = 1$

11. $4 = |x| - 2$

12. $3 = |x| - 5$

13. $|x - 2| = 5$

14. $|x + 1| = 2$

15. $|a - 4| = \dfrac{5}{3}$

16. $|a + 2| = \dfrac{7}{5}$

17. $1 = |3 - x|$

18. $2 = |4 - x|$

19. $\left|\dfrac{3}{5}a + \dfrac{1}{2}\right| = 1$

20. $\left|\dfrac{2}{7}a + \dfrac{3}{4}\right| = 1$

21. $60 = |20x - 40|$

22. $800 = |400x - 200|$

23. $|2x + 1| = -3$

24. $|2x - 5| = -7$

25. $\left|\dfrac{3}{4}x - 6\right| = 9$

26. $\left|\dfrac{4}{5}x - 5\right| = 15$

27. $\left|1 - \dfrac{1}{2}a\right| = 3$

28. $\left|2 - \dfrac{1}{3}a\right| = 10$

Solve each equation. [Examples 3–5]

29. $|3x + 4| + 1 = 7$

30. $|5x - 3| - 4 = 3$

31. $|3 - 2y| + 4 = 3$

32. $|8 - 7y| + 9 = 1$

33. $3 + |4t - 1| = 8$

34. $2 + |2t - 6| = 10$

35. $\left|9 - \dfrac{3}{5}x\right| + 6 = 12$

36. $\left|4 - \dfrac{2}{7}x\right| + 2 = 14$

37. $5 = \left|\dfrac{2x}{7} + \dfrac{4}{7}\right| - 3$

38. $7 = \left|\dfrac{3x}{5} + \dfrac{1}{5}\right| + 2$

39. $2 = -8 + \left|4 - \dfrac{1}{2}y\right|$

40. $1 = -3 + \left|2 - \dfrac{1}{4}y\right|$

41. $|3a + 1| = |2a - 4|$

42. $|5a + 2| = |4a + 7|$

43. $\left|x - \dfrac{1}{3}\right| = \left|\dfrac{1}{2}x + \dfrac{1}{6}\right|$

44. $\left|\dfrac{1}{10}x - \dfrac{1}{2}\right| = \left|\dfrac{1}{5}x + \dfrac{1}{10}\right|$

45. $|y - 2| = |y + 3|$

46. $|y - 5| = |y - 4|$

47. $|3x - 1| = |3x + 1|$

48. $|5x - 8| = |5x + 8|$

49. $|3 - m| = |m + 4|$

50. $|5 - m| = |m + 8|$

51. $|0.03 - 0.01x| = |0.04 + 0.05x|$

52. $|0.07 - 0.01x| = |0.08 - 0.02x|$

53. $|x - 2| = |2 - x|$

54. $|x - 4| = |4 - x|$

55. $\left|\dfrac{x}{5} - 1\right| = \left|1 - \dfrac{x}{5}\right|$

56. $\left|\dfrac{x}{3} - 1\right| = \left|1 - \dfrac{x}{3}\right|$

57. Work each problem according to the instructions.

   a. Solve: $4x - 5 = 0$.

   b. Solve: $|4x - 5| = 0$.

   c. Solve: $4x - 5 = 3$.

   d. Solve: $|4x - 5| = 3$.

   e. Solve: $|4x - 5| = |2x + 3|$.

58. Work each problem according to the instructions.

   a. Solve: $3x + 6 = 0$.

   b. Solve: $|3x + 6| = 0$.

   c. Solve: $3x + 6 = 4$.

   d. Solve: $|3x + 6| = 4$.

   e. Solve: $|3x + 6| = |7x + 4|$.

Applying the Concepts

59. Bridges The chart shows the longest suspension bridges in the United States. Write an expression using absolute value to find the difference in length between the George Washington Bridge and the Golden Gate Bridge.

Bridging the Gap

Bridge	Length
Verrazano-Narrows Bridge, NY	4,260 ft
Golden Gate Bridge, CA	4,200 ft
Mackinac Bridge, MI	3,800 ft
George Washington Bridge, NY	3,500 ft

Source: infoplease.com

60. Bridges Using the chart in the previous problem, write an expression using absolute value to find the difference in length between the Mackinac Bridge and the Verrazano Narrows Bridge.

Maintaining Your Skills

Graph the following inequalities.

61. $x < -2$ or $x > 8$

62. $1 < x < 4$

63. $-2 \leq x \leq 1$

64. $x \leq -\dfrac{3}{2}$ or $x \geq 3$

Simplify each expression.

65. $\dfrac{38}{30}$

66. $\dfrac{10}{25}$

67. $\dfrac{240}{6}$

68. $\dfrac{39}{13}$

69. $\dfrac{0+6}{0-3}$

70. $\dfrac{6+6}{6-3}$

71. $\dfrac{4-4}{4-2}$

72. $\dfrac{3+6}{3-3}$

Getting Ready for the Next Section

To understand all of the explanations and examples in the next section, you must be able to work the problems below.
Solve each inequality. Do not graph the solution set.

73. $2x - 5 < 3$

74. $-3 < 2x - 5$

75. $-4 \le 3a + 7$

76. $3a + 7 \le 4$

77. $4t - 3 \le -9$

78. $4t - 3 \ge 9$

Solve each inequality for x. Assume a, b, and c are positive.

79. $|x - a| = b$

80. $|x + a| - b = 0$

81. $|ax + b| = c$

82. $|ax - b| - c = 0$

83. $\left|\dfrac{x}{a} + \dfrac{y}{b}\right| = 1$

84. $\left|\dfrac{x}{a} + \dfrac{y}{b}\right| = c$

Inequalities Involving Absolute Value

2.6

OBJECTIVES

A   Solve inequalities with absolute value and graph the solution set.

TICKET TO SUCCESS

Keep these questions in mind as you read through the section. Then respond in your own words and in complete sentences.

1. Write an inequality containing absolute value, the solution to which is all the numbers between -5 and 5 on the number line.

2. Translate $|x| < 3$ into words using the definition of absolute value.

3. Explain in words what the inequality $|x - 5| < 2$ means with respect to distance on the number line.

4. Why is there no solution to the inequality $|2x - 3| < 0$?

The chart below shows the heights of the three highest buildings in the world.

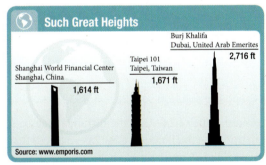

Source: www.emporis.com

The relative size of objects can be compared by extending the use of absolute value to include linear inequalities. In this section, we will again apply the definition of absolute value to solve inequalities involving absolute value. You will see how to compare the heights of these three buildings.

A Solving Inequalities with Absolute Value

The absolute value of x, which is $|x|$, represents the distance that x is from 0 on the number line. We will begin by considering three absolute value equations/inequalities and their English translations.

In Symbols	In Words
$|x| = 7$	x is exactly 7 units from 0 on the number line.
$|a| < 5$	a is less than 5 units from 0 on the number line.
$|y| \geq 4$	y is greater than or equal to 4 units from 0 on the number line.

Once we have translated the expression into words, we can use the translation to graph the original equation or inequality. The graph then is used to write a final equation or inequality that does not involve absolute value.

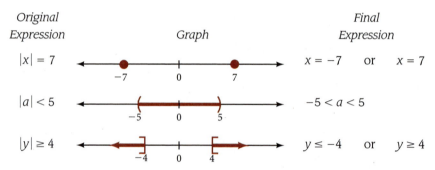

Original Expression	Graph	Final Expression
$|x| = 7$		$x = -7$ or $x = 7$
$|a| < 5$		$-5 < a < 5$
$|y| \geq 4$		$y \leq -4$ or $y \geq 4$

Although we will not always write out the English translation of an absolute value inequality, it is important that we understand the translation. Our second expression, $|a| < 5$, means a is within 5 units of 0 on the number line. The graph of this relationship is

which can be written with the following continued inequality:

$$-5 < a < 5$$

We can follow this same kind of reasoning to solve more complicated absolute value inequalities.

PRACTICE PROBLEMS

1. Graph the solution set.
$$|2x + 1| \leq 7$$

EXAMPLE 1   Graph the solution set: $|2x - 5| < 3$.

SOLUTION   The absolute value of $2x - 5$ is the distance that $2x - 5$ is from 0 on the number line. We can translate the inequality as "$2x - 5$ is less than 3 units from 0 on the number line"; that is, $2x - 5$ must appear between -3 and 3 on the number line. A picture of this relationship is

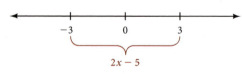

Using the picture, we can write an inequality without absolute value that describes the situation:

$$-3 < 2x - 5 < 3$$

Answers

1–8. See Solutions Section.

Next, we solve the continued inequality by first adding 5 to all three members and then multiplying all three by $\frac{1}{2}$.

$$-3 < 2x - 5 < 3$$

$$2 < 2x < 8 \qquad \text{Add 5 to all three members.}$$

$$1 < x < 4 \qquad \text{Multiply each member by } \frac{1}{2}.$$

The graph of the solution set is

We can see from the solution that for the absolute value of $2x - 5$ to be within 3 units of 0 on the number line, x must be between 1 and 4. ■

EXAMPLE 2   Solve and graph $|3a + 7| \leq 4$.

SOLUTION   We can read the inequality as, "The distance between $3a + 7$ and 0 is less than or equal to 4." Or, "$3a + 7$ is within 4 units of 0 on the number line." This relationship can be written without absolute value as

$$-4 \leq 3a + 7 \leq 4$$

Solving as usual, we have

$$-4 \leq 3a + 7 \leq 4$$

$$-11 \leq \quad 3a \quad \leq -3 \qquad \text{Add } -7 \text{ to all three members.}$$

$$-\frac{11}{3} \leq \quad a \quad \leq -1 \qquad \text{Multiply each member by } \frac{1}{3}.$$

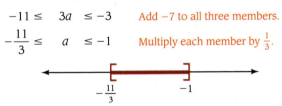

We can see from Examples 1 and 2 that to solve an inequality involving absolute value, we must be able to write an equivalent expression that does not involve absolute value.

EXAMPLE 3   Solve $|x - 3| > 5$ and graph the solution.

SOLUTION   We interpret the absolute value inequality to mean that $x - 3$ is more than 5 units from 0 on the number line. The quantity $x - 3$ must be either above 5 or below -5. Here is a picture of the relationship:

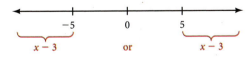

An inequality without absolute value that also describes this situation is

$$x - 3 < -5 \qquad \text{or} \qquad x - 3 > 5$$

Adding 3 to both sides of each inequality we have

$$x < -2 \qquad \text{or} \qquad x > 8$$

The graph of which is

■

2. Solve and graph $|4a - 3| < 5$.

3. Solve $|x + 2| > 7$ and graph the solution.

4. Graph the solution set.

$|3t - 6| \geq 3$

EXAMPLE 4   Graph the solution set: $|4t - 3| \geq 9$

SOLUTION   The quantity $4t - 3$ is greater than or equal to 9 units from 0. It must be either above 9 or below -9.

$$4t - 3 \leq -9 \qquad \text{or} \qquad 4t - 3 \geq 9$$

$$4t \leq -6 \qquad \text{or} \qquad 4t \geq 12 \qquad \textcolor{red}{\text{Add 3.}}$$

$$t \leq -\frac{6}{4} \qquad \text{or} \qquad t \geq \frac{12}{4} \qquad \textcolor{red}{\text{Multiply by } \tfrac{1}{4}.}$$

$$t \leq -\frac{3}{2} \qquad \text{or} \qquad t \geq 3$$

We can use the results of our first few examples and the material in the previous section to summarize the information we have related to absolute value equations and inequalities.

Rule Rewriting Absolute Value Equations and Inequalities

If c is a positive real number, then each of the following statements on the left is equivalent to the corresponding statement on the right.

With Absolute Value	**Without Absolute Value**
$\lvert x \rvert = c$	$x = -c \qquad \text{or} \qquad x = c$
$\lvert x \rvert < c$	$-c < x < c$
$\lvert x \rvert > c$	$x < -c \qquad \text{or} \qquad x > c$
$\lvert ax + b \rvert = c$	$ax + b = -c \qquad \text{or} \qquad ax + b = c$
$\lvert ax + b \rvert < c$	$-c < ax + b < c$
$\lvert ax + b \rvert > c$	$ax + b < -c \qquad \text{or} \qquad ax + b > c$

---

5. Solve and graph.

$|2x + 5| - 2 < 9$

EXAMPLE 5   Solve and graph $|2x + 3| + 4 < 9$.

SOLUTION   Before we can apply the method of solution we used in the previous examples, we must isolate the absolute value on one side of the inequality. To do so, we add -4 to each side.

$$|2x + 3| + 4 < 9$$

$$|2x + 3| + 4 + (\mathbf{-4}) < 9 + (\mathbf{-4})$$

$$|2x + 3| < 5$$

From this last line we know that $2x + 3$ must be between -5 and 5.

$$-5 < 2x + 3 < 5$$

$$-8 < 2x < 2 \qquad \textcolor{red}{\text{Add } -3 \text{ to each member.}}$$

$$-4 < x < 1 \qquad \textcolor{red}{\text{Multiply each member by } \tfrac{1}{2}.}$$

The graph is

EXAMPLE 6   Solve and graph $|4 - 2t| > 2$.

SOLUTION   The inequality indicates that $4 - 2t$ is less than -2 or greater than 2. Writing this without absolute value symbols, we have

$$4 - 2t < -2 \quad \text{or} \quad 4 - 2t > 2$$

To solve these inequalities, we begin by adding -4 to each side.

$$4 + (-4) - 2t < -2 + (-4) \quad \text{or} \quad 4 + (-4) - 2t > 2 + (-4)$$

$$-2t < -6 \qquad \text{or} \qquad -2t > -2$$

Next we must multiply both sides of each inequality by $-\frac{1}{2}$. When we do so, we must also reverse the direction of each inequality symbol.

$$-2t < -6 \qquad \text{or} \qquad -2t > -2$$

$$-\frac{1}{2}(-2t) > -\frac{1}{2}(-6) \quad \text{or} \quad -\frac{1}{2}(-2t) < -\frac{1}{2}(-2)$$

$$t > 3 \qquad \text{or} \qquad t < 1$$

Although, in situations like this, we are used to seeing the "less than" symbol written first, the meaning of the solution is clear. We want to graph all real numbers that are either greater than 3 or less than 1. Here is the graph.

Because absolute value always results in a nonnegative quantity, we sometimes come across special solution sets when a negative number appears on the right side of an absolute value inequality.

EXAMPLE 7   Solve $|7y - 1| < -2$.

SOLUTION   The *left* side is never negative because it is an absolute value. The *right* side is negative. We have a positive quantity (or zero) less than a negative quantity, which is impossible. The solution set is the empty set, $\varnothing$. There is no real number to substitute for y to make this inequality a true statement. ∎

EXAMPLE 8   Solve $|6x + 2| > -5$.

SOLUTION   This is the opposite case from that in Example 7. No matter what real number we use for x on the *left* side, the result will always be positive, or zero. The *right* side is negative. We have a positive quantity (or zero) greater than a negative quantity. Every real number we choose for x gives us a true statement. The solution set is the set of all real numbers. ∎

6. Solve and graph:
$|5 - 2t| > 3$.

NOTE
Remember, the multiplication property for inequalities requires that we reverse the direction of the inequality symbol every time we multiply both sides of an inequality by a negative number.

7. Solve $|8y + 3| \leq -3$.

8. Solve $|2x - 9| \geq -7$.

Problem Set 2.6

A Solve each of the following inequalities using the definition of absolute value. Graph the solution set in each case. [Examples 1–4, 7–8]

1. $|x| < 3$

2. $|x| \leq 7$

3. $|x| \geq 2$

4. $|x| > 4$

5. $|x| + 2 < 5$

6. $|x| - 3 < -1$

7. $|t| - 3 > 4$

8. $|t| + 5 > 8$

9. $|y| < -5$

10. $|y| > -3$

11. $|x| \geq -2$

12. $|x| \leq -4$

13. $|x - 3| < 7$

14. $|x + 4| < 2$

15. $|a + 5| \geq 4$

16. $|a - 6| \geq 3$

17. $|a - 1| < -3$

18. $|a + 2| \geq -5$

19. $|2x - 4| < 6$

20. $|2x + 6| < 2$

21. $|3y + 9| \geq 6$

22. $|5y - 1| \geq 4$

23. $|2k + 3| \geq 7$

24. $|2k - 5| \geq 3$

25. $|x - 3| + 2 < 6$

26. $|x + 4| - 3 < -1$

27. $|2a + 1| + 4 \geq 7$

28. $|2a - 6| - 1 \geq 2$

29. $|3x + 5| - 8 < 5$

30. $|6x - 1| - 4 \leq 2$

Solve each inequality, and graph the solution set. Keep in mind that if you multiply or divide both sides of an inequality by a negative number, you must reverse the inequality sign. [Example 6]

31. $|5 - x| > 3$

32. $|7 - x| > 2$

33. $\left|3 - \dfrac{2}{3}x\right| \geq 5$

34. $\left|3 - \dfrac{3}{4}x\right| \geq 9$

35. $\left|2 - \dfrac{1}{2}x\right| > 1$

36. $\left|3 - \dfrac{1}{3}x\right| > 1$

Solve each inequality.

37. $|x - 1| < 0.01$

38. $|x + 1| < 0.01$

39. $|2x + 1| \geq \dfrac{1}{5}$

40. $|2x - 1| \geq \dfrac{1}{8}$

41. $\left|\dfrac{3x - 2}{5}\right| \leq \dfrac{1}{2}$

42. $\left|\dfrac{4x - 3}{2}\right| \leq \dfrac{1}{3}$

43. $\left|2x - \dfrac{1}{5}\right| < 0.3$

44. $\left|3x - \dfrac{3}{5}\right| < 0.2$

45. Write $-1 \leq x - 5 \leq 1$ as a single inequality involving absolute value.

46. Write $-3 \leq x + 2 \leq 3$ as a single inequality involving absolute value.

47. Is 0 a solution to $|5x + 3| > 7$?

48. Is 0 a solution to $|-2x - 5| > 1$?

49. Solve: $|5x + 3| > 7$.

50. Solve: $|-2x - 5| > 1$.

Maintaining Your Skills

Simplify each expression as much as possible.

51. $-9 \div \dfrac{3}{2}$

52. $-\dfrac{4}{5} \div (-4)$

53. $3 - 7(-6 - 3)$

54. $(3 - 7)(-6 - 3)$

55. $-4(-2)^3 - 5(-3)^2$

56. $4(2 - 5)^3 - 3(4 - 5)^5$

57. $\dfrac{2(-3) - 5(-6)}{-1 - 2 - 3}$

58. $\dfrac{4 - 8(3 - 5)}{2 - 4(3 - 5)}$

59. $6(1) - 1(-5) + 1(2)$

60. $-2(-14) + 3(-4) - 1(-10)$

61. $1(0)(1) + 3(1)(4) + (-2)(2)(-1)$

62. $-3(-1 - 1) + 4(-2 + 2) - 5[2 - (-2)]$

63. $4(-1)^2 + 3(-1) - 2$

64. $5 \cdot 2^3 - 3 \cdot 2^2 + 4 \cdot 2 - 5$

Extending the Concepts

Solve each inequality for x. (Assume a, b, and c are all positive.)

65. $|x - a| < b$

66. $|x - a| > b$

67. $|ax - b| > c$

68. $|ax - b| < c$

69. $|ax + b| \leq c$

70. $|ax + b| \geq c$

Chapter 2 Summary

EXAMPLES

■ Addition Property of Equality [2.1]

1. We can solve

$$x + 3 = 5$$

by adding -3 to both sides:

$$x + 3 + (-3) = 5 + (-3)$$
$$x = 2$$

For algebraic expressions A, B, and C,

$$\text{if} \qquad A = B$$
$$\text{then} \qquad A + C = B + C$$

This property states that we can add the same quantity to both sides of an equation without changing the solution set.

■ Multiplication Property of Equality [2.1]

2. We can solve $3x = 12$ by multiplying both sides by $\frac{1}{3}$.

$$3x = 12$$
$$\frac{1}{3}(3x) = \frac{1}{3}(12)$$
$$x = 4$$

For algebraic expressions A, B, and C,

$$\text{if} \qquad A = B$$
$$\text{then} \qquad AC = BC \quad (C \neq 0)$$

Multiplying both sides of an equation by the same nonzero quantity never changes the solution set.

■ Strategy for Solving Linear Equations in One Variable [2.1]

3. Solve: $3(2x - 1) = 9$.

$$3(2x - 1) = 9$$
$$6x - 3 = 9$$
$$6x - 3 + 3 = 9 + 3$$
$$6x = 12$$
$$\frac{1}{6}(6x) = \frac{1}{6}(12)$$
$$x = 2$$

Step 1: **a.** Use the distributive property to separate terms, if necessary.

b. If fractions are present, consider multiplying both sides by the LCD to eliminate the fractions. If decimals are present, consider multiplying both sides by a power of 10 to clear the equation of decimals.

c. Combine similar terms on each side of the equation.

Step 2: Use the addition property of equality to get all variable terms on one side of the equation and all constant terms on the other side. A variable term is a term that contains the variable (for example, $5x$). A constant term is a term that does not contain the variable (the number 3, for example).

Step 3: Use the multiplication property of equality to get the variable by itself on one side of the equation.

Step 4: Check your solution in the original equation to be sure that you have not made a mistake in the solution process.

■ Formulas [2.2]

4. Solve for w.

$$P = 2l + 2w$$
$$P - 2l = 2w$$
$$\frac{P - 2l}{2} = w$$

A *formula* in algebra is an equation involving more than one variable. To solve a formula for one of its variables, simply isolate that variable on one side of the equation.

◼ Blueprint for Problem Solving [2.3]

5. The perimeter of a rectangle is 32 inches. If the length is 3 times the width, find the dimensions.

Step 1: This step is done mentally.

Step 2: Let x = the width. Then the length is $3x$.

Step 3: The perimeter is 32; therefore

$$2x + 2(3x) = 32$$

Step 4: $\qquad 8x = 32$
$\qquad\qquad x = 4$

Step 5: The width is 4 inches. The length is $3(4) = 12$ inches.

Step 6: The perimeter is $2(4) + 2(12)$, which is 32. The length is 3 times the width.

Step 1: **Read** the problem, and then mentally **list** the items that are known and the items that are unknown.

Step 2: **Assign a variable** to one of the unknown items. (In most cases this will amount to letting x = the item that is asked for in the problem.) Then **translate** the other **information** in the problem to expressions involving the variable.

Step 3: **Reread** the problem, and then **write an equation,** using the items and variables listed in steps 1 and 2, that describes the situation.

Step 4: **Solve the equation** found in step 3.

Step 5: **Write your answer** using a complete sentence.

Step 6: **Reread** the problem, and **check** your solution with the original words in the problem.

◼ Addition Property for Inequalities [2.4]

6. Adding 5 to both sides of the inequality $x - 5 < -2$ gives

$$x - 5 + \mathbf{5} < -2 + \mathbf{5}$$
$$x < 3$$

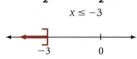

For expressions A, B, and C,

$$\text{if} \qquad\qquad A < B$$

$$\text{then} \qquad A + C < B + C$$

Adding the same quantity to both sides of an inequality never changes the solution set.

◼ Multiplication Property for Inequalities [2.4]

7. Multiplying both sides of $-2x \geq 6$ by $-\frac{1}{2}$ gives

$$-2x \geq 6$$
$$-\tfrac{1}{2}(-2x) \leq -\tfrac{1}{2}(6)$$
$$x \leq -3$$

For expressions A, B, and C,

$$\text{if} \qquad\qquad A < B$$

$$\text{then} \qquad AC < BC \quad \text{if} \quad C > 0 \ (C \text{ is positive})$$

$$\text{or} \qquad AC > BC \quad \text{if} \quad C < 0 \ (C \text{ is negative})$$

We can multiply both sides of an inequality by the same nonzero number without changing the solution set as long as each time we multiply by a negative number we also reverse the direction of the inequality symbol.

Absolute Value Equations [2.5]

8. To solve

$$|2x - 1| + 2 = 7$$

we first isolate the absolute value on the left side by adding -2 to each side to obtain

$$|2x - 1| = 5$$
$$
\begin{array}{lll}
2x - 1 = 5 & \text{or} & 2x - 1 = -5 \\
2x = 6 & \text{or} & 2x = -4 \\
x = 3 & \text{or} & x = -2
\end{array}
$$

To solve an equation that involves absolute value, we isolate the absolute value on one side of the equation and then rewrite the absolute value equation as two separate equations that do not involve absolute value. In general, if b is a positive real number, then

$$|a| = b \quad \text{is equivalent to} \quad a = b \quad \text{or} \quad a = -b$$

Absolute Value Inequalities [2.6]

9. To solve

$$|x - 3| + 2 < 6$$

we first add -2 to both sides to obtain

$$|x - 3| < 4$$

which is equivalent to

$$
\begin{array}{l}
-4 < x - 3 < 4 \\
-1 < x \;\;\;\; < 7
\end{array}
$$

To solve an inequality that involves absolute value, we first isolate the absolute value on the left side of the inequality symbol. Then we rewrite the absolute value inequality as an equivalent continued or compound inequality that does not contain absolute value symbols. In general, if b is a positive real number, then

$$|a| < b \quad \text{is equivalent to} \quad -b < a < b$$

and

$$|a| > b \quad \text{is equivalent to} \quad a < -b \quad \text{or} \quad a > b$$

🚫 **COMMON MISTAKES**

A very common mistake in solving inequalities is to forget to reverse the direction of the inequality symbol when multiplying both sides by a negative number. When this mistake occurs, the graph of the solution set is always drawn on the wrong side of the endpoint.

Chapter 2 Review

Solve each equation. [2.1]

1. $x - 3 = 7$

2. $5x - 2 = 8$

3. $400 - 100a = 200$

4. $5 - \dfrac{2}{3}a = 7$

5. $4x - 2 = 7x + 7$

6. $\dfrac{3}{2}x - \dfrac{1}{6} = -\dfrac{7}{6}x - \dfrac{1}{6}$

7. $7y - 5 - 2y = 2y - 3$

8. $\dfrac{3y}{4} - \dfrac{1}{2} + \dfrac{3y}{2} = 2 - y$

9. $3(2x + 1) = 18$

10. $-\dfrac{1}{2}(4x - 2) = -x$

11. $8 - 3(2t + 1) = 5(t + 2)$

12. $8 + 4(1 - 3t) = -3(t - 4) + 2$

Substitute the given values in each formula and then solve for the variable that does not have a numerical replacement. [2.2]

13. $P = 2b + 2h$: $P = 40$, $b = 3$

14. $A = P + Prt$: $A = 2,000$, $P = 1,000$, $r = 0.05$

Solve each formula for the indicated variable. [2.2]

15. $I = prt$ for p

16. $y = mx + b$ for x

17. $4x - 3y = 12$ for y

18. $d = vt + 16t^2$ for v

Solve each formula for y. [2.2]

19. $5x + 3y - 6 = 0$

20. $\dfrac{x}{3} + \dfrac{y}{2} = 1$

21. $y + 3 = -2(x - 1)$

22. $\dfrac{y + 2}{x - 1} = -3$

Solve each application. In each case, be sure to show the equation that describes the situation. [2.3]

23. **Geometry** The length of a rectangle is 3 times the width. The perimeter is 32 feet. Find the length and width.

24. **Geometry** The three sides of a triangle are given by three consecutive integers. If the perimeter is 12 meters, find the length of each side.

25. **Salary** A teacher has a salary of $25,920 for her second year on the job. If this is 4.2% more than her first year salary, how much did she earn her first year?

26. **Brick Laying** The formula $N = 7 \cdot L \cdot H$ gives the number of standard bricks needed in a wall of L feet long and H feet high and is called the *bricklayer's formula*.

    a. How many bricks would be required in a wall 45 feet long by 12 feet high?

    b. An 8-foot-high wall is to be built from a load of 35,000 bricks. What length wall can be built?

Solve each inequality. Write your answer using interval notation. [2.4]

27. $-8a > -4$

28. $6 - a \geq -2$

29. $\dfrac{3}{4}x + 1 \leq 10$

30. $800 - 200x < 1,000$

31. $\dfrac{1}{3} \leq \dfrac{1}{6}x \leq 1$

32. $-0.01 \leq 0.02x - 0.01 \leq 0.01$

33. $5t + 1 \leq 3t - 2$ or $-7t \leq -21$

34. $3(x + 1) < 2(x + 2)$ or $2(x - 1) \geq x + 2$

Solve each equation. [2.5]

35. $|x| = 2$

36. $|a| - 3 = 1$

37. $|x - 3| = 1$

38. $|2y - 3| = 5$

39. $|4x - 3| + 2 = 11$

40. $\left|\dfrac{7}{3} - \dfrac{x}{3}\right| + \dfrac{4}{3} = 2$

41. $|5t - 3| = |3t - 5|$

42. $\left|\dfrac{1}{2} - x\right| = \left|x + \dfrac{1}{2}\right|$

Solve each inequality and graph the solution set. [2.6]

43. $|x| < 5$

44. $|0.01a| \geq 5$

45. $|x| < 0$

46. $|2t + 1| - 3 < 2$

Solve each equation or inequality, if possible. [2.1, 2.4, 2.5, 2.6]

47. $2x - 3 = 2(x - 3)$

48. $3\left(5x - \dfrac{1}{2}\right) = 15x + 2$

49. $|4y + 8| = -1$

50. $|x| > 0$

51. $|5 - 8t| + 4 \leq 1$

52. $|2x + 1| \geq -4$

Simplify each of the following.

1. 7^2

2. $|-3|$

3. $-2 \cdot 5 + 4$

4. $-2(5 + 4)$

5. $6 + 2 \cdot 5 - 3$

6. $(6 + 2)(5 - 3)$

7. $3 + 2(7 + 4)$

8. $35 - 15 \div 3 + 6$

9. $-(6 - 4) - (3 - 8)$

10. $3(-5)^2 - 6(-2)^3$

11. $\dfrac{-7 - 7}{-3 - 4}$

12. $\dfrac{5(-3) + 6}{8(-3) + 7(3)}$

Reduce the following fractions to lowest terms.

13. $\dfrac{630}{735}$

14. $\dfrac{294}{392}$

Add the following fractions.

15. $\dfrac{11}{30} + \dfrac{13}{42}$

16. $\dfrac{7}{24} + \dfrac{5}{42}$

17. Find the difference of -3 and 6.

18. Add -4 to the product of 5 and -6.

Let $A = \{3, 6, 9\}$; $B = \{4, 5, 6, 7\}$; $C = \{2, 4, 6, 8\}$. Find the following.

19. $A \cup C$

20. $A \cap B$

21. $\{x \mid x \in B \text{ and } x \le 5\}$

22. $\{x \mid x \in C \text{ and } x > 6\}$

For the set $\{-13, -6.7, -\sqrt{5}, 0, \frac{1}{2}, 2, \frac{5}{2}, \pi, \sqrt{13}\}$, list all the numbers that are in each of the following sets.

23. Rational numbers

24. Integers

25. Irrational numbers

26. Whole numbers

Find the next number in the following sequences.

27. $2, -6, 18, \ldots$

28. $-\dfrac{1}{2}, \dfrac{1}{4}, -\dfrac{1}{8}, \ldots$

Simplify each of the following expressions.

29. $6\left(\dfrac{x}{3} - 2\right)$

30. $5(2x + 4) + 6$

31. $x\left(1 - \dfrac{4}{x}\right)$

32. $100(0.07x + 0.06y)$

Solve each of the following equations.

33. $7 - 3y = 14$

34. $5(4 + x) = 3(5 + 2x)$

35. $3 - \dfrac{4}{5}a = -5$

36. $100x^3 = 500x^2$

37. $|y| - 2 = 3$

38. $|3y - 2| = 7$

Substitute the given values in each formula, and then solve for the variable that does not have a numerical replacement.

39. $A = P + Prt$: $A = 1{,}000$, $P = 500$, $r = 0.1$

40. $A = a + (n - 1)d$: $A = 40$, $a = 4$, $d = 9$

Solve each formula for the indicated variable.

41. $4x - 3y = 12$ for x

42. $F = \dfrac{9}{5}C + 32$ for C

43. $A = \dfrac{1}{2}bh$ for h

44. $ax + 6 = bx + 4$ for x

Solve each application. In each case, be sure to show the equation that describes the situation.

45. Angles Two angles are complementary. If the larger angle is 15° more than twice the smaller angle, find the measure of each angle.

46. Rectangle A rectangle is twice as long as it is wide. The perimeter is 78 feet. Find the dimensions.

Solve each inequality. Write your answer using interval notation.

47. $600 - 300x < 900$

48. $-\dfrac{1}{2} \le \dfrac{1}{6}x \le \dfrac{1}{3}$

Solve each inequality and graph the solution set.

49. $|5x - 1| > 3$

50. $|2t + 1| - 1 < 5$

The illustration shows the average price for a gallon of gas in May 2011. Use this information to answer Problems 51 and 52.

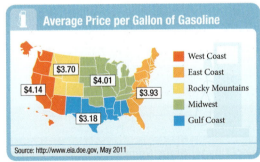

Average Price per Gallon of Gasoline

$3.70
$4.01
$4.14
$3.93
$3.18

West Coast
East Coast
Rocky Mountains
Midwest
Gulf Coast

Source: http://www.eia.doe.gov, May 2011

51. Write an inequality that describes the gas prices in the Gulf Coast relative to the prices in the Midwest.

52. Find the set A using the chart that A contains regions of the United states and $A = \{x \mid x < \$4.00 \text{ and } x > \$3.50\}$?

Chapter 2 Test

Solve the following equations. [2.1]

1. $x - 5 = 7$ **2.** $3y = -4$

3. $5(x - 1) - 2(2x + 3) = 5x - 4$

4. $0.07 - 0.02(3x + 1) = -0.04x + 0.01$

Use the formula $4x - 6y = 12$ to find y when [2.2]

5. $x = 0$ **6.** $x = 3$

Use the formula $-2x + 4y = -8$ to find x when [2.2]

7. $y = 0$ **8.** $y = 2$

Solve for the indicated variable. [2.2]

9. $P = 2l + 2w$ for w **10.** $A = \dfrac{1}{2}h(b + B)$ for B

Solve for y. [2.2]

11. $3x - 2y = -4$ **12.** $5x + 3y = 6$

13. $y - 3 = \dfrac{2}{3}(x + 6)$ **14.** $y + 4 = \dfrac{1}{2}(6x - 4)$

Solve each percent problem. [2.2]

15. 15 is what percent of 25?

16. What percent of 45 is 20?

17. What is 25% of 220?

18. What is 130% of 20?

Solve each geometry problem. [2.3]

19. Find the dimensions of the rectangle below if its length is four times the width and the perimeter is 60 cm.

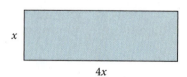

20. The perimeter of the triangle below is 35 meters. Find the value of x.

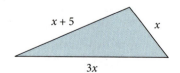

21. Find the measures of the two supplementary angles.

22. Find the measures of the two acute angles in the given right triangle.

Solve each of the following. [2.3]

23. Sales Tax At the beginning of the day, the cash register at a coffee shop contains $75. At the end of the day, it contains $881.25. If the sales tax rate is 7.5%, how much of the total is sales tax?

24. Suppose you invest a certain amount of money in an account that pays 11% interest annually, and $4,000 more than that in an account that pays 12% annually. How much money do you have in each account if the total interest for a year is $940?

	Dollars Invested at 11%	Dollars Invested at 12%
Dollars		x
Interest		

Solve the following inequalities. Write the solution set using interval notation, then graph the solution set. [2.4]

25. $-5t \le 30$ **26.** $5 - \dfrac{3}{2}x > -1$

27. $1.6x - 2 < 0.8x + 2.8$ **28.** $3(2y + 4) \ge 5(y - 8)$

Solve the following equations. [2.5]

29. $\left|\dfrac{1}{4}x - 1\right| = \dfrac{1}{2}$ **30.** $\left|\dfrac{2}{3}a + 4\right| = 6$

31. $|3 - 2x| + 5 = 2$ **32.** $5 = |3y + 6| - 4$

Solve the following inequalities and graph the solutions. [2.6]

33. $|6x - 1| > 7$ **34.** $|3x - 5| - 4 \le 3$

35. $|5 - 4x| \ge -7$ **36.** $|4t - 1| < -3$

EQUATIONS AND INEQUALITIES IN ONE VARIABLE

Finding the Maximum Height of a Model Rocket

Number of People 3

Time Needed 20 minutes

Equipment Paper and pencil

Background In this chapter, we used formulas to do some table building. Once we have a table, it is sometimes possible to use just the table information to extend what we know about the situation described by the table. In this project, we take some basic information from a table and then look for patterns among the table entries. Once we have established the patterns, we continue them and, in so doing, solve a realistic application problem.

Procedure A model rocket is launched into the air. Table 1 gives the height of the rocket every second after takeoff for the first 5 seconds. Figure 1 is a graphical representation of the information in Table 1.

TABLE 1			
HEIGHT OF A MODEL ROCKET			
Time (seconds)	Height (feet)	First Differences	Second Differences
0	0		
1	176		
2	320		
3	432		
4	512		
5	560		

TABLE 1	
HEIGHT OF A MODEL ROCKET	
Time (seconds)	Height (feet)
0	0
1	176
2	320
3	432
4	512
5	560

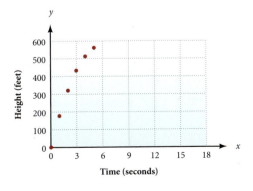

FIGURE 1

1. Table 1 is shown again above with two new columns. Fill in the first five entries in the First Differences column by finding the difference of consecutive heights. For example, the second entry in the First Differences column will be the difference of 320 and 176, which is 144.

2. Start filling in the Second Differences column by finding the differences of the First Differences.

3. Once you see the pattern in the Second Differences table, fill in the rest of the entries.

4. Now, using the results in the Second Differences table, go back and complete the First Differences table.

5. Now, using the results in the First Differences table, go back and complete the Heights column in the original table.

6. Plot the rest of the points from Table 1 on the graph in Figure 1.

7. What is the maximum height of the rocket?

8. How long was the rocket in the air?

The Equals Sign

We have been using the equals sign, =, for some time now. The first published use of the symbol was in 1557, with the publication of *The Whetstone of Witte* by the English mathematician and physician Robert Recorde. Research the first use of the symbols we use for addition, subtraction, multiplication, and division and then write an essay on the subject from your results.

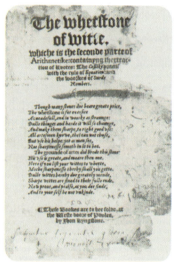

Courtesy of Smithsonian Libraries, Washington DC

TAKING A VACATION

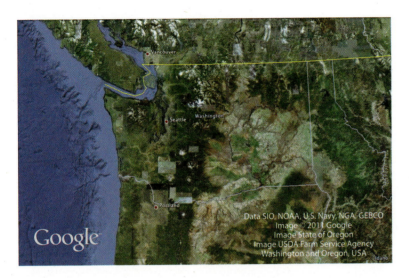

Amtrak provides train service to nearly 30,000 people each year along its 21,000 miles of track. If you've ever taken a road trip, you may have considered riding the train instead of driving your car. Which mode of transportation do you suppose would get you to your destination faster? Let's compare.

STEP 1   Use Amtrak's website at www.amtrak.com to determine the length of time it would take to ride a train from the train station in Seattle, WA (SEA) to the train station in Portland, OR (PDX).

    a. Enter your starting point and destination for a one-way ticket on the home screen.

    b. Select tomorrow's date for your departure date

    c. Select "Anytime" for your departure time.

    d. Click "GO!"

    e. Your screen will show the shortest trip time. Record this time.

STEP 2   Now use Google Earth to get driving directions from Seattle, WA to Portland, OR. Note the total distance in miles and the total estimated driving time.

STEP 3   Suppose a train and a car leave Seattle, WA at the same time and make no stops along the way to Portland, OR. Assuming all rates of speed are averaged, use the rate equation to calculate the speed of the car and the speed of the train. Round all answers to one decimal place. Do the car and the train arrive at the same time?

Equations and Inequalities in Two Variables

3

The University of Bologna, shown in the above Google Earth photo, is the oldest continuously operating university in the world. Many advances in science and astronomy have been made there. Maria Gaetana Agnesi is recognized as the first woman to publish a book of mathematics, and went on to become the first female professor of mathematics at the university.

Born in Milan in 1718 to a wealthy and literate family, Maria seemed destined for such academic surroundings. The Agnesi home was a gathering place of the most recognized intellectual minds of the day. Maria participated in these gatherings until the death of her mother, when she took over management of the household and the education of her siblings. The texts she wrote for this task were the beginnings of her published work, *Analytical Institutions*. This work discusses several curves that can be drawn on a rectangular coordinate system. The rectangular coordinate system is the foundation upon which we build our study of graphing in the this chapter.

Preview

Key Words	Definition
Ordered Pair	A pair of numbers enclosed by parentheses and separated by a comma
***x*-intercept**	The *x*-coordinate of a point where the graph of an equation crosses the *x*-axis
***y*-intercept**	The *y*-coordinate of the point where the graph of an equation crosses the *y*-axis
Function	A rule that pairs each element in a set with exactly one element in a second set
Relation	A rule that pairs each element in a set with one or more elements from a second set

Chapter Outline

3.1 Paired Data and the Rectangular Coordinate System

A Graph ordered pairs on a rectangular coordinate system.

B Graph linear equations by finding intercepts or by making a table.

C Graph horizontal and vertical lines.

3.2 The Slope of a Line

A Find the slope of a line from its graph.

B Find the slope of a line given two points on the line.

3.3 The Equation of a Line

A Find the equation of a line given its slope and *y*-intercept.

B Find the slope and *y*-intercept from the equation of a line.

C Find the equation of a line given the slope and a point on the line.

D Find the equation of a line given two points on the line.

3.4 Linear Inequalities in Two Variables

A Graph linear inequalities in two variables.

3.5 Introduction to Functions

A Construct a table or a graph from a function rule.

B Identify the domain and range of a function or a relation.

C Determine whether a relation is also a function.

D Use the vertical line test to determine whether a graph is a function.

3.6 Function Notation

A Use function notation to find the value of a function for a given value of the variable.

B Use graphs to visualize the relationship between a function and a variable.

C Use function notation in formulas.

3.7 Algebra and Composition with Functions

A Find the sum, difference, product, and quotient of two functions.

B For two functions f and g, find $f(g(x))$ and $g(f(x))$.

3.8 Variation

A Set up and solve problems with direct variation.

B Set up and solve problems with inverse variation.

C Set up and solve problems with joint variation.

Paired Data and the Rectangular Coordinate System

OBJECTIVES

A Graph ordered pairs on a rectangular coordinate system.

B Graph linear equations by finding intercepts or by making a table.

C Graph horizontal and vertical lines.

TICKET TO SUCCESS

Keep these questions in mind as you read through the section. Then respond in your own words and in complete sentences.

1. Explain how you would construct a rectangular coordinate system from two real number lines.
2. What is the definition of a linear equation in two variables?
3. How can you tell if an ordered pair is a solution to the equation $y = \frac{1}{3}x + 5$?
4. What are the x- and y-intercepts of the graph of an equation?

Imagno/Getty Images

In the 17th century, mathematicians and scientists treated algebra and geometry as separate subjects. In 1637, the French philosopher and mathematician Rene Descartes discovered a distinct link between algebra and geometry by associating geometric shapes with algebraic equations. He plotted this connection on a graph using an x-and a y-axis drawn perpendicular to each other, later named the Cartesian coordinate system after him. Our ability to graph equations is due to Descartes' invention of the Cartesian coordinate system. In this section we will begin our work with graphing equations by examining how data can be paired together.

A   Ordered Pairs

Paired data plays an important role in equations that contain two variables. Working with these equations is easier if we standardize the terminology and notation associated with paired data. So here is a definition that will do just that.

> **Definition**
>
> A pair of numbers enclosed in parentheses and separated by a comma, such as $(-2, 1)$, is called an ordered pair of numbers. The first number in the pair is called the **x-coordinate** of the ordered pair; the second number is called the **y-coordinate**. For the ordered pair $(-2, 1)$, the x-coordinate is -2 and the y-coordinate is 1.

To standardize the way in which we display paired data visually, we use a rectangular coordinate system. A *rectangular coordinate system* is made by drawing two real number lines at right angles to each other. The two number lines, called *axes,* cross each other at 0. This point is called the *origin.* Positive directions are to the right and up. Negative directions are down and to the left. The rectangular coordinate system is shown in Figure 3.

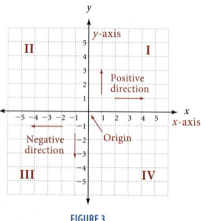

FIGURE 3

The horizontal number line is called the *x-axis* and the vertical number line is called the *y-axis.* The two number lines divide the coordinate system into four quadrants, which we number I through IV in a counterclockwise direction. Points on the axes are not considered as being in any quadrant.

To graph the ordered pair (a, b) on a rectangular system, we start at the origin and move a units right or left (right if a is positive, left if a is negative). Then we move b units up or down (up if b is positive and down if b is negative). The point where we end up is the graph of the ordered pair (a, b).

EXAMPLE 1   Plot (graph) the ordered pairs $(2, 5)$, $(-2, 5)$, $(-2, -5)$, and $(2, -5)$.

SOLUTION   To graph the ordered pair $(2, 5)$, we start at the origin and move 2 units to the right, then 5 units up. We are now at the point with coordinates $(2, 5)$. We graph the other three ordered pairs in a similar manner (see Figure 4).

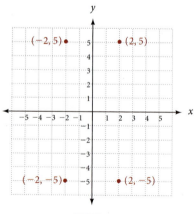

FIGURE 4

PRACTICE PROBLEMS

1. Plot the ordered pairs $(1, 3)$, $(-1, 3)$, $(-1, -3)$, and $(1, -3)$.

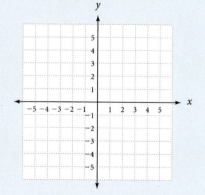

Answers
See Solutions to Selected Practice Problems for all answers in this section.

EXAMPLE 2 Graph the ordered pairs $(1, -3)$, $(0, 5)$, $(3, 0)$, $(0, -2)$, $(-1, 0)$, and $\left(\frac{1}{2}, 2\right)$.

SOLUTION

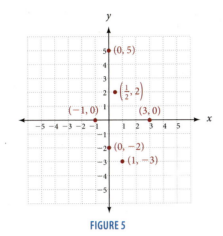

FIGURE 5

From Figure 5, we see that any point on the x-axis has a y-coordinate of 0 (with no vertical displacement), and any point on the y-axis has an x-coordinate of 0 (with no horizontal displacement). ■

B Linear Equations

We can plot a single point from an ordered pair, but to draw a line, we need two points or an equation in two variables.

> **Definition**
>
> Any equation that can be put in the form $ax + by = c$, where a, b, and c are real numbers and a and b are not both 0, is called a **linear equation in two variables.** The graph of any equation of this form is a straight line (that is why these equations are called "linear"). The form $ax + by = c$ is called **standard form.**

To graph a linear equation in two variables, we simply graph its solution set; that is, we draw a line through all the points with coordinates that satisfy the equation.

EXAMPLE 3 Graph the equation $y = -\frac{1}{3}x + 2$.

SOLUTION We need to find three ordered pairs that satisfy the equation. To do so, we can let x equal any numbers we choose and find corresponding values of y. But since every value of x we substitute into the equation is going to be multiplied by $-\frac{1}{3}$, let's use numbers for x that are divisible by 3, like -3, 0, and 3. That way, when we multiply them by $-\frac{1}{3}$, the result will be an integer.

2. Plot the ordered pairs $(4, -2)$, $\left(-\frac{1}{2}, 3\right)$, $(1, 0)$, $(0, -5)$, $(-6, 0)$, and $(0, 4)$.

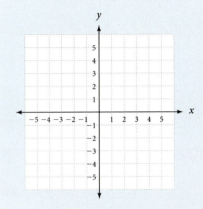

3. Graph $y = \frac{1}{2}x + 3$.

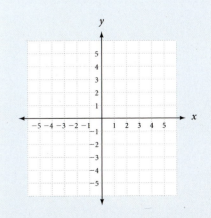

$$\text{Let } x = -3; \quad y = -\frac{1}{3}(-3) + 2$$

$$y = 1 + 2$$

$$y = 3$$

The ordered pair $(-3, 3)$ is one solution.

$$\text{Let } x = 0; \quad y = -\frac{1}{3}(0) + 2$$

$$y = 0 + 2$$

$$y = 2$$

The ordered pair $(0, 2)$ is a second solution.

$$\text{Let } x = 3; \quad y = -\frac{1}{3}(3) + 2$$

$$y = -1 + 2$$

$$y = 1$$

The ordered pair $(3, 1)$ is a third solution.

In table form

x	y
-3	3
0	2
3	1

Plotting the ordered pairs $(-3, 3)$, $(0, 2)$, and $(3, 1)$ and drawing a straight line through their graphs, we have the graph of the equation $y = -\frac{1}{3}x + 2$, as shown in Figure 6.

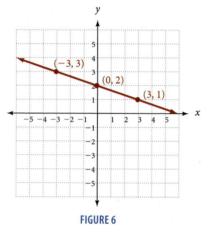

FIGURE 6

> **NOTE**
> It takes only two points to determine a straight line. We have included a third point for "insurance." If all three points do not line up in a straight line, we have made a mistake.

Example 3 illustrates again the connection between algebra and geometry that we mentioned earlier in this section. Descartes' rectangular coordinate system allows us to associate the equation $y = -\frac{1}{3}x + 2$ (an algebraic concept) with a specific straight line (a geometric concept). The study of the relationship between equations in algebra and their associated geometric figures is called *analytic geometry*.

Intercepts

Two important points on the graph of a straight line, if they exist, are the points where the graph crosses the axes.

Because any point on the x-axis has a y-coordinate of 0, we can find the x-intercept by letting $y = 0$ and solving the equation for x. We find the y-intercept by letting $x = 0$ and solving for y.

> **Definition**
>
> The **x-intercept** of the graph of an equation is the x-coordinate of the point where the graph crosses the x-axis. The **y-intercept** is defined similarly.

EXAMPLE 4 Find the x- and y-intercepts for $2x + 3y = 6$; then graph the solution set.

SOLUTION To find the y-intercept we let $x = 0$.

When → $\qquad\qquad x = 0$

we have → $\qquad 2(0) + 3y = 6$

$$3y = 6$$

$$y = 2$$

The y-intercept is 2, and the graph crosses the y-axis at the point $(0, 2)$.

When → $\qquad\qquad y = 0$

we have → $\qquad 2x + 3(0) = 6$

$$2x = 6$$

$$x = 3$$

The x-intercept is 3, so the graph crosses the x-axis at the point $(3, 0)$. We use these results to graph the solution set for $2x + 3y = 6$. The graph is shown in Figure 7.

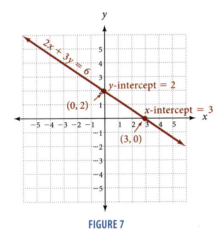

FIGURE 7

C Horizontal and Vertical Lines

EXAMPLE 5 Graph each of the following lines.

a. $y = \dfrac{1}{2}x$ $\quad$ **b.** $x = 3$ $\quad$ **c.** $y = -2$

SOLUTION

a. The line $y = \dfrac{1}{2}x$ passes through the origin because $(0, 0)$ satisfies the equation. To sketch the graph we need at least one more point on the line. When x is 2, we obtain the point $(2, 1)$, and when x is -4, we obtain the point $(-4, -2)$. The graph of $y = \dfrac{1}{2}x$ is shown in Figure 8A.

b. The line $x = 3$ is the set of all points whose x-coordinate is 3. The variable y does not appear in the equation, so the y-coordinate can be any number. Note that we can write our equation as a linear equation in two variables by writing it as $x + 0y = 3$. Because the product of 0 and y will always be 0, y can be any number. The graph of $x = 3$ is the vertical line shown in Figure 8B.

4. Find the x- and y-intercepts for $2x - 3y = 6$, and then graph the solution set.

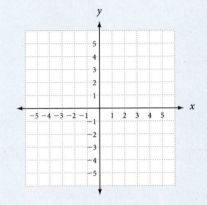

NOTE
Graphing straight lines by finding the intercepts works best when the coefficients of x and y are factors of the constant term.

5. Graph the line $x = -1$ and the line $y = 4$.

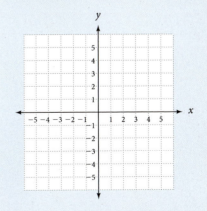

c. The line $y = -2$ is the set of all points whose y-coordinate is -2. The variable x does not appear in the equation, so the x-coordinate can be any number. Again, we can write our equation as a linear equation in two variables by writing it as $0x + y = -2$. Because the product of 0 and x will always be 0, x can be any number. The graph of $y = -2$ is the horizontal line shown in Figure 8C.

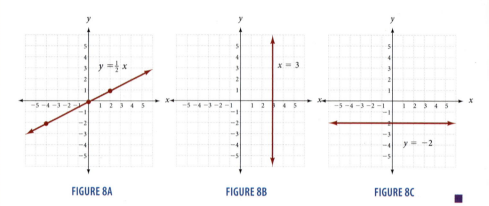

FIGURE 8A   FIGURE 8B   FIGURE 8C

FACTS FROM GEOMETRY   **Special Equations and Their Graphs**

For the equations below, m, a, and b are real numbers.

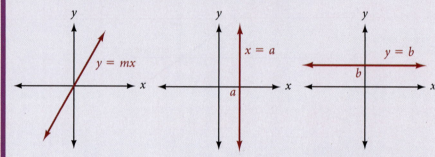

FIGURE 9A   Any equation of the form $y = mx$ has a graph that passes through the origin.

FIGURE 9B   Any equation of the form $x = a$ has a vertical line for its graph.

FIGURE 9C   Any equation of the form $y = b$ has a horizontal line for its graph.

Problem Set 3.1

Moving Toward Success

"One secret of success in life is for a man to be ready for his opportunity when it comes."

—Benjamin Disraeli, 1804–1881 British statesman and literary figure

1. Are you using the study time set aside for this class effectively? If not, how can you improve?

2. What are things you can do to make sure you study with intention?

A Graph each of the following ordered pairs on a rectangular coordinate system. [Examples 1–2]

1. a. (1, 2)
  b. (−1, −2)
  c. (5, 0)
  d. (0, 2)
  e. (−5, −5)
  f. $\left(\frac{1}{2}, 2\right)$

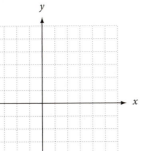

2. a. (−1, 2)
  b. (1, −2)
  c. (0, −3)
  d. (4, 0)
  e. (−4, −1)
  f. $\left(3, \frac{1}{4}\right)$

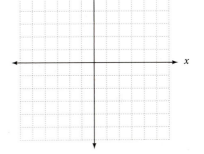

A Give the coordinates of each point.

3.

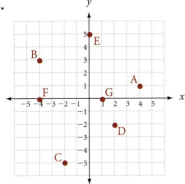

4.

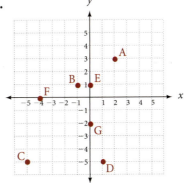

B Graph each of the following linear equations by first finding the intercepts. [Examples 3–4]

5. $2x - 3y = 6$

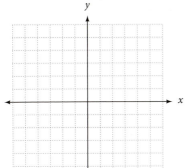

6. $y - 2x = 4$

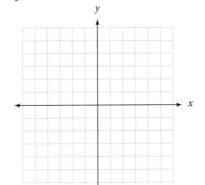

7. $4x - 5y = 20$

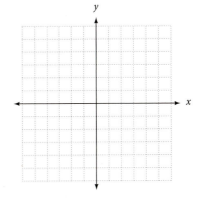

8. $5x - 3y - 15 = 0$

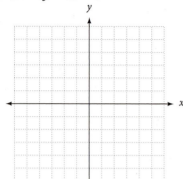

9. $y = 2x + 3$

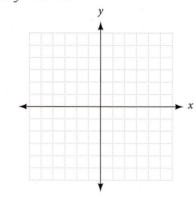

10. $y = 3x - 2$

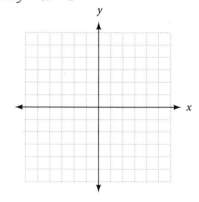

11. $-3x + 2y = 12$

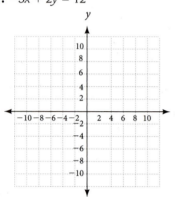

12. $5x - 7y = -35$

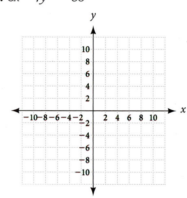

13. $6x - 5y - 20 = 0$

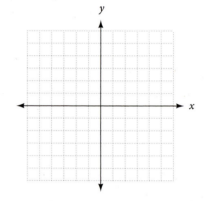

14. $-4x - 6y + 15 = 0$

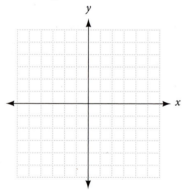

15. $y = 3x - 5$

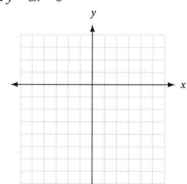

16. $y = -4x + 1$

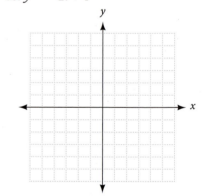

17. $\dfrac{x}{2} + \dfrac{y}{3} = 1$

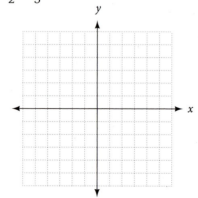

18. $\dfrac{x}{4} - \dfrac{y}{7} = 1$

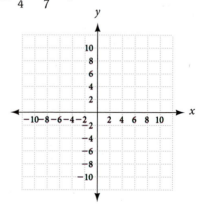

19. Which of the following tables could be produced from the equation $y = 2x - 6$?

a.

x	y
0	6
1	4
2	2
3	0

b.

x	y
0	-6
1	-4
2	-2
3	0

c.

x	y
0	-6
1	-5
2	-4
3	-3

20. Which of the following tables could be produced from the equation $3x - 5y = 15$?

a.

x	y
0	5
-3	0
10	3

b.

x	y
0	-3
5	0
10	3

c.

x	y
0	-3
-5	0
10	-3

B Graph each of the following lines. [Example 3, 4]

21. $y = \dfrac{1}{3}x$

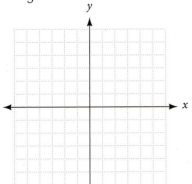

22. $y = \dfrac{1}{2}x$

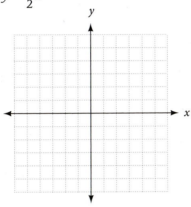

23. $-2x + y = -3$

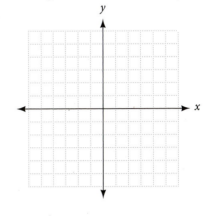

24. $-3x + y = -2$

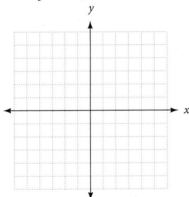

25. $y = -\dfrac{2}{3}x + 1$

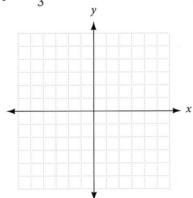

26. $y = -\dfrac{2}{3}x - 1$

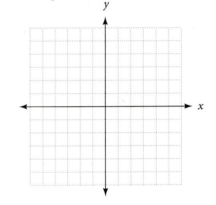

27. $\dfrac{x}{3} + \dfrac{y}{4} = 1$

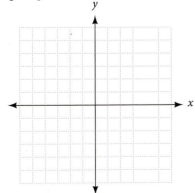

28. $\dfrac{x}{-2} + \dfrac{y}{3} = 1$

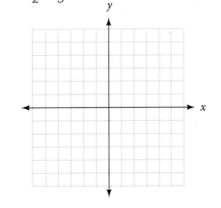

29. The graph shown here is the graph of which of the following equations?

   a. $3x - 2y = 6$

   b. $2x - 3y = 6$

   c. $2x + 3y = 6$

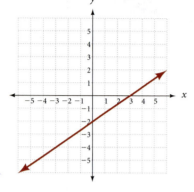

30. The graph shown here is the graph of which of the following equations?

   a. $3x - 2y = 8$

   b. $2x - 3y = 8$

   c. $2x + 3y = 8$

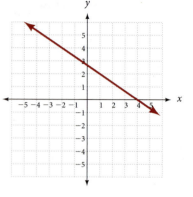

The next two problems are intended to give you practice reading, and paying attention to, the instructions that accompany the problems you are working. Working these problems is an excellent way to get ready for a test or a quiz.

31. Work each problem according to the instructions.

   a. Solve: $4x + 12 = -16$.

   b. Find x when y is 0: $4x + 12y = -16$.

   c. Find y when x is 0: $4x + 12y = -16$.

   d. Graph: $4x + 12y = -16$.

   e. Solve for y: $4x + 12y = -16$.

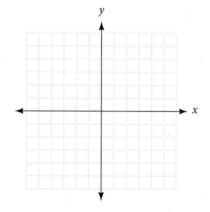

32. Work each problem according to the instructions.

   a. Solve: $3x - 8 = -12$.

   b. Find x when y is 0: $3x - 8y = -12$.

   c. Find y when x is 0: $3x - 8y = -12$.

   d. Graph: $3x - 8y = -12$.

   e. Solve for y: $3x - 8y = -12$.

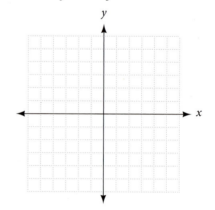

C Graph each of the following lines. [Example 5]

33. **a.** $y = 2x$ **b.** $x = -3$ **c.** $y = 2$

a.

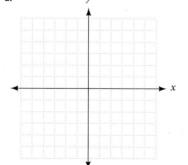

b.

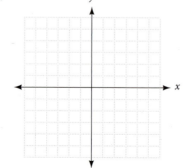

c.

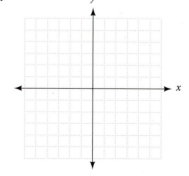

34. Graph each of the following lines.

    a. $y = 3x$ **b.** $x = -2$ **c.** $y = 4$

    a.

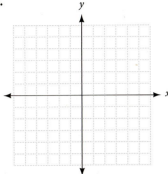

    b.

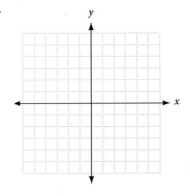

    c.

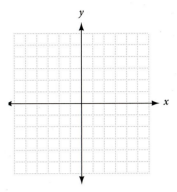

35. Graph each of the following lines.

    a. $y = -\dfrac{1}{2}x$ **b.** $x = 4$ **c.** $y = -3$

    a.

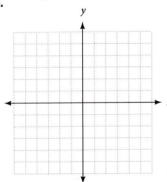

    b.

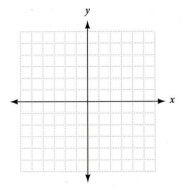

    c.

36. Graph each of the following lines.

    a. $y = -\dfrac{1}{3}x$ **b.** $x = 1$ **c.** $y = -5$

    a.

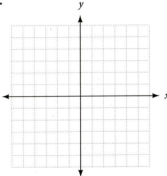

    b.

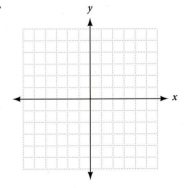

    c.

37. Graph the line $0.02x + 0.03y = 0.06$.

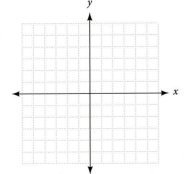

38. Graph the line $0.05x - 0.03y = 0.15$.

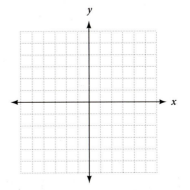

39. The ordered pairs that satisfy the equation $y = 3x$ all have the form $(x, 3x)$ because y is always 3 times x. Graph all ordered pairs of the form $(x, 3x)$.

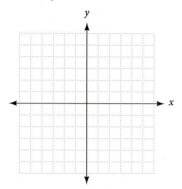

40. Graph all ordered pairs of the form $(x, -3x)$.

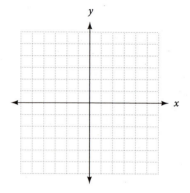

Applying the Concepts

41. Hourly Wages Jane takes a job at the local Marcy's department store. Her job pays $11.00 per hour. The graph shows how much Jane earns for working from 0 to 40 hours in a week.

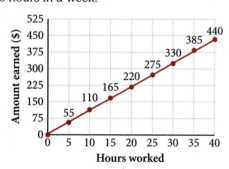

a. List three ordered pairs that lie on the line graph.

b. How much will she earn for working 40 hours?

c. If her check for one week is $275, how many hours did she work?

d. She works 35 hours one week, but her paycheck before deductions are subtracted out is for $350. Is this correct? Explain.

42. Hourly Wages Judy takes a job at Gigi's boutique. Her job pays $9.00 per hour plus $50 per week in commission. The graph shows how much Judy earns for working from 0 to 40 hours in a week.

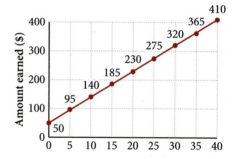

a. List three ordered pairs that lie on the line graph.

b. How much will she earn for working 40 hours?

c. If her check for one week is $320, how many hours did she work?

d. She works 35 hours one week, but her paycheck before deductions are subtracted out is for $365. Is this correct? Explain.

43. Non-Camera Phone Sales The table and bar chart shown here show the projected sales of non-camera phones. Use the information from the table and chart to construct a line graph.

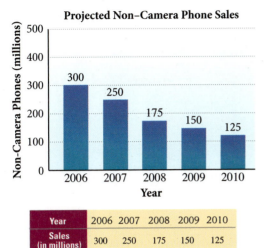

Projected Non–Camera Phone Sales

Year	2006	2007	2008	2009	2010
Sales (in millions)	300	250	175	150	125

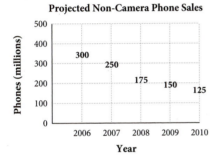

Projected Non-Camera Phone Sales

44. Camera Phone Sales The table and bar chart shown here show the projected sales of camera phones from 2006 to 2010. Use the information from the table and chart to construct a line graph.

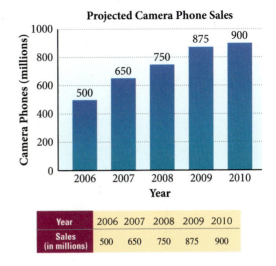

Projected Camera Phone Sales

Year	2006	2007	2008	2009	2010
Sales (in millions)	500	650	750	875	900

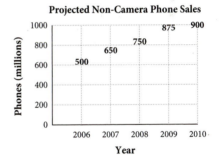

Projected Non-Camera Phone Sales

45. Laptops The chart shows the increasing percentage of adults with laptop computers. If x represents the year in question, and y represents the percentage of adults with laptops, write five ordered pairs that describe the information in the graph.

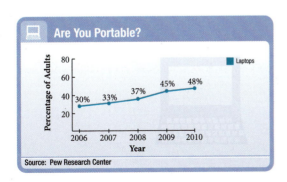

Are You Portable?

Laptops

Source: Pew Research Center

46. Social Networking The chart shows the increase in social networking users globally. Use the graph to list the ordered pairs that describe the information in the chart.

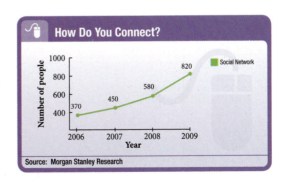

How Do You Connect?

Social Network

Source: Morgan Stanley Research

47. Social Networking The graph used in the previous problem shows the increase in social networking users globally. Use the chart to make a table of x and y values, where x is the year and y is the number of users in millions. Round to the nearest ten million.

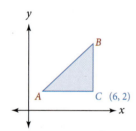

x (year)	y (age)

48. Camera Phones The graph shows what percentage of different generations have cell phones. The x-axis shows the youngest age in the generation. Use the chart to estimate the ordered pairs that describe the information in the chart.

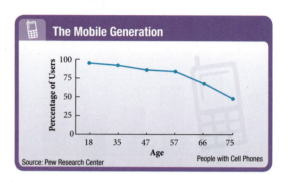

Source: Pew Research Center

49. In the figure below, right triangle ABC has legs of length 5. Find the coordinates of points A and B.

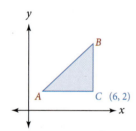

50. In the figure below, right triangle ABC has legs of length 5. Find the coordinates of points A and B.

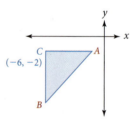

51. In the figure below, rectangle $ABCD$ has a length of 5 and a width of 3. Find points A, B, and C.

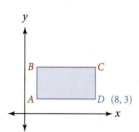

52. In the figure below, rectangle $ABCD$ has a length of 5 and a width of 3. Find points A, B, and C.

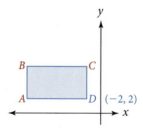

53. **Internet** The chart shows the decline in dial-up internet users. Use the chart to answer the following questions.

   a. Could the graph contain the point (2001, 36)?

   b. Could the graph contain the point (2004, 35)?

   c. Could the graph contain the point (2010, 20)?

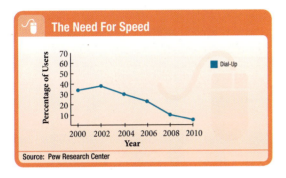

54. **Health Care Costs** The graph shows the projected rise in the cost of health care from 2002 to 2014. Using years as x and billions of dollars as y, write five ordered pairs that describe the information in the graph.

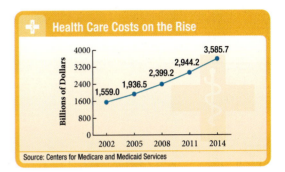

Maintaining Your Skills

The problems that follow review some of the more important skills you have learned in previous sections and chapters.

Solve each equation.

55. $5x - 4 = -3x + 12$

56. $\dfrac{1}{2} - \dfrac{y}{5} = -\dfrac{9}{10} + \dfrac{y}{2}$

57. $\dfrac{1}{2} - \dfrac{1}{8}(3t - 4) = -\dfrac{7}{8}t$

58. $3(5t - 1) - (3 - 2t) = 5t - 8$

59. $50 = \dfrac{K}{24}$

60. $3.5d = 16(3.5)^2$

61. $2(1) + y = 4$

62. $2(2y + 6) + 3y = 5$

63. $4\left(\dfrac{19}{15}\right) - 2y = 4$

64. $2x - 3\left(-\dfrac{5}{11}\right) = 4$

Getting Ready for the Next Section

65. Write -0.06 as a fracton with a denominator 100.

66. Write -0.07 as a fraction with a denominator 100.

67. If $y = 2x - 3$, find y when $x = 2$.

68. If $y = 2x - 3$, find x when $y = 5$.

Simplify.

69. $\dfrac{1 - (-3)}{-5 - (-2)}$

70. $\dfrac{-3 - 1}{-2 - (-5)}$

71. $\dfrac{-1 - 4}{3 - 3}$

72. $\dfrac{-3 - (-3)}{2 - (-1)}$

73. The product of $\dfrac{2}{3}$ and what number will result in

   a. 1?

   b. -1?

74. The product of 3 and what number will result in

   a. 1?

   b. -1?

Extending the Concepts

Find the x- and y-intercepts.

75. $ax + by = c$

76. $ax - by = c$

77. $\dfrac{x}{a} + \dfrac{y}{b} = 1$

78. $y = ax + b$

The Slope of a Line

OBJECTIVES

A Find the slope of a line from its graph.

B Find the slope of a line given two points on the line.

TICKET TO SUCCESS

Keep these questions in mind as you read through the section. Then respond in your own words and in complete sentences.

1. How would you geometrically define the slope of a line?
2. Describe the behavior of a line with a negative slope.
3. Use symbols to define the slope of the line between the points (x_1, y_1) and (x_2, y_2).
4. How are the slopes of two perpendicular lines different than the slopes of two parallel lines?

B & T Media Group Inc./Shutterstock.com

During Stage 4 of the Amgen Tour of California 2011, cyclists traveled nearly 82 miles from Livermore, CA to San Jose, CA. Although not the longest stage of the race, cyclists faced Sierra Road, which was one of the steepest parts of the tour route. The 3.5-mile section of the race logged a grueling 9.4% grade. This grade means that for every 100 feet a cyclist moved horizontally, the change in elevation was 9.4 feet. In mathematics, we say this change in elevation in relation to the change in horizontal distance is called the *slope*. For Sierra Road, the slope can be shown as follows:

$$0.094 = \frac{94}{1000} = \frac{47}{500}$$

In other words, the slope of $\frac{47}{500}$ is the ratio of the vertical change to the accompanying horizontal change for that road.

In defining the slope of a straight line, we are looking for a number to associate with that line that does two things. First, we want the slope of a line to measure the "steepness" of the line; that is, in comparing two lines, the slope of the steeper line should have the larger numerical absolute value. Second, we want a line that *rises* going from left to right to have a *positive* slope. We want a line that *falls* going from left to right to have a *negative* slope. (A line that neither rises nor falls going from left to right must, therefore, have 0 slope.)

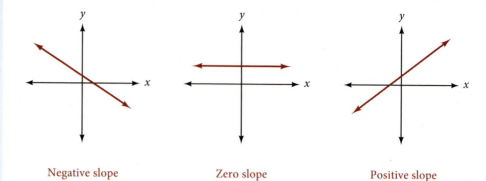

Negative slope   Zero slope   Positive slope

Geometrically, we can define the *slope* of a line as the ratio of the vertical change to the horizontal change encountered when moving from one point to another on the line. The vertical change is sometimes called the *rise*. The horizontal change is called the *run*.

A Finding the Slope from Graphs

PRACTICE PROBLEMS

1. Graph the line $y = 3x - 5$ and then find the slope.

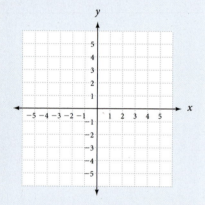

EXAMPLE 1   Find the slope of the line $y = 2x - 3$.

SOLUTION   To use our geometric definition, we first graph $y = 2x - 3$ (Figure 1). We then pick any two convenient points and find the ratio of rise to run. By convenient points we mean points with integer coordinates. If we let $x = 2$ in the equation, then $y = 1$. Likewise if we let $x = 4$, then $y = 5$.

Our line has a slope of 2.   ∎

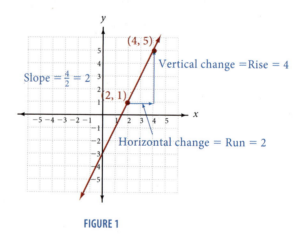

FIGURE 1

Notice that we can measure the vertical change (rise) by subtracting the y-coordinates of the two points shown in Figure 1: $5 - 1 = 4$. The horizontal change (run) is the difference of the x-coordinates: $4 - 2 = 2$. This gives us a second way of defining the slope of a line.

Answer
1. 3

B Finding Slopes Using Two Points

Definition

The **slope** of the line between two points (x_1, y_1) and (x_2, y_2) is given by

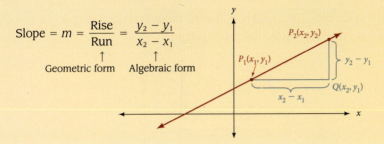

$$\text{Slope} = m = \frac{\text{Rise}}{\text{Run}} = \frac{y_2 - y_1}{x_2 - x_1}$$

↑   ↑
Geometric form   Algebraic form

EXAMPLE 2 Find the slope of the line through $(-2, -3)$ and $(-5, 1)$.

SOLUTION $m = \dfrac{y_2 - y_1}{x_2 - x_1} = \dfrac{1 - (-3)}{-5 - (-2)} = \dfrac{4}{-3} = -\dfrac{4}{3}$

Looking at the graph of the line through the two points (Figure 2), we can see that our geometric approach does not conflict with our algebraic approach. ■

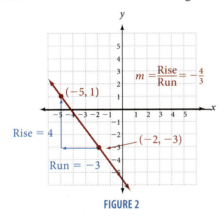

FIGURE 2

We should note here that it does not matter which ordered pair we call (x_1, y_1) and which we call (x_2, y_2). If we were to reverse the order of subtraction of both the x- and y-coordinates in the preceding example, we would have

$$m = \frac{-3 - 1}{-2 - (-5)} = \frac{-4}{3} = -\frac{4}{3}$$

which is the same as our previous result.

⊘ COMMON MISTAKES

NOTE The two most common mistakes students make when first working with the formula for the slope of a line are

1. Putting the difference of the x-coordinates over the difference of the y-coordinates.

2. Subtracting in one order in the numerator and then subtracting in the opposite order in the denominator. You would make this mistake in Example 2 if you wrote $1 - (-3)$ in the numerator and then $-2 - (-5)$ in the denominator.

2. Find the slope of the line through $(3, 4)$ and $(1, -2)$.

Answer
2. 3

3. Find the slope of the line through $(2, -3)$ and $(-1, -3)$.

EXAMPLE 3 Find the slope of the line containing $(3, -1)$ and $(3, 4)$.

SOLUTION Using the definition for slope, we have

$$m = \frac{-1 - 4}{3 - 3} = \frac{-5}{0}$$

The expression $\frac{-5}{0}$ is undefined; that is, there is no real number to associate with it. In this case, we say the line *has no slope*.

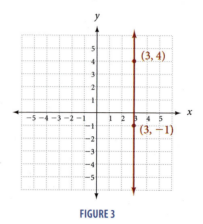

FIGURE 3

The graph of our line is shown in Figure 3. Our line with no slope is a vertical line. All vertical lines have no slope. (And all horizontal lines, as we mentioned earlier, have 0 slope.) ∎

Slopes of Parallel and Perpendicular Lines

In geometry, we call lines in the same plane that never intersect parallel. For two lines to be nonintersecting, they must rise or fall at the same rate. In other words, two lines are *parallel* if and only if they have the *same slope.*

Although it is not as obvious, it is also true that two nonvertical lines are *perpendicular* if and only if the *product of their slopes is* -1. This is the same as saying their slopes are negative reciprocals. We can state these facts with symbols as follows:

> If line l_1 has slope m_1 and line l_2 has slope m_2, then
>
> $$l_1 \text{ is } \textbf{parallel} \text{ to } l_2 \Leftrightarrow m_1 = m_2$$
>
> and
>
> $$l_1 \text{ is } \textbf{perpendicular} \text{ to } l_2 \Leftrightarrow m_1 \cdot m_2 = -1 \left(\text{or } m_1 = \frac{-1}{m_2} \right)$$

For example, if a line has a slope of $\frac{2}{3}$, then any line parallel to it has a slope of $\frac{2}{3}$. Any line perpendicular to it has a slope of $-\frac{3}{2}$ (the negative reciprocal of $\frac{2}{3}$).

Although we cannot give a formal proof of the relationship between the slopes of perpendicular lines at this level of mathematics, we can offer some justification for the relationship. Figure 4 shows the graphs of two lines. One of the lines has a slope of $\frac{2}{3}$; the other has a slope of $-\frac{3}{2}$. As you can see, the lines are perpendicular.

Answer

3. 0

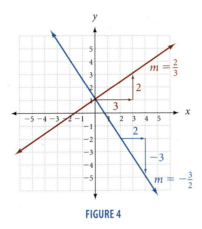

FIGURE 4

Slope and Rate of Change

So far, the slopes we have worked with represent the ratio of the change in y to a corresponding change in x; or, on the graph of the line, the ratio of vertical change to horizontal change in moving from one point on the line to another. However, when our variables represent quantities from the world around us, slope can have additional interpretations.

EXAMPLE 4 On the chart below, find the slope of the line connecting the first point (1955, 0.29) with the highest point (2005, 2.93). Explain the significance of the result.

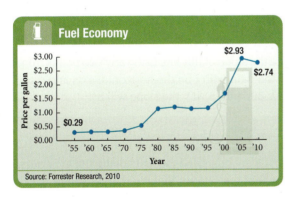

Source: Forrester Research, 2010

SOLUTION The slope of the line connecting the first point (1955, 0.29) with the highest point (2005, 2.93), is

$$m = \frac{2.93 - 0.29}{2005 - 1955} = \frac{2.64}{50} = 0.0528$$

The units are dollars/year. If we write this in terms of cents we have

$$m = 5.28 \text{ cents/year}$$

which is the average change in the price of a gallon of gasoline over a 50-year period of time.

Likewise, if we connect the points (1995, 1.10) and (2005, 2.93), the line that results has a slope of

$$m = \frac{2.93 - 1.10}{2005 - 1995} = \frac{1.83}{10} = 0.183 \text{ dollars/year} = 18.3 \text{ cents/year}$$

4. On the chart, find the slope of the line connecting the point (1985, 1.25) with the highest point (2005, 2.93). Explain in words what your result represents.

Answer
4. 8.4 cents/year, which is the average rate of change of the price of a gallon of gasoline over a 20-year period of time

which is the average change in the price of a gallon of gasoline over a 10-year period. As you can imagine by looking at the chart, the line connecting the first and last point is not as steep as the line connecting the points from 1995 and 2005, and this is what we are seeing numerically with our slope calculations. If we were summarizing this information for an article in the newspaper, we could say, although the price of a gallon of gasoline has increased only 5.28 cents per year over the last 50 years, in the last 10 years the average annual rate of increase was more than triple that, at 18.3 cents per year.   ■

Slope and Average Speed

Previously we introduced the rate equation $d = rt$. Suppose that a boat is traveling at a constant speed of 15 miles per hour in still water. The following table shows the distance the boat will have traveled in the specified number of hours. The graph of these data is shown in Figure 5. Notice that the points all lie along a line.

t (hours)	d (miles)
0	0
1	15
2	30
3	45
4	60
5	75

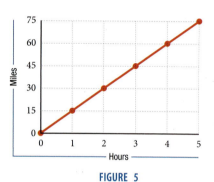

FIGURE 5

We can calculate the slope of this line using any two points from the table. Notice we have graphed the data with t on the horizontal axis and d on the vertical axis. Using the points (2, 30) and (3, 45), the slope will be

$$m = \frac{\text{rise}}{\text{run}} = \frac{45 - 30}{3 - 2} = \frac{15}{1} = 15$$

The units of the rise are miles and the units of the run are hours, so the slope will be in units of miles per hour. We see that the slope is simply the change in distance divided by the change in time, which is how we compute the average speed. Since the speed is constant, the slope of the line represents the speed of 15 miles per hour.

EXAMPLE 5   A car is traveling at a constant speed. A graph (Figure 6) of the distance the car has traveled over time is shown below. Use the graph to find the speed of the car.

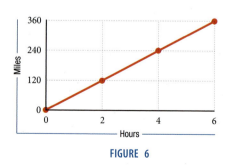

FIGURE 6

5. A car is traveling at a constant speed. A graph (Figure 7) of the distance the car has traveled over time is shown below. Use the graph to find the speed of the car.

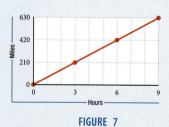

FIGURE 7

Answer
5. 70 miles per hour

SOLUTION Using the second and third points, we see the rise is $240 - 120 = 120$ miles, and the run is $4 - 2 = 2$ hours. The speed is given by the slope, which is

$$m = \frac{\text{rise}}{\text{run}}$$

$$= \frac{120 \text{ miles}}{2 \text{ hours}}$$

$$= 60 \text{ miles per hour} \qquad \blacksquare$$

USING TECHNOLOGY

Families of Curves

We can use a graphing calculator to investigate the effects of the numbers a and b on the graph of $y = ax + b$. To see how the number b affects the graph, we can hold a constant and let b vary. Doing so will give us a *family* of curves. Suppose we set $a = 1$ and then let b take on integer values from -3 to 3. The equations we obtain are

$y = x - 3$
$y = x - 2$
$y = x - 1$
$y = x$
$y = x + 1$
$y = x + 2$
$y = x + 3$

On the next page, we will give three methods of graphing this set of equations on a graphing calculator.

Method 1: Y-Variables List

To use the Y-variables list, enter each equation at one of the Y variables, set the graph window, then graph. The calculator will graph the equations in order, starting with Y_1 and ending with Y_7. Following is the Y-variables list, an appropriate window, and a sample of the type of graph obtained (Figure 7).

$Y_1 = X - 3$

$Y_2 = X - 2$

$Y_3 = X - 1$

$Y_4 = X$

$Y_5 = X + 1$

$Y_6 = X + 2$

$Y_7 = X + 3$

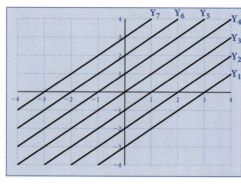

FIGURE 7

Window: X from -4 to 4, Y from -4 to 4

Method 2: Programming

The same result can be obtained by programming your calculator to graph $y = x + b$ for $b = -3, -2, -1, 0, 1, 2,$ and 3. Here is an outline of a program that will do this. Check the manual that came with your calculator to find the commands for your calculator.

Step 1: Clear screen
Step 2: Set window for X from -4 to 4 and Y from -4 to 4
Step 3: $-3 \rightarrow B$
Step 4: Label 1
Step 5: Graph Y = X + B
Step 6: $B + 1 \rightarrow B$
Step 7: If $B < 4$, go to 1
Step 8: End

Method 3: Using Lists

On the TI-83/84 you can set Y_1 as follows

$$Y_1 = X + \{-3, -2, -1, 0, 1, 2, 3\}$$

When you press $\boxed{\text{GRAPH}}$, the calculator will graph each line from
$$y = x + (-3) \text{ to } y = x + 3.$$

Each of the three methods will produce graphs similar to those in Figure 7.

Problem Set 3.2

Moving Toward Success

"Life is ten percent what happens to you and ninety percent how you respond to it."

—Lou Holtz, 1937–present, American college football coach and motivational speaker

1. Do you think it is important to take notes as you read this book? Explain.

2. Why is it helpful to put the material you learn in your own words when you take notes or answer homework questions?

A Find the slope of each of the following lines from the given graph. [Example 1]

1.

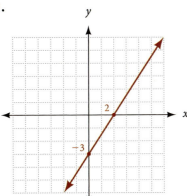

2.

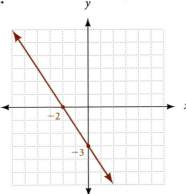

3.

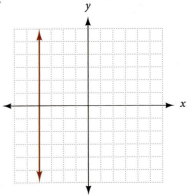

4.

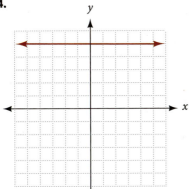

5.

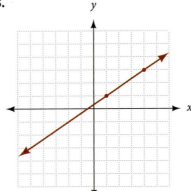

6.

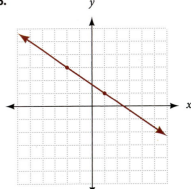

B Find the slope of the line through each of the following pairs of points. Then, plot each pair of points, draw a line through them, and indicate the rise and run in the graph in the manner shown in Example 2. [Examples 2–3]

7. (2, 1), (4, 4)

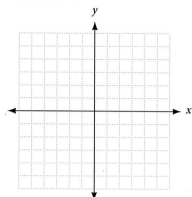

8. (3, 1), (5, 4)

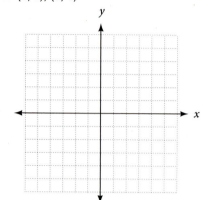

9. (1, 4), (5, 2)

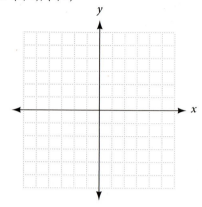

10. (1, 3), (5, 2)

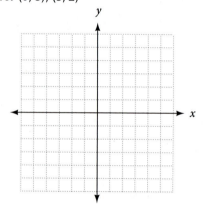

11. (1, −3), (4, 2)

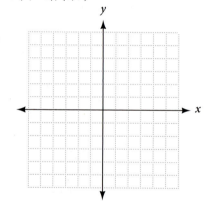

12. (2, −3), (5, 2)

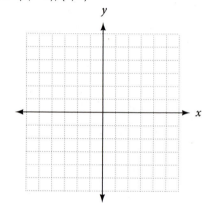

13. (−3, −2), (1, 3)

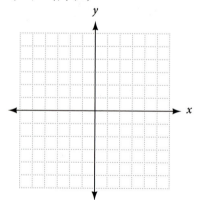

14. (−3, −1), (1, 4)

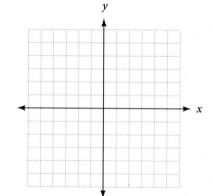

15. (−3, 2), (3, −2)

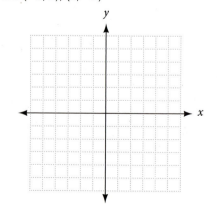

16. $(-3, 3), (3, -1)$

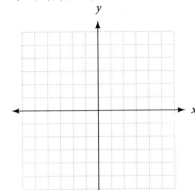

17. $(2, -5), (3, -2)$

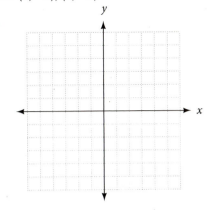

18. $(2, -4), (3, -1)$

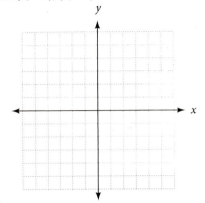

Solve for the indicated variable if the line through the two given points has the given slope.

19. $(a, 3)$ and $(2, 6)$, $m = -1$

20. $(a, -2)$ and $(4, -6)$, $m = -3$

21. $(2, b)$ and $(-1, 4b)$, $m = -2$

22. $(-4, y)$ and $(-1, 6y)$, $m = 2$

For each equation below, complete the table, and then use the results to find the slope of the graph of the equation.

23. $2x + 3y = 6$

x	y
0	
	0

24. $3x - 2y = 6$

x	y
0	
	0

25. $y = \dfrac{2}{3}x - 5$

x	y
0	
3	

26. $y = -\dfrac{3}{4}x + 2$

x	y
0	
4	

27. Finding Slope from Intercepts Graph the line with x-intercept -4 and y-intercept -2. What is the slope of this line?

28. Parallel Lines Find the slope of a line parallel to the line through $(2, 3)$ and $(-8, 1)$.

29. Parallel Lines Find the slope of a line parallel to the line through $(2, 5)$ and $(5, -3)$.

30. Perpendicular Lines Line l contains the points $(5, -6)$ and $(5, 2)$. Give the slope of any line perpendicular to l.

31. Perpendicular Lines Line *l* contains the points (3, 4) and (−3, 1). Give the slope of any line perpendicular to *l*.

32. Parallel Lines Line *l* contains the points (−2, 1) and (4, −5). Find the slope of any line parallel to *l*.

33. Parallel Lines Line *l* contains the points (3, −4) and (−2, −6). Find the slope of any line parallel to *l*.

34. Perpendicular Lines Line *l* contains the points (−2, −5) and (1, −3). Find the slope of any line perpendicular to *l*.

35. Perpendicular Lines Line *l* contains the points (6, −3) and (−2, 7). Find the slope of any line perpendicular to *l*.

36. Graph the line with *x*-intercept 4 and *y*-intercept 2. What is the slope of this line?

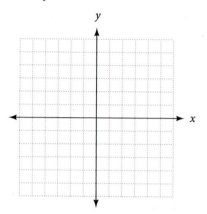

37. Determine if each of the following tables could represent ordered pairs from an equation of a line.

a.

x	y
0	5
1	7
2	9
3	11

b.

x	y
−2	−5
0	−2
2	0
4	1

38. The following lines have slope 2, $\frac{1}{2}$, 0, and −1. Match each line to its slope value.

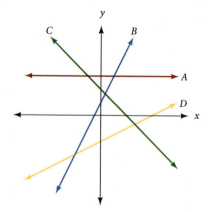

39. Line *l* has a slope of $\frac{2}{3}$. A horizontal change of 12 will always be accompanied by how much of a vertical change?

40. For any line with slope $\frac{4}{5}$, a vertical change of 8 is always accompanied by how much of a horizontal change?

Applying the Concepts

41. An object is traveling at a constant speed. The distance and time data are shown on the given graphs. Use the graphs to find the speed of the object.

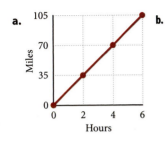

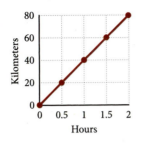

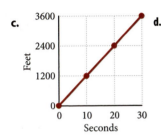

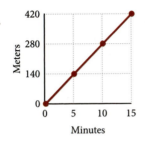

42. Cycling A cyclist is traveling at a constant speed. The graph shows the distance the cyclist travels over time when there is no wind present, when she travels against the wind, and when she travels with the wind.

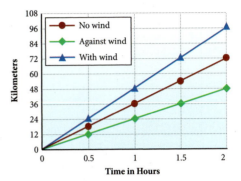

a. Use the concept of slope to find the speed of the cyclist under each of the three conditions.

b. Compare the speed of the cyclist when she is traveling without any wind to when she is riding against the wind. How do the two speeds differ?

c. Compare the speed of the cyclist when she is traveling without any wind to when she is riding with the wind. How do the two speeds differ?

d. Assuming no other forces are involved, what is the speed of the wind?

43. Non-Camera Phone Sales The table and line graph here each show the projected non-camera phone sales each year through 2010. Find the slope of each of the three line segments A, B, and C.

Year	2006	2007	2008	2009	2010
Sales (in millions)	300	250	175	150	125

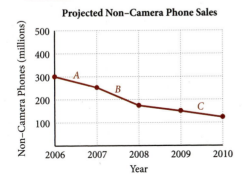

44. Camera Phone Sales The table and line graph here show the projected sales of camera phones from 2006 to 2010. Find the slopes of line segments A, B, and C.

Year	2006	2007	2008	2009	2010
Sales (in millions)	500	650	750	875	900

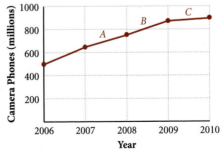

45. Slope of a Highway A sign at the top of the Cuesta Grade, outside of San Luis Obispo, CA reads "7% downgrade next 3 miles." The diagram shown here is a model of the Cuesta Grade that takes into account the information on that sign.

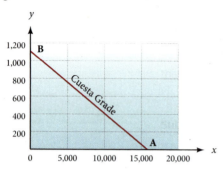

a. At point B, the graph crosses the *y*-axis at 1,106 feet. How far is Point A from the origin?

b. What is the slope of the Cuesta Grade?

46. Cell Phones The graph shows what percentage of people by income level, think that cell phones are a necessity. What is the slope of the line segment from (0, 65) to (30, 70)?

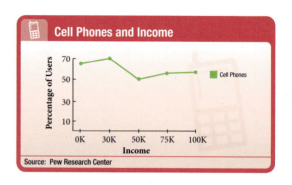

47. Solar Energy The graph below shows the annual shipments of solar thermal collectors in the United States. Using the graph below, find the slope of the line connecting the first (1997, 8,000) and last (2008, 17,000) endpoints and then explain in words what the slope represents.

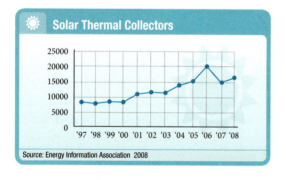

48. Age of New Mothers The graph shows the increase in the average age of first time mothers in the U.S. since 1970. Find the slope of the line that connects the points (1975, 21.75) and (1990, 24.25). Round to the nearest hundredth. Explain in words what the slope represents.

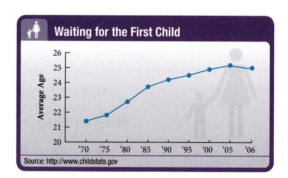

49. Slope Some of the steepest roads in the United States are found in the Bighorn Mountains in Colorado. These roads can have up to a 15% downgrade. What is the slope of these roads?

50. Slope San Francisco's Filbert Street has an incline of 31.5%. It is one of the steepest drivable streets in the Western Hemisphere. What is the slope of Filbert Street?

51. Horse Racing The graph shows the amount of money bet on horse racing from 1985 to 2010. Use the chart to work the following problems involving slope.

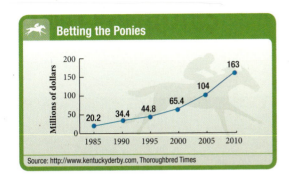

Betting the Ponies

Source: http://www.kentuckyderby.com, Thoroughbred Times

a. Find the slope of the line from 1985 to 1990, and then explain in words what the slope represents.

b. Find the slope of the line from 2000 to 2005, and then explain in words what the slope represents.

52. Light Bulbs The chart shows a comparison of power usage between incandescent and LED light bulbs. Use the chart to work the following problems involving slope.

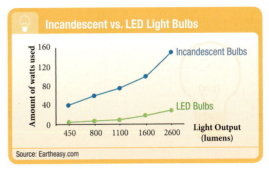

Incandescent vs. LED Light Bulbs

Source: Eartheasy.com

a. Find the slope of the line for the incandescent bulb from the two endpoints (450, 40) and (2,600, 160) and then explain in words what the slope represents.

b. Find the slope of the line for the LED bulb from the two endpoints (450, 1) and (2,600, 20) and then explain in words what the slope represents.

c. Which light bulb is more efficient? Why?

Maintaining Your Skills

The problems that follow review some of the more important skills you have learned in previous sections and chapters.

53. If $3x + 2y = 12$, find y when x is 4.

54. If $y = 3x - 1$, find x when y is 0.

55. Solve the formula $3x + 2y = 12$ for y.

56. Solve the formula $y = 3x - 1$ for x.

57. Solve the formula $A = P + Prt$ for t.

58. Solve the formula $S = \pi r^2 + 2\pi rh$ for h.

Getting Ready for the Next Section

Simplify.

59. $2\left(-\dfrac{1}{2}\right)$

60. $\dfrac{3 - (-1)}{-3 - 3}$

61. $\dfrac{5 - (-3)}{2 - 6}$

62. $3\left(-\dfrac{2}{3}x + 1\right)$

Solve for y.

63. $\dfrac{y - b}{x - 0} = m$

64. $2x + 3y = 6$

65. $y - 3 = -2(x + 4)$

66. $y + 1 = -\dfrac{2}{3}(x - 3)$

67. If $y = -\dfrac{4}{3}x + 5$, find y when x is 0.

68. If $y = -\dfrac{4}{3}x + 5$, find y when x is 3.

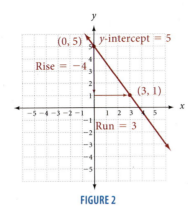

FIGURE 2

B Finding Slopes and Intercepts

EXAMPLE 2   Give the slope and y-intercept for the line $2x - 3y = 5$.

SOLUTION   To use the slope-intercept form, we must solve the equation for y in terms of x.

$$2x - 3y = 5$$

$$-3y = -2x + 5 \qquad \text{Add } -2x \text{ to both sides.}$$

$$y = \frac{2}{3}x - \frac{5}{3} \qquad \text{Divide by } -3.$$

The last equation has the form $y = mx + b$. The slope must be $m = \frac{2}{3}$, and the y-intercept is $b = -\frac{5}{3}$. ∎

EXAMPLE 3   Graph the equation $2x + 3y = 6$ using the slope and y-intercept.

SOLUTION   Although we could graph this equation by finding ordered pairs that are solutions to the equation and drawing a line through their graphs, it is sometimes easier to graph a line using the slope-intercept form of the equation.

Solving the equation for y, we have

$$2x + 3y = 6$$

$$3y = -2x + 6 \qquad \text{Add } -2x \text{ to both sides.}$$

$$y = -\frac{2}{3}x + 2 \qquad \text{Divide by 3.}$$

The slope is $m = -\frac{2}{3}$ and the y-intercept is $b = 2$. Therefore, the point $(0, 2)$ is on the graph, and the ratio of rise to run going from $(0, 2)$ to any other point on the line is $-\frac{2}{3}$. If we start at $(0, 2)$ and move 2 units up (that's a rise of 2) and 3 units to the left (a run of -3), we will be at another point on the graph. (We could also go down 2 units and right 3 units and still be assured of ending up at another point on the line because $\frac{2}{-3}$ is the same as $\frac{-2}{3}$.)

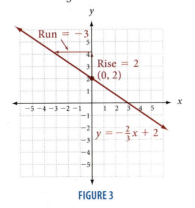

FIGURE 3 ∎

---

2. Give the slope and y-intercept for the line $4x - 5y = 7$.

3. Graph the line $-3x + 2y = -6$ on the following coordinate system. Use the method shown in Example 3.

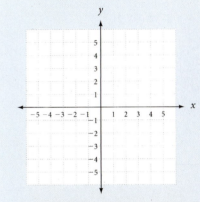

NOTE
As we mentioned earlier, the rectangular coordinate system is the tool we use to connect algebra and geometry. Example 3 illustrates this connection, as do the many other examples in this chapter. In Example 3, Descartes's rectangular coordinate system allows us to associate the equation $2x + 3y = 6$ (an algebraic concept) with the straight line (a geometric concept) shown in Figure 3.

Answers
2. $m = \frac{4}{5}, b = -\frac{7}{5}$
3. See Solutions Section.

C Finding the Line Using the Slope and a Point

A second useful form of the equation of a straight line is the *point-slope form*.

Let line *l* contain the point (x_1, y_1) and have slope *m*. If (x, y) is any other point on *l*, then by the definition of slope we have

$$\frac{y - y_1}{x - x_1} = m$$

Multiplying both sides by $(x - x_1)$ gives us

$$(x - x_1) \cdot \frac{y - y_1}{x - x_1} = m(x - x_1)$$

$$y - y_1 = m(x - x_1)$$

This last equation is known as the *point-slope form* of the equation of a straight line.

Point-Slope Form

The equation of the line through (x_1, y_1) with slope *m* is given by
$$y - y_1 = m(x - x_1)$$

This form of the equation of a straight line is used to find the equation of a line, either given one point on the line and the slope, or given two points on the line.

EXAMPLE 4   Find the equation of the line with slope -2 that contains the point $(-4, 3)$. Write the answer in slope-intercept form.

SOLUTION

$$\begin{array}{lll}
\text{Using} & (x_1, y_1) = (-4, 3) \text{ and } m = -2 & \\
\text{in} & y - y_1 = m(x - x_1) & \text{\color{red}Point-slope form} \\
\text{gives us} & y - 3 = -2(x + 4) & \text{\color{red}Note: } x - (-4) = x + 4 \\
& y - 3 = -2x - 8 & \text{\color{red}Multiply out right side.} \\
& y = -2x - 5 & \text{\color{red}Add 3 to each side.}
\end{array}$$

Figure 4 is the graph of the line that contains $(-4, 3)$ and has a slope of -2. Notice that the *y*-intercept on the graph matches that of the equation we found.

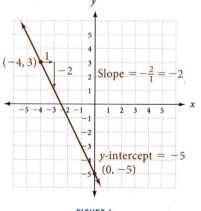

FIGURE 4

4. Find the equation of the line with slope 3 that contains the point $(-1, 2)$.

Answer

4. $y = 3x + 5$

D Finding the Line Using Points

EXAMPLE 5   Find the equation of the line that passes through the points $(-3, 3)$ and $(3, -1)$.

SOLUTION   We begin by finding the slope of the line.

$$m = \frac{3 - (-1)}{-3 - 3} = \frac{4}{-6} = -\frac{2}{3}$$

Using $(x_1, y_1) = (3, -1)$ and $m = -\frac{2}{3}$ in $y - y_1 = m(x - x_1)$ yields

$$y + 1 = -\frac{2}{3}(x - 3)$$

$$y + 1 = -\frac{2}{3}x + 2 \qquad \text{Multiply out right side.}$$

$$y = -\frac{2}{3}x + 1 \qquad \text{Add } -1 \text{ to each side.}$$

Figure 5 shows the graph of the line that passes through the points $(-3, 3)$ and $(3, -1)$. As you can see, the slope and y-intercept are $-\frac{2}{3}$ and 1, respectively.

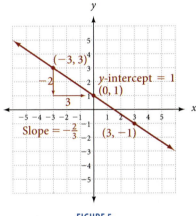

FIGURE 5   ■

The last form of the equation of a line that we will consider in this section is called *standard form*. It is used mainly to write equations in a form that is free of fractions and is easy to compare with other equations.

> **Standard Form**
>
> If a, b, and c are integers, then the equation of a line is in standard form when it has the form
> $$ax + by = c$$

If we were to write the equation

$$y = -\frac{2}{3}x + 1$$

in standard form, we would first multiply both sides by 3 to obtain

$$3y = -2x + 3$$

Then we would add $2x$ to each side, yielding

$$2x + 3y = 3$$

which is a linear equation in standard form.

5. Find the equation of the line through $(2, 5)$, and $(6, -3)$.

NOTE
In Example 5, we could have used the point $(-3, 3)$ instead of $(3, -1)$ and obtained the same equation; that is, using $(x_1, y_1) = (-3, 3)$ and $m = -\frac{2}{3}$ in $y - y_1 = m(x - x_1)$ gives us
$$y - 3 = -\frac{2}{3}(x + 3)$$
$$y - 3 = -\frac{2}{3}x - 2$$
$$y = -\frac{2}{3}x + 1$$
which is the same result we obtained using $(3, -1)$.

Answer
5. $y = -2x + 9$

6. Find the equation of the line through (3, 2) that is perpendicular to the graph of $3x - y = 2$. Write your answer in standard form.

EXAMPLE 6 Give the equation of the line through $(-1, 4)$ whose graph is perpendicular to the graph of $2x - y = -3$. Write the answer in standard form.

SOLUTION To find the slope of $2x - y = -3$, we solve for y.

$$2x - y = -3$$
$$y = 2x + 3$$

The slope of this line is 2. The line we are interested in is perpendicular to the line with slope 2 and must, therefore, have a slope of $-\frac{1}{2}$.

Using $(x_1, y_1) = (-1, 4)$ and $m = -\frac{1}{2}$, we have

$$y - y_1 = m(x - x_1)$$
$$y - 4 = -\frac{1}{2}(x + 1)$$

Because we want our answer in standard form, we multiply each side by 2.

$$2y - 8 = -1(x + 1)$$
$$2y - 8 = -x - 1$$
$$x + 2y - 8 = -1$$
$$x + 2y = 7$$

The last equation is in standard form. ∎

As a final note, all horizontal lines have equations of the form $y = b$ and slopes of 0. Because they cross the y-axis at b, the y-intercept is b; there is no x-intercept. Vertical lines have no slope and equations of the form $x = a$. Each will have an x-intercept at a and no y-intercept. Finally, equations of the form $y = mx$ have graphs that pass through the origin. The slope is always m and both the x-intercept and the y-intercept are 0.

USING TECHNOLOGY

Graphing Calculators

One advantage of using a graphing calculator to graph lines is that a calculator does not care whether the equation has been simplified or not. To illustrate, in Example 5 we found that the equation of the line with slope $-\frac{2}{3}$ that passes through the point $(3, -1)$ is

$$y + 1 = -\frac{2}{3}(x - 3)$$

Normally, to graph this equation we would simplify it first. With a graphing calculator, we add -1 to each side and enter the equation this way:

$$Y_1 = -(2/3)(X - 3) - 1$$

Answer
6. $x + 3y = 9$

Problem Set 3.3

Moving Toward Success

"There is no chance, no destiny, no fate, that can hinder or control the firm resolve of a determined soul."

—Ella Wheeler Wilcox, 1850–1919, American author and poet

1. Do you have rules for yourself while you study? Do they yield a quality studying experience? Why or why not?

2. How can you improve your studying routine?

A Write the equation of the line with the given slope and *y*-intercept. [Example 1]

1. $m = 2, b = 3$

2. $m = -4, b = 2$

3. $m = 1, b = -5$

4. $m = -5, b = -3$

5. $m = \dfrac{1}{2}, b = \dfrac{3}{2}$

6. $m = \dfrac{2}{3}, b = \dfrac{5}{6}$

7. $m = 0, b = 4$

8. $m = 0, b = -2$

B Give the slope and *y*-intercept for each of the following equations. Sketch the graph using the slope and *y*-intercept. Give the slope of any line perpendicular to the given line. [Examples 2, 3, 6]

9. $y = 3x - 2$

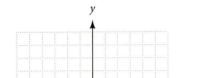

10. $y = 2x + 3$

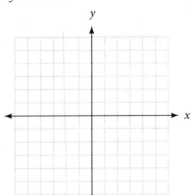

11. $2x - 3y = 12$

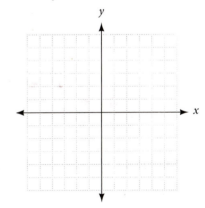

12. $3x - 2y = 12$

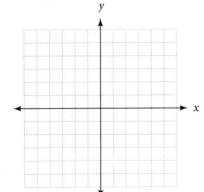

13. $4x + 5y = 20$

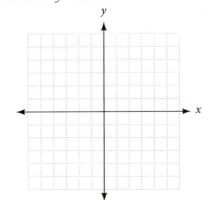

14. $5x - 4y = 20$

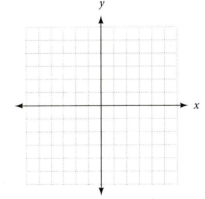

For each of the following lines, name the slope and *y*-intercept. Then write the equation of the line in slope-intercept form.

15.

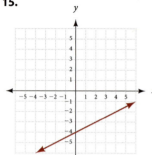

16.

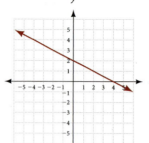

17.

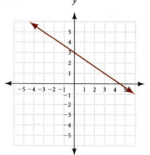

18.

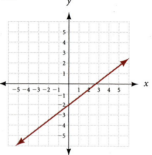

C For each of the following problems, the slope and one point on a line are given. In each case, find the equation of that line. (Write the equation for each line in slope-intercept form.) [Example 4]

19. $(-2, -5)$; $m = 2$

20. $(-1, -5)$; $m = 2$

21. $(-4, 1)$; $m = -\dfrac{1}{2}$

22. $(-2, 1)$; $m = -\dfrac{1}{2}$

23. $\left(-\dfrac{1}{3}, 2\right)$; $m = -3$

24. $\left(-\dfrac{2}{3}, 5\right)$; $m = -3$

25. $(-4, -2)$, $m = \dfrac{2}{3}$

26. $(3, 4)$, $m = -\dfrac{1}{3}$

27. $(-5, 2)$, $m = -\dfrac{1}{4}$

28. $(-4, 3)$, $m = \dfrac{1}{6}$

D Find the equation of the line that passes through each pair of points. Write your answers in standard form. [Example 5]

29. $(-2, -4)$, $(1, -1)$

30. $(2, 4)$, $(-3, -1)$

31. $(-1, -5)$, $(2, 1)$

32. $(-1, 6)$, $(1, 2)$

33. $\left(\dfrac{1}{3}, -\dfrac{1}{5}\right)$, $\left(-\dfrac{1}{3}, -1\right)$

34. $\left(-\dfrac{1}{2}, -\dfrac{1}{2}\right)$, $\left(\dfrac{1}{2}, \dfrac{1}{10}\right)$

D Find the slope of a line (a) parallel and (b) perpendicular to the given line. [Example 6]

35. $y = 3x - 4$

36. $y = -4x + 1$

37. $3x + y = -2$

38. $2x - y = -4$

39. $2x + 5y = -11$

40. $3x - 5y = -4$

41. The equation $3x - 2y = 10$ is a linear equation in standard form. From this equation, answer the following:

  a. Find the x and y intercepts.

  b. Find a solution to this equation other than the intercepts in part a.

  c. Write this equation in slope-intercept form.

  d. Is the point $(2, 2)$ a solution to the equation?

42. The equation $4x + 3y = 8$ is a linear equation in standard form. From this equation, answer the following:

  a. Find the x and y intercepts.

  b. Find a solution to this equation other than the intercepts in part a.

  c. Write this equation in slope-intercept form.

  d. Is the point $(-3, 2)$ a solution to the equation?

43. The equation $\frac{3x}{4} - \frac{y}{2} = 1$ is a linear equation in standard form. From this equation, answer the following:

  a. Find the x and y intercepts.

  b. Find a solution to this equation other than the intercepts in part a.

  c. Write this equation in slope-intercept form.

  d. Is the point $(1, 2)$ a solution to the equation?

44. The equation $\frac{3x}{5} + \frac{2y}{3} = 1$ is a linear equation in standard form. From this equation, answer the following:

  a. Find the x and y intercepts.

  b. Find a solution to this equation other than the intercepts in part a.

  c. Write this equation in slope-intercept form.

  d. Is the point $(-5, 3)$ a solution to the equation?

The next two problems are intended to give you practice reading, and paying attention to the instructions that accompany the problems you are working. Working these problems is an excellent way to get ready for a test or a quiz.

45. Work each problem according to the instructions.

    a. Solve: $-2x + 1 = -3$.

    b. Find x when y is 0: $-2x + y = -3$.

    c. Find y when x is 0: $-2x + y = -3$.

    d. Graph: $-2x + y = -3$.

    e. Solve for y: $-2x + y = -3$.

46. Work each problem according to the instructions.

    a. Solve: $\frac{x}{3} + \frac{1}{4} = 1$.

    b. Find x when y is 0: $\frac{x}{3} + \frac{y}{4} = 1$.

    c. Find y when x is 0: $\frac{x}{3} + \frac{y}{4} = 1$.

    d. Graph: $\frac{x}{3} + \frac{y}{4} = 1$.

    e. Solve for y: $\frac{x}{3} + \frac{y}{4} = 1$.

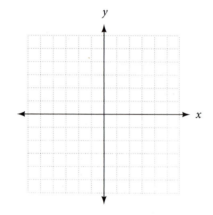

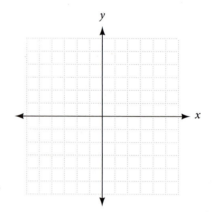

For each of the following lines, name the coordinates of any two points on the line. Then use those two points to find the equation of the line.

47.

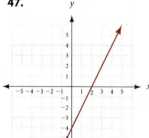

48.

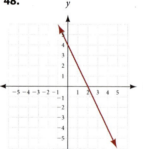

49.

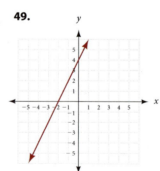

50.

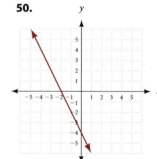

51. Give the slope and y-intercept of $y = -2$. Sketch the graph.

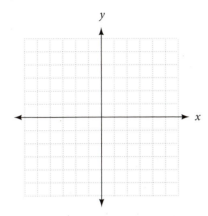

52. For the line $x = -3$, sketch the graph, give the slope, and name any intercepts.

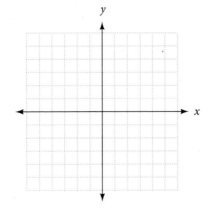

53. Find the equation of the line parallel to the graph of $3x - y = 5$ that contains the point $(-1, 4)$.

54. Find the equation of the line parallel to the graph of $2x - 4y = 5$ that contains the point $(0, 3)$.

55. Line l is perpendicular to the graph of the equation $2x - 5y = 10$ and contains the point $(-4, -3)$. Find the equation for l.

56. Line l is perpendicular to the graph of the equation $-3x - 5y = 2$ and contains the point $(2, -6)$. Find the equation for l.

57. Give the equation of the line perpendicular to the graph of $y = -4x + 2$ that has an x-intercept of -1.

58. Write the equation of the line parallel to the graph of $7x - 2y = 14$ that has an x-intercept of 5.

59. Find the equation of the line with x-intercept 3 and y-intercept 2.

60. Find the equation of the line with x-intercept 2 and y-intercept 3.

Applying the Concepts

61. Internet The chart shows the percentage of adults who use broadband to access the internet. Find an equation for the line segment from 2000 to 2002. Write your answer in slope-intercept form.

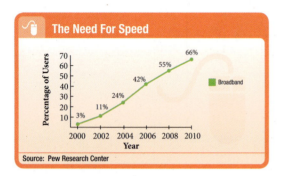

Source: Pew Research Center

62. Cell Phones The graph shows the percentage of teens that use cell phones to make calls. Find an equation for the line that connects (12,17) and (17,60). Write your answer in slope-intercept form.

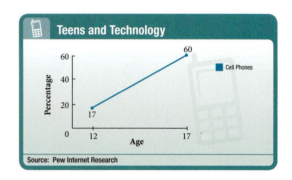

Source: Pew Internet Research

63. Deriving the Temperature Equation The table from the introduction to this section is repeated here. The rows of the table give us ordered pairs (C, F).

Degrees Celsius	Degrees Fahrenheit
C	F
0	32
25	77
50	122
75	167
100	212

a. Use any two of the ordered pairs from the table to derive the equation $F = \frac{9}{5}C + 32$.

b. Use the equation from part (a) to find the Fahrenheit temperature that corresponds to a Celsius temperature of 30°.

64. Maximum Heart Rate The table gives the maximum heart rate for adults 30, 40, 50, and 60 years old. Each row of the table gives us an ordered pair (A, M).

Age (years)	Maximum Heart Rate (beats per minute)
A	M
30	190
40	180
50	170
60	160

Source: webmd.com

a. Use any two of the ordered pairs from the table to derive the equation $M = 220 - A$, which gives the maximum heart rate M for an adult whose age is A.

b. Use the equation from part (a) to find the maximum heart rate for a 25-year-old adult.

65. Daniel starts a new job in which he sells TVs. He earns $1,000 a month plus a certain amount for each TV he sells. The graph below shows the amount per month Daniel will make based on how many TVs he sells.

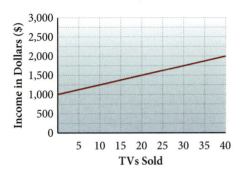

a. How much will he earn if he sells 20 TVs?

b. How much will he earn if he sells 40 TVs?

c. What is the slope of the line?

d. How much does he make for each TV he sells?

e. Find the equation of the line where y is his income and x is TV sales.

66. Mark gets a new job where he sells hats. He will earn $1,250 a month plus a certain amount for each hat he sells. The graph below shows the amount Mark will make based on how many hats he sells.

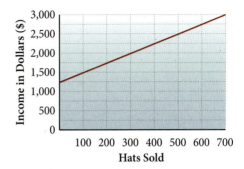

a. How much will he earn if he sells 300 hats?

b. How much does he have to sell to earn $2,500?

c. What is the slope of the line?

d. How much does he make for each hat he sells?

e. Find the equation of the line.

Improving Your Quantitative Literacy

67. Courtney buys a new car. It cost $20,000 and will decrease in value each year. The graph below shows the value of the car after five years.

a. How much is the car worth after five years?

b. Find the equation of the line connecting the endpoints.

c. Using your equation, how much will the car be worth in ten years? What's wrong with the answer?

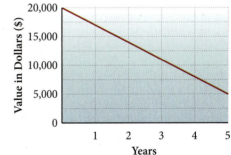

68. Definitions

a. Use an online dictionary to find definitions for the following words: *interpolation* and *extrapolation*

b. Which of the two words describes what you did in problem 69 part (c) above.

Maintaining Your Skills

69. The length of a rectangle is 3 inches more than 4 times the width. The perimeter is 56 inches. Find the length and width.

70. One angle is 10 degrees less than four times another. Find the measure of each angle if

a. the two angles are complementary.

b. the two angles are supplementary.

71. The cash register in a candy shop contains $66 at the beginning of the day. At the end of the day, it contains $732.50. If the sales tax rate is 7.5%, how much of the total is sales tax?

72. The third angle in an isosceles triangle is 20 degrees less than twice as large as each of the two base angles. Find the measure of each angle.

Getting Ready for the Next Section

73. Which of the following are solutions to $x + y \leq 4$?

$(0, 0)$ $(4, 0)$ $(2, 3)$

74. Which of the following are solutions to $y < 2x - 3$?

$(0, 0)$ $(3, -2)$ $(-3, 2)$

75. Which of the following are solutions to $y \leq \frac{1}{2}x$?

$(0, 0)$ $(2, 0)$ $(-2, 0)$

76. Which of the following are solutions to $y > -2x$?

$(0, 0)$ $(2, 0)$ $(-2, 0)$

Extending the Concepts

The midpoint M of two points (x_1, y_1) and (x_2, y_2) is defined to be the average of each of their coordinates, so

$$M = \left(\frac{x_1 + x_2}{2}, \frac{y_1 + y_2}{2} \right)$$

For example, the midpoint of $(-2, 3)$ and $(6, 8)$ is given by

$$\left(\frac{-2 + 6}{2}, \frac{3 + 8}{2} \right) = \left(2, \frac{11}{2} \right)$$

For each given pair of points, find the equation of the line that is perpendicular to the line through these points and that passes through their midpoint. Answer using slope-intercept form.

Note: This line is called the perpendicular bisector of the line segment connecting the two points.

77. $(1, 4)$ and $(7, 8)$

78. $(-2, 1)$ and $(6, 7)$

79. $(-5, 1)$ and $(-1, 4)$

80. $(-6, -2)$ and $(2, 1)$

Linear Inequalities in Two Variables

TICKET TO SUCCESS

Keep these questions in mind as you read through the section. Then respond in your own words and in complete sentences.

1. When graphing a linear inequality in two variables, how do you find the equation of the boundary line?
2. What is the significance of a broken line in the graph of an inequality?
3. When graphing a linear inequality in two variables, how do you know which side of the boundary line to shade?
4. Does the graph of $x + y < 4$ include the boundary line? Explain.

Artfran/Shutterstock.com

The King County Water Taxi operates a passenger-only ferry service from Downtown Seattle, Washington to Vashon Island. The one-way fare to ride the ferry is higher for adults than it is for seniors age 65 and older. The ferry also has a maximum passenger capacity of 150 people. Therefore, it is important to know the different combinations of adults and seniors that can ride the ferry at any one time. The shaded region in Figure 1 contains all the capacity combinations. The line $x + y = 150$ shows the combinations for a full ferry. The y-intercept corresponds to a ferry full of adults, and the x-intercept corresponds to a ferry full of seniors. In the shaded region below the line $x + y = 150$ are the combinations that occur if the ferry is not full.

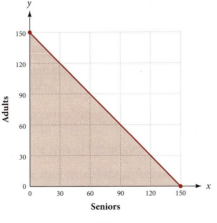

FIGURE 1

A Graphing Inequalities

A *linear inequality in two variables* is any expression that can be put in the form

$$ax + by < c$$

where *a*, *b*, and *c* are real numbers (*a* and *b* not both 0). The inequality symbol can be any one of the following four: $<, \le, >, \ge$.

Some examples of linear inequalities are

$$2x + 3y < 6 \qquad y \ge 2x + 1 \qquad x - y \le 0$$

Although not all of these examples have the form $ax + by < c$, each one can be put in that form.

The solution set for a linear inequality is a *section of the coordinate plane*. The *boundary* for the section is found by replacing the inequality symbol with an equal sign and graphing the resulting equation. The boundary is included in the solution set (and represented with a *solid line*) if the inequality symbol originally used is $\le$ or $\ge$. The boundary is not included (and is represented with a *broken line*) if the original symbol is $<$ or $>$.

PRACTICE PROBLEMS

1. Graph the solution set for $x - y \ge 3$. Follow Example 1 carefully. First graph the boundary. Then shade in the correct region after testing a point not on the boundary.

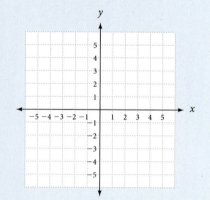

EXAMPLE 1 Graph the solution set for $x + y \le 4$.

SOLUTION The boundary for the graph is the graph of $x + y = 4$. The boundary is included in the solution set because the inequality symbol is $\le$.

Figure 2 is the graph of the boundary.

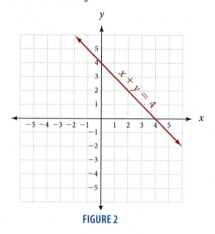

FIGURE 2

The boundary separates the coordinate plane into two regions: the region above the boundary and the region below it. The solution set for $x + y \le 4$ is one of these two regions along with the boundary. To find the correct region, we simply choose any convenient point that is *not* on the boundary. Then we substitute the coordinates of the point into the original inequality, $x + y \le 4$. If the point we choose satisfies the inequality, then it is a member of the solution set, and we can assume that all points on the same side of the boundary as the chosen point are also in the solution set. If the coordinates of our point do not satisfy the original inequality, then the solution set lies on the other side of the boundary.

In this example, a convenient point that is not on the boundary is the origin.

Substituting →   (0, 0)

into → $x + y \le 4$

gives us → $0 + 0 \le 4$

$0 \le 4$ A true statement

Answers
See Solutions to Selected Practice Problems for all answers in this section.

Because the origin is a solution to the inequality $x + y \leq 4$ and the origin is below the boundary, all other points below the boundary are also solutions.

Figure 3 is the graph of $x + y \leq 4$.

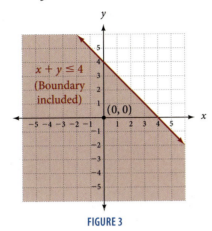

FIGURE 3

The region above the boundary is described by the inequality $x + y > 4$. ■

Here is a list of steps to follow when graphing the solution set for a linear inequality in two variables.

Strategy   To Graph a Linear Inequalities in Two Variables

Step 1 Replace the inequality symbol with an equal sign. The resulting equation represents the boundary for the solution set.

Step 2 Graph the boundary found in step 1 using a solid line if the boundary is included in the solution set; that is, if the original inequality symbol was either $\leq$ or $\geq$. Use a broken line to graph the boundary if it is not included in the solution set. (It is not included if the original inequality was either $<$ or $>$.)

Step 3 Choose any convenient point not on the boundary and substitute the coordinates into the original inequality. If the resulting statement is true, the graph lies on the same side of the boundary as the chosen point. If the resulting statement is false, the solution set lies on the opposite side of the boundary.

2. Graph the solution set for
$$y < \frac{1}{2}x + 3.$$

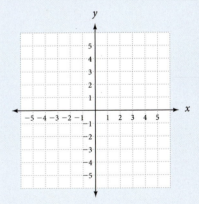

EXAMPLE 2   Graph the solution set for $y < 2x - 3$.

SOLUTION   The boundary is the graph of $y = 2x - 3$, a line with slope 2 and y-intercept -3. The boundary is not included because the original inequality symbol is $<$. We therefore use a broken line to represent the boundary in Figure 4.

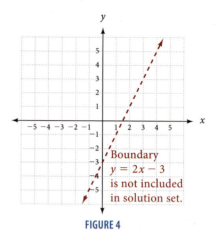

FIGURE 4

A convenient test point is again the origin.

$$\text{Using} \rightarrow \quad (0, 0)$$

$$\text{in} \rightarrow \quad y < 2x - 3$$

$$\text{we have} \rightarrow \quad 0 < 2(0) - 3$$

$$0 < -3 \qquad \text{A false statement}$$

Because our test point gives us a false statement and it lies above the boundary, the solution set must lie on the other side of the boundary (Figure 5).

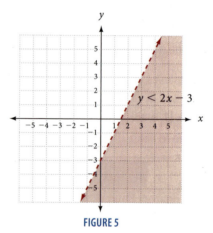

FIGURE 5

USING TECHNOLOGY

Graphing Calculators

Most graphing calculators have a Shade command that allows a portion of a graphing screen to be shaded. With this command we can visualize the solution sets to linear inequalities in two variables. Because most graphing calculators cannot draw a dotted line, however, we are not actually "graphing" the solution set, only visualizing it.

Strategy   Visualizing a Linear Inequality in Two Variables on a Graphic Calculator

Step 1   Solve the inequality for y.

Step 2   Replace the inequality symbol with an equal sign. The resulting equation represents the boundary for the solution set.

Step 3   Graph the equation in an appropriate viewing window.

Step 4   Use the Shade command to indicate the solution set.

For inequalities having the $<$ or $\leq$ sign, use Shade(Ymin, Y_1).

For inequalities having the $>$ or $\geq$ sign, use Shade(Y_1, Ymax).

Note: On the TI-83/84, step 4 can be done by manipulating the icons in the left column in the list of Y variables.

Figures 6 and 7 show the graphing calculator screens that help us visualize the solution set to the inequality $y < 2x - 3$ that we graphed in Example 2.

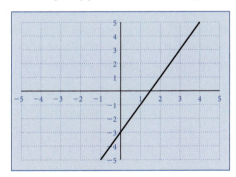

FIGURE 6   Y1 = 2X − 3

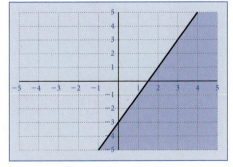

FIGURE 7   Shade (Xmin, Y_1)

Windows: X from −5 to 5, Y from −5 to 5

3. Graph $y > -2$.

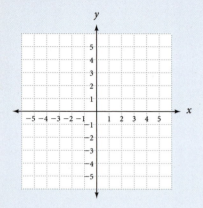

4. Graph the solution set for $y > -\frac{2}{3}x$.

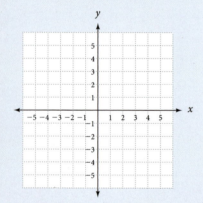

EXAMPLE 3   Graph the solution set for $x \le 5$.

SOLUTION   The boundary is $x = 5$, which is a vertical line. All points in Figure 8 to the left of the boundary have x-coordinates less than 5 and all points to the right have x-coordinates greater than 5.

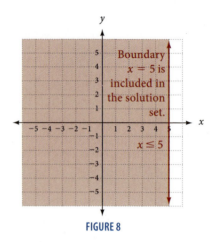

Boundary $x = 5$ is included in the solution set.

$x \le 5$

FIGURE 8   ■

EXAMPLE 4   Graph the solution set for $y > \frac{1}{4}x$.

SOLUTION   The boundary is the line $y = \frac{1}{4}x$, which has a slope of $\frac{1}{4}$ and passes through the origin. The graph of the boundary line is shown in Figure 9. Since the boundary passes through the origin, we cannot use the origin as our test point. Remember, the test point cannot be on the boundary line. Let's use the point $(0, -4)$ as our test point. It lies below the boundary line. When we substitute the coordinates into our original inequality, the result is the false statement $-4 > 0$. This tells us that the solution set is on the other side of the boundary line. The solution set for our original inequality is shown in Figure 10.

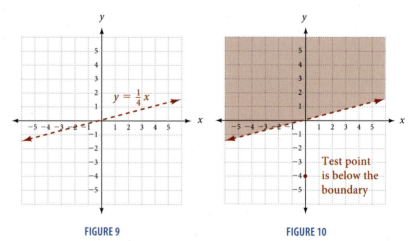

$y = \frac{1}{4}x$

Test point is below the boundary

FIGURE 9   **FIGURE 10**   ■

Problem Set 3.4

Moving Toward Success

"If there were dreams to sell, what would you buy?"

—Thomas Lovell Beddoes, 1803–1849, British poet

1. Why are objectives and goals important to set for your study group?

2. Why is it important to come prepared to a study group session?

A   Graph the solution set for each of the following. [Examples 1–4]

1. $x + y < 5$

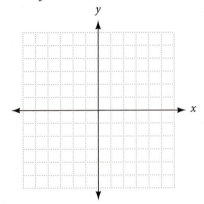

2. $x + y \leq 5$

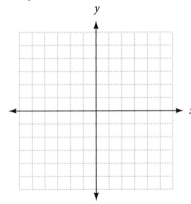

3. $x - y \geq -3$

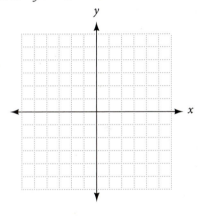

4. $x - y > -3$

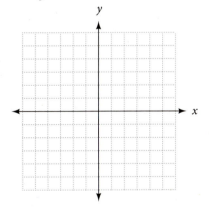

5. $2x + 3y < 6$

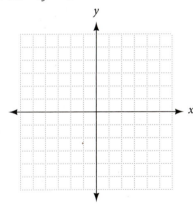

6. $2x - 3y > -6$

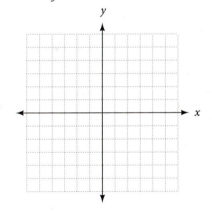

7. $-x + 2y > -4$

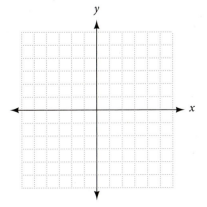

8. $-x - 2y < 4$

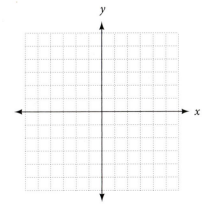

9. $2x + y < 5$

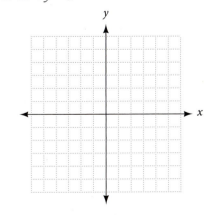

10. $2x + y < -5$

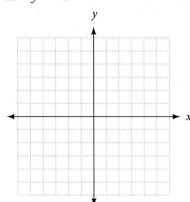

11. $y < 2x - 1$

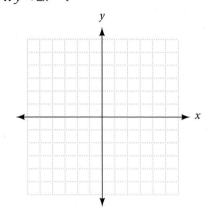

12. $y \leq 2x - 1$

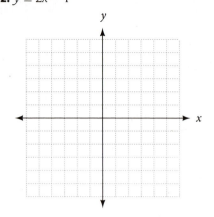

13. $3x - 4y < 12$

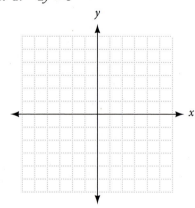

14. $-2x + 3y < 6$

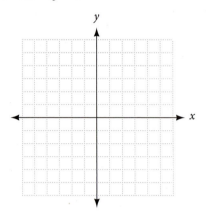

15. $-5x + 2y \leq 10$

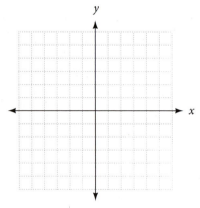

16. $4x - 2y \leq 8$

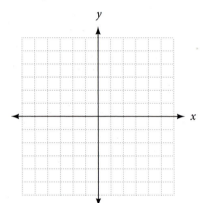

17. $x \geq 3$

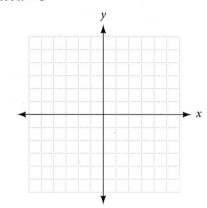

18. $x > -2$

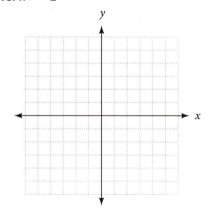

19. $y \leq 4$

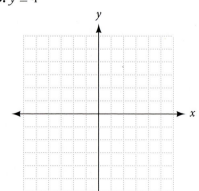

20. $y > -5$

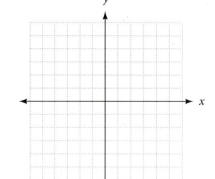

21. $y < 2x$

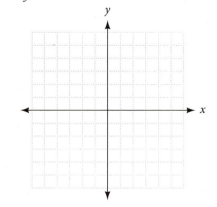

22. $y > -3x$

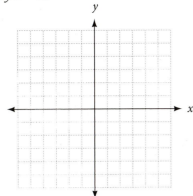

23. $y \geq \frac{1}{2}x$

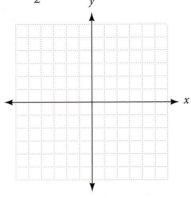

24. $y \leq \frac{1}{3}x$

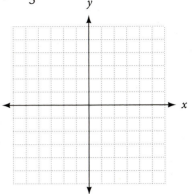

25. $y \geq \frac{3}{4}x - 2$

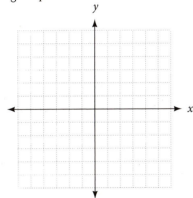

26. $\frac{x}{3} + \frac{y}{2} > 1$

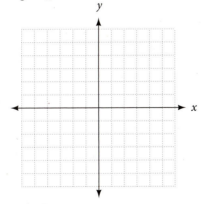

27. $y > -\frac{2}{3}x + 3$

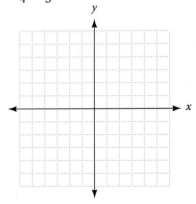

28. $\frac{x}{5} + \frac{y}{4} < 1$

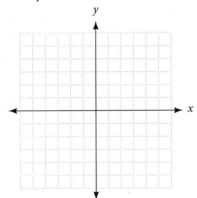

29. $\frac{x}{3} - \frac{y}{2} > 1$

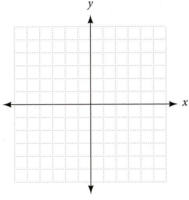

30. $-\frac{x}{4} - \frac{y}{3} > 1$

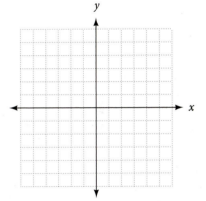

31. $y \leq -\frac{2}{3}x$

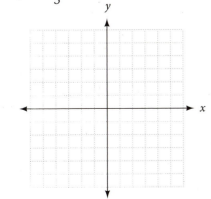

32. $y \geq \frac{1}{4}x$

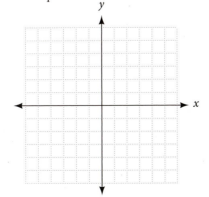

33. $5x - 3y < 0$

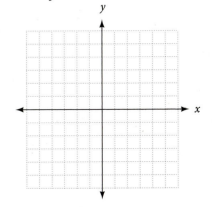

34. $2x + 3y < 0$

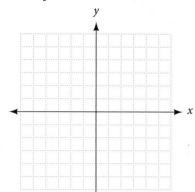

35. $\dfrac{x}{4} + \dfrac{y}{5} \le 1$

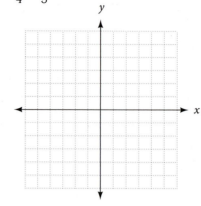

36. $\dfrac{x}{2} + \dfrac{y}{3} < 1$

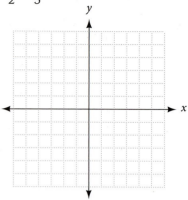

For each graph shown here, name the linear inequality in two variables that is represented by the shaded region.

37.

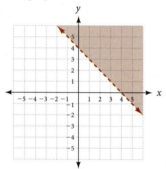

38.

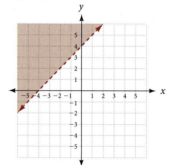

39.

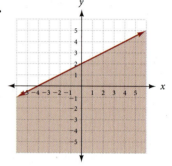

40.

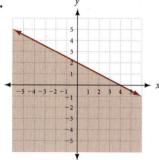

41.

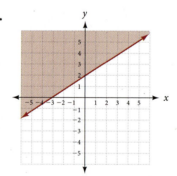

42.

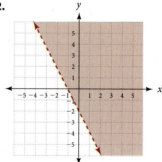

The next two problems are intended to give you practice reading, and paying attention to, the instructions that accompany the problems you are working.

43. Work each problem according to the instructions.

  a. Solve: $\frac{1}{3} + \frac{y}{2} < 1$.

  b. Solve: $\frac{1}{3} - \frac{y}{2} < 1$.

  c. Solve for y: $\frac{x}{3} + \frac{y}{2} = 1$.

  d. Graph: $y < -\frac{4}{3}x + 4$.

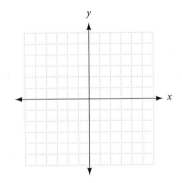

44. Work each problem according to the instructions.

  a. Solve: $3x + 4 \geq -8$.

  b. Solve: $-3x + 4 \geq -8$.

  c. Solve for y: $-3x + 4y = -8$.

  d. Graph: $3x + 4y \geq -8$.

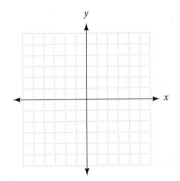

Applying the Concepts

45. Number of People in a Dance Club A dance club holds a maximum of 200 people. The club charges one price for students and a higher price for nonstudents. If the number of students in the club at any time is x and the number of nonstudents is y, sketch a graph representing the people in the club. Then, shade the region in the first quadrant that contains all combinations of students and nonstudents that are in the club at any time.

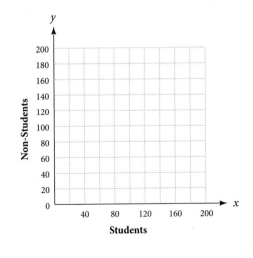

46. Many Perimeters Suppose you have 500 feet of fencing that you will use to build a rectangular livestock pen. Let x represent the length of the pen and y represent the width. Sketch a graph of the situation and shade the region in the first quadrant that contains all possible values of x and y that will give you a rectangle from 500 feet of fencing. (You don't have to use all of the fencing, so the perimeter of the pen could be less than 500 feet.)

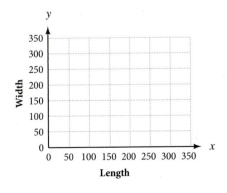

47. Gas Mileage You have two cars. The first car travels an average of 12 miles on a gallon of gasoline, and the second averages 22 miles per gallon. Suppose you can afford to buy up to 30 gallons of gasoline this month. If the first car is driven x miles this month, and the second car is driven y miles this month, draw a graph and shade the region in the first quadrant that gives all the possible values of x and y that will keep you from buying more than 30 gallons of gasoline this month.

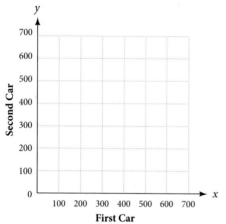

48. Student Loan Payments When considering how much debt to incur in student loans, it is advisable to keep your student loan payment after graduation to 8% or less of your starting monthly income. Let x represent your starting monthly salary and let y represent your monthly student loan payment, and write an inequality that describes this situation. Graph the inequality and shade the region in the first quadrant that is a solution to your inequality.

49. Solar and Wind Energy The chart shows the cost to install either solar panels or a wind turbine. A company that specializes in alternate energy sources is placing an order for more cables for both wind and solar energy solutions. If the total bill is some amount less than $11,440, graph the inequality and shade the region in the first quadrant that contains all the possible combinations of products.

50. Gas Prices CJ has two vehicles, a gasoline car and a diesel truck. Suppose that the price of gas and the price of diesel are both $4.20 per gallon. Assume the price of crude oil accounts for 68% of the price of gasoline and 62% of the price of diesel. The amount of gasoline and diesel CJ buys comes to less than 15 gallons. Graph the inequality and shade the region in the first quadrant that contains all possible combination of money he paid for crude oil. If necessary, round to the nearest cent.

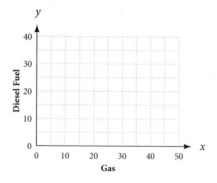

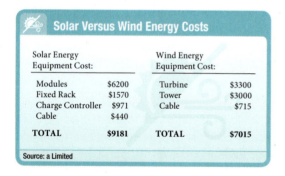

Solar Energy Equipment Cost:		Wind Energy Equipment Cost:	
Modules	$6200	Turbine	$3300
Fixed Rack	$1570	Tower	$3000
Charge Controller	$971	Cable	$715
Cable	$440		
TOTAL	**$9181**	**TOTAL**	**$7015**

Source: a Limited

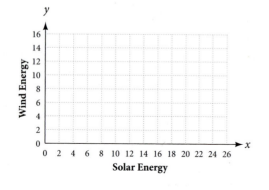

Maintaining Your Skills

51. $\dfrac{1}{3} + \dfrac{y}{5} \leq \dfrac{26}{15}$

52. $-\dfrac{1}{3} \geq \dfrac{1}{6} - \dfrac{y}{2}$

53. $5t - 4 > 3t - 8$

54. $-3(t - 2) < 6 - 5(t + 1)$

55. $-9 < -4 + 5t < 6$

56. $-3 < 2t + 1 < 3$

Getting Ready for the Next Section

Complete each table using the given equation.

57. $y = 7.5x$

x	y
0	
10	
20	

58. $h = 32t - 16t^2$

t	h
0	
$\frac{1}{2}$	
1	

59. $x = y^2$

x	y
	0
	1
	-1

60. $y = |x|$

x	y
-3	
0	
3	

Extending the Concepts

Graph each inequality.

61. $y < |x + 2|$
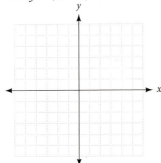

62. $y > |x - 2|$
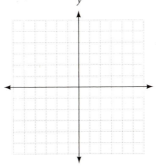

63. $y > |x - 3|$
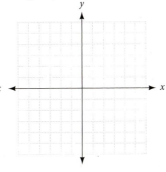

64. $y < |x + 3|$
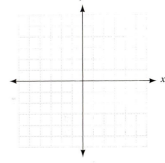

65. $y \leq |x - 1|$
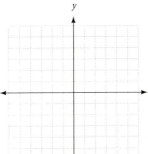

66. $y \geq |x + 1|$
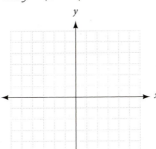

67. $y < x^2$
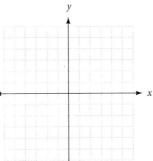

68. $y > x^2 - 2$
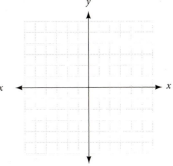

69. The Associated Students organization holds a *Night at the Movies* fundraiser. Student tickets are $1.00 each and nonstudent tickets are $2.00 each. The theater holds a maximum of 200 people. The club needs to collect at least $100 to make money. Sketch a graph of the inequality and shade the region in the first quadrant that contains all combinations of students and nonstudents that could attend the movie night so the club makes money.

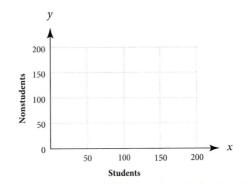

Introduction to Functions

OBJECTIVES

A Construct a table or a graph from a function rule.

B Identify the domain and range of a function or a relation.

C Determine whether a relation is also a function.

D Use the vertical line test to determine whether a graph is a function.

TICKET TO SUCCESS

Keep these questions in mind as you read through the section. Then respond in your own words and in complete sentences.

1. Define function using the terms domain and range.
2. Which variable is usually associated with the domain of a function?
3. What is a relation?
4. Explain the vertical line test.

ilker canikligil/Shutterstock.com

International Laws of the Game for soccer state that the length of the playing field can be between 100 and 130 yards. Suppose you are laying out a field at a local park. The width of the field must be 60 yards due to some landscaping. How would you decide what the area of the field could be? In this section, we will learn how to find the area y as a function of the length x to lay out your regulation soccer field.

An Informal Look at Functions

The ad shown in the margin appeared in the help wanted section of the local newspaper the day I was writing this section of the book. We can use the information in the ad to continue an informal discussion of functions.

To begin, suppose you have a job that pays $7.50 per hour and that you work anywhere from 0 to 40 hours per week. The amount of money you make in one week depends on the number of hours you work that week. In mathematics, we say that your weekly earnings are a *function* of the number of hours you work. If we let the variable x represent hours and the variable y represent the money you make, then the relationship between x and y can be written as

$$y = 7.5x \quad \text{for} \quad 0 \le x \le 40$$

A Constructing Tables and Graphs

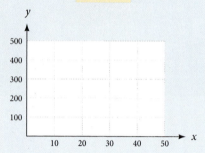

EXAMPLE 1 Construct a table and graph for the function.

$$y = 9.5x \quad \text{for} \quad 0 \le x \le 40$$

SOLUTION Table 1 gives some of the paired data that satisfy the equation $y = 9.5x$. Figure 1 is the graph of the equation with the restriction $0 \le x \le 40$.

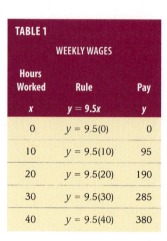

TABLE 1

WEEKLY WAGES

Hours Worked	Rule	Pay
x	$y = 9.5x$	y
0	$y = 9.5(0)$	0
10	$y = 9.5(10)$	95
20	$y = 9.5(20)$	190
30	$y = 9.5(30)$	285
40	$y = 9.5(40)$	380

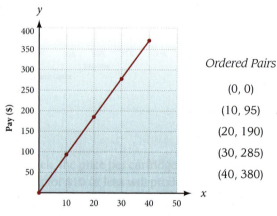

FIGURE 1 Weekly wages at $9.50 per hour

Ordered Pairs

(0, 0)

(10, 95)

(20, 190)

(30, 285)

(40, 380)

The equation $y = 9.5x$ with the restriction $0 \le x \le 40$, Table 1, and Figure 1 are three ways to describe the same relationship between the number of hours you work in one week and your gross pay for that week. In all three, we *input* values of x, and then use the function rule to *output* values of y. ■

B Domain and Range of a Function

We began this discussion by saying that the number of hours worked during the week was from 0 to 40, so these are the values that x can assume. From the line graph in Figure 1, we see that the values of y range from 0 to 380. We call the complete set of values that x can assume the *domain* of the function. The values that are assigned to y are called the *range* of the function.

The Function Rule

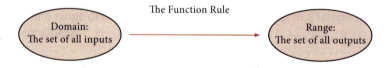

EXAMPLE 2 State the domain and range for the function.

$$y = 9.5x, \quad 0 \le x \le 40$$

SOLUTION From the previous discussion, we have

$$\text{Domain} = \{x \mid 0 \le x \le 40\}$$

$$\text{Range} = \{y \mid 0 \le y \le 380\}$$ ■

Function Maps

Another way to visualize the relationship between x and y is with the diagram in Figure 2, which we call a *function map*:

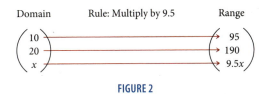

FIGURE 2

Although Figure 2 does not show all the values that x and y can assume, it does give us a visual description of how x and y are related. It shows that values of y in the range come from values of x in the domain according to a specific rule (multiply by 9.5 each time).

A Formal Look at Functions

What is apparent from the preceding discussion is that we are working with paired data. The solutions to the equation $y = 9.5x$ are pairs of numbers; the points on the line graph in Figure 1 come from paired data; and the diagram in Figure 2 pairs numbers in the domain with numbers in the range. We are now ready for the formal definition of a function.

> **Definition**
>
> A **function** is a rule that pairs each element in one set, called the **domain**, with exactly one element from a second set, called the **range**.

In other words, a function is a rule for which each input is paired with exactly one output.

Furthermore, the function rule $y = 9.5x$ from Example 1 produces ordered pairs of numbers (x, y). The same thing happens with all functions: The function rule produces ordered pairs of numbers. We use this result to write an alternative definition for a function.

> **Alternate Definition**
>
> A **function** is a set of ordered pairs in which no two different ordered pairs have the same first coordinate. The set of all first coordinates is called the **domain** of the function. The set of all second coordinates is called the **range** of the function.

The restriction on first coordinates in the alternative definition keeps us from assigning a number in the domain to more than one number in the range.

C A Relationship That Is Not a Function

You may be wondering if any sets of paired data fail to qualify as functions. The answer is yes, as the next example reveals.

3. With respect to Table 2 and Figure 3, how many outputs are paired with each of the following inputs?
a. 2008
b. 2007

EXAMPLE 3   Table 2 shows the prices of used Ford Mustangs that were listed in the local newspaper. The diagram in Figure 3 is called a *scatter diagram*. It gives a visual representation of the data in Table 2. Why are these data not a function?

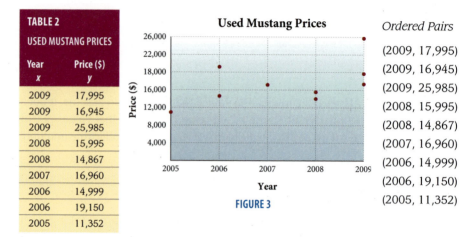

TABLE 2	
USED MUSTANG PRICES	
Year *x*	Price ($) *y*
2009	17,995
2009	16,945
2009	25,985
2008	15,995
2008	14,867
2007	16,960
2006	14,999
2006	19,150
2005	11,352

FIGURE 3

Ordered Pairs

(2009, 17,995)
(2009, 16,945)
(2009, 25,985)
(2008, 15,995)
(2008, 14,867)
(2007, 16,960)
(2006, 14,999)
(2006, 19,150)
(2005, 11,352)

SOLUTION   In Table 2, the year 2009 is paired with three different prices: $17,995, $16,945, and $25,985. That is enough to disqualify the data from belonging to a function. For a set of paired data to be considered a function, each number in the domain must be paired with exactly one number in the range. ∎

Still, there is a relationship between the first coordinates and second coordinates in the used-car data in Example 3. It is not a function relationship, but it is a relationship. To classify all relationships specified by ordered pairs, whether they are functions or not, we include the following two definitions.

> **Definition**
>
> A **relation** is a rule that pairs each element in one set, called the **domain**, with one or more elements from a second set, called the **range**.

> **Alternate Definition**
>
> A **relation** is a set of ordered pairs. The set of all first coordinates is the **domain** of the relation. The set of all second coordinates is the **range** of the relation.

Here are some facts that will help clarify the distinction between relations and functions.

1. Any rule that assigns numbers from one set to numbers in another set is a relation. If that rule makes the assignment so that no input has more than one output, then it is also a function.

2. Any set of ordered pairs is a relation. If none of the first coordinates of those ordered pairs is repeated, the set of ordered pairs is also a function.

3. Every function is a relation.

4. Not every relation is a function.

Answers
3. a. 2 **b.** 1

Graphing Relations and Functions

To give ourselves a wider perspective on functions and relations, we consider some equations whose graphs are not straight lines.

EXAMPLE 4 Kendra is tossing a softball into the air with an underhand motion. The distance of the ball above her hand at any time is given by the function

$$h = 32t - 16t^2 \quad \text{for} \quad 0 \le t \le 2$$

where h is the height of the ball in feet and t is the time in seconds. Construct a table that gives the height of the ball at quarter-second intervals, starting with $t = 0$ and ending with $t = 2$. Construct a graph from the table.

SOLUTION We construct Table 3 using the following values of t: $0, \frac{1}{4}, \frac{1}{2}, \frac{3}{4},$ $1, \frac{5}{4}, \frac{3}{2}, \frac{7}{4}, 2$. The values of h come from substituting these values of t into the equation $h = 32t - 16t^2$. (This equation comes from physics. If you take a physics class, you will learn how to derive this equation.) Then we construct the graph in Figure 4 from the table. The graph appears only in the first quadrant because neither t nor h can be negative.

TABLE 3

TOSSING A SOFTBALL INTO THE AIR

Time (sec) t	Function Rule $h = 32t - 16t^2$	Distance (ft) h
0	$h = 32(0) - 16(0)^2 = 0 - 0 = 0$	0
$\frac{1}{4}$	$h = 32\left(\frac{1}{4}\right) - 16\left(\frac{1}{4}\right)^2 = 8 - 1 = 7$	7
$\frac{1}{2}$	$h = 32\left(\frac{1}{2}\right) - 16\left(\frac{1}{2}\right)^2 = 16 - 4 = 12$	12
$\frac{3}{4}$	$h = 32\left(\frac{3}{4}\right) - 16\left(\frac{3}{4}\right)^2 = 24 - 9 = 15$	15
1	$h = 32(1) - 16(1)^2 = 32 - 16 = 16$	16
$\frac{5}{4}$	$h = 32\left(\frac{5}{4}\right) - 16\left(\frac{5}{4}\right)^2 = 40 - 25 = 15$	15
$\frac{3}{2}$	$h = 32\left(\frac{3}{2}\right) - 16\left(\frac{3}{2}\right)^2 = 48 - 36 = 12$	12
$\frac{7}{4}$	$h = 32\left(\frac{7}{4}\right) - 16\left(\frac{7}{4}\right)^2 = 56 - 49 = 7$	7
2	$h = 32(2) - 16(2)^2 = 64 - 64 = 0$	0

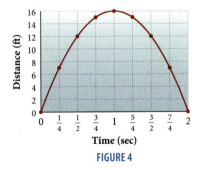

FIGURE 4

Here is a summary of what we know about functions as it applies to this example: We input values of t and output values of h according to the function rule

$$h = 32t - 16t^2 \quad \text{for} \quad 0 \le t \le 2$$

4. Kendra is tossing a softball into the air so that the distance h the ball is above her hand t seconds after she begins the toss is given by

$$h = 48t - 16t^2, \quad 0 \le t \le 3$$

Construct a table and line graph for this function.

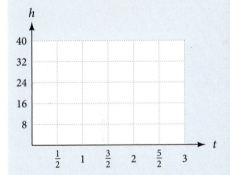

t	h
0	
$\frac{1}{2}$	
1	
$\frac{3}{2}$	
2	
$\frac{5}{2}$	
3	

Answer
4. See Solutions Section.

The domain is given by the inequality that follows the equation; it is

$$\text{Domain} = \{t \mid 0 \le t \le 2\}$$

The range is the set of all outputs that are possible by substituting the values of t from the domain into the equation. From our table and graph, it seems that the range is

$$\text{Range} = \{h \mid 0 \le h \le 16\}$$ ■

USING TECHNOLOGY

More About Example 4

Most graphing calculators can easily produce the information in Table 3. Simply set Y_1 equal to $32X - 16X^2$. Then set up the table so it starts at 0 and increases by an increment of 0.25 each time. (On a TI-83/84, use the TBLSET key to set up the table.)

Table Setup

Table minimum = 0
Table increment = .25
Dependent variable: Auto
Independent variable: Auto

Y-Variables Setup

$Y_1 = 32X - 16X^2$

The table will look like this:

X	Y_1
0.00	0
0.25	7
0.50	12
0.75	15
1.00	16
1.25	15
1.50	12

5. Use the equation

$$x = y^2 - 4$$

to fill in the following table. Then use the table to sketch the graph

x	y
	-3
	-2
	-1
	0
	1
	2
	3

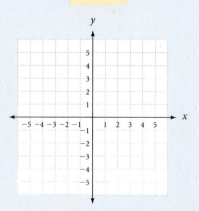

Answer

5. See Solutions Section.

EXAMPLE 5

Sketch the graph of $x = y^2$.

SOLUTION Without going into much detail, we graph the equation $x = y^2$ by finding a number of ordered pairs that satisfy the equation, plotting these points, and then drawing a smooth curve that connects them. A table of values for x and y that satisfy the equation follows, along with the graph of $x = y^2$ shown in Figure 5.

x	y
0	0
1	1
1	-1
4	2
4	-2
9	3
9	-3

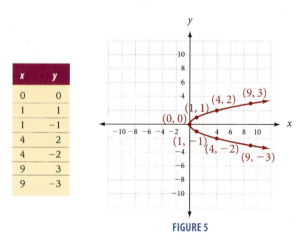

FIGURE 5

As you can see from looking at the table and the graph in Figure 5, several ordered pairs whose graphs lie on the curve have repeated first coordinates. For instance, (1, 1) and (1, −1), (4, 2) and (4, −2), and (9, 3) and (9, −3). The graph is therefore not the graph of a function. ∎

D Vertical Line Test

Look back at the scatter diagram for used Mustang prices shown in Figure 3. Notice that some of the points on the diagram lie above and below each other along vertical lines. This is an indication that the data do not constitute a function. Two data points that lie on the same vertical line must have come from two ordered pairs with the same first coordinates.

Now, look at the graph shown in Figure 5. The reason this graph is the graph of a relation, but not of a function, is that some points on the graph have the same first coordinates—for example, the points (4, 2) and (4, −2). Furthermore, any time two points on a graph have the same first coordinates, those points must lie on a vertical line. [To convince yourself, connect the points (4, 2) and (4, −2) with a straight line. You will see that it must be a vertical line.] This allows us to write the following test that uses the graph to determine whether a relation is also a function.

Vertical Line Test

If a vertical line crosses the graph of a relation in more than one place, the relation cannot be a function. If no vertical line can be found that crosses a graph in more than one place, then the graph is the graph of a function.

If we look back to the graph of $h = 32t - 16t^2$ as shown in Figure 4, we see that no vertical line can be found that crosses this graph in more than one place. The graph shown in Figure 4 is therefore the graph of a function.

EXAMPLE 6 Graph $y = |x|$. Use the graph to determine whether we have the graph of a function. State the domain and range.

SOLUTION We let x take on values of −4, −3, −2, −1, 0, 1, 2, 3, and 4. The corresponding values of y are shown in the table. The graph is shown in Figure 6.

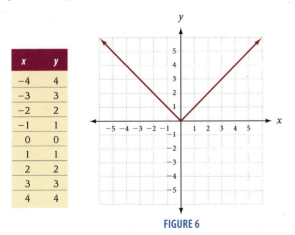

x	y
−4	4
−3	3
−2	2
−1	1
0	0
1	1
2	2
3	3
4	4

FIGURE 6

Because no vertical line can be found that crosses the graph in more than one place, $y = |x|$ is a function. The domain is all real numbers. The range is $\{y \mid y \geq 0\}$. ∎

6. Graph $y = |x| - 3$.

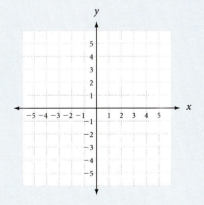

Problem Set 3.5

Moving Toward Success

"The art of living lies less in eliminating our troubles than in growing with them."

—Bernard M. Baruch, 1870–1965, American financier and statesman

1. Why should you mentally list the items that are known and unknown in a word problem?

2. Why is making a list of difficult problems important?

B For each of the following relations, give the domain and range, and indicate which are also functions. [Example 2]

1. $\{(1, 3), (2, 5), (4, 1)\}$

2. $\{(3, 1), (5, 7), (2, 3)\}$

3. $\{(-1, 3), (1, 3), (2, -5)\}$

4. $\{(3, -4), (-1, 5), (3, 2)\}$

5. $\{(7, -1), (3, -1), (7, 4)\}$

6. $\{(5, -2), (3, -2), (5, -1)\}$

7. $\{(a, 3), (b, 4), (c, 3), (d, 5)\}$

8. $\{(a, 5), (b, 5), (c, 4), (d, 5)\}$

9. $\{(a, 1), (a, 2), (a, 3), (a, 4)\}$

10. $\{(a, 1), (b, 1), (c, 1), (d, 1)\}$

B Determine the domain and range of the following functions. Assume the entire function is shown. [Example 2]

11.
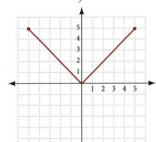

12.
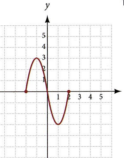

13.
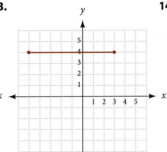

14.
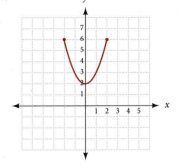

C **D** Graph each of the following relations. In each case, use the graph to find the domain and range, and indicate whether the graph is the graph of a function. [Example 6]

15. $y = x^2 - 1$

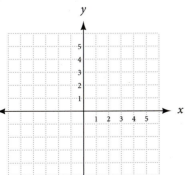

16. $y = x^2 + 1$

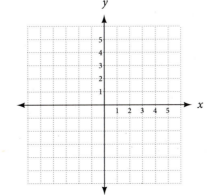

17. $y = x^2 + 4$

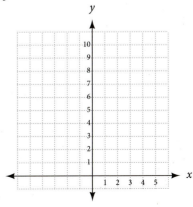

18. $y = x^2 - 9$

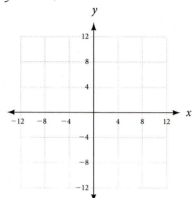

19. $x = y^2 - 1$

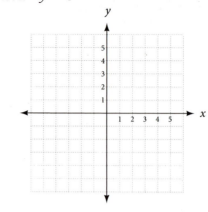

20. $x = y^2 + 1$

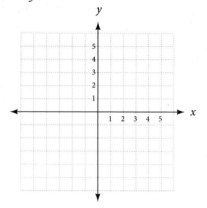

21. $x = y^2 + 4$

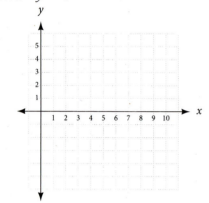

22. $x = y^2 - 9$

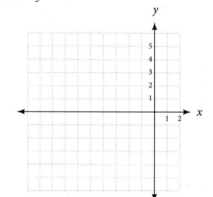

23. $y = |x - 2|$

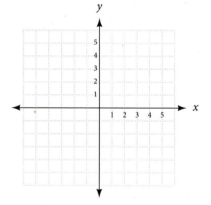

24. $y = |x + 2|$

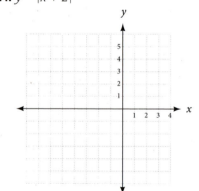

25. $y = |x| - 2$

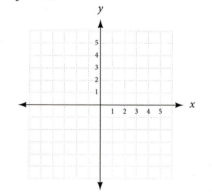

26. $y = |x| + 2$

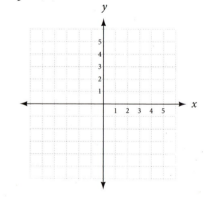

D State whether each of the following graphs represents a function. [Example 6]

27.
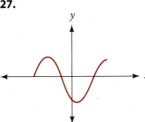

28.
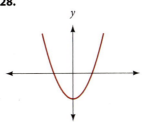

29.
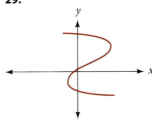

30.
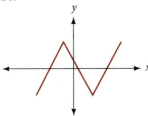

31.
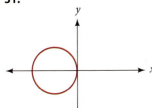

32.
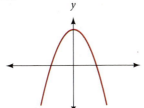

33.
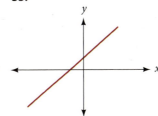

34.
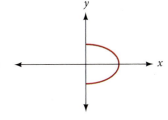

35.
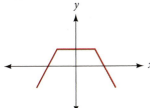

36.
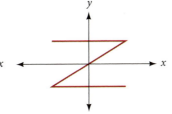

Applying the Concepts

37. **Wind Energy** The chart shows the electricity generated by wind energy throughout the world. Use the chart and list the domain and range that describes the data.

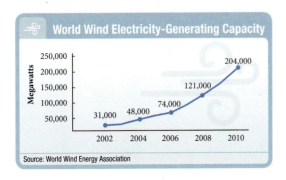

38. **Phones** The chart shows what percentage of people, by income level, who use land lines and cell phones. Use the chart to state the domain and range of the data for people who use cell phones.

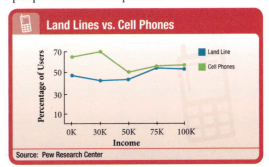

39. Weekly Wages Suppose you have a job that pays $8.50 per hour and you work anywhere from 10 to 40 hours per week.

a. Write an equation, with a restriction on the variable *x*, that gives the amount of money, *y*, you will earn for working *x* hours in one week.

b. Use the function rule to complete the table.

c. Use the data from the table to graph the function.

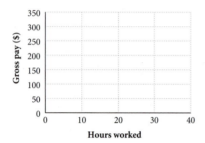

WEEKLY WAGES		
Hours Worked	Function Rule	Gross Pay ($)
x		*y*
10		
20		
30		
40		

d. State the domain and range of this function.

e. What is the minimum amount you can earn in a week with this job? What is the maximum amount?

40. Weekly Wages The ad shown here was in the local newspaper. Suppose you are hired for the job described in the ad.

a. If *x* is the number of hours you work per week and *y* is your weekly gross pay, write the equation for *y*. (Be sure to include any restrictions on the variable *x* that are given in the ad.)

b. Use the function rule to complete the table.

c. Use the data from the table to graph the function.

WEEKLY WAGES		
Hours Worked	Function Rule	Gross Pay ($)
x		*y*
15		
20		
25		
30		

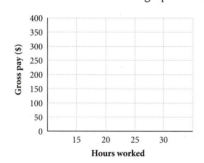

d. State the domain and range of this function.

e. What is the minimum amount you can earn in a week with this job? What is the maximum amount?

41. Tossing a Coin Hali is tossing a quarter into the air with an underhand motion. The height the quarter is above her hand at any time is given by the function

$$h = 16t - 16t^2 \quad \text{for} \quad 0 \le t \le 1$$

where h is the height of the quarter in feet, and t is the time in seconds.

a. Fill in the table.

b. State the domain and range of this function.

c. Use the data from the table to graph the function.

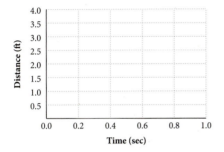

Time (sec)	Function Rule	Height (ft)
t	$h = 16t - 16t^2$	h
0		
0.1		
0.2		
0.3		
0.4		
0.5		
0.6		
0.7		
0.8		
0.9		
1		

42. Intensity of Light The formula below gives the intensity of light that falls on a surface at various distances from a 100-watt light bulb:

$$I = \frac{120}{d^2} \quad \text{for} \quad d > 0$$

where I is the intensity of light (in lumens per square foot), and d is the distance (in feet) from the light bulb to the surface.

a. Fill in the table.

b. Use the data from the table to graph the function.

Distance (ft)	Function Rule	Intensity
d	$I = \dfrac{120}{d^2}$	I
1		
2		
3		
4		
5		
6		

43. Area of a Circle The formula for the area A of a circle with radius r is given by $A = \pi r^2$. The formula shows that A is a function of r.

a. Graph the function $A = \pi r^2$ for $0 \le r \le 3$. (On the graph, let the horizontal axis be the r-axis, and let the vertical axis be the A-axis.)

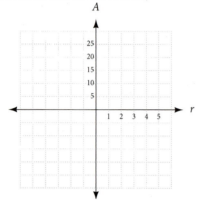

b. State the domain and range of the function $A = \pi r^2$, $0 \le r \le 3$.

44. Area and Perimeter of a Rectangle A rectangle is 2 inches longer than it is wide. Let x = the width, P = the perimeter, and A = the area of the rectangle.

a. Write an equation that will give the perimeter P in terms of the width x of the rectangle. Are there any restrictions on the values that x can assume?

b. Graph the relationship between P and x.

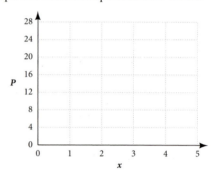

45. Tossing a Ball A ball is thrown straight up into the air from ground level. The relationship between the height h of the ball at any time t is illustrated by the following graph:

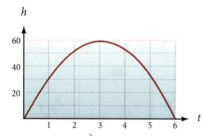

The horizontal axis represents time t, and the vertical axis represents height h.

a. Is this graph the graph of a function?

b. State the domain and range.

c. At what time does the ball reach its maximum height?

d. What is the maximum height of the ball?

e. At what time does the ball hit the ground?

46. Company Profits The amount of profit a company earns is based on the number of items it sells. The relationship between the profit P and number of items it sells x, is illustrated by the following graph:

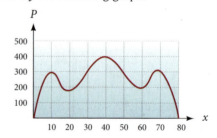

The horizontal axis represents items sold, x, and the vertical axis represents the profit, P.

a. Is this graph the graph of a function?

b. State the domain and range.

c. How many items must the company sell to make their maximum profit?

d. What is their maximum profit?

47. Profits Match each of the following statements to the appropriate graph indicated by labels I-IV.

  a. Sarah works 25 hours to earn $250.

  b. Justin works 35 hours to earn $560.

  c. Rosemary works 30 hours to earn $360.

  d. Marcus works 40 hours to earn $320.

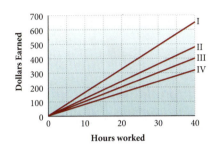

48. Find an equation for each of the functions shown in the graph for problem 47. Show dollars earned, E, as a function of hours worked, t. Then indicate the domain and range of each function.

  a. Graph I: $E =$
    Domain $= \{t \mid \qquad \}$
    Range $= \{E \mid \qquad \}$

  b. Graph II: $E =$
    Domain $= \{t \mid \qquad \}$
    Range $= \{E \mid \qquad \}$

  c. Graph III: $E =$
    Domain $= \{t \mid \qquad \}$
    Range $= \{E \mid \qquad \}$

  d. Graph IV: $E =$
    Domain $= \{t \mid \qquad \}$
    Range $= \{E \mid \qquad \}$

Maintaining Your Skills

The problems that follow review some of the more important skills you have learned in previous sections and chapters.

For the equation $y = 3x - 2$

49. find y if x is 4.

50. find y if x is 0.

51. find y if x is -4.

52. find y if x is -2.

For the equation $y = x^2 - 3$

53. find y if x is 2.

54. find y if x is -2.

55. find y if x is 0.

56. find y if x is -4.

57. If $x - 2y = 4$ and $x = \dfrac{8}{5}$, find y.

58. If $5x - 10y = 15$, find y when x is 3.

59. Let $x = 0$ and $y = 0$ in $y = a(x - 80)^2 + 70$ and solve for a.

60. Find R if $p = 2.5$ and $R = (900 - 300p)p$.

Getting Ready for the Next Section

Simplify. Round to the nearest whole number if necessary.

61. $7.5(20)$

62. $60 \div 7.5$

63. $4(3.14)(9)$

64. $\dfrac{4}{3}(3.14) \cdot 3^3$

65. $4(-2) - 1$

66. $3(3)^2 + 2(3) - 1$

67. If $s = \dfrac{60}{t}$, find s when

    a. $t = 10$ **b.** $t = 8$

68. If $y = 3x^2 + 2x - 1$, find y when

    a. $x = 0$ **b.** $x = -2$

69. Find the value of $x^2 + 2$ for

    a. $x = 5$ **b.** $x = -2$

70. Find the value of $125 \cdot 2^t$ for

    a. $t = 0$ **b.** $t = 1$

Extending the Concepts

Graph each of the following relations. In each case, use the graph to find the domain and range, and indicate whether the graph is the graph of a function.

71. $y = 5 - |x|$

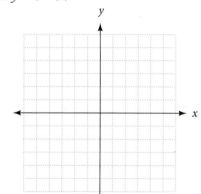

72. $y = |x| - 3$

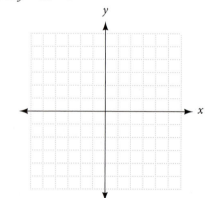

73. $x = |y| + 3$

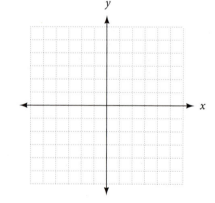

74. $x = 2 - |y|$

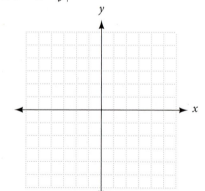

75. $|x| + |y| = 4$

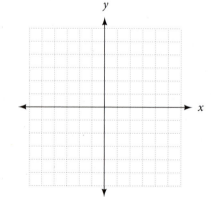

76. $2|x| + |y| = 6$

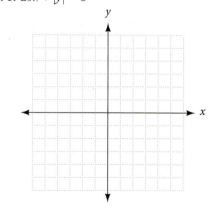

Function Notation

3.6

OBJECTIVES

A Use function notation to fnd the value of a function for a given value of the variable.

B Use graphs to visualize the relationship between a function and a variable.

C Use function notation in formulas.

TICKET TO SUCCESS

Keep these questions in mind as you read through the section. Then respond in your own words and in complete sentences.

1. Explain what you are calculating when you find $f(2)$ for a given function f.
2. If $s(t) = \frac{60}{t}$, how do you find $s(10)$?
3. If $f(2) = 3$ for a function f, what is the relationship between the numbers 2 and 3 and the graph of f?
4. If $f(6) = 0$ for a particular function f, then you can immediately graph one of the intercepts. Explain.

Rob Wilson/Shutterstock.com

Let's return to the discussion that introduced us to functions. If the width of the field 60 yards is multiplied by the length x to get the area y, it can be said that area is a function of length. The exact relationship between x and y is written

$$y = 60x$$

Because the area depends on the length of the touchline, we call y the dependent variable and x the independent variable. If we let f represent all the ordered pairs produced by the equation, then we can write

$$f = \{(x, y) \mid y = 60x \text{ and } 100 \le x \le 130\}$$

A Evaluate a Function at a Point

If a job pays $9.50 per hour for working from 0 to 40 hours a week, then the amount of money y earned in one week is a function of the number of hours worked x. The exact relationship between x and y is written

$$y = 9.5x \quad \text{for} \quad 0 \le x \le 40$$

Because the amount of money earned, y, depends on the number of hours worked, x, we call y the *dependent variable* and x the *independent variable*. Furthermore, if we let f represent all the ordered pairs produced by the equation, then we can write

$$f = \{(x, y) \mid y = 9.5x \text{ and } 0 \le x \le 40\}$$

Once we have named a function with a letter, we can use an alternative notation to represent the dependent variable y. The alternative notation for y is $f(x)$. It is read "f of x" and can be used instead of the variable y when working with functions. The notation y and the notation $f(x)$ are equivalent; that is,

$$y = 9.5x \Leftrightarrow f(x) = 9.5x$$

When we use the notation $f(x)$ we are using *function notation*. The benefit of using function notation is that we can write more information with fewer symbols than we can by using just the variable y. For example, asking how much money a person will make for working 20 hours is simply a matter of asking for $f(20)$. Without function notation, we would have to say, "find the value of y that corresponds to a value of $x = 20$." To illustrate further, using the variable y, we can say, "y is 190 when x is 20." Using the notation $f(x)$, we simply say, "$f(20) = 190$." Each expression indicates that you will earn $190 for working 20 hours.

PRACTICE PROBLEMS

1. If $f(x) = 8x$, find
   a. $f(0)$
   b. $f(5)$
   c. $f(10.5)$

> **NOTE**
> Some students like to think of functions as machines. Values of x are put into the machine, which transforms them into values of $f(x)$, which then are output by the machine.
>
>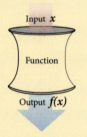

> **EXAMPLE 1** If $f(x) = 9.5x$, find $f(0)$, $f(10)$, and $f(20)$.

SOLUTION   To find $f(0)$ we substitute 0 for x in the expression $9.5x$ and simplify. We find $f(10)$ and $f(20)$ in a similar manner—by substitution.

$$\text{If} \rightarrow \qquad f(x) = 9.5x$$

$$\text{then} \rightarrow \qquad f(\mathbf{0}) = 9.5(\mathbf{0}) = 0$$

$$f(\mathbf{10}) = 9.5(\mathbf{10}) = 95$$

$$f(\mathbf{20}) = 9.5(\mathbf{20}) = 190$$ ∎

If we changed the example in the discussion that opened this section so that the hourly wage was $8.50 per hour, we would have a new equation to work with.

$$y = 8.5x \qquad \text{for} \qquad 0 \le x \le 40$$

Suppose we name this new function with the letter g. Then

$$g = \{(x, y) \mid y = 8.5x \text{ and } 0 \le x \le 40\}$$

and

$$g(x) = 8.5x$$

If we want to talk about both functions in the same discussion, having two different letters, f and g, makes it easy to distinguish between them. For example, because $f(x) = 9.5x$ and $g(x) = 8.5x$, asking how much money a person makes for working 20 hours is simply a matter of asking for $f(20)$ or $g(20)$, avoiding any confusion over which hourly wage we are talking about.

The diagrams shown in Figure 1 further illustrate the similarities and differences between the two functions we have been discussing.

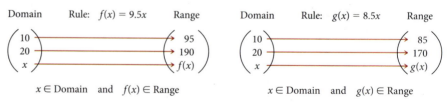

FIGURE 1 Function maps

B Function Notation and Graphs

We can visualize the relationship between x and $f(x)$ or $g(x)$ on the graphs of the two functions. Figure 2 shows the graph of $f(x) = 9.5x$ along with two additional line segments. The horizontal line segment corresponds to $x = 20$, and the vertical line segment corresponds to $f(20)$. Figure 3 shows the graph of $g(x) = 8.5x$ along with the horizontal line segment that corresponds to $x = 20$, and the vertical line segment that corresponds to $g(20)$. (Note that the domain in each case is restricted to $0 \le x \le 40$.)

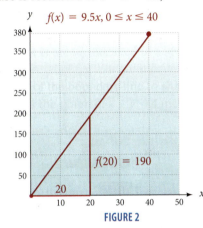

FIGURE 2

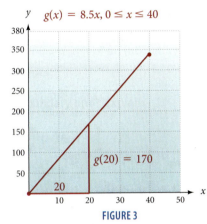

FIGURE 3

C Using Function Notation

The remaining examples in this section show a variety of ways to use and interpret function notation.

EXAMPLE 2   If it takes Lorena t minutes to run a mile, then her average speed $s(t)$ in miles per hour is given by the formula

$$s(t) = \frac{60}{t} \quad \text{for} \quad t > 0$$

Find $s(10)$ and $s(8)$, and then explain what they mean.

SOLUTION   To find $s(10)$, we substitute 10 for t in the equation and simplify.

$$s(\mathbf{10}) = \frac{60}{\mathbf{10}} = 6$$

In words: When Lorena runs a mile in 10 minutes, her average speed is 6 miles per hour.

We calculate $s(8)$ by substituting 8 for t in the equation. Doing so gives us

$$s(\mathbf{8}) = \frac{60}{\mathbf{8}} = 7.5$$

In words: Running a mile in 8 minutes is running at a rate of 7.5 miles per hour. ∎

EXAMPLE 3   A painting is purchased as an investment for $125. If its value increases continuously so that it doubles every 5 years, then its value is given by the function

$$V(t) = 125 \cdot 2^{t/5} \quad \text{for} \quad t \ge 0$$

where t is the number of years since the painting was purchased, and $V(t)$ is its value (in dollars) at time t. Find $V(5)$ and $V(10)$, and explain what they mean.

SOLUTION   The expression $V(5)$ is the value of the painting when $t = 5$ (5 years after it is purchased). We calculate $V(5)$ by substituting 5 for t in the equation $V(t) = 125 \cdot 2^{t/5}$. Here is our work:

$$V(\mathbf{5}) = 125 \cdot 2^{\mathbf{5}/5} = 125 \cdot 2^1 = 125 \cdot 2 = 250$$

2. When Lorena runs a mile in t minutes, then her average speed in feet per seconds is given by $s(t) = \frac{88}{t}$, $t > 0$.
  a. Find $s(8)$ and explain what it means.
  b. Find $s(11)$ and explain what it means.

3. A medication has a half-life of 5 days. If the concentration of the medication in a patient's system is 80 ng/mL, and the patient stops taking it, then t days later the concentration will be

$$C(t) = 80\left(\frac{1}{2}\right)^{t/5}$$

Find each of the following, and explain what they mean.
  a. $C(5)$
  b. $C(10)$

Answers
2. a. $s(8) = 11$; runs a mile in 8 minutes, average speed is 11 feet per second.
  b. $s(11) = 8$; runs a mile in 11 minutes, average speed is 8 feet per second.
3. a. $C(5) = 40$ ng/mL; after 5 days the concentration is 40 ng per mL.
  b. $C(10) = 20$ ng/mL; after 10 days the concentration is 20 ng per mL.

In words: After 5 years, the painting is worth $250.

The expression $V(10)$ is the value of the painting after 10 years. To find this number, we substitute 10 for t in the equation.

$$V(\mathbf{10}) = 125 \cdot 2^{10/5} = 125 \cdot 2^2 = 125 \cdot 4 = 500$$

In words: The value of the painting 10 years after it is purchased is $500.   ∎

4. The following formulas give the circumference and area of a circle with a radius of r. Use the formulas to find the circumference and area of a circular plate if the radius is 5 inches.

$$C(r) = 2\pi r$$
$$A(r) = \pi r^2$$

EXAMPLE 4   A balloon has the shape of a sphere with a radius of 3 inches. Use the following formulas to find the volume and surface area of the balloon.

$$V(r) = \frac{4}{3}\pi r^3 \qquad S(r) = 4\pi r^2$$

SOLUTION   As you can see, we have used function notation to write the two formulas for volume and surface area because each quantity is a function of the radius. To find these quantities when the radius is 3 inches, we evaluate $V(3)$ and $S(3)$.

$$V(\mathbf{3}) = \frac{4}{3}\pi \mathbf{3}^3 = \frac{4}{3}\pi 27 = 36\pi \text{ cubic inches, or 113 cubic inches}$$ Round to the nearest whole number.

$$S(\mathbf{3}) = 4\pi \mathbf{3}^2 = 36\pi \text{ square inches, or 113 square inches}$$ Round to the nearest whole number.

The fact that $V(3) = 36\pi$ means that the ordered pair $(3, 36\pi)$ belongs to the function V. Likewise, the fact that $S(3) = 36\pi$ tells us that the ordered pair $(3, 36\pi)$ is a member of function S.   ∎

We can generalize the discussion at the end of Example 4 this way:

$$(a, b) \in f \quad \text{if and only if} \quad f(a) = b$$

5. If $f(x) = 4x^2 - 3$, find
   a. $f(0)$
   b. $f(3)$
   c. $f(-2)$

EXAMPLE 5   If $f(x) = 3x^2 + 2x - 1$, find $f(0)$, $f(3)$, and $f(-2)$.

SOLUTION   Because $f(x) = 3x^2 + 2x - 1$, we have

$$f(\mathbf{0}) = 3(\mathbf{0})^2 + 2(\mathbf{0}) - 1 \qquad = 0 + 0 - 1 = -1$$

$$f(\mathbf{3}) = 3(\mathbf{3})^2 + 2(\mathbf{3}) - 1 \qquad = 27 + 6 - 1 = 32$$

$$f(-\mathbf{2}) = 3(-\mathbf{2})^2 + 2(-\mathbf{2}) - 1 = 12 - 4 - 1 = 7$$ ∎

6. If $f(x) = 2x + 1$ and $g(x) = x^2 - 3$, find
   a. $f(5)$
   b. $g(5)$
   c. $f(-2)$
   d. $g(-2)$
   e. $f(a)$
   f. $g(a)$

In Example 5, the function f is defined by the equation $f(x) = 3x^2 + 2x - 1$. We could just as easily have said $y = 3x^2 + 2x - 1$; that is, $y = f(x)$. Saying $f(-2) = 7$ is exactly the same as saying y is 7 when x is -2.

7. If $f = \{(-4, 1), (2, -3), (7, 9)\}$, find
   a. $f(-4)$
   b. $f(2)$
   c. $f(7)$

EXAMPLE 6   If $f(x) = 4x - 1$ and $g(x) = x^2 + 2$, then

$$f(\mathbf{5}) = 4(\mathbf{5}) - 1 = 19 \qquad \text{and} \qquad g(\mathbf{5}) = \mathbf{5}^2 + 2 = 27$$

$$f(-\mathbf{2}) = 4(-\mathbf{2}) - 1 = -9 \qquad \text{and} \qquad g(-\mathbf{2}) = (-\mathbf{2})^2 + 2 = 6$$

$$f(\mathbf{0}) = 4(\mathbf{0}) - 1 = -1 \qquad \text{and} \qquad g(\mathbf{0}) = \mathbf{0}^2 + 2 = 2$$

$$f(z) = 4z - 1 \qquad \text{and} \qquad g(z) = z^2 + 2$$

$$f(a) = 4a - 1 \qquad \text{and} \qquad g(a) = a^2 + 2$$ ∎

Answers
4. $C(5) = 10\pi \approx 31.4$ inches
   $A(5) = 25\pi \approx 78.5$ in^2
5. a. -3 **b.** 33   **c.** 13
6. a. 11 **b.** 22   **c.** -3
   d. 1   **e.** $2a + 1$ **f.** $a^2 - 3$
7. a. 1   **b.** -3 **c.** 9

EXAMPLE 7   If the function f is given by

$$f = \{(-2, 0), (3, -1), (2, 4), (7, 5)\}$$

then $f(-2) = 0$, $f(3) = -1$, $f(2) = 4$, and $f(7) = 5$.   ∎

Problem Set 3.6

Moving Toward Success

"Nothing is to be rated higher than the value of the day."

—Johann Wolfgang von Goethe, 1749–1832,
German dramatist and novelist

1. When should you review material for the next exam?

   a. Only at the last minute before an exam

   b. After you see it on the exam

   c. Each day

   d. Only after you learn it

2. Should you work problems when you review material? Why or why not?

A C Let $f(x) = 2x - 5$ and $g(x) = x^2 + 3x + 4$. Evaluate the following. [Examples 1, 5, 6]

1. $f(2)$

2. $f(3)$

3. $f(-3)$

4. $g(-2)$

5. $g(-1)$

6. $f(-4)$

7. $g(-3)$

8. $g(2)$

9. $g(4) + f(4)$

10. $f(2) - g(3)$

11. $f(3) - g(2)$

12. $g(-1) + f(-1)$

A C Let $f(x) = 3x^2 - 4x + 1$ and $g(x) = 2x - 1$. Evaluate the following. [Examples 1, 2]

13. $f(0)$

14. $g(0)$

15. $g(-4)$

16. $f(1)$

17. $f(-1)$

18. $g(-1)$

19. $g(10)$

20. $f(10)$

21. $f(3)$

22. $g(3)$

23. $g\left(\dfrac{1}{2}\right)$

24. $g\left(\dfrac{1}{4}\right)$

25. $f(a)$

26. $g(b)$

A **C** Let $f(x) = 2x^2 - 8$ and $g(x) = \frac{1}{2}x + 1$. Evaluate each of the following. [Examples 1, 5–6]

27. $f(0)$

28. $g(0)$

29. $g(-4)$

30. $f(1)$

31. $f(a)$

32. $g(z)$

33. $f(b)$

34. $g(t)$

B

35. Graph the function $f(x) = \frac{1}{2}x + 2$. Then draw and label the line segments that represent $x = 4$ and $f(4)$.

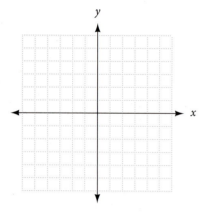

36. Graph the function $f(x) = -\frac{1}{2}x + 6$. Then draw and label the line segments that represent $x = 4$ and $f(4)$.

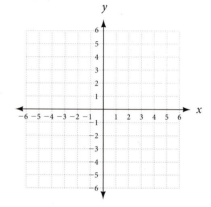

37. For the function $f(x) = \frac{1}{2}x + 2$, find the value of x for which $f(x) = x$.

38. For the function $f(x) = -\frac{1}{2}x + 6$, find the value of x for which $f(x) = x$.

39. Graph the function $f(x) = x^2$. Then draw and label the line segments that represent $x = 1$ and $f(1)$, $x = 2$ and $f(2)$, and, finally, $x = 3$ and $f(3)$.

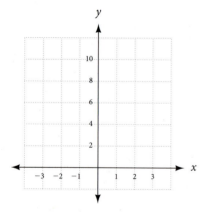

40. Graph the function $f(x) = x^2 - 2$. Then draw and label the line segments that represent $x = 2$ and $f(2)$, and the line segments corresponding to $x = 3$ and $f(3)$.

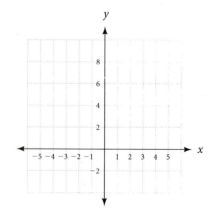

C If $f = \{(1, 4), (-2, 0), \left(3, \frac{1}{2}\right), (\pi, 0)\}$ and $g = \{(1, 1), (-2, 2), \left(\frac{1}{2}, 0\right)\}$, find each of the following values of f and g. [Example 7]

41. $f(1)$

42. $g(1)$

43. $g\left(\frac{1}{2}\right)$

44. $f(3)$

45. $g(-2)$

46. $f(\pi)$

Applying the Concepts

47. Investing in Art A painting is purchased as an investment for $150. If its value increases continuously so that it doubles every 3 years, then its value is given by the function

$$V(t) = 150 \cdot 2^{t/3} \quad \text{for} \quad t \geq 0$$

where t is the number of years since the painting was purchased, and $V(t)$ is its value (in dollars) at time t. Find $V(3)$ and $V(6)$, and then explain what they mean.

48. Average Speed If it takes Minke t minutes to run a mile, then her average speed $s(t)$, in miles per hour, is given by the formula

$$s(t) = \frac{60}{t} \quad \text{for} \quad t > 0$$

Find $s(4)$ and $s(5)$, and then explain what they mean.

49. Social Networking The chart shows the percentage of people, by age, who are using various social networking sites.

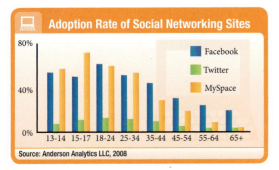

Adoption Rate of Social Networking Sites

Source: Anderson Analytics LLC, 2008

Suppose the following:

m is the percentage of people in an age group that use Myspace.

f is the percentage of people in an age group that use Facebook.

t is the percentage of people in an age group that use Twitter.

Use this information to determine if each if the following statements are true or false.

a. $f(15–17) > f(55–64)$

b. $t(65+) > t(18–24)$

c. $m(25–34) < m(15–17)$

50. Internet The graph shows the percentage of people who use broadband or dial-up to connect to the internet.

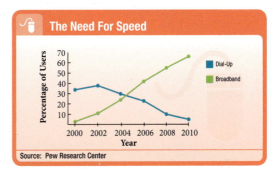

The Need For Speed

Source: Pew Research Center

Suppose the following:

f is the function for broadband.

g is the function for dial-up.

Use this information to estimate the following:

a. $f(2002) - g(2008)$

b. $f(2008) - f(2000)$

c. $g(2002) - f(2008)$

51. Social Networking Refer to the figure in number 49. Suppose x represents one of the age groups in the chart. Suppose further that we have three functions m, f, and t that do the following:

> m is the percentage of people in an age group that use Myspace.

> f is the percentage of people in an age group that use Facebook.

> t is the percentage of people in an age group that use Twitter.

For each statement below, indicate whether the statement is true or false.

a. $m(55–64) < m(18–24)$

b. $m(15–17) > f(15–17)$

c. $t(25–34) > 20\%$

d. $t(18–24) > t(35–44) > t(65+)$

52. Mobile Phone Sales Suppose x represents one of the years in the chart. Suppose further that we have three functions f, g, and h that do the following:

> f pairs each year with the number of camera phones sold that year.

> g pairs each year with the number of non-camera phones sold that year.

> h is such that $h(x) = f(x) + g(x)$.

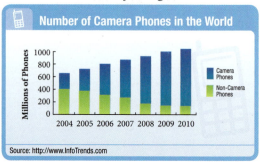

Source: http://www.InfoTrends.com

For each statement below, indicate whether the statement is true or false.

a. The domain of f is {2004, 2005, 2006, 2007, 2008, 2009, 2010}

b. $h(2007) = 741,000$

c. $f(2009) > g(2009)$

d. $f(2004) < f(2005)$

e. $h(2010) > h(2007) > h(2004)$

Straight-Line Depreciation Straight-line depreciation is an accounting method used to help spread the cost of new equipment over a number of years. It takes into account both the cost when new and the salvage value, which is the value of the equipment at the time it gets replaced.

53. Value of a Copy Machine The function $V(t) = -3,300t + 18,000$, where V is value and t is time in years, can be used to find the value of a large copy machine during the first 5 years of use.

a. What is the value of the copier after 3 years and 9 months?

b. What is the salvage value of this copier if it is replaced after 5 years?

c. State the domain of this function.

d. Sketch the graph of this function.

e. What is the range of this function?

f. After how many years will the copier be worth only $10,000?

54. Value of a Forklift The function $V(t) = -16{,}500t + 125{,}000$, where V is value and t is time in years, can be used to find the value of an electric forklift during the first 6 years of use.

a. What is the value of the forklift after 2 years and 3 months?

b. What is the salvage value of this forklift if it is replaced after 6 years?

c. State the domain of this function.

d. Sketch the graph of this function.

e. What is the range of this function?

f. After how many years will the forklift be worth only $45,000?

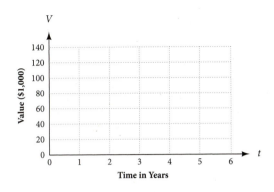

Getting Ready for the Next Section

Simplify.

55. $(35x - 0.1x^2) - (8x + 500)$

56. $70 + 0.6(M - 70)$

57. $(4x^2 + 3x + 2) + (2x^2 - 5x - 6)$

58. $(4x - 3) + (4x^2 - 7x + 3)$

59. $(4x^2 + 3x + 2) - (2x^2 - 5x - 6)$

60. $(x + 5)^2 - 2(x + 5)$

Multiply.

61. $0.6(M - 70)$

62. $x(35 - 0.1x)$

63. $(4x - 3)(x - 1)$

64. $(4x - 3)(4x^2 - 7x + 3)$

Maintaining Your Skills

The problems that follow review some of the more important skills you have learned in previous sections and chapters.

Solve each equation.

65. $|3x - 5| = 7$

66. $|0.04 - 0.03x| = 0.02$

67. $|4y + 2| - 8 = -2$

68. $4 = |3 - 2y| - 5$

69. $5 + |6t + 2| = 3$

70. $7 + \left|3 - \frac{3}{4}t\right| = 10$

Extending the Concepts

The graphs of two functions are shown in Figures 4 and 5. Use the graphs to find the following values in Problems 71 and 72.

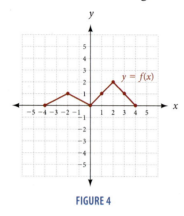

FIGURE 4

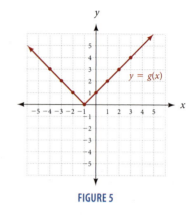

FIGURE 5

71. a. $f(2)$

    b. $f(-4)$

    c. $g(0)$

    d. $g(3)$

72. a. $g(2) - f(2)$

    b. $f(1) + g(1)$

    c. $f(1) - g(3)$

    d. $g(2) - f(3)$

73. Step Function Figure 6 shows the graph of the step function C that was used to calculate the first-class postage on a large envelop weighing x ounces in 2010. Use this graph to answer questions a through d.

    a. Fill in the following table:

Weight (ounces)	0.6	1.0	1.1	2.5	3.0	4.8	5.0	5.3
Cost (cents)								

    b. If a large envelope costs 122 cents to mail, how much does it weigh? State your answer in words and as an inequality.

    c. If the entire function is shown in Figure 6, state the domain.

    d. State the range of the function shown in Figure 6.

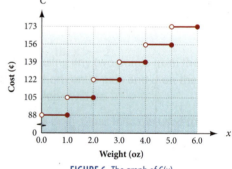

FIGURE 6 The graph of C(x)

74. Step Function A taxi ride in Boston is \$2.60 for the first $\frac{1}{7}$ mile, and then \$0.40 for each additional $\frac{1}{7}$ of a mile. The following graph shows how much you will pay for a taxi ride of 1 mile or less.

    a. What is the most you will pay for this taxi ride?

    b. How much does it cost to ride the taxi for $\frac{5}{7}$ of a mile?

    c. Find the values of A and B on the horizontal axis.

    d. If a taxi ride costs \$4.60, what distance was the ride?

    e. If the complete function is shown in Figure 7, find the domain and range of the function.

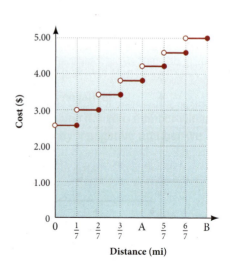

FIGURE 7

Algebra and Composition with Functions

TICKET TO SUCCESS

Keep these questions in mind as you read through the section. Then respond in your own words and in complete sentences.

1. Use function notation to show how profit, revenue, and cost are related.
2. Explain in words the definition for $(f + g)(x) = f(x) + g(x)$.
3. What is the composition of functions?
4. How would you find maximum heart rate using a composition of functions?

OBJECTIVES

A Find the sum, difference, product, and quotient of two functions.

B For two functions f and g, find $f(g(x))$ and $f(g(x))$.

Khomula Anna/Shutterstock.com

A company produces and sells popular apps for smartphones. The price the company charges for each app is related to the number of downloads sold by the demand function

$$p(x) = 35 - 0.1x$$

We find the revenue for this business by multiplying the number of downloads sold by the price per download. When we do so, we are forming a new function by combining two existing functions; that is, if $n(x) = x$ is the number of downloads sold and $p(x) = 35 - 0.1x$ is the price per download, then revenue is

$$R(x) = n(x) \cdot p(x) = x(35 - 0.1x) = 35x - 0.1x^2$$

In this case, the revenue function is the product of two functions. When we combine functions in this manner, we are applying our rules for algebra to functions.

To carry this situation further, we know that the profit function is the difference between two functions. If the cost function for producing x applications is $C(x) = 8x + 500$, then the profit function is

$$P(x) = R(x) - C(x) = (35x - 0.1x^2) - (8x + 500) = -500 + 27x - 0.1x^2$$

The relationship between these last three functions is shown visually in Figure 1.

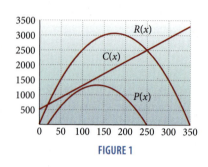

FIGURE 1

Again, when we combine functions in the manner shown, we are applying our rules for algebra to functions. To begin this section, we take a formal look at addition, subtraction, multiplication, and division with functions.

A Algebra with Functions

If we are given two functions f and g with a common domain, we can define four other functions as follows:

Definition

$(f + g)(x) = f(x) + g(x)$ The function $f + g$ is the sum of the functions f and g.

$(f - g)(x) = f(x) - g(x)$ The function $f - g$ is the difference of the functions f and g.

$(fg)(x) = f(x)g(x)$ The function fg is the product of the functions f and g.

$\left(\dfrac{f}{g}\right)(x) = \dfrac{f(x)}{g(x)}$ The function $\dfrac{f}{g}$ is the quotient of the functions f and g, where $g(x) \neq 0$.

PRACTICE PROBLEMS

1. If $f(x) = x^2 - 4$ and $g(x) = x + 2$, find formulas for
  a. $f + g$
  b. $f - g$

EXAMPLE 1   If $f(x) = 4x^2 + 3x$ and $g(x) = -5x - 6$, write the formula for the functions $f + g$ and $f - g$.

SOLUTION   The function $f + g$ is defined by

$$(f + g)(x) = f(x) + g(x)$$

$$= (4x^2 + 3x) + (-5x - 6)$$

$$= 4x^2 - 2x - 6$$

The function $f - g$ is defined by

$$(f - g)(x) = f(x) - g(x)$$

$$= (4x^2 + 3x) - (-5x - 6)$$

$$= 4x^2 + 3x + 5x + 6$$

$$= 4x^2 + 8x + 6$$

∎

Answers

1. a. $x^2 + x - 2$ **b.** $x^2 - x - 6$

EXAMPLE 2 Let $f(x) = 4x - 3$, and $g(x) = 4x^2$. Find $f + g$, fg, and $\frac{g}{f}$.

SOLUTION The function $f + g$, the sum of functions f and g, is defined by

$$(f + g)(x) = f(x) + g(x)$$

$$= (4x - 3) + 4x^2$$

$$= 4x^2 + 4x - 3$$

The product of the functions f and g, fg, is given by

$$(fg)(x) = f(x)g(x)$$

$$= (4x - 3)(4x^2)$$

$$= 16x^3 - 12x^2$$

The quotient of the functions g and f, $\frac{g}{f}$, is defined as

$$\left(\frac{g}{f}\right)(x) = \frac{g(x)}{f(x)}$$

$$= \frac{4x^2}{4x - 3}$$ ∎

B Composition of Functions

In addition to the four operations used to combine functions shown so far in this section, there is a fifth way to combine two functions to obtain a new function. It is called *composition of functions*. To illustrate the concept, recall the definition of training heart rate: Training heart rate, in beats per minute, is resting heart rate plus 60% of the difference between maximum heart rate and resting heart rate. If your resting heart rate is 70 beats per minute, then your training heart rate is a function of your maximum heart rate M.

$$T(M) = 70 + 0.6(M - 70) = 70 + 0.6M - 42 = 28 + 0.6M$$

But your maximum heart rate is found by subtracting your age in years from 220. So, if x represents your age in years, then your maximum heart rate is

$$M(x) = 220 - x$$

Therefore, if your resting heart rate is 70 beats per minute and your age in years is x, then your training heart rate can be written as a function of x.

$$T(x) = 28 + 0.6(220 - x)$$

This last line is the composition of functions T and M. We input x into function M, which outputs $M(x)$. Then we input $M(x)$ into function T, which outputs $T(M(x))$. This is the training heart rate as a function of age x. Here is a diagram of the situation, as a function map.

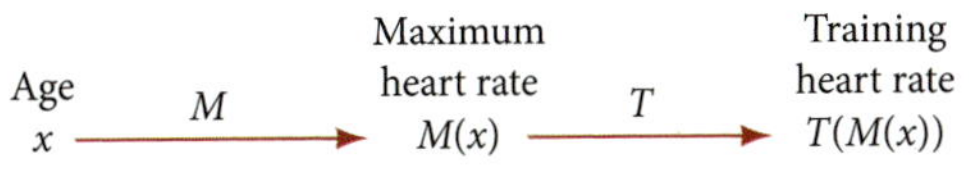

	Maximum		Training
Age	M	heart rate	T	heart rate
x	→	$M(x)$	→	$T(M(x))$

FIGURE 2

2. Let $f(x) = 3x + 2$,
$g(x) = 3x^2 - 10x - 8$, and
$h(x) = x - 4$. Find

a. $f + g$

b. fg

c. $\frac{g}{f}$

Answers
2. a. $3x^2 - 7x - 6$
   b. $9x^3 - 24x^2 - 44x - 16$
   c. $h(x)$

Now let's generalize the preceding ideas into a formal development of composition of functions. To find the composition of two functions f and g, we first require that the range of g have numbers in common with the domain of f. Then the composition of f with g, denoted $f \circ g$, is defined this way:

$$(f \circ g)(x) = f(g(x))$$

To understand this new function, we begin with a number x, and we operate on it with g, giving us $g(x)$. Then we take $g(x)$ and operate on it with f, giving us $f(g(x))$. The only numbers we can use for the domain of the composition of f with g are numbers x in the domain of g, for which $g(x)$ is in the domain of f. The diagrams in Figure 3 illustrate the composition of f with g.

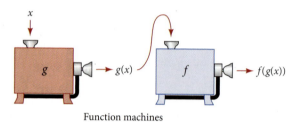

Function machines

FIGURE 3

Composition of functions is not commutative. The composition of f with g, $f \circ g$, may therefore be different from the composition of g with f, $g \circ f$.

$$(g \circ f)(x) = g(f(x))$$

Again, the only numbers we can use for the domain of the composition of g with f are numbers in the domain of f, for which $f(x)$ is in the domain of g. The diagrams in Figure 4 illustrate the composition of g with f.

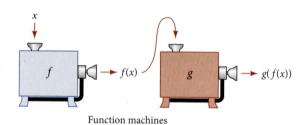

Function machines

FIGURE 4

EXAMPLE 3 If $f(x) = x + 5$ and $g(x) = 2x$, find $(f \circ g)(x)$.

SOLUTION The composition of f with g is

$$(f \circ g)(x) = f(g(x))$$
$$= f(2x)$$
$$= 2x + 5$$

The composition of g with f is

$$(g \circ f)(x) = g(f(x))$$
$$= g(x + 5)$$
$$= 2(x + 5)$$
$$= 2x + 10$$ ∎

3. If $f(x) = x - 4$ and $g(x) = x^2 + 3x$, find

  a. $(f \circ g)(x)$

  b. $(g \circ f)(x)$

Answers

3. a. $x^2 + 3x - 4$ **b.** $x^2 - 5x + 4$

Problem Set 3.7

Moving Toward Success

"A strong mind always hopes, and has always cause to hope."

—Thomas Carlyle, 1795–1881, Scottish philosopher and author

1. Do you feel anxiety when you are taking a test? Why or why not?

2. What might you do to remain calm and boost your confidence if you feel anxious during a test?

A Let $f(x) = 4x - 3$ and $g(x) = 2x + 5$. Write a formula for each of the following functions. [Example 1]

1. $f + g$ **2.** $f - g$ **3.** $g - f$ **4.** $g + f$

A If the functions f, g, and h are defined by $f(x) = 3x - 5$, $g(x) = x - 2$, and $h(x) = 3x^2$, write a formula for each of the following functions. [Examples 1, 2]

5. $g + f$ **6.** $f + h$ **7.** $g + h$

8. $f - g$ **9.** $g - f$ **10.** $h - g$

11. fh **12.** gh **13.** $\dfrac{h}{f}$

14. $\dfrac{h}{g}$ **15.** $\dfrac{f}{h}$ **16.** $\dfrac{g}{h}$

17. $f + g + h$ **18.** $h - g + f$

A Let $f(x) = 2x + 1$, $g(x) = 4x + 2$, and $h(x) = 4x^2 + 4x + 1$, and find the following. [Examples 1, 2]

19. $(f + g)(2)$ **20.** $(f - g)(-1)$ **21.** $(fg)(3)$

22. $\left(\dfrac{f}{g}\right)(-3)$

23. $\left(\dfrac{h}{g}\right)(1)$

24. $(hg)(1)$

25. $(fh)(0)$

26. $(h - g)(-4)$

27. $(f + g + h)(2)$

28. $(h - f + g)(0)$

29. $(h + fg)(3)$

30. $(h - fg)(5)$

B [Example 3]

31. Let $f(x) = x^2$ and $g(x) = x + 4$, and find
   a. $(f \circ g)(5)$

   b. $(g \circ f)(5)$

   c. $(f \circ g)(x)$

   d. $(g \circ f)(x)$

32. Let $f(x) = 3 - x$ and $g(x) = x^3 - 1$, and find
   a. $(f \circ g)(0)$

   b. $(g \circ f)(0)$

   c. $(f \circ g)(x)$

33. Let $f(x) = x^2 + 3x$ and $g(x) = 4x - 1$, and find
   a. $(f \circ g)(0)$

   b. $(g \circ f)(0)$

   c. $(g \circ f)(x)$

34. Let $f(x) = (x - 2)^2$ and $g(x) = x + 1$, and find the following
   a. $(f \circ g)(-1)$

   b. $(g \circ f)(-1)$

For each of the following pairs of functions f and g, show that $(f \circ g)(x) = (g \circ f)(x) = x$.

35. $f(x) = 5x - 4$ and $g(x) = \dfrac{x + 4}{5}$

36. $f(x) = \dfrac{x}{6} - 2$ and $g(x) = 6x + 12$

Use the graph to answer the following problems.

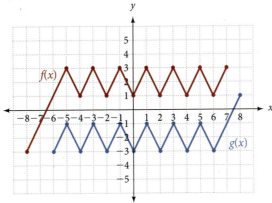

Evaluate.

37. $f(2) + 5$

38. $g(-2) - 5$

39. $f(-3) + g(-3)$

40. $f(5) - g(5)$

41. $(f \circ g)(0)$

42. $(g \circ f)(0)$

43. Find x if $f(x) = -3$

44. Find x if $g(x) = 1$

45. Profit, Revenue, and Cost A company manufactures and sells prerecorded DVDs. Here are the equations they use in connection with their business.

Number of DVDs sold each day: $n(x) = x$

Selling price for each DVD: $p(x) = 11.5 - 0.05x$

Daily fixed costs: $f(x) = 200$

Daily variable costs: $v(x) = 2x$

Find the following functions.

  a. Revenue $= R(x) =$ the product of the number of DVDs sold each day and the selling price of each DVD.

  b. Cost $= C(x) =$ the sum of the fixed costs and the variable costs.

  c. Profit $= P(x) =$ the difference between revenue and cost.

  d. Average cost $= \overline{C}(x) =$ the quotient of cost and the number of DVDs sold each day.

46. Profit, Revenue, and Cost A company manufactures and sells flash drives for home computers. Here are the equations they use in connection with their business.

Number of flash drives sold each day: $n(x) = x$

Selling price for each flash drive: $p(x) = 3 - \frac{1}{300}x$

Daily fixed costs: $f(x) = 200$

Daily variable costs: $v(x) = 2x$

Find the following functions.

  a. Revenue $= R(x) =$ the product of the number of flash drives sold each day and the selling price of each flash drive.

  b. Cost $= C(x) =$ the sum of the fixed costs and the variable costs.

  c. Profit $= P(x) =$ the difference between revenue and cost.

  d. Average cost $= \overline{C}(x) =$ the quotient of cost and the number of flash drives sold each day.

47. Training Heart Rate Find the training heart rate function, $T(M)$, for a person with a resting heart rate of 62 beats per minute, then find the following.

   a. Find the maximum heart rate function, $M(x)$, for a person x years of age.

   b. What is the maximum heart rate for a 24-year-old person?

   c. What is the training heart rate for a 24-year-old person with a resting heart rate of 62 beats per minute? Round to the nearest whole number.

   d. What is the training heart rate for a 36-year-old person with a resting heart rate of 62 beats per minute? Round to the nearest whole number.

   e. What is the training heart rate for a 48-year-old person with a resting heart rate of 62 beats per minute? Round to the nearest whole number.

48. Training Heart Rate Find the training heart rate function, $T(M)$, for a person with a resting heart rate of 72 beats per minute, then find the following to the nearest whole number.

   a. Find the maximum heart rate function, $M(x)$, for a person x years of age.

   b. What is the maximum heart rate for a 20-year-old person?

   c. What is the training heart rate for a 20-year-old person with a resting heart rate of 72 beats per minute? Round to the nearest whole number.

   d. What is the training heart rate for a 30-year-old person with a resting heart rate of 72 beats per minute? Round to the nearest whole number.

   e. What is the training heart rate for a 40-year-old person with a resting heart rate of 72 beats per minute? Round to the nearest whole number.

Maintaining Your Skills

49. $|x - 3| < 1$

50. $|x - 3| > 1$

51. $|6 - x| > 2$

52. $\left|1 - \frac{1}{2}x\right| > 2$

53. $|7x - 1| \le 6$

54. $|7x - 1| \ge 6$

Getting Ready for the Next Section

55. $16(3.5)^2$

56. $\dfrac{2{,}400}{100}$

57. $\dfrac{180}{45}$

58. $4(2)(4)^2$

59. $\dfrac{0.0005(200)}{(0.25)^2}$

60. $\dfrac{0.2(0.5)^2}{100}$

61. If $y = Kx$, find K if $x = 5$ and $y = 15$.

62. If $d = Kt^2$, find K if $t = 2$ and $d = 64$.

63. If $V = \dfrac{K}{P}$, find K if $P = 48$ and $V = 50$.

64. If $y = Kxz^2$, find K if $x = 5$, $z = 3$, and $y = 180$.

Applying the Concepts

37. Length of a Spring The length a spring stretches is directly proportional to the force applied. If a force of 5 pounds stretches a spring 3 inches, how much force is necessary to stretch the same spring 10 inches?

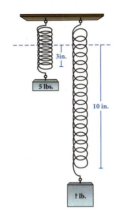

38. Weight and Surface Area The weight of a certain material varies directly with the surface area of that material. If 8 square feet weighs half a pound, how much will 10 square feet weigh?

Cars The chart shows the fastest cars in the world. Use the chart to answer Problems 39 and 40

39. The speed of a car in miles per hour is directly proportional to its speed in feet per second. If the speed of the SSC Ultimate Aero in feet per second is $400\frac{2}{5}$, what is the speed of the Bugatti in feet per second?

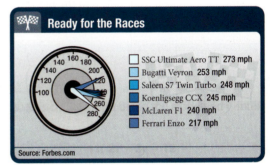

Ready for the Races

☐ SSC Ultimate Aero TT **273 mph**
☐ Bugatti Veyron **253 mph**
☐ Saleen S7 Twin Turbo **248 mph**
☐ Koenligsegg CCX **245 mph**
☐ McLaren F1 **240 mph**
☐ Ferrari Enzo **217 mph**

Source: Forbes.com

40. The time that it takes a car to travel a given distance varies inversely with the speed of the car. If a McLaren F1 traveled a certain distance in 2 hours, how long would it take the Saleen S7 Twin Turbo to travel the same distance? Round to the nearest hundredth.

41. Pressure and Temperature The temperature of a gas, T, varies directly with its pressure, P. A temperature of 200 K (Kelvin) produces a pressure of 50 pounds per square inch.

  a. Find the equation that relates pressure and temperature.

  b. Graph the equation from part a in the first quadrant only.

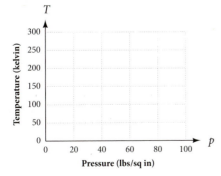

  c. What pressure will the gas have at 280° K?

42. Circumference and Diameter The circumference of a wheel is directly proportional to its diameter. A wheel has a circumference of 8.5 feet and a diameter of 2.7 feet.

  a. Find the equation that relates circumference and diameter.

  b. Graph the equation from part a in the first quadrant only.

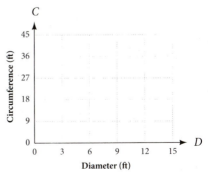

  c. What is the circumference of a wheel that has a diameter of 11.3 feet? Round your answer to the nearest tenth.

43. Volume and Pressure The volume of a gas is inversely proportional to the pressure. If a pressure of 36 pounds per square inch corresponds to a volume of 25 cubic feet, what pressure is needed to produce a volume of 75 cubic feet?

44. Wave Frequency The frequency of an electromagnetic wave varies inversely with the wavelength. If a wavelength of 200 meters has a frequency of 800 kilocycles per second, what frequency will be associated with a wavelength of 500 meters?

45. *f*-Stop and Aperture Diameter The relative aperture or *f*-stop for a camera lens is inversely proportional to the diameter of the aperture. An *f*-stop of 2 corresponds to an aperture diameter of 40 millimeters for the lens on a camera.

  a. Find the equation that relates *f*-stop and diameter.

  b. Graph the equation from part a in the first quadrant only.

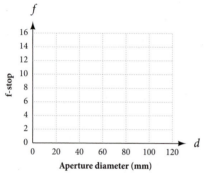

  c. What is the *f*-stop of this camera when the aperture diameter is 10 millimeters?

46. *f*-Stop and Aperture Diameter The relative aperture or *f*-stop for a camera lens is inversely proportional to the diameter of the aperture. An *f*-stop of 2.8 corresponds to an aperture diameter of 75 millimeters for a certain telephoto lens.

  a. Find the equation that relates *f*-stop and diameter.

  b. Graph the equation from part a in the first quadrant only.

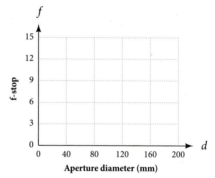

  c. What aperture diameter corresponds to an *f*-stop of 5.6?

47. Surface Area of a Cylinder The surface area of a hollow cylinder varies jointly with the height and radius of the cylinder. If a cylinder with radius 3 inches and height 5 inches has a surface area of 94 square inches, what is the surface area of a cylinder with radius 2 inches and height 8 inches?

48. Capacity of a Cylinder The capacity of a cylinder varies jointly with its height and the square of its radius. If a cylinder with a radius of 3 centimeters and a height of 6 centimeters has a capacity of 169.56 cubic centimeters, what will be the capacity of a cylinder with radius 4 centimeters and height 9 centimeters?

49. Electrical Resistance The resistance of a wire varies directly with its length and inversely with the square of its diameter. If 100 feet of wire with diameter 0.01 inch has a resistance of 10 ohms, what is the resistance of 60 feet of the same type of wire if its diameter is 0.02 inch?

50. Volume and Temperature The volume of a gas varies directly with its temperature and inversely with the pressure. If the volume of a certain gas is 30 cubic feet at a temperature of 300 K and a pressure of 20 pounds per square inch, what is the volume of the same gas at 340 K when the pressure is 30 pounds per square inch?

51. Music A musical tone's pitch varies inversely with its wavelength. If one tone has a pitch of 420 vibrations each second and a wavelength of 2.2 meters, find the wavelength of a tone that has a pitch of 720 vibrations each second. Round your answer to the nearest hundredth.

52. Hooke's Law Hooke's law states that the stress (force per unit area) placed on a solid object varies directly with the strain (deformation) produced.

 a. Using the variables S_1 for stress and S_2 for strain, state this law in algebraic form.

 b. Find the constant, K, if for one type of material $S_1 = 24$ and $S_2 = 72$.

53. Gravity In Book Three of his *Principia,* Isaac Newton (depicted on the postage stamp) states that there is a single force in the universe that holds everything together, called the force of universal gravity. Newton stated that the force of universal gravity, F, is directly proportional with the product of two masses, m_1 and m_2, and inversely proportional with the square of the distance d between them. Write the equation for Newton's force of universal gravity, using the symbol G as the constant of proportionality.

54. Boyle's Law and Charles's Law Boyle's law states that for low pressures, the pressure of an ideal gas kept at a constant temperature varies inversely with the volume of the gas. Charles's law states that for low pressures, the density of an ideal gas kept at a constant pressure varies inversely with the absolute temperature of the gas.

 a. State Boyle's law as an equation using the symbols P, K, and V.

 b. State Charles's law as an equation using the symbols D, K, and T.

Maintaining Your Skills

The problems that follow review some of the more important skills you have learned in previous sections and chapters.

Solve the following equations.

55. $x - 5 = 7$

56. $3y = -4$

57. $5 - \dfrac{4}{7}a = -11$

58. $\dfrac{1}{5}x - \dfrac{1}{2} - \dfrac{1}{10}x + \dfrac{2}{5} = \dfrac{3}{10}x + \dfrac{1}{2}$

59. $5(x - 1) - 2(2x + 3) = 5x - 4$

60. $0.07 - 0.02(3x + 1) = -0.04x + 0.01$

Solve for the indicated variable.

61. $P = 2l + 2w$ for w

62. $A = \dfrac{1}{2}h(b + B)$ for B

Solve the following inequalities.

63. $-5t \le 30$

64. $5 - \dfrac{3}{2}x > -1$

65. $1.6x - 2 < 0.8x + 2.8$

66. $3(2y + 4) \ge 5(y - 8)$

Solve the following equations.

67. $\left| \dfrac{1}{4}x - 1 \right| = \dfrac{1}{2}$

68. $\left| \dfrac{2}{3}a + 4 \right| = 6$

69. $|3 - 2x| + 5 = 2$

70. $5 = |3y + 6| - 4$

Solve each inequality and graph the solution set.

71. $\left| \dfrac{x}{5} + 1 \right| \ge \dfrac{4}{5}$

72. $|2 - 6t| < -5$

73. $|3 - 4t| > -5$

74. $|6y - 1| - 4 \le 2$

Extending the Concepts

75. Human Cannonball A circus company is deciding where to position the net for the human cannonball so that he will land safely during the act. They do this by firing a 100-pound sack of potatoes out of the cannon at different speeds and then measuring how far from the cannon the sack lands. The results are shown in the table.

Speed in Miles/Hour	Distance in Feet
40	108
50	169
60	243
70	331

a. Does distance vary directly with the speed, or directly with the square of the speed?

b. Write the equation that describes the relationship between speed and distance.

c. If the cannon will fire a human safely at 55 miles/hour, where should they position the net so the cannonball has a safe landing?

d. How much farther will he land if his speed out of the cannon is 56 miles/hour?

Chapter 3 Summary

Linear Equations in Two Variables [3.1, 3.3]

1. The equation $3x + 2y = 6$ is an example of a linear equation in two variables.

A *linear equation in two variables* is any equation that can be put in *standard form* $ax + by = c$. The graph of every linear equation is a straight line.

Intercepts [3.1]

2. To find the x-intercept for $3x + 2y = 6$, we let $y = 0$ and get

$$3x = 6$$
$$x = 2$$

In this case the x-intercept is 2, and the graph crosses the x-axis at $(2, 0)$.

The x-intercept of an equation is the x-coordinate of the point where the graph crosses the x-axis. The y-intercept is the y-coordinate of the point where the graph crosses the y-axis. We find the y-intercept by substituting $x = 0$ into the equation and solving for y. The x-intercept is found by letting $y = 0$ and solving for x.

The Slope of a Line [3.2]

3. The slope of the line through $(6, 9)$ and $(1, -1)$ is

$$m = \frac{9 - (-1)}{6 - 1}$$
$$= \frac{10}{5}$$
$$= 2$$

The *slope* of the line containing points (x_1, y_1) and (x_2, y_2) is given by

$$\text{Slope} = m = \frac{\text{Rise}}{\text{Run}} = \frac{y_2 - y_1}{x_2 - x_1}$$

Horizontal lines have 0 slope, and vertical lines have no slope.
Parallel lines have equal slopes, and perpendicular lines have slopes that are negative reciprocals.

The Slope-Intercept Form of a Line [3.3]

4. The equation of the line with slope 5 and y-intercept 3 is

$$y = 5x + 3$$

The equation of a line with slope m and y-intercept b is given by

$$y = mx + b$$

The Point-Slope Form of a Line [3.3]

5. The equation of the line through $(3, 2)$ with slope -4 is

$$y - 2 = -4(x - 3)$$

which can be simplified to

$$y = -4x + 14$$

The equation of the line through (x_1, y_1) that has slope m can be written as

$$y - y_1 = m(x - x_1)$$

■ Linear Inequalities in Two Variables [3.4]

6. The graph of

$$x - y \le 3 \text{ is}$$

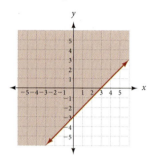

An inequality of the form $ax + by < c$ is a *linear inequality in two variables.* The equation for the boundary of the solution set is given by $ax + by = c$. (This equation is found by simply replacing the inequality symbol with an equal sign.)

   To graph a linear inequality, first graph the boundary using a solid line if the boundary is included in the solution set and a broken line if the boundary is not included in the solution set. Next, choose any point not on the boundary and substitute its coordinates into the original inequality. If the resulting statement is true, the graph lies on the same side of the boundary as the test point. A false statement indicates that the solution set lies on the other side of the boundary.

■ Functions and Relations [3.5]

7. The relation

$$\{(8, 1), (6, 1), (-3, 0)\}$$

is also a function because no ordered pairs have the same first coordinates. The domain is $\{8, 6, -3\}$ and the range is $\{1, 0\}$.

A *function* is a rule that pairs each element in one set, called the *domain,* with exactly one element from a second set, called the *range.*

   A *relation* is any set of ordered pairs. The set of all first coordinates is called the *domain* of the relation, and the set of all second coordinates is the *range* of the relation. A function is a relation in which no two different ordered pairs have the same first coordinates.

■ Vertical Line Test [3.5]

8. The graph of $x = y^2$ shown in Figure 5 in Section 3.5 fails the vertical line test. It is not the graph of a function.

If a vertical line crosses the graph of a relation in more than one place, the relation cannot be a function. If no vertical line can be found that crosses the graph in more than one place, the relation must be a function.

■ Function Notation [3.6]

9. If $f(x) = 5x - 3$ then
$$f(0) = 5(0) - 3$$
$$= -3$$
$$f(1) = 5(1) - 3$$
$$= 2$$
$$f(-2) = 5(-2) - 3$$
$$= -13$$
$$f(a) = 5a - 3$$

The alternative notation for y is $f(x)$. It is read "f of x" and can be used instead of the variable y when working with functions. The notation y and the notation $f(x)$ are equivalent; that is, $y = f(x)$.

■ Algebra with Functions [3.7]

If f and g are any two functions with a common domain, then

$(f + g)(x) = f(x) + g(x)$ The function $f + g$ is the sum of the functions f and g.

$(f - g)(x) = f(x) - g(x)$ The function $f - g$ is the difference of the functions f and g.

$(fg)(x) = f(x)g(x)$ The function fg is the product of the functions f and g.

$\dfrac{f}{g}(x) = \dfrac{f(x)}{g(x)}$ The function $\dfrac{f}{g}$ is the quotient of the functions f and g, where $g(x) \ne 0$

Composition of Functions [3.7]

If f and g are two functions for which the range of each has numbers in common with the domain of the other, then we have the following definitions:

$$\text{The composition of } f \text{ with } g: (f \circ g)(x) = f(g(x))$$

$$\text{The composition of } g \text{ with } f: (g \circ f)(x) = g(f(x))$$

Variation [3.8]

10. If y varies directly with x, then

$$y = Kx$$

Then if y is 18 when x is 6,

$$18 = K \cdot 6$$

or

$$K = 3$$

So the equation can be written more specifically as

$$y = 3x$$

If we want to know what y is when x is 4, we simply substitute:

$$y = 3 \cdot 4$$
$$y = 12$$

If y varies *directly* with x (y is directly proportional to x), then

$$y = Kx$$

If y varies *inversely* with x (y is inversely proportional to x), then

$$y = \frac{K}{x}$$

If z varies *jointly* with x and y (z is directly proportional to both x and y), then

$$z = Kxy$$

In each case, K is called the *constant of variation*.

🚫 COMMON MISTAKES

1. When graphing ordered pairs, the most common mistake is to associate the first coordinate with the y-axis and the second with the x-axis. If you make this mistake you would graph (3, 1) by going up 3 and to the right 1, which is just the reverse of what you should do. Remember, the first coordinate is always associated with the horizontal axis, and the second coordinate is always associated with the vertical axis.

2. The two most common mistakes students make when first working with the formula for the slope of a line are the following:
   a. Putting the difference of the x-coordinates over the difference of the y-coordinates.
   b. Subtracting in one order in the numerator and then subtracting in the opposite order in the denominator.

3. When graphing linear inequalities in two variables, remember to graph the boundary with a broken line when the inequality symbol is $<$ or $>$. The only time you use a solid line for the boundary is when the inequality symbol is $\leq$ or $\geq$.

Graph each line. [3.1]

1. $3x + 2y = 6$

2. $y = -\dfrac{3}{2}x + 1$

3. $x = 3$

Find the slope of the line through the following pairs of points. [3.2]

4. $(5, 2), (3, 6)$

5. $(-4, 2), (3, 2)$

Find x if the line through the two given points has the given slope. [3.2]

6. $(4, x), (1, -3); m = 2$

7. $(-4, 7), (2, x); m = -\dfrac{1}{3}$

8. Find the slope of any line parallel to the line through $(3, 8)$ and $(5, -2)$. [3.2]

9. The line through $(5, 3y)$ and $(2, y)$ is parallel to a line with slope 4. What is the value of y? [3.2]

Give the equation of the line with the following slope and y-intercept. [3.3]

10. $m = 3, b = 5$

11. $m = -2, b = 0$

Give the slope and y-intercept of each equation. [3.3]

12. $3x - y = 6$

13. $2x - 3y = 9$

Find the equation of the line that contains the given point and has the given slope. [3.3]

14. $(2, 4), m = 2$

15. $(-3, 1), m = -\dfrac{1}{3}$

Find the equation of the line that contains the given pair of points. [3.3]

16. $(2, 5), (-3, -5)$

17. $(-3, 7), (4, 7)$

18. $(-5, -1), (-3, -4)$

19. Find the equation of the line that is parallel to $2x - y = 4$ and contains the point $(2, -3)$. [3.3]

20. Find the equation of the line perpendicular to $y = -3x + 1$ that has an x-intercept of 2. [3.3]

Graph each linear inequality. [3.4]

21. $y \le 2x - 3$

22. $x \ge -1$

State the domain and range of each relation, and then indicate which relations are also functions. [3.5]

23. $\{(2, 4), (3, 3), (4, 2)\}$

24. $\{(6, 3), (-4, 3), (-2, 0)\}$

If $f = \{(2, -1), (-3, 0), \left(4, \dfrac{1}{2}\right), (\pi, 2)\}$ and $g = \{(2, 2), (-1, 4), (0, 0)\}$, find the following. [3.6]

25. $f(-3)$

26. $f(2) + g(2)$

Let $f(x) = 2x^2 - 4x + 1$ and $g(x) = 3x + 2$, and evaluate each of the following. [3.6]

27. $f(0)$

28. $g(a)$

29. $f(g(0))$

30. $f(g(1))$

For the following problems, y varies directly with x. [3.8]

31. If y is 6 when x is 2, find y when x is 8.

32. If y is -3 when x is 5, find y when x is -10.

For the following problems, y varies inversely with the square of x. [3.8]

33. If y is 9 when x is 2, find y when x is 3.

34. If y is 4 when x is 5, find y when x is 2.

Solve each application problem. [3.8]

35. Tension in a Spring The tension t in a spring varies directly with the distance d the spring is stretched. If the tension is 42 pounds when the spring is stretched 2 inches, find the tension when the spring is stretched twice as far.

36. Light Intensity The intensity of a light source, measured in foot-candles varies inversely with the square of the distance from the source. Four feet from the source the intensity is 9 foot-candles. What is the intensity 3 feet from the source?

Simplify each of the following.

1. -5^2

2. $-|-6|$

3. $5^2 + 7^2$

4. $(5 + 7)^2$

5. $48 \div 8 \cdot 6$

6. $75 \div 15 \cdot 5$

7. $30 - 15 \div 3 + 2$

8. $30 - 15 \div (3 + 2)$

Find the value of each expression when x is 4.

9. $x^2 - 10x + 25$

10. $x^2 - 25$

Reduce the following fractions to lowest terms.

11. $\dfrac{336}{432}$

12. $\dfrac{721}{927}$

Subtract.

13. $\dfrac{3}{15} - \dfrac{2}{20}$

14. $\dfrac{6}{28} - \dfrac{5}{42}$

15. Add $-\dfrac{3}{4}$ to the product of -3 and $\dfrac{5}{12}$.

16. Subtract $\dfrac{2}{5}$ from the product of -3 and $\dfrac{6}{15}$.

Simplify each of the following expressions.

17. $8\left(\dfrac{3}{4}x - \dfrac{1}{2}y\right)$

18. $12\left(\dfrac{5}{6}x + \dfrac{3}{4}y\right)$

Solve the following equations.

19. $\dfrac{2}{3}a - 4 = 6$

20. $\dfrac{3}{4}(8x - 3) + \dfrac{3}{4} = 2$

21. $-4 + 3(3x + 2) = 7$

22. $3x - (x + 2) = -6$

23. $|2x - 3| - 7 = 1$

24. $|2y + 3| = 2y - 7$

Solve the following expressions for y and simplify.

25. $y - 5 = -3(x + 2)$

26. $y + 3 = \dfrac{1}{2}(3x - 4)$

Solve each inequality and graph the solution set.

27. $|3x - 2| \le 7$

28. $|3x + 4| - 7 > 3$

Graph on a rectangular coordinate system.

29. $6x - 5y = 30$

30. $x = 2$

31. $y = \dfrac{1}{2}x - 1$

32. $2x - y \le -3$

33. Find the slope of the line through $(-3, -2)$ and $(-5, 3)$.

34. Find y if the line through $(-1, 2)$ and $(-3, y)$ has a slope of -2.

35. Find the slope of the line $y = 3$.

36. Give the equation of a line with slope $-\dfrac{3}{4}$ and y-intercept $= 2$.

37. Find the slope and y-intercept of $4x - 5y = 20$.

38. Find the equation of the line with x-intercept -2 and y-intercept -1.

39. Find the equation of the line that is parallel to $5x + 2y = -1$ and contains the point $(-4, -2)$.

40. Find the equation of a line with slope $\dfrac{7}{3}$ that contains the point $(6, 8)$. Write the answer in slope-intercept form.

If $f(x) = x^2 - 3x$, $g(x) = x - 1$ and $h(x) = 3x - 1$, find the following.

41. $h(0)$

42. $f(-1) + g(4)$

43. $(g \circ f)(-2)$

44. $(h - g)(x)$

45. Specify the domain and range for the relation $\{(-1, 3), (2, -1), (3, 3)\}$. Is the relation also a function?

46. y varies inversely with the square of x. If $y = 4$ when $x = 4$; find y when $x = 6$.

47. Zac left a $6 tip for a $30 lunch with his girlfriend. What percent of the cost of lunch was the tip?

48. If the sales tax is 7.25% of the purchase price, how much sales tax will Hailey pay on a $130 dress?

Use the information in the illustration to work Problems 49–50.

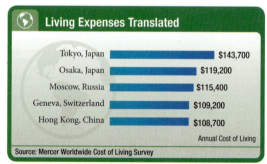

Living Expenses Translated

Tokyo, Japan	$143,700
Osaka, Japan	$119,200
Moscow, Russia	$115,400
Geneva, Switzerland	$109,200
Hong Kong, China	$108,700

Annual Cost of Living

Source: Mercer Worldwide Cost of Living Survey

49. How much more does it cost, expressed as a percent, to live in Tokyo than in Moscow?

50. How much less does it cost, expressed as a percent, to live in Geneva than in Osaka?

For each of the following straight lines, identify the
x-intercept, y-intercept, and slope, and sketch the graph.
[3.1–3.3]

1. $2x + y = 6$

2. $y = -2x - 3$

3. $y = \dfrac{3}{2}x + 4$

4. $x = -2$

16. $y = x^2 - 9$

17. $\{(0, 0), (1, 3), (2, 5)\}$

18. $y = 3x - 5$

Use slope-intercept form to write the equation of each line
below. [3.3]

5.
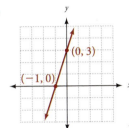

6.
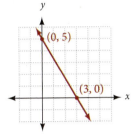

Determine the domain and range of the following
functions. Assume the entire function is shown. [3.5]

19.
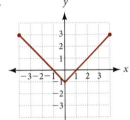

20.
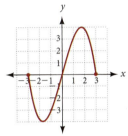

Find the equation for each line. [3.3]

7. The line through $(-1, 3)$ that has slope $m = 2$

8. The line through $(-3, 2)$ and $(4, -1)$

9. The line which contains the point $(5, -3)$ and is
parallel to the line of $2x - 5y = 10$

10. The line which contains the point $(-1, -2)$ and is
perpendicular to the line of $y = 3x - 2$

For each graph below, state the inequality represented by
the shaded region. [3.4]

11.
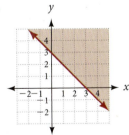

12.
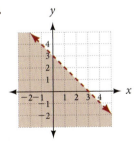

Graph the following linear inequalities. [3.4]

13. $3x - 4y < 12$

14. $y \le -x + 2$

State the domain and range for the following relations, and
indicate which relations are also functions. [3.5]

15. $\{(-2, 0), (-3, 0), (-2, 1)\}$

Let $f(x) = x - 2$, $g(x) = 3x + 4$ and $h(x) = 3x^2 - 2x - 8$, and
find the following. [3.6]

21. $f(3) + g(2)$

22. $h(0) + g(0)$

23. $f(g(2))$

24. $g(f(2))$

Use the graph to answer questions 25–28.

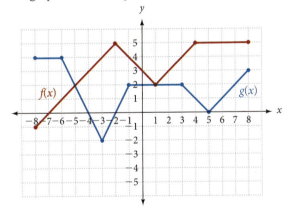

Evaluate. [3.7]

25. $f(-6) + g(-6)$

26. $(f - g)(1)$

27. $(f \circ g)(5)$

28. $(g \circ f)(5)$

Solve the following variation problems. [3.8]

29. Direct Variation Quantity y varies directly with the
square of x. If y is 50 when x is 5, find y when x is 3.

30. Joint Variation Quantity z varies jointly with x and the
cube of y. If z is 15 when x is 5 and y is 2, find z when
x is 2 and y is 3.

EQUATIONS AND INEQUALITIES IN TWO VARIABLES

Group Project

Light Intensity

Number of People	2–3
Time Needed	15 minutes
Equipment	Paper and pencil
Background	I found the following diagram while shopping for some track lighting for my home. I was impressed by the diagram because it displays a lot of useful information in a very efficient manner. As the diagram indicates, the amount of light that falls on a surface depends on how far above the surface the light is placed and how much the light spreads out on the surface. Assume that this light illuminates a circle on a flat surface, and work the following problems.

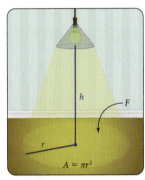

$$A = \pi r^2$$

b. Use the templates in Figures 1 and 2 to construct line graphs from the data in the tables.

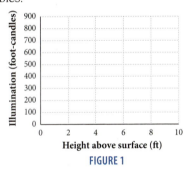

FIGURE 1

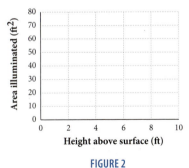

FIGURE 2

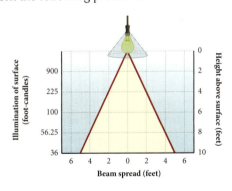

Procedure **a.** Fill in each table.

Height Above Surface (ft)	Illumination (foot-candles)
2	
4	
6	
8	
10	

Height Above Surface (ft)	Area of Illuminated Region (ft²)
2	
4	
6	
8	
10	

c. Which of the relationships is direct variation, and which is inverse variation?

d. Let F represent the number of foot-candles that fall on the surface, h the distance the light source is above the surface, and A the area of the illuminated region. Write an equation that shows the relationship between A and h, then write another equation that gives the relationship between F and h.

Descartes and Pascal

René Descartes, the inventor of the rectangular coordinate system, is the person who made the statement, "I think, therefore I am." Blaise Pascal, another French philosopher, is responsible for the statement, "The heart has its reasons which reason does not know." Although Pascal and Descartes were contemporaries, the philosophies of the two men differed greatly. Research the philosophy of both Descartes and Pascal, and then write an essay that gives the main points of each man's philosophy. In the essay, show how the quotations given here fit in with the philosophy of the man responsible for the quotation.

René Descartes, 1596–1650

Blaise Pascal, 1623–1662

Systems of Linear Equations and Inequalities

4

Image © 2010 Google
Hartsfield-Jackson International Airport
Atlanta, Georgia

The Airports Council International determines an airport as "busy" using three factors: (1) passenger traffic, (2) cargo traffic, and (3) number of takeoffs and landings. By these standards, Hartsfield-Jackson International Airport in Atlanta, Georgia is labeled the busiest airport in the world. In 2007, Hartsfield-Jackson set a world record with 994,346 takeoffs and landings. With such a busy airport, it is important to have skilled and alert traffic controllers to manage all those flights.

Bryan Busovicki/Shutterstock.com

Picture two airplanes, each with a straight flight path, approaching the airport. We could assign the path of each plane a linear equation in two variables, using rate of speed, distance, and time as variables. Then we could graph those equations to determine when or if the planes would intersect. Those two equations when considered together are called a system of linear equations in two variables, which is the primary focus of this chapter.

Key Words	Definition
System of Linear Equations	Two or more equations considered together
Inconsistent Systems	A system of equations that have no solution in common
Dependent Equations	A system of equations that have all solutions in common

Chapter Outline

4.1 Systems of Linear Equations in Two Variables

  A Solve systems of linear equations in two variables by graphing.

  B Solve systems of linear equations in two variables by the addition method.

  C Solve systems of linear equations in two variables by the substitution method.

4.2 Systems of Linear Equations in Three Variables

  A Solve systems of linear equations in three variables.

4.3 Applications of Linear Systems

  A Solve application problems whose solutions are found through systems of linear equations.

4.4 Matrix Solutions to Linear Systems

  A Solve a system of linear equations using an augmented matrix.

4.5 Systems of Linear Inequalities

  A Graph the solution to a system of linear inequalities in two variables.

Systems of Linear Equations in Two Variables

OBJECTIVES

A Solve systems of linear equations in two variables by graphing.

B Solve systems of linear equations in two variables by the addition method.

C Solve systems of linear equations in two variables by the substitution method.

TICKET TO SUCCESS

Keep these questions in mind as you read through the section. Then respond in your own words and in complete sentences.

1. How would you define a solution for a linear system of equations?
2. How would you use the addition method to solve a system of linear equations?
3. When would the substitution method be more efficient than the addition method in solving a system of linear equations?
4. Explain what an inconsistent system of linear equations looks like graphically and what would result algebraically when attempting to solve the system.

Graa Victoria/Shutterstock.com

Suppose you and a friend want to order burritos and tacos for lunch. The restaurant's lighted menu, however, is broken and you can't read it. You order 2 burritos and 3 tacos and are charged a total of $12. Your friend orders 1 burrito and 4 tacos for a total of $11. How much is each burrito? How much is each taco? In this section, we will begin our work with systems of equations in two variables and use these systems to answer your lunch question.

A Solve Systems by Graphing

Previously, we found the graph of an equation of the form $ax + by = c$ to be a straight line. Since the graph is a straight line, the equation is said to be a linear equation. Two linear equations considered together form a *linear system* of equations. For example,

$$3x - 2y = 6$$
$$2x + 4y = 20$$

is a linear system. The solution set to the system is the set of all ordered pairs that satisfy both equations. If we graph each equation on the same set of axes, we can see the solution set (Figure 1).

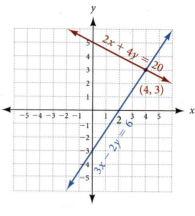

FIGURE 1

The point (4, 3) lies on both lines and therefore must satisfy both equations. It is obvious from the graph that it is the only point that does so. The solution set for the system is {(4, 3)}.

More generally, if $a_1 x + b_1 y = c_1$ and $a_2 x + b_2 y = c_2$ are linear equations, then the solution set for the system

$$a_1 x + b_1 y = c_1$$
$$a_2 x + b_2 y = c_2$$

can be illustrated through one of the graphs in Figure 2.

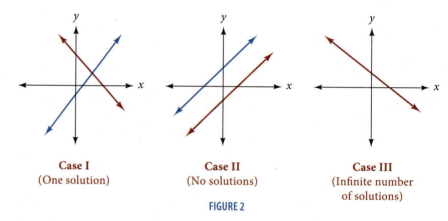

Case I
(One solution)

Case II
(No solutions)

Case III
(Infinite number
of solutions)

FIGURE 2

Case I The two lines intersect at one and only one point. The coordinates of the point give the solution to the system. This is what usually happens.

Case II The lines are parallel and therefore have no points in common. The solution set to the system is the empty set, $\varnothing$. In this case, we say the system is *inconsistent*.

Case III The lines coincide; that is, their graphs represent the same line. The solution set consists of all ordered pairs that satisfy either equation. In this case, the equations are said to be *dependent*.

In the beginning of this section we found the solution set for the system

$$3x - 2y = 6$$
$$2x + 4y = 20$$

by graphing each equation and then reading the solution set from the graph. Solving a system of linear equations by graphing is the least accurate method. If the coordinates of the point of intersection are not integers, it can be very difficult to read the solution set from the graph. There is another method of solving a linear system that does not depend on the graph. It is called the *addition method*.

B The Addition Method

EXAMPLE 1   Solve the system.

$$4x + 3y = 10$$
$$2x + y = 4$$

SOLUTION   If we multiply the bottom equation by -3, the coefficients of y in the resulting equation and the top equation will be opposites.

$$
\begin{array}{l}
4x + 3y = 10 \xrightarrow{\text{No change}} 4x + 3y = 10 \\
2x + y = 4 \xrightarrow{\text{Multiply by } -3.} -6x - 3y = -12
\end{array}
$$

Adding the left and right sides of the resulting equations, we have

$$
\begin{array}{r}
4x + 3y = 10 \\
-6x - 3y = -12 \\
\hline
-2x = -2
\end{array}
$$

The result is a linear equation in one variable. We have eliminated the variable y from the equations by addition. (It is for this reason we call this method of solving a linear system the *addition method*.) Solving $-2x = -2$ for x, we have

$$x = 1$$

This is the x-coordinate of the solution to our system. To find the y-coordinate, we substitute $x = 1$ into any of the equations containing both the variables x and y. Let's try the second equation in our original system.

$$2(1) + y = 4$$
$$2 + y = 4$$
$$y = 2$$

This is the y-coordinate of the solution to our system. The ordered pair $(1, 2)$ is the solution to the system.

CHECKING SOLUTIONS   We can check our solution by substituting it into both of our equations.

Substituting $x = 1$ and $y = 2$ into $4x + 3y = 10$, we have

$$4(1) + 3(2) \overset{?}{=} 10$$
$$4 + 6 \overset{?}{=} 10$$
$$10 = 10 \quad \text{A true statement}$$

Substituting $x = 1$ and $y = 2$ into $2x + y = 4$, we have

$$2(1) + 2 \overset{?}{=} 4$$
$$2 + 2 \overset{?}{=} 4$$
$$4 = 4 \quad \text{A true statement}$$

Our solution satisfies both equations; therefore, it is a solution to our system of equations. ∎

EXAMPLE 2   Solve the system.

$$3x - 5y = -2$$
$$2x - 3y = 1$$

SOLUTION   We can eliminate either variable. Let's decide to eliminate the variable x. We can do so by multiplying the top equation by 2 and the bottom equation by -3, and then adding the left and right sides of the resulting equations.

$$\begin{array}{ll} & \text{Multiply by 2.} \\ 3x - 5y = -2 \longrightarrow & 6x - 10y = -4 \\ 2x - 3y = 1 \longrightarrow & \underline{-6x + 9y = -3} \\ & \text{Multiply by } -3. \qquad -y = -7 \\ & y = 7 \end{array}$$

The y-coordinate of the solution to the system is 7. Substituting this value of y into any of the equations with both x- and y-variables gives $x = 11$. The solution to the system is $(11, 7)$. It is the only ordered pair that satisfies both equations.

CHECKING SOLUTIONS   Checking $(11, 7)$ in each equation looks like this:

Substituting $x = 11$ and $y = 7$ into   Substituting $x = 11$ and $y = 7$ into

$3x - 5y = -2$, we have $2x - 3y = 1$, we have

$\qquad 3(11) - 5(7) \overset{?}{=} -2 \qquad\qquad\qquad 2(11) - 3(7) \overset{?}{=} 1$

$\qquad\quad 33 - 35 \overset{?}{=} -2 \qquad\qquad\qquad\qquad 22 - 21 \overset{?}{=} 1$

$\qquad\qquad\qquad -2 = -2$ A true statement $\qquad\qquad 1 = 1$ A true statement

Our solution satisfies both equations; therefore, $(11, 7)$ is a solution to our system. ■

3. Solve the system.
$3x - 5y = 2$
$2x + 4y = 1$

EXAMPLE 3   Solve the system.

$$2x - 3y = 4$$
$$4x + 5y = 3$$

SOLUTION   We can eliminate x by multiplying the top equation by -2 and adding it to the bottom equation.

$$\begin{array}{ll} & \text{Multiply by } -2. \\ 2x - 3y = 4 \longrightarrow & -4x + 6y = -8 \\ 4x + 5y = 3 \longrightarrow & \underline{4x + 5y = 3} \\ & \text{No change} \qquad 11y = -5 \\ & y = -\dfrac{5}{11} \end{array}$$

The y-coordinate of our solution is $-\dfrac{5}{11}$. If we were to substitute this value of y back into either of our original equations, we would find the arithmetic necessary to solve for x cumbersome. For this reason, it is probably best to go back to the original system and solve it a second time—for x instead of y. Here is how we do that:

$$\begin{array}{ll} & \text{Multiply by 5.} \\ 2x - 3y = 4 \longrightarrow & 10x - 15y = 20 \\ 4x + 5y = 3 \longrightarrow & \underline{12x + 15y = 9} \\ & \text{Multiply by 3.} \quad 22x = 29 \\ & x = \dfrac{29}{22} \end{array}$$

The solution to our system is $\left(\dfrac{29}{22}, -\dfrac{5}{11}\right)$. ■

The main idea in solving a system of linear equations by the addition method is to use the multiplication property of equality on one or both of the original equations, if necessary, to make the coefficients of either variable opposites. The following box shows some steps to follow when solving a system of linear equations by the addition method.

Answer

3. $\left(\dfrac{13}{22}, -\dfrac{1}{22}\right)$

Strategy Solving a System of Linear Equations by the Addition Method

Step 1 Decide which variable to eliminate. (In some cases, one variable will be easier to eliminate than the other. With some practice, you will notice which one it is.)

Step 2 Use the multiplication property of equality on each equation separately to make the coefficients of the variable that is to be eliminated opposites.

Step 3 Add the respective left and right sides of the system together.

Step 4 Solve for the remaining variable.

Step 5 Substitute the value of the variable from step 4 into an equation containing both variables and solve for the other variable. (Or repeat steps 2–4 to eliminate the other variable.)

Step 6 Check your solution in both equations, if necessary.

EXAMPLE 4 Solve the system.
$$5x - 2y = 5$$
$$-10x + 4y = 15$$

SOLUTION We can eliminate y by multiplying the first equation by 2 and adding the result to the second equation.

$$
\begin{array}{lcl}
& \text{Multiply by 2.} & \\
5x - 2y = 5 & \longrightarrow & 10x - 4y = 10 \\
-10x + 4y = 15 & \longrightarrow & \underline{-10x + 4y = 15} \\
& \text{No change} & \ 0 = 25
\end{array}
$$

The result is the false statement $0 = 25$, which indicates there is no solution to the system. If we were to graph the two lines, we would find that they are parallel as shown in Figure 3. In a case like this, we say the system is *inconsistent*. Whenever both variables have been eliminated and the resulting statement is false, the solution set for the system will be the empty set, $\varnothing$.

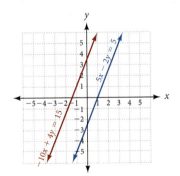

FIGURE 3 A visual representation of the situation in Example 4—the two lines are parallel.

4. Solve the system.
$$2x + 7y = 3$$
$$4x + 14y = 1$$

Answer
4. $\varnothing$

5. Solve the system.
$$2x + 7y = 3$$
$$4x + 14y = 6$$

Solve the system.

$$4x + 3y = 2$$
$$8x + 6y = 4$$

SOLUTION Multiplying the top equation by -2 and adding, we can eliminate the variable x.

Multiply by -2.

$4x + 3y = 2$ $\xrightarrow{\hspace{2cm}}$ $-8x - 6y = -4$

$8x + 6y = 4$ $\xrightarrow{\hspace{2cm}}$ $\underline{8x + 6y = 4}$

No change

$0 = 0$

Both variables have been eliminated and the resulting statement $0 = 0$ is true. In this case, the lines coincide as shown in Figure 4 and the equations are said to be *dependent*. The solution set consists of all ordered pairs that satisfy either equation. We can write the solution set as $\{(x, y) \mid 4x + 3y = 2\}$ or $\{(x, y) \mid 8x + 6y = 4\}$.

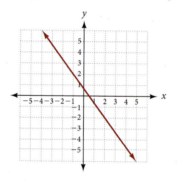

FIGURE 4   A visual representation of the situation in Example 5—both equations produce the same graph.

Special Cases

The previous two examples illustrate the two special cases in which the graphs of the equations in the system either coincide or are parallel. In both cases, the left-hand sides of the equations were multiples of each other. In the case of the dependent equations, the right-hand sides were also multiples. The following generalizes these observations for the system

$$a_1 x + b_1 y = c_1$$
$$a_2 x + b_2 y = c_2$$

SPECIAL CASES FOR SYSTEMS OF LINEAR EQUATIONS		
What happens	**Geometric Interpretation**	**Algebraic Interpretation**
Inconsistent System		
Both variables are eliminated, and the resulting statement is false.	The lines are parallel, and there is no solution to the system.	$\dfrac{a_1}{a_2} = \dfrac{b_1}{b_2} \neq \dfrac{c_1}{c_2}$
Dependent Equations		
Both variables are eliminated, and the resulting statement is true.	The lines coincide, and there are an infinite number of solutions to the system.	$\dfrac{a_1}{a_2} = \dfrac{b_1}{b_2} = \dfrac{c_1}{c_2}$

Answer

5. $\{(x, y) \mid 2x + 7y = 3\}$

EXAMPLE 6 Solve the system.

$$\frac{1}{2}x - \frac{1}{3}y = 2$$
$$\frac{1}{4}x + \frac{2}{3}y = 6$$

SOLUTION Although we could solve this system without clearing the equations of fractions, there is probably less chance for error if we have only integer coefficients to work with. So let's begin by multiplying both sides of the top equation by 6 and both sides of the bottom equation by 12 to clear each equation of fractions.

Multiply by 6.
$$\frac{1}{2}x - \frac{1}{3}y = 2 \longrightarrow 3x - 2y = 12$$

$$\frac{1}{4}x + \frac{2}{3}y = 6 \longrightarrow 3x + 8y = 72$$
Multiply by 12.

Now we can eliminate x by multiplying the top equation by -1 and leaving the bottom equation unchanged.

Multiply by -1.
$$3x - 2y = 12 \longrightarrow \quad -3x + 2y = -12$$
$$3x + 8y = 72 \longrightarrow \quad \underline{3x + 8y = \quad 72}$$
No change $\qquad 10y = \quad 60$
$$y = \quad 6$$

We can substitute $y = 6$ into any equation that contains both x and y. Let's use $3x - 2y = 12$.

$$3x - 2(6) = 12$$
$$3x - 12 = 12$$
$$3x = 24$$
$$x = 8$$

The solution to the system is (8, 6). ∎

C The Substitution Method

EXAMPLE 7 Solve the system.

$$2x - 3y = -6$$
$$y = 3x - 5$$

SOLUTION The second equation tells us y is $3x - 5$. Substituting the expression $3x - 5$ for y in the first equation, we have

$$2x - 3(3x - 5) = -6$$

The result of the substitution is the elimination of the variable y. Solving the resulting linear equation for x as usual, we have

$$2x - 9x + 15 = -6$$
$$-7x + 15 = -6$$
$$-7x = -21$$
$$x = 3$$

6. Solve the system.
$$\frac{1}{3}x + \frac{1}{2}y = 4$$
$$\frac{2}{3}x - \frac{1}{4}y = 3$$

7. Solve the system
$$4x - 2y = -2$$
$$y = x + 3$$
by substituting the expression for y given in the second equation into the first equation.

Answers
6. (6, 4)
7. (2, 5)

Putting $x = 3$ into the second equation in the original system, we have

$$y = 3(3) - 5$$
$$= 9 - 5$$
$$= 4$$

The solution to the system is (3, 4).

CHECKING SOLUTIONS Checking (3, 4) in each equation looks like this:

Substituting $x = 3$ and $y = 4$ into

$2x - 3y = -6$, we have

$2(3) - 3(4) \stackrel{?}{=} -6$

$6 - 12 \stackrel{?}{=} -6$

$-6 = -6$ A true statement

Substituting $x = 3$ and $y = 4$ into

$y = 3x - 5$, we have

$4 \stackrel{?}{=} 3(3) - 5$

$4 \stackrel{?}{=} 9 - 5$

$4 = 4$ A true statement

Our solution satisfies both equations; therefore, (3, 4) is a solution to our system. ∎

Strategy   **Solving a System of Equations by the Substitution Method**

Step 1 Solve either one of the equations for x or y. (This step is not necessary if one of the equations is already in the correct form, as in Example 7.)

Step 2 Substitute the expression for the variable obtained in step 1 into the other equation and solve it.

Step 3 Substitute the solution for step 2 into any equation in the system that contains both variables and solve it.

Step 4 Check your results, if necessary.

---

8. Solve by substitution.

$$5x - 3y = -4$$
$$x + 2y = 7$$

NOTE
Both the substitution method and the addition method can be used to solve any system of linear equations in two variables. Systems like the one in Example 7, however, are easier to solve using the substitution method because one of the variables is already written in terms of the other. A system like the one in Example 2 is easier to solve using the addition method because solving for one of the variables would lead to an expression involving fractions. The system in Example 8 could be solved easily by either method because solving the second equation for x is a one-step process.

Answer
8. (1, 3)

EXAMPLE 8   Solve by substitution.

$$2x + 3y = 5$$
$$x - 2y = 6$$

SOLUTION   To use the substitution method, we must solve one of the two equations for x or y. We can solve for x in the second equation by adding $2y$ to both sides.

$$x - 2y = 6$$
$$x = 2y + 6 \quad \text{Add } 2y \text{ to both sides.}$$

Substituting the expression $2y + 6$ for x in the first equation of our system, we have

$$2(2y + 6) + 3y = 5$$
$$4y + 12 + 3y = 5$$
$$7y + 12 = 5$$
$$7y = -7$$
$$y = -1$$

Using $y = -1$ in either equation in the original system, we get $x = 4$. The solution is (4, −1). ∎

Problem Set 4.1

Moving Toward Success

"When one has a great deal to put into it, a day has a hundred pockets."

—Friedrich Nietzsche, 1844–1900, German philosopher

1. How would keeping a calendar of your daily schedule help you lay out a successful study plan?

2. Why should you keep a daily to do list and write quiz and exam dates on your calendar?

A Solve each system by graphing both equations on the same set of axes and then reading the solution from the graph.

1. $3x - 2y = 6$
   $x - y = 1$

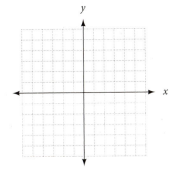

2. $5x - 2y = 10$
   $x - y = -1$

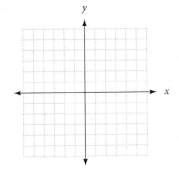

3. $y = \dfrac{3}{5}x - 3$
   $2x - y = -4$

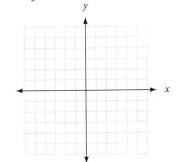

4. $y = \dfrac{1}{2}x - 2$
   $2x - y = -1$

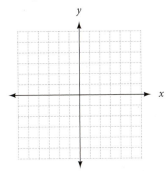

5. $y = \dfrac{1}{2}x$
   $y = -\dfrac{3}{4}x + 5$

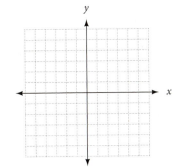

6. $y = \dfrac{2}{3}x$
   $y = -\dfrac{1}{3}x + 6$

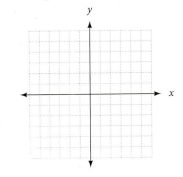

7. $3x + 3y = -2$
   $y = -x + 4$

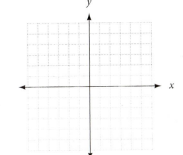

8. $2x - 2y = 6$
   $y = x - 3$

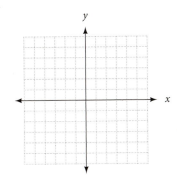

9. $2x - y = 5$
   $y = 2x - 5$

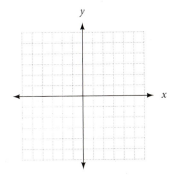

10. $x + 2y = 5$

$y = -\dfrac{1}{2}x + 3$

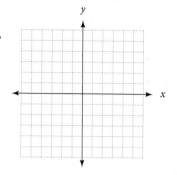

B Solve each of the following systems by the addition method. [Examples 1–6]

11. $x + y = 5$
$3x - y = 3$

12. $x - y = 4$
$-x + 2y = -3$

13. $3x + y = 4$
$4x + y = 5$

14. $6x - 2y = -10$
$6x + 3y = -15$

15. $3x - 2y = 6$
$6x - 4y = 12$

16. $4x + 5y = -3$
$-8x - 10y = 3$

17. $x + 2y = 0$
$2x - 6y = 5$

18. $x + 3y = 3$
$2x - 9y = 1$

19. $2x - 5y = 16$
$4x - 3y = 11$

20. $5x - 3y = -11$
$7x + 6y = -12$

21. $6x + 3y = -1$
$9x + 5y = 1$

22. $5x + 4y = -1$
$7x + 6y = -2$

23. $4x + 3y = 14$
$9x - 2y = 14$

24. $7x - 6y = 13$
$6x - 5y = 11$

25. $2x - 5y = 3$
$-4x + 10y = 3$

26. $3x - 2y = 1$
$-6x + 4y = -2$

27. $\dfrac{1}{4}x - \dfrac{1}{6}y = -2$

$-\dfrac{1}{6}x + \dfrac{1}{5}y = 4$

28. $-\dfrac{1}{3}x + \dfrac{1}{4}y = 0$

$\dfrac{1}{5}x - \dfrac{1}{10}y = 1$

29. $\dfrac{1}{2}x + \dfrac{1}{3}y = 13$

$\dfrac{2}{5}x + \dfrac{1}{4}y = 10$

30. $\dfrac{1}{2}x + \dfrac{1}{3}y = \dfrac{2}{3}$

$\dfrac{2}{3}x + \dfrac{2}{5}y = \dfrac{14}{15}$

C Solve each of the following systems by the substitution method. [Examples 7–8]

31. $7x - y = 24$
$x = 2y + 9$

32. $3x - y = -8$
$y = 6x + 3$

33. $6x - y = 10$
$y = -\dfrac{3}{4}x - 1$

34. $2x - y = 6$
$y = -\dfrac{4}{3}x + 1$

35. $3y + 4z = 23$
$6y + z = 32$

36. $2x - y = 650$
$3.5x - y = 1,400$

37. $y = 3x - 2$
$y = 4x - 4$

38. $y = 5x - 2$
$y = -2x + 5$

39. $2x - y = 5$
$4x - 2y = 10$

40. $-10x + 8y = -6$
$y = \dfrac{5}{4}x$

41. $\dfrac{1}{3}x - \dfrac{1}{2}y = 0$
$x = \dfrac{3}{2}y$

42. $\dfrac{2}{5}x - \dfrac{2}{3}y = 0$
$y = \dfrac{3}{5}x$

You may want to read Example 3 again before solving the systems that follow.

43. $4x - 7y = 3$
$5x + 2y = -3$

44. $3x - 4y = 7$
$6x - 3y = 5$

45. $9x - 8y = 4$
$2x + 3y = 6$

46. $4x - 7y = 10$
$-3x + 2y = -9$

47. $3x - 5y = 2$
$7x + 2y = 1$

48. $4x - 3y = -1$
$5x + 8y = 2$

Solve each of the following systems by using either the addition or substitution method. Choose the method that is most appropriate for the problem.

49. $x - 3y = 7$
$2x + y = -6$

50. $2x - y = 9$
$x + 2y = -11$

51. $y = \dfrac{1}{2}x + \dfrac{1}{3}$
$y = -\dfrac{1}{3}x + 2$

52. $y = \dfrac{3}{4}x - \dfrac{4}{5}$
$y = \dfrac{1}{2}x - \dfrac{1}{2}$

53. $3x - 4y = 12$
$x = \dfrac{2}{3}y - 4$

54. $-5x + 3y = -15$
$x = \dfrac{4}{5}y - 2$

55. $4x - 3y = -7$
$-8x + 6y = -11$

56. $3x - 4y = 8$
$y = \dfrac{3}{4}x - 2$

57. $3y + z = 17$
$5y + 20z = 65$

58. $x + y = 850$
$1.5x + y = 1,100$

59. $\dfrac{3}{4}x - \dfrac{1}{3}y = 1$
$y = \dfrac{1}{4}x$

60. $-\dfrac{2}{3}x - \dfrac{1}{2}y = -1$
$y = -\dfrac{1}{3}x$

61. $\dfrac{1}{4}x - \dfrac{1}{2}y = \dfrac{1}{3}$
$\dfrac{1}{3}x - \dfrac{1}{4}y = \dfrac{2}{3}$

62. $\dfrac{1}{5}x - \dfrac{1}{10}y = -\dfrac{1}{3}$
$\dfrac{2}{3}x - \dfrac{1}{2}y = -\dfrac{1}{6}$

63. $\dfrac{3}{4}x + \dfrac{1}{3}y = 2$
$\dfrac{1}{2}x - y = 0$

64. $\dfrac{5}{6}x - \dfrac{1}{2}y = 4$
$x - \dfrac{1}{3}y = 2$

65. $\dfrac{3}{5}x - \dfrac{1}{2}y = \dfrac{7}{10}$
$\dfrac{1}{6}x - \dfrac{2}{3}y = -\dfrac{1}{2}$

66. $\dfrac{5}{3}x - \dfrac{1}{2}y = -\dfrac{4}{3}$
$\dfrac{1}{2}x + \dfrac{3}{4}y = \dfrac{1}{4}$

The next two problems are intended to give you practice reading, and paying attention to, the instructions that accompany the problems you are working.

67. Work each problem according to the instructions.

   a. Simplify: $(3x - 4y) - 3(x - y)$.

   b. Find y when x is 0 in $3x - 4y = 8$.

   c. Find the y-intercept: $3x - 4y = 8$

   d. Graph: $3x - 4y = 8$

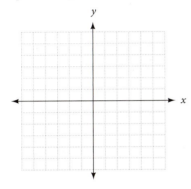

   e. Find the point where the graphs of $3x - 4y = 8$ and $x - y = 2$ cross.

68. Work each problem according to the instructions.

   a. Solve: $4x - 5 = 20$.

   b. Solve for y: $4x - 5y = 20$.

   c. Solve for x: $x - y = 5$.

   d. Solve: $4x - 5y = 20$
             $x - y = 5$

69. Multiply both sides of the second equation in the following system by 100, and then solve as usual.

$$x + y = 10{,}000$$
$$0.06x + 0.05y = 560$$

70. Multiply both sides of the second equation in the following system by 10, and then solve as usual.

$$x + y = 12$$
$$0.20x + 0.50y = 0.30(12)$$

71. What value of c will make the following system a dependent system (one in which the lines coincide)?

$$6x - 9y = 3$$
$$4x - 6y = c$$

72. What value of c will make the following system a dependent system?

$$5x - 7y = c$$
$$-15x + 21y = 9$$

73. Where do the graphs of the lines $x + y = 4$ and $x - 2y = 4$ intersect?

74. Where do the graphs of the line $x = -1$ and $x - 2y = 4$ intersect?

75. Lunch Suppose you and your friend make the order mentioned in the section opener. If you ordered 2 burritos and 3 tacos for $12, and your friend ordered 1 burrito and 4 tacos for $11, how much does each taco cost? How much does each burrito cost?

76. Lunch You and a friend are at lunch. You order 3 mini hamburgers and 1 side for $9. Your friend orders 1 mini hamburger and 2 sides for $8. What would be the cost for 1 mini hamburger? How much would 1 side cost?

Maintaining Your Skills

77. Find the slope of the line that contains $(-4, -1)$ and $(-2, 5)$.

78. A line has a slope of $\frac{2}{3}$. Find the slope of any line
   a. parallel to it.

   b. perpendicular to it.

79. Give the slope and y-intercept of the line $2x - 3y = 6$.

80. Give the equation of the line with slope -3 and y-intercept 5.

81. Find the equation of the line with slope $\frac{2}{3}$ that contains the point $(-6, 2)$.

82. Find the equation of the line through $(1, 3)$ and $(-1, -5)$.

83. Find the equation of the line with x-intercept 3 and y-intercept -2.

84. Find the equation of the line through $(-1, 4)$ whose graph is perpendicular to the graph of $y = 2x + 3$.

85. Find the equation of the line through $(-2, 3)$ whose graph is parallel to the graph of $4x - 3y = 6$.

86. Find the equation of the line through $(-3, 3)$ whose graph is perpendicular to the graph of the line through $(1, 1)$ and $(-2, -8)$.

Getting Ready for the Next Section

Simplify.

87. $2 - 2(6)$

88. $2(1) - 2 + 3$

89. $(x + 3y) - 1(x - 2z)$

90. $(x + y + z) + (2x - y + z)$

Solve.

91. $-9y = -9$

92. $30x = 38$

93. $3(1) + 2z = 9$

94. $4\left(\dfrac{19}{15}\right) - 2y = 4$

Apply the distributive property, then simplify if possible.

95. $2(5x - z)$

96. $-1(x - 2z)$

97. $3(3x + y - 2z)$

98. $2(2x - y + z)$

Extending the Concepts

99. Find a and b so that the line $ax + by = 7$ passes through the points $(1, -2)$ and $(3, 1)$.

100. Find a and b so that the line $ax + by = 2$ passes through the points $(2, 2)$ and $(6, 7)$.

101. Find m and b so that the line $y = mx + b$ passes through the points $(-6, 0)$ and $(3, 6)$.

102. Find m and b so that the line $y = mx + b$ passes through the points $(4, 1)$ and $(-2, -8)$.

Systems of Linear Equations in Three Variables

TICKET TO SUCCESS

Keep these questions in mind as you read through the section. Then respond in your own words and in complete sentences.

1. What is an ordered triple of numbers?
2. Explain what it means for (1, 2, 3) to be a solution to a system of linear equations in three variables.
3. Explain in a general way the procedure you would use to solve a system of three linear equations in three variables.
4. How would you know when a system of linear equations in three variables has no single ordered triple for a solution?

Alexey Stiop/Shutterstock.com

In the previous section we were working with systems of equations in two variables. However, imagine you are in charge of running a concession stand selling sodas, hot dogs, and chips for a fundraiser. The price for each soda is $.50, each hot dog is $2.00, and each bag of chips is $1.00. You have not kept track of your inventory, but you know you have sold a total of 85 items for a total of $100.00. You also know that you sold twice as many sodas as hot dogs. How many sodas, hot dogs, and chips did you sell? We will now begin to solve systems of equations with three unknowns like this one.

A Solve Systems in Three Variables

A solution to an equation in three variables, such as

$$2x + y - 3z = 6$$

is an *ordered triple* of numbers (x, y, z). For example, the ordered triples $(0, 0, -2)$, $(2, 2, 0)$, and $(0, 9, 1)$ are solutions to the equation $2x + y - 3z = 6$ since they produce a true statement when their coordinates are substituted for $x, y,$ and z in the equation.

> **Definition**
>
> The **solution set** for a system of three linear equations in three variables is the set of ordered triples that satisfy all three equations.

PRACTICE PROBLEMS

1. Solve the system.
$$x + 2y + z = 2$$
$$x + y - z = 6$$
$$x - y + 2z = -7$$

EXAMPLE 1   Solve the system.

$$x + y + z = 6 \qquad (1)$$
$$2x - y + z = 3 \qquad (2)$$
$$x + 2y - 3z = -4 \qquad (3)$$

SOLUTION   We want to find the ordered triple (x, y, z) that satisfies all three equations. We have numbered the equations so it will be easier to keep track of where they are and what we are doing.

There are many ways to proceed. The main idea is to take two different pairs of equations and eliminate the same variable from each pair. We begin by adding equations (1) and (2) to eliminate the variable y. The resulting equation is numbered (4):

NOTE
Solving a system in three variables can be overwhelming at first. Don't get discouraged. Remember to read the solution's explanation slowly, and you may need to read it more than once to fully grasp each step.

$$x + y + z = 6 \qquad (1)$$
$$\underline{2x - y + z = 3} \qquad (2)$$
$$3x + 2z = 9 \qquad (4)$$

Adding twice equation (2) to equation (3) will also eliminate the variable y. The resulting equation is numbered (5):

$$4x - 2y + 2z = 6 \qquad \text{Twice (2)}$$
$$\underline{x + 2y - 3z = -4} \qquad (3)$$
$$5x - z = 2 \qquad (5)$$

Equations (4) and (5) form a linear system in two variables. By multiplying equation (5) by 2 and adding the result to equation (4), we succeed in eliminating the variable z from the new pair of equations:

$$3x + 2z = 9 \qquad (4)$$
$$\underline{10x - 2z = 4} \qquad \text{Twice (5)}$$
$$13x = 13$$
$$x = 1$$

Substituting $x = 1$ into equation (4), we have

$$3(1) + 2z = 9$$
$$2z = 6$$
$$z = 3$$

Using $x = 1$ and $z = 3$ in equation (1) gives us

$$1 + y + 3 = 6$$
$$y + 4 = 6$$
$$y = 2$$

The solution is the ordered triple $(1, 2, 3)$. ∎

Answer
1. $(1, 2, -3)$

EXAMPLE 2 Solve the system.

$$2x + y - z = 3 \quad (1)$$
$$3x + 4y + z = 6 \quad (2)$$
$$2x - 3y + z = 1 \quad (3)$$

SOLUTION It is easiest to eliminate z from the equations using the addition method. The equation produced by adding (1) and (2) is

$$5x + 5y = 9 \quad (4)$$

The equation that results from adding (1) and (3) is

$$4x - 2y = 4 \quad (5)$$

Equations (4) and (5) form a linear system in two variables. We can eliminate the variable y from this system as follows:

$$
\begin{array}{ll}
& \text{Multiply by 2.} \\
5x + 5y = 9 \xrightarrow{\hspace{2cm}} & 10x + 10y = 18 \\
4x - 2y = 4 \xrightarrow{\hspace{2cm}} & \underline{20x - 10y = 20} \\
& \text{Multiply by 5.} \quad 30x \hspace{1.1cm} = 38
\end{array}
$$

$$x = \frac{38}{30}$$

$$= \frac{19}{15}$$

Substituting $x = \frac{19}{15}$ into equation (5) or equation (4) and solving for y gives

$$y = \frac{8}{15}$$

Using $x = \frac{19}{15}$ and $y = \frac{8}{15}$ in equation (1), (2), or (3) and solving for z results in

$$z = \frac{1}{15}$$

The ordered triple that satisfies all three equations is $\left(\frac{19}{15}, \frac{8}{15}, \frac{1}{15}\right)$. ■

EXAMPLE 3 Solve the system.

$$2x + 3y - z = 5 \quad (1)$$
$$4x + 6y - 2z = 10 \quad (2)$$
$$x - 4y + 3z = 5 \quad (3)$$

SOLUTION Multiplying equation (1) by -2 and adding the result to equation (2) looks like this:

$$
\begin{array}{ll}
-4x - 6y + 2z = -10 & -2 \text{ times (1)} \\
\underline{4x + 6y - 2z = 10} & (2) \\
 0 = 0 &
\end{array}
$$

All three variables have been eliminated, and we are left with a true statement. As was the case in the previous section, this implies that the two equations are dependent. With a system of three equations in three variables, however, a system such as this one can have no solution or an infinite number of solutions. In either case, we have no unique solutions, meaning there is no single ordered triple that is the only solution to the system. ■

2. Solve the system.
$$3x - 2y + z = 2$$
$$3x + y + 3z = 7$$
$$x + 4y - z = 4$$

3. Solve the system.
$$3x + 5y - 2z = 1$$
$$6x + 10y - 4z = 2$$
$$x - 8y + z = 4$$

Answers
2. $(1, 1, 1)$
3. No unique solution; it is a dependent system.

4. Solve the system.
$$3x - y + 2z = 4$$
$$6x - 2y + 4z = 2$$
$$5x - 3y + 7z = 5$$

EXAMPLE 4 Solve the system.

$$x - 5y + 4z = 8 \quad (1)$$
$$3x + y - 2z = 7 \quad (2)$$
$$-9x - 3y + 6z = 5 \quad (3)$$

SOLUTION Multiplying equation (2) by 3 and adding the result to equation (3) produces

$$9x + 3y - 6z = 21 \quad \text{3 times (2)}$$
$$\underline{-9x - 3y + 6z = 5} \quad (3)$$
$$0 = 26$$

In this case, all three variables have been eliminated, and we are left with a false statement. The system is inconsistent: there are no ordered triples that satisfy both equations. The solution set for the system is the empty set, $\varnothing$. If equations (2) and (3) have no ordered triples in common, then certainly (1), (2), and (3) do not either. ■

5. Solve the system.
$$x + 2y = 0$$
$$3y + z = -3$$
$$2x - z = 5$$

EXAMPLE 5 Solve the system.

$$x + 3y = 5 \quad (1)$$
$$6y + z = 12 \quad (2)$$
$$x - 2z = -10 \quad (3)$$

SOLUTION It may be helpful to rewrite the system as

$$x + 3y = 5 \quad (1)$$
$$6y + z = 12 \quad (2)$$
$$x - 2z = -10 \quad (3)$$

Equation (2) does not contain the variable x. If we multiply equation (3) by -1 and add the result to equation (1), we will be left with another equation that does not contain the variable x.

$$x + 3y = 5 \quad (1)$$
$$\underline{-x + 2z = 10} \quad \text{-1 times (3)}$$
$$3y + 2z = 15 \quad (4)$$

Equations (2) and (4) form a linear system in two variables. Multiplying equation (2) by -2 and adding the result to equation (4) eliminates the variable z.

$$\text{Multiply by -2.}$$

$$6y + z = 12 \longrightarrow -12y - 2z = -24$$
$$3y + 2z = 15 \longrightarrow \underline{3y + 2z = 15}$$
$$\text{No change} \qquad -9y = -9$$
$$y = 1$$

Using $y = 1$ in equation (4) and solving for z, we have

$$z = 6$$

Substituting $y = 1$ into equation (1) gives

$$x = 2$$

The ordered triple that satisfies all three equations is $(2, 1, 6)$. ■

Answers
4. No solution; it is an inconsistent system.
5. $(4, -2, 3)$

The Geometry Behind Equations in Three Variables

We can graph an ordered triple on a coordinate system with three axes. The graph will be a point in space. The coordinate system is drawn in perspective; you have to imagine that the x-axis comes out of the paper and is perpendicular to both the y-axis and the z-axis. To graph the point $(3, 4, 5)$, we move 3 units in the x-direction, 4 units in the y-direction, and then 5 units in the z-direction, as shown in Figure 1.

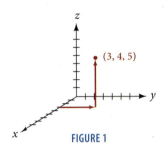

FIGURE 1

Although in actual practice, it is sometimes difficult to graph equations in three variables, if we have to graph a linear equation in three variables, we would find that the graph was a plane in space. A system of three equations in three variables is represented by three planes in space.

There are a number of possible ways in which these three planes can intersect, some of which are shown in the margin on this page. There are still other possibilities that are not among those shown in the margin.

In Example 3, we found that equations 1 and 2 were dependent equations. They represent the same plane; that is, they have all their points in common. But the system of equations that they came from has either no solution or an infinite number of solutions. It all depends on the third plane. If the third plane coincides with the first two, then the solution to the system is a plane. If the third plane is distinct from but parallel to the first two, then there is no solution to the system. And, finally, if the third plane intersects the first two, but does not coincide with them, then the solution to the system is that line of intersection.

In Example 4, we found that trying to eliminate a variable from the second and third equations resulted in a false statement. This means that the two planes represented by these equations are parallel. It makes no difference where the third plane is; there is no solution to the system in Example 4. (If we were to graph the three planes from Example 4, we would obtain a diagram similar to Case 4 in the margin.)

If, in the process of solving a system of linear equations in three variables, we eliminate all the variables from a pair of equations and are left with a false statement, we will say the system is inconsistent. If we eliminate all the variables and are left with a true statement, then we will say the system has no unique solution.

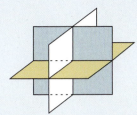

Case 1 The three planes have exactly one point in common. In this case we get one solution to our system, as in Examples 1, 2, and 5.

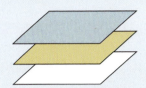

Case 2 The three planes have no points in common because they are all parallel to each other. The system they represent is an inconsistent system.

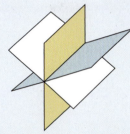

Case 3 The three planes intersect in a line. Any point on the line is a solution to the system.

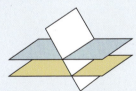

Case 4 In this case the three planes have no points in common. There is no solution to the system; it is an inconsistent system.

Problem Set 4.2

Moving Toward Success

"Even a mistake may turn out to be the one thing necessary to a worthwhile achievement."

—Henry Ford, 1863–1947, American industrialist and founder of Ford Motor Company

1. Why is making mistakes important to the process of learning mathematics?

2. What will you do if you notice you are repeatedly making mistakes on certain types of problems?

A Solve the following systems. [Examples 1–5]

1. $x + y + z = 4$
$x - y + 2z = 1$
$x - y - 3z = -4$

2. $x - y - 2z = -1$
$x + y + z = 6$
$x + y - z = 4$

3. $x + y + z = 6$
$x - y + 2z = 7$
$2x - y - 4z = -9$

4. $x + y + z = 0$
$x + y - z = 6$
$x - y + 2z = -7$

5. $x + 2y + z = 3$
$2x - y + 2z = 6$
$3x + y - z = 5$

6. $2x + y - 3z = -14$
$x - 3y + 4z = 22$
$3x + 2y + z = 0$

7. $2x + 3y - 2z = 4$
$x + 3y - 3z = 4$
$3x - 6y + z = -3$

8. $4x + y - 2z = 0$
$2x - 3y + 3z = 9$
$-6x - 2y + z = 0$

9. $-x + 4y - 3z = 2$
$2x - 8y + 6z = 1$
$3x - y + z = 0$

10. $4x + 6y - 8z = 1$
$-6x - 9y + 12z = 0$
$x - 2y - 2z = 3$

11. $\frac{1}{2}x - y + z = 0$
$2x + \frac{1}{3}y + z = 2$
$x + y + z = -4$

12. $\frac{1}{3}x + \frac{1}{2}y + z = -1$
$x - y + \frac{1}{5}z = 1$
$x + y + z = 5$

13. $2x - y - 3z = 1$
$x + 2y + 4z = 3$
$4x - 2y - 6z = 2$

14. $3x + 2y + z = 3$
$x - 3y + z = 4$
$-6x - 4y - 2z = 1$

15. $2x - y + 3z = 4$
$x + 2y - z = -3$
$4x + 3y + 2z = -5$

16. $6x - 2y + z = 5$
$3x + y + 3z = 7$
$x + 4y - z = 4$

17. $x + y = 9$
$y + z = 7$
$x - z = 2$

18. $x - y = -3$
$x + z = 2$
$y - z = 7$

19. $2x + y = 2$
$y + z = 3$
$4x - z = 0$

20. $2x + y = 6$
$3y - 2z = -8$
$x + z = 5$

21. $2x - 3y = 0$
$6y - 4z = 1$
$x + 2z = 1$

22. $3x + 2y = 3$
$y + 2z = 2$
$6x - 4z = 1$

23. $x + y - z = 2$
$2x + y + 3z = 4$
$x - 2y + 2z = 6$

24. $x + 2y - 2z = 4$
$3x + 4y - z = -2$
$2x + 3y - 3z = -5$

25. $2x + 3y = -\dfrac{1}{2}$

$4x + 8z = 2$

$3y + 2z = -\dfrac{3}{4}$

26. $3x - 5y = 2$

$4x + 6z = \dfrac{1}{3}$

$5y - 7z = \dfrac{1}{6}$

27. $\dfrac{1}{3}x + \dfrac{1}{2}y - \dfrac{1}{6}z = 4$

$\dfrac{1}{4}x - \dfrac{3}{4}y + \dfrac{1}{2}z = \dfrac{3}{2}$

$\dfrac{1}{2}x - \dfrac{2}{3}y - \dfrac{1}{4}z = -\dfrac{16}{3}$

28. $-\dfrac{1}{4}x + \dfrac{3}{8}y + \dfrac{1}{2}z = -1$

$\dfrac{2}{3}x - \dfrac{1}{6}y - \dfrac{1}{2}z = 2$

$\dfrac{3}{4}x - \dfrac{1}{2}y - \dfrac{1}{8}z = 1$

29. $x - \dfrac{1}{2}y - \dfrac{1}{3}z = -\dfrac{4}{3}$

$\dfrac{1}{3}x + \ y - \dfrac{1}{2}z = 5$

$-\dfrac{1}{4}x + \dfrac{2}{3}y - z = -\dfrac{3}{4}$

30. $x + \dfrac{1}{3}y - \dfrac{1}{2}z = -\dfrac{3}{2}$

$\dfrac{1}{2}x - y + \dfrac{1}{3}z = 8$

$\dfrac{1}{3}x - \dfrac{1}{4}y - z = -\dfrac{5}{6}$

31. $\dfrac{1}{2}x + \dfrac{2}{3}y = \dfrac{5}{2}$

$\dfrac{1}{5}x - \dfrac{1}{2}z = -\dfrac{3}{10}$

$\dfrac{1}{3}y - \dfrac{1}{4}z = \dfrac{3}{4}$

32. $\dfrac{1}{2}x - \dfrac{1}{3}y = \dfrac{1}{6}$

$\dfrac{1}{3}y - \dfrac{1}{3}z = 1$

$\dfrac{1}{5}x - \dfrac{1}{2}z = -\dfrac{4}{5}$

33. $\dfrac{1}{2}x - \dfrac{1}{4}y + \dfrac{1}{2}z = -2$

$\dfrac{1}{4}x - \dfrac{1}{12}y - \dfrac{1}{3}z = \dfrac{1}{4}$

$\dfrac{1}{6}x + \dfrac{1}{3}y - \dfrac{1}{2}z = \dfrac{3}{2}$

34. $\dfrac{1}{2}x + \dfrac{1}{2}y + z = \dfrac{1}{2}$

$\dfrac{1}{2}x - \dfrac{1}{4}y - \dfrac{1}{4}z = 0$

$\dfrac{1}{4}x + \dfrac{1}{12}y + \dfrac{1}{6}z = \dfrac{1}{6}$

Applying the Concepts

35. Electric Current In the following diagram of an electrical circuit, x, y, and z represent the amount of current (in amperes) flowing across the 5-ohm, 20-ohm, and 10-ohm resistors, respectively. (In circuit diagrams resistors are represented by ‑∧∧∧‑ and potential differences by ‑|⊢.)

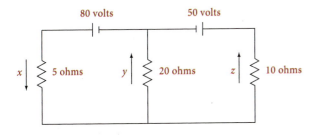

The system of equations used to find the three currents x, y, and z is

$$x - y - z = 0$$
$$5x + 20y = 80$$
$$20y - 10z = 50$$

Solve the system for all variables.

36. Electric Current In the following diagram of an electrical circuit, x, y, and z represent the amount of current (in amperes) flowing across the 10-ohm, 15-ohm, and 5-ohm resistors, respectively.

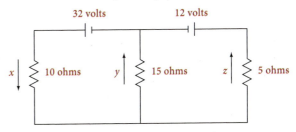

The system of equations used to find the three currents x, y, and z is

$$x - y - z = 0$$
$$10x + 15y = 32$$
$$15y - 5z = 12$$

Solve the system for all variables.

Snacks A concession stand sells sodas for $0.50, hot dogs for $2.00, and chips for $1.00. Use this information to answer the following questions.

37. If 85 items were sold for a total of $100 with the same number of sodas sold as hot dogs, then how many of each item were sold?

38. If 95 items were sold for a total of $110 with twice as many sodas sold as chips, then how many of each item were sold?

Getting Ready for the Next Section

Simplify.

39. $1(4) - 3(2)$

40. $3(7) - (-2)(5)$

41. $1(1) - 3(-2) + (-2)(-2)$

42. $-4(0)(-2) - (-1)(1)(1) - 1(2)(3)$

43. $-3(-1 - 1) + 4(-2 + 2) - 5[2 - (-2)]$

44. $12 + 4 - (-1) - 6$

Solve.

45. $-5x = 20$

46. $4x - 2x = 8$

Maintaining Your Skills

47. If y varies directly with the square of x, and y is 75 when x is 5, find y when x is 7.

48. Suppose y varies directly with the cube of x. If y is 16 when x is 2, find y when x is 3.

49. Suppose y varies inversely with x. If y is 10 when x is 25, find x when y is 5.

50. If y varies inversely with the cube of x, and y is 2 when x is 2, find y when x is 4.

51. Suppose z varies jointly with x and the square of y. If z is 40 when x is 5 and y is 2, find z when x is 2 and y is 5.

52. Suppose z varies jointly with x and the cube of y. If z is 48 when x is 3 and y is 2, find z when x is 4 and y is $\frac{1}{2}$.

Extending the Concepts

Solve each system for the solution (x, y, z, w).

53.
$$x + y + z + w = 10$$
$$x + 2y - z + w = 6$$
$$x - y - z + 2w = 4$$
$$x - 2y + z - 3w = -12$$

54.
$$x + y + z + w = 16$$
$$x - y + 2z - w = 1$$
$$x + 3y - z - w = -2$$
$$x - 3y - 2z + 2w = -4$$

Applications of Linear Systems

OBJECTIVES

A Solve application problems whose solutions are found through systems of linear equations.

TICKET TO SUCCESS

Keep these questions in mind as you read through the section. Then respond in your own words and in complete sentences.

1. To apply the Blueprint for Problem Solving to the examples in this section, what is the first step?

2. When would you write a system of equations while working a problem in this section using the Blueprint for Problem Solving?

3. When working application problems involving boats moving in rivers, how does the current of the river affect the speed of the boat?

4. Write an application problem for which the solution depends on solving the system

$$x + y = 1,000$$
$$0.05x + 0.06y = 55$$

Dmitry Kalinovsky/Shutterstock.comt

Suppose you want to be able to quantify your weekly driving habits. You put 25 gallons of gas in your car, then drive 425 miles in a week, using all the gas. You know your car gets 25 miles per gallon on the highway and 15 miles per gallon in the city. How many gallons were used in the highway and how many were used in the city? In this section, we will put our knowledge of systems of equations to use solving different kinds of applications.

Many times word problems involve more than one unknown quantity. If a problem is stated in terms of two unknowns and we represent each unknown quantity with a different variable, then we must write the relationships between the variables with two equations. The two equations written in terms of the two variables form a system of linear equations that we solve using the methods developed in this chapter. If we find a problem that relates three unknown quantities, then we need three equations to form a linear system we can solve.

Here is our Blueprint for Problem Solving, modified to fit the application problems that you will find in this section.

A Application Problems

> **Strategy**   **Problem Solving Using a System of Equations**
>
> **Step 1** **Read** the problem, and then mentally **list** the items that are known and the items that are unknown.
>
> **Step 2** **Assign variables** to each of the unknown items; that is, let x = one of the unknown items and y = the other unknown item (and z = the third unknown item, if there is a third one). Then **translate** the other **information** in the problem to expressions involving the two (or three) variables.
>
> **Step 3** **Reread** the problem, and then **write a system of equations,** using the items and variables listed in steps 1 and 2, that describes the situation.
>
> **Step 4** **Solve the system** found in step 3.
>
> **Step 5** **Write your answers** using complete sentences.
>
> **Step 6** **Reread** the problem, and **check** your solution with the original words in the problem.

PRACTICE PROBLEMS

1. A number is 1 less than twice another. Their sum is 14. Find the two numbers.

EXAMPLE 1   One number is 2 more than 3 times another. Their sum is 26. Find the two numbers.

SOLUTION   Applying the steps from our Blueprint, we have

Step 1:   **Read and list.**
We know that we have two numbers, whose sum is 26. One of them is 2 more than 3 times the other. The unknown quantities are the two numbers.

Step 2:   **Assign variables and translate information.**
Let x = one of the numbers and y = the other number.

Step 3:   **Write a system of equations.**
The first sentence in the problem translates into $y = 3x + 2$. The second sentence gives us a second equation: $x + y = 26$. Together, these two equations give us the following system of equations:

$$x + y = 26$$
$$y = 3x + 2$$

Step 4:   **Solve the system.**
Substituting the expression for y from the second equation into the first and solving for x yields

$$x + (3x + 2) = 26$$
$$4x + 2 = 26$$
$$4x = 24$$
$$x = 6$$

Using $x = 6$ in $y = 3x + 2$ gives the second number.

$$y = 3(6) + 2$$
$$y = 20$$

Answers

1. 5, 9

Step 5: ***Write answers.***
The two numbers are 6 and 20.

Step 6: ***Reread and check.***
The sum of 6 and 20 is 26, and 20 is 2 more than 3 times 6.

∎

EXAMPLE 2   Suppose 850 tickets were sold for a game for a total of $1,100. If adult tickets cost $1.50 and children's tickets cost $1.00, how many of each kind of ticket were sold?

SOLUTION   **Step 1:** ***Read and list.***
The total number of tickets sold is 850. The total income from tickets is $1,100. Adult tickets are $1.50 each. Children's tickets are $1.00 each. We don't know how many of each type of ticket have been sold.

Step 2: ***Assign variables and translate information.***
We let x = the number of adult tickets and y = the number of children's tickets.

Step 3: ***Write a system of equations.***
The total number of tickets sold is 850, giving us our first equation.

$$x + y = 850$$

Because each adult ticket costs $1.50 and each children's ticket costs $1.00 and the total amount of money paid for tickets was $1,100, a second equation is

$$1.50x + 1.00y = 1,100$$

The same information can also be obtained by summarizing the problem with a table. One such table follows. Notice that the two equations we obtained previously are given by the two rows of the table.

	Adult Tickets	Children's Tickets	Total
Number	x	y	850
Value	1.50x	1.00y	1,100

Whether we use a table to summarize the information in the problem or just talk our way through the problem, the system of equations that describes the situation is

$$x + y = 850$$

$$1.50x + 1.00y = 1,100$$

Step 4: ***Solve the system.***
If we multiply the second equation by 10 to clear it of decimals, we have the system

$$x + y = 850$$
$$15x + 10y = 11,000$$

Multiplying the first equation by -10 and adding the result to the second equation eliminates the variable y from the system.

2. There were 750 tickets sold for a basketball game for a total of $1,090. If adult tickets cost $2.00 and children's tickets cost $1.00, how many of each kind were sold?

Answers
2. 340 adult tickets;
   410 children's tickets

$$-10x - 10y = -8,500$$
$$\underline{15x + 10y = 11,000}$$
$$5x \quad\quad = 2,500$$
$$x = 500$$

The number of adult tickets sold was 500. To find the number of children's tickets, we substitute $x = 500$ into $x + y = 850$ to get

$$500 + y = 850$$
$$y = 350$$

Step 5: **Write answers.**

The number of children's tickets is 350, and the number of adult tickets is 500.

Step 6: **Reread and check.**

The total number of tickets is $350 + 500 = 850$. The amount of money from selling the two types of tickets is

350 children's tickets at $1.00 each is $350(1.00) = \$350$
500 adult tickets at $1.50 each is $500(1.50) = \$750$

The total income from ticket sales is $1,100.

■

EXAMPLE 3 Suppose a person invests a total of $10,000 in two accounts. One account earns 8% annually, and the other earns 9% annually. If the total interest earned from both accounts in a year is $860, how much was invested in each account?

SOLUTION **Step 1:** **Read and list.**

The total investment is $10,000 split between two accounts. One account earns 8% annually, and the other earns 9% annually. The interest from both accounts is $860 in 1 year. We don't know how much is in each account.

Step 2: **Assign variables and translate information.**

We let x equal the amount invested at 9% and y be the amount invested at 8%.

Step 3: **Write a system of equations.**

Because the total investment is $10,000, one relationship between x and y can be written as

$$x + y = 10,000$$

The total interest earned from both accounts is $860. The amount of interest earned on x dollars at 9% is $0.09x$, whereas the amount of interest earned on y dollars at 8% is $0.08y$. This relationship is represented by the equation

$$0.09x + 0.08y = 860$$

The two equations we have just written can also be found by first summarizing the information from the problem in a table. Again, the two rows of the table yield the two equations just written. Here is the table:

	Dollars at 9%	Dollars at 8%	Total
Number	x	y	10,000
Interest	0.09x	0.08y	860

3. A person invests $12,000 in two accounts. One account earns 6% annually and the other earns 7%. If the total interest in a year is $790, how much was invested in each account?

The system of equations that describes this situation is given by

$$x + y = 10{,}000$$
$$0.09x + 0.08y = 860$$

Step 4: Solve the system.

Multiplying the second equation by 100 will clear it of decimals. The system that results after doing so is

$$x + y = 10{,}000$$
$$9x + 8y = 86{,}000$$

We can eliminate y from this system by multiplying the first equation by -8 and adding the result to the second equation.

$$-8x - 8y = -80{,}000$$
$$\underline{9x + 8y = 86{,}000}$$
$$x = 6{,}000$$

The amount of money invested at 9% is $6,000. Because the total investment was $10,000, the amount invested at 8% must be $4,000.

Step 5: Write answers.

The amount invested at 8% is $4,000, and the amount invested at 9% is $6,000.

Step 6: Reread and check.

The total investment is $4,000 + $6,000 = $10,000. The amount of interest earned from the two accounts is

In 1 year, $4,000 invested at 8% earns $0.08(4{,}000) = \$320$

$\underline{\text{In 1 year, }\$6{,}000\text{ invested at 9\% earns }0.09(6{,}000) = \$540}$

The total interest from the two accounts is $860.

■

EXAMPLE 4 How much 20% alcohol solution and 50% alcohol solution must be mixed to get 12 gallons of 30% alcohol solution?

SOLUTION To solve this problem, we must first understand that a 20% alcohol solution is 20% alcohol and 80% water.

Step 1: Read and list.

We will mix two solutions to obtain 12 gallons of solution that is 30% alcohol. One of the solutions is 20% alcohol and the other is 50% alcohol. We don't know how much of each solution we need.

Step 2: Assign variables and translate information.

Let x = the number of gallons of 20% alcohol solution needed and y = the number of gallons of 50% alcohol solution needed.

Step 3: Write a system of equations.

Because we must end up with a total of 12 gallons of solution, one equation for the system is

$$x + y = 12$$

4. How much 30% alcohol solution and 70% alcohol solution must be mixed to get 16 gallons of 60% solution?

The amount of alcohol in the x gallons of 20% solution is $0.20x$, whereas the amount of alcohol in the y gallons of 50% solution is $0.50y$. Because the total amount of alcohol in the 20% and 50% solutions must add up to the amount of alcohol in the 12 gallons of 30% solution, the second equation in our system can be written as

$$0.20x + 0.50y = 0.30(12)$$

Again, let's make a table that summarizes the information we have to this point in the problem.

	20% Solution	50% Solution	Final Solution
Total number of gallons	x	y	12
Gallons of alcohol	$0.20x$	$0.50y$	$0.30(12)$

Our system of equations is

$$x + y = 12$$
$$0.20x + 0.50y = 0.30(12) = 3.6$$

Step 4:   Solve the system.

Multiplying the second equation by 10 gives us an equivalent system.

$$x + y = 12$$
$$2x + 5y = 36$$

Multiplying the top equation by -2 to eliminate the x-variable, we have

$$\begin{array}{rcl} -2x - 2y &=& -24 \\ 2x + 5y &=& 36 \\ \hline 3y &=& 12 \\ y &=& 4 \end{array}$$

Substituting $y = 4$ into $x + y = 12$, we solve for x.

$$x + 4 = 12$$
$$x = 8$$

Step 5:   Write answers.

It takes 8 gallons of 20% alcohol solution and 4 gallons of 50% alcohol solution to produce 12 gallons of 30% alcohol solution.

Step 6:   Reread and check.

If we mix 8 gallons of 20% solution and 4 gallons of 50% solution, we end up with a total of 12 gallons of solution. To check the percentages, we look for the total amount of alcohol in the two initial solutions and in the final solution.

In the initial solutions

The amount of alcohol in 8 gallons of 20% solution is $0.20(8) = 1.6$ gallons

The amount of alcohol in 4 gallons of 50% solution is $0.50(4) = 2.0$ gallons

The total amount of alcohol in the initial solutions is 3.6 gallons.

In the final solution

The amount of alcohol in 12 gallons of 30% solution is $0.30(12) = 3.6$ gallons.

∎

EXAMPLE 5 It takes 2 hours for a boat to travel 28 miles downstream (with the current). The same boat can travel 18 miles upstream (against the current) in 3 hours. What is the speed of the boat in still water, and what is the speed of the current of the river?

Current

SOLUTION

Step 1: Read and list.
A boat travels 18 miles upstream and 28 miles downstream. The trip upstream takes 3 hours. The trip downstream takes 2 hours. We don't know the speed of the boat or the speed of the current.

Step 2: Assign variables and translate information.
Let x = the speed of the boat in still water and let y = the speed of the current. The average speed (rate) of the boat upstream is $x - y$ because it is traveling against the current. The rate of the boat downstream is $x + y$ because the boat is traveling with the current.

Step 3: Write a system of equations.
Putting the information into a table, we have

	d (distance, miles)	r (rate, mph)	t (time, h)
Upstream	18	$x - y$	3
Downstream	28	$x + y$	2

The formula for the relationship between distance d, rate r, and time t is $d = rt$ (the rate equation). Because $d = r \cdot t$, the system we need to solve the problem is

$$18 = (x - y) \cdot 3$$
$$28 = (x + y) \cdot 2$$

which is equivalent to

$$6 = x - y$$
$$14 = x + y$$

Step 4: Solve the system.
Adding the two equations, we have

$$20 = 2x$$
$$x = 10$$

Substituting $x = 10$ into $14 = x + y$, we see that

$$y = 4$$

Step 5: Write answers.
The speed of the boat in still water is 10 miles per hour; the speed of the current is 4 miles per hour.

Step 6: Reread and check.
The boat travels at $10 + 4 = 14$ miles per hour downstream, so in 2 hours it will travel $14 \cdot 2 = 28$ miles. The boat travels at $10 - 4 = 6$ miles per hour upstream, so in 3 hours it will travel $6 \cdot 3 = 18$ miles. ■

5. A boat can travel 20 miles downstream in 2 hours. The same boat can travel 18 miles upstream in 3 hours. What is the speed of the boat in still water, and what is the speed of the current?

Answers
5. Boat 8 miles per hour; current 2 miles per hour

6. A collection of nickels, dimes, and quarters consists of 15 coins with a total value of $1.10. If the number of nickels is one less than 4 times the number of dimes, how many of each coin is contained in the collection?

EXAMPLE 6 A coin collection consists of 14 coins with a total value of $1.35. If the coins are nickels, dimes, and quarters, and the number of nickels is 3 less than twice the number of dimes, how many of each coin is there in the collection?

SOLUTION This problem will require three variables and three equations.

Step 1: Read and list.

We have 14 coins with a total value of $1.35. The coins are nickels, dimes, and quarters. The number of nickels is 3 less than twice the number of dimes. We do not know how many of each coin we have.

Step 2: Assign variables and translate information.

Since we have three types of coins, we will have to use three variables. Let's let x = the number of nickels, y = the number of dimes, and z = the number of quarters.

Step 3: Write a system of equations.

Because the total number of coins is 14, our first equation is

$$x + y + z = 14$$

Because the number of nickels is 3 less than twice the number of dimes, a second equation is

$$x = 2y - 3 \qquad \text{which is equivalent to} \qquad x - 2y = -3$$

Our last equation is obtained by considering the value of each coin and the total value of the collection. Let's write the equation in terms of cents, so we won't have to clear it of decimals later.

$$5x + 10y + 25z = 135$$

Here is our system, with the equations numbered for reference:

$$
\begin{array}{rl}
x + y + z = 14 & \quad (1) \\
x - 2y = -3 & \quad (2) \\
5x + 10y + 25z = 135 & \quad (3)
\end{array}
$$

Step 4: Solve the system.

Let's begin by eliminating x from the first and second equations and the first and third equations. Adding -1 times the second equation to the first equation gives us an equation in only y and z. We call this equation (4).

$$3y + z = 17 \qquad (4)$$

Adding -5 times equation (1) to equation (3) gives us

$$5y + 20z = 65 \qquad (5)$$

We can eliminate z from equations (4) and (5) by adding -20 times (4) to (5). Here is the result:

$$-55y = -275$$
$$y = 5$$

Substituting $y = 5$ into equation (4) gives us $z = 2$. Substituting $y = 5$ and $z = 2$ into equation (1) gives us $x = 7$.

Answers

6. 11 nickels, 3 dimes, 1 quarter

Step 5: ***Write answers.***

The collection consists of 7 nickels, 5 dimes, and 2 quarters.

Step 6: ***Reread and check.***

The total number of coins is $7 + 5 + 2 = 14$. The number of nickels, 7, is 3 less than twice the number of dimes, 5. To find the total value of the collection, we have

The value of the 7 nickels is $7(0.05) = \$0.35$

The value of the 5 dimes is $5(0.10) = \$0.50$

The value of the 2 quarters is $2(0.25) = \$0.50$

The total value of the collection is $1.35.

∎

If you go on to take a chemistry class, you may see the next example (or one much like it).

EXAMPLE 7   In a chemistry lab, students record the temperature of water at room temperature and find that it is 77° on the Fahrenheit temperature scale and 25° on the Celsius temperature scale. The water is then heated until it boils. The temperature of the boiling water is 212°F and 100°C. Assume that the relationship between the two temperature scales is a linear one. Then, use the preceding data to find the formula that gives the Celsius temperature C in terms of the Fahrenheit temperature F.

7. Suppose the students mentioned in Example 7 take two additional temperature measurements while they are heating the water and find that when $F = 95°$, $C = 35°$, and when $F = 167°$, $C = 75°$. Assume that the relationship between the two temperature scales is linear, and then derive the formula that gives C in terms of F using these new data points.

SOLUTION   The data are summarized in the following table.

CORRESPONDING TEMPERATURES	
In Degrees Fahrenheit	In Degrees Celsius
77	25
212	100

If we assume the relationship is linear, then the formula that relates the two temperature scales can be written in slope-intercept form as

$$C = mF + b$$

Substituting $C = 25$ and $F = 77$ into this formula gives us

$$25 = 77m + b$$

Substituting $C = 100$ and $F = 212$ into the formula yields

$$100 = 212m + b$$

Answer

7. $C = \dfrac{5}{9}F - \dfrac{160}{9}$

Together, the two equations form a system of equations, which we can solve using the addition method.

$$\begin{array}{l}
25 = 77m + b \xrightarrow{\text{Multiply by } -1.} -25 = -77m - b \\
100 = 212m + b \xrightarrow{\phantom{\text{Multiply by } -1.}} \underline{100 = 212m + b} \\
 \text{No change} \quad 75 = 135m
\end{array}$$

$$m = \frac{75}{135} = \frac{5}{9}$$

To find the value of b, we substitute $m = \dfrac{5}{9}$ into $25 = 77m + b$ and solve for b.

$$25 = 77\left(\frac{5}{9}\right) + b$$

$$25 = \frac{385}{9} + b$$

$$b = 25 - \frac{385}{9} = \frac{225}{9} - \frac{385}{9} = -\frac{160}{9}$$

The equation that gives C in terms of F is

$$C = \frac{5}{9}F - \frac{160}{9}$$ ■

USING TECHNOLOGY

Graphing Calculators: Solving a System That Intersects at Exactly One Point

A graphing calculator can be used to solve a system of equations in two variables if the equations intersect at exactly one point. To solve the system shown, we first solve each equation for y. Here is the result:

$$2x - 3y = 4 \quad \text{becomes} \quad y = \frac{4 - 2x}{-3}$$

$$4x + 5y = 3 \quad \text{becomes} \quad y = \frac{3 - 4x}{5}$$

Graphing these two functions on the calculator gives a diagram similar to the one in Figure 1.

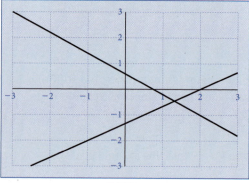

FIGURE 1

Using the Trace and Zoom features, we find that the two lines intersect at $x = 1.32$ and $y = -0.45$.

Problem Set 4.3

Moving Toward Success

"Quality means doing it right when no one is looking."

—Henry Ford, 1863–1947, American industrialist
and founder of Ford Motor Company

1. Why is it important to show all your work when you solve problems?
2. How do you make sure your answers are correct and that you fully understand how you calculated them?

Number Problems

1. One number is 3 more than twice another. The sum of the numbers is 18. Find the two numbers.

2. The sum of two numbers is 32. One of the numbers is 4 less than 5 times the other. Find the two numbers.

3. The difference of two numbers is 6. Twice the smaller is 4 more than the larger. Find the two numbers.

4. The larger of two numbers is 5 more than twice the smaller. If the smaller is subtracted from the larger, the result is 12. Find the two numbers.

5. The sum of three numbers is 8. Twice the smallest is 2 less than the largest, and the sum of the largest and smallest is 5. Use a linear system in three variables to find the three numbers.

6. The sum of three numbers is 14. The largest is 4 times the smallest, the sum of the smallest and twice the largest is 18. Use a linear system in three variables to find the three numbers.

Ticket and Interest Problems

7. A total of 925 tickets was sold for a game for a total of $1,150. If adult tickets sold for $2.00 and children's tickets sold for $1.00, how many of each kind of ticket were sold?

8. If tickets for a show cost $2.00 for adults and $1.50 for children, how many of each kind of ticket were sold if a total of 300 tickets were sold for $525?

9. Mr. Jones has $20,000 to invest. He invests part at 6% and the rest at 7%. If he earns $1,280 in interest after 1 year, how much did he invest at each rate?

10. A man invests $17,000 in two accounts. One account earns 5% interest per year and the other earns 6.5%. If his total interest after one year is $970, how much did he invest at each rate?

11. Susan invests twice as much money at 7.5% as she does at 6%. If her total interest after a year is $840, how much does she have invested at each rate?

12. A woman earns $1,350 in interest from two accounts in a year. If she has three times as much invested at 7% as she does at 6%, how much does she have in each account?

13. A man invests $2,200 in three accounts that pay 6%, 8%, and 9% in annual interest, respectively. He has three times as much invested at 9% as he does at 6%. If his total interest for the year is $178, how much is invested at each rate?

14. A student has money in three accounts that pay 5%, 7%, and 8% in annual interest. She has three times as much invested at 8% as she does at 5%. If the total amount she has invested is $1,600 and her interest for the year comes to $115, how much money does she have in each account?

15. Martin invests money in three accounts that pay 6%, 8%, and 12%. He has twice as much money invested at 8% than as 6%. If the total amount he has invested is $1,300 and his interest for the year comes to $114, how much money does he have in each account?

16. Albert invests money in three accounts that pay 4%, 10%, an 15%. He has three times as much money invested at 10% than he does at 15%. If the total amount he has invested is $5,550 and his interest for the year comes to $570, how much money does he have in each account?

Mixture Problems

17. How many gallons of 20% alcohol solution and 50% alcohol solution must be mixed to get 9 gallons of 30% alcohol solution?

18. How many ounces of 30% hydrochloric acid solution and 80% hydrochloric acid solution must be mixed to get 10 ounces of 50% hydrochloric acid solution?

19. A mixture of 16% disinfectant solution is to be made from 20% and 14% disinfectant solutions. How much of each solution should be used if 15 gallons of the 16% solution are needed?

20. Paul mixes nuts worth $1.55 per pound with oats worth $1.35 per pound to get 25 pounds of trail mix worth $1.45 per pound. How many pounds of nuts and how many pounds of oats did he use?

21. Metal workers solve systems of equations when forming metal alloys. If a certain metal alloy is 40% copper and another alloy is 60% copper, then a system of equations may be written to determine the amount of each alloy necessary to make 50 pounds of a metal alloy that is 55% copper. Write the system and determine this amount.

22. A chemist has three different acid solutions. The first acid solution contains 20% acid, the second contains 40%, and the third contains 60%. He wants to use all three solutions to obtain a mixture of 60 liters containing 50% acid, using twice as much of the 60% solution as the 40% solution. How many liters of each solution should be used?

Rate Problems

23. It takes about 2 hours to travel 24 miles downstream and 3 hours to travel 18 miles upstream. What is the speed of the boat in still water? What is the speed of the current of the river?

24. A boat on a river travels 20 miles downstream in only 2 hours. It takes the same boat 6 hours to travel 12 miles upstream. What are the speed of the boat and the speed of the current?

18 miles in 3 hrs.

24 miles in 2 hrs.

25. An airplane flying with the wind can cover a certain distance in 2 hours. The return trip against the wind takes $2\frac{1}{2}$ hours. How fast is the plane and what is the speed of the wind, if the one-way distance is 600 miles?

26. An airplane covers a distance of 1,500 miles in 3 hours when it flies with the wind and $3\frac{1}{3}$ hours when it flies against the wind. What is the speed of the plane in still air?

2 hour trip

$2\frac{1}{2}$ hour trip

Wind

600 mi.

Coin Problems

27. Bob has 20 coins totaling $1.40. If he has only dimes and nickels, how many of each coin does he have?

28. If Amy has 15 coins totaling $2.70, and the coins are quarters and dimes, how many of each coin does she have?

29. A collection of nickels, dimes, and quarters consists of 9 coins with a total value of $1.20. If the number of dimes is equal to the number of nickels, find the number of each type of coin.

30. A coin collection consists of 12 coins with a total value of $1.20. If the collection consists only of nickels, dimes, and quarters, and the number of dimes is two more than twice the number of nickels, how many of each type of coin are in the collection?

31. A collection of nickels, dimes, and quarters amounts to $10.00. If there are 140 coins in all and there are twice as many dimes as there are quarters, find the number of nickels.

32. A cash register contains a total of 95 coins consisting of pennies, nickels, dimes, and quarters. There are only 5 pennies and the total value of the coins is $12.05. Also, there are 5 more quarters than dimes. How many of each coin is in the cash register?

33. Justin has $5.25 in quarters, dimes, and nickels. If he has a total of 31 coins and he has twice as many dimes as nickels, how many of each coin does he have?

34. Russ has $46.00 in 1, 5, and 10 dollar bills. If he has 11 bills in his wallet and he has twice as many $1 bills as $10 bills, how many of each bill does he have?

Additional Problems

35. Price and Demand A manufacturing company finds that they can sell 300 items if the price per item is $2.00, and 400 items if the price is $1.50 per item. If the relationship between the number of items sold x and the price per item p is a linear one, find a formula that gives x in terms of p. Then use the formula to find the number of items they will sell if the price per item is $3.00.

36. Price and Demand A company manufactures and sells bracelets. They have found from experience that they can sell 300 bracelets each week if the price per bracelet is $2.00, but only 150 bracelets are sold if the price is $2.50 per bracelet. If the relationship between the number of bracelets sold x and the price per bracelet p is a linear one, find a formula that gives x in terms of p. Then use the formula to find the number of bracelets they will sell at $3.00 each.

37. Height of a Ball A ball is tossed into the air so that the height after 1, 3, and 5 seconds is as given in the following table. If the relationship between the height of the ball h and the time t is quadratic, then the relationship can be written as

$$h = at^2 + bt + c$$

Use the information in the table to write a system of three equations in three variables a, b, and c. Solve the system to find the exact relationship between h and t.

38. Height of a Ball A ball is tossed into the air and its height above the ground after 1, 3, and 4 seconds is recorded as shown in the following table. The relationship between the height of the ball h and the time t is quadratic and can be written as

$$h = at^2 + bt + c$$

Use the information in the table to write a system of three equations in three variables a, b, and c. Solve the system to find the exact relationship between the variables h and t.

t (sec)	h (ft)
1	128
3	128
5	0

t (sec)	h (ft)
1	96
3	64
4	0

39. Fuel Economy Suppose your car gets 25 miles per gallon on the highway and 15 miles per gallon in the city. If you drive 425 miles on 25 gallons, how many gallons did you use for highway driving and how many used for city driving?

40. Fuel Economy The standard Chevy Camaro gets 17 miles per gallon in the city and 29 miles per gallon on the freeway. If the car can travel 467 miles on 19 gallons of gasoline, how many gallons were used for highway driving and how many gallons were used for city driving?

Getting Ready for the Next Section

41. Does the graph of $x + y < 4$ include the boundary line?

42. Does the graph of $-x + y \leq 3$ include the boundary line?

43. Where do the graphs of the lines $x + y = 4$ and $x - 2y = 4$ intersect?

44. Where do the graphs of the line $x = -1$ and $x - 2y = 4$ intersect?

Solve.

45. $20x + 9{,}300 > 18{,}000$

46. $20x + 4{,}800 > 18{,}000$

Maintaining Your Skills

Graph each inequality.

47. $2x + 3y < 6$

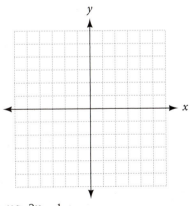

48. $2x + y < -5$

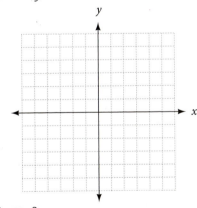

49. $y \geq -3x - 4$

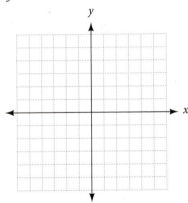

50. $y \geq 2x - 1$

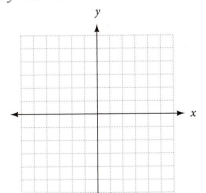

51. $x \geq 3$

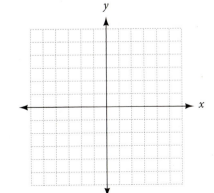

52. $y > -5$

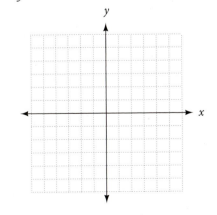

Extending the Concepts

High School Dropout Rate The high school dropout rates for males and females over the years 1965 to 2007 are shown in the following table.

Year	Female Dropout Rate (%)	Male Dropout Rate (%)
1970	16	14
1980	13	15
1985	12	13
1990	12	12
1995	12	12
2000	10	12
2005	8	11
2008	8	10

Source: infoplease.com

53. Plotting the years along the horizontal axis and the dropout rates along the vertical axis, draw a line graph for these data. Draw the female line dashed and the male line solid for easier reading and comparison.

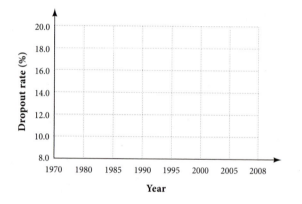

54. Refer to the data in the preceding exercise to answer the following questions.

  a. Using the slope, determine the time intervals when the decline in the female dropout rate was the steepest.

  b. Using the slope, determine the time interval when the increase in the male dropout rate was the steepest.

  c. What appears unusual about the time period from 1970 to 1980?

  d. Are the dropout rates for males and females generally increasing or generally decreasing?

55. Refer to the data about high school dropout rates for males and females.

  a. Write a linear equation for the dropout rate of males M based on the year x for the years 2000 and 2005.

  b. Write a linear equation for the dropout rate of females, F, based on the year, x, for the years 2000 and 2005.

  c. Using your results from parts a and b, determine when the dropout rates for males and females were the same.

Matrix Solutions to Linear Systems

4.4

OBJECTIVES

A Solve a system of linear equations using an augmented matrix.

TICKET TO SUCCESS

Keep these questions in mind as you read through the section. Then respond in your own words and in complete sentences.

1. What is an augmented matrix?
2. What does the vertical line in a matrix represent?
3. Explain the three row operations for an augmented matrix?
4. Briefly explain how you would transform an augmented matrix into a matrix that has 1's down the diagonal of the coefficient matrix, and 0's below it.

IDAK/Shutterstock.com

In mathematics, a matrix is a rectangular array of elements considered as a whole. We can use matrices to represent systems of linear equations. The Chinese, between 200 BC and 100 BC, in a text called *Nine Chapters on the Mathematical Art*, written during the Han Dynasty, gives the first known example of the use of matrices. The following problem appears in that text:

> *There are three types of corn, of which three bundles of the first, two of the second, and one of the third make 39 measures. Two of the first, three of the second and one of the third make 24 measures. And one of the first, two of the second and three of the third make 26 measures. How many measures of corn are contained in one bundle of each type?*

The author creates linear equations in three variables using that information and sets up the coefficients of the three unknowns as a table on a *counting board*.

$$\begin{bmatrix} 1 & 2 & 3 & | & 26 \\ 2 & 3 & 1 & | & 24 \\ 3 & 2 & 1 & | & 39 \end{bmatrix}$$

The counting board shown above is an example of a matrix, which is the focus of this section. We will begin our work with matrices by writing the coefficients of the variables and the constant terms in the same position in the matrix as they occur in the system of equations. To show where the coefficients end and the constant terms begin, we use vertical lines instead of equal signs. For example, the system

$$2x + 5y = -4$$
$$x - 3y = 9$$

can be represented by the matrix

$$\begin{bmatrix} 2 & 5 & \vline & -4 \\ 1 & -3 & \vline & 9 \end{bmatrix}$$

which is called an *augmented matrix* because it includes both the coefficients of the variables and the constant terms.

To solve a system of linear equations by using the augmented matrix for that system, we need the following row operations as the tools of that solution process. The row operations tell us what we can do to an augmented matrix that may change the numbers in the matrix, but will always produce a matrix that represents a system of equations with the same solution as that of our original system.

A Row Operations

1. We can interchange any two rows of a matrix.
2. We can multiply any row by a nonzero constant.
3. We can add to any row a constant multiple of another row.

The three row operations are simply a list of the properties we use to solve systems of linear equations, translated to fit an augmented matrix. For instance, the second operation in our list is actually just another way to state the multiplication property of equality.

We solve a system of linear equations by first transforming the augmented matrix into a matrix that has 1's down the diagonal of the coefficient matrix, and 0's below it. For instance, we will solve the system

$$2x + 5y = -4$$
$$x - 3y = 9$$

by transforming the matrix

$$\begin{bmatrix} 2 & 5 & \vline & -4 \\ 1 & -3 & \vline & 9 \end{bmatrix}$$

using the row operations listed earlier to get a matrix of the form

$$\begin{bmatrix} 1 & - & \vline & - \\ 0 & 1 & \vline & - \end{bmatrix}$$

To accomplish this, we begin with the first column and try to produce a 1 in the first position and a 0 below it. Interchanging rows 1 and 2 gives us a 1 in the top position of the first column:

$$\begin{bmatrix} 1 & -3 & \vline & 9 \\ 2 & 5 & \vline & -4 \end{bmatrix} \qquad \text{Interchange rows 1 and 2.}$$

Multiplying row 1 by −2 and adding the result to row 2 gives us a 0 where we want it.

$$\begin{bmatrix} 1 & -3 & \vline & 9 \\ 0 & 11 & \vline & -22 \end{bmatrix} \qquad \text{Multiply row 1 by −2 and add the result to row 2.}$$

Continue to produce 1's down the diagonal by multiplying row 2 by $\dfrac{1}{11}$.

$$\begin{bmatrix} 1 & -3 & \vline & 9 \\ 0 & 1 & \vline & -2 \end{bmatrix} \qquad \text{Multiply row 2 by } \tfrac{1}{11}.$$

Taking this last matrix and writing the system of equations it represents, we have

$$x - 3y = 9$$
$$y = -2$$

Substituting -2 for y in the top equation gives us

$$x = 3$$

The solution to our system is $(3, -2)$.

EXAMPLE 1 Solve the following system using an augmented matrix:

$$x + y - z = 2$$
$$2x + 3y - z = 7$$
$$3x - 2y + z = 9$$

SOLUTION We begin by writing the system in terms of an augmented matrix.

$$\left[\begin{array}{ccc|c} 1 & 1 & -1 & 2 \\ 2 & 3 & -1 & 7 \\ 3 & -2 & 1 & 9 \end{array}\right]$$

Next, we want to produce 0's in the second two postions of column 1:

$$\left[\begin{array}{ccc|c} 1 & 1 & -1 & 2 \\ 0 & 1 & 1 & 3 \\ 3 & -2 & 1 & 9 \end{array}\right]$$ Multiply row 1 by -2 and add the result to row 2.

$$\left[\begin{array}{ccc|c} 1 & 1 & -1 & 2 \\ 0 & 1 & 1 & 3 \\ 0 & -2 & 4 & 3 \end{array}\right]$$ Multiply row 1 by -3 and add the result to row 3.

Note that we could have done these two steps in one single step. As you become more familiar with this method of solving systems of equations, you will do just that.

$$\left[\begin{array}{ccc|c} 1 & 1 & -1 & 2 \\ 0 & 1 & 1 & 3 \\ 0 & 0 & 9 & 18 \end{array}\right]$$ Multiply row 2 by 5 and add the result to row 3.

$$\left[\begin{array}{ccc|c} 1 & 1 & -1 & 2 \\ 0 & 1 & 1 & 3 \\ 0 & 0 & 1 & 2 \end{array}\right]$$ Multiply row 3 by $\frac{1}{9}$.

Converting back to a system of equations, we have

$$x + y - z = 2$$
$$y + z = 3$$
$$z = 2$$

This system is equivalent to our first one, but much easier to solve.

Substituting $z = 2$ into the second equation, we have

$$y = 1$$

Substituting $z = 2$ and $y = 1$ into the first equation, we have

$$x = 3$$

The solution to our original system is $(3, 1, 2)$. It satisfies each of our original equations. You can check this, if you want. ■

Problem Set 4.4

Moving Toward Success

"One's first step in wisdom is to question everything— and one's last is to come to terms with everything."

—Georg Christoph Lichtenberg, 1742–1799,
German scientist and philosopher

1. Why is it important to recognize the different types of instructions for a problem?
2. Should you still add to your list of difficult problems this far into the class? Why or why not?

A Solve the following systems of equations by using matrices. [Example 1]

1. $x + y = 5$
$3x - y = 3$

2. $x + y = -2$
$2x - y = -10$

3. $3x - 5y = 7$
$-x + y = -1$

4. $2x - y = 4$
$x + 3y = 9$

5. $2x - 8y = 6$
$3x - 8y = 13$

6. $3x - 6y = 3$
$-2x + 3y = -4$

7. $2x - y = -10$
$4x + 3y = 0$

8. $3x - 7y = 36$
$5x - 4y = 14$

9. $5x - 3y = 27$
$6x + 2y = -18$

10. $3x + 4y = 2$
$5x + 3y = 29$

11. $5x + 2y = -14$
$y = 2x + 11$

12. $3x + 5y = 3$
$x = 4y + 1$

13. $2x + 3y = 11$
$-x - y = -2$

14. $5x + 2y = -25$
$-3x + 2y = -1$

15. $3x - 2y = 16$
$4x + 3y = -24$

16. $6x + y = 3$
$ x = 4y + 13$

17. $3x - 2y = 16$
$ y = 2x - 12$

18. $4x - 3y = 28$
$ y = -x - 7$

19. $x + y + z = 4$
$ x - y + 2z = 1$
$ x - y - z = -2$

20. $x - y - 2z = -1$
$ x + y + z = 6$
$ x + y - z = 4$

21. $x + 2y + z = 3$
$ 2x - y + 2z = 6$
$ 3x + y - z = 5$

22. $x - 3y + 4z = -4$
$ 2x + y - 3z = 14$
$ 3x + 2y + z = 10$

23. $ x - 2y + z = -4$
$ 2x + y - 3z = 7$
$ 5x - 3y + z = -5$

24. $3x - 2y + 3z = -3$
$ x + y + z = 4$
$ x - 4y + 2z = -9$

25. $5x - 3y + z = 10$
$ x - 2y - z = 0$
$ 3x - y + 2z = 10$

26. $ 2x - y - z = 1$
$ x + 3y + 2z = 13$
$ 4x + y - z = 7$

27. $2x - 5y + 3z = 2$
$ 3x - 7y + z = 0$
$ x + y + 2z = 5$

28. $3x - 4y + 2z = -2$
$ 2x + y + 3z = 13$
$ x - 3y + 2z = -3$

29. $x + 2y = 3$
$ y + z = 3$
$ 4x - z = 2$

30. $ x + y = 2$
$ 3y - 2z = -8$
$ x + z = 5$

31. $x + 3y = 7$
$3x - 4z = -8$
$5y - 2z = -5$

32. $x + 4y = 13$
$2x - 5z = -3$
$4y - 3z = 9$

33. $x + 2y = 13$
$x - 3z = 11$
$3y + 4z = 4$

34. $x - 2y = 5$
$4x + 3z = 11$
$5y + 4z = -12$

35. $x - 2y + z = -5$
$2x + 3y - 2z = -9$
$2x - y + 2z = -1$

36. $-4x - 3y - z = -7$
$3x + 2y + 2z = 7$
$-x - y + 2z = 2$

37. $4x - 2y - z = -5$
$x + 3y - 4z = 13$
$3x - y - 3z = 0$

38. $3x - 5y + z = 15$
$2x + 6y - 4z = 10$
$x - 5y - 3z = -5$

39. $5y + z = 11$
$7x - 2y = 1$
$5x + 2z = -3$

40. $x - 2y - z = 1$
$3x - 2y + 3z = 3$
$2x + y + 4z = 5$

Solve each system using matrices. Remember, multiplying a row by a nonzero constant will not change the solution to a system.

41. $\frac{1}{3}x + \frac{1}{5}y = 2$

   $\frac{1}{3}x - \frac{1}{2}y = -\frac{1}{3}$

42. $\frac{1}{2}x + \frac{1}{3}y = 13$

   $\frac{1}{5}x + \frac{1}{8}y = 5$

43. $\frac{1}{3}x - \frac{1}{4}y = 1$

   $\frac{1}{3}x + \frac{1}{4}y = 3$

44. $\frac{1}{3}x - \frac{5}{6}y = 16$

   $-\frac{1}{2}x + \frac{3}{4}y = -18$

The systems that follow are inconsistent systems. In both cases, the lines are parallel. Try solving each system using matrices and see what happens.

45. $2x - 3y = 4$

   $4x + 6y = 4$

46. $10x - 15y = 5$

   $-4x + 6y = -4$

The systems that follow are dependent systems. In each case, the lines coincide. Try solving each system using matrices and see what happens

47. $-6x + 4y = 8$

   $-3x + 2y = 4$

48. $x + 2y = 5$

   $-x - 2y = -5$

Getting Ready for the Next Section

Graph the following equations.

49. $y = 3x - 5$

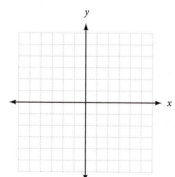

50. $y = -\frac{3}{4}x$

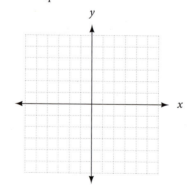

51. $x = -3$

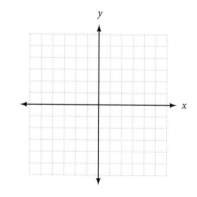

Systems of Linear Inequalities

TICKET TO SUCCESS

Keep these questions in mind as you read through the section. Then respond in your own words and in complete sentences.

1. What is the solution set to a system of inequalities?
2. For the boundary lines of a system of linear inequalities, when would you use a dotted line rather than a solid line?
3. Once you have graphed the solution set for each inequality in a system, how would you determine the region to shade for the solution to the system of inequalities?
4. How would you find the solution set by graphing a system that contained three linear inequalities?

Marie C. Fiedis\Shutterstock.com

Imagine you work at a local coffee house and your boss wants you to create a new house blend using the three most popular coffees. The dark roast coffee you are to use costs $20.00 per pound, the medium roast costs $14.50 per pound, and the flavored coffee costs $16.00 per pound. Your boss wants the cost of the new blend to be at most $17.00 per pound. She will want to see possible combinations of the coffee to keep the price at most $17.00. As discussed in Chapter 1, words like "at most" imply the use of inequalities. Working through this section will help us visualize solutions to inequalities in two and three variables.

A Graphing Solutions to Linear Inequalities

In the previous chapter, we graphed linear inequalities in two variables. To review, we graph the boundary line using a solid line if the boundary is part of the solution set and a broken line if the boundary is not part of the solution set. Then we test any point that is not on the boundary line in the original inequality. A true statement tells us that the point lies in the solution set; a false statement tells us the solution set is the other region.

Figure 1 shows the graph of the inequality $x + y < 4$. Note that the boundary is not included in the solution set and is therefore drawn with a broken line.

Figure 2 shows the graph of $-x + y \leq 3$. Note that the boundary is drawn with a solid line because it is part of the solution set.

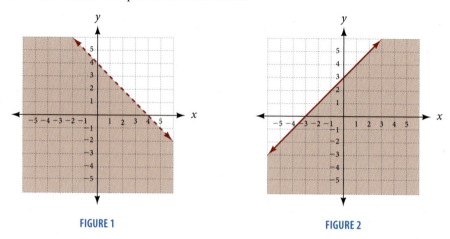

FIGURE 1   FIGURE 2

If we form a system of inequalities with the two inequalities, the solution set will be all the points common to both solution sets shown in the two figures above; it is the intersection of the two solution sets. Therefore, the solution set for the system of inequalities

$$x + y < 4$$
$$-x + y \leq 3$$

is all the ordered pairs that satisfy both inequalities. It is the set of points that are below the line $x + y = 4$ and also below (and including) the line $-x + y = 3$. The graph of the solution set to this system is shown in Figure 3. We have written the system in Figure 3 with the word *and* just to remind you that the solution set to a system of equations or inequalities is all the points that satisfy both equations or inequalities.

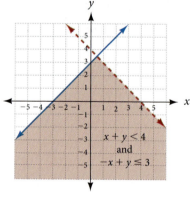

$x + y < 4$
and
$-x + y \leq 3$

FIGURE 3

EXAMPLE 1   Graph the solution to the system of linear inequalities.

$$y < \frac{1}{2}x + 3$$

$$y \geq \frac{1}{2}x - 2$$

SOLUTION   Figures 4 and 5 show the solution set for each of the inequalities separately.

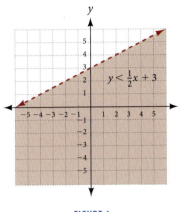

FIGURE 4

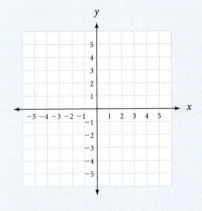

FIGURE 5

Figure 6 is the solution set to the system of inequalities. It is the region consisting of points whose coordinates satisfy both inequalities

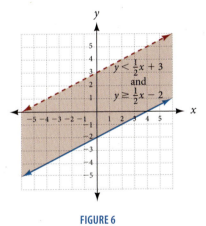

FIGURE 6

■

EXAMPLE 2   Graph the solution to the system of linear inequalities.

$$x + y < 4$$

$$x \geq 0$$

$$y \geq 0$$

SOLUTION   We graphed the first inequality, $x + y < 4$, in Figure 1 at the beginning of this section. The solution set to the inequality $x \geq 0$, shown in Figure 7, is all the points to the right of the y-axis; that is, all the points with x-coordinates that are greater than or equal to 0. Figure 8 shows the graph of $y \geq 0$. It consists of all points with y-coordinates greater than or equal to 0; that is, all points from the x-axis up.

PRACTICE PROBLEMS

1. Graph the solution set to the system.

$$y \leq \frac{1}{3}x + 1$$

$$y > \frac{1}{3}x - 3$$

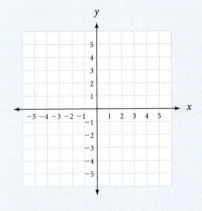

2. Graph the solution set to the system.

$$x + y < 3$$

$$x \geq 0$$

$$y \geq 0$$

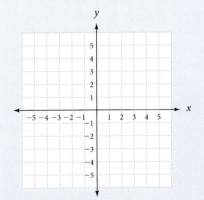

Answers
1–4. See Solutions Section.

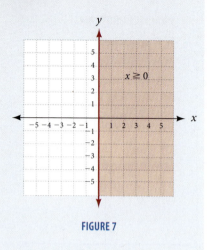

FIGURE 7

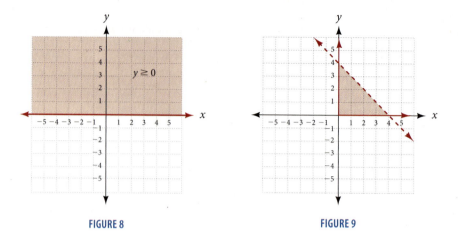

FIGURE 8

FIGURE 9

The regions shown in Figures 7 and 8 overlap in the first quadrant. Therefore, putting all three regions together we have the points in the first quadrant that are below the line $x + y = 4$. This region is shown in Figure 9, and it is the solution to our system of inequalities.

Extending the discussion in Example 2, we can name the points in each of the four quadrants using systems of inequalities.

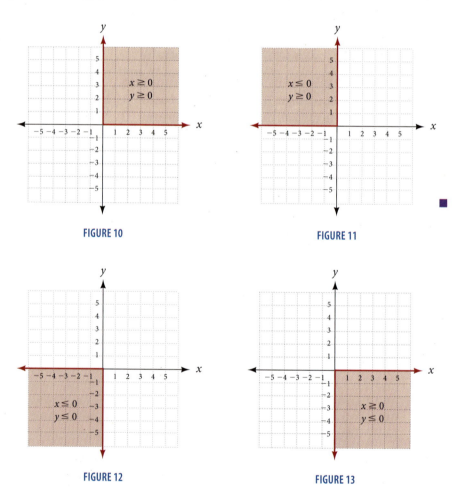

FIGURE 10

FIGURE 11

FIGURE 12

FIGURE 13

EXAMPLE 3 Graph the solution to the system of linear inequalities.

$$x \leq 4$$

$$y \geq -3$$

SOLUTION The solution to this system will consist of all points to the left of and including the vertical line $x = 4$ that intersect with all points above and including the horizontal line $y = -3$. The solution set is shown in Figure 14.

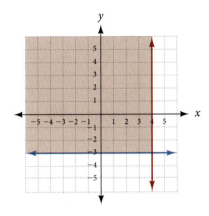

FIGURE 14

■

EXAMPLE 4 Graph the solution set for the following system.

$$x - 2y \leq 4$$

$$x + y \leq 4$$

$$x \geq -1$$

SOLUTION We have three linear inequalities, representing three sections of the coordinate plane. The graph of the solution set for this system will be the intersection of these three sections. The graph of $x - 2y \leq 4$ is the section above and including the boundary $x - 2y = 4$. The graph of $x + y \leq 4$ is the section below and including the boundary line $x + y = 4$. The graph of $x \geq -1$ is all the points to the right of, and including, the vertical line $x = -1$. The intersection of these three graphs is shown in Figure 15.

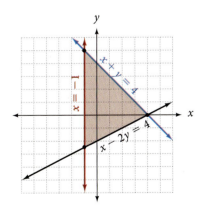

FIGURE 15

3. Graph the solution to the system.

$$x > -3$$

$$y < 4$$

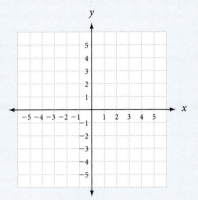

4. Graph the solution set for the following system.

$$x - y < 5$$

$$x + y < 5$$

$$x > 1$$

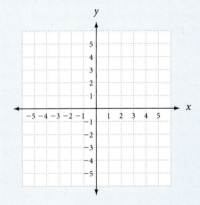

5. How would the graph in Figure 16 change if they had reserved only 300 tickets at the $15 rate?

EXAMPLE 5   A college basketball arena plans on charging $20 for certain seats and $15 for others. They want to bring in more than $18,000 from all ticket sales and they have reserved at least 500 tickets at the $15 rate. Find a system of inequalities describing all possibilities and sketch the graph. If 620 tickets are sold for $15, at least how many tickets are sold for $20?

SOLUTION   Let x = the number of $20 tickets and y = the number of $15 tickets. We need to write a list of inequalities that describe this situation. That list will form our system of inequalities. First of all, we note that we cannot use negative numbers for either x or y. So, we have our first inequalities:

$$x \geq 0$$
$$y \geq 0$$

Next, we note that they are selling at least 500 tickets for $15, so we can replace our second inequality with $y \geq 500$. Now our system is

$$x \geq 0$$
$$y \geq 500$$

Now, the amount of money brought in by selling $20 tickets is $20x$, and the amount of money brought in by selling $15 tickets is $15y$. If the total income from ticket sales is to be more than $18,000, then $20x + 15y$ must be greater than 18,000. This gives us our last inequality and completes our system.

$$20x + 15y > 18,000$$
$$x \geq 0$$
$$y \geq 500$$

We have used all the information in the problem to arrive at this system of inequalities. The solution set contains all the values of x and y that satisfy all the conditions given in the problem. Here is the graph of the solution set.

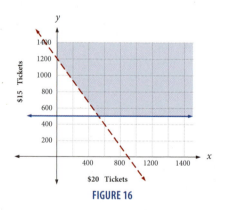

FIGURE 16

If 620 tickets are sold for $15, then we substitute 620 for y in our first inequality to obtain

$20x + 15(620) > 18,000$	Substitute 620 for y.
$20x + 9,300 > 18,000$	Multiply.
$20x > 8,700$	Add $-9,300$ to each side.
$x > 435$	Divide each side by 20.

If they sell 620 tickets for $15 each, then they need to sell more than 435 tickets at $20 each to bring in more than $18,000. ∎

Problem Set 4.5

Moving Toward Success

"There is no comparison between that which is lost by not succeeding and that which is lost by not trying."

—Francis Bacon, 1561–1626, English philosopher and statesman

1. Do you take notes in class? Why or why not?
2. Research some note taking techniques online. What are some tips that you should employ when taking notes for this class?

A Graph the solution set for each system of linear inequalities. [Examples 1–4]

1. $x + y < 5$
$2x - y > 4$

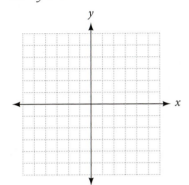

2. $x + y < 5$
$2x - y < 4$

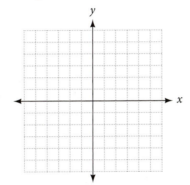

3. $y < \frac{1}{3}x + 4$
$y \geq \frac{1}{3}x - 3$

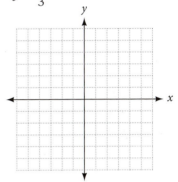

4. $y < 2x + 4$
$y \geq 2x - 3$

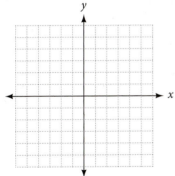

5. $3x + 4y > 12$
$2x - y \leq 4$

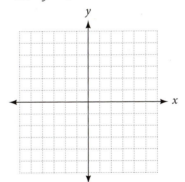

6. $x - 4y \geq 4$
$2x + 5y < -10$

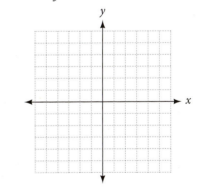

7. $x \geq 3y - 6$
$y < \frac{2}{3}x + 1$

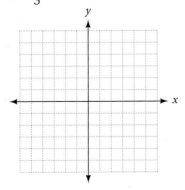

8. $y > -\frac{1}{3}x + 2$
$y \leq 4x - 3$

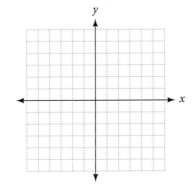

9. $x \geq -3$
$y < 2$

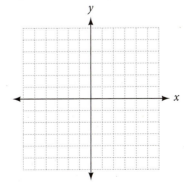

10. $x \le 4$
  $y \ge -2$

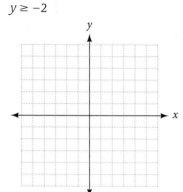

11. $1 \le x \le 3$
  $2 \le y \le 4$

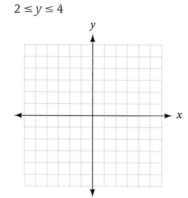

12. $-4 \le x \le -2$
  $1 \le y \le 3$

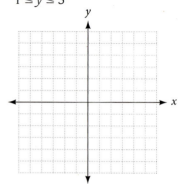

13. $x + y \le 4$
  $x \ge 0$
  $y \ge 0$

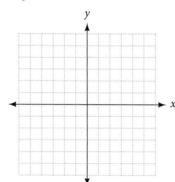

14. $x - y \le 2$
  $x \ge 0$
  $y \le 0$

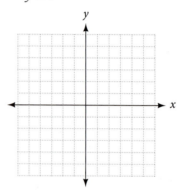

15. $-3 \le x \le 2$
  $y > \dfrac{2}{3}x - 1$
  $y \le -\dfrac{3}{4}x + 3$

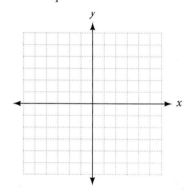

16. $-2 \le y \le 4$
   $y \ge -3x + 1$
  $12x + 4y \le 24$

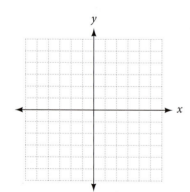

17. $-1 < x < 2$
  $x \le -\dfrac{2}{3}y - 2$
  $y < \dfrac{2}{3}x - 2$

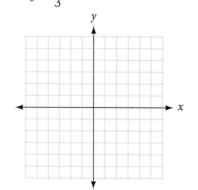

18. $-2 \le x \le 4$
   $y \ge \dfrac{3}{2}x - 1$
  $2x + y \le 4$

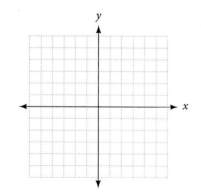

19. $x + y \le 3$
$x - 3y \le 3$
$x \ge -2$

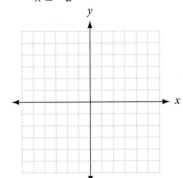

20. $x - y \le 4$
$x + 2y \le 4$
$x \ge -1$

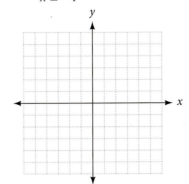

21. $x + y \le 2$
$-x + y \le 2$
$y \ge -2$

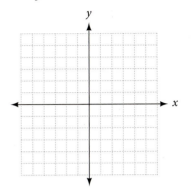

22. $x - y \le 3$
$-x - y \le 3$
$y \le -1$

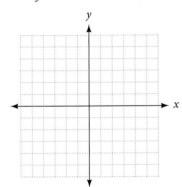

23. $x + y < 5$
$y > x$
$y \ge 0$

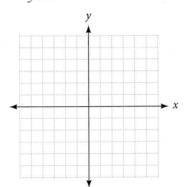

24. $x + y < 5$
$y > x$
$x \ge 0$

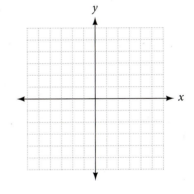

25. $2x + 3y \le 6$
$x \ge 0$
$y \ge 0$

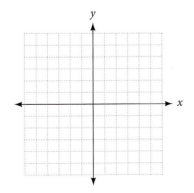

26. $x + 2y \le 10$
$3x + y \le 12$
$x \ge 0$
$y \ge 0$

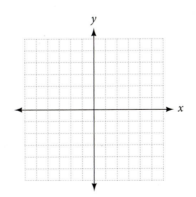

For each graph below, find a system of inequalities that describes the shaded region.

27.

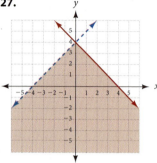

28.

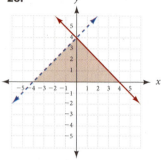

29.

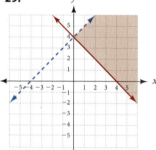

30.

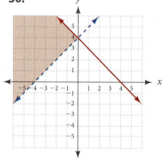

Applying the Concepts

31. Office Supplies An office worker wants to purchase some $0.55 postage stamps and also some $0.65 postage stamps totaling no more than $40. He also wants to have at least twice as many $0.55 stamps and more than 15 of the $0.55 stamps.

a. Find a system of inequalities describing all the possibilities and sketch the graph.

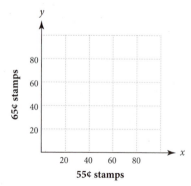

b. If he purchases 20 of the $0.55 stamps, what is the maximum number of $0.65 stamps he can purchase?

32. Inventory A store sells two brands of DVD players. Customer demand indicates that it is necessary to stock at least twice as many DVD players of brand A as of brand B. At least 30 of brand A and 15 of brand B must be on hand. There is room for not more than 100 DVD players in the store.

a. Find a system of inequalities describing all possibilities, then sketch the graph.

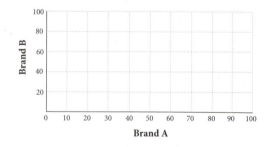

b. If there are 35 DVD players of brand A, what is the maximum number of brand B DVD players on hand?

33. Coffee A coffee house wants to make a special holiday blend. They will blend dark roast coffee that costs $20.00 per pound and a medium roast at $14.50 per pound. They will make a batch of less than 25 pounds that will cost at most $17.00 per pound.

a. Find a system of inequalities describing all the possibilities and sketch the graph.

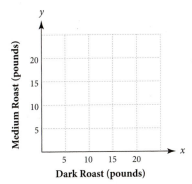

b. If the coffee house makes the blend with 8 pounds of dark roast, what is the maximum amount of medium roast they can use?

34. Coffee The coffee house in problem 33 wants to try out a new blend using the dark and medium roasts. They want the blend to cost at most $16.00. They will make no more than 20 pounds and they want to use at least 4 more pounds of medium roast than dark roast.

a. Find a system of inequalities describing all the possibilities and sketch the graph.

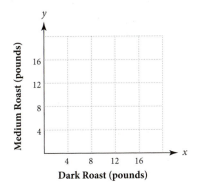

b. If the coffee house uses 3 pounds of the dark roast, what is the minimum and maximum number of pounds of medium roast that can be used?

Extending the Concepts

Use the table to answer the following questions.

ESPRESSO BEANS	
Blend	**Cost Per Pound**
Grand Espresso	$7.75
Super Crema Espresso	$9.25
Gran Riserva Espresso	$13.00

35. Coffee A coffee house wants to try a new blend containing Grand Espresso and Gran Riserva Espresso. They want to make at least 35 pounds, and keep the cost below $10.00 per pound. They want to use at least 6 pounds more of the Grand Espresso. Find a system of inequalities describing all the possibilities and sketch the graph.

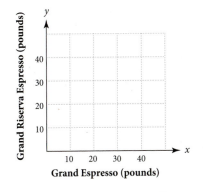

36. Coffee A coffee house is going to make a special edition blend that contains 5 pounds of Gran Riserva and the total amount of the blend made will be no more than 35 pounds and cost less than $10.00 per pound. Find a system of inequalities describing all the possibilities and sketch the graph.

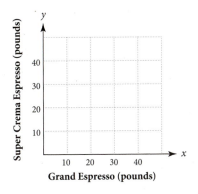

For each of the following straight lines, identify the x-intercept, y-intercept, and slope, and sketch the graph.

37. $2x + y = 6$

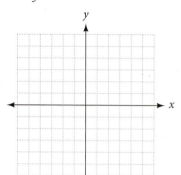

38. $y = \dfrac{3}{2}x + 4$

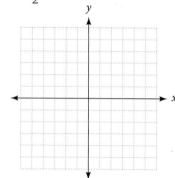

39. $x = -2$

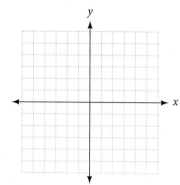

Find the equation for each line.

40. Give the equation of the line through $(-1, 3)$ that has slope $m = 2$.

41. Give the equation of the line through $(-3, 2)$ and $(4, -1)$.

42. Line l contains the point $(5, -3)$ and has a graph parallel to the graph of $2x - 5y = 10$. Find the equation for l.

43. Give the equation of the vertical line through $(4, -7)$.

State the domain and range for the following relations, and indicate which relations are also functions.

44. $\{(-2, 0), (-3, 0), (-2, 1)\}$

45. $y = x^2 - 9$

Let $f(x) = x - 2$, $g(x) = 3x + 4$ and $h(x) = 3x^2 - 2x - 8$, and find the following.

46. $f(3) + g(2)$

47. $h(0) + g(0)$

48. $f(g(2))$

49. $g(f(2))$

Solve the following variation problems.

50. Direct Variation Quantity y varies directly with the square of x. If y is 50 when x is 5, find y when x is 3.

51. Joint Variation Quantity z varies jointly with x and the cube of y. If z is 15 when x is 5 and y is 2, find z when x is 2 and y is 3.

Chapter 4 Summary

EXAMPLES

Systems of Linear Equations [4.1, 4.2]

1. The solution to the system

$$x + 2y = 4$$
$$x - y = 1$$

is the ordered pair (2, 1). It is the only ordered pair that satisfies both equations.

A system of linear equations consists of two or more linear equations considered simultaneously. The solution set to a linear system in two variables is the set of ordered pairs that satisfy both equations. The solution set to a linear system in three variables consists of all the ordered triples that satisfy each equation in the system.

To Solve a System by the Addition Method [4.1]

2. We can eliminate the y-variable from the system in Example 1 by multiplying both sides of the second equation by 2 and adding the result to the first equation.

$$
\begin{array}{ll}
x + 2y = 4 \xrightarrow{\ \text{No change}\ } & x + 2y = 4 \\
x - y = 1 \xrightarrow{\ \text{Multiply by 2.}\ } & \underline{2x - 2y = 2} \\
& 3x \qquad = 6 \\
& x = 2
\end{array}
$$

Substituting $x = 2$ into either of the original two equations gives $y = 1$. The solution is (2, 1).

Step 1: Look the system over to decide which variable will be easier to eliminate.

Step 2: Use the multiplication property of equality on each equation separately, if necessary, to ensure that the coefficients of the variable to be eliminated are opposites.

Step 3: Add the left and right sides of the system produced in step 2, and solve the resulting equation.

Step 4: Substitute the solution from step 3 back into any equation with both x- and y-variables, and solve.

Step 5: Check your solution in both equations if necessary.

To Solve a System by the Substitution Method [4.1]

3. We can apply the substitution method to the system in Example 1 by first solving the second equation for x to get

$$x = y + 1$$

Substituting this expression for x into the first equation we have

$$(y + 1) + 2y = 4$$
$$3y + 1 = 4$$
$$3y = 3$$
$$y = 1$$

Using $y = 1$ in either of the original equations gives $x = 2$.

Step 1: Solve either of the equations for one of the variables (this step is not necessary if one of the equations has the correct form already).

Step 2: Substitute the results of step 1 into the other equation, and solve.

Step 3: Substitute the results of step 2 into an equation with both x- and y-variables, and solve. (The equation produced in step 1 is usually a good one to use.)

Step 4: Check your solution if necessary.

Inconsistent and Dependent Equations [4.1, 4.2]

4. If the two lines are parallel, then the system will be inconsistent and the solution is $\varnothing$. If the two lines coincide, then the equations are dependent.

A system of two linear equations that have no solutions in common is said to be an *inconsistent* system, whereas two linear equations that have all their solutions in common are said to be *dependent* equations.

5. One number is 2 more than 3 times another. Their sum is 26. Find the numbers.

Step 1: Read and List. Known items: two numbers, whose sum is 26. One is 2 more than 3 times the other. Unknown items are the two numbers.

Step 2: Assign variables. Let x = one of the numbers. Then y = the other number.

Step 3: Write a system of equations.

$$x + y = 26$$
$$y = 3x + 2$$

Step 4: Solve the system. Substituting the expression for y from the second equation into the first and solving for x yields

$$x + (3x + 2) = 26$$
$$4x + 2 = 26$$
$$4x = 24$$
$$x = 6$$

Using $x = 6$ in $y = 3x + 2$ gives the second number: $y = 20$.

Step 5: Write answers
The two numbers are 6 and 20.

Step 6: Reread and check.
The sum of 6 and 20 is 26, and 20 is 2 more than 3 times 6.

■ Applications of Linear Systems [4.3]

Strategy **Problem Solving Using a System of Equations**

Step 1 **Read** the problem, and then mentally **list** the items that are known and the items that are unknown.

Step 2 **Assign variables** to each of the unknown items; that is, let x = one of the unknown items and y = the other unknown item (and z = the third unknown item, if there is a third one). Then **translate** the other **information** in the problem to expressions involving the two (or three) variables.

Step 3 **Reread** the problem, and then **write a system of equations,** using the items and variables listed in steps 1 and 2, that describes the situation.

Step 4 **Solve the system** found in step 3.

Step 5 **Write your answers** using complete sentences.

Step 6 **Reread** the problem, and **check** your solution with the original words in the problem.

■ Matrix Solutions [4.4]

A matrix is a rectangular array of elements considered as a whole. We can use matrices to represent systems of linear equations. An augmented matrix includes both the coefficients of the variables and the constant terms of the linear system.

To solve a system of linear equations using an augmented matrix, we must follow row operations.

1. We can interchange any two rows of a matrix.

2. We can multiply any row by a nonzero constant.

3. We can add to any row a constant multiple of another row.

■ Systems of Linear Inequalities [4.5]

6. The solution set for the system

$$x + y < 4$$
$$-x + y \leq 3.$$

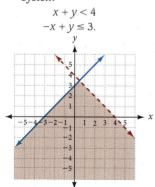

A system of linear inequalities is two or more linear inequalities considered at the same time. To find the solution set to the system, we graph each of the inequalities on the same coordinate system. The solution set is the region that is common to all the regions graphed.

Solve each system using the addition method. [4.1]

1. $x + y = 4$
$2x - y = 14$

2. $3x + y = 2$
$2x + y = 0$

3. $2x - 4y = 5$
$-x + 2y = 3$

4. $5x - 2y = 7$
$3x + y = 2$

5. $6x - 5y = -5$
$3x + y = 1$

6. $6x + 4y = 8$
$9x + 6y = 12$

7. $3x - 7y = 2$
$-4x + 6y = -6$

8. $6x + 5y = 9$
$4x + 3y = 6$

9. $-7x + 4y = -1$
$5x - 3y = 0$

10. $\frac{1}{2}x - \frac{3}{4}y = -4$
$\frac{1}{4}x + \frac{3}{2}y = 13$

11. $\frac{2}{3}x - \frac{1}{6}y = 0$
$\frac{4}{3}x + \frac{5}{6}y = 14$

12. $-\frac{1}{2}x + \frac{1}{3}y = -\frac{13}{6}$
$\frac{4}{5}x + \frac{3}{4}y = \frac{9}{10}$

Solve each system by the substitution method. [4.1]

13. $x + y = 2$
$y = x - 1$

14. $2x - 3y = 5$
$y = 2x - 7$

15. $x + y = 4$
$2x + 5y = 2$

16. $x + y = 3$
$2x + 5y = -6$

17. $3x + 7y = 6$
$x = -3y + 4$

18. $5x - y = 4$
$y = 5x - 3$

Solve each system. [4.2]

19. $x + y + z = 6$
$x - y - 3z = -8$
$x + y - 2z = -6$

20. $3x + 2y + z = 4$
$2x - 4y + z = -1$
$x + 6y + 3z = -4$

21. $5x + 8y - 4z = -7$
$7x + 4y + 2z = -2$
$3x - 2y + 8z = 8$

22. $5x - 3y - 6z = 5$
$4x - 6y - 3z = 4$
$-x + 9y + 9z = 7$

23. $5x - 2y + z = 6$
$-3x + 4y - z = 2$
$6x - 8y + 2z = -4$

24. $4x - 6y + 8z = 4$
$5x + y - 2z = 4$
$6x - 9y + 12z = 6$

25. $2x - y = 5$
$3x - 2z = -2$
$5y + z = -1$

26. $x - y = 2$
$y - z = -3$
$x - z = -1$

Use systems of equations to solve each application problem. In each case, be sure to show the system used. [4.3]

27. Ticket Prices Tickets for the show cost $2.00 for adults and $1.50 for children. How many adult tickets and how many children's tickets were sold if a total of 127 tickets was sold for $214?

28. Coin Collection John has 20 coins totaling $3.20. If he has only dimes and quarters, how many of each coin does he have?

29. Investments Ms. Jones invests money in two accounts, one of which pays 12% per year, and the other pays 15% per year. If her total investment is $12,000 and the interest after 1 year is $1,650, how much is invested in each account?

30. Speed It takes a boat on a river 2 hours to travel 28 miles downstream and 3 hours to travel 30 miles upstream. What is the speed of the boat and the current of the river?

Solve the following systems of equations by using matrices. [4.4]

31. $5x - 3y = 13$
$2x + 6y = -2$

32. $2x + 3y = 6$
$x - 2y = -11$

33. $5x + 3y + z = 6$
$x - 4y - 3z = 4$
$3x + y - 5z = -6$

34. $4x - 3y + 4z = -2$
$x - y + 2z = 1$
$2x + y + 3z = -5$

Graph the solution set for each system of linear inequalities. [4.5]

35. $3x + 4y < 12$
$-3x + 2y \leq 6$

36. $3x + 4y < 12$
$-3x + 2y \leq 6$
$y \geq 0$

37. $3x + 4y < 12$
$x \geq 0$
$y \geq 0$

38. $x > -3$
$y > -2$

Simplify each of the following.

1. $4^2 - 8^2$

2. $(4 - 8)^2$

3. $15 - 12 \div 4 - 3 \cdot 2$

4. $6(11 - 13)^3 - 5(8 - 11)^2$

5. $8\left(\dfrac{3}{2}\right) - 24\left(\dfrac{3}{2}\right)^2$

6. $16\left(\dfrac{3}{8}\right) - 32\left(\dfrac{3}{4}\right)^2$

Find the value of each expression when x is -3.

7. $x^2 + 12x - 36$

8. $(x - 6)(x + 6)$

Simplify each of the following expressions.

9. $4(3x - 2) + 3(2x + 5)$

10. $5 - 3[2x - 4(x - 2)]$

11. $-4\left(\dfrac{3}{4}x - \dfrac{3}{2}y\right)$

12. $-6\left(\dfrac{5}{3}x + \dfrac{3}{6}y\right)$

Solve.

13. $-4y - 2 = 6y + 8$

14. $-6 + 2(2x + 3) = 0$

15. $|x + 5| + 3 = 2$

16. $|2x - 3| + 7 = 1$

Solve the following equations for y.

17. $y - 3 = -5(x - 2)$

18. $y - 3 = \dfrac{2}{3}(x - 6)$

Solve the following equations for x.

19. $ax - 5 = bx - 3$

20. $ax + 7 = cx - 2$

Solve each inequality and write your answers using interval notation.

21. $-3t \geq 12$

22. $|2x - 1| \geq 5$

Let $A = \{0, 3, 6, 9\}$ and $B = \{2, 4, 6\}$. Find the following.

23. $A \cup B$

24. $A \cap B$

For the set $\{-1, 0, \sqrt{2}, 2.35, \sqrt{3}, 4\}$ list all the elements in the following sets.

25. Rational numbers

26. Integers

Solve each of the following systems.

27. $\begin{aligned} 2x - 5y &= -7 \\ -3x + 4y &= 0 \end{aligned}$

28. $\begin{aligned} 8x + 6y &= 4 \\ 12x + 9y &= 8 \end{aligned}$

29. $\begin{aligned} 2x + y &= 3 \\ y &= -2x + 3 \end{aligned}$

30. $\begin{aligned} 2x + 2y &= -8z \\ 4y - 2z &= -9 \\ 3x - 2z &= -6 \end{aligned}$

Graph on a rectangular coordinate system.

31. $0.03x + 0.04y = 0.04$

32. $3x - y < -2$

33. Graph the solution set to the system.

$$\begin{aligned} x - y &< -3 \\ y &< 5 \end{aligned}$$

34. Find the slope of the line through $\left(\dfrac{2}{3}, -\dfrac{1}{2}\right)$ and $\left(\dfrac{1}{2}, -\dfrac{5}{6}\right)$.

35. Find the slope and y-intercept of $3x - 5y = 15$.

36. Give the equation of a line with slope $-\dfrac{2}{3}$ and y-intercept $= -3$.

37. Find the equation of the line that is perpendicular to $2x - 5y = 10$ and contains the point $(0, 1)$.

38. Find the equation of the line through $(1, 4)$ and $(-1, -2)$.

Let $f(x) = 2x + 3$ and $g(x) = x^2 - 5$. Find the following.

39. $f(-2)$

40. $g(-2) + 6$

41. $(f - g)(x)$

42. $(g \circ f)(x)$

Use the graph below to work Problems 43–46.

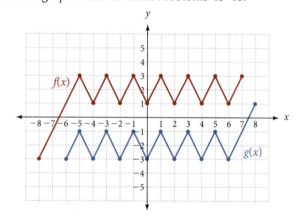

43. Find $f(3)$.

44. Find $(f + g)(-1)$.

45. Find $(g \circ f)(6)$.

46. Find x if $f(x) = -1$.

Solve the following systems by the addition method. [4.1]

1. $2x - 5y = -8$
$3x + y = 5$

2. $\dfrac{1}{3}x - \dfrac{1}{6}y = 3$
$-\dfrac{1}{5}x + \dfrac{1}{4}y = 0$

Solve the following systems by the substitution method. [4.1]

3. $2x - 5y = 14$
$y = 3x + 8$

4. $6x - 3y = 0$
$x + 2y = 5$

Solve the system. [4.2]

5. $2x - y + z = 9$
$x + y - 3z = -2$
$3x + y - z = 6$

6. Electric Current In the following diagram of an electrical circuit, x, y, and z represent the amount of current (in amperes) flowing across the 5-ohm, 20-ohm, and 10-ohm resistors, respectively. (In circuit diagrams, resistors are represented by ⟶⋀⋀⋀⟶ and potential differences by ⊣⊢.)

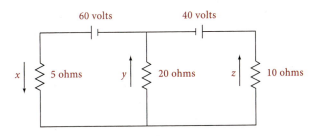

60 volts   40 volts

x 5 ohms y 20 ohms z 10 ohms

The system of equations used to find the three currents x, y, and z is

$$x - y - z = 0$$
$$5x + 20y = 60$$
$$20y - 10z = 40$$

Solve the system for all variables.

Solve each word problem. [4.3]

7. Number Problem A number is 1 less than twice another. Their sum is 14. Find the two numbers.

8. Investing John invests twice as much money at 6% as he does at 5%. If his investments earn a total of $680 in 1 year, how much does he have invested at each rate?

9. Ticket Cost There were 750 tickets sold for a basketball game for a total of $1,090. If adult tickets costs $2.00 and children's tickets cost $1.00, how many of each kind was sold?

10. Speed of a Boat A boat can travel 20 miles downstream in 2 hours. The same boat can travel 18 miles upstream in 3 hours. What is the speed of the boat in still water, and what is the speed of the current?

11. Coin Problem A collection of nickels, dimes, and quarters consists of 15 coins with a total value of $1.10. If the number of nickels is one less than 4 times the number of dimes, how many of each coin are contained in the collection?

Solve the following systems of equations by using matrices. [4.4]

12. $2x - y = -13$
$y = 4x + 23$

13. $3x - 4y - 3z = -3$
$x + 3y + 4z = 19$
$2y - 5z = -20$

Graph the solution set for each system of linear inequalities. [4.5]

14. $x + 4y \le 4$
$-3x + 2y > -12$

15. $y < -\dfrac{1}{2}x + 4$
$x \ge 0$
$y \ge 0$

Find a system of inequalities that describes the shaded region. [4.5]

16.

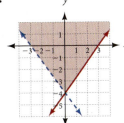

17.

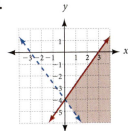

SYSTEMS OF LINEAR EQUATIONS AND INEQUALITIES

Group Project

Break-Even Point

Number of People	2 or 3
Time Needed	10–15 minutes
Equipment	Pencil and paper
Background	The break-even point for a company occurs when the revenue from sales of a product equals the cost of producing the product. This group project is designed to give you more insight into revenue, cost, and the break-even point.
Procedure	A company is planning to open a factory to manufacture calculators.

1. It costs them $120,000 to open the factory, and it will cost $10 for each calculator they make. What is the expression for $C(x)$, the cost of making x calculators?

2. They can sell the calculators for $50 each. What is the expression for $R(x)$, their revenue from selling x calculators? Remember that $R = px$, where p is the price per calculator.

3. Graph both the cost equation $C(x)$ and the revenue equation $R(x)$ on a coordinate system like the one below.

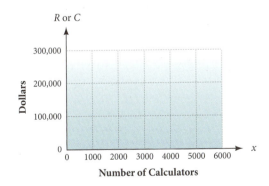

4. The break-even point is the value of x (the number of calculators) for which the revenue is equal to the cost. Where is the break-even point on the graph you produced in Part 3? Estimate the break-even point from the graph.

5. Set the cost equal to the revenue and solve for x to find the exact value of the break-even point. How many calculators do they need to make and sell to exactly break even? What will be their revenue and their cost for that many calculators?

6. Write an inequality that gives the values of x that will produce a profit for the company. (A profit occurs when the revenue is larger than the cost.)

7. Write an inequality that gives the values of x that will produce a loss for the company. (A loss occurs when the cost is larger than the revenue.)

8. Profit is the difference between revenue and cost, or $P(x) = R(x) - C(x)$. Write the equation for profit and then graph it on a coordinate system like the one below.

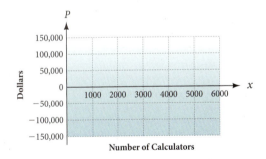

9. How do you recognize the break-even point and the regions of loss and profit on the graph you produced above?

Research Project

Zeno's Paradoxes

Zeno of Elea was born at about the same time that Pythagoras (of the Pythagorean theorem) died. He is responsible for three paradoxes that have come to be known as Zeno's paradoxes. One of the three has to do with a race between Achilles and a tortoise. Achilles is much faster than the tortoise, but the tortoise has a head start. According to Zeno's method of reasoning, Achilles can never pass the tortoise because each time he reaches the place where the tortoise was, the tortoise is gone. Research Zeno's paradox concerning Achilles and the tortoise. Put your findings into essay form that begins with a definition for the word "paradox." Then use Zeno's method of reasoning to describe a race between Achilles and the tortoise — if Achilles runs at 10 miles per hour, the tortoise runs at 1 mile per hour, and the tortoise has a 1-mile head start. Next, use the methods shown in this chapter to find the distance and the time at which Achilles reaches the tortoise. Conclude your essay by summarizing what you have done and showing how the two results you have obtained form a paradox.

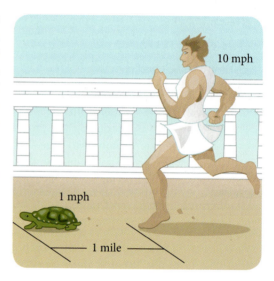

Exponents and Polynomials

Gray Buildings © 2008 Sanborn
Image © 2011 DigitalGlobe
Data SIO, NOAA, U.S. Navy, NGA, GEBCO
Empire State Building, New York, New York

The Empire State Building is a popular tourist destination in New York City. It hosts one of the most spectacular views of the city from its observation deck nearly a quarter of a mile in the sky. Since its grand opening in 1931, the building has welcomed more than 110,000,000 visitors. The building even has its own invitational athletic event called the Empire State Building Run Up. Every year, hundreds of athletes from around the world compete in this event by racing up 1,576 steps of the building's 86 flights. In 2011, German student Thomas Dold won the event with a time of 10 minutes and 10 seconds.

jovannig/Shutterstock.com

Let's examine a few of the numbers we presented about the Empire State Building. The number of steps in the building and even Thomas Dold's winning time are manageable numbers if we were to apply them to a math problem. But what about the number of visitors? In this chapter, we will work more with exponents, including a way to write large numbers, such as 110,000,000, in a more manageable form using exponents. This form is called scientific notation. Here is an example:

$$110{,}000{,}000 = \underline{1.1 \times 10^8}$$

Scientific notation

Preview

Key Words	Definition
Scientific Notation	A number written as the product of a number between 1 and 10 and an integer power of 10
Term (Monomial)	A constant or the product of a constant and one or more variables raised to a whole-number power
Polynomial	Any finite sum of terms
Degree of a Polynomial	The highest power to which the variable is raised in any one term
Similar Terms	Two or more terms that differ only in the numerical coefficient
Greatest Common Factor	The largest monomial that evenly divides each term of the polynomial
Quadratic Equation	Any equation that can be put in the form $ax^2 + bx + c = 0$ where a is not 0

Chapter Outline

5.1 Properties of Exponents
A Simplify expression using the properties of exponents.
B Convert back and forth between scientific notation and expanded form.
C Multiply and divide expressions written in scientific notation.

5.2 Polynomials, Sums, and Differences
A Give the degree of a polynomial.
B Add and subtract polynomials.
C Evaluate a polynomial for a given value of its variable.

5.3 Multiplication of Polynomials
A Multiply polynomials.
B Multiply binomials using the FOIL method.
C Find the square of a binomial.
D Multiply binomials to find the difference of two squares.

5.4 The Greatest Common Factor and Factoring by Grouping
A Factor out the greatest common factor.
B Factor by grouping.

5.5 Factoring Trinomials
A Factoring trinomials in which the leading coefficient is 1.
B Factor trinomials in which the leading coefficient is a number other than 1.

5.6 Special Factoring
A Factor perfect square trinomials.
B Factor the difference of two squares.
C Factor the sum or difference of two cubes.

5.7 Factoring: A General Review
A Factor a variety of polynomials.

5.8 Solving Equations by Factoring
A Solve equations by factoring.
B Apply the Blueprint for Problem Solving to solve application problems whose solutions involve quadratic equations.
C Solve problems that contain formulas that are quadratic.

Properties of Exponents

OBJECTIVES

A Simplify expressions using the properties of exponent.

B Convert back and forth between scientific notation and expanded form.

C Multiply and divide expressions written in scientific notation.

TICKET TO SUCCESS

Keep these questions in mind as you read through the section. Then respond in your own words and in complete sentences.

1. Using symbols, explain the product property for exponents.
2. Compare and contrast the power and the distributive property for exponents.
3. How does a negative integer exponent affect the base number?
4. What is scientific notation?

nanadou/Shutterstock.com

The figure shows a square and a cube, each with a side of length 2 centimeters. To find the area of the square, we raise 2 to the second power: 2^2. To find the volume of the cube, we raise 2 to the third power: 2^3.

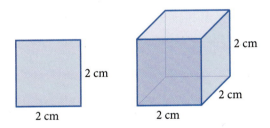

2 cm

2 cm

2 cm

2 cm

2 cm

Because the area of the square is 2^2, we say second powers are *squares;* that is, x^2 is read "x squared." Likewise, since the volume of the cube is 2^3, we say third powers are *cubes,* that is, x^3 is read "x cubed." Exponents and the vocabulary associated with them are topics we will study in this section.

A Properties of Exponents

In this section, we will be concerned with the simplification of expressions that involve exponents. We begin by making some generalizations about exponents, based on specific examples.

PRACTICE PROBLEMS

1. Write the product $x^5 \cdot x^4$ with a single exponent.

EXAMPLE 1 Write the product $x^3 \cdot x^4$ with a single exponent.

SOLUTION

$$x^3 \cdot x^4 = (x \cdot x \cdot x)(x \cdot x \cdot x \cdot x)$$
$$= (x \cdot x \cdot x \cdot x \cdot x \cdot x \cdot x)$$
$$= x^7 \qquad \text{Note: } 3 + 4 = 7 \qquad\blacksquare$$

We can generalize this result into the first property of exponents.

Product Property for Exponents

If a is a real number and r and s are integers, then

$$a^r \cdot a^s = a^{r+s}$$

2. Simplify $(4^5)^2$ by writing an equivalent expression with a single exponent.

EXAMPLE 2 Write $(5^3)^2$ with a single exponent.

SOLUTION

$$(5^3)^2 = 5^3 \cdot 5^3$$
$$= 5^6 \qquad \text{Note: } 3 \cdot 2 = 6 \qquad\blacksquare$$

Generalizing this result, we have a second property of exponents.

Power Property for Exponents

If a is a real number and r and s are integers, then

$$(a^r)^s = a^{r \cdot s}$$

A third property of exponents arises when we have the product of two or more numbers raised to an integer power.

3. Expand $(2x)^3$ and then multiply.

EXAMPLE 3 Expand $(3x)^4$ and then multiply.

SOLUTION

$$(3x)^4 = (3x)(3x)(3x)(3x)$$
$$= (3 \cdot 3 \cdot 3 \cdot 3)(x \cdot x \cdot x \cdot x)$$
$$= 3^4 \cdot x^4 \qquad \text{Note: The exponent 4 distributes over the product } 3x.$$
$$= 81x^4 \qquad\blacksquare$$

Generalizing Example 3, we have the third property for exponents.

Distributive Property for Exponents

If a and b are any two real numbers and r is an integer, then

$$(ab)^r = a^r \cdot b^r$$

Here are some examples that use combinations of the first three properties of exponents to simplify expressions involving exponents.

Simplify each expression.
4. $(-2x^3)(4x^5)$

EXAMPLE 4 Simplify $(-3x^2)(5x^4)$ using the properties of exponents.

SOLUTION

$$(-3x^2)(5x^4) = -3(5)(x^2 \cdot x^4) \qquad \text{Commutative and associative}$$
$$= -15x^6 \qquad \text{Product property} \qquad\blacksquare$$

Answers
1. x^9
2. 4^{10}
3. $8x^3$
4. $-8x^8$

EXAMPLE 5 Simplify $(-2x^2)^3(4x^5)$ using the properties of exponents.

SOLUTION

$$\begin{aligned}
(-2x^2)^3(4x^5) &= (-2)^3(x^2)^3(4x^5) && \text{Distributive property}\\
&= -8x^6 \cdot (4x^5) && \text{Power property}\\
&= (-8 \cdot 4)(x^6 \cdot x^5) && \text{Commutative and associative}\\
&= -32x^{11} && \text{Product property}
\end{aligned}$$ ∎

EXAMPLE 6 Simplify $(x^2)^4(x^2y^3)^2(y^4)^3$ using the properties of exponents.

SOLUTION

$$\begin{aligned}
(x^2)^4(x^2y^3)^2(y^4)^3 &= x^8 \cdot x^4 \cdot y^6 \cdot y^{12} && \text{Power and distributive properties}\\
&= x^{12}y^{18} && \text{Product property}
\end{aligned}$$ ∎

The next property of exponents deals with negative integer exponents.

> **Negative Exponent Property**
>
> If a is any nonzero real number and r is a positive integer, then
>
> $$a^{-r} = \frac{1}{a^r}$$

EXAMPLE 7 Write 5^{-2} with positive exponents, then simplify.

SOLUTION

$$5^{-2} = \frac{1}{5^2} = \frac{1}{25}$$ ∎

EXAMPLE 8 Write $(-2)^{-3}$ with positive exponents, then simplify.

SOLUTION

$$(-2)^{-3} = \frac{1}{(-2)^3} = \frac{1}{-8} = -\frac{1}{8}$$ ∎

EXAMPLE 9 Write $\left(\frac{3}{4}\right)^{-2}$ with positive exponents, then simplify.

SOLUTION

$$\left(\frac{3}{4}\right)^{-2} = \frac{1}{\left(\frac{3}{4}\right)^2} = \frac{1}{\frac{9}{16}} = \frac{16}{9}$$ ∎

If we generalize the result in Example 9, we have the following extension of the negative exponent property:

$$\left(\frac{a}{b}\right)^{-r} = \left(\frac{b}{a}\right)^r$$

which indicates that raising a fraction to a negative power is equivalent to raising the reciprocal of the fraction to the positive power.

The distributive property for exponents indicated that exponents distribute over products. Since division is defined in terms of multiplication, we can expect that exponents will distribute over quotients as well. The fifth property of exponents is the formal statement of this fact.

> **Expanded Distributive Property for Exponents**
>
> If a and b are any two real numbers with $b \neq 0$, and r is an integer, then
>
> $$\left(\frac{a}{b}\right)^r = \frac{a^r}{b^r}$$

Simplify each expression.

5. $(-3y^3)^3(2y^6)$

6. $(a^2)^3(a^3b^4)^2(b^5)^2$

> **NOTE**
> This property is actually a definition; that is, we are defining negative-integer exponents as indicating reciprocals. Doing so gives us a way to write an expression with a negative exponent as an equivalent expression with a positive exponent.

Write with positive exponents, then simplify.

7. 3^{-2}

8. $(-5)^{-3}$

Write with positive exponents, then simplify.

9. $\left(\frac{2}{3}\right)^{-2}$

Answers

5. $-54y^{15}$

6. $a^{12}b^{18}$

7. $\frac{1}{9}$

8. $-\frac{1}{125}$

9. $\frac{9}{4}$

Proof of the Expanded Distributive Property for Exponents

$$\left(\frac{a}{b}\right)^r = \underbrace{\left(\frac{a}{b}\right)\left(\frac{a}{b}\right)\left(\frac{a}{b}\right)\cdots\left(\frac{a}{b}\right)}_{r\ \text{factors}}$$

$$= \frac{a \cdot a \cdot a \cdots a}{b \cdot b \cdot b \cdots b} \quad \begin{array}{l}\leftarrow r\ \text{factors} \\ \leftarrow r\ \text{factors}\end{array}$$

$$= \frac{a^r}{b^r}$$

Since multiplication with the same base resulted in addition of exponents, it seems reasonable to expect division with the same base to result in subtraction of exponents.

> **Quotient Property for Exponents**
>
> If a is any nonzero real number, and r and s are any two integers, then
>
> $$\frac{a^r}{a^s} = a^{r-s}$$

Notice again that we have specified r and s to be any integers. Our definition of negative exponents is such that the properties of exponents hold for all integer exponents, whether positive or negative integers. Here is proof of the sixth property of exponents.

Proof of the Quotient Property for Exponents

Our proof is centered on the fact that division by a number is equivalent to multiplication by the reciprocal of the number.

$$\frac{a^r}{a^s} = a^r \cdot \frac{1}{a^s} \qquad \text{Dividing by } a^s \text{ is equivalent to multiplying by } \frac{1}{a^s}.$$

$$= a^r a^{-s} \qquad \text{Negative exponent property}$$

$$= a^{r+(-s)} \qquad \text{Product property}$$

$$= a^{r-s} \qquad \text{Definition of subtraction}$$

10. Apply the quotient property for exponents to each expression, and then simplify. All answers should contain positive exponents only.

a. $\dfrac{3^7}{3^4}$

b. $\dfrac{x^3}{x^{10}}$

c. $\dfrac{a^5}{a^{-7}}$

d. $\dfrac{m^{-3}}{m^{-5}}$

EXAMPLE 10 Apply the quotient property for exponents to each expression, and then simplify the result. All answers that contain exponents should contain positive exponents only.

SOLUTION

a. $\dfrac{2^8}{2^3} = 2^{8-3} = 2^5 = 32$

b. $\dfrac{x^2}{x^{18}} = x^{2-18} = x^{-16} = \dfrac{1}{x^{16}}$

c. $\dfrac{a^6}{a^{-8}} = a^{6-(-8)} = a^{14}$

d. $\dfrac{m^{-5}}{m^{-7}} = m^{-5-(-7)} = m^2$ ∎

Let's complete our list of properties by looking at how the numbers 0 and 1 behave when used as exponents.

We can use the original definition for exponents when the number 1 is used as an exponent.

$$a^1 = a \leftarrow 1 \text{ factor}$$

For 0 as an exponent, consider the expression $\dfrac{3^4}{3^4}$. Since $3^4 = 81$, we have

$$\frac{3^4}{3^4} = \frac{81}{81} = 1$$

Answers

10. a. 27 **b.** $\dfrac{1}{x^7}$ **c.** a^{12} **d.** m^2

However, because we have the quotient of two expressions with the same base, we can subtract exponents.

$$\frac{3^4}{3^4} = 3^{4-4} = 3^0$$

Hence, 3^0 must be the same as 1.

Summarizing these results, we have our last properties for exponents.

Identity Property for Exponents

If a is any real number, then

$$a^1 = a$$

Zero Exponent Property

$$a^0 = 1 \text{ if } a \neq 0$$

EXAMPLE 11 Simplify.

SOLUTION **a.** $(2x^2y^4)^0 = 1$

b. $(2x^2y^4)^1 = 2x^2y^4$ ∎

Here are some examples that use many of the properties of exponents. There are a number of ways to proceed on problems like these. You should use the method that works best for you.

EXAMPLE 12 Simplify $\dfrac{(x^3)^{-2}(x^4)^5}{(x^{-2})^7}$.

SOLUTION

$$\frac{(x^3)^{-2}(x^4)^5}{(x^{-2})^7} = \frac{x^{-6}x^{20}}{x^{-14}} \qquad \text{Power property}$$

$$= \frac{x^{14}}{x^{-14}} \qquad \text{Product property}$$

$$= x^{28} \qquad \text{Quotient property: } x^{14-(-14)} = x^{28}$$ ∎

EXAMPLE 13 Simplify $\dfrac{6a^5b^{-6}}{12a^3b^{-9}}$.

SOLUTION

$$\frac{6a^5b^{-6}}{12a^3b^{-9}} = \frac{6}{12} \cdot \frac{a^5}{a^3} \cdot \frac{b^{-6}}{b^{-9}} \qquad \text{Write as separate fractions.}$$

$$= \frac{1}{2}a^2b^3 \qquad \text{Quotient property}$$

Note: This last answer also can be written as $\dfrac{a^2b^3}{2}$. Either answer is correct. ∎

EXAMPLE 14 Simplify $\dfrac{(4x^{-5}y^3)^2}{(x^4y^{-6})^{-3}}$.

SOLUTION

$$\frac{(4x^{-5}y^3)^2}{(x^4y^{-6})^{-3}} = \frac{16x^{-10}y^6}{x^{-12}y^{18}} \qquad \text{Power and distributive properties}$$

$$= 16x^2y^{-12} \qquad \text{Quotient property}$$

$$= 16x^2 \cdot \frac{1}{y^{12}} \qquad \text{Negative exponent property}$$

$$= \frac{16x^2}{y^{12}} \qquad \text{Multiply.}$$ ∎

11. Simplify.
  a. $(3xy^5)^0$
  b. $(3xy^5)^1$

12. Simplify.
$$\frac{(x^2)^{-4}(x^3)^6}{(x^{-3})^5}$$

13. Simplify.
$$\frac{18a^7b^{-4}}{36a^2b^{-8}}$$

14. Simplify.
$$\frac{(3x^{-2}y^7)^2}{(x^5y^{-2})^{-4}}$$

Answers
11. a. 1 **b.** $3xy^5$
12. x^{25}
13. $\dfrac{a^5b^4}{2}$
14. $9x^{16}y^6$

B Scientific Notation

A black hole is a region in space that is so dense not even light can escape its gravitational pull. Astronomers have discovered a large bubble in space fueled by a black hole. This bubble is making its way through a nearby galaxy called NGC 7793, which is 12 million light years away. Large numbers, like 12 million, are sometimes difficult to work with because of their size.

We will now introduce a way in which to write very large or very small numbers in a more manageable form. This form is called *scientific notation*. Here is the definition.

> **Definition**
>
> A number is written in **scientific notation** if it is written as the product of a number between 1 and 10 and an integer power of 10. A number written in scientific notation has the form
>
> $$n \times 10^r$$
>
> where $1 \leq n < 10$ and $r =$ an integer.

15. Write 27,400 in scientific notation.

EXAMPLE 15   Write 376,000 in scientific notation.

SOLUTION   We must rewrite 376,000 as the product of a number between 1 and 10 and a power of 10. To do so, we move the decimal point five places to the left so that it appears between the 3 and the 7. Then we multiply this number by 10^5. The number that results has the same value as our original number and is written in scientific notation.   ■

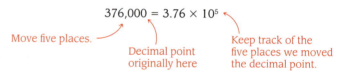

$$376{,}000 = 3.76 \times 10^5$$

Move five places.
Decimal point originally here
Keep track of the five places we moved the decimal point.

If a number written in expanded form is greater than or equal to 10, then when the number is written in scientific notation the exponent on 10 will be positive. A number that is less than 1 will have a negative exponent when written in scientific notation.

16. Write 3.91×10^4 in expanded form.

EXAMPLE 16   Write 4.52×10^3 in expanded form.

SOLUTION   Since 10^3 is 1,000, we can think of this as simply a multiplication problem; that is,

$$4.52 \times 10^3 = 4.52 \times 1{,}000 = 4{,}520$$

However, we can think of the exponent 3 as indicating the number of places we need to move the decimal point to write our number in expanded form. Since our exponent is positive 3, we move the decimal point three places to the right.

$$4.52 \times 10^3 = 4{,}520$$ ■

Answers

15. 2.74×10^4

16. 39,100

The following table lists some additional examples of numbers written in expanded form and in scientific notation. In each case, note the relationship between the number of places the decimal point is moved and the exponent on 10.

Number Written in Expanded Form		Number Written Again in Scientific Notation
376,000	=	3.76×10^5
49,500	=	4.95×10^4
3,200	=	3.2×10^3
591	=	5.91×10^2
46	=	4.6×10^1
8	=	8×10^0
0.47	=	4.7×10^{-1}
0.093	=	9.3×10^{-2}
0.00688	=	6.88×10^{-3}
0.0002	=	2×10^{-4}
0.000098	=	9.8×10^{-5}

USING TECHNOLOGY

Calculator Note Many calculators have a key that allows you to enter numbers in scientific notation. The key is labeled

$$\boxed{\textbf{EXP}} \text{ or } \boxed{\textbf{EE}} \text{ or } \boxed{\textbf{SCI}}$$

To enter the number 3.45×10^6, you first enter the decimal number, then press the scientific notation key, and finally enter the exponent.

$$3.45 \boxed{\textbf{EXP}} 6$$

We can use our properties of exponents to do arithmetic with numbers written in scientific notation. Here are some examples.

C Simplifying Expressions with Scientific Notation

EXAMPLE 17 Simplify each expression, and write all answers in scientific notation.

SOLUTION **a.** $(2 \times 10^8)(3 \times 10^{-3}) = (2)(3) \times (10^8)(10^{-3})$

$$= 6 \times 10^5$$

b. $\dfrac{4.8 \times 10^9}{2.4 \times 10^{-3}} = \dfrac{4.8}{2.4} \times \dfrac{10^9}{10^{-3}}$

$$= 2 \times 10^{9-(-3)}$$

$$= 2 \times 10^{12}$$

c. $\dfrac{(6.8 \times 10^5)(3.9 \times 10^{-7})}{7.8 \times 10^{-4}} = \dfrac{(6.8)(3.9)}{7.8} \times \dfrac{(10^5)(10^{-7})}{10^{-4}}$

$$= 3.4 \times 10^2 \qquad ■$$

USING TECHNOLOGY

Calculator Note On a scientific calculator with a scientific notation key, you would use the following sequence of keys to do Example 17 (b):

$$4.8 \boxed{\textbf{EXP}} 9 \boxed{÷} 2.4 \boxed{\textbf{EXP}} 3 \boxed{+/-} \boxed{=}$$

17. Simplify and write all answers in scientific notation.

   a. $(3 \times 10^7)(2 \times 10^{-4})$

   b. $\dfrac{3.9 \times 10^6}{1.3 \times 10^{-4}}$

   c. $\dfrac{(2.4 \times 10^6)(1.8 \times 10^{-4})}{1.2 \times 10^{-3}}$

NOTE

Remember, on some calculators the scientific notation key may be labeled $\boxed{\textbf{EE}}$ or $\boxed{\textbf{SCI}}$.

Answers

17. a. 6×10^3 **b.** 3×10^{10}

   c. 3.6×10^5

Problem Set 5.1

Moving Toward Success

"Vision without action is a daydream. Action without vision is a nightmare."

—Japanese proverb

1. Why do you think it is important to create pictures in your head as you learn mathematics?
2. Why can it be helpful to sometimes read this book or your notes out loud as you study?

Evaluate each of the following.

1. 4^2
2. $(-4)^2$
3. -4^2
4. $-(-4)^2$
5. -0.3^3
6. $(-0.3)^3$

7. 2^5
8. 2^4
9. $\left(\frac{1}{2}\right)^3$
10. $\left(\frac{3}{4}\right)^2$
11. $-\left(\frac{5}{6}\right)^2$
12. $-\left(\frac{7}{8}\right)^2$

A Use the properties of exponents to simplify each of the following as much as possible. [Examples 1–6]

13. $x^5 \cdot x^4$
14. $x^6 \cdot x^3$
15. $(2^3)^2$
16. $(3^2)^2$

17. $\left(-\frac{2}{3}x^2\right)^3$
18. $\left(-\frac{3}{5}x^4\right)^3$
19. $-3a^2(2a^4)$
20. $5a^7(-4a^6)$

A Write each of the following with positive exponents. Then simplify as much as possible. [Examples 7–9]

21. 3^{-2}
22. $(-5)^{-2}$
23. $(-2)^{-5}$
24. 2^{-5}

25. $\left(\frac{3}{4}\right)^{-2}$
26. $\left(\frac{3}{5}\right)^{-2}$
27. $\left(\frac{1}{3}\right)^{-2} + \left(\frac{1}{2}\right)^{-3}$
28. $\left(\frac{1}{2}\right)^{-2} + \left(\frac{1}{3}\right)^{-3}$

A Simplify each expression. Write all answers with positive exponents only. (Assume all variables are nonzero.)

29. $x^{-4}x^7$
30. $x^{-3}x^8$
31. $(a^2b^{-5})^3$

32. $(a^4b^{-3})^3$
33. $(5y^4)^{-3}(2y^{-2})^3$
34. $(3y^5)^{-2}(2y^{-4})^3$

35. $\left(\dfrac{1}{2}x^3\right)\left(\dfrac{2}{3}x^4\right)\left(\dfrac{3}{5}x^{-7}\right)$

36. $\left(\dfrac{1}{7}x^{-3}\right)\left(\dfrac{7}{8}x^{-5}\right)\left(\dfrac{8}{9}x^8\right)$

37. $(4a^5b^2)(2b^{-5}c^2)(3a^7c^4)$

38. $(3a^{-2}c^3)(5b^{-6}c^5)(4a^6b^{-2})$

39. $(2x^2y^{-5})^3(3x^{-4}y^2)^{-4}$

40. $(4x^{-4}y^9)^{-2}(5x^4y^{-3})^2$

A Use the properties of exponents to simplify each expression. Write all answers with positive exponents only. (Assume all variables are nonzero.) [Examples 10–14]

41. $\dfrac{x^{-1}}{x^9}$

42. $\dfrac{x^{-3}}{x^5}$

43. $\dfrac{a^4}{a^{-6}}$

44. $\dfrac{a^5}{a^{-2}}$

45. $\dfrac{t^{-10}}{t^{-4}}$

46. $\dfrac{t^{-8}}{t^{-5}}$

47. $\left(\dfrac{x^5}{x^3}\right)^6$

48. $\left(\dfrac{x^7}{x^4}\right)^5$

49. $\dfrac{(x^5)^6}{(x^3)^4}$

50. $\dfrac{(x^7)^3}{(x^4)^5}$

51. $\dfrac{(x^{-2})^3(x^3)^{-2}}{x^{10}}$

52. $\dfrac{(x^{-4})^3(x^3)^{-4}}{x^{10}}$

53. $\dfrac{5a^8b^3}{20a^5b^{-4}}$

54. $\dfrac{7a^6b^{-2}}{21a^2b^{-5}}$

55. $\dfrac{(3x^{-2}y^8)^4}{(9x^4y^{-3})^2}$

56. $\dfrac{(6x^{-3}y^{-5})^2}{(3x^{-4}y^{-3})^4}$

57. $\left(\dfrac{8x^2y}{4x^4y^{-3}}\right)^4$

58. $\left(\dfrac{5x^4y^5}{10xy^{-2}}\right)^3$

59. $\left(\dfrac{x^{-5}y^2}{x^{-3}y^5}\right)^{-2}$

60. $\left(\dfrac{x^{-8}y^{-3}}{x^{-5}y^6}\right)^{-1}$

Write each expression as a perfect square.

61. $x^4y^2 = (\quad)^2$

62. $x^8y^6 = (\quad)^2$

63. $9a^2b^4 = (\quad)^2$

64. $225x^6y^{12} = (\quad)^2$

Write each expression as a perfect cube.

65. $8a^3 = ($ $)^3$

66. $27b^3 = ($ $)^3$

67. $64x^3y^{12} = ($ $)^3$

68. $216x^{15}y^{21} = ($ $)^3$

69. Let $x = 2$ in each of the following expressions and simplify.

  a. x^3x^2

  b. $(x^3)^2$

  c. x^5

  d. x^6

70. Let $x = -1$ in each of the following expressions and simplify.

  a. x^3x^4

  b. $(x^3)^4$

  c. x^7

  d. x^{12}

71. Let $x = 2$ in each of the following expressions and simplify.

  a. $\dfrac{x^5}{x^2}$

  b. x^3

  c. $\dfrac{x^2}{x^6}$

  d. x^{-4}

72. Let $x = -1$ in each of the following expressions and simplify.

  a. $\dfrac{x^{14}}{x^9}$

  b. x^5

  c. $\dfrac{x^{13}}{x^9}$

  d. x^4

73. Write each expression as a perfect square.

  a. $\dfrac{1}{49} = ($ $)^2$

  b. $\dfrac{1}{121} = ($ $)^2$

  c. $\dfrac{1}{4x^2} = ($ $)^2$

  d. $\dfrac{1}{64x^4} = ($ $)^2$

74. Write each expression as a perfect cube.

  a. $\dfrac{1}{125x^3} = ($ $)^3$

  b. $\dfrac{1}{64y^{12}} = ($ $)^3$

  c. $\dfrac{x^6}{216y^9} = ($ $)^3$

  d. $\dfrac{8a^9}{27b^{15}} = ($ $)^3$

Simplify.

75. $2 \cdot 2^{n-1}$

76. $3 \cdot 3^{n-1}$

77. $\dfrac{ar^6}{ar^3}$

78. $\dfrac{ar^7}{ar^4}$

B Write each number in scientific notation. [Example 15]

79. 378,000

80. 3,780,000

81. 4,900

82. 490

83. 0.00037

84. 0.000037

85. 0.00495

86. 0.0495

B Write each number in expanded form. [Example 16]

87. 5.34×10^3

88. 5.34×10^2

89. 7.8×10^6

90. 7.8×10^4

91. 3.44×10^{-3}

92. 3.44×10^{-5}

93. 4.9×10^{-1}

94. 4.9×10^{-2}

C Use the properties of exponents to simplify each of the following expressions. Write all answers in scientific notation. [Example 17]

95. $(4 \times 10^{10})(2 \times 10^{-6})$

96. $(3 \times 10^{-12})(3 \times 10^4)$

97. $\dfrac{8 \times 10^{14}}{4 \times 10^5}$

98. $\dfrac{6 \times 10^8}{2 \times 10^3}$

99. $\dfrac{(5 \times 10^6)(4 \times 10^{-8})}{8 \times 10^4}$

100. $\dfrac{(6 \times 10^{-7})(3 \times 10^9)}{5 \times 10^6}$

Polynomials, Sums, and Differences

OBJECTIVES

A Give the degree of a polynomial.

B Add and subtract polynomials.

C Evaluate a polynomial for a given value of its variable.

TICKET TO SUCCESS

Keep these questions in mind as you read through the section. Then respond in your own words and in complete sentences.

1. What is a monomial?
2. What is the degree of a polynomial?
3. Why do you have to consider similar terms when adding or subtracting polynomials?
4. Write a problem that contains a polynomial and then restate it using function notation.

Quang Ho/Shutterstock.com

Suppose you have authored your first novel and are considering having an online self-publishing company print your work. The company's initial publishing package will cost $599. You've also chosen a royalty package that will pay you $3.00 for every printed copy sold. This $3.00 can be considered your personal revenue. Furthermore, you will need to sell at least 200 copies of your book to recoup your initial publishing cost. Any additional copies sold would generate a profit. In other words, profit is the difference between revenue and cost.

The relationship between profit, revenue, and cost is one application of the polynomials we will study in this section. Let's begin with a definition that we will use to build polynomials.

Polynomials in General

> **Definition**
>
> A **term**, or **monomial**, is a constant or the product of a constant and one or more variables raised to whole-number exponents.

The following are monomials, or terms:

$$-16 \qquad 3x^2y \qquad -\frac{2}{5}a^3b^2c \qquad xy^2z$$

The numerical part of each monomial is called the *numerical coefficient,* or just *coefficient*. For the preceding terms, the coefficients are -16, 3, $-\frac{2}{5}$, and 1. Notice that the coefficient for xy^2z is understood to be 1.

> **Definition**
>
> A **polynomial** is any finite sum of terms. Because subtraction can be written in terms of addition, finite differences are also included in this definition.

The following are polynomials:

$$2x^2 - 6x + 3 \qquad -5x^2y + 2xy^2 \qquad 4a - 5b + 6c + 7d$$

Polynomials can be classified further according to the number of terms present. If a polynomial consists of two terms, it is said to be a *binomial*. If it has three terms, it is called a *trinomial*. And, as stated, a polynomial with only one term is said to be a *monomial*.

A Degree of a Polynomial

> **Definition**
>
> The **degree** of a polynomial with one variable is the highest power to which the variable is raised in any one term.

PRACTICE PROBLEMS

1. Identify each expression as monomial, binomial, or trinomial, and give the degree of each.

a. $3x + 1$

b. $4x^2 + 2x + 5$

c. -17

d. $4x^5 - 7x^3$

e. $4x^3 - 5x^2 + 2x$

EXAMPLE 1   Identify each expression as monomial, binomial, or trinomial, and give the degree of each.

a.	$6x^2 + 2x - 1$	A trinomial of degree 2
b.	$5x - 3$	A binomial of degree 1
c.	$7x^6 - 5x^3 + 2x - 4$	A polynomial of degree 6
d.	$-7x^4$	A monomial of degree 4
e.	15	A monomial of degree 0 ∎

Polynomials in one variable are usually written in decreasing powers of the variable. When this is the case, the coefficient of the first term is called the *leading coefficient*. In Example 1a, the leading coefficient is 6. In Example 1b, it is 5. The leading coefficient in Example 1c is 7.

> **Definition**
>
> Two or more terms that differ only in their numerical coefficients are called **similar**, or **like**, terms. Since similar terms can differ only in their coefficients, they have identical variable parts.

B Addition and Subtraction of Polynomials

To add two polynomials, we simply apply the commutative and associative properties to group similar terms together and then use the distributive property as we have in the following example.

Answers

1. a. Binomial, 1

   b. Trinomial, 2

   c. Monomial, 0

   d. Binomial, 5

   e. Trinomial, 3

EXAMPLE 2 Add $5x^2 - 4x + 2$ and $3x^2 + 9x - 6$.

SOLUTION $(5x^2 - 4x + 2) + (3x^2 + 9x - 6)$

$$= (5x^2 + 3x^2) + (-4x + 9x) + (2 - 6)$$

Commutative and associative properties

$$= (5 + 3)x^2 + (-4 + 9)x + (2 - 6)$$

Distributive property

$$= 8x^2 + 5x + (-4)$$

$$= 8x^2 + 5x - 4$$ ∎

2. Add $3x^2 + 2x - 5$ and $2x^2 - 7x + 3$.

NOTE
In practice it is not necessary to show all the steps shown in Example 2. It is important to understand that addition of polynomials is equivalent to combining similar terms.

EXAMPLE 3 Find the sum of $-8x^3 + 7x^2 - 6x + 5$ and $10x^3 + 3x^2 - 2x - 6$.

SOLUTION We can add the two polynomials using the method in Example 2, or we can arrange similar terms in columns and add vertically. Using the column method, we have

$$\begin{array}{r} -8x^3 + 7x^2 - 6x + 5 \\ 10x^3 + 3x^2 - 2x - 6 \\ \hline 2x^3 + 10x^2 - 8x - 1 \end{array}$$ ∎

3. Find the sum of $x^3 + 7x^2 + 3x + 2$ and $-3x^3 - 2x^2 + 3x - 1$.

To find the difference of two polynomials, we need to use the fact that the opposite of a sum is the sum of the opposites; that is,

$$-(a + b) = -a + (-b)$$

One way to remember this is to observe that $-(a + b)$ is equivalent to $-1(a + b) = (-1)a + (-1)b = -a + (-b)$.

If a negative sign directly precedes the parentheses surrounding a polynomial, we may remove the parentheses and the preceding negative sign by changing the sign of each term within the parentheses. For example,

$$-(3x + 4) = -3x + (-4) = -3x - 4$$

$$-(5x^2 - 6x + 9) = -5x^2 + 6x - 9$$

$$-(-x^2 + 7x - 3) = x^2 - 7x + 3$$

EXAMPLE 4 Subtract $(9x^2 - 3x + 5) - (4x^2 + 2x - 3)$.

SOLUTION We subtract by adding the opposite of each term in the polynomial that follows the subtraction sign.

$$(9x^2 - 3x + 5) - (4x^2 + 2x - 3)$$

$$= 9x^2 - 3x + 5 + (-4x^2) + (-2x) + 3$$

The opposite of a sum is the sum of the opposites.

$$= (9x^2 - 4x^2) + (-3x - 2x) + (5 + 3)$$

Commutative and associative properties

$$= 5x^2 - 5x + 8$$

Combine similar terms. ∎

4. Subtract.
$(4x^2 - 2x + 7) - (7x^2 - 3x + 1)$

EXAMPLE 5 Subtract $4x^2 - 9x + 1$ from $-3x^2 + 5x - 2$.

SOLUTION Again, to subtract, we add the opposite.

$$(-3x^2 + 5x - 2) - (4x^2 - 9x + 1)$$

$$= -3x^2 + 5x - 2 - 4x^2 + 9x - 1$$

$$= (-3x^2 - 4x^2) + (5x + 9x) + (-2 - 1)$$

$$= -7x^2 + 14x - 3$$ ∎

5. Subtract $3x + 5$ from $7x - 4$.

Answers
2. $5x^2 - 5x - 2$
3. $-2x^3 + 5x^2 + 6x + 1$
4. $-3x^2 + x + 6$
5. $4x - 9$

6. Simplify $2x - 4[6 - (5x + 3)]$.

EXAMPLE 6

Simplify $4x - 3[2 - (3x + 4)]$.

SOLUTION Removing the innermost parentheses first, we have

$$4x - 3[2 - (3x + 4)] = 4x - 3(2 - 3x - 4)$$
$$= 4x - 3(-3x - 2)$$
$$= 4x + 9x + 6$$
$$= 13x + 6 \qquad \blacksquare$$

7. Simplify.
$(9x - 4) - [(2x + 5) - (x + 3)]$

EXAMPLE 7

Simplify $(2x + 3) - [(3x + 1) - (x - 7)]$.

SOLUTION

$$(2x + 3) - [(3x + 1) - (x - 7)] = (2x + 3) - (3x + 1 - x + 7)$$
$$= (2x + 3) - (2x + 8)$$
$$= 2x + 3 - 2x - 8$$
$$= -5 \qquad \blacksquare$$

C Evaluating Polynomials

In the example that follows, we will find the value of a polynomial for a given value of the variable.

8. Find the value of
$2x^3 - 3x^2 + 4x - 8$ when x is -2.

EXAMPLE 8

Find the value of $5x^3 - 3x^2 + 4x - 5$ when x is 2.

SOLUTION We begin by substituting 2 for x in the original polynomial:

When → $\qquad x = 2$

the polynomial → $5x^3 - 3x^2 + 4x - 5$

becomes → $\quad 5 \cdot 2^3 - 3 \cdot 2^2 + 4 \cdot 2 - 5 = 5 \cdot 8 - 3 \cdot 4 + 4 \cdot 2 - 5$

$$= 40 - 12 + 8 - 5$$
$$= 31 \qquad \blacksquare$$

Polynomials and Function Notation

Example 8 can be restated using function notation by calling the polynomial $P(x)$ and asking for $P(2)$. The solution would look like this:

If $\qquad P(x) = 5x^3 - 3x^2 + 4x - 5$

then $\qquad P(2) = 5 \cdot 2^3 - 3 \cdot 2^2 + 4 \cdot 2 - 5$

$$= 31$$

Our next example is stated in terms of function notation.

Three functions that occur very frequently in business and economics classes are profit, revenue, and cost functions. If a company manufactures and sells x items, then the revenue $R(x)$ is the total amount of money obtained by selling all x items. The cost $C(x)$ is the total amount of money it costs the company to manufacture the x items. The profit $P(x)$ obtained by selling all x items is the difference between the revenue and the cost and is given by the equation

$$P(x) = R(x) - C(x)$$

Answers
6. $22x - 12$
7. $8x - 6$
8. -44

Problem Set 5.2

Moving Toward Success

"To the man who only has a hammer in the toolkit, every problem looks like a nail."

—Abraham Maslow, 1908–1970, American psychologist

1. A mnemonic device is a mental tool that uses an acronym or a short verse to help a person remember. Do you think a mnemonic device is a helpful learning tool? Why?

2. Have you used a mnemonic device to help you study mathematics? If so, what was it? If not, create one for this section.

A   Identify those of the following that are monomials, binomials, or trinomials. Give the degree of each, and name the leading coefficient. [Example 1]

1. $5x^2 - 3x + 2$

2. $2x^2 + 4x - 1$

3. $3x - 5$

4. $5y + 3$

5. $8a^2 + 3a - 5$

6. $9a^2 - 8a - 4$

7. $4x^3 - 6x^2 + 5x - 3$

8. $9x^4 + 4x^3 - 2x^2 + x$

9. $-\dfrac{3}{4}$

10. -16

11. $4x - 5 + 6x^3$

12. $9x + 2 + 3x^3$

B   Simplify each of the following by combining similar terms. [Examples 2–5]

13. $(4x + 2) + (3x - 1)$

14. $(8x - 5) + (-5x + 4)$

15. $2x^2 - 3x + 10x - 15$

16. $6x^2 - 4x - 15x + 10$

17. $12a^2 + 8ab - 15ab - 10b^2$

18. $28a^2 - 8ab + 7ab - 2b^2$

19. $(5x^2 - 6x + 1) - (4x^2 + 7x - 2)$

20. $(11x^2 - 8x) - (4x^2 - 2x - 7)$

21. $\left(\frac{1}{2}x^2 - \frac{1}{3}x - \frac{1}{6}\right) - \left(\frac{1}{4}x^2 + \frac{7}{12}x\right) + \left(\frac{1}{3}x - \frac{1}{12}\right)$

22. $\left(\frac{2}{3}x^2 - \frac{1}{2}x\right) - \left(\frac{1}{4}x^2 + \frac{1}{6}x\right) + \frac{1}{12} - \left(\frac{1}{2}x^2 + \frac{1}{4}\right)$

23. $(y^3 - 2y^2 - 3y + 4) - (2y^3 - y^2 + y - 3)$

24. $(8y^3 - 3y^2 + 7y + 2) - (-4y^3 + 6y^2 - 5y - 8)$

25. $(5x^3 - 4x^2) - (3x + 4) + (5x^2 - 7) - (3x^3 + 6)$

26. $(x^3 - x) - (x^2 + x) + (x^3 - 1) - (-3x + 2)$

27. $\left(\frac{4}{7}x^2 - \frac{1}{7}xy + \frac{1}{14}y^2\right) - \left(\frac{1}{2}x^2 - \frac{2}{7}xy - \frac{9}{14}y^2\right)$

28. $\left(\frac{1}{5}x^2 - \frac{1}{2}xy + \frac{1}{10}y^2\right) - \left(-\frac{3}{10}x^2 + \frac{2}{5}xy - \frac{1}{2}y^2\right)$

29. $(3a^3 + 2a^2b + ab^2 - b^3) - (6a^3 - 4a^2b + 6ab^2 - b^3)$

30. $(a^3 - 3a^2b + 3ab^2 - b^3) - (a^3 + 3a^2b + 3ab^2 + b^3)$

31. Subtract $2x^2 - 4x$ from $2x^2 - 7x$.

32. Subtract $-3x + 6$ from $-3x + 9$.

33. Find the sum of $x^2 - 6xy + y^2$ and $2x^2 - 6xy - y^2$.

34. Find the sum of $9x^3 - 6x^2 + 2$ and $3x^2 - 5x + 4$.

35. Subtract $-8x^5 - 4x^3 + 6$ from $9x^5 - 4x^3 - 6$.

36. Subtract $4x^4 - 3x^3 - 2x^2$ from $2x^4 + 3x^3 + 4x^2$.

37. Find the sum of $11a^2 + 3ab + 2b^2$, $9a^2 - 2ab + b^2$, and $-6a^2 - 3ab + 5b^2$.

38. Find the sum of $a^2 - ab - b^2$, $a^2 + ab - b^2$, and $a^2 + 2ab + b^2$.

B Simplify each of the following. Begin by working within the innermost parentheses. [Examples 6, 7]

39. $-[2 - (4 - x)]$

40. $-[-3 - (x - 6)]$

41. $-5[-(x - 3) - (x + 2)]$

42. $-6[(2x - 5) - 3(8x - 2)]$

43. $4x - 5[3 - (x - 4)]$

44. $x - 7[3x - (2 - x)]$

45. $-(3x - 4y) - [(4x + 2y) - (3x + 7y)]$

46. $(8x - y) - [-(2x + y) - (-3x - 6y)]$

47. $4a - (3a + 2[a - 5(a + 1) + 4])$

48. $6a - (-2a - 6[2a + 3(a - 1) - 6])$

C [Example 8]

49. Find the value of $2x^2 - 3x - 4$ when x is 2.

50. Find the value of $4x^2 + 3x - 2$ when x is -1.

51. If $P(x) = \dfrac{3}{2}x^2 - \dfrac{3}{4}x + 1$, find

   a. $P(12)$

   b. $P(-8)$

52. If $P(x) = \dfrac{2}{5}x^2 - \dfrac{1}{10}x + 2$, find

   a. $P(10)$

   b. $P(-10)$

53. If $Q(x) = x^3 - x^2 + x - 1$, find

   a. $Q(4)$

   b. $Q(-2)$

54. If $Q(x) = x^3 + x^2 + x - 1$, find

   a. $Q(5)$

   b. $Q(-2)$

55. If $R(x) = 11.5x - 0.05x^2$, find

   a. $R(10)$

   b. $R(-10)$

56. If $R(x) = 11.5x - 0.01x^2$, find

   a. $R(10)$

   b. $R(-10)$

57. If $P(x) = 600 + 1{,}000x - 100x^2$, find

   a. $P(-4)$

   b. $P(4)$

58. If $P(x) = 500 + 800x - 100x^2$, find

   a. $P(-6)$

   b. $P(8)$

Applying the Concepts

59. Education The chart shows the average income for people with different levels of education. In a high school's graduating class, x students plan to get their Bachelor's Degree and y students plan to go on and get their Master's Degree. The next year, twice as many students plan to get their Bachelor's Degree than the year before but the number of students that plan to go on and get their Master's Degree stays the same as the year before. Write an expression that describes the total income of each year and then find the total income for both years in terms of x and y.

60. Energy Usage The chart shows how much energy is used by different gaming systems. If twice as many Wiis are being played as PS3s and one PC is being played, write an expression that describes the energy used in this situation. Then find the value if 2 PS3s were played.

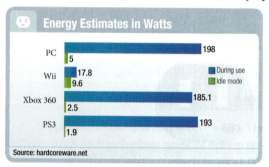

C Problems 61–66 may be solved using a graphing calculator.

61. Height of an Object If an object is thrown straight up into the air with a velocity of 128 feet/second, then its height $h(t)$ above the ground t seconds later is given by the formula

$$h(t) = -16t^2 + 128t$$

Find the height after 3 seconds and after 5 seconds. [Find $h(3)$ and $h(5)$.]

62. Height of an Object The formula for the height of an object that has been thrown straight up with a velocity of 64 feet/second is

$$h(t) = -16t^2 + 64t$$

Find the height after 1 second and after 3 seconds. [Find $h(1)$ and $h(3)$.]

63. Profits The total cost (in dollars) for a company to manufacture and sell x items per week is $C(x) = 60x + 300$. If the revenue brought in by selling all x items is $R(x) = 100x - 0.5x^2$, find the weekly profit. How much profit will be made by producing and selling 60 items each week?

64. Profits The total cost (in dollars) for a company to produce and sell x items per week is $C(x) = 200x + 1{,}600$. If the revenue brought in by selling all x items is $R(x) = 300x - 0.6x^2$, find the weekly profit. How much profit will be made by producing and selling 50 items each week?

65. Profits Suppose it costs a company selling patterns $C(x) = 800 + 6.5x$ dollars to produce and sell x patterns a month. If the revenue obtained by selling x patterns is $R(x) = 10x - 0.002x^2$, what is the profit equation? How much profit will be made if 1,000 patterns are produced and sold in May?

66. Profits Suppose a company manufactures and sells x picture frames each month with a total cost of $C(x) = 1{,}200 + 3.5x$ dollars. If the revenue obtained by selling x frames is $R(x) = 9x - 0.003x^2$, find the profit equation. How much profit will be made if 1,000 frames are manufactured and sold in June?

Getting Ready for the Next Section

Simplify.

67. $2x^2 - 3x + 10x - 15$

68. $12a^2 + 8ab - 15ab - 10b^2$

69. $(6x^3 - 2x^2y + 8xy^2) + (-9x^2y + 3xy^2 - 12y^3)$

70. $(3x^3 - 15x^2 + 18x) + (2x^2 - 10x + 12)$

71. $4x^3(-3x)$

72. $5x^2(-4x)$

73. $4x^3(5x^2)$

74. $5x^2(3x^2)$

75. $(a^3)^2$

76. $(a^4)^2$

77. $11.5(130) - 0.05(130)^2$

78. $-0.05(130)^2 + 9.5(130) - 200$

Maintaining Your Skills

Simplify each expression.

79. $-1(5 - x)$

80. $-1(a - b)$

81. $-1(7 - x)$

82. $-1(6 - y)$

83. $5\left(x - \dfrac{1}{5}\right)$

84. $7\left(x + \dfrac{1}{7}\right)$

85. $x\left(1 - \dfrac{1}{x}\right)$

86. $a\left(1 + \dfrac{1}{a}\right)$

87. $12\left(\dfrac{1}{4}x + \dfrac{2}{3}y\right)$

88. $20\left(\dfrac{2}{5}x + \dfrac{1}{4}y\right)$

Extending the Concepts

89. The graphs of two polynomial functions are given in Figures 1 and 2. Use the graphs to find the following.

a. $f(-3)$ **b.** $f(0)$ **c.** $f(1)$ **d.** $g(-1)$

e. $g(0)$ **f.** $g(2)$ **g.** $f(g(2))$ **h.** $g(f(2))$

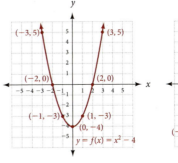

FIGURE 1

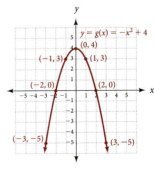

FIGURE 2

Multiplication of Polynomials

TICKET TO SUCCESS

Keep these questions in mind as you read through the section. Then respond in your own words and in complete sentences.

1. In words, state the rule for multiplying polynomials.
2. How would you use the FOIL method to multiply two binomials?
3. Write a problem that demonstrates how to expand a binomial square.
4. Discuss the rule for the difference of two squares.

OBJECTIVES

A Multiply polynomials.

B Multiply binomials using the FOIL method.

C Find the square of a binomial.

D Multiply binomials to find the difference of two squares.

Abel Tumik/Shutterstock.com

Imagine you play for your university's club volleyball team. Your club has decided to sell t-shirts adorned with the club's logo to help pay for new uniforms. Each t-shirt costs \$15. In order to calculate the total revenue generated from t-shirt sales, you must multiply the number of shirts sold x by the price of each shirt p. This relationship can be given by the formula

$$R = xp$$

This formula is extremely common in the business world. Many times, x and p are polynomials, which means that the expression xp is the product of two polynomials. In this section, we learn how to multiply polynomials, and in so doing, increase our understanding of the equations and formulas that describe business applications.

A Multiplying Polynomials

EXAMPLE 1 Find the product of $4x^3$ and $5x^2 - 3x + 1$.

SOLUTION To multiply, we apply the distributive property.

$$4x^3(5x^2 - 3x + 1)$$
$$= 4x^3(5x^2) + 4x^3(-3x) + 4x^3(1) \qquad \text{Distributive property}$$
$$= 20x^5 - 12x^4 + 4x^3$$

Notice that we multiply coefficients and add exponents. ∎

PRACTICE PROBLEMS

1. Multiply $5x^2(3x^2 - 4x + 2)$.

Answer
1. $15x^4 - 20x^3 + 10x^2$

2. Multiply $4x - 1$ and $x + 3$.

EXAMPLE 2   Multiply $2x - 3$ and $x + 5$.

SOLUTION   Distributing the $2x - 3$ across the sum $x + 5$ gives us

$$(2x - 3)(x + 5)$$
$$= (2x - 3)x + (2x - 3)5 \qquad \text{\color{red}Distributive property}$$
$$= 2x(x) + (-3)x + 2x(5) + (-3)5 \qquad \text{\color{red}Distributive property}$$
$$= 2x^2 - 3x + 10x - 15$$
$$= 2x^2 + 7x - 15 \qquad \text{\color{red}Combine like terms.}$$

Notice the third line in this example. It consists of all possible products of terms in the first binomial and those of the second binomial. We can generalize this into a rule for multiplying two polynomials. ■

> **Rule Multiplying Polynomials**
>
> To multiply two polynomials, multiply each term in the first polynomial by each term in the second polynomial.

Multiplying polynomials can be accomplished by a method that looks very similar to long multiplication with whole numbers.

3. Multiply vertically.
$(3x + 2)(x^2 - 5x + 6)$

EXAMPLE 3   Multiply $(2x - 3y)$ and $(3x^2 - xy + 4y^2)$ vertically.

SOLUTION

$$
\begin{array}{r}
3x^2 - \quad xy + \quad 4y^2 \\
2x - \quad 3y \\
\hline
6x^3 - \quad 2x^2y + \quad 8xy^2 \qquad \text{\color{red}Multiply } (3x^2 - xy + 4y^2) \text{ by } 2x. \\
- \quad 9x^2y + \quad 3xy^2 - 12y^3 \qquad \text{\color{red}Multiply } (3x^2 - xy + 4y^2) \text{ by } -3y. \\
\hline
6x^3 - 11x^2y + 11xy^2 - 12y^3 \qquad \text{\color{red}Add similar terms.}
\end{array}
$$

■

> **NOTE**
> The vertical method of multiplying polynomials does not directly show the use of the distributive property. It is, however, very useful since it always gives the correct result and is easy to remember.

B Multiplying Binomials—The FOIL Method

Consider the product of $(2x - 5)$ and $(3x - 2)$. Distributing $(3x - 2)$ over $2x$ and -5, we have

$$(2x - 5)(3x - 2) = (2x)(3x - 2) + (-5)(3x - 2)$$
$$= (2x)(3x) + (2x)(-2) + (-5)(3x) + (-5)(-2)$$
$$= 6x^2 - 4x - 15x + 10$$
$$= 6x^2 - 19x + 10$$

Looking closely at the second and third lines, we notice the following:

1. $6x^2$ comes from multiplying the *first* terms in each binomial:

$$(2x - 5)(3x - 2) \qquad 2x(3x) = 6x^2 \qquad \text{\color{red}\textit{First} terms}$$

2. $-4x$ comes from multiplying the *outside* terms in the product:

$$(2x - 5)(3x - 2) \qquad 2x(-2) = -4x \qquad \text{\color{red}\textit{Outside} terms}$$

3. $-15x$ comes from multiplying the *inside* terms in the product:

$$(2x - 5)(3x - 2) \qquad -5(3x) = -15x \qquad \text{\color{red}\textit{Inside} terms}$$

> **NOTE**
> The FOIL method does not show the properties used in multiplying two binomials. It is simply a way of finding products of binomials quickly. Remember, the FOIL method applies only to products of two binomials. The vertical method applies to all products of polynomials with two or more terms.

Answers

2. $4x^2 + 11x - 3$

3. $3x^3 - 13x^2 + 8x + 12$

4. 10 comes from multiplying the *last* two terms in the product:

$$(2x - 5)(3x - 2) \qquad -5(-2) = 10 \qquad \textit{Last terms}$$

Once we know where the terms in the answer come from, we can reduce the number of steps used in finding the product:

$$(2x - 5)(3x - 2) = 6x^2 - 4x - 15x + 10 = 6x^2 - 19x + 10$$

First Outside Inside Last

EXAMPLE 4 Multiply $(4a - 5b)(3a + 2b)$ using the FOIL method.

SOLUTION $(4a - 5b)(3a + 2b) = 12a^2 + 8ab - 15ab - 10b^2$

F   O   I   L

$$= 12a^2 - 7ab - 10b^2 \qquad \blacksquare$$

EXAMPLE 5 Multiply $(3 - 2t)(4 + 7t)$ using the FOIL method.

SOLUTION $(3 - 2t)(4 + 7t) = 12 + 21t - 8t - 14t^2$

F   O   I   L

$$= 12 + 13t - 14t^2 \qquad \blacksquare$$

EXAMPLE 6 Multiply $\left(2x + \dfrac{1}{2}\right)\left(4x - \dfrac{1}{2}\right)$ using the FOIL method.

SOLUTION $\left(2x + \dfrac{1}{2}\right)\left(4x - \dfrac{1}{2}\right) = 8x^2 - x + 2x - \dfrac{1}{4} = 8x^2 + x - \dfrac{1}{4}$

F   O   I   L $\qquad \blacksquare$

EXAMPLE 7 Multiply $(a^5 + 3)(a^5 - 7)$ using the FOIL method.

SOLUTION $(a^5 + 3)(a^5 - 7) = a^{10} - 7a^5 + 3a^5 - 21$

F   O   I   L

$$= a^{10} - 4a^5 - 21 \qquad \blacksquare$$

EXAMPLE 8 Multiply $(2x + 3)(5y - 4)$ using the FOIL method.

SOLUTION $(2x + 3)(5y - 4) = 10xy - 8x + 15y - 12$

F   O   I   L $\qquad \blacksquare$

C The Square of a Binomial

EXAMPLE 9 Find $(4x - 6)^2$.

SOLUTION Applying the definition of exponents and then the FOIL method, we have

$$(4x - 6)^2 = (4x - 6)(4x - 6)$$

$$= 16x^2 - 24x - 24x + 36$$

F   O   I   L

$$= 16x^2 - 48x + 36 \qquad \blacksquare$$

Multiply using the FOIL method.

4. $(2a - 3b)(5a - b)$

5. $(6 - 3t)(2 + 5t)$

6. $\left(3x + \dfrac{1}{4}\right)\left(4x - \dfrac{1}{3}\right)$

7. $(a^4 - 2)(a^4 + 5)$

8. $(7x - 2)(3y + 8)$

9. Expand and multiply $(3x - 2)^2$.

Answers

4. $10a^2 - 17ab + 3b^2$

5. $12 + 24t - 15t^2$

6. $12x^2 - \dfrac{1}{12}$

7. $a^8 + 3a^4 - 10$

8. $21xy + 56x - 6y - 16$

9. $9x^2 - 12x + 4$

This example is the square of a binomial. This type of product occurs frequently enough in algebra that we have special formulas for it. Here are the formulas for binomial squares:

$$(a + b)^2 = (a + b)(a + b) = a^2 + ab + ab + b^2 = a^2 + 2ab + b^2$$

$$(a - b)^2 = (a - b)(a - b) = a^2 - ab - ab + b^2 = a^2 - 2ab + b^2$$

Observing the results in both cases, we have the following rule.

> **Rule Squaring a Binomial**
>
> The square of a binomial is the sum of the square of the first term, twice the product of the two terms, and the square of the last term. Or:
>
> $$(a + b)^2 = \quad a^2 \quad + \quad 2ab \quad + \quad b^2$$
>
> | Square of first term | Twice the product of the two terms | Square of last term |
>
> $$(a - b)^2 = \quad a^2 \quad - \quad 2ab \quad + \quad b^2$$

10. Expand and simplify.

a. $(x - y)^2$

b. $(4t + 5)^2$

c. $(5x - 3y)^2$

d. $(6 - a^4)^2$

EXAMPLE 10   Use the preceding formulas to expand each binomial square.

a. $(x + 7)^2 = x^2 + 2(x)(7) + 7^2 = x^2 + 14x + 49$

b. $(3t - 5)^2 = (3t)^2 - 2(3t)(5) + 5^2 = 9t^2 - 30t + 25$

c. $(4x + 2y)^2 = (4x)^2 + 2(4x)(2y) + (2y)^2 = 16x^2 + 16xy + 4y^2$

d. $(5 - a^3)^2 = 5^2 - 2(5)(a^3) + (a^3)^2 = 25 - 10a^3 + a^6$ ■

D Products Resulting in the Difference of Two Squares

Another frequently occurring kind of product is found when multiplying two binomials that differ only in the sign between their terms.

11. Multiply $(4x - 3)$ and $(4x + 3)$.

EXAMPLE 11   Multiply $(3x - 5)$ and $(3x + 5)$.

SOLUTION   Applying the FOIL method, we have

$$(3x - 5)(3x + 5) = 9x^2 + 15x - 15x - 25 \qquad \text{Two middle terms add to 0}$$

$$ \text{F} \qquad \text{O} \qquad \text{I} \qquad \text{L}$$

$$= 9x^2 - 25 \qquad ■$$

The outside and inside products in Example 11 are opposites and therefore add to 0. Here it is in general:

$$(a - b)(a + b) = a^2 + ab - ab - b^2 \qquad \text{Two middle terms add to 0}$$

$$= a^2 - b^2$$

> **Rule Difference of Two Squares**
>
> To multiply two binomials that differ only in the sign between their two terms, simply subtract the square of the second term from the square of the first term:
>
> $$(a + b)(a - b) = a^2 - b^2$$

The expression $a^2 - b^2$ is called the *difference of two squares*.

Answers

10. a. $x^2 - 2xy + y^2$

b. $16t^2 + 40t + 25$

c. $25x^2 - 30xy + 9y^2$

d. $36 - 12a^4 + a^8$

11. $16x^2 - 9$

EXAMPLE 12 Find the following products.

    a. $(x - 5)(x + 5) = x^2 - 25$

    b. $(2a - 3)(2a + 3) = 4a^2 - 9$

    c. $(x^2 + 4)(x^2 - 4) = x^4 - 16$

    d. $(x^3 - 2a)(x^3 + 2a) = x^6 - 4a^2$ ■

12. Multiply.
    a. $(x + 2)(x - 2)$
    b. $(5a + 7)(5a - 7)$
    c. $(x^3 + 4)(x^3 - 4)$
    d. $(x^4 - 5a)(x^4 + 5a)$

More About Function Notation

Recall that the revenue obtained from selling x items at p dollars per item is

$$R = \text{Revenue} = xp \qquad \text{(The number of items} \times \text{price per item)}$$

For example, if a store sells 100 items at \$4.50 per item, the revenue is $100(4.50) = \$450$. If we have an equation that gives the relationship between x and p, then we can write the revenue in terms of x or in terms of p. With function notation, we would write the revenue as either $R(x)$ or $R(p)$, where

    $R(x)$ is the revenue function that gives the revenue R in terms of the number of items x, and

    $R(p)$ is the revenue function that gives the revenue R in terms of the price per item p.

With function notation we can see exactly which variables we want our formulas written in terms of.

    In the next two examples, we will use function notation to combine a number of problems we have worked previously.

EXAMPLE 13 A company manufactures and sells DVDs. They find that they can sell x DVDs each day at p dollars per disc, according to the equation $x = 230 - 20p$. Find $R(x)$ and $R(p)$.

13. Repeat Example 13 using the equation
$$x = 250 - 25p$$

SOLUTION The notation $R(p)$ tells us we are to write the revenue equation in terms of the variable p. To do so, we use the formula $R(p) = xp$ and substitute $230 - 20p$ for x to obtain

$$R(p) = xp = (230 - 20p)p = 230p - 20p^2$$

The notation $R(x)$ indicates that we are to write the revenue in terms of the variable x. We need to solve the equation $x = 230 - 20p$ for p. Let's begin by interchanging the two sides of the equation:

$$230 - 20p = x$$

    $-20p = -230 + x$ Add -230 to each side.

    $p = \dfrac{-230 + x}{-20}$ Divide each side by -20.

    $p = 11.5 - 0.05x$ $\frac{230}{20} = 11.5$ and $\frac{1}{20} = 0.05$

Now we can find $R(x)$ by substituting $11.5 - 0.05x$ for p in the formula $R(x) = xp$:

$$R(x) = xp = x(11.5 - 0.05x) = 11.5x - 0.05x^2$$

Answers
12. a. $x^2 - 4$
    b. $25a^2 - 49$
    c. $x^6 - 16$
    d. $x^8 - 25a^2$
13. $R(p) = 250p - 25p^2$
    $R(x) = 10x - 0.04x^2$

Our two revenue functions are actually equivalent. To offer some justification for this, suppose that the company decides to sell each disc for $5. The equation $x = 230 - 20p$ indicates that, at $5 per disc, they will sell $x = 230 - 20(5) = 230 - 100 = 130$ discs per day. To find the revenue from selling the discs for $5 each, we use $R(p)$ with $p = 5$.

$$\text{If} \qquad p = 5$$
$$\text{then} \qquad R(p) = R(5)$$
$$= 230(5) - 20(5)^2$$
$$= 1{,}150 - 500$$
$$= \$650$$

However, to find the revenue from selling 130 discs, we use $R(x)$ with $x = 130$.

$$\text{If} \qquad x = 130$$
$$\text{then} \qquad R(x) = R(130)$$
$$= 11.5(130) - 0.05(130)^2$$
$$= 1{,}495 - 845$$
$$= \$650 \qquad \blacksquare$$

EXAMPLE 14 Suppose the daily cost function for the DVDs in Example 13 is $C(x) = 200 + 2x$. Find the profit function $P(x)$ and then find $P(130)$.

SOLUTION Since profit is equal to the difference of the revenue and the cost, we have

$$P(x) = R(x) - C(x)$$
$$= 11.5x - 0.05x^2 - (200 + 2x)$$
$$= -0.05x^2 + 9.5x - 200$$

Notice that we used the formula for $R(x)$ from Example 13 instead of the formula for $R(p)$. We did so because we were asked to find $P(x)$, meaning we want the profit P only in terms of the variable x.

Next, we use the formula we just obtained to find $P(130)$.

$$P(130) = -0.05(130)^2 + 9.5(130) - 200$$
$$= -0.05(16{,}900) + 9.5(130) - 200$$
$$= -845 + 1{,}235 - 200$$
$$= \$190$$

Because $P(130) = \$190$, the company will make a profit of $190 per day by selling 130 discs per day. $\blacksquare$

14. Find $P(x)$ and $P(130)$ if
$$x = 250 - 25p$$
and
$$C(x) = 210 + 1.5x$$

Answers
14. $P(x) = -0.04x^2 + 8.5x - 210$;
$P(130) = \$219$

Problem Set 5.3

Moving Toward Success

"When what we are is what we want to be, that's happiness."

—Malcolm S. Forbes, 1917–1990, American publisher

1. What traits do you like in a study partner?
2. Have you possessed those traits when studying for this class? Why or why not?

A Multiply the following by applying the distributive property. [Examples 1, 2]

1. $2x(6x^2 - 5x + 4)$

2. $-3x(5x^2 - 6x - 4)$

3. $-3a^2(a^3 - 6a^2 + 7)$

4. $4a^3(3a^2 - a + 1)$

5. $2a^2b(a^3 - ab + b^3)$

6. $5a^2b^2(8a^2 - 2ab + b^2)$

A Multiply the following vertically. [Example 3]

7. $(x - 5)(x + 3)$

8. $(x + 4)(x + 6)$

9. $(2x^2 - 3)(3x^2 - 5)$

10. $(3x^2 + 4)(2x^2 - 5)$

11. $(x + 3)(x^2 + 6x + 5)$

12. $(x - 2)(x^2 - 5x + 7)$

13. $(a - b)(a^2 + ab + b^2)$

14. $(a + b)(a^2 - ab + b^2)$

15. $(2x + y)(4x^2 - 2xy + y^2)$

16. $(x - 3y)(x^2 + 3xy + 9y^2)$

17. $(2a - 3b)(a^2 + ab + b^2)$

18. $(5a - 2b)(a^2 - ab - b^2)$

B Multiply the following using the FOIL method. [Examples 4–8]

19. $(x - 2)(x + 3)$

20. $(x + 2)(x - 3)$

21. $(2a + 3)(3a + 2)$

22. $(5a - 4)(2a + 1)$

23. $(5 - 3t)(4 + 2t)$

24. $(7 - t)(6 - 3t)$

25. $(x^3 + 3)(x^3 - 5)$

26. $(x^3 + 4)(x^3 - 7)$

27. $(5x - 6y)(4x + 3y)$

28. $(6x - 5y)(2x - 3y)$

29. $\left(3t + \dfrac{1}{3}\right)\left(6t - \dfrac{2}{3}\right)$

30. $\left(5t - \dfrac{1}{5}\right)\left(10t + \dfrac{3}{5}\right)$

C D Find the following special products. [Examples 9–12]

31. $(5x + 2y)^2$

32. $(3x - 4y)^2$

33. $(5 - 3t^3)^2$

34. $(7 - 2t^4)^2$

35. $(2a + 3b)(2a - 3b)$

36. $(6a - 1)(6a + 1)$

37. $(3r^2 + 7s)(3r^2 - 7s)$

38. $(5r^2 - 2s)(5r^2 + 2s)$

39. $\left(y + \dfrac{3}{2}\right)^2$

40. $\left(y - \dfrac{7}{2}\right)^2$

41. $\left(a - \dfrac{1}{2}\right)^2$

42. $\left(a - \dfrac{5}{2}\right)^2$

43. $\left(x + \dfrac{1}{4}\right)^2$

44. $\left(x - \dfrac{3}{8}\right)^2$

45. $\left(t + \dfrac{1}{3}\right)^2$

46. $\left(t - \dfrac{2}{5}\right)^2$

47. $\left(\dfrac{1}{3}x - \dfrac{2}{5}\right)\left(\dfrac{1}{3}x + \dfrac{2}{5}\right)$

48. $\left(\dfrac{3}{4}x - \dfrac{1}{7}\right)\left(\dfrac{3}{4}x + \dfrac{1}{7}\right)$

Find the following products.

49. $(x - 2)^3$

50. $(4x + 1)^3$

51. $\left(x - \dfrac{1}{2}\right)^3$

52. $\left(x + \dfrac{1}{4}\right)^3$

53. $3(x - 1)(x - 2)(x - 3)$

54. $2(x + 1)(x + 2)(x + 3)$

55. $(b^2 + 8)(a^2 + 1)$

56. $(b^2 + 1)(a^4 - 5)$

57. $(x + 1)^2 + (x + 2)^2 + (x + 3)^2$

58. $(x - 1)^2 + (x - 2)^2 + (x - 3)^2$

59. $(2x + 3)^2 - (2x - 3)^2$

60. $(x - 3)^3 - (x + 3)^3$

Applying the Concepts

Solar and Wind Energy The chart shows the cost to install either solar panels or a wind turbine. Use the chart to answer Problems 69 and 70.

61. A homeowner is buying a certain number of solar panel modules. He is going to get a discount on each module that is equal to 25 dollars for each module he buys. Write an equation that describes this situation, then simplify and find the cost if he buys 3 modules.

Solar Versus Wind Energy Costs

Solar Energy Equipment Cost:		Wind Energy Equipment Cost:	
Modules	$6200	Turbine	$3300
Fixed Rack	$1570	Tower	$3000
Charge Controller	$971	Cable	$715
Cable	$440		
TOTAL	**$9181**	**TOTAL**	**$7015**

Source: a Limited

62. A farmer is replacing several turbines in his field. He is going to get a discount that is equal to 50 dollars for each turbine he buys. Write an expression that describes this situation, then simplify and find the cost if he replaces 5 turbines.

63. Revenue A store selling art supplies finds that it can sell x sketch pads per week at p dollars each, according to the formula $x = 900 - 300p$. Write formulas for $R(p)$ and $R(x)$. Then find the revenue obtained by selling the pads for $1.60 each.

64. Revenue A company selling CDs finds that it can sell x CDs per day at p dollars per CD, according to the formula $x = 800 - 100p$. Write formulas for $R(p)$ and $R(x)$. Then find the revenue obtained by selling the CDs for $3.80 each.

65. Revenue A company sells an inexpensive accounting program for home computers. If it can sell x programs per week at p dollars per program, according to the formula $x = 350 - 10p$, find formulas for $R(p)$ and $R(x)$. How much will the weekly revenue be if it sells 65 programs?

66. Revenue A company sells boxes of greeting cards through the mail. It finds that it can sell x boxes of cards each week at p dollars per box, according to the formula $x = 1,475 - 250p$. Write formulas for $R(p)$ and $R(x)$. What revenue will it bring in each week if it sells 200 boxes of cards?

67. Profit If the cost to produce the x programs in Problem 73 is $C(x) = 5x + 500$, find $P(x)$ and $P(60)$.

68. Profit If the cost to produce the x CDs in Problem 72 is $C(x) = 2x + 200$, find $P(x)$ and $P(40)$.

69. Interest If you deposit $100 in an account with an interest rate r that is compounded annually, then the amount of money in that account at the end of 4 years is given by the formula $A = 100(1 + r)^4$. Expand the right side of this formula.

70. Interest If you deposit P dollars in an account with an annual interest rate r that is compounded twice a year, then at the end of a year the amount of money in that account is given by the formula

$$A = P\left(1 + \frac{r}{2}\right)^2$$

Expand the right side of this formula.

Getting Ready for the Next Section

71. $\dfrac{8a^3}{a}$

72. $\dfrac{-8a^2}{a}$

73. $\dfrac{-48a}{a}$

74. $\dfrac{-32a}{a}$

75. $\dfrac{16a^5b^4}{8a^2b^3}$

76. $\dfrac{12x^4y^5}{3x^3y^3}$

77. $\dfrac{-24a^5b^5}{8a^5b^3}$

78. $\dfrac{-15x^5y^3}{3x^3y^3}$

79. $\dfrac{x^3y^4}{-x^3}$

80. $\dfrac{x^2y^2}{-x^2}$

Maintaining Your Skills

Solve.

81. $x + y + z = 6$
$2x - y + z = 3$
$x + 2y - 3z = -4$

82. $x + y + z = 6$
$x - y + 2z = 7$
$2x - y - z = 0$

83. $3x + 4y = 15$
$2x - 5z = -3$
$4y - 3z = 9$

84. $x + 3y = 5$
$6y + z = 12$
$x - 2z = -10$

Extending the Concepts

85. Multiply $(x + y - 4)(x + y + 5)$ by first writing it like this:

$$[(x + y) - 4][(x + y) + 5]$$

and then applying the FOIL method.

86. Multiply $(x - 5 - y)(x - 5 + y)$ by first writing it like this:

$$[(x - 5) - y][(x - 5) + y]$$

and then applying the FOIL method.

Assume n is a positive integer and multiply.

87. $(x^n - 2)(x^n - 3)$

88. $(x^{2n} + 3)(x^{2n} - 3)$

89. $(2x^n + 3)(5x^n - 1)$

90. $(4x^n - 3)(7x^n + 2)$

91. $(x^n + 5)^2$

92. $(x^n - 2)^2$

93. $(x^n + 1)(x^{2n} - x^n + 1)$

94. $(x^{3n} - 3)(x^{6n} + 3x^{3n} + 9)$

The Greatest Common Factor and Factoring by Grouping

TICKET TO SUCCESS

Keep these questions in mind as you read through the section. Then respond in your own words and in complete sentences.

1. What is the relationship between factoring and multiplication?
2. What is the greatest common factor for a polynomial?
3. After factoring a polynomial, how can you check your result?
4. When would you try to factor by grouping?

Alexandru Cristian Ciobanu/Shutterstock.com

Picture a football player throwing a ball straight up in the air. The distance the ball is from the ground can be represented by the polynomial equations $d = -16t^2 + 80t$ where t is the time the ball is in the air. If we needed to solve this equation, our first step would be to factor out the greatest common factor.

In this section, we will begin our work factoring polynomials. In general, factoring is the reverse of multiplication. The diagram here illustrates the relationship between factoring and multiplication. Reading from left to right, we say the product of 3 and 7 is 21. Reading in the other direction, from right to left, we say 21 factors into 3 times 7. Or, 3 and 7 are factors of 21.

Multiplication

Factors → 3 · 7 = 21 ← Product

Factoring

A Greatest Common Factor

> **Definition**
>
> The **greatest common factor** for a polynomial is the largest monomial that divides (is a factor of) each term of the polynomial.

PRACTICE PROBLEMS

Factor the greatest common factor from the following polynomials.
1. $15a^7 - 25a^5 + 30a^3$

2. $12x^4y^5 - 9x^3y^4 - 15x^5y^3$

3. $4(a + b)^4 - 6(a + b)^3 + 16(a + b)^2$

4. If the total weekly revenue for the company mentioned in Example 4 is
$$R(x) = 80.5x - 0.35x^2$$
find the formula that gives the price p in terms of x.

The greatest common factor for the polynomial $25x^5 + 20x^4 - 30x^3$ is $5x^3$ since it is the largest monomial that is a factor of each term. We can apply the distributive property and write

$$25x^5 + 20x^4 - 30x^3 = 5x^3(5x^2) + 5x^3(4x) - 5x^3(6)$$
$$= 5x^3(5x^2 + 4x - 6)$$

The last line is written in factored form.

EXAMPLE 1   Factor the greatest common factor from $8a^3 - 8a^2 - 48a$.

SOLUTION   The greatest common factor is $8a$. It is the largest monomial that divides each term of our polynomial. We can write each term in our polynomial as the product of $8a$ and another monomial. Then, we apply the distributive property to factor $8a$ from each term.

$$8a^3 - 8a^2 - 48a = 8a(a^2) - 8a(a) - 8a(6)$$
$$= 8a(a^2 - a - 6)$$ ∎

EXAMPLE 2   Factor the greatest common factor from
$$16a^5b^4 - 24a^2b^5 - 8a^3b^3.$$

SOLUTION   The largest monomial that divides each term is $8a^2b^3$. We write each term of the original polynomial in terms of $8a^2b^3$ and apply the distributive property to write the polynomial in factored form.

$$16a^5b^4 - 24a^2b^5 - 8a^3b^3 = 8a^2b^3(2a^3b) - 8a^2b^3(3b^2) - 8a^2b^3(a)$$
$$= 8a^2b^3(2a^3b - 3b^2 - a)$$ ∎

EXAMPLE 3   Factor the greatest common factor from
$$5x^2(a + b) - 6x(a + b) - 7(a + b).$$

SOLUTION   The greatest common factor is $a + b$. Factoring it from each term, we have

$$5x^2(a + b) - 6x(a + b) - 7(a + b) = (a + b)(5x^2 - 6x - 7)$$ ∎

EXAMPLE 4   A company manufacturing DVDs finds that the total daily revenue for selling x DVDs is given by

$$R(x) = 11.5x - 0.05x^2$$

Factor x from each term on the right side of the equation to find the formula that gives the price p in terms of x.

SOLUTION   We begin by factoring x from the right side of the equation.

If $R(x) = 11.5x - 0.05x^2$

then $R(x) = x(11.5 - 0.05x)$

Because R is always xp, the quantity in parentheses must be p. The price it should charge if it wants to sell x items per day is therefore

$$p = 11.5 - 0.05x$$ ∎

Answers
1. $5a^3(3a^4 - 5a^2 + 6)$
2. $3x^3y^3(4xy^2 - 3y - 5x^2)$
3. $2(a + b)^2[2(a + b)^2 - 3(a + b) + 8]$
4. $p = 80.5 - 0.35x$

B Factoring by Grouping

The polynomial $5x + 5y + x^2 + xy$ can be factored by noticing that the first two terms have a 5 in common, whereas the last two have an x in common. Applying the distributive property, we have

$$5x + 5y + x^2 + xy = 5(x + y) + x(x + y)$$

This last expression can be thought of as having two terms, $5(x + y)$ and $x(x + y)$, each of which has a common factor $(x + y)$. We apply the distributive property again to factor $(x + y)$ from each term.

$$5(x + y) + x(x + y) = (x + y)(5 + x)$$

EXAMPLE 5 Factor $a^2b^2 + b^2 + 8a^2 + 8$.

SOLUTION The first two terms have b^2 in common; the last two have 8 in common.

$$a^2b^2 + b^2 + 8a^2 + 8 = b^2(a^2 + 1) + 8(a^2 + 1)$$

$$= (a^2 + 1)(b^2 + 8) ∎$$

EXAMPLE 6 Factor $15 - 5y^4 - 3x^3 + x^3y^4$.

SOLUTION Let's try factoring a 5 from the first two terms and an $-x^3$ from the last two terms.

$$15 - 5y^4 - 3x^3 + x^3y^4 = 5(3 - y^4) - x^3(3 - y^4)$$
$$= (3 - y^4)(5 - x^3) ∎$$

EXAMPLE 7 Factor $x^3 + 2x^2 + 9x + 18$.

SOLUTION We begin by factoring x^2 from the first two terms and 9 from the second two terms.

$$x^3 + 2x^2 + 9x + 18 = x^2(x + 2) + 9(x + 2)$$
$$= (x + 2)(x^2 + 9) ∎$$

Factor each polynomial by grouping.
5. $ab^3 + b^3 + 6a + 6$

6. $15 - 3y^2 - 5x^2 + x^2y^2$

7. $x^3 + 5x^2 + 3x + 15$

Answers
5. $(a + 1)(b^3 + 6)$
6. $(5 - y^2)(3 - x^2)$
7. $(x + 5)(x^2 + 3)$

Problem Set 5.4

Moving Toward Success

"Kind words do not cost much. Yet they accomplish much."

—Blaise Pascal, 1623–1662, French mathematician

1. What would you do if your study partner is constantly complaining about the class or wanting to quit?

2. What are some other methods you can use to prevent negative thoughts toward this class?

A Factor the greatest common factor from each of the following. [Examples 1–4]

1. $10x^3 - 15x^2$

2. $12x^5 + 18x^7$

3. $9y^6 + 18y^3$

4. $24y^4 - 8y^2$

5. $9a^2b - 6ab^2$

6. $30a^3b^4 + 20a^4b^3$

7. $21xy^4 + 7x^2y^2$

8. $14x^6y^3 - 6x^2y^4$

9. $3a^2 - 21a + 30$

10. $3a^2 - 3a - 6$

11. $4x^3 - 16x^2 - 20x$

12. $2x^3 - 14x^2 + 20x$

13. $10x^4y^2 + 20x^3y^3 - 30x^2y^4$

14. $6x^4y^2 + 18x^3y^3 - 24x^2y^4$

15. $-x^2y + xy^2 - x^2y^2$

16. $-x^3y^2 - x^2y^3 - x^2y^2$

17. $4x^3y^2z - 8x^2y^2z^2 + 6xy^2z^3$

18. $7x^4y^3z^2 - 21x^2y^2z^2 - 14x^2y^3z^4$

19. $20a^2b^2c^2 - 30ab^2c + 25a^2bc^2$

20. $8a^3bc^5 - 48a^2b^4c + 16ab^3c^5$

21. $5x(a - 2b) - 3y(a - 2b)$

22. $3a(x - y) - 7b(x - y)$

23. $3x^2(x + y)^2 - 6y^2(x + y)^2$

24. $10x^3(2x - 3y) - 15x^2(2x - 3y)$

25. $2x^2(x + 5) + 7x(x + 5) + 6(x + 5)$

26. $2x^2(x + 2) + 13x(x + 2) + 15(x + 2)$

B Factor each of the following by grouping. [Examples 5–7]

27. $3xy + 3y + 2ax + 2a$

28. $5xy^2 + 5y^2 + 3ax + 3a$

29. $x^2y + x + 3xy + 3$

30. $x^3y^3 + 2x^3 + 5x^2y^3 + 10x^2$

31. $3xy^2 - 6y^2 + 4x - 8$

32. $8x^2y - 4x^2 + 6y - 3$

33. $x^2 - ax - bx + ab$

34. $ax - x^2 - bx + ab$

35. $ab + 5a - b - 5$

36. $x^2 - xy - ax + ay$

37. $a^4b^2 + a^4 - 5b^2 - 5$

38. $2a^2 - a^2b - bc^2 + 2c^2$

39. $x^3 + 3x^2 - 4x - 12$

40. $x^3 + 5x^2 - 4x - 20$

41. $x^3 + 2x^2 - 25x - 50$

42. $x^3 + 4x^2 - 9x - 36$

43. $2x^3 + 3x^2 - 8x - 12$

44. $3x^3 + 2x^2 - 27x - 18$

45. $4x^3 + 12x^2 - 9x - 27$

46. $9x^3 + 18x^2 - 4x - 8$

47. The greatest common factor of the binomial $3x - 9$ is 3. The greatest common factor of the binomial $6x - 2$ is 2. What is the greatest common factor of their product, $(3x - 9)(6x - 2)$, when it has been multiplied out?

48. The greatest common factors of the binomials $5x - 10$ and $2x + 4$ are 5 and 2, respectively. What is the greatest common factor of their product, $(5x - 10)(2x + 4)$, when it has been multiplied out?

Applying the Concepts

49. Investing If P dollars are placed in a savings account in which the rate of interest r is compounded yearly, then at the end of one year the amount of money in the account can be written as $P + Pr$. At the end of two years the amount of money in the account is

$$P + Pr + (P + Pr)r$$

Use factoring by grouping to show that this last expression can be written as $P(1 + r)^2$.

50. Investing At the end of 3 years, the amount of money in the savings account in Problem 49 will be

$$P(1 + r)^2 + P(1 + r)^2 r$$

Use factoring to show that this last expression can be written as $P(1 + r)^3$.

Use Example 4 as a guide in solving the next four problems.

51. Price A manufacturing company that produces prerecorded DVDs finds that the total daily revenue R for selling x items at p dollars per item is given by

$$R(x) = 11.5x - 0.05x^2$$

Factor x from each term on the right side of the equation to find the formula that gives the price p in terms of x. Then, use it to find the price they should charge if they want to sell 125 DVDs per day.

52. Price A company producing CDs finds that the total daily revenue for selling x items at p dollars per item is given by

$$R(x) = 8x - 0.01x^2$$

Use the fact that $R = xp$ and your knowledge of factoring to find a formula that gives the price p in terms of x. Then, use it to find the price they should charge if they want to sell 420 CDs per day.

53. Price The weekly revenue equation for a company selling an inexpensive accounting program for home computers is given by the equation

$$R(x) = 35x - 0.1x^2$$

where x is the number of programs they sell per week. What price p should they charge if they want to sell 65 programs per week?

54. Price The weekly revenue equation for a small mail-order company selling boxes of greeting cards is

$$R(x) = 5.9x - 0.004x^2$$

where x is the number of boxes they sell per week. What price p should they charge if they want to sell 200 boxes each week?

Getting Ready for the Next Section

Factor out the greatest common factor.

55. $3x^4 - 9x^3y - 18x^2y^2$

56. $5x^2 + 10x + 30$

57. $2x^2(x - 3) - 4x(x - 3) - 3(x - 3)$

58. $3x^2(x - 2) - 8x(x - 2) + 2(x - 2)$

Multiply.

59. $(x + 2)(3x - 1)$

60. $(x - 2)(3x + 1)$

61. $(x - 1)(3x - 2)$

62. $(x + 1)(3x + 2)$

63. $(x + 2)(x + 3)$

64. $(x - 2)(x - 3)$

65. $(2y + 5)(3y - 7)$

66. $(2y - 5)(3y + 7)$

67. $(4 - 3a)(5 - a)$

68. $(4 - 3a)(5 + a)$

Maintaining Your Skills

Solve the following systems of equations.

69. $2x + 5y = 4$
$3x - 2y = -13$

70. $6x + 3y = -15$
$3x - 9y = 17$

71. $2x + 6y = -12$
$y = x - 14$

72. $4x - 6y = 2$
$x = 5y - 10$

Factoring Trinomials

5.5

OBJECTIVES

A Factor trinomials in which the leading coefficient is 1.

B Factor trinomials in which the leading coefficient is a number other than 1.

TICKET TO SUCCESS

Keep these questions in mind as you read through the section. Then respond in your own words and in complete sentences.

1. What is a prime polynomial?
2. What is the first step in factoring a polynomial with a lead coefficient other than 1?
3. When would you factor a polynomial using trial and error?
4. Briefly explain the method of factoring a polynomial by grouping.

Condor/Shutterstock.com

Suppose you toss a large shell upward with an initial velocity of 16 feet per second and it lands into the ocean 32 feet below. The equation that gives the height of the shell at any time is $h = 32 + 16t - 16t^2$. How would you find the height of the shell at any given time? In this section, we will expand our work with factoring to look at how to factor several different kinds of trinomials.

A Factoring Trinomials with a Leading Coefficient of 1

Earlier in this chapter, we multiplied binomials.

$$(x - 2)(x + 3) = x^2 + x - 6$$

$$(x + 5)(x + 2) = x^2 + 7x + 10$$

In each case, the product of two binomials is a trinomial. The first term in the resulting trinomial is obtained by multiplying the first term in each binomial. The middle term comes from adding the product of the two inside terms with the product of the two outside terms. The last term is the product of the last terms in each binomial.

In general,

$$(x + a)(x + b) = x^2 + ax + bx + ab$$

$$= x^2 + (a + b)x + ab$$

Writing this as a factoring problem, we have

$$x^2 + (a + b)x + ab = (x + a)(x + b)$$

To factor a trinomial with a leading coefficient of 1, we simply find the two numbers a and b whose sum is the coefficient of the middle term and whose product is the constant term.

EXAMPLE 1 Factor $x^2 + 2x - 15$.

SOLUTION Since the leading coefficient is 1, we need two integers whose product is -15 and whose sum is 2. The integers are 5 and -3.

$$x^2 + 2x - 15 = (x + 5)(x - 3)$$

In the preceding example, we found factors of $x + 5$ and $x - 3$. These are the only two such factors for $x^2 + 2x - 15$. There is no other pair of binomials $x + a$ and $x + b$ whose product is $x^2 + 2x - 15$. ∎

EXAMPLE 2 Factor $x^2 - xy - 12y^2$.

SOLUTION We need two expressions whose product is $-12y^2$ and whose sum is $-y$. The expressions are $-4y$ and $3y$.

$$x^2 - xy - 12y^2 = (x - 4y)(x + 3y)$$

Checking this result gives

$$(x - 4y)(x + 3y) = x^2 + 3xy - 4xy - 12y^2$$
$$= x^2 - xy - 12y^2$$ ∎

EXAMPLE 3 Factor $x^2 - 8x + 6$.

SOLUTION Since there is no pair of integers whose product is 6 and whose sum is -8, the trinomial $x^2 - 8x + 6$ is not factorable. We say it is a *prime polynomial*. ∎

B Factoring When the Lead Coefficient Is Not 1

EXAMPLE 4 Factor $3x^4 - 15x^3y - 18x^2y^2$.

SOLUTION The leading coefficient is not 1. Each term is divisible by $3x^2$, however. Factoring this out to begin with we have

$$3x^4 - 15x^3y - 18x^2y^2 = 3x^2(x^2 - 5xy - 6y^2)$$

Factoring the resulting trinomial as in the previous examples gives

$$3x^2(x^2 - 5xy - 6y^2) = 3x^2(x - 6y)(x + y)$$ ∎

Factoring Other Trinomials by Trial and Error

We want to turn our attention now to trinomials with leading coefficients other than 1 and with no greatest common factor other than 1.

Suppose we want to factor $3x^2 - x - 2$. The factors will be a pair of binomials. The product of the first terms will be $3x^2$, and the product of the last terms will be -2. We can list all the possible factors along with their products as follows.

PRACTICE PROBLEMS

1. Factor $x^2 - x - 12$.

2. Factor $x^2 + 2xy - 15y^2$.

3. Factor $x^2 + x + 1$.

4. Factor $5x^2 + 25x + 30$.

NOTE
As a general rule, it is best to factor out the greatest common factor first.

Answers
1. $(x - 4)(x + 3)$
2. $(x + 5y)(x - 3y)$
3. Does not factor.
4. $5(x + 3)(x + 2)$

Possible Factors	First Term	Middle Term	Last Term
$(x + 2)(3x - 1)$	$3x^2$	$+5x$	-2
$(x - 2)(3x + 1)$	$3x^2$	$-5x$	-2
$(x + 1)(3x - 2)$	$3x^2$	$+x$	-2
$(x - 1)(3x + 2)$	$3x^2$	$-x$	-2

From the last line we see that the factors of $3x^2 - x - 2$ are $(x - 1)(3x + 2)$. That is,

$$3x^2 - x - 2 = (x - 1)(3x + 2)$$

To factor trinomials with leading coefficients other than 1, when the greatest common factor is 1, we must use trial and error or list all the possible factors. In either case the idea is this: look only at pairs of binomials whose products give the correct first and last terms, then look for the combination that will give the correct middle term.

EXAMPLE 5 Factor $2x^2 + 13xy + 15y^2$.

SOLUTION Listing all possible factors the product of whose first terms is $2x^2$ and the product of whose last terms is $15y^2$ yields

Possible Factors	Middle Term of Product
$(2x - 5y)(x - 3y)$	$-11xy$
$(2x - 3y)(x - 5y)$	$-13xy$
$(2x + 5y)(x + 3y)$	$+11xy$
$(2x + 3y)(x + 5y)$	$+13xy$
$(2x + 15y)(x + y)$	$+17xy$
$(2x - 15y)(x - y)$	$-17xy$
$(2x + y)(x + 15y)$	$+31xy$
$(2x - y)(x - 15y)$	$-31xy$

The fourth line has the correct middle term.

$$2x^2 + 13xy + 15y^2 = (2x + 3y)(x + 5y)$$

Actually, we did not need to check the first two pairs of possible factors in the preceding list. Because all the signs in the trinomial $2x^2 + 13xy + 15y^2$ are positive, the binomial factors must be of the form $(ax + b)(cx + d)$, where a, b, c, and d are all positive. ▪

There are other ways to reduce the number of possible factors to consider. For example, if we were to factor the trinomial $2x^2 - 11x + 12$, we would not have to consider the pair of possible factors $(2x - 4)(x - 3)$. If the original trinomial has no greatest common factor other than 1, then neither of its binomial factors will either. The trinomial $2x^2 - 11x + 12$ has a greatest common factor of 1, but the possible factor $2x - 4$ has a greatest common factor of 2: $2x - 4 = 2(x - 2)$. Therefore, we do not need to consider $2x - 4$ as a possible factor.

EXAMPLE 6 Factor $12x^4 + 17x^2 + 6$.

SOLUTION This is a trinomial in x^2:

$$12x^4 + 17x^2 + 6 = (4x^2 + 3)(3x^2 + 2)$$ ▪

5. Factor $3x^2 - x - 2$.

6. Factor $15x^4 + x^2 - 2$.

Answers

5. $(3x + 2)(x - 1)$

6. $(5x^2 + 2)(3x^2 - 1)$

7. Factor
$3x^2(x - 2) - 7x(x - 2) + 2(x - 2)$.

Factor $2x^2(x - 3) - 5x(x - 3) - 3(x - 3)$.

SOLUTION   We begin by factoring out the greatest common factor $(x - 3)$. Then we factor the trinomial that remains.

$$2x^2(x - 3) - 5x(x - 3) - 3(x - 3) = (x - 3)(2x^2 - 5x - 3)$$

$$= (x - 3)(2x + 1)(x - 3)$$

$$= (x - 3)^2(2x + 1) \quad \blacksquare$$

Another Method of Factoring Trinomials

As an alternative to the trial-and-error method of factoring trinomials, we present the following method. The new method does not require as much trial and error. To use this new method, we must rewrite our original trinomial in such a way that the factoring by grouping method can be applied.

Here are the steps we use to factor $ax^2 + bx + c$.

Step 1: Form the product ac.
Step 2: Find a pair of numbers whose product is ac and whose sum is b.
Step 3: Rewrite the polynomial to be factored so that the middle term bx is written as the sum of two terms whose coefficients are the two numbers found in step 2.
Step 4: Factor by grouping.

8. Factor $8x^2 - 5x - 3$.

Factor $3x^2 - 10x - 8$ using these steps.

SOLUTION   The trinomial $3x^2 - 10x - 8$ has the form $ax^2 + bx + c$, where $a = 3$, $b = -10$, and $c = -8$.

Step 1: The product ac is $3(-8) = -24$.
Step 2: We need to find two numbers whose product is -24 and whose sum is -10. Let's list all the pairs of numbers whose product is -24 to find the pair whose sum is -10.

Product	Sum
$1(-24) = -24$	$1 + (-24) = -23$
$-1(24) = -24$	$-1 + 24 = 23$
$2(-12) = -24$	$2 + (-12) = -10$
$-2(12) = -24$	$-2 + 12 = 10$
$3(-8) = -24$	$3 + (-8) = -5$
$-3(8) = -24$	$-3 + 8 = 5$
$4(-6) = -24$	$4 + (-6) = -2$
$-4(6) = -24$	$-4 + 6 = 2$

As you can see, of all the pairs of numbers whose product is -24, only 2 and -12 have a sum of -10.

Step 3: We now rewrite our original trinomial so the middle term $-10x$ is written as the sum of $-12x$ and $2x$:

$$3x^2 - 10x - 8 = 3x^2 - 12x + 2x - 8$$

Step 4: Factoring by grouping, we have

$$3x^2 - 12x + 2x - 8 = 3x(x - 4) + 2(x - 4)$$

$$= (x - 4)(3x + 2)$$

You can see that this method works by multiplying $x - 4$ and $3x + 2$ to get

$$3x^2 - 10x - 8 \quad \blacksquare$$

Answers
7. $(x - 2)^2(3x - 1)$
8. $(x - 1)(8x + 3)$

Problem Set 5.5

Moving Toward Success

"Human beings, by changing the inner attitudes of their minds, can change the outer aspects of their lives."

—William James, 1842–1910, American psychologist and author

1. Thoughts like "I can't do this problem" or "I'm going to get a bad grade so why try" are called negative self talk. How does this type of thinking hurt your success in this class?

2. What are some positive things you can say to yourself to combat any negative self talk? Why would these things help relieve stress in this class?

A Factor each of the following trinomials. [Examples 1–3]

1. $x^2 + 7x + 12$
2. $x^2 - 7x + 12$
3. $x^2 - x - 12$
4. $x^2 + x - 12$

5. $y^2 + y - 6$
6. $y^2 - y - 6$
7. $16 - 6x - x^2$
8. $3 + 2x - x^2$

9. $12 + 8x + x^2$
10. $15 - 2x - x^2$
11. $x^2 + 3xy + 2y^2$
12. $x^2 - 5xy - 24y^2$

13. $a^2 + 3ab - 18b^2$
14. $a^2 - 8ab - 9b^2$
15. $x^2 - 2xa - 48a^2$
16. $x^2 + 14xa + 48a^2$

17. $x^2 - 12xb + 36b^2$
18. $x^2 + 10xb + 25b^2$

B Factor completely by first factoring out the greatest common factor and then factoring the trinomial that remains. [Example 4]

19. $3a^2 - 21a + 30$
20. $3a^2 - 3a - 6$
21. $4x^3 - 16x^2 - 20x$
22. $2x^3 - 14x^2 + 20x$

C Factor completely. Be sure to factor out the greatest common factor first if it is a number other than 1. [Examples 4–6]

23. $3x^2 - 6xy - 9y^2$

24. $5x^2 + 25xy + 20y^2$

25. $2a^5 + 4a^4b + 4a^3b^2$

26. $3a^4 - 18a^3b + 27a^2b^2$

27. $10x^4y^2 + 20x^3y^3 - 30x^2y^4$

28. $6x^4y^2 + 18x^3y^3 - 24x^2y^4$

29. $2x^2 + 7x - 15$

30. $2x^2 - 7x - 15$

31. $2x^2 + x - 15$

32. $2x^2 - x - 15$

33. $2x^2 - 13x + 15$

34. $2x^2 + 13x + 15$

35. $2x^2 - 11x + 15$

36. $2x^2 + 11x + 15$

37. $2x^2 + 7x + 15$

38. $2x^2 + x + 15$

39. $2 + 7a + 6a^2$

40. $2 - 7a + 6a^2$

41. $60y^2 - 15y - 45$

42. $72y^2 + 60y - 72$

43. $6x^4 - x^3 - 2x^2$

44. $3x^4 + 2x^3 - 5x^2$

45. $40r^3 - 120r^2 + 90r$

46. $40r^3 + 200r^2 + 250r$

47. $4x^2 - 11xy - 3y^2$

48. $3x^2 + 19xy - 14y^2$

49. $10x^2 - 3xa - 18a^2$

50. $9x^2 + 9xa - 10a^2$

51. $18a^2 + 3ab - 28b^2$

52. $6a^2 - 7ab - 5b^2$

53. $600 + 800t - 800t^2$

54. $200 - 600t - 350t^2$

55. $9y^4 + 9y^3 - 10y^2$

56. $4y^5 + 7y^4 - 2y^3$

57. $24a^2 - 2a^3 - 12a^4$

58. $60a^2 + 65a^3 - 20a^4$

59. $8x^4y^2 - 2x^3y^3 - 6x^2y^4$

60. $8x^4y^2 - 47x^3y^3 - 6x^2y^4$

61. $300x^4 + 1,000x^2 + 300$

62. $600x^4 - 100x^2 - 200$

63. $20a^4 + 37a^2 + 15$

64. $20a^4 + 13a^2 - 15$

65. $9 + 3r^2 - 12r^4$

66. $2 - 4r^2 - 30r^4$

D Factor each of the following by first factoring out the greatest common factor and then factoring the trinomial that remains. [Example 7]

67. $2x^2(x + 5) + 7x(x + 5) + 6(x + 5)$

68. $2x^2(x + 2) + 13x(x + 2) + 15(x + 2)$

69. $x^2(2x + 3) + 7x(2x + 3) + 10(2x + 3)$

70. $2x^2(x + 1) + 7x(x + 1) + 6(x + 1)$

71. $3x^2(x - 3) + 7x(x - 3) - 20(x - 3)$

72. $4x^2(x + 6) + 23x(x + 6) + 15(x + 6)$

73. $6x^2(x - 2) - 17x(x - 2) + 12(x - 2)$

74. $10x^2(x + 4) - 33x(x + 4) - 7(x + 4)$

75. $12x^2(x + 3) + 7x(x + 3) - 45(x + 3)$

76. $24x^2(x - 6) + 38x(x - 6) + 15(x - 6)$

77. $6x^2(5x - 2) - 11x(5x - 2) - 10(5x - 2)$

78. $14x^2(3x + 4) - 39x(3x + 4) + 10(3x + 4)$

79. $20x^2(2x + 3) + 47x(2x + 3) + 21(2x + 3)$

80. $15x^2(4x - 5) - 2x(4x - 5) - 24(4x - 5)$

81. What polynomial gives $(3x + 5y)(3x - 5y)$ when factored?

82. What polynomial gives $(7x + 2y)(7x - 2y)$ when factored?

83. One factor of the trinomial $a^2 + 260a + 2{,}500$ is $a + 10$. What is the other factor?

84. One factor of the trinomial $a^2 - 75a - 2{,}500$ is $a + 25$. What is the other factor?

85. One factor of the trinomial $12x^2 - 107x + 210$ is $x - 6$. What is the other factor?

86. One factor of the trinomial $36x^2 + 134x - 40$ is $2x + 8$. What is the other factor?

87. One factor of the trinomial $54x^2 + 111x + 56$ is $6x + 7$. What is the other factor?

88. One factor of the trinomial $63x^2 + 110x + 48$ is $7x + 6$. What is the other factor?

89. One factor of the trinomial $35x^2 + 19x - 24$ is $5x - 3$. What is the other factor?

90. One factor of the trinomial $36x^2 + 43x - 35$ is $4x + 7$. What is the other factor?

91. Factor the right side of the equation $y = 4x^2 + 18x - 10$, and then use the result to find y when x is $\frac{1}{2}$, when x is -5, and when x is 2.

92. Factor the right side of the equation $y = 9x^2 + 33x - 12$, and use the result to find y when x is $\frac{1}{3}$, when x is -4, and when x is 3.

Maintaining Your Skills

Multiply.

93. $(2x - 3)(2x + 3)$

94. $(4 - 5x)(4 + 5x)$

95. $(2x - 3)^2$

96. $(4 - 5x)^2$

97. $(2x - 3)(4x^2 + 6x + 9)$

98. $(2x + 3)(4x^2 - 6x + 9)$

Getting Ready for the Next Section

For each problem below, place a number or expression inside the parentheses so that the resulting statement is true.

99. $\dfrac{25}{64} = (\quad)^2$

100. $\dfrac{4}{9} = (\quad)^2$

101. $x^6 = (\quad)^2$

102. $x^8 = (\quad)^2$

103. $16x^4 = (\quad)^2$

104. $81y^4 = (\quad)^2$

Write as a perfect cube.

105. $\dfrac{1}{8} = (\quad)^3$

106. $\dfrac{1}{27} = (\quad)^3$

107. $x^6 = (\quad)^3$

108. $x^{12} = (\quad)^3$

109. $27x^3 = (\quad)^3$

110. $125y^3 = (\quad)^3$

111. $8y^3 = (\quad)^3$

112. $1000x^3 = (\quad)^3$

Extending the Concepts

Factor completely.

113. $8x^6 + 26x^3y^2 + 15y^4$

114. $24x^4 + 6x^2y^3 - 45y^6$

115. $3x^2 + 295x - 500$

116. $3x^2 + 594x - 1{,}200$

117. $\dfrac{1}{8}x^2 + x + 2$

118. $\dfrac{1}{9}x^2 + x + 2$

119. $2x^2 + 1.5x + 0.25$

120. $6x^2 + 2x + 0.16$

Special Factoring

TICKET TO SUCCESS

Keep these questions in mind as you read through the section. Then respond in your own words and in complete sentences.

1. What is a perfect square trinomial?
2. Is it possible to factor the sum of two squares?
3. What is the formula for factoring the sum of two cubes?
4. Write a problem that uses the formula for the difference of two cubes.

Pixelbliss/Shutterstock.com

Suppose a smoothie shop can make x smoothies for a total cost of $C = 10x^2 + 100x + 250$. If we begin by factoring out the greatest common factor of 10, we find that the resulting equation looks like this:

$$C = 10(x^2 + 10x + 25)$$

This is a special kind of trinomial called a perfect square trinomial. We will explore these and other kinds of special trinomials in this section.

To find the area of the large square in the margin, we can square the length of its side, giving us $(a + b)^2$. However, we can add the areas of the four smaller figures to arrive at the same result.

Since the area of the large square is the same whether we find it by squaring a side or by adding the four smaller areas, we can write the following relationship:

$$(a + b)^2 = a^2 + 2ab + b^2$$

This is the formula for factoring the square of a binomial. The figure gives us a geometric interpretation for one of the special multiplication formulas. We begin this section by looking at the special multiplication formulas from a factoring perspective.

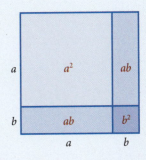

A Perfect Square Trinomials

We previously listed some special products found in multiplying polynomials. Two of the formulas looked like this:

$$(a + b)^2 = a^2 + 2ab + b^2$$
$$(a - b)^2 = a^2 - 2ab + b^2$$

If we exchange the left and right sides of each formula, we have two special formulas for factoring:

$$a^2 + 2ab + b^2 = (a + b)^2$$
$$a^2 - 2ab + b^2 = (a - b)^2$$

The left side of each formula is called a *perfect square trinomial*. The right sides are binomial squares. Perfect square trinomials can always be factored using the usual methods for factoring trinomials. However, if we notice that the first and last terms of a trinomial are perfect squares, it is wise to see whether the trinomial factors as a binomial square before attempting to factor by the usual method.

EXAMPLE 1   Factor $x^2 - 6x + 9$.

SOLUTION   Since the first and last terms are perfect squares, we attempt to factor according to the preceding formulas.

$$x^2 - 6x + 9 = (x - 3)^2$$

If we expand $(x - 3)^2$, we have $x^2 - 6x + 9$, indicating we have factored correctly. ∎

EXAMPLE 2   Factor each of the following perfect square trinomials.

SOLUTION
  a. $16a^2 + 40ab + 25b^2 = (4a + 5b)^2$
  b. $49 - 14t + t^2 = (7 - t)^2$
  c. $9x^4 - 12x^2 + 4 = (3x^2 - 2)^2$
  d. $(y + 3)^2 + 10(y + 3) + 25 = [(y + 3) + 5]^2 = (y + 8)^2$ ∎

EXAMPLE 3   Factor $8x^2 - 24xy + 18y^2$.

SOLUTION   We begin by factoring the greatest common factor 2 from each term.

$$8x^2 - 24xy + 18y^2 = 2(4x^2 - 12xy + 9y^2)$$
$$= 2(2x - 3y)^2$$ ∎

B The Difference of Two Squares

Recall the formula that results in the difference of two squares:

$$(a + b)(a - b) = a^2 - b^2.$$

Writing this as a factoring formula, we have

$$a^2 - b^2 = (a + b)(a - b)$$

EXAMPLE 4   Each of the following is the difference of two squares. Use the formula $a^2 - b^2 = (a + b)(a - b)$ to factor each one.

SOLUTION   **a.** $x^2 - 25 = x^2 - 5^2 = (x + 5)(x - 5)$

         b. $49 - t^2 = 7^2 - t^2 = (7 + t)(7 - t)$

         c. $81a^2 - 25b^2 = (9a)^2 - (5b)^2 = (9a + 5b)(9a - 5b)$

         d. $4x^6 - 1 = (2x^3)^2 - 1^2 = (2x^3 + 1)(2x^3 - 1)$

         e. $x^2 - \dfrac{4}{9} = x^2 - \left(\dfrac{2}{3}\right)^2 = \left(x + \dfrac{2}{3}\right)\left(x - \dfrac{2}{3}\right)$ ■

As our next example shows, the difference of two fourth powers can be factored as the difference of two squares.

EXAMPLE 5   Factor $16x^4 - 81y^4$.

SOLUTION   The first and last terms are perfect squares. We factor according to the preceding formula.

$$16x^4 - 81y^4 = (4x^2)^2 - (9y^2)^2$$

$$= (4x^2 + 9y^2)(4x^2 - 9y^2)$$

Notice that the second factor is also the difference of two squares. Factoring completely, we have

$$16x^4 - 81y^4 = (4x^2 + 9y^2)(2x + 3y)(2x - 3y)$$ ■

Here is another example of the difference of two squares.

EXAMPLE 6   Factor $(x - 3)^2 - 25$.

SOLUTION   This example has the form $a^2 - b^2$, where a is $x - 3$ and b is 5. We factor it according to the formula for the difference of two squares.

$$(x - 3)^2 - 25 = (x - 3)^2 - 5^2 \qquad \text{Write 25 as } 5^2.$$

$$= [(x - 3) + 5][(x - 3) - 5] \qquad \text{Factor.}$$

$$= (x + 2)(x - 8) \qquad \text{Simplify.}$$

Notice in this example we could have expanded $(x - 3)^2$, subtracted 25, and then factored to obtain the same result.

$$(x - 3)^2 - 25 = x^2 - 6x + 9 - 25 \qquad \text{Expand } (x - 3)^2.$$

$$= x^2 - 6x - 16 \qquad \text{Simplify.}$$

$$= (x - 8)(x + 2) \qquad \text{Factor.}$$ ■

4. Factor.

   a. $x^2 - 16$

   b. $64 - t^2$

   c. $25x^2 - 36y^2$

   d. $9x^6 - 1$

   e. $x^2 - \dfrac{25}{64}$

5. Factor $x^4 - 81$.

NOTE
The sum of two squares never factors into the product of two binomials; that is, if we were to attempt to factor $(4x^2 + 9y^2)$ in Example 5, we would be unable to find two binomials (or any other polynomials) whose product is $4x^2 + 9y^2$. The factors do not exist as polynomials.

6. Factor $(x - 4)^2 - 9$.

Answers

4. a. $(x + 4)(x - 4)$

   b. $(8 + t)(8 - t)$

   c. $(5x + 6y)(5x - 6y)$

   d. $(3x^3 + 1)(3x^3 - 1)$

   e. $\left(x + \dfrac{5}{8}\right)\left(x - \dfrac{5}{8}\right)$

5. $(x^2 + 9)(x + 3)(x - 3)$

6. $(x - 7)(x - 1)$

7. Factor $x^2 - 6x + 9 - y^2$.

EXAMPLE 7 Factor $x^2 - 10x + 25 - y^2$.

SOLUTION Notice the first three terms form a perfect square trinomial; that is, $x^2 - 10x + 25 = (x - 5)^2$. If we replace the first three terms by $(x - 5)^2$, the expression that results has the form $a^2 - b^2$. We can factor as we did in Example 6.

$$x^2 - 10x + 25 - y^2 = (x^2 - 10x + 25) - y^2 \qquad \text{Group first three terms together.}$$

$$= (x - 5)^2 - y^2 \qquad \text{This has the form } a^2 - b^2.$$

$$= [(x - 5) + y][(x - 5) - y] \qquad \begin{array}{l}\text{Factor according to} \\ \text{the formula} \\ a^2 - b^2 = (a + b)(a - b).\end{array}$$

$$= (x - 5 + y)(x - 5 - y) \qquad \text{Simplify.}$$

We could check this result by multiplying the two factors together. (You may want to do that to convince yourself that we have the correct result.) ∎

8. Factor completely.
$$x^3 + 5x^2 - 4x - 20$$

EXAMPLE 8 Factor $x^3 + 2x^2 - 9x - 18$ completely.

SOLUTION We use factoring by grouping to begin and then factor the difference of two squares.

$$x^3 + 2x^2 - 9x - 18 = x^2(x + 2) - 9(x + 2)$$

$$= (x + 2)(x^2 - 9)$$

$$= (x + 2)(x + 3)(x - 3) \qquad ∎$$

C The Sum and Difference of Two Cubes

Here are the formulas for factoring the sum and difference of two cubes:

$$a^3 + b^3 = (a + b)(a^2 - ab + b^2) \qquad \text{Sum of two cubes}$$

$$a^3 - b^3 = (a - b)(a^2 + ab + b^2) \qquad \text{Difference of two cubes}$$

Since these formulas are unfamiliar, it is important that we verify them.

9. Multiply $(x - 3)(x^2 + 3x + 9)$.

EXAMPLE 9 Verify the two formulas.

SOLUTION We verify the formulas by multiplying the right sides and comparing the results with the left sides.

$$
\begin{array}{r}
a^2 - ab\ \ + b^2 \\
a\ \ + b \\
\hline
a^3 - a^2b + ab^2 \\
a^2b - ab^2 + b^3 \\
\hline
a^3 \qquad\qquad + b^3
\end{array}
$$

The first formula is correct.

$$
\begin{array}{r}
a^2 + ab\ \ + b^2 \\
a\ \ - b \\
\hline
a^3 + a^2b + ab^2 \\
-a^2b - ab^2 - b^3 \\
\hline
a^3 \qquad\qquad - b^3
\end{array}
$$

The second formula is correct. ∎

Answers

7. $(x - 3 + y)(x - 3 - y)$

8. $(x + 5)(x + 2)(x - 2)$

9. $x^3 - 27$

Here are some examples using the formulas for factoring the sum and difference of two cubes.

Factor.

10. $27 + x^3$

EXAMPLE 10 Factor $64 + t^3$.

SOLUTION The first term is the cube of 4 and the second term is the cube of t.

$$64 + t^3 = 4^3 + t^3$$

$$= (4 + t)(16 - 4t + t^2) \qquad \blacksquare$$

EXAMPLE 11 Factor $27x^3 + 125y^3$.

11. $8x^3 + y^3$

SOLUTION Writing both terms as perfect cubes, we have

$$27x^3 + 125y^3 = (3x)^3 + (5y)^3$$

$$= (3x + 5y)(9x^2 - 15xy + 25y^2) \qquad \blacksquare$$

EXAMPLE 12 Factor $a^3 - \dfrac{1}{8}$.

12. $a^3 - \dfrac{1}{27}$

SOLUTION The first term is the cube of a and the second term is the cube of $\frac{1}{2}$.

$$a^3 - \frac{1}{8} = a^3 - \left(\frac{1}{2}\right)^3$$

$$= \left(a - \frac{1}{2}\right)\left(a^2 + \frac{1}{2}a + \frac{1}{4}\right) \qquad \blacksquare$$

EXAMPLE 13 Factor $x^6 - y^6$.

13. $x^6 - 1$

SOLUTION We have a choice of how we want to write the two terms to begin. We can write the expression as the difference of two squares, $(x^3)^2 - (y^3)^2$, or as the difference of two cubes, $(x^2)^3 - (y^2)^3$. It is better to use the difference of two squares if we have a choice.

$$x^6 - y^6 = (x^3)^2 - (y^3)^2$$

$$= (x^3 - y^3)(x^3 + y^3)$$

$$= (x - y)(x^2 + xy + y^2)(x + y)(x^2 - xy + y^2)$$

Try this example again writing the first line as the difference of two cubes instead of the difference of two squares. It will become apparent why it is better to use the difference of two squares. $\qquad \blacksquare$

Answers

10. $(3 + x)(9 - 3x + x^2)$

11. $(2x + y)(4x^2 - 2xy + y^2)$

12. $\left(a - \dfrac{1}{3}\right)\left(a^2 + \dfrac{1}{3}a + \dfrac{1}{9}\right)$

13. $(x + 1)(x^2 - x + 1)(x - 1)(x^2 + x + 1)$

Problem Set 5.6

Moving Toward Success

"No one can cheat you out of ultimate success but yourself."
—Ralph Waldo Emerson, 1803–1882, American poet and essayist

1. Placing blame on others if you do poorly will only hurt your success in this class. Why?
2. What is one thing you can do today to perform better in this class?

A Factor each perfect square trinomial. [Example 1–3]

1. $x^2 - 6x + 9$

2. $x^2 + 10x + 25$

3. $a^2 - 12a + 36$

4. $36 - 12a + a^2$

5. $25 - 10t + t^2$

6. $64 + 16t + t^2$

7. $\dfrac{1}{9}x^2 + 2x + 9$

8. $\dfrac{1}{4}x^2 - 2x + 4$

9. $4y^4 - 12y^2 + 9$

10. $9y^4 + 12y^2 + 4$

11. $16a^2 + 40ab + 25b^2$

12. $25a^2 - 40ab + 16b^2$

13. $\dfrac{1}{25} + \dfrac{1}{10}t^2 + \dfrac{1}{16}t^4$

14. $\dfrac{1}{9} - \dfrac{1}{3}t^3 + \dfrac{1}{4}t^6$

15. $y^2 + 3y + \dfrac{9}{4}$

16. $y^2 - 7y + \dfrac{49}{4}$

17. $a^2 - a + \dfrac{1}{4}$

18. $a^2 - 5a + \dfrac{25}{4}$

19. $x^2 - \dfrac{1}{2}x + \dfrac{1}{16}$

20. $x^2 - \dfrac{3}{4}x + \dfrac{9}{64}$

21. $t^2 + \dfrac{2}{3}t + \dfrac{1}{9}$

22. $t^2 - \dfrac{4}{5}t + \dfrac{4}{25}$

23. $16x^2 - 48x + 36$

24. $36x^2 + 48x + 16$

25. $75a^3 + 30a^2 + 3a$

26. $45a^4 - 30a^3 + 5a^2$

27. $(x + 2)^2 + 6(x + 2) + 9$

28. $(x + 5)^2 + 4(x + 5) + 4$

B Factor each as the difference of two squares. Be sure to factor completely. [Example 4]

29. $x^2 - 9$

30. $x^2 - 16$

31. $49x^2 - 64y^2$

32. $81x^2 - 49y^2$

33. $4a^2 - \dfrac{1}{4}$

34. $25a^2 - \dfrac{1}{25}$

35. $x^2 - \dfrac{9}{25}$

36. $x^2 - \dfrac{25}{36}$

37. $9x^2 - 16y^2$

38. $25x^2 - 49y^2$

39. $250 - 10t^2$

40. $640 - 10t^2$

B Factor each as the difference of two squares. Be sure to factor completely. [Example 5]

41. $x^4 - 81$

42. $x^4 - 16$

43. $9x^6 - 1$

44. $25x^6 - 1$

45. $16a^4 - 81$

46. $81a^4 - 16b^4$

47. $\dfrac{1}{81} - \dfrac{y^4}{16}$

48. $\dfrac{1}{25} - \dfrac{y^4}{64}$

B Factor completely. [Examples 6–8]

49. $x^6 - y^6$

50. $x^6 - 1$

51. $2a^7 - 128a$

52. $128a^8 - 2a^2$

53. $(x - 2)^2 - 9$

54. $(x + 2)^2 - 9$

55. $(y + 4)^2 - 16$

56. $(y - 4)^2 - 16$

57. $x^2 - 10x + 25 - y^2$

58. $x^2 - 6x + 9 - y^2$

59. $a^2 + 8a + 16 - b^2$

60. $a^2 + 12a + 36 - b^2$

61. $x^2 + 2xy + y^2 - a^2$

62. $a^2 + 2ab + b^2 - y^2$

63. $x^3 + 3x^2 - 4x - 12$

64. $x^3 + 5x^2 - 4x - 20$

65. $x^3 + 2x^2 - 25x - 50$

66. $x^3 + 4x^2 - 9x - 36$

67. $2x^3 + 3x^2 - 8x - 12$

68. $3x^3 + 2x^2 - 27x - 18$

69. $4x^3 + 12x^2 - 9x - 27$

70. $9x^3 + 18x^2 - 4x - 8$

71. $(2x - 5)^2 - 100$

72. $(7a + 5)^2 - 64$

73. $(a - 3)^2 - (4b)^2$

74. $(2x - 5)^2 - (6y)^2$

75. $a^2 - 6a + 9 - 16b^2$

76. $x^2 - 10x + 25 - 9y^2$

77. $x^2(x + 4) - 6x(x + 4) + 9(x + 4)$

78. $x^2(x - 6) + 8x(x - 6) + 16(x - 6)$

C Factor each of the following as the sum or difference of two cubes. [Examples 9–13]

79. $x^3 - y^3$

80. $x^3 + y^3$

81. $a^3 + 8$

82. $a^3 - 8$

83. $27 + x^3$

84. $27 - x^3$

85. $y^3 - 1$

86. $y^3 + 1$

87. $10r^3 - 1{,}250$

88. $10r^3 + 1{,}250$

89. $64 + 27a^3$

90. $27 - 64a^3$

91. $8x^3 - 27y^3$

92. $27x^3 - 8y^3$

93. $t^3 + \dfrac{1}{27}$

94. $t^3 - \dfrac{1}{27}$

95. $27x^3 - \dfrac{1}{27}$

96. $8x^3 + \dfrac{1}{8}$

97. $64a^3 + 125b^3$

98. $125a^3 - 27b^3$

99. Find two values of b that will make $9x^2 + bx + 25$ a perfect square trinomial.

100. Find a value of c that will make $49x^2 - 42x + c$ a perfect square trinomial.

Getting Ready for the Next Section

Factor out the greatest common factor.

101. $y^3 + 25y$

102. $y^4 + 36y^2$

103. $2ab^5 + 8ab^4 + 2ab^3$

104. $3a^2b^3 + 6a^2b^2 - 3a^2b$

Factor by grouping.

105. $4x^2 - 6x + 2ax - 3a$

106. $6x^2 - 4x + 3ax - 2a$

Factor the difference of squares.

107. $x^2 - 4$

108. $x^2 - 9$

Factor the perfect square trinomial.

109. $x^2 - 6x + 9$

110. $x^2 - 10x + 25$

Factor.

111. $6a^2 - 11a + 4$

112. $6x^2 - x - 15$

Factor the sum or difference of cubes.

113. $x^3 + 8$

114. $x^3 - 27$

Maintaining Your Skills

Solve each system by using matrices.

115. $2x - 4y = -2$
$3x - 2y = 9$

116. $3x + 5y = 14$
$5x + 2y = -2$

117. $3x + 4y = 15$
$2x - 5z = -3$
$4y - 3z = 9$

118. $x + 3y = 5$
$6y + z = 12$
$x - 2z = -10$

Extending the Concepts

Factor completely.

119. $a^2 - b^2 + 6b - 9$

120. $a^2 - b^2 - 18b - 81$

121. $(x - 3)^2 - (y + 5)^2$

122. $(a + 7)^2 - (b - 9)^2$

Find k such that each trinomial becomes a perfect square trinomial.

123. $kx^2 - 168xy + 49y^2$

124. $kx^2 + 110xy + 121y^2$

125. $49x^2 + kx + 81$

126. $64x^2 + kx + 169$

Factoring: A General Review

TICKET TO SUCCESS

Keep these questions in mind as you read through the section. Then respond in your own words and in complete sentences.

1. How do you know when you have factored completely?
2. What is the first step in the strategy for factoring a polynomial?
3. Why is it important to not skip the last step in the strategy for factoring a polynomial?
4. What method of factoring would you use to factor a polynomial with four terms?

kmiragaya/Shutterstock.com

The process of factoring has been used by mathematicians for thousands of years. As early as 2000 BC, the Babylonians were factoring polynomials by carving their numeric characters into stone tablets.

A   Factoring Review

In this section, we will review the different methods of factoring that we have presented in the previous sections of this chapter. This section is important because it will give you an opportunity to factor a variety of polynomials.

Here are some examples illustrating how we use the steps in our list. There are no new factoring problems in this section. The problems here are all similar to the problems you have seen before. What is different is that they are not all of the same type.

PRACTICE PROBLEMS

1. Factor $3x^8 - 27x^6$.

2. Factor $4x^4 + 40x^3 + 100x^2$.

3. Factor $y^4 + 36y^2$.

4. Factor $6x^2 - x - 15$.

5. Factor $3x^5 - 81x^2$.

6. Factor $3a^2b^3 + 6a^2b^2 - 3a^2b$.

EXAMPLE 1   Factor $2x^5 - 8x^3$.

SOLUTION   First we check to see if the greatest common factor is other than 1. Since the greatest common factor is $2x^3$, we begin by factoring it out. Once we have done so, we notice that the binomial that remains is the difference of two squares, which we factor according to the formula $a^2 - b^2 = (a + b)(a - b)$.

$$2x^5 - 8x^3 = 2x^3(x^2 - 4) \qquad \text{Factor out the greatest common factor, } 2x^3.$$
$$= 2x^3(x + 2)(x - 2) \quad \text{Factor the difference of two squares.} \qquad ■$$

EXAMPLE 2   Factor $3x^4 - 18x^3 + 27x^2$.

SOLUTION   Step 1 is to factor out the greatest common factor $3x^2$. After we have done so, we notice that the trinomial that remains is a perfect square trinomial, which will factor as the square of a binomial.

$$3x^4 - 18x^3 + 27x^2 = 3x^2(x^2 - 6x + 9) \qquad \text{Factor out } 3x^2.$$
$$= 3x^2(x - 3)^2 \qquad x^2 - 6x + 9 \text{ is the square of } x - 3. \qquad ■$$

EXAMPLE 3   Factor $y^3 + 25y$.

SOLUTION   We begin by factoring out the y that is common to both terms. The binomial that remains after we have done so is the sum of two squares, which does not factor, so after the first step, we are finished.

$$y^3 + 25y = y(y^2 + 25) \qquad ■$$

EXAMPLE 4   Factor $6a^2 - 11a + 4$.

SOLUTION   Here we have a trinomial that does not have a greatest common factor other than 1. Since it is not a perfect square trinomial, we factor it by trial and error. Without showing all the different possibilities, here is the answer:

$$6a^2 - 11a + 4 = (3a - 4)(2a - 1) \qquad ■$$

EXAMPLE 5   Factor $2x^4 + 16x$.

SOLUTION   This binomial has a greatest common factor of $2x$. The binomial that remains after the $2x$ has been factored from each term is the sum of two cubes, which we factor according to the formula $a^3 + b^3 = (a + b)(a^2 - ab + b^2)$.

$$2x^4 + 16x = 2x(x^3 + 8) \qquad \text{Factor } 2x \text{ from each term.}$$
$$= 2x(x + 2)(x^2 - 2x + 4) \qquad \text{The sum of two cubes} \qquad ■$$

EXAMPLE 6   Factor $2ab^5 + 8ab^4 + 2ab^3$.

SOLUTION   The greatest common factor is $2ab^3$. We begin by factoring it from each term. After that we find that the trinomial that remains cannot be factored further.

$$2ab^5 + 8ab^4 + 2ab^3 = 2ab^3(b^2 + 4b + 1) \qquad ■$$

Answers
1. $3x^6(x + 3)(x - 3)$
2. $4x^2(x + 5)^2$
3. $y^2(y^2 + 36)$
4. $(3x - 5)(2x + 3)$
5. $3x^2(x - 3)(x^2 + 3x + 9)$
6. $3a^2b(b^2 + 2b - 1)$

Problem Set 5.7

Moving Toward Success

"If you don't like something, change it. If you can't change it, change your attitude. Don't complain."

— Maya Angelou, 1928–present, American author and poet

1. How do you plan on staying positive before, during, and after a test?

2. If you receive a poor grade on a test, what do you plan to do to perform better on the next test?

A Factor each of the following polynomials completely. Once you are finished factoring, none of the factors you obtain should be factorable. Also, note that the even-numbered problems are not necessarily similar to the odd-numbered problems that precede them in this problem set. [Examples 1–6]

1. $x^2 - 81$

2. $x^2 - 18x + 81$

3. $x^2 + 2x - 15$

4. $15x^2 + 13x - 6$

5. $x^2(x + 2) + 6x(x + 2) + 9(x + 2)$

6. $12x^2 - 11x + 2$

7. $x^2y^2 + 2y^2 + x^2 + 2$

8. $21y^2 - 25y - 4$

9. $2a^3b + 6a^2b + 2ab$

10. $6a^2 - ab - 15b^2$

11. $x^2 + x + 1$

12. $x^2y + 3y + 2x^2 + 6$

13. $12a^2 - 75$

14. $18a^2 - 50$

15. $9x^2 - 12xy + 4y^2$

16. $x^3 - x^2$

17. $25 - 10t + t^2$

18. $t^2 + 4t + 4 - y^2$

19. $4x^3 + 16xy^2$

20. $16x^2 + 49y^2$

21. $2y^3 + 20y^2 + 50y$

22. $x^2 + 5bx - 2ax - 10ab$

23. $a^7 + 8a^4b^3$

24. $5a^2 - 45b^2$

25. $t^2 + 6t + 9 - x^2$

26. $36 + 12t + t^2$

27. $x^3 + 5x^2 - 9x - 45$

28. $x^3 + 5x^2 - 16x - 80$

29. $5a^2 + 10ab + 5b^2$

30. $3a^3b^2 + 15a^2b^2 + 3ab^2$

31. $x^2 + 49$

32. $16 - x^4$

33. $3x^2 + 15xy + 18y^2$

34. $3x^2 + 27xy + 54y^2$

35. $9a^2 + 2a + \dfrac{1}{9}$

36. $18 - 2a^2$

37. $x^2(x - 3) - 14x(x - 3) + 49(x - 3)$

38. $x^2 + 3ax - 2bx - 6ab$

39. $x^2 - 64$

40. $9x^2 - 4$

41. $8 - 14x - 15x^2$

42. $5x^4 + 14x^2 - 3$

43. $49a^7 - 9a^5$

44. $a^6 - b^6$

45. $r^2 - \dfrac{1}{25}$

46. $27 - r^3$

47. $49x^2 + 9y^2$

48. $12x^4 - 62x^3 + 70x^2$

49. $100x^2 - 100x - 600$

50. $100x^2 - 100x - 1{,}200$

51. $25a^3 + 20a^2 + 3a$

52. $16a^5 - 54a^2$

53. $3x^4 - 14x^2 - 5$

54. $8 - 2x - 15x^2$

55. $24a^5b - 3a^2b$

56. $18a^4b^2 - 24a^3b^3 + 8a^2b^4$

57. $64 - r^3$

Solving Equations by Factoring

OBJECTIVES

A Solve equations by factoring.

B Apply the Blueprint for Problem Solving to solve application problems whose solutions involve quadratic equations.

C Solve problems that contain formulas that are quadratic.

TICKET TO SUCCESS

Keep these questions in mind as you read through the section. Then respond in your own words and in complete sentences.

1. What is standard form?
2. What is the zero-factor property?
3. Briefly explain the strategy to solve an equation by factoring.
4. Explain the Pythagorean theorem in words.

GLUE STOCK/Shutterstock.com

A producer of novelty sunglasses finds that the number of glasses sold N is related to the retail price p by the formula $N = 20p - p^2$. What price should they charge if they want to sell the 96 glasses they have in stock? Solving equations like this requires the use of our new factoring skills. Let's put those skills to work.

In this section, we will use our knowledge of factoring to solve equations. Most of the equations we will solve in this section are *quadratic equations.* Here is the definition of a quadratic equation.

> **Definition**
>
> Any equation that can be written in the form
>
> $$ax^2 + bx + c = 0$$
>
> where a, b, and c are constants and a is not 0 ($a \neq 0$) is called a **quadratic equation**. The form $ax^2 + bx + c = 0$ is called **standard form** for quadratic equations.

Each of the following is a quadratic equation:

$$2x^2 = 5x + 3 \qquad 5x^2 = 75 \qquad 4x^2 - 3x + 2 = 0$$

Notation For a quadratic equation written in standard form, the first term ax^2 is called the *quadratic term;* the second term bx is the *linear term;* and the third term c is called the *constant term.*

> **NOTE**
> The third equation is clearly a quadratic equation since it is in standard form. (Notice that a is 4, b is -3, and c is 2.) The first two equations are also quadratic because they could be put in the form $ax^2 + bx + c = 0$ by using the addition property of equality.

In the past we have noticed that the number 0 is a special number. There is another property of 0 that is the key to solving quadratic equations. It is called the *zero-factor property*.

> **Zero-Factor Property**
> For all real numbers r and s,
>
> $$r \cdot s = 0 \quad \text{if and only if} \quad r = 0 \quad \text{or} \quad s = 0 \quad \text{(or both)}$$

A Solving Equations by Factoring

EXAMPLE 1 Solve $x^2 - 2x - 24 = 0$.

SOLUTION We begin by factoring the left side as $(x - 6)(x + 4)$ and get

$$(x - 6)(x + 4) = 0$$

Now both $(x - 6)$ and $(x + 4)$ represent real numbers. We notice that their product is 0. By the zero-factor property, one or both of them must be 0.

$$x - 6 = 0 \quad \text{or} \quad x + 4 = 0$$

We have used factoring and the zero-factor property to rewrite our original second-degree equation as two first-degree equations connected by the word *or*. Completing the solution, we solve the two first-degree equations.

$$x - 6 = 0 \quad \text{or} \quad x + 4 = 0$$
$$x = 6 \quad \text{or} \quad x = -4$$

We check our solutions in the original equation as follows:

Check $x = 6$	Check $x = -4$
$6^2 - 2(6) - 24 \stackrel{?}{=} 0$	$(-4)^2 - 2(-4) - 24 \stackrel{?}{=} 0$
$36 - 12 - 24 \stackrel{?}{=} 0$	$16 + 8 - 24 \stackrel{?}{=} 0$
$0 = 0$	$0 = 0$

In both cases the result is a true statement, which means that both 6 and -4 are solutions to the original equation. ∎

Although the next equation is not quadratic, the method we use is similar.

EXAMPLE 2 Solve $\frac{1}{3}x^3 = \frac{5}{6}x^2 + \frac{1}{2}x$.

SOLUTION We can simplify our work if we clear the equation of fractions. Multiplying both sides by the LCD, 6, we have

$$\mathbf{6} \cdot \frac{1}{3}x^3 = \mathbf{6} \cdot \frac{5}{6}x^2 + \mathbf{6} \cdot \frac{1}{2}x$$
$$2x^3 = 5x^2 + 3x$$

Next we add $-5x^2$ and $-3x$ to each side so that the right side will become 0.

$$2x^3 - 5x^2 - 3x = 0 \qquad \text{Standard form}$$

We factor the left side and then use the zero-factor property to set each factor to 0.

$$x(2x^2 - 5x - 3) = 0 \qquad \text{Factor out the greatest common factor.}$$

$$x(2x + 1)(x - 3) = 0 \qquad \text{Continue factoring.}$$

$$x = 0 \quad \text{or} \quad 2x + 1 = 0 \quad \text{or} \quad x - 3 = 0 \qquad \text{Zero-factor property}$$

Solving each of the resulting equations, we have

$$x = 0 \quad \text{or} \quad x = -\frac{1}{2} \quad \text{or} \quad x = 3 \qquad \blacksquare$$

To generalize the preceding example, here are the steps used in solving a quadratic equation by factoring.

Strategy　**To Solve a Quadratic Equation by Factoring**

Step 1　Write the equation in standard form.

Step 2　Factor the left side.

Step 3　Use the zero-factor property to set each factor equal to 0.

Step 4　Solve the resulting linear equations.

EXAMPLE 3　Solve $100x^2 = 300x$.

3. Solve $100x^2 = 500x$.

SOLUTION　We begin by writing the equation in standard form and factoring:

$$100x^2 = 300x$$

$$100x^2 - 300x = 0 \qquad \text{Standard form}$$

$$100x(x - 3) = 0 \qquad \text{Factor.}$$

Using the zero-factor property to set each factor equal to 0, we have

$$100x = 0 \quad \text{or} \quad x - 3 = 0$$

$$x = 0 \qquad\qquad x = 3$$

The two solutions are 0 and 3. $\qquad \blacksquare$

EXAMPLE 4　Solve $(x - 2)(x + 1) = 4$.

4. Solve $(x + 1)(x + 2) = 12$.

SOLUTION　We begin by multiplying the two factors on the left side. (Notice that it would be incorrect to set each of the factors on the left side equal to 4. The fact that the product is 4 does not imply that either of the factors must be 4.)

$$(x - 2)(x + 1) = 4$$

$$x^2 - x - 2 = 4 \qquad \text{Multiply the left side.}$$

$$x^2 - x - 6 = 0 \qquad \text{Standard form}$$

$$(x - 3)(x + 2) = 0 \qquad \text{Factor.}$$

$$x - 3 = 0 \quad \text{or} \quad x + 2 = 0 \qquad \text{Zero-factor property}$$

$$x = 3 \qquad\qquad x = -2 \qquad \blacksquare$$

Answers
3. 0, 5
4. −5, 2

5. Solve $x^3 + 5x^2 - 4x - 20 = 0$.

EXAMPLE 5   Solve $x^3 + 2x^2 - 9x - 18 = 0$ for x.

SOLUTION   We start with factoring by grouping.

$$x^3 + 2x^2 - 9x - 18 = 0$$

$$x^2(x + 2) - 9(x + 2) = 0$$

$$(x + 2)(x^2 - 9) = 0$$

$$(x + 2)(x - 3)(x + 3) = 0 \qquad \text{The difference of two squares}$$

$$x + 2 = 0 \quad \text{or} \quad x - 3 = 0 \quad \text{or} \quad x + 3 = 0 \qquad \text{Set factors equal to 0.}$$

$$x = -2 \qquad\qquad x = 3 \qquad\qquad x = -3$$

We have three solutions: -2, 3, and -3. ■

B Applications

6. One integer is 3 more than another. The sum of their squares is 29. Find the two integers.

EXAMPLE 6   The sum of the squares of two consecutive integers is 25. Find the two integers.

SOLUTION   We apply the Blueprint for Problem Solving to solve this application problem. Remember, step 1 in the blueprint is done mentally.

Step 1: Read and list.
   Known items: Two consecutive integers. If we add their squares, the result is 25.
   Unknown items: The two integers

Step 2: Assign a variable and translate information.
   Let $x =$ the first integer; then $x + 1 =$ the next consecutive integer.

Step 3: Reread and write an equation.
   Since the sum of the squares of the two integers is 25, the equation that describes the situation is

$$x^2 + (x + 1)^2 = 25$$

Step 4: Solve the equation.

$$x^2 + (x + 1)^2 = 25$$

$$x^2 + (x^2 + 2x + 1) = 25$$

$$2x^2 + 2x - 24 = 0$$

$$x^2 + x - 12 = 0$$

$$(x + 4)(x - 3) = 0$$

$$x = -4 \quad \text{or} \quad x = 3$$

NOTE
The common factor can be divided out, since 0 divided by any number is 0.

Step 5: Write the answer.
   If $x = -4$, then $x + 1 = -3$. If $x = 3$, then $x + 1 = 4$. The two integers are -4 and -3, or the two integers are 3 and 4.

Step 6: Reread and check.
   The two integers in each pair are consecutive integers, and the sum of the squares of either pair is 25. ■

Answers
5. $-5, -2, 2$
6. $-5, -2,$ or $2, 5$

Another application of quadratic equations involves the Pythagorean theorem, an important theorem from geometry. The theorem gives the relationship between the sides of any right triangle (a triangle with a 90-degree angle). We state it here without proof.

Pythagorean Theorem

In any right triangle, the square of the length of the longest side (hypotenuse) is equal to the sum of the squares of the length of the other two sides (legs).

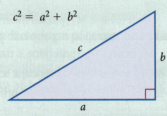

$$c^2 = a^2 + b^2$$

EXAMPLE 7   The lengths of the three sides of a right triangle are given by three consecutive integers. Find the lengths of the three sides.

SOLUTION   **Step 1:   Read and list.**
Known items: A right triangle. The three sides are three consecutive integers.
Unknown items: The three sides

Step 2:   Assign a variable and translate information.
Let x = first integer (shortest side).
Then $x + 1$ = next consecutive integer
$x + 2$ = last consecutive integer (longest side)

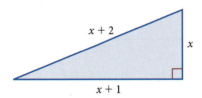

Step 3:   Reread and write an equation.
By the Pythagorean theorem, we have

$$(x + 2)^2 = (x + 1)^2 + x^2$$

Step 4:   Solve the equation.

$$x^2 + 4x + 4 = x^2 + 2x + 1 + x^2$$

$$x^2 - 2x - 3 = 0$$

$$(x - 3)(x + 1) = 0$$

$$x = 3 \quad \text{or} \quad x = -1$$

Step 5:   Write the answer.
Since x is the length of a side in a triangle, it must be a positive number. Therefore, $x = -1$ cannot be used.
The shortest side is 3. The other two sides are 4 and 5.

7. The longest side of a right triangle is 4 more than the shortest side. The third side is 2 more than the shortest side. Find the length of each side.

Answers
7. 6, 8, 10

Step 6: Reread and check.

The three sides are given by consecutive integers. The square of the longest side is equal to the sum of the squares of the two shorter sides.   ■

8. Two boats leave from an island port at the same time. One travels due east at a speed of 5 miles per hour, and the other travels due south at a speed of 12 miles per hour. How long until the distance between the boats is 52 miles?

EXAMPLE 8   Two boats leave from an island port at the same time. One travels due north at a speed of 12 miles per hour, and the other travels due west at a speed of 16 miles per hour. How long until the distance between the boats is 60 miles?

SOLUTION   **Step 1: Read and list.**

Known items: The speed and direction of both boats. The distance between the boats.

Unknown items: The distance traveled by each boat, and the time.

Step 2: Assign a variable and translate information.

Let t = the time.

Then $12t$ = the distance traveled by boat going north

$16t$ = the distance traveled by boat going west

If we draw a diagram for the problem, we see that the distances traveled by the two boats form the legs of a right triangle. The hypotenuse of the triangle will be the distance between the boats, which is 60 miles.

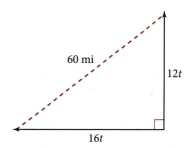

Step 3: Reread and write an equation.

By the Pythagorean theorem, we have

$$(16t)^2 + (12t)^2 = 60^2$$

Step 4: Solve the equation.

$$256t^2 + 144t^2 = 3600$$

$$400t^2 = 3600$$

$$400t^2 - 3600 = 0$$

$$t^2 - 9 = 0$$

$$(t + 3)(t - 3) = 0$$

$$t = -3 \quad \text{or} \quad t = 3$$

Step 5: Write the answer.

Because t is measuring time, it must be a positive number. Therefore, $t = -3$ cannot be used.

The two boats will be 60 miles apart after 3 hours.

Answer

8. 4 hours

Step 6: ***Reread and check.***

The boat going north will travel $12 \cdot 3 = 36$ miles in 3 hours, and the boat going west will travel $16 \cdot 3 = 48$ miles. The distance between them after 3 hours will be 60 miles ($48^2 + 36^2 = 60^2$). ■

C Formulas

Our next two examples involve formulas that are quadratic.

EXAMPLE 9 If an object is projected into the air with an initial vertical velocity of v feet/second, its height h, in feet, above the ground after t seconds will be given by

$$h = vt - 16t^2$$

Find t if $v = 64$ feet/second and $h = 48$ feet.

SOLUTION Substituting $v = 64$ and $h = 48$ into the preceding formula, we have

$$48 = 64t - 16t^2$$

which is a quadratic equation. We write it in standard form and solve by factoring.

$$16t^2 - 64t + 48 = 0$$

$$t^2 - 4t + 3 = 0 \qquad \text{Divide each side by 16.}$$

$$(t - 1)(t - 3) = 0$$

$$t - 1 = 0 \quad \text{or} \quad t - 3 = 0$$

$$t = 1 \qquad\qquad t = 3$$

Here is how we interpret our results: If an object is projected upward with an initial vertical velocity of 64 feet/second, it will be 48 feet above the ground after 1 second and after 3 seconds; that is, it passes 48 feet going up and also coming down. ■

EXAMPLE 10 A manufacturer of headphones knows that the number of headphones she can sell each week is related to the price of the headphones by the equation $x = 1,300 - 100p$, where x is the number of headphones and p is the price per set. What price should she charge for each set of headphones if she wants the weekly revenue to be $4,000?

SOLUTION The formula for total revenue is $R = xp$. Since we want R in terms of p, we substitute $1,300 - 100p$ for x in the equation $R = xp$.

$$\text{If} \qquad R = xp$$

$$\text{and} \qquad x = 1,300 - 100p$$

$$\text{then} \qquad R = (1,300 - 100p)p$$

We want to find p when R is 4,000. Substituting 4,000 for R in the formula gives us

$$4,000 = (1,300 - 100p)p$$

$$4,000 = 1,300p - 100p^2$$

which is a quadratic equation. To write it in standard form, we add $100p^2$ and $-1,300p$ to each side, giving us

9. Use the formula in Example 9 to find t if $v = 64$ feet/second and $h = 64$ feet.

10. Use the information in Example 10 to find the price that should be charged if the weekly revenue is to be $3,600.

Answers
9. 2 seconds
10. $4 or $9

$$100p^2 - 1,300p + 4,000 = 0$$

$$p^2 - 13p + 40 = 0 \qquad \text{Divide each side by 100.}$$

$$(p - 5)(p - 8) = 0$$

$$p - 5 = 0 \quad \text{or} \quad p - 8 = 0$$

$$p = 5 \qquad\qquad p = 8$$

If she sells the headphones for $5 each or for $8 each she will have a weekly revenue of $4,000. ∎

Problem Set 5.8

Moving Toward Success

"You don't save a pitcher for tomorrow. Tomorrow it may rain."
—Leo Durocher, 1905–1991, Hall of Fame American baseball manager and player

1. How can distraction impede your success in this class?

2. How do questions like "Why am I taking this class?" and "When am I ever going to use this stuff?" distract you?

A Solve each equation. [Examples 1–5]

1. $x^2 - 5x - 6 = 0$

2. $x^2 + 5x - 6 = 0$

3. $x^3 - 5x^2 + 6x = 0$

4. $x^3 + 5x^2 + 6x = 0$

5. $3y^2 + 11y - 4 = 0$

6. $3y^2 - y - 4 = 0$

7. $60x^2 - 130x + 60 = 0$

8. $90x^2 + 60x - 80 = 0$

9. $\dfrac{1}{10}t^2 - \dfrac{5}{2} = 0$

10. $\dfrac{2}{7}t^2 - \dfrac{7}{2} = 0$

11. $100x^4 = 400x^3 + 2{,}100x^2$

12. $100x^4 = -400x^3 + 2{,}100x^2$

13. $\dfrac{1}{5}y^2 - 2 = -\dfrac{3}{10}y$

14. $\dfrac{1}{2}y^2 + \dfrac{5}{3} = \dfrac{17}{6}y$

15. $9x^2 - 12x = 0$

16. $4x^2 + 4x = 0$

17. $0.02r + 0.01 = 0.15r^2$

18. $0.02r - 0.01 = -0.08r^2$

19. $9a^3 = 16a$

20. $16a^3 = 25a$

21. $-100x = 10x^2$

22. $800x = 100x^2$

23. $(x + 6)(x - 2) = -7$

24. $(x - 7)(x + 5) = -20$

25. $(y - 4)(y + 1) = -6$

26. $(y - 6)(y + 1) = -12$

27. $(x + 1)^2 = 3x + 7$

28. $(x + 2)^2 = 9x$

29. $(2r + 3)(2r - 1) = -(3r + 1)$

30. $(3r + 2)(r - 1) = -(7r - 7)$

31. $x^3 + 3x^2 - 4x - 12 = 0$

32. $x^3 + 5x^2 - 4x - 20 = 0$

33. $x^3 + 2x^2 - 25x - 50 = 0$

34. $x^3 + 4x^2 - 9x - 36 = 0$

35. $2x^3 + 3x^2 - 8x - 12 = 0$

36. $3x^3 + 2x^2 - 27x - 18 = 0$

37. $4x^3 + 12x^2 - 9x - 27 = 0$

38. $9x^3 + 18x^2 - 4x - 8 = 0$

A Problems 39–48 are problems you will see later in the book. Solve each equation. [Examples 1–5]

39. $3x^2 + x = 10$

40. $y^2 + y - 20 = 2y$

41. $12(x + 3) + 12(x - 3) = 3(x^2 - 9)$

42. $8(x + 2) + 8(x - 2) = 3(x^2 - 4)$

43. $(y + 3)^2 + y^2 = 9$

44. $(2y + 4)^2 + y^2 = 4$

45. $(x + 3)^2 + 1^2 = 2$

46. $(x - 3)^2 + (-1)^2 = 10$

47. $(x + 2)(x) = 2^3$

48. $(x + 3)(x) = 2^2$

49. Let $f(x) = \left(x + \dfrac{3}{2}\right)^2$. Find all values for the variable x, for which $f(x) = 0$.

50. Let $f(x) = \left(x - \dfrac{5}{2}\right)^2$. Find all values for the variable x, for which $f(x) = 0$.

51. Let $f(x) = (x - 3)^2 - 25$. Find all values for the variable x, for which $f(x) = 0$.

52. Let $f(x) = 9x^3 + 18x^2 - 4x - 8$. Find all values for the variable x, for which $f(x) = 0$.

Let $f(x) = x^2 + 6x + 3$. Find all values for the variable x, for which $f(x) = g(x)$.

53. $g(x) = -6$ **54.** $g(x) = 19$ **55.** $g(x) = 10$ **56.** $g(x) = -2$

Let $h(x) = x^2 - 5x$. Find all values for the variable x, for which $h(x) = f(x)$.

57. $f(x) = 0$ **58.** $f(x) = -6$ **59.** $f(x) = 2x + 8$ **60.** $f(x) = -2x + 10$

61. Solve each equation.

  a. $9x - 25 = 0$

  b. $9x^2 - 25 = 0$

  c. $9x^2 - 25 = 56$

  d. $9x^2 - 25 = 30x - 50$

62. Solve each equation.

  a. $5x - 6 = 0$

  b. $(5x - 6)^2 = 0$

  c. $25x^2 - 36 = 0$

  d. $25x^2 - 36 = 28$

Applying the Concepts

63. Distance Two cyclists leave from an intersection at the same time. One travels due north at a speed of 15 miles per hour, and the other travels due east at a speed of 20 miles per hour. How long until the distance between the two cyclists is 75 miles?

64. Distance Two airplanes leave from an airport at the same time. One travels due south at a speed of 480 miles per hour, and the other travels due west at a speed of 360 miles per hour. How long until the distance between the two airplanes is 2400 miles?

65. Consecutive Integers The square of the sum of two consecutive integers is 81. Find the two integers.

66. Consecutive Integers Find two consecutive even integers whose sum squared is 100.

67. Right Triangle A 25-foot ladder is leaning against a building. The base of the ladder is 7 feet from the side of the building. How high does the ladder reach along the side of the building?

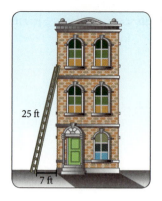

25 ft

7 ft

68. Right Triangle Noreen wants to place a 13-foot ramp against the side of her house so that the top of the ramp rests on a ledge that is 5 feet above the ground. How far will the base of the ramp be from the house?

69. Right Triangle The lengths of the three sides of a right triangle are given by three consecutive even integers. Find the lengths of the three sides.

70. Right Triangle The longest side of a right triangle is 3 less than twice the shortest side. The third side measures 12 inches. Find the length of the shortest side.

71. Geometry The length of a rectangle is 2 feet more than 3 times the width. If the area is 16 square feet, find the width and the length.

72. Geometry The length of a rectangle is 4 yards more than twice the width. If the area is 70 square yards, find the width and the length.

73. Geometry The base of a triangle is 2 inches more than 4 times the height. If the area is 36 square inches, find the base and the height.

74. Geometry The height of a triangle is 4 feet less than twice the base. If the area is 48 square feet, find the base and the height.

75. Projectile Motion If an object is thrown straight up into the air with an initial velocity of 32 feet per second, then its height above the ground at any time t is given by the formula $h = 32t - 16t^2$. Find the times at which the object is on the ground by letting $h = 0$ in the equation and solving for t.

76. Projectile Motion An object is projected into the air with an initial velocity of 64 feet per second. Its height at any time t is given by the formula $h = 64t - 16t^2$. Find the times at which the object is on the ground.

C   The formula $h = vt - 16t^2$ gives the height h, in feet, of an object projected into the air with an initial vertical velocity v, in feet per second, after t seconds. [Examples 9, 10]

77. Projectile Motion If an object is projected upward with an initial velocity of 48 feet per second, at what times will it reach a height of 32 feet above the ground?

78. Projectile Motion If an object is projected upward into the air with an initial velocity of 80 feet per second, at what times will it reach a height of 64 feet above the ground?

79. Projectile Motion An object is projected into the air with a vertical velocity of 24 feet per second. At what times will the object be on the ground? (It is on the ground when h is 0.)

80. Projectile Motion An object is projected into the air with a vertical velocity of 20 feet per second. At what times will the object be on the ground?

81. Height of a Bullet A bullet is fired into the air with an initial upward velocity of 80 feet per second from the top of a building 96 feet high. The equation that gives the height of the bullet at any time t is $h = 96 + 80t - 16t^2$. At what times will the bullet be 192 feet in the air?

82. Height of an Arrow An arrow is shot into the air with an upward velocity of 48 feet per second from a hill 32 feet high. The equation that gives the height of the arrow at any time t is $h = 32 + 48t - 16t^2$. Find the times at which the arrow will be 64 feet above the ground.

83. Price and Revenue A company that manufactures hair ribbons knows that the number of ribbons x it can sell each week is related to the price per ribbon p by the equation $x = 1,200 - 100p$. At what price should it sell the ribbons if it wants the weekly revenue to be \$3,200? (*Remember:* The equation for revenue is $R = xp$.)

84. Price and Revenue A company manufactures CDs for home computers. It knows from past experience that the number of CDs x it can sell each day is related to the price per CD p by the equation $x = 800 - 100p$. At what price should it sell its CDs if it wants the daily revenue to be \$1,200?

85. Price and Revenue The relationship between the number of calculators x a company sells per day and the price of each calculator p is given by the equation $x = 1,700 - 100p$. At what price should the calculators be sold if the daily revenue is to be \$7,000?

86. Price and Revenue The relationship between the number of pencil sharpeners x a company can sell each week and the price of each sharpener p is given by the equation $x = 1,800 - 100p$. At what price should the sharpeners be sold if the weekly revenue is to be \$7,200?

Maintaining Your Skills

Solve each system.

87. $2x - 5y = -8$
$3x + y = 5$

88. $4x - 7y = -2$
$-5x + 6y = -3$

89. $\dfrac{1}{3}x - \dfrac{1}{6}y = 3$
$-\dfrac{1}{5}x + \dfrac{1}{4}y = 0$

90. $2x - 5y = 14$
$y = 3x + 8$

91. $2x - y + z = 9$
$x + y - 3z = -2$
$3x + y - z = 6$

92. Number Problem A number is 2 less than three times another. Their sum is 22. Find the two numbers.

93. Investing John invests twice as much money at 6% as he does at 5%. If his investments earn a total of $680 in 1 year, how much does he have invested at each rate?

94. Speed of a Boat A boat can travel 24 miles downstream in 3 hours. The same boat can travel 16 miles upstream in 4 hours. What is the speed of the boat in still water, and what is the speed of the current?

Graph the solution set for each system.

95. $3x + 2y < 6$
$-2x + 3y < 6$

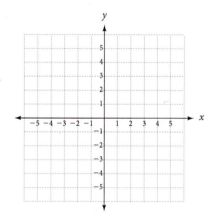

96. $y \leq x + 3$
$y > x - 4$

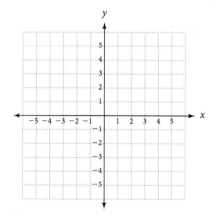

97. $x \leq 4$
$y < 2$

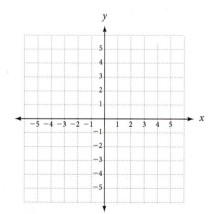

98. $2x + y < 4$
$x \geq 0$
$y \geq 0$

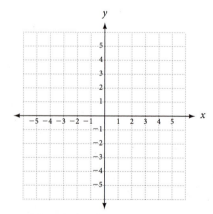

Chapter 5 Summary

EXAMPLES

■ Properties of Exponents [5.1]

1. These expressions illustrate the properties of exponents.

a. $x^2 \cdot x^3 = x^{2+3} = x^5$

b. $(x^2)^3 = x^{2 \cdot 3} = x^6$

c. $(3x)^2 = 3^2 \cdot x^2 = 9x^2$

d. $2^{-3} = \dfrac{1}{2^3} = \dfrac{1}{8}$

e. $\left(\dfrac{x}{5}\right)^2 = \dfrac{x^2}{5^2} = \dfrac{x^2}{25}$

f. $\dfrac{x^7}{x^5} = x^{7-5} = x^2$

g. $3^1 = 3$

$3^0 = 1$

If a and b represent real numbers and r and s represent integers, then

1. $a^r \cdot a^s = a^{r+s}$ Product property

2. $(a^r)^s = a^{r \cdot s}$ Power property

3. $(ab)^r = a^r \cdot b^r$ Distributive property

4. $a^{-r} = \dfrac{1}{a^r}$ $(a \neq 0)$ Negative exponent property

5. $\left(\dfrac{a}{b}\right)^r = \dfrac{a^r}{b^r}$ $(b \neq 0)$ Expanded distributive property

6. $\dfrac{a^r}{a^s} = a^{r-s}$ $(a \neq 0)$ Quotient property

7. $a^1 = a$ Identity property

8. $a^0 = 1$ $(a \neq 0)$ Zero exponent property

■ Scientific Notation [5.1]

2. $49{,}800{,}000 = 4.98 \times 10^7$

$0.00462 = 4.62 \times 10^{-3}$

A number is written in scientific notation when it is written as the product of a number between 1 and 10 and an integer power of 10; that is, when it has the form

$$n \times 10^r$$

where $1 \leq n < 10$ and $r =$ an integer.

■ Addition of Polynomials [5.2]

3. $(3x^2 + 2x - 5) + (4x^2 - 7x + 2)$

$= 7x^2 - 5x - 3$

To add two polynomials, simply combine the coefficients of similar terms.

■ Negative Signs Preceding Parentheses [5.2]

4. $-(2x^2 - 8x - 9)$

$= -2x^2 + 8x + 9$

If there is a negative sign directly preceding the parentheses surrounding a polynomial, we may remove the parentheses and preceding negative sign by changing the sign of each term within the parentheses. (This procedure is actually just another application of the distributive property.)

■ Multiplication of Polynomials [5.3]

5. $(3x - 5)(x + 2)$

$= 3x^2 + 6x - 5x - 10$

$= 3x^2 + x - 10$

To multiply two polynomials, multiply each term in the first by each term in the second.

Special Products [5.3]

6. The following are examples of the three special products:

$(x + 3)^2 = x^2 + 6x + 9$

$(5 - x)^2 = 25 - 10x + x^2$

$(x + 7)(x - 7) = x^2 - 49$

$$(a + b)^2 = a^2 + 2ab + b^2$$

$$(a - b)^2 = a^2 - 2ab + b^2$$

$$(a + b)(a - b) = a^2 - b^2$$

Business Applications [5.2, 5.3, 5.4]

7. A company makes x items each week and sells them for p dollars each, according to the equation $p = 35 - 0.1x$. Then, the revenue is

$R = x(35 - 0.1x) = 35x - 0.1x^2$

If the total cost to make all x items is $C = 8x + 500$, then the profit gained by selling the x items is

$P = 35x - 0.1x^2 - (8x + 500)$

$= -500 + 27x - 0.1x^2$

If a company manufacturers and sells x items at p dollars per item, then the revenue R is given by the formula

$$R = xp$$

If the total cost to manufacture all x items is C, then the profit obtained from selling all x items is

$$P = R - C$$

Greatest Common Factor [5.4]

8. The greatest common factor of $10x^5 - 15x^4 + 30x^3$ is $5x^3$. Factoring it out of each term, we have

$5x^3(2x^2 - 3x + 6)$

The greatest common factor of a polynomial is the largest monomial (the monomial with the largest coefficient and highest exponent) that divides each term of the polynomial. The first step in factoring a polynomial is to factor the greatest common factor (if it is other than 1) out of each term.

Factoring Trinomials [5.5]

9. $x^2 + 5x + 6 = (x + 2)(x + 3)$

$x^2 - 5x + 6 = (x - 2)(x - 3)$

$x^2 + x - 6 = (x - 2)(x + 3)$

$x^2 - x - 6 = (x + 2)(x - 3)$

We factor a trinomial by writing it as the product of two binomials. (This refers to trinomials whose greatest common factor is 1.) Each factorable trinomial has a unique set of factors. Finding the factors is sometimes a matter of trial and error.

Special Factoring [5.6]

10. Here are some binomials that have been factored this way.

$x^2 + 6x + 9 = (x + 3)^2$

$x^2 - 6x + 9 = (x - 3)^2$

$x^2 - 9 = (x + 3)(x - 3)$

$x^3 - 27 = (x - 3)(x^2 + 3x + 9)$

$x^3 + 27 = (x + 3)(x^2 - 3x + 9)$

$$a^2 + 2ab + b^2 = (a + b)^2$$ Perfect square trinomials

$$a^2 - 2ab + b^2 = (a - b)^2$$

$$a^2 - b^2 = (a - b)(a + b)$$ Difference of two squares

$$a^3 - b^3 = (a - b)(a^2 + ab + b^2)$$ Difference of two cubes

$$a^3 + b^3 = (a + b)(a^2 - ab + b^2)$$ Sum of two cubes

■ To Factor Polynomials in General [5.7]

11. Factor completely.

  a. $3x^3 - 6x^2 = 3x^2(x - 2)$

  b. $x^2 - 9 = (x + 3)(x - 3)$
    $x^3 - 8 = (x - 2)(x^2 + 2x + 4)$
    $x^3 + 27 = (x + 3)(x^2 - 3x + 9)$

  c. $x^2 - 6x + 9 = (x - 3)^2$
    $6x^2 - 7x - 5 = (2x + 1)(3x - 5)$

  d. $x^2 + ax + bx + ab$
    $= x(x + a) + b(x + a)$
    $= (x + a)(x + b)$

Step 1: If the polynomial has a greatest common factor other than 1, then factor out the greatest common factor.

Step 2: If the polynomial has two terms (it is a binomial), then see if it is the difference of two squares, or the sum or difference of two cubes, and then factor accordingly. Remember, if it is the sum of two squares it will not factor.

Step 3: If the polynomial has three terms (a trinomial), then it is either a perfect square trinomial, which will factor into the square of a binomial, or it is not a perfect square trinomial, in which case you use one of the methods developed in Section 5.5.

Step 4: If the polynomial has more than three terms, then try to factor it by grouping.

Step 5: As a final check, see if any of the factors you have written can be factored further. If you have overlooked a common factor, you can catch it here.

■ To Solve an Equation by Factoring [5.8]

12. Solve $x^2 - 5x = -6$.
    $x^2 - 5x + 6 = 0$
    $(x - 3)(x - 2) = 0$
$x - 3 = 0$ or $x - 2 = 0$
   $x = 3$ $x = 2$

Step 1: Write the equation in standard form.

Step 2: Factor the left side.

Step 3: Use the zero-factor property to set each factor equal to zero.

Step 4: Solve the resulting linear equations.

🚫 COMMON MISTAKES

When we subtract one polynomial from another, it is common to forget to add the opposite of each term in the second polynomial. For example

$$(6x - 5) - (3x + 4) = 6x - 5 - 3x + 4 \qquad \text{Mistake}$$

This mistake occurs if the negative sign outside the second set of parentheses is not distributed over all terms inside the parentheses. To avoid this mistake, remember: the opposite of a sum is the sum of the opposites, or,

$$-(3x + 4) = -3x + (-4)$$

Chapter 5 Review

Simplify each of the following. [5.1]

1. $x^3 \cdot x^7$

2. $(5x^3)^2$

3. $(2x^3y)^2(-2x^4y^2)^3$

Write with positive exponents, and then simplify. [5.1]

4. 2^{-3}

5. $\left(\dfrac{2}{3}\right)^{-2}$

6. $2^{-2} + 4^{-1}$

Write in scientific notation. [5.1]

7. 34,500,000

8. 0.00357

Write in expanded form. [5.1]

9. 4.45×10^4

10. 4.45×10^{-4}

Simplify each expression. All answers should contain positive exponents only. (Assume all variables are nonnegative.) [5.1]

11. $\dfrac{a^{-4}}{a^5}$

12. $\dfrac{(4x^2)(-3x^3)^2}{(12x^{-2})^2}$

13. $\dfrac{x^n x^{3n}}{x^{4n-2}}$

Simplify each expression as much as possible. Write all answers in scientific notation. [5.1]

14. $(2 \times 10^3)(4 \times 10^{-5})$

15. $\dfrac{(600,000)(0.000008)}{(4,000)(3,000,000)}$

Simplify by combining similar terms. [5.2]

16. $(6x^2 - 3x + 2) - (4x^2 + 2x - 5)$

17. $(x^3 - x) - (x^2 + x) + (x^3 - 3) - (x^2 + 1)$

18. Subtract $2x^2 - 3x + 1$ from $3x^2 - 5x - 2$.

19. Simplify $-3[2x - 4(3x + 1)]$.

20. Find the value of $2x^2 - 3x + 1$ when x is -2.

Multiply. [5.3]

21. $3x(4x^2 - 2x + 1)$

22. $2a^2b^3(a^2 + 2ab + b^2)$

23. $(6 - y)(3 - y)$

24. $(2x^2 - 1)(3x^2 + 4)$

25. $(2x - 3)(4x^2 + 6x + 9)$

26. $(a^2 - 2)^2$

27. $\left(x - \dfrac{1}{3}\right)\left(x + \dfrac{1}{3}\right)$

28. $(2a + b)(2a - b)$

Factor out the greatest common factor. [5.4]

29. $6x^4y - 9xy^4 + 18x^3y^3$

30. $4x^2(x + y)^2 - 8y^2(x + y)^2$

Factor by grouping. [5.4, 5.6]

31. $8x^2 + 10 - 4x^2y - 5y$

32. $x^3 + 8b^2 - x^3y^2 - 8y^2b^2$

Factor completely. [5.4, 5.5]

33. $x^2 - 5x + 6$

34. $2x^3 + 4x^2 - 30x$

35. $20a^2 - 41ab + 20b^2$

36. $6x^4 - 11x^3 - 10x^2$

37. $24x^2y - 6xy - 45y$

Factor completely. [5.6]

38. $x^4 - 16$

39. $3a^4 + 18a^2 + 27$

40. $a^3 - 8$

41. $5x^3 + 30x^2y + 45xy^2$

42. $36 - 25a^2$

43. $x^3 + 4x^2 - 9x - 36$

Solve each equation. [5.8]

44. $x^2 + 5x + 6 = 0$

45. $\dfrac{5}{6}y^2 = \dfrac{1}{4}y + \dfrac{1}{3}$

46. $9x^2 - 25 = 0$

47. $5x^2 = -10x$

48. $(x + 2)(x - 5) = 8$

49. $x^3 + 4x^2 - 9x - 36 = 0$

Solve each application. In each case, be sure to show the equation used. [5.8]

50. **Consecutive Numbers** The product of two consecutive even integers is 80. Find the two integers.

51. **Consecutive Numbers** The sum of the squares of two consecutive integers is 41. Find the two integers.

52. **Geometry** The lengths of the three sides of a right triangle are given by three consecutive integers. Find the three sides.

53. **Geometry** The lengths of three sides of a right triangle are given by three consecutive even integers. Find the three sides.

Simplify each of the following.

1. $(-5)^3$

2. $(-5)^{-3}$

3. $|-32 - 41| - 13$

4. $2^3 + 2(6^2 - 4^2)$

5. $96 \div 8 \cdot 4$

6. $55 \div 5 \cdot 11$

7. $\dfrac{3^3 + 5}{-2(5^2 - 3^2)}$

8. $\dfrac{2^3(17 - 2 \cdot 4)}{7^2 - 13}$

Let $P(x) = 11.5x - 0.01x^2$. Find the following.

9. $P(1)$

10. $P(-1)$

Let $Q(x) = \dfrac{3}{2}x^2 + \dfrac{3}{4}x - 5$. Find the following.

11. $Q(4)$

12. $Q(-4)$

Simplify each of the following expressions.

13. $6\left(x - \dfrac{1}{6}\right)$

14. $15\left(\dfrac{4}{3}x - \dfrac{3}{5}y\right)$

15. $(3x - 2) - (5x + 2)$

16. $3a - 2[4 - (a + 3)]$

17. $(3x^3 - 2) - (5x^2 + 2) + (5x^3 - 6) - (2x^2 + 4)$

18. $\left(\dfrac{5}{6}x^2 - \dfrac{2}{3}x\right) - \left(\dfrac{5}{3}x^2 + \dfrac{1}{2}x\right) + \left(\dfrac{5}{2}x^2 - \dfrac{1}{6}x\right) - \left(\dfrac{4}{3}x^2 + \dfrac{3}{2}x\right)$

Multiply or divide as indicated.

19. $(3x^2y)^4$

20. $(-3a^2b)(5a^3b^4)$

21. $\dfrac{32x^2y^6}{-16xy^2}$

22. $\dfrac{(-3a^2b)^4}{(a^3b^4)^2}$

23. $(3x - 2)(5x + 2)$

24. $5x(3x^2 + 2x + 7)$

25. $\left(x - \dfrac{2}{3}\right)^2$

26. $\left(t + \dfrac{2}{3}\right)^3$

27. $(4 \times 10^3)(7 \times 10^6)$

28. $\dfrac{8 \times 10^6}{2 \times 10^4}$

Factor each of the following expressions.

29. $x^2 - xy - ax + ay$

30. $6a^2 + a - 35$

31. $(x + 3)^2 - 25$

32. $\dfrac{1}{8} + t^3$

Solve the following equations.

33. $3x + 2 = 8$

34. $|2x - 4| + 7 = 13$

35. $\dfrac{3}{4}x - 7 = -34$

36. $-\dfrac{x}{2} = -\dfrac{3}{4} - \dfrac{3}{2}x$

37. $25a^3 = 36a$

38. $(x + 3)^2 = -2x - 7$

Solve each system.

39. $x - y = -4$
$x + 3y = 4$

40. $y = \dfrac{2}{3}x - 3$
$y = -\dfrac{3}{2}x + 23$

Find the slope of the line through the given points.

41. $(-2, 1), (5, -4)$

42. $(-10, -6), (-4, -4)$

Let $f(x) = \dfrac{3x - 5}{2}$ and $g(x) = \dfrac{2x + 5}{3}$. Find the following.

43. $f(-7)$

44. $g(-13)$

45. $(g \circ f)(2)$

46. $(f \circ g)(3)$

47. y varies inversely with the square of x. If $y = 4$ when $x = 4$, find y when $x = 6$.

48. z varies jointly with x and the cube of y. If $z = -48$ when $x = 3$ and $y = 2$, find z when $x = 2$ and $y = 3$.

49. Geometry The height of a triangle is 5 feet less than 2 times the base. If the area is 75 square feet, find the base and height.

50. Geometry Find all three angles in a triangle if the smallest angle is one-sixth the largest angle and the remaining angle is 20 degrees more than the smallest angle.

The results of a survey of 80 internet users is displayed here as a Venn diagram. Use this information to answer Questions 51 and 52.

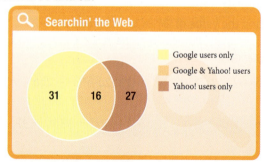

51. How many of the internet users surveyed use either Yahoo or Google?

52. How many of the internet users surveyed use neither Yahoo nor Google?

Simplify. All answers should contain positive exponents only. (Assume all variables are nonzero.) [5.1]

1. $\left(\dfrac{3}{4}\right)^{-2}$

2. 2^{-5}

3. $x^4 \cdot x^7 \cdot x^{-3}$

4. $\dfrac{a^{-5}}{a^{-7}}$

5. $(2x^2y)^3(2x^3y^4)^2$

6. $\dfrac{(2ab^3)^{-2}}{(a^{-4}b^3)^4}$

Write each number in scientific notation. [5.1]

7. 6,530,000

8. 0.00087

Perform the indicated operations, and write your answers in scientific notation. [5.1]

9. $(2.9 \times 10^{12})(3 \times 10^{-5})$

10. $\dfrac{(6 \times 10^{-4})(4 \times 10^9)}{8 \times 10^{-3}}$

Let $P(x) = 15x - 0.01x^2$ and find the following.

11. $P(10)$

12. $P(-10)$

Simplify the following expressions. [5.2]

13. $\left(\dfrac{3}{4}x^3 - x^2 - \dfrac{3}{2}\right) - \left(\dfrac{1}{4}x^2 + 2x - \dfrac{1}{2}\right)$

14. $3 - 4[2x - 3(x + 6)]$

Profit, Revenue, and Cost A company making ceramic coffee cups finds that it can sell x cups per week at p dollars each, according to the formula $p = 25 - 0.2x$. If the total cost to produce and sell x coffee cups is $C = 2x + 100$, find [5.2, 5.3]

15. an equation for the revenue that gives the revenue in terms of x.

16. the profit equation.

17. the revenue brought in by selling 100 coffee cups.

18. the cost of producing 100 coffee cups.

19. the profit obtained by making and selling 100 coffee cups.

Multiply. [5.3]

20. $(3y - 7)(2y + 5)$

21. $(2x - 5)(x^2 + 4x - 3)$

22. $(8 - 3t^3)^2$

23. $(1 - 6y)(1 + 6y)$

24. $2x(x - 3)(2x + 5)$

25. $\left(5t^2 - \dfrac{1}{2}\right)\left(2t^2 + \dfrac{1}{5}\right)$

Factor the following expressions. [5.4, 5.5, 5.6, 5.7]

26. $x^2 + x - 12$

27. $12x^4 + 26x^2 - 10$

28. $16a^4 - 81y^4$

29. $7ax^2 - 14ay - b^2x^2 + 2b^2y$

30. $t^3 + \dfrac{1}{8}$

31. $4a^5b - 24a^4b^2 - 64a^3b^3$

32. $x^2 - 10x + 25 - b^2$

33. $81 - x^4$

Solve each equation. [5.8]

34. $\dfrac{1}{5}x^2 = \dfrac{1}{3}x + \dfrac{2}{15}$

35. $100x^3 = 500x^2$

36. $(x + 1)(x + 2) = 12$

37. $x^3 + 2x^2 - 16x - 32 = 0$

Let $f(x) = x^2 - 5x + 6$. Find all values for the variable x for which $f(x) = g(x)$. [5.8]

38. $g(x) = 0$

39. $g(x) = x - 2$

40. Find the value of the variable x in the following figure. [5.8]

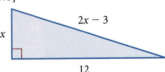

41. The area of the figure below is 35 square inches. Find the value of x. [5.8]

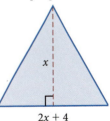

42. The area of the figure below is 48 square centimeters. Find the value of x. [5.8]

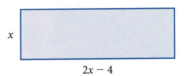

EXPONENTS AND POLYNOMIALS

Group Project

Discovering Pascal's Triangle

Number of People	3
Time Needed	20 minutes
Equipment	Paper and pencils
Background	The triangular array of numbers shown here is known as Pascal's triangle, after the French philosopher Blaise Pascal (1623–1662).

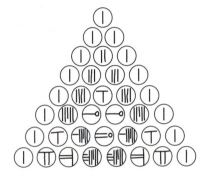

Pascal's triangle in Japanese (1781)

```
              1
            1   1
          1   2   1
        1   3   3   1
      1   4   6   4   1
    1   5  10  10   5   1
```

Procedure Look at Pascal's triangle and discover how the numbers in each row of the triangle are obtained from the numbers in the row above it.

1. Once you have discovered how to extend the triangle, write the next two rows.

2. Pascal's triangle can be linked to the Fibonacci sequence by rewriting Pascal's triangle so that the 1's on the left side of the triangle line up under one another and the other columns are equally spaced to the right of the first column. Rewrite Pascal's triangle as indicated and then look along the diagonals of the new array until you discover how the Fibonacci sequence can be obtained from it.

3. The diagram above shows Pascal's triangle as written in Japanese in 1781. Use your knowledge of Pascal's triangle to translate the numbers written in Japanese into our number system. Then write down the Japanese numbers from 1 to 20.

Binomial Expansions

The title on the following diagram is *Binomial Expansions* because each line gives the expansion of the binomial $x + y$ raised to a whole-number power.

Binomial Expansions

$$(x + y)^0 = \qquad\qquad 1$$

$$(x + y)^1 = \qquad\qquad x + y$$

$$(x + y)^2 = \qquad\qquad x^2 + 2xy + y^2$$

$$(x + y)^3 = \qquad x^3 + 3x^2y + 3xy^2 + y^3$$

$$(x + y)^4 =$$

$$(x + y)^5 =$$

The fourth row in the diagram was completed by expanding $(x + y)^3$ using the methods developed in this chapter. Next, complete the diagram by expanding the binomials $(x + y)^4$ and $(x + y)^5$ using the multiplication procedures you have learned in this chapter. Finally, study the completed diagram until you see patterns that will allow you to continue the diagram one more row without using multiplication. (One pattern that you will see is Pascal's triangle, which we mentioned in the preceding group project.) When you are finished, write an essay in which you describe what you have done and the results you have obtained.

THE GREAT PYRAMIDS

Image © 2011 GeoEye
The Great Pyramid of Giza
Giza, Egypt

We've worked a great deal with area. Now we can use what we've learned about area and apply it to finding the volume of a pyramid. The shape of the Great Pyramid of Giza is known as a square pyramid, having a square base and a vertex V directly above the center of the base. There are two types of square pyramids: one has sides the shape of an equilateral triangle, and the other has sides the shape of an isosceles triangle. Let's travel to Egypt and find the volume of the Great Pyramid of Giza, and also determine if it has equilateral or isosceles triangles for its sides.

STEP 1   Use Google Earth to find the Great Pyramid of Giza in Egypt. Use the **Ruler** tool set to **Centimeters** to measure the length of the pyramid's base. Calculate the area b of the base. Write your answer in scientific notation.

STEP 2   Use the **Ruler** tool set to **Centimeters** to measure the edges (from the corners of the base to the vertex) of the pyramid. What shape sides does the Great Pyramid of Giza have?

STEP 3   Now use your cursor to find the elevation of the ground at the base of the pyramid. Next, find the elevation of the vertex of the pyramid. Subtract the ground's elevation from the vertex's elevation. This calculation is equal to the height of the vertex. Notice Google Earth uses meters to measure elevation. Convert your elevation measurement to centimeters and write it in scientific notation.

STEP 4   Using the area of the base and the height measurement written in scientific notation, find the volume V of the pyramid. The volume of a pyramid can be found using the formula

$$V = \frac{1}{3}bh$$

where b is the area of the base and h is the height of the pyramid at its vertex.

Rational Expressions and Rational Functions

6

The Mazda Raceway Laguna Seca is a world-renowned paved-road racetrack near Monterey, CA. Every year, the track hosts the United States Grad Prix, a premier opportunity for Formula One racers to test their skills. We can use the rate equation that gives a relationship between rate of speed, distance, and time to examine the performance of racecars during this event. One form of the rate equation is given by a ratio of distance d to time t, which equals the rate of speed r:

$$r = \frac{d}{t}$$

Therefore, if we know two of the three variables, we can calculate the remaining variable. For instance, an official lap record of 1'07.722 for this track was set by Helio Castroneves in 2000. Since we know the track is 2.238 miles long, we can use $r = \frac{d}{t}$ to calculate the average speed of Castroneves's car.

$$r = \frac{d}{t} = \frac{2.238 \text{ mi}}{1 \text{ min } 7.722 \text{ sec}}$$

$$= 118.969 \text{ mph}$$

The ratio of d to t is called a rational expression, making the equation $r = \frac{d}{t}$ a rational function. We will work more with rational expressions and rational functions in this chapter.

Key Words	Definition
Rational Function	Any function that can be written in the form $f(x) = \frac{P(x)}{Q(x)}$ where $P(x)$ and $Q(x)$ are polynomials
Least Common Denominator	The smallest expression that is divisible by each of the denominators
Complex Fraction	A rational expression that contains in its numerator or denominator other rational expressions

Chapter Outline

Basic Properties and Reducing to Lowest Terms

TICKET TO SUCCESS

Keep these questions in mind as you read through the section. Then respond in your own words and in complete sentences.

1. What is a rational expression?
2. Explain how to determine if a rational expression is in "lowest terms."
3. What is a rational function?
4. When is a rational expression undefined?

sylvaine thomas/Shutterstock.com

The Kentucky Derby is often referred to as "the fastest two minutes in sports." The race is 1.25 miles long, and the average speed of a horse in the race is given by the rational function

$$s(t) = \frac{1.25}{t}$$

where $s(t)$ is the speed in miles per minute and t is the duration in minutes.

In this section, we will begin our work with rational expressions and functions, which will help you decide if it is indeed a two minute race.

We will begin this section with the definition of a rational expression. We then will state the two basic properties associated with rational expressions, and go on to apply one of the properties to reduce rational expressions to lowest terms.

Recall from Chapter 1 that a *rational number* is any number that can be expressed as the ratio of two integers.

$$\text{Rational numbers} = \left\{ \frac{a}{b} \,\middle|\, a \text{ and } b \text{ are integers, } b \neq 0 \right\}$$

A rational expression is defined similarly as any expression that can be written as the ratio of two polynomials.

$$\text{Rational expressions} = \left\{ \frac{P}{Q} \,\middle|\, P \text{ and } Q \text{ are polynomials, } Q \neq 0 \right\}$$

Some examples of rational expressions are

$$\frac{2x - 3}{x + 5} \qquad \frac{x^2 - 5x - 6}{x^2 - 1} \qquad \frac{a - b}{b - a}$$

Basic Properties

For rational expressions, multiplying the numerator and denominator by the same nonzero expression may change the form of the rational expression, but it will always produce an expression equivalent to the original one. The same is true when dividing the numerator and denominator by the same nonzero quantity.

> **Property**
>
> If P, Q, and K are polynomials with $Q \neq 0$ and $K \neq 0$, then
>
> $$\frac{P}{Q} = \frac{PK}{QK} \quad \text{and} \quad \frac{P}{Q} = \frac{\frac{P}{K}}{\frac{Q}{K}}$$

A   Reducing to Lowest Terms

The fraction $\frac{6}{8}$ can be written in lowest terms as $\frac{3}{4}$. The process is shown here:

$$\frac{6}{8} = \frac{3 \cdot \overset{1}{\cancel{2}}}{4 \cdot \underset{1}{\cancel{2}}} = \frac{3}{4}$$

Reducing $\frac{6}{8}$ to $\frac{3}{4}$ involves dividing the numerator and denominator by 2, the factor they have in common. Before dividing out the common factor 2, we must notice that the common factor is 2. (This may not be as obvious with other numbers. Since we are very familiar with the numbers 6 and 8 we do not have to put much thought into finding what number divides both of them.)

We reduce rational expressions to lowest terms by first factoring the numerator and denominator and then dividing both numerator and denominator by any factors they have in common.

EXAMPLE 1   Reduce $\dfrac{x^2 - 9}{x - 3}$ to lowest terms.

SOLUTION   Factoring, we have

$$\frac{x^2 - 9}{x - 3} = \frac{(x + 3)(x - 3)}{x - 3}$$

The numerator and denominator have the factor $x - 3$ in common. Dividing the numerator and denominator by $x - 3$, we have

$$\frac{(x + 3)\cancel{(x - 3)}}{\underset{1}{\cancel{x - 3}}} = \frac{x + 3}{1} = x + 3 \qquad \blacksquare$$

Notice in Example 1, there is an implied restriction on the variable x: it cannot be 3. If x were 3, the expression $\frac{(x^2 - 9)}{(x - 3)}$ would become $\frac{0}{0}$, an expression that we cannot associate with a real number. For all problems involving rational expressions, we restrict the variable to only those values that result in a nonzero denominator. When we state the relationship

$$\frac{x^2 - 9}{x - 3} = x + 3$$

we are assuming that it is true for all values of x except $x = 3$.

Here are some other examples of reducing rational expressions to lowest terms.

EXAMPLE 2 Reduce $\dfrac{y^2 - 5y - 6}{y^2 - 1}$ to lowest terms.

SOLUTION Factoring, we have

$$\frac{y^2 - 5y - 6}{y^2 - 1} = \frac{(y - 6)(y + 1)}{(y - 1)(y + 1)}$$ Factor numerator and denominator.

$$= \frac{y - 6}{y - 1}$$ Divide out common factor $y + 1$. ∎

Reduce to lowest terms.

2. $\dfrac{y^2 - y - 6}{y^2 - 4}$

EXAMPLE 3 Reduce $\dfrac{2a^3 - 16}{4a^2 - 12a + 8}$ to lowest terms.

SOLUTION Factoring, we have

$$\frac{2a^3 - 16}{4a^2 - 12a + 8} = \frac{2(a^3 - 8)}{4(a^2 - 3a + 2)}$$ Factor numerator.

$$= \frac{2(a - 2)(a^2 + 2a + 4)}{4(a - 2)(a - 1)}$$ and denominator

$$= \frac{a^2 + 2a + 4}{2(a - 1)}$$ Divide out common factor $2(a - 2)$. ∎

3. $\dfrac{3a^3 + 3}{6a^2 - 6a + 6}$

EXAMPLE 4 Reduce $\dfrac{x^2 - 3x + ax - 3a}{x^2 - ax - 3x + 3a}$ to lowest terms.

SOLUTION Factoring, we have

$$\frac{x^2 - 3x + ax - 3a}{x^2 - ax - 3x + 3a} = \frac{x(x - 3) + a(x - 3)}{x(x - a) - 3(x - a)}$$ Factor numerator and denominator.

$$= \frac{(x - 3)(x + a)}{(x - a)(x - 3)}$$

$$= \frac{x + a}{x - a}$$ Divide out common factor $x - 3$. ∎

4. $\dfrac{x^2 + 4x + ax + 4a}{x^2 + ax + 4x + 4a}$

The answer to Example 4 is $\frac{(x + a)}{(x - a)}$. The problem cannot be reduced further. It is a fairly common mistake to attempt to divide out an x or an a in this last expression. Remember, we can divide out only the factors common to the numerator and denominator of a rational expression. For the last expression in Example 4, neither the numerator nor the denominator can be factored further; x is not a factor of the numerator or the denominator, and neither is a. The expression is in lowest terms.

The next example involves what we call a trick. The trick is to reverse the order of the terms in a difference by factoring -1 from each term. The next examples illustrate how this is done.

EXAMPLE 5 Reduce to lowest terms: $\dfrac{a - b}{b - a}$.

SOLUTION The relationship between $a - b$ and $b - a$ is that they are opposites. We can show this fact by factoring -1 from each term in the numerator.

$$\frac{a - b}{b - a} = \frac{-1(-a + b)}{b - a}$$ Factor -1 from each term in the numerator.

$$= \frac{-1(b - a)}{b - a}$$ Reverse the order of the terms in the numerator.

$$= -1$$ Divide out common factor $b - a$. ∎

5. Replace a with 7 and b with 4 and then simplify.

$$\frac{a - b}{b - a}$$

Answers

2. $\dfrac{y - 3}{y - 2}$

3. $\dfrac{a + 1}{2}$

4. 1

5. -1

6. Reduce to lowest terms.

$$\frac{7 - x}{x^2 - 49}$$

EXAMPLE 6   Reduce to lowest terms: $\dfrac{x^2 - 25}{5 - x}$.

SOLUTION   Begin by factoring the numerator.

$$\frac{x^2 - 25}{5 - x} = \frac{(x - 5)(x + 5)}{5 - x}$$

The factors $x - 5$ and $5 - x$ are similar but are not exactly the same. We can reverse the order of either by factoring -1 from it; that is

$$5 - x = -1(-5 + x) = -1(x - 5).$$

$$\frac{(x - 5)(x + 5)}{5 - x} = \frac{\cancel{(x - 5)}(x + 5)}{-1\cancel{(x - 5)}}$$

$$= \frac{x + 5}{-1}$$

$$= -(x + 5) \qquad \blacksquare$$

B   Ratios

You may recall from previous math classes that the ratio of a to b is the same as the fraction $\frac{a}{b}$. Here are two ratios that are used frequently in mathematics:

1. The number π is defined as the ratio of the circumference of a circle to the diameter of a circle; that is

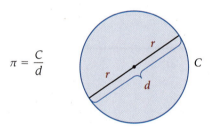

$$\pi = \frac{C}{d}$$

Multiplying both sides of this formula by d, we have the more common form $C = \pi d$.

2. The *average speed* of a moving object is defined to be the ratio of distance to time. If you drive your car for 5 hours and travel a distance of 200 miles, then your average rate of speed is

$$\text{Average speed} = \frac{200 \text{ miles}}{5 \text{ hours}} = 40 \text{ miles per hour}$$

The formula we use for the relationship between average speed r, distance d, and time t is

$$r = \frac{d}{t}$$

The formula is sometimes called the *rate equation*. Multiplying both sides by t, we have an equivalent form of the rate equation, $d = rt$.

Answer

6. $\dfrac{-1}{x + 7}$

C Rational Functions

Our next example involves both the formula for the circumference of a circle and the rate equation.

EXAMPLE 7 The first Ferris wheel was designed and built by George Ferris in 1893. The diameter of the wheel was 250 feet. It had 36 carriages, equally spaced around the wheel, each of which held a maximum of 40 people. One trip around the wheel took 20 minutes. Find the average speed of a rider on the first Ferris wheel. (Use 3.14 as an approximation for π.)

Circumference
250 ft

SOLUTION The distance traveled is the circumference of the wheel, which is

$$C = 250\pi = 250(3.14) = 785 \text{ feet}$$

To find the average speed, we divide the distance traveled by the amount of time it took to go once around the wheel.

$$r = \frac{d}{t} = \frac{785 \text{ feet}}{20 \text{ minutes}} = 39.3 \text{ feet per minute} \qquad \text{To the nearest tenth}$$

Later in this chapter, we will convert this ratio into an equivalent ratio that gives the speed of the rider in miles per hour. ∎

The Ferris wheel in Example 7 has a circumference of 785 feet. In the next example, we graph the relationship between the average speed of a person riding this wheel and the amount of time it takes the wheel to complete one revolution.

EXAMPLE 8 A Ferris wheel has a circumference of 785 feet. If one complete revolution of the wheel takes 10 to 30 minutes, then the relationship between the average speed of a rider on the wheel and the amount of time it takes the wheel to complete one revolution is given by the function

$$r(t) = \frac{785}{t} \qquad 10 \le t \le 30$$

where $r(t)$ is the average speed (in feet per minute) and t is the amount of time (in minutes) it takes the wheel to complete one revolution. Graph the function.

SOLUTION Since the variables r and t represent speed and time, both must be positive quantities. Therefore, the graph of this function will lie in the first quadrant only. The following table displays the values of t and $r(t)$ found from the function, along with the graph of the function (Figure 1). (Some of the numbers in the table have been rounded to the nearest tenth.)

7. A Ferris wheel was built in St. Louis in 1986. It is named *Colossus*. The diameter of the wheel is 165 feet. It has 40 cars, each of which holds 6 passengers. A trip around the wheel takes 40 seconds. Find the average speed of a rider on *Colossus*. (Use 3.14 as an approximation for π, and round your answer to the nearest tenth.)

8. Extend the table in Example 8 by finding $r(35)$. Then explain your result in words.

Answers
7. 13.0 feet per second
8. $r(35) = 22.4$ to the nearest tenth. If one trip around the wheel takes 35 minutes, then the rider is moving at approximately 22.4 feet per minute.

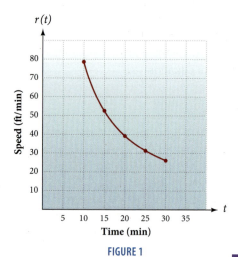

Time to Complete One Revolution t	Speed (ft/min) $r(t)$
10	78.5
15	52.3
20	39.3
25	31.4
30	26.2

FIGURE 1   ■

USING TECHNOLOGY

More About Example 8

If we use a graphing calculator to graph the equation in Example 8, it is not necessary to construct the table first. In fact, if we graph

$$Y_1 = 785/X \qquad \text{Window:} \qquad \text{X from 0 to 40, Y from 0 to 90}$$

we can use the Trace and Zoom features together to produce the numbers in the table next to Figure 1. Graph the preceding equation, and zoom in on the point with x-coordinate 20 until you are convinced that the table values for x and y are correct.

The function shown in Example 8 is called a *rational function* because the right side, $\frac{785}{t}$, is a rational expression (the numerator, 785, is a polynomial of degree 0). We can extend our knowledge of rational expressions to functions with the following definition:

Definition

A **rational function** is any function that can be written in the form

$$f(x) = \frac{P(x)}{Q(x)}$$

where $P(x)$ and $Q(x)$ are polynomials and $Q(x) \neq 0$.

---

9. If $f(x) = \dfrac{x+6}{x-3}$, find
  a. $f(0)$
  b. $f(-6)$
  c. $f(6)$
  d. $f(3)$

EXAMPLE 9   For the rational function $f(x) = \dfrac{x-4}{x-2}$, find $f(0), f(-4), f(4), f(-2),$ and $f(2)$.

SOLUTION   To find these function values, we substitute the given value of x into the rational expression, and then simplify if possible.

$$f(0) = \frac{0-4}{0-2} = \frac{-4}{-2} = 2 \qquad\qquad f(-2) = \frac{-2-4}{-2-2} = \frac{-6}{-4} = \frac{3}{2}$$

$$f(-4) = \frac{-4-4}{-4-2} = \frac{-8}{-6} = \frac{4}{3} \qquad\qquad f(2) = \frac{2-4}{2-2} = \frac{-2}{0} \quad \text{Undefined}$$

$$f(4) = \frac{4-4}{4-2} = \frac{0}{2} = 0$$

■

Answers
9. a. −2 **b.** 0 **c.** 4 **d.** undefined

Because the rational function in Example 9 is not defined when x is 2, the domain of that function does not include 2. We have more to say about the domain of a rational function next.

The Domain of a Rational Function

In Example 8, the domain of the rational function is specified as $10 \leq t \leq 30$, and the function is defined for all values of t in that domain. If the domain of a rational function is not specified, it is assumed to be all real numbers for which the function is defined; that is, the domain of the rational function

$$f(x) = \frac{P(x)}{Q(x)}$$

is all x for which $Q(x)$ is nonzero. For example,

The domain for $r(t) = \dfrac{785}{t}$, $10 \leq t \leq 30$, is $\{t \mid 10 \leq t \leq 30\}$.

The domain for $f(x) = \dfrac{x - 4}{x - 2}$ is $\{x \mid x \neq 2\}$.

The domain for $g(x) = \dfrac{x^2 + 5}{x + 1}$ is $\{x \mid x \neq -1\}$.

The domain for $h(x) = \dfrac{x}{x^2 - 9}$ is $\{x \mid x \neq -3, x \neq 3\}$.

Notice that, for these functions, $f(2)$, $g(-1)$, $h(-3)$, and $h(3)$ are all undefined, and that is why the domains are written as shown.

EXAMPLE 10 Graph the equation $y = \dfrac{x^2 - 9}{x - 3}$. How is this graph different from the graph of $y = x + 3$?

SOLUTION We know from the discussion in Example 1 that

$$y = \frac{x^2 - 9}{x - 3} = \frac{(x + 3)(x - 3)}{x - 3} = x + 3$$

This relationship is true for all x except $x = 3$ because the rational expression with $x - 3$ in the denominator is undefined when x is 3. However, for all other values of x, the expressions

$$\frac{x^2 - 9}{x - 3} \quad \text{and} \quad x + 3$$

are equal. Therefore, the graphs of

$$y = \frac{x^2 - 9}{x - 3} \quad \text{and} \quad y = x + 3$$

will be the same except when x is 3. In the first equation, there is no value of y to correspond to $x = 3$. In the second equation, $y = x + 3$, so y is 6 when x is 3.

10. Graph the equation $y = \dfrac{x^2 - 1}{x + 1}$ and explain how this graph differs from the graph of $y = x - 1$

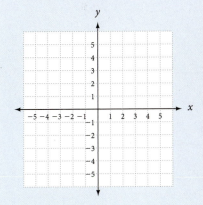

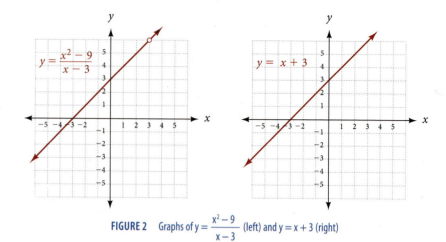

FIGURE 2   Graphs of $y = \dfrac{x^2 - 9}{x - 3}$ (left) and $y = x + 3$ (right)

Now you can see the difference in the graphs of the two equations. To show that there is no y value for $x = 3$ in the graph on the left in Figure 2, we draw an open circle at that point on the line.   ■

Notice that the two graphs shown in Figure 2 are both graphs of functions. Suppose we use function notation to designate them as follows:

$$f(x) = \frac{x^2 - 9}{x - 3} \qquad \text{and} \qquad g(x) = x + 3$$

The two functions, f and g, are equivalent except when $x = 3$ because $f(3)$ is undefined, while $g(3) = 6$. The domain of the function f is all real numbers except $x = 3$, while the domain for g is all real numbers, with no restrictions.

Problem Set 6.1

Moving Toward Success

"Time is what we want most, but what we use worst."

—William Penn, 1644–1718, founder of Pennsylvania

1. What are the three main ways you waste your time?

2. How can you prevent these three things from interrupting your study time?

A

1. Simplify each expression. State any restrictions on the variable. [Examples 1–4]

a. $\dfrac{6 + 1}{36 - 1}$ b. $\dfrac{x + 3}{x^2 - 9}$ c. $\dfrac{x^2 - 3x}{x^2 - 9}$ d. $\dfrac{x^3 - 27}{x^2 - 9}$

2. Simplify each expression. State any restrictions on the variable.

a. $\dfrac{64 - 80 + 25}{64 - 25}$ b. $\dfrac{x^2 - 10x + 25}{x^2 - 25}$ c. $\dfrac{x^2 - 26x + 25}{x^2 - 25x}$ d. $\dfrac{x^2 + 5x + ax + 5a}{x^2 - 25}$

3. If $h(t) = \dfrac{t - 3}{t + 1}$, find $h(0)$, $h(-3)$, $h(3)$, $h(-1)$, and $h(1)$, if possible.

4. If $h(t) = \dfrac{t - 2}{t + 1}$, find $h(0)$, $h(-2)$, $h(2)$, $h(-1)$, and $h(1)$, if possible.

A Reduce each rational expression to lowest terms. [Examples 1–4]

5. $\dfrac{x^2 - 16}{6x + 24}$ **6.** $\dfrac{5x + 25}{x^2 - 25}$ **7.** $\dfrac{12x - 9y}{3x^2 + 3xy}$ **8.** $\dfrac{x^3 - xy^2}{4x + 4y}$

9. $\dfrac{a^4 - 81}{a - 3}$ **10.** $\dfrac{a + 4}{a^2 - 16}$ **11.** $\dfrac{a^2 - 4a - 12}{a^2 + 8a + 12}$ **12.** $\dfrac{a^2 - 7a + 12}{a^2 - 9a + 20}$

13. $\dfrac{20y^2 - 45}{10y^2 - 5y - 15}$ **14.** $\dfrac{54y^2 - 6}{18y^2 - 60y + 18}$ **15.** $\dfrac{a^3 + b^3}{a^2 - b^2}$ **16.** $\dfrac{a^2 - b^2}{a^3 - b^3}$

Reduce to lowest terms. [Examples 1–4]

17. $\dfrac{8x^4 - 8x}{4x^4 + 4x^3 + 4x^2}$

18. $\dfrac{6x^5 - 48x^2}{12x^3 + 24x^2 + 48x}$

19. $\dfrac{6x^2 + 7xy - 3y^2}{6x^2 + xy - y^2}$

20. $\dfrac{4x^2 - y^2}{4x^2 - 8xy - 5y^2}$

21. $\dfrac{ax + 2x + 3a + 6}{ay + 2y - 4a - 8}$

22. $\dfrac{ax - x - 5a + 5}{ax + x - 5a - 5}$

23. $\dfrac{x^2 + bx - 3x - 3b}{x^2 - 2bx - 3x + 6b}$

24. $\dfrac{x^2 - 3ax - 2x + 6a}{x^2 - 3ax + 2x - 6a}$

25. $\dfrac{x^3 + 3x^2 - 4x - 12}{x^2 + x - 6}$

26. $\dfrac{x^3 + 5x^2 - 4x - 20}{x^2 + 7x + 10}$

27. $\dfrac{4x^4 - 25}{6x^3 - 4x^2 + 15x - 10}$

28. $\dfrac{16x^4 - 49}{8x^3 - 12x^2 + 14x - 21}$

29. $\dfrac{x^3 - 8}{x^2 - 4}$

30. $\dfrac{y^2 - 9}{y^3 - 27}$

31. $\dfrac{64 + t^3}{16 - 4t + t^2}$

32. $\dfrac{25 + 5a + a^2}{125 - a^3}$

33. $\dfrac{8x^3 - 27}{4x^2 - 9}$

34. $\dfrac{25y^2 - 4}{125y^3 + 8}$

Reduce the following to lowest terms. [Example 5, 6]

35. $\dfrac{x - 4}{4 - x}$

36. $\dfrac{6 - x}{x - 6}$

37. $\dfrac{y^2 - 36}{6 - y}$

38. $\dfrac{1 - y}{y^2 - 1}$

39. $\dfrac{1 - 9a^2}{9a^2 - 6a + 1}$

40. $\dfrac{1 - a^2}{a^2 - 2a + 1}$

Simplify each expression.

41. $\dfrac{(3x - 5) - (3a - 5)}{x - a}$

42. $\dfrac{(2x + 3) - (2a + 3)}{x - a}$

43. $\dfrac{(x^2 - 4) - (a^2 - 4)}{x - a}$

44. $\dfrac{(x^2 - 1) - (a^2 - 1)}{x - a}$

C State the domain for each rational function.

45. $f(x) = \dfrac{x - 3}{x - 1}$

46. $g(x) = \dfrac{x^2 - 4}{x - 2}$

47. $h(t) = \dfrac{t - 4}{t^2 - 16}$

48. $h(t) = \dfrac{t - 5}{t^2 - 25}$

49. $f(x) = \dfrac{3(x^2 - 25)}{3x - 15}$

50. $g(x) = -\dfrac{4x - 36}{4x + 28}$

51. $f(x) = -\dfrac{x^2 + 25}{10}$

52. $g(x) = -\dfrac{x - 9}{45}$

53. $f(x) = \dfrac{2(x^2 + 49)}{7x}$

54. $g(x) = -\dfrac{x^3 - 27}{4x}$

55. $h(x) = \dfrac{x^3 - 8}{x^2 - x - 20}$

56. $f(x) = \dfrac{x + \pi - 5}{x^2 + x - 12}$

C Let $f(x) = \dfrac{x^2 - 4}{x - 2}$ and $g(x) = x + 2$, and evaluate the following expressions, if possible. [Example 9]

57. $f(0)$ and $g(0)$

58. $f(1)$ and $g(1)$

59. $f(2)$ and $g(2)$

60. $f(3)$ and $g(3)$

Let $f(x) = \dfrac{x^2 - 1}{x - 1}$ and $g(x) = x + 1$, and evaluate the following expressions, if possible.

61. $f(0)$ and $g(0)$

62. $f(1)$ and $g(1)$

63. $f(2)$ and $g(2)$

64. $f(-1)$ and $g(-1)$

Applying the Concepts

65. Diet If you have ever been on a weight loss diet you know that you lose more weight in the beginning than you do later. The quantity $W(x)$ is your weight (in pounds) after x weeks of dieting. Use the function to fill in the table, rounding to the nearest pound.

$$W(x) = \frac{80(2x + 15)}{x + 6}$$

Weeks x	Weight (pounds) W (x)
0	
1	
4	
12	
24	

66. Drag Racing The following rational function gives the speed $V(x)$, in miles per hour, of a dragster at each second x during a quarter-mile race. Use the function to fill in the table, rounding to the nearest tenth.

$$V(x) = \frac{340x}{x + 3}$$

Weeks x	Weight (pounds) W (x)
0	
1	
2	
3	
4	
5	
6	

Average Speed For Problems 67 and 68, use 3.14 as an approximation for π. Round answers to the nearest tenth.

67. A person riding a Ferris wheel with a diameter of 65 feet travels once around the wheel in 30 seconds. What is the average speed of the rider in feet per second?

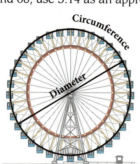

68. A person riding a Ferris wheel with a diameter of 102 feet travels once around the wheel in 3.5 minutes. What is the average speed of the rider in feet per minute?

69. Average Speed A Ferris wheel has a circumference of 204 feet (to the nearest foot). If a ride on the wheel takes from 20 to 50 seconds, then the relationship between the average speed of a rider and the amount of time it takes to complete one revolution is given by the function

$$r(t) = \frac{204}{t} \qquad 20 \le t \le 50$$

where $r(t)$ is in feet per second and t is in seconds.

a. State the domain for this function.

b. Graph this function.

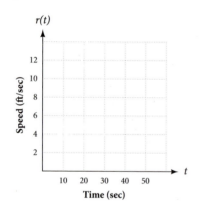

70. Average Speed The section opener explained that the average speed of a horse in the Kentucky Derby is given by the rational function

$$s(t) = \frac{1.25}{t} \qquad 0.5 \le t \le 2$$

where $s(t)$ is the speed in miles per minute and t is the duration in minutes.

a. State the domain for this function.

b. Graph this function.

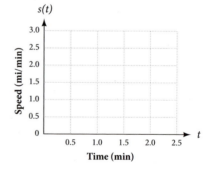

71. Intensity of Light

The relationship between the intensity of light that falls on a surface from a 100-watt light bulb and the distance from that surface is given by the rational function

$$I(d) = \frac{120}{d^2}$$

for $1 \le d \le 6$

where $I(d)$ is the intensity of light (in lumens per square foot) and d is the distance (in feet) from the light bulb to the surface.

a. State the domain for this function.

b. Graph this function.

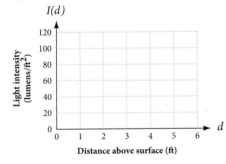

72. Average Speed If it takes Maria t minutes to run a mile, then her average speed $s(t)$ is given by the rational function

$$s(t) = \frac{60}{t}$$

for $6 \le t \le 12$

where $s(t)$ is in miles per hour and t is in minutes.

a. State the domain for this function.

b. Graph this function.

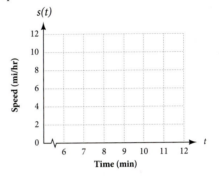

Average Speed The abbreviation "rpm" stands for revolutions per minute. If a point on a circle rotates at 300 rpm, then it rotates through one complete revolution 300 times every minute. The length of time it takes to rotate once around the circle is $\frac{1}{300}$ minute. Use 3.14 as an approximation for π.

73. An audio CD, when placed in a CD player, rotates at 300 rpm (1 revolution takes $\frac{1}{300}$ minute). Find the average speed of a point 2 inches from the center of the CD. Then find the average speed of a point 1.5 inches from the center of the CD.

74. A power generating turbine rotates at 3,600 rpm. Find the average speed of a point 2 feet from the center of the turbine. Then find the average speed of a point 1.5 feet from the center.

Maintaining Your Skills

Subtract as indicated.

75. Subtract $x^2 + 2x + 1$ from $4x^2 - 5x + 5$.

76. Subtract $3x^2 - 5x + 2$ from $7x^2 + 6x + 4$.

77. Subtract $10x - 20$ from $10x - 11$.

78. Subtract $-6x - 18$ from $-6x + 5$.

79. Subtract $4x^3 - 8x^2$ from $4x^3$.

80. Subtract $2x^2 + 6x$ from $2x^2$.

Solve the linear systems using any method.

81. $5x + 2y = -3$
$\quad\ x - 3y = -21$

82. $3x - 2y = 10$
$\quad\ 5x + 3y = -15$

83. $\dfrac{1}{4}x + \dfrac{1}{2}y = \dfrac{3}{4}$
$\quad\ \dfrac{1}{3}x + \dfrac{4}{5}y = \dfrac{2}{15}$

84. $\dfrac{2}{3}x - \dfrac{1}{2}y = \dfrac{5}{6}$
$\quad\ \dfrac{3}{2}x + \dfrac{3}{4}y = -\dfrac{5}{8}$

85. $2x - 5y = 2$
$\quad\quad\ y = x + 2$

86. $3x - 4y = 24$
$\quad\ 2x = -4y - 4$

87. $x - 2y - 2z = 4$
$\quad\ 3x + 3y + z = 2$
$\quad\quad x + y - z = 6$

88. $\ x - 3y = 16$
$\quad\ x + 2z = 9$
$\quad\ 2y + 4z = 6$

Getting Ready for the Next Section

Divide.

89. $\dfrac{10x^5}{5x^2}$

90. $\dfrac{-15x^4}{5x^2}$

91. $\dfrac{4x^4y^3}{-2x^2y}$

92. $\dfrac{10a^4b^2}{4a^2b^2}$

93. $4{,}628 \div 25$

94. $7{,}546 \div 35$

Multiply.

95. $2x^2(2x - 4)$

96. $3x^2(x - 2)$

97. $(2x - 4)(2x^2 + 4x + 5)$

98. $(x - 2)(3x^2 + 6x + 15)$

Division of Polynomials, and Difference Quotients

OBJECTIVES

A Divide polynomials by factoring.

B Evaluate difference quotients.

C Divide polynomials using long division.

TICKET TO SUCCESS

Keep these questions in mind as you read through the section. Then respond in your own words and in complete sentences.

1. What property of real numbers is the key to dividing a polynomial by a monomial?
2. What is a difference quotient?
3. What are the four steps used in long division with polynomials?
4. When must long division be performed, and when can factoring be used to divide polynomials?

Dmitrijs Dmitrijevs/Shutterstock.com

Suppose the front and back yards of your rental house are extensively overgrown. You decide to hire a landscaper who charges a $50 service fee plus $15 per hour. If the landscaper works x hours, the total work cost is given by the formula $C(x) = 50 + 15x$. From this formula, we see that the more hours the landscaper works, the more we pay him. But it is also true that the more hours he works, the lower the cost per hour. To find the cost per hour, we use the average cost function, which divides the total cost by the number of hours worked.

$$\text{Average Cost} = \overline{C}(x) = \frac{C(x)}{x} = \frac{50 + 15x}{x}$$

This last expression gives us the average cost per hour for each of the x hours the landscaper works. To work with this last expression, we need to know something about division with polynomials, and that is what we will cover in this section.

We begin by considering division of a polynomial by a monomial. This is the simplest kind of polynomial division. The rest of the section is devoted to division of a polynomial by a polynomial. This kind of division is similar to long division with whole numbers.

A Dividing a Polynomial by a Monomial

To divide a polynomial by a monomial, we use the definition of division and apply the distributive property. The following example illustrates the procedure.

EXAMPLE 1   Divide $\dfrac{10x^5 - 15x^4 + 20x^3}{5x^2}$.

SOLUTION

$$= (10x^5 - 15x^4 + 20x^3) \cdot \frac{1}{5x^2}$$

Dividing by $5x^2$ is the same as multiplying by $\frac{1}{5x^2}$.

$$= 10x^5 \cdot \frac{1}{5x^2} - 15x^4 \cdot \frac{1}{5x^2} + 20x^3 \cdot \frac{1}{5x^2}$$

Distributive property

$$= \frac{10x^5}{5x^2} - \frac{15x^4}{5x^2} + \frac{20x^3}{5x^2}$$

Multiplying by $\frac{1}{5x^2}$ is the same as dividing by $5x^2$.

$$= 2x^3 - 3x^2 + 4x$$

Divide coefficients, subtract exponents ∎

Notice that division of a polynomial by a monomial is accomplished by dividing each term of the polynomial by the monomial. The first two steps are usually not shown in a problem like this. They are part of Example 1 to justify distributing $5x^2$ under all three terms of the polynomial $10x^5 - 15x^4 + 20x^3$.

Here are some more examples of this kind of division.

EXAMPLE 2   Divide. Write all results with positive exponents.

SOLUTION

$$\frac{8x^3y^5 - 16x^2y^2 + 4x^4y^3}{-2x^2y} = \frac{8x^3y^5}{-2x^2y} + \frac{-16x^2y^2}{-2x^2y} + \frac{4x^4y^3}{-2x^2y}$$

$$= -4xy^4 + 8y - 2x^2y^2$$ ∎

EXAMPLE 3   Divide. Write all results with positive exponents.

SOLUTION

$$\frac{10a^4b^2 + 8ab^3 - 12a^3b + 6ab}{4a^2b^2} = \frac{10a^4b^2}{4a^2b^2} + \frac{8ab^3}{4a^2b^2} - \frac{12a^3b}{4a^2b^2} + \frac{6ab}{4a^2b^2}$$

$$= \frac{5a^2}{2} + \frac{2b}{a} - \frac{3a}{b} + \frac{3}{2ab}$$ ∎

Notice in Example 3 that the result is not a polynomial because of the last three terms. If we were to write each term as a product, some of the variables would have negative exponents. For example, the second term would be

$$\frac{2b}{a} = 2a^{-1}b$$

The divisor in each of the previous examples was a monomial. We now want to turn our attention to division of polynomials in which the divisor has two or more terms.

EXAMPLE 4   Divide $\dfrac{x^2 - 6xy - 7y^2}{x + y}$.

SOLUTION   In this case, we can factor the numerator and perform our division by simply dividing out common factors, just like we did in the previous section.

$$\frac{x^2 - 6xy - 7y^2}{x + y} = \frac{\cancel{(x + y)}(x - 7y)}{\cancel{x + y}}$$

$$= x - 7y$$ ∎

B Difference Quotients

Figure 1 is an important diagram from calculus. Although it may look complicated, the point of it is simple: the slope m of the line passing through the points P and Q is given by the formula

$$\text{Slope of line through } PQ = m = \frac{f(x) - f(a)}{x - a}$$

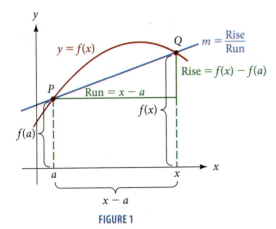

FIGURE 1

The expression $\frac{f(x) - f(a)}{x - a}$ is called a difference quotient. When $f(x)$ is a polynomial, the difference quotient will be a rational expression.

EXAMPLE 5 If $f(x) = 3x - 5$, find $\frac{f(x) - f(a)}{x - a}$.

SOLUTION

$$\frac{f(x) - f(a)}{x - a} = \frac{(3x - 5) - (3a - 5)}{x - a}$$

$$= \frac{3x - 5 - 3a + 5}{x - a}$$

$$= \frac{3x - 3a}{x - a}$$

$$= \frac{3(x - a)}{x - a}$$

$$= 3 \qquad ■$$

EXAMPLE 6 If $f(x) = x^2 - 4$, find $\frac{f(x) - f(a)}{x - a}$ and simplify.

SOLUTION Because $f(x) = x^2 - 4$ and $f(a) = a^2 - 4$, we have

$$\frac{f(x) - f(a)}{x - a} = \frac{(x^2 - 4) - (a^2 - 4)}{x - a}$$

$$= \frac{x^2 - 4 - a^2 + 4}{x - a}$$

$$= \frac{x^2 - a^2}{x - a}$$

$$= \frac{(x + a)(x - a)}{x - a} \qquad \text{Factor and divide out common factor.}$$

$$= x + a \qquad ■$$

5. If $f(x) = 2x + 3$, find $\frac{f(x) - f(a)}{x - a}$.

6. If $f(x) = x^2 - 1$, find $\frac{f(x) - f(a)}{x - a}$.

Answers
5. 2
6. $x + a$

The diagram in Figure 2 is similar to the one in Figure 1. The main difference is in how we label the points. From Figure 2, we can see another difference quotient that gives us the slope of the line through the points P and Q.

$$\text{Slope of line through } PQ = m = \frac{f(x + h) - f(x)}{h}$$

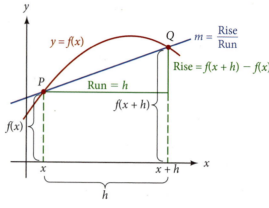

FIGURE 2

Examples 7 and 8 use the same functions used in Examples 5 and 6, but this time the new difference quotient is used.

7. If $f(x) = 2x + 3$,

find $\dfrac{f(x + h) - f(x)}{h}$.

EXAMPLE 7 If $f(x) = 3x - 5$, find $\dfrac{f(x + h) - f(x)}{h}$.

SOLUTION The expression $f(x + h)$ is given by

$$f(x + h) = 3(x + h) - 5$$
$$= 3x + 3h - 5$$

Using this result gives us

$$\frac{f(x + h) - f(x)}{h} = \frac{(3x + 3h - 5) - (3x - 5)}{h}$$

$$= \frac{3h}{h}$$

$$= 3 \qquad\blacksquare$$

8. If $f(x) = x^2 - 1$,

find $\dfrac{f(x + h) - f(x)}{h}$.

EXAMPLE 8 If $f(x) = x^2 - 4$, find $\dfrac{f(x + h) - f(x)}{h}$.

SOLUTION The expression $f(x + h)$ is given by

$$f(x + h) = (x + h)^2 - 4$$
$$= x^2 + 2xh + h^2 - 4$$

Using this result gives us

$$\frac{f(x + h) + f(x)}{h} = \frac{(x^2 + 2xh + h^2 - 4) - (x^2 - 4)}{h}$$

$$= \frac{2xh + h^2}{h}$$

$$= \frac{h(2x + h)}{h}$$

$$= 2x + h \qquad\blacksquare$$

Answers
7. 2
8. $2x + h$

C Long Division

For the type of division shown in Examples 4 through 6, the denominator must be a factor of the numerator. When the denominator is not a factor of the numerator, or in the case where we can't factor the numerator, the method used in Examples 4 through 6 won't work. We need to develop a new method for these cases. Since this new method is very similar to long division with whole numbers, we will review it here.

EXAMPLE 9 Divide $25\overline{)4628}$.

9. Divide $35\overline{)7546}$.

SOLUTION

$$
\begin{array}{r}
1 \\
25\overline{)4628} \\
\underline{25} \\
21
\end{array}
$$

Estimate 25 into 46.

Multiply $1 \times 25 = 25$.

Subtract $46 - 25 = 21$.

$$
\begin{array}{r}
1 \\
25\overline{)4628} \\
\underline{25\downarrow} \\
212
\end{array}
$$

Bring down the 2.

These are the four basic steps in long division: estimate, multiply, subtract, and bring down the next term. To complete the problem, we simply perform the same four steps:

$$
\begin{array}{r}
18 \\
25\overline{)4628} \\
\underline{25\downarrow} \\
212 \\
\underline{200} \\
128
\end{array}
$$

8 is the estimate.

Multiply to get 200.

Subtract to get 12, then bring down the 8.

One more time:

$$
\begin{array}{r}
185 \\
25\overline{)4628} \\
\underline{25\downarrow} \\
212 \\
\underline{200} \\
128 \\
\underline{125} \\
3
\end{array}
$$

5 is the estimate.

Multiply to get 125.

Subtract to get 3.

Since 3 is less than 25 and we have no more terms to bring down, we have our answer:

$$
\frac{4628}{25} = 185 + \frac{3}{25}
$$

To check our answer, we multiply 185 by 25 and then add 3 to the result.

$$
25(185) + 3 = 4{,}625 + 3 = 4{,}628 \qquad \blacksquare
$$

Long division with polynomials is very similar to long division with whole numbers. Both use the same four basic steps: estimate, multiply, subtract, and bring down the next term. We use long division with polynomials when the denominator has two or more terms and is not a factor of the numerator. Here is an example.

> **NOTE**
> You may realize when looking over this example that you don't have a very good idea why you proceed as you do with the steps in long division. What you do know is the process always works. We are going to approach the explanation for division of two polynomials with this in mind; that is, we won't always be sure why the steps we use are important, only that they always produce the correct result.

Answer

9. $215 + \dfrac{3}{5}$

Divide.

10. $\dfrac{3x^2 - 8x - 1}{x - 3}$

EXAMPLE 10 Divide $\dfrac{2x^2 - 7x + 9}{x - 2}$.

SOLUTION

$$
\begin{array}{r}
2x \\
x - 2\overline{)2x^2 - 7x + 9} \\
\cancel{2x^2} \; \cancel{4x} \\
\hline
-3x
\end{array}
$$

Estimate $2x^2 \div x = 2x$.

Multiply $2x(x - 2) = 2x^2 - 4x$.

Subtract $(2x^2 - 7x) - (2x^2 - 4x) = -3x$.

$$
\begin{array}{r}
2x \\
x - 2\overline{)2x^2 - 7x + 9} \\
\cancel{2x^2} \; \cancel{4x} \\
\hline
-3x + 9
\end{array}
$$

Bring down the 9.

Notice we change the signs on $2x^2 - 4x$ and add in the subtraction step. Subtracting a polynomial is equivalent to adding its opposite. We repeat the four steps.

$$
\begin{array}{r}
2x - 3 \\
x - 2\overline{)2x^2 - 7x + 9} \\
\cancel{2x^2} \; \cancel{4x} \\
\hline
-3x + 9 \\
\cancel{3x} \; \cancel{6} \\
\hline
3
\end{array}
$$

-3 is the estimate: $-3x \div x = -3$.

Multiply $-3(x - 2) = -3x + 6$.

Subtract $(-3x + 9) - (-3x + 6) = 3$.

Since we have no other term to bring down, we have our answer:

$$
\frac{2x^2 - 7x + 9}{x - 2} = 2x - 3 + \frac{3}{x - 2}
$$

To check, we multiply $(2x - 3)(x - 2)$ to get $2x^2 - 7x + 6$; then, adding the remainder 3 to this result, we have $2x^2 - 7x + 9$. ∎

In setting up a division problem involving two polynomials, we must remember two things: both polynomials should be in decreasing powers of the variable, and neither should skip any powers from the highest power down to the constant term. If there are any missing terms, they can be filled in using a coefficient of 0.

11. Divide $x - 2\overline{)3x^3 + 3x + 1}$.

EXAMPLE 11 Divide $2x - 4\overline{)4x^3 - 6x - 11}$.

SOLUTION Since the trinomial is missing a term in x^2, we can fill it in with $0x^2$.

$$
4x^3 - 6x - 11 = 4x^3 + 0x^2 - 6x - 11
$$

Adding $0x^2$ does not change our original problem.

$$
\begin{array}{r}
2x^2 + 4x + 5 \\
2x - 4\overline{)4x^3 + 0x^2 - 6x - 11} \\
\cancel{4x^3} \; \cancel{8x^2} \\
\hline
+ 8x^2 - 6x \\
\cancel{8x^2} \; \cancel{16x} \\
\hline
+ 10x - 11 \\
\cancel{10x} \; \cancel{20} \\
\hline
+ 9
\end{array}
$$

Note: Adding the $0x^2$ term gives us a column in which to write $-8x^2$.

$$
\frac{4x^3 - 6x - 11}{2x - 4} = 2x^2 + 4x + 5 + \frac{9}{2x - 4}
$$

Answers

10. $3x + 1 + \dfrac{2}{x - 3}$

11. $3x^2 + 6x + 15 + \dfrac{31}{x - 2}$

To check this result, we multiply $2x - 4$ and $2x^2 + 4x + 5$.

$$
\begin{array}{r}
2x^2 + 4x + 5 \\
2x - 4 \\
\hline
4x^3 + 8x^2 + 10x \\
-8x^2 - 16x - 20 \\
\hline
4x^3 \qquad -6x - 20
\end{array}
$$

Adding 9 (the remainder) to this result gives us the polynomial $4x^3 - 6x - 11$. Our answer checks. ∎

Let's do Example 4 again, but this time use long division.

EXAMPLE 12   Divide $\dfrac{x^2 - 6xy - 7y^2}{x + y}$.

SOLUTION

$$
\begin{array}{r}
x - 7y \\
x + y \overline{)\; x^2 - 6xy - 7y^2} \\
x^2 + xy \\
\hline
-7xy - 7y^2 \\
7xy + 7y^2 \\
\hline
0
\end{array}
$$

In this case, the remainder is 0 and we have

$$
\frac{x^2 - 6xy - 7y^2}{x + y} = x - 7y
$$

which is easy to check since

$$
(x + y)(x - 7y) = x^2 - 6xy - 7y^2 \qquad \blacksquare
$$

EXAMPLE 13   Factor $x^3 + 9x^2 + 26x + 24$ completely if $x + 2$ is one of its factors.

SOLUTION   Because $x + 2$ is one of the factors of the polynomial we are trying to factor, it must divide that polynomial evenly — that is, without a remainder. Therefore, we begin by dividing the polynomial by $x + 2$:

$$
\begin{array}{r}
x^2 + 7x + 12 \\
x + 2 \overline{)\; x^3 + 9x^2 + 26x + 24} \\
x^3 + 2x^2 \qquad\qquad \downarrow \\
\hline
+7x^2 + 26x \\
7x^2 + 14x \qquad \downarrow \\
\hline
+12x + 24 \\
12x + 24 \\
\hline
0
\end{array}
$$

Now we know that the polynomial we are trying to factor is equal to the product of $x + 2$ and $x^2 + 7x + 12$. To factor completely, we simply factor $x^2 + 7x + 12$:

$$
x^3 + 9x^2 + 26x + 24 = (x + 2)(x^2 + 7x + 12)
$$

$$
= (x + 2)(x + 3)(x + 4) \qquad \blacksquare
$$

Divide.

12. $\dfrac{2x^2 - 5xy + 3y^2}{x - y}$

13. Factor $x^3 + 6x^2 + 11x + 6$ completely if $x + 1$ is one of its factors.

Answers

12. $2x - 3y$

13. $(x + 1)(x + 2)(x + 3)$

Problem Set 6.2

Moving Toward Success

"Just as iron rusts from disuse, even so does inaction spoil the intellect."

—Leonardo da Vinci, 1452–1519, Italian painter and sculptor

1. What time during the day do you feel productive? Why?

2. How can you incorporate studying for this class during those productive hours?

A Find the following quotients. [Examples 1–3]

1. $\dfrac{4x^3 - 8x^2 + 6x}{2x}$

2. $\dfrac{6x^3 + 12x^2 - 9x}{3x}$

3. $\dfrac{10x^4 + 15x^3 - 20x^2}{-5x^2}$

4. $\dfrac{12x^5 - 18x^4 - 6x^3}{6x^3}$

5. $\dfrac{8y^5 + 10y^3 - 6y}{4y^3}$

6. $\dfrac{6y^4 - 3y^3 + 18y^2}{9y^2}$

7. $\dfrac{5x^3 - 8x^2 - 6x}{-2x^2}$

8. $\dfrac{-9x^5 + 10x^3 - 12x}{-6x^4}$

9. $\dfrac{28a^3b^5 + 42a^4b^3}{7a^2b^2}$

10. $\dfrac{a^2b + ab^2}{ab}$

11. $\dfrac{10x^3y^2 - 20x^2y^3 - 30x^3y^3}{-10x^2y}$

12. $\dfrac{9x^4y^4 + 18x^3y^4 - 27x^2y^4}{-9xy^3}$

A Divide by factoring numerators and then dividing out common factors. [Example 4]

13. $\dfrac{x^2 - x - 6}{x - 3}$

14. $\dfrac{x^2 - x - 6}{x + 2}$

15. $\dfrac{2a^2 - 3a - 9}{2a + 3}$

16. $\dfrac{2a^2 + 3a - 9}{2a - 3}$

17. $\dfrac{5x^2 - 14xy - 24y^2}{x - 4y}$

18. $\dfrac{5x^2 - 26xy - 24y^2}{5x + 4y}$

19. $\dfrac{x^3 - y^3}{x - y}$

20. $\dfrac{x^3 + 8}{x + 2}$

21. $\dfrac{y^4 - 16}{y - 2}$

22. $\dfrac{y^4 - 81}{y - 3}$

23. $\dfrac{x^3 + 2x^2 - 25x - 50}{x - 5}$

24. $\dfrac{x^3 + 2x^2 - 25x - 50}{x + 5}$

B For the functions below, evaluate [Examples 5–8]

a. $\dfrac{f(x + h) - f(x)}{h}$ **b.** $\dfrac{f(x) - f(a)}{x - a}$

25. $f(x) = 4x$

26. $f(x) = -3x$

27. $f(x) = 5x + 3$

28. $f(x) = 6x - 5$

29. $f(x) = x^2$

30. $f(x) = 3x^2$

31. $f(x) = x^2 + 1$

32. $f(x) = x^2 - 3$

33. $f(x) = x^2 - 3x + 4$

34. $f(x) = x^2 + 4x - 7$

35. $f(x) = 2x^2 + 3x - 4$

36. $f(x) = 5x^2 + 3x - 7$

37. $f(x) = -3x^2 + 2x - 1$

38. $f(x) = -2x^2 + 5x + 6$

C Divide using the long division method. [Examples 10–13]

39. $\dfrac{x^2 - 5x - 7}{x + 2}$

40. $\dfrac{x^2 + 4x - 8}{x - 3}$

41. $\dfrac{2x^3 - 3x^2 - 4x + 5}{x + 1}$

42. $\dfrac{3x^3 - 5x^2 + 2x - 1}{x - 2}$

43. $\dfrac{2y^3 - 9y^2 - 17y + 39}{2y - 3}$

44. $\dfrac{3y^3 - 19y^2 + 17y + 4}{3y - 4}$

45. $\dfrac{6y^3 - 8y + 5}{2y - 4}$

46. $\dfrac{9y^3 - 6y^2 + 8}{3y - 3}$

47. $\dfrac{a^4 - 2a + 5}{a - 2}$

48. $\dfrac{a^4 + a^3 - 1}{a + 2}$

49. $\dfrac{y^4 - 16}{y - 2}$

50. $\dfrac{y^4 - 81}{y - 3}$

51. Let $f(x) = x^2 - 36$ and $g(x) = 4x - 24$. If $h(x) = \dfrac{f(x)}{g(x)}$, find $h(x)$, then state the domain.

52. Let $f(x) = x^2 - 49$ and $g(x) = 2x + 14$. If $h(x) = \dfrac{f(x)}{g(x)}$, find $h(x)$, then state the domain.

53. Let $f(x) = x^2 - 16x + 64$ and $g(x) = x^2 - 4x - 32$. If $h(x) = \dfrac{f(x)}{g(x)}$, find $h(x)$, then state the domain.

54. Let $f(x) = x^2 + 20x + 100$ and $g(x) = x^2 + 5x - 50$. If $h(x) = \dfrac{f(x)}{g(x)}$, find $h(x)$, then state the domain.

55. Let $f(x) = x^3 - 27$ and $g(x) = x - 3$. If $h(x) = \dfrac{f(x)}{g(x)}$, find $h(x)$, then state the domain.

56. Let $f(x) = x^3 + 125$ and $g(x) = x + 5$. If $h(x) = \dfrac{f(x)}{g(x)}$, find $h(x)$, then state the domain.

57. The Factor Theorem The factor theorem of algebra states that if $x - a$ is a factor of a polynomial, $P(x)$, then $P(a) = 0$. Verify that

a. $x - 2$ is a factor of $P(x) = x^3 - 3x^2 + 5x - 6$, and that $P(2) = 0$.

b. $x - 5$ is a factor of $P(x) = x^4 - 5x^3 - x^2 + 6x - 5$, and that $P(5) = 0$.

58. The Remainder Theorem The remainder theorem of algebra states that if a polynomial, $P(x)$, is divided by $x - a$, then the remainder is $P(a)$. Verify the remainder theorem by showing that when $P(x) = x^2 - x + 3$ is divided by $x - 2$ the remainder is 5 and that $P(2) = 5$.

59. One factor of $x^3 + 10x^2 + 29x + 20$ is $x + 4$.

a. Factor $x^3 + 10x^2 + 29x + 20$ completely.

b. Reduce $\dfrac{x^3 + 10x^2 + 29x + 20}{x + 4}$.

60. One factor of $x^3 + 5x^2 - 2x - 24$ is $x + 3$.

a. Factor $x^3 + 5x^2 - 2x - 24$ completely.

b. Reduce $\dfrac{x^3 + 5x^2 - 2x - 24}{x + 3}$.

61. One factor of $x^3 + 3x^2 - 10x - 24$ is $x + 2$.

a. Factor $x^3 + 3x^2 - 10x - 24$ completely.

b. Reduce $\dfrac{x^3 + 3x^2 - 10x - 24}{x + 2}$.

62. One factor of $x^3 + 6x^2 + 11x + 6$ is $x + 3$.

a. Factor $x^3 + 6x^2 + 11x + 6$ completely.

b. Reduce $\dfrac{x^3 + 6x^2 + 11x + 6}{x + 3}$.

63. Find $P(3)$ if $P(x) = x^2 + 4x - 8$. Compare it with the remainder in problem 40.

64. Find $P(-2)$ if $P(x) = x^2 - 5x - 7$. Compare it with the remainder in problem 39.

Multiplication and Division of Rational Expressions

TICKET TO SUCCESS

Keep these questions in mind as you read through the section. Then respond in your own words and in complete sentences.

1. Summarize the steps used to multiply fractions.

2. What is the first step in multiplying two rational expressions?

3. Why is factoring important when multiplying and dividing rational expressions?

4. How is division with rational expressions different than multiplication of rational expressions?

Wally Stemberger/Shutterstock.com

Suppose you and some friends buy tickets to a concert. The following rational expression represents the price of each ticket:

$$\frac{y}{(x + 1)}$$

where y is the total cost of tickets and x is the number of friends accompanying you to the concert. However, one day later, the tickets go on sale for two-thirds the original price. Here is the rational expression that represents the new ticket price:

$$\frac{y}{x + 1} \cdot \frac{2}{3}$$

To simplify this rational expression, we need to use multiplication, which is one of the topics we cover in this section.

In the first section of this chapter, we found the process of reducing rational expressions to lowest terms to be the same process used in reducing fractions to lowest terms. The similarity also holds for the process of multiplication or division of rational expressions.

A Multiplying and Dividing Rational Expressions

Multiplication with fractions is the simplest of the four basic operations. To multiply two fractions we simply multiply numerators and multiply denominators; that is, if a, b, c, and d are real numbers, with $b \neq 0$ and $d \neq 0$, then

$$\frac{a}{b} \cdot \frac{c}{d} = \frac{ac}{bd}$$

PRACTICE PROBLEMS

1. Multiply $\frac{3}{4} \cdot \frac{12}{27}$.

EXAMPLE 1 Multiply $\frac{6}{7} \cdot \frac{14}{18}$.

SOLUTION

$$\frac{6}{7} \cdot \frac{14}{18} = \frac{6(14)}{7(18)}$$ Multiply numerators and denominators.

$$= \frac{2 \cdot 3(2 \cdot 7)}{7(2 \cdot 3 \cdot 3)}$$ Factor.

$$= \frac{2}{3}$$ Divide out common factors. ∎

Our next example is similar to some of the problems we worked in an earlier chapter. We multiply fractions whose numerators and denominators are monomials by multiplying numerators and multiplying denominators and then reducing to lowest terms. Here is how it looks.

2. Multiply $\frac{6x^4}{4y^9} \cdot \frac{12y^5}{3x^2}$.

EXAMPLE 2 Multiply $\frac{8x^3}{27y^8} \cdot \frac{9y^3}{12x^2}$.

SOLUTION We multiply numerators and denominators without actually carrying out the multiplication.

$$\frac{8x^3}{27y^8} \cdot \frac{9y^3}{12x^2} = \frac{8 \cdot 9x^3y^3}{27 \cdot 12x^2y^8}$$ Multiply numerators.
Multiply denominators.

$$= \frac{4 \cdot 2 \cdot 9x^3y^3}{9 \cdot 3 \cdot 4 \cdot 3x^2y^8}$$ Factor coefficients.

$$= \frac{2x}{9y^5}$$ Divide out common factors. ∎

> **NOTE**
> Notice how we factor the coefficients just enough so that we can see the factors they have in common. If you want to show this step without showing the factoring, it would look like this:
> $$\frac{\overset{2}{\cancel{8}} \cdot \overset{1}{\cancel{9}}x^3y^3}{\underset{3}{\cancel{27}} \cdot \underset{3}{\cancel{12}}x^2y^8}$$

Once again, we should mention that the little slashes we have drawn through the factors are used to denote the factors we have divided out of the numerator and denominator.

The process of multiplying rational expressions is the same as the process of multiplying fractions. The product of two rational expressions is the product of their numerators over the product of their denominators.

3. Multiply $\frac{x+5}{x^2-25} \cdot \frac{x-5}{x^2-10x+25}$.

EXAMPLE 3 Multiply $\frac{x-3}{x^2-4} \cdot \frac{x+2}{x^2-6x+9}$.

SOLUTION We begin by multiplying numerators and denominators. We then factor all polynomials and divide out factors common to the numerator and denominator.

$$\frac{x-3}{x^2-4} \cdot \frac{x+2}{x^2-6x+9} = \frac{(x-3)(x+2)}{(x^2-4)(x^2-6x+9)}$$ Multiply.

$$= \frac{(x-3)(x+2)}{(x+2)(x-2)(x-3)(x-3)}$$ Factor.

$$= \frac{1}{(x-2)(x-3)}$$ Divide out common factors. ∎

Answers
1. $\frac{1}{3}$
2. $\frac{6x^2}{y^4}$
3. $\frac{1}{(x-5)^2}$

The first two steps can be combined to save time. We can perform the multiplication and factoring steps together.

EXAMPLE 4   Multiply $\dfrac{2y^2 - 4y}{2y^2 - 2} \cdot \dfrac{y^2 - 2y - 3}{y^2 - 5y + 6}$.

SOLUTION

$$\frac{2y^2 - 4y}{2y^2 - 2} \cdot \frac{y^2 - 2y - 3}{y^2 - 5y + 6} = \frac{2y(y-2)(y-3)(y+1)}{2(y+1)(y-1)(y-3)(y-2)}$$

$$= \frac{y}{y - 1} \qquad \blacksquare$$

Notice in both of the preceding examples that we did not actually multiply the polynomials as we did in the chapter on exponents and polynomials. It would be senseless to do that since we would then have to factor each of the resulting products to reduce them to lowest terms.

The quotient of two rational expressions is the product of the first and the reciprocal of the second; that is, we find the quotient of two rational expressions the same way we find the quotient of two fractions.

To divide one rational expression by another, we use the definition of division to multiply by the reciprocal of the expression that follows the division symbol.

EXAMPLE 5   Divide $\dfrac{8x^3}{5y^2} \div \dfrac{4x^2}{10y^6}$.

SOLUTION   First, we rewrite the problem in terms of multiplication. Then we multiply.

$$\frac{8x^3}{5y^2} \div \frac{4x^2}{10y^6} = \frac{8x^3}{5y^2} \cdot \frac{10y^6}{4x^2}$$

$$= \frac{\overset{2}{8} \cdot \overset{2}{10} x^3 y^6}{\underset{1}{4} \cdot \underset{1}{5} x^2 y^2}$$

$$= 4xy^4 \qquad \blacksquare$$

EXAMPLE 6   Divide $\dfrac{x^2 - y^2}{x^2 - 2xy + y^2} \div \dfrac{x^3 + y^3}{x^3 - x^2 y}$.

SOLUTION   We begin by writing the problem as the product of the first and the reciprocal of the second and then proceed as in the previous example.

$$\frac{x^2 - y^2}{x^2 - 2xy + y^2} \div \frac{x^3 + y^3}{x^3 - x^2 y} = \frac{x^2 - y^2}{x^2 - 2xy + y^2} \cdot \frac{x^3 - x^2 y}{x^3 + y^3}$$

Multiply by the reciprocal of the divisor.

$$= \frac{(x-y)(x+y)(x^2)(x-y)}{(x-y)(x-y)(x+y)(x^2 - xy + y^2)}$$

Factor and multiply.

$$= \frac{x^2}{x^2 - xy + y^2}$$

Divide out common factors.

$\qquad \blacksquare$

4. Multiply $\dfrac{3y^2 - 3y}{3y - 12} \cdot \dfrac{y^2 - 2y - 8}{y^2 + 3y + 2}$.

5. Divide $\dfrac{9x^4}{4y^3} \div \dfrac{3x^2}{8y^6}$.

6. Divide. $\dfrac{xy^2 - y^3}{x^2 - y^2} \div \dfrac{x^3 + y^3}{x^2 + 2xy + y^2}$.

Answers

4. $\dfrac{y(y - 1)}{y + 1}$

5. $6x^2 y^2$

6. $\dfrac{y^2}{x^2 - xy + y^2}$

Here are some more examples of multiplication and division with rational expressions.

7. Perform the indicated operations.

$$\frac{a^2 + 3a - 4}{a - 4} \cdot \frac{a + 3}{a^2 - 4a + 3} \div \frac{a + 1}{a^2 - 2a - 3}$$

EXAMPLE 7 Perform the indicated operations.

$$\frac{a^2 - 8a + 15}{a + 4} \cdot \frac{a + 2}{a^2 - 5a + 6} \div \frac{a^2 - 3a - 10}{a^2 + 2a - 8}$$

SOLUTION First we rewrite the division as multiplication by the reciprocal. Then we proceed as usual.

$$\frac{a^2 - 8a + 15}{a + 4} \cdot \frac{a + 2}{a^2 - 5a + 6} \div \frac{a^2 - 3a - 10}{a^2 + 2a - 8}$$

Change division to multiplication by the reciprocal.

$$= \frac{(a^2 - 8a + 15)(a + 2)(a^2 + 2a - 8)}{(a + 4)(a^2 - 5a + 6)(a^2 - 3a - 10)}$$

$$= \frac{(a - 5)(a - 3)(a + 2)(a + 4)(a - 2)}{(a + 4)(a - 3)(a - 2)(a - 5)(a + 2)}$$

Factor.

$$= 1$$

Divide out common factors.

Our next example involves factoring by grouping. As you may have noticed, working the problems in this chapter gives you a very detailed review of factoring.

8. Multiply.

$$\frac{xa + xb - ya - yb}{xa + 2x + ya + 2y} \cdot \frac{xa + 2x + ya + 2y}{xa + xb + ya + yb}$$

EXAMPLE 8 Multiply $\dfrac{xa + xb + ya + yb}{xa - xb - ya + yb} \cdot \dfrac{xa + xb - ya - yb}{xa - xb + ya - yb}$

SOLUTION We will factor each polynomial by grouping, which takes two steps.

$$\frac{xa + xb + ya + yb}{xa - xb - ya + yb} \cdot \frac{xa + xb - ya - yb}{xa - xb + ya - yb}$$

$$= \frac{x(a + b) + y(a + b)}{x(a - b) - y(a - b)} \cdot \frac{x(a + b) - y(a + b)}{x(a - b) + y(a - b)}$$

$$= \frac{(a + b)(x + y)(a + b)(x - y)}{(a - b)(x - y)(a - b)(x + y)}$$

Factor by grouping.

$$= \frac{(a + b)^2}{(a - b)^2}$$

9. Multiply.

$$(5x^2 - 45) \cdot \frac{3}{5x - 15}$$

EXAMPLE 9 Multiply $(4x^2 - 36) \cdot \dfrac{12}{4x + 12}$.

SOLUTION We can think of $4x^2 - 36$ as having a denominator of 1. Thinking of it in this way allows us to proceed as we did in the previous examples.

$$(4x^2 - 36) \cdot \frac{12}{4x + 12} = \frac{4x^2 - 36}{1} \cdot \frac{12}{4x + 12}$$

Write $4x^2 - 36$ with denominator 1.

$$= \frac{4(x - 3)(x + 3)12}{4(x + 3)}$$

Factor.

$$= 12(x - 3)$$

Divide out common factors.

Answers

7. $\dfrac{(a + 4)(a + 3)}{a - 4}$

8. $\dfrac{x - y}{x + y}$

9. $3(x + 3)$

Problem Set 6.3

Moving Toward Success

"It is common sense to take a method and try it; if it fails, admit it frankly and try another. But above all, try something."

—Franklin D. Roosevelt, 1882–1945, 32nd President of the United States

1. Why should you be actively involved in an word problem?

2. Why is making the decision to be successful important to solving word problems?

A Perform the indicated operations involving fractions. [Examples 1, 2, 5]

1. $\dfrac{2}{9} \cdot \dfrac{3}{4}$

2. $\dfrac{5}{6} \cdot \dfrac{7}{8}$

3. $\dfrac{3}{4} \div \dfrac{1}{3}$

4. $\dfrac{3}{8} \div \dfrac{5}{4}$

5. $\dfrac{3}{7} \cdot \dfrac{14}{24} \div \dfrac{1}{2}$

6. $\dfrac{6}{5} \cdot \dfrac{10}{36} \div \dfrac{3}{4}$

7. $\dfrac{10x^2}{5y^2} \cdot \dfrac{15y^3}{2x^4}$

8. $\dfrac{8x^3}{7y^4} \cdot \dfrac{14y^6}{16x^2}$

9. $\dfrac{11a^2b}{5ab^2} \div \dfrac{22a^3b^2}{10ab^4}$

10. $\dfrac{8ab^3}{9a^2b} \div \dfrac{16a^2b^2}{18ab^3}$

11. $\dfrac{6x^2}{5y^3} \cdot \dfrac{11z^2}{2x^2} \div \dfrac{33z^5}{10y^8}$

12. $\dfrac{4x^3}{7y^2} \cdot \dfrac{6z^5}{5x^6} \div \dfrac{24z^2}{35x^6}$

A Perform the indicated operations. Be sure to write all answers in lowest terms. [Examples 3, 4]

13. $\dfrac{x^2 - 9}{x^2 - 4} \cdot \dfrac{x - 2}{x - 3}$

14. $\dfrac{x^2 - 16}{x^2 - 25} \cdot \dfrac{x - 5}{x - 4}$

15. $\dfrac{y^2 - 1}{y + 2} \cdot \dfrac{y^2 + 5y + 6}{y^2 + 2y - 3}$

16. $\dfrac{y - 1}{y^2 - y - 6} \cdot \dfrac{y^2 + 5y + 6}{y^2 - 1}$

17. $\dfrac{3x - 12}{x^2 - 4} \cdot \dfrac{x^2 + 6x + 8}{x - 4}$

18. $\dfrac{x^2 + 5x + 1}{4x - 4} \cdot \dfrac{x - 1}{x^2 + 5x + 1}$

A Perform the indicated operations. Be sure to write all answers in lowest terms. [Examples 3, 4, 6–9]

19. $\dfrac{5x + 2y}{25x^2 - 5xy - 6y^2} \cdot \dfrac{20x^2 - 7xy - 3y^2}{4x + y}$

20. $\dfrac{7x + 3y}{42x^2 - 17xy - 15y^2} \cdot \dfrac{12x^2 - 4xy - 5y^2}{2x + y}$

21. $\dfrac{a^2 - 5a + 6}{a^2 - 2a - 3} \div \dfrac{a - 5}{a^2 + 3a + 2}$

22. $\dfrac{a^2 + 7a + 12}{a - 5} \div \dfrac{a^2 + 9a + 18}{a^2 - 7a + 10}$

23. $\dfrac{4t^2 - 1}{6t^2 + t - 2} \div \dfrac{8t^3 + 1}{27t^3 + 8}$

24. $\dfrac{9t^2 - 1}{6t^2 + 7t - 3} \div \dfrac{27t^3 + 1}{8t^3 + 27}$

25. $\dfrac{2x^2 - 5x - 12}{4x^2 + 8x + 3} \div \dfrac{x^2 - 16}{2x^2 + 7x + 3}$

26. $\dfrac{x^2 - 2x + 1}{3x^2 + 7x - 20} \div \dfrac{x^2 + 3x - 4}{3x^2 - 2x - 5}$

27. $\dfrac{6a^2b + 2ab^2 - 20b^3}{4a^2b - 16b^3} \cdot \dfrac{10a^2 - 22ab + 4b^2}{27a^3 - 125b^3}$

28. $\dfrac{12a^2b - 3ab^2 - 42b^3}{9a^2 - 36b^2} \cdot \dfrac{6a^2 - 15ab + 6b^2}{8a^3b - b^4}$

29. $\dfrac{360x^3 - 490x}{36x^2 + 84x + 49} \cdot \dfrac{30x^2 + 83x + 56}{150x^3 + 65x^2 - 280x}$

30. $\dfrac{490x^2 - 640}{49x^2 - 112x + 64} \cdot \dfrac{28x^2 - 95x + 72}{56x^3 - 62x^2 - 144x}$

31. $\dfrac{x^5 - x^2}{5x^5 - 5x} \cdot \dfrac{10x^4 - 10x^2}{2x^4 + 2x^3 + 2x^2}$

32. $\dfrac{2x^4 - 16x}{3x^6 - 48x^2} \cdot \dfrac{6x^5 + 24x^3}{4x^4 + 8x^3 + 16x^2}$

33. $\dfrac{a^2 - 16b^2}{a^2 - 8ab + 16b^2} \cdot \dfrac{a^2 - 9ab + 20b^2}{a^2 - 7ab + 12b^2} \div \dfrac{a^2 - 25b^2}{a^2 - 6ab + 9b^2}$

34. $\dfrac{a^2 - 6ab + 9b^2}{a^2 - 4b^2} \cdot \dfrac{a^2 - 5ab + 6b^2}{(a - 3b)^2} \div \dfrac{a^2 - 9b^2}{a^2 - ab - 6b^2}$

35. $\dfrac{2y^2 - 7y - 15}{42y^2 - 29y - 5} \cdot \dfrac{12y^2 - 16y + 5}{7y^2 - 36y + 5} \div \dfrac{4y^2 - 9}{49y^2 - 1}$

36. $\dfrac{8y^2 + 18y - 5}{21y^2 - 16y + 3} \cdot \dfrac{35y^2 - 22y + 3}{6y^2 + 17y + 5} \div \dfrac{16y^2 - 1}{9y^2 - 1}$

37. $\dfrac{xy - 2x + 3y - 6}{xy + 2x - 4y - 8} \cdot \dfrac{xy + x - 4y - 4}{xy - x + 3y - 3}$

38. $\dfrac{ax + bx + 2a + 2b}{ax - 3a + bx - 3b} \cdot \dfrac{ax - bx - 3a + 3b}{ax - bx - 2a + 2b}$

39. $\dfrac{xy^2 - y^2 + 4xy - 4y}{xy - 3y + 4x - 12} \div \dfrac{xy^3 + 2xy^2 + y^3 + 2y^2}{xy^2 - 3y^2 + 2xy - 6y}$

40. $\dfrac{4xb - 8b + 12x - 24}{xb^2 + 3b^2 + 3xb + 9b} \div \dfrac{4xb - 8b - 8x + 16}{xb^2 + 3b^2 - 2xb - 6b}$

41. $\dfrac{2x^3 + 10x^2 - 8x - 40}{x^3 + 4x^2 - 9x - 36} \cdot \dfrac{x^2 + x - 12}{2x^2 + 14x + 20}$

42. $\dfrac{x^3 + 2x^2 - 9x - 18}{x^4 + 3x^3 - 4x^2 - 12x} \cdot \dfrac{x^3 + 5x^2 + 6x}{x^2 - x - 6}$

The next two problems are intended to give you practice reading, and paying attention to, the instructions that accompany the problems you are working. Working these problems is an excellent way to get ready for a test or a quiz.

43. Work each problem according to the instructions.

    a. Simplify: $\dfrac{16 - 1}{64 - 1}$

    b. Reduce: $\dfrac{25x^2 - 9}{125x^3 - 27}$

    c. Multiply: $\dfrac{25x^2 - 9}{125x^3 - 27} \cdot \dfrac{5x - 3}{5x + 3}$

    d. Divide: $\dfrac{25x^2 - 9}{125x^3 - 27} \div \dfrac{5x - 3}{25x^2 + 15x + 9}$

44. Work each problem according to the instructions.

    a. Simplify: $\dfrac{64 - 49}{64 + 112 + 49}$

    b. Reduce: $\dfrac{9x^2 - 49}{9x^2 + 42x + 49}$

    c. Multiply: $\dfrac{9x^2 - 49}{9x^2 + 42x + 49} \cdot \dfrac{3x + 7}{3x - 7}$

    d. Divide: $\dfrac{9x^2 - 49}{9x^2 + 42x + 49} \div \dfrac{3x + 7}{3x - 7}$

The work you did with the algebra of functions will help with these.

45. Let $f(x) = \dfrac{x^2 - x - 6}{x - 1}$ and $g(x) = \dfrac{x + 2}{x^2 - 4x + 3}$. Find

    a. $f(x) \cdot g(x)$

    b. $f(x) \div g(x)$

46. Let $f(x) = \dfrac{x^2 - x - 12}{x^2 - 4x + 3}$ and $g(x) = \dfrac{x^2 - x - 12}{x^2 - 5x + 4}$. Find

    a. $f(x) \cdot g(x)$

    b. $f(x) \div g(x)$

47. Let $f(x) = \dfrac{x^3 - 9x^2 - 3x + 27}{4x^2 - 12}$ and $g(x) = \dfrac{x^2 - 2x - 8}{x^2 - 81}$ find $f(x) \cdot g(x)$.

48. Let $f(x) = \dfrac{x^2 - 7x + 12}{x^2 - 16}$ and $g(x) = \dfrac{x^2 - 4x + 3}{x^2 - 6x + 9}$ find $f(x) \cdot g(x)$.

49. Let $f(x) = \dfrac{x^3 - 3x^2 - 4x + 12}{x + 2}$ and $g(x) = \dfrac{x^2 + 7x + 12}{x^2 - 5x + 6}$ find $f(x) \cdot g(x)$.

50. Let $f(x) = 2x^2 - 9x + 9$ and $g(x) = \dfrac{2x + 8}{x^2 + x - 12}$ find $f(x) \cdot g(x)$.

Find the following products. [Example 9]

51. $(3x - 6) \cdot \dfrac{x}{x - 2}$

52. $(4x + 8) \cdot \dfrac{x}{x + 2}$

53. $(x^2 - 25) \cdot \dfrac{2}{x - 5}$

54. $(x^2 - 49) \cdot \dfrac{5}{x + 7}$

55. $(x^2 - 3x + 2) \cdot \dfrac{3}{3x - 3}$

56. $(x^2 - 3x + 2) \cdot \dfrac{-1}{x - 2}$

57. $(y - 3)(y - 4)(y + 3) \cdot \dfrac{-1}{y^2 - 9}$

58. $(y + 1)(y + 4)(y - 1) \cdot \dfrac{3}{y^2 - 1}$

59. $a(a + 5)(a - 5) \cdot \dfrac{a + 1}{a^2 + 5a}$

60. $a(a + 3)(a - 3) \cdot \dfrac{a - 1}{a^2 - 3a}$

Divide.

61. $(x^2 - 2x - 8) \div \dfrac{x^2 - x - 6}{x - 4}$

62. $(2x^2 + 7x - 15) \div \dfrac{6x^2 + 21x - 45}{2x - 3}$

63. $(3 - x) \div \dfrac{x^2 - 9}{x - 1}$

64. $(-x^2 + 3x - 2) \div \dfrac{x^2 + 2x - 8}{x^2 + 3x - 4}$

65. $(xy - 2x - 3y + 6) \div \dfrac{x^2 - 2x - 3}{x^2 - 6x - 7}$

66. $(x^3 - x^2 - 9x + 9) \div \dfrac{x^3 - 3x^2 - 4x + 12}{6x - 12}$

Divide.

67. $\dfrac{x^2(x + 2) + 6x(x + 2) + 9(x + 2)}{x^2 - 2x - 8}$

68. $\dfrac{x^2(x - 3) - 14x(x - 3) + 49(x - 3)}{x^2 - 3x - 28}$

69. $\dfrac{2x^2(x + 3) - 5x(x + 3) + 3(x + 3)}{x^2 + x - 6}$

70. $\dfrac{3x^2(x + 1) - 4x(x + 1) + (x + 1)}{x^3 + 5x^2 - x - 5}$

Applying the Concepts

The following demand function gives the price (in dollars) per DVD $p(x)$ a company charges for making x copies. Use this function to work Problems 71–74.

$$p(x) = \frac{2(x + 60)}{x + 5}$$

71. Demand Equation Use the demand equation to fill in the table.

Number of Copies	Price per Copy ($)
1	
10	
20	
50	
100	

72. Demand Equation To find the revenue for selling 50 copies of a DVD, we multiply the price per DVD by 50. Find the revenue for selling 50 DVDs.

73. Revenue Find the revenue for selling 100 DVDs.

74. Revenue Find the revenue equation $R(x)$.

Use the demand equation $p(x) = \dfrac{3(x + 40)}{x + 5}$ to work problems 75–78.

75. Demand Equation Use the demand equation to fill in the table.

Number of Copies	Price per Copy ($)
1	
10	
20	
50	
100	

76. Revenue Find the revenue for selling 20 copies.

77. Revenue Find the revenue for selling 100 copies.

78. Revenue Find the revenue equation $R(x)$.

79. Area The following box has a square top. The front face of the box has an area of $A = x^3 - 2x^2 - 2x - 3$. The height of the box is $h = x^2 + x + 1$. Find a formula for the area of the top square in terms of x.

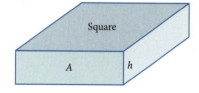

Square

A h

80. Surface Area of a Cylinder The surface area of the cylinder in the figure is defined as the area of its two circular bases and the lateral, or side, area. The surface area may be found by the formula

$$A = 2\pi r^2 + 2\pi r h$$

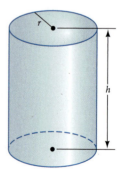

If the surface area is 6π, and $h = 2$, find r.

Maintaining Your Skills

Multiply.

81. $2x^2(5x^3 + 4x - 3)$

82. $3x^3(7x^2 - 4x - 8)$

83. $(3a - 1)(4a + 5)$

84. $(6a - 3)(2a + 1)$

85. $(3x + 7)(4y - 2)$

86. $(x + 2a)(2 - 3b)$

87. $(3 - t^2)^2$

88. $(2 - t^3)^2$

89. $3(x + 1)(x + 2)(x + 3)$

90. $4(x - 1)(x - 2)(x - 3)$

Getting Ready for the Next Section

Combine.

91. $\dfrac{4}{9} + \dfrac{2}{9}$

92. $\dfrac{3}{8} + \dfrac{1}{8}$

93. $\dfrac{3}{14} + \dfrac{7}{30}$

94. $\dfrac{3}{10} + \dfrac{11}{42}$

Multiply.

95. $-1(7 - x)$

96. $-1(3 - x)$

Factor.

97. $x^2 - 1$

98. $x^2 - 2x - 3$

99. $2x + 10$

100. $x^2 + 4x + 3$

101. $a^3 - b^3$

102. $8y^3 - 27$

103. $16x^3 + 54y^3$

104. $3x^4 - 24x$

Extending the Concepts

105. $\dfrac{x^6 + y^6}{x^4 + 4x^2y^2 + 3y^4} \div \dfrac{x^4 + 3x^2y^2 + 2y^4}{x^4 + 5x^2y^2 + 6y^4}$

106. $\dfrac{x^2 + 9xy + 8y^2}{x^2 + 7xy - 8y^2} \div \dfrac{x^2 - y^2}{x^2 + 5xy - 6y^2}$

107. $\dfrac{a^2(2a + b) + 6a(2a + b) + 5(2a + b)}{3a^2(2a + b) - 2a(2a + b) + (2a + b)} \div \dfrac{a + 1}{a - 1}$

108. $\dfrac{2x^2(x - 3z) - 5x(x - 3z) + 2(x - 3z)}{4x^2(x - 3z) - 11x(x - 3z) + 6(x - 3z)} \div \dfrac{4x - 3}{4x + 1}$

109. $\dfrac{a^3 - a^2b}{ac - a} \div \left(\dfrac{a - b}{c - 1} \right)^2$

110. $\dfrac{p^3 + q^3}{q - p} \div \dfrac{(p + q)^2}{p^2 - q^2}$

111. $\dfrac{x^3 - x^2y}{(x - y)^2} \div \dfrac{x^2 + xy}{x^2 - y^2}$

112. $\dfrac{m^3 + n^3}{n - m} \cdot \dfrac{m^3 - n^2m}{m^2 + 2mn + n^2}$

Addition and Subtraction of Rational Expressions

OBJECTIVES

A Add and subtract rational expressions with the same denominator.

B Add and subtract rational expressions with different denominators.

TICKET TO SUCCESS

Keep these questions in mind as you read through the section. Then respond in your own words and in complete sentences.

1. How would you use the distributive property to add two rational expressions that have the same denominator?
2. What is the definition of the least common denominator?
3. Why is factoring important in finding a least common denominator?
4. What is the last step in adding or subtracting two rational expressions?

afaizal/Shutterstock.com

Imagine you are going for a 5-mile ride on your skateboard. The first half of your journey is down a steep hill and the second half is flat. If your downhill speed is a constant 25 miles per hour and your speed on the flat part is a constant 10 miles per hour, how can you find your average velocity of your skateboard trip? In this section, we will be concerned with addition and subtraction of rational expressions, which are the tools you need to answer this question.

In the first part of this section we will look at addition of expressions that have the same denominator. In the second part we will look at addition of expressions that have different denominators.

A Addition and Subtraction with the Same Denominator

To add two expressions that have the same denominator, we simply add numerators and put the sum over the common denominator. Since the process we use to add and subtract rational expressions is the same process used to add and subtract fractions, we will begin with an example involving fractions.

EXAMPLE 1   Add $\dfrac{4}{9} + \dfrac{2}{9}$.

SOLUTION   We add fractions with the same denominator by using the distributive property. Here is a detailed look at the steps involved.

$$\frac{4}{9} + \frac{2}{9} = 4\left(\frac{1}{9}\right) + 2\left(\frac{1}{9}\right)$$

$$= (4 + 2)\left(\frac{1}{9}\right) \qquad \text{Distributive property}$$

$$= 6\left(\frac{1}{9}\right)$$

$$= \frac{6}{9}$$

$$= \frac{2}{3} \qquad \text{Divide numerator and denominator} \atop \text{by common factor 3.}$$

Note that the important thing about the fractions in this example is that they each have a denominator of 9. If they did not have the same denominator, we could not have written them as two terms with a factor of $\dfrac{1}{9}$ in common. Without the $\dfrac{1}{9}$ common to each term, we couldn't apply the distributive property. And without the distributive property, we would not have been able to add the two fractions.   ■

In the examples that follow, we will not show all the steps we showed in Example 1. The steps were shown in Example 1 so that you could see why both fractions must have the same denominator before we can add them. In practice, we simply add numerators and place the result over the common denominator.

We add and subtract rational expressions with the same denominator by combining numerators and writing the result over the common denominator. Then we reduce the result to lowest terms, if possible. Example 2 shows this process in detail. If you see the similarities between operations on rational numbers and operations on rational expressions, this chapter will look like an extension of rational numbers rather than a completely new set of topics.

EXAMPLE 2   Add $\dfrac{x}{x^2 - 1} + \dfrac{1}{x^2 - 1}$.

SOLUTION   Since the denominators are the same, we simply add numerators.

$$\frac{x}{x^2 - 1} + \frac{1}{x^2 - 1} = \frac{x + 1}{x^2 - 1} \qquad \text{Add numerators.}$$

$$= \frac{\cancel{x + 1}}{(x - 1)\cancel{(x + 1)}} \qquad \text{Factor denominator.}$$

$$= \frac{1}{x - 1} \qquad \text{Divide out common factor } x + 1. \quad ■$$

Our next example involves subtraction of rational expressions. Pay careful attention to what happens to the signs of the terms in the numerator of the second expression when we subtract it from the first expression.

EXAMPLE 3   Subtract $\dfrac{2x - 5}{x - 2} - \dfrac{x - 3}{x - 2}$.

SOLUTION   Since each expression has the same denominator, we simply subtract the numerator in the second expression from the numerator in the first expression and write the difference over the common denominator $x - 2$. We must be careful, however, that we subtract both terms in the second numerator. To ensure that we do, we will enclose that numerator in parentheses.

$$\frac{2x-5}{x-2} - \frac{x-3}{x-2} = \frac{2x-5-(x-3)}{x-2} \qquad \text{Subtract numerators.}$$

$$= \frac{2x-5-x+3}{x-2} \qquad \text{Remove parentheses.}$$

$$= \frac{x-2}{x-2} \qquad \text{Combine similar terms in the numerator.}$$

$$= 1 \qquad \text{Reduce (or divide).}$$

Note the 3 in the numerator of the second step. It is a very common mistake to write that as −3 by forgetting to subtract both terms in the numerator of the second expression. Whenever the expression we are subtracting has two or more terms in its numerator, we have to watch for this mistake. ∎

Next we consider addition and subtraction of fractions and rational expressions that have different denominators.

B Addition and Subtraction with Different Denominators

Before we look at an example of addition of fractions with different denominators, we need to review the definition for the least common denominator.

> **Definition**
>
> The **least common denominator**, abbreviated LCD, for a set of denominators is the smallest expression that is divisible by each of the denominators.

The first step in combining two fractions is to find the LCD. Once we have the common denominator, we rewrite each fraction as an equivalent fraction with the common denominator. After that, we simply add or subtract as we did in our first three examples.

Example 4 is a review of the step-by-step procedure used to add two fractions with different denominators.

EXAMPLE 4   Add $\frac{3}{14} + \frac{7}{30}$.

SOLUTION

Step 1: Find the LCD.

To do this, we first factor both denominators into prime factors.

$$\text{Factor 14:} \qquad 14 = 2 \cdot 7$$

$$\text{Factor 30:} \qquad 30 = 2 \cdot 3 \cdot 5$$

Since the LCD must be divisible by 14, it must have factors of $2 \cdot 7$. It must also be divisible by 30 and, therefore, have factors of $2 \cdot 3 \cdot 5$. We do not need to repeat the 2 that appears in both the factors of 14 and those of 30. Therefore,

$$\text{LCD} = 2 \cdot 3 \cdot 5 \cdot 7 = 210$$

Step 2: Change to equivalent fractions.

Since we want each fraction to have a denominator of 210 and at the same time keep its original value, we multiply each by 1 in the appropriate form.

4. Add $\frac{3}{10} + \frac{11}{42}$.

Answer

4. $\frac{59}{105}$

NOTE

When we multiply $\frac{3}{14}$ by $\frac{15}{15}$ we obtain a fraction with the same value as $\frac{3}{14}$ (because we multiplied by 1) but with the common denominator 210.

Change $\frac{3}{14}$ to a fraction with denominator 210.

$$\frac{3}{14} \cdot \frac{15}{15} = \frac{45}{210}$$

Change $\frac{7}{30}$ to a fraction with denominator 210.

$$\frac{7}{30} \cdot \frac{7}{7} = \frac{49}{210}$$

Step 3: Add numerators of equivalent fractions found in step 2.

$$\frac{45}{210} + \frac{49}{210} = \frac{94}{210}$$

Step 4: Reduce to lowest terms if necessary.

$$\frac{94}{210} = \frac{47}{105} \qquad \blacksquare$$

The main idea in adding fractions is to write each fraction again with the LCD for a denominator. In doing so, we must be sure not to change the value of either of the original fractions.

5. Add $\dfrac{-3}{x^2 - 2x - 8} + \dfrac{4}{x^2 - 16}$.

EXAMPLE 5 Add $\dfrac{-2}{x^2 - 2x - 3} + \dfrac{3}{x^2 - 9}$.

SOLUTION

Step 1: Factor each denominator and build the LCD from the factors.

$$\left.\begin{array}{l} x^2 - 2x - 3 = (x-3)(x+1) \\ x^2 - 9 \quad\ = (x-3)(x+3) \end{array}\right\} \text{LCD} = (x-3)(x+3)(x+1)$$

Step 2: Change each rational expression to an equivalent expression that has the LCD for a denominator.

$$\frac{-2}{x^2 - 2x - 3} = \frac{-2}{(x-3)(x+1)} \cdot \frac{(x+3)}{(x+3)} = \frac{-2x-6}{(x-3)(x+3)(x+1)}$$

$$\frac{3}{x^2 - 9} = \frac{3}{(x-3)(x+3)} \cdot \frac{(x+1)}{(x+1)} = \frac{3x+3}{(x-3)(x+3)(x+1)}$$

Step 3: Add numerators of the rational expressions found in step 2.

$$\frac{-2x-6}{(x-3)(x+3)(x+1)} + \frac{3x+3}{(x-3)(x+3)(x+1)} = \frac{x-3}{(x-3)(x+3)(x+1)}$$

Step 4: Reduce to lowest terms by dividing out the common factor $x - 3$.

$$= \frac{1}{(x+3)(x+1)} \qquad \blacksquare$$

6. Add $\dfrac{x-4}{2x-6} + \dfrac{3}{x^2-9}$.

EXAMPLE 6 Subtract $\dfrac{x+4}{2x+10} - \dfrac{5}{x^2 - 25}$.

SOLUTION We begin by factoring each denominator.

$$\frac{x+4}{2x+10} - \frac{5}{x^2 - 25} = \frac{x+4}{2(x+5)} - \frac{5}{(x+5)(x-5)}$$

Answers

5. $\dfrac{1}{(x+4)(x+2)}$

6. $\dfrac{x+2}{2(x+3)}$

The LCD is $2(x + 5)(x - 5)$. Completing the solution we have

$$= \frac{x + 4}{2(x + 5)} \cdot \frac{(x - 5)}{(x - 5)} - \frac{5}{(x + 5)(x - 5)} \cdot \frac{2}{2}$$

$$= \frac{x^2 - x - 20}{2(x + 5)(x - 5)} - \frac{10}{2(x + 5)(x - 5)}$$

$$= \frac{x^2 - x - 30}{2(x + 5)(x - 5)}$$

To see if this expression will reduce, we factor the numerator into $(x - 6)(x + 5)$.

$$= \frac{(x - 6)\cancel{(x + 5)}}{2\cancel{(x + 5)}(x - 5)}$$

$$= \frac{x - 6}{2(x - 5)} \quad \blacksquare$$

EXAMPLE 7 Subtract $\dfrac{2x - 2}{x^2 + 4x + 3} - \dfrac{x - 1}{x^2 + 5x + 6}$.

SOLUTION We factor each denominator and build the LCD from those factors.

$$\frac{2x - 2}{x^2 + 4x + 3} - \frac{x - 1}{x^2 + 5x + 6}$$

$$= \frac{2x - 2}{(x + 3)(x + 1)} - \frac{x - 1}{(x + 3)(x + 2)}$$

$$= \frac{2x - 2}{(x + 3)(x + 1)} \cdot \frac{(x + 2)}{(x + 2)} - \frac{x - 1}{(x + 3)(x + 2)} \cdot \frac{(x + 1)}{(x + 1)} \qquad \text{Build the LCD.}$$

$$= \frac{2x^2 + 2x - 4}{(x + 1)(x + 2)(x + 3)} - \frac{x^2 - 1}{(x + 1)(x + 2)(x + 3)} \qquad \begin{array}{l}\text{Multiply out} \\ \text{each numerator.}\end{array}$$

$$= \frac{(2x^2 + 2x - 4) - (x^2 - 1)}{(x + 1)(x + 2)(x + 3)} \qquad \text{Subtract numerators.}$$

$$= \frac{x^2 + 2x - 3}{(x + 1)(x + 2)(x + 3)}$$

$$= \frac{\cancel{(x + 3)}(x - 1)}{(x + 1)(x + 2)\cancel{(x + 3)}} \qquad \begin{array}{l}\text{Factor numerator to see if we can} \\ \text{reduce.}\end{array}$$

$$= \frac{x - 1}{(x + 1)(x + 2)} \qquad \text{Reduce.} \qquad \blacksquare$$

EXAMPLE 8 Add $\dfrac{x^2}{x - 7} + \dfrac{6x + 7}{7 - x}$.

SOLUTION In the first section of this chapter, we were able to reverse the terms in a factor such as $7 - x$ by factoring -1 from each term. In a problem like this, the same result can be obtained by multiplying the numerator and denominator by -1.

$$\frac{x^2}{x - 7} + \frac{6x + 7}{7 - x} \cdot \frac{-1}{-1} = \frac{x^2}{x - 7} + \frac{-6x - 7}{x - 7}$$

$$= \frac{x^2 - 6x - 7}{x - 7} \qquad \text{Add numerators.}$$

$$= \frac{\cancel{(x - 7)}(x + 1)}{\cancel{(x - 7)}} \qquad \text{Factor numerator.}$$

$$= x + 1 \qquad \text{Divide out } x - 7. \qquad \blacksquare$$

For our next example, we will look at a problem in which we combine a whole number and a rational expression.

7. Subtract.

$$\frac{2x - 4}{x^2 + 5x + 4} - \frac{x - 4}{x^2 + 6x + 8}$$

8. Add $\dfrac{x^2}{x - 4} + \dfrac{x + 12}{4 - x}$.

Answers

7. $\dfrac{x - 1}{(x + 1)(x + 2)}$

8. $x + 3$

9. Add $2 + \dfrac{25}{5x - 1}$.

EXAMPLE 9 Subtract $2 - \dfrac{9}{3x + 1}$.

SOLUTION To subtract these two expressions, we think of 2 as a rational expression with a denominator of 1.

$$2 - \frac{9}{3x + 1} = \frac{2}{1} - \frac{9}{3x + 1}$$

The LCD is $3x + 1$. Multiplying the numerator and denominator of the first expression by $3x + 1$ gives us a rational expression equivalent to 2 but with a denominator of $3x + 1$.

$$\frac{2}{1} \cdot \frac{(3x + 1)}{(3x + 1)} - \frac{9}{3x + 1} = \frac{6x + 2 - 9}{3x + 1}$$

$$= \frac{6x - 7}{3x + 1}$$

The numerator and denominator of this last expression do not have any factors in common other than 1, so the expression is in lowest terms. ■

10. One number is three times another. Write an expression for the sum of the reciprocals of the two numbers. Then simplify that expression.

EXAMPLE 10 Write an expression for the sum of a number and twice its reciprocal. Then, simplify that expression.

SOLUTION If x is the number, then its reciprocal is $\dfrac{1}{x}$. Twice its reciprocal is $\dfrac{2}{x}$. The sum of the number and twice its reciprocal is

$$x + \frac{2}{x}$$

To combine these two expressions, we think of the first term x as a rational expression with a denominator of 1. The least common denominator is x.

$$x + \frac{2}{x} = \frac{x}{1} + \frac{2}{x}$$

$$= \frac{x}{1} \cdot \frac{x}{x} + \frac{2}{x}$$

$$= \frac{x^2 + 2}{x}$$

■

Problem Set 6.4

Moving Toward Success

"Courage is what it takes to stand up and speak. Courage is also what it takes to sit down and listen."

—Winston Churchill, 1874–1965, English politician and author

1. Why will increasing the effectiveness of the time you spend learning help you?

2. Should you still add to your list of difficult work problems this far into the class? Why or why not?

A Combine the following rational expressions. Reduce all answers to lowest terms. [Examples 2, 3]

1. $\dfrac{x}{x+3} + \dfrac{3}{x+3}$

2. $\dfrac{5x}{5x+2} + \dfrac{2}{5x+2}$

3. $\dfrac{4}{y-4} - \dfrac{y}{y-4}$

4. $\dfrac{8}{y+8} + \dfrac{y}{y+8}$

5. $\dfrac{x}{x^2-y^2} - \dfrac{y}{x^2-y^2}$

6. $\dfrac{x}{x^2-y^2} + \dfrac{y}{x^2-y^2}$

7. $\dfrac{2x-3}{x-2} - \dfrac{x-1}{x-2}$

8. $\dfrac{2x-4}{x+2} - \dfrac{x-6}{x+2}$

9. $\dfrac{1}{a} + \dfrac{2}{a^2} - \dfrac{3}{a^3}$

10. $\dfrac{3}{a} + \dfrac{2}{a^2} - \dfrac{1}{a^3}$

11. $\dfrac{7x-2}{2x+1} - \dfrac{5x-3}{2x+1}$

12. $\dfrac{7x-1}{3x+2} - \dfrac{4x-3}{3x+2}$

13. Work each problem according to the instructions.

   a. Multiply: $\dfrac{3}{8} \cdot \dfrac{1}{6}$.

   b. Divide: $\dfrac{3}{8} \div \dfrac{1}{6}$.

   c. Add: $\dfrac{3}{8} + \dfrac{1}{6}$.

   d. Multiply: $\dfrac{x+3}{x-3} \cdot \dfrac{5x+15}{x^2-9}$.

   e. Divide: $\dfrac{x+3}{x-3} \div \dfrac{5x+15}{x^2-9}$.

   f. Subtract: $\dfrac{x+3}{x-3} - \dfrac{5x+15}{x^2-9}$.

14. Work each problem according to the instructions.

   a. Multiply: $\dfrac{16}{49} \cdot \dfrac{1}{28}$.

   b. Divide: $\dfrac{16}{49} \div \dfrac{1}{28}$.

   c. Add: $\dfrac{16}{49} - \dfrac{1}{28}$.

   d. Multiply: $\dfrac{3x-2}{3x+2} \cdot \dfrac{15x+6}{9x^2-4}$.

   e. Divide: $\dfrac{3x-2}{3x+2} \div \dfrac{15x+6}{9x^2-4}$.

   f. Subtract: $\dfrac{3x-2}{3x+2} - \dfrac{15x+6}{9x^2-4}$.

B Combine the following fractions. [Example 4]

15. $\dfrac{3}{4} + \dfrac{1}{2}$

16. $\dfrac{5}{6} + \dfrac{1}{3}$

17. $\dfrac{2}{5} - \dfrac{1}{15}$

18. $\dfrac{5}{8} - \dfrac{1}{4}$

19. $\dfrac{5}{6} + \dfrac{7}{8}$

20. $\dfrac{3}{4} + \dfrac{2}{3}$

21. $\dfrac{9}{48} - \dfrac{3}{54}$

22. $\dfrac{6}{28} - \dfrac{5}{42}$

23. $\dfrac{3}{4} - \dfrac{1}{8} + \dfrac{2}{3}$

24. $\dfrac{1}{3} - \dfrac{5}{6} + \dfrac{5}{12}$

B Combine the following rational expressions. Reduce all answers to lowest terms. [Examples 5–10]

25. $\dfrac{2}{t^2} - \dfrac{3}{2t}$

26. $\dfrac{5}{3t} - \dfrac{4}{t^2}$

27. $\dfrac{3x + 1}{2x - 6} - \dfrac{x + 2}{x - 3}$

28. $\dfrac{x + 1}{x - 2} - \dfrac{4x + 7}{5x - 10}$

29. $\dfrac{x + 1}{2x - 2} - \dfrac{2}{x^2 - 1}$

30. $\dfrac{x + 7}{2x + 12} + \dfrac{6}{x^2 - 36}$

31. $\dfrac{1}{a - b} - \dfrac{3ab}{a^3 - b^3}$

32. $\dfrac{1}{a + b} + \dfrac{3ab}{a^3 + b^3}$

33. $\dfrac{1}{2y - 3} - \dfrac{18y}{8y^3 - 27}$

34. $\dfrac{1}{3y - 2} - \dfrac{18y}{27y^3 - 8}$

35. $\dfrac{x}{x^2 - 5x + 6} - \dfrac{3}{3 - x}$

36. $\dfrac{x}{x^2 + 4x + 4} - \dfrac{2}{2 + x}$

37. $\dfrac{2}{4t - 5} + \dfrac{9}{8t^2 - 38t + 35}$

38. $\dfrac{3}{2t - 5} + \dfrac{21}{8t^2 - 14t - 15}$

39. $\dfrac{1}{a^2 - 5a + 6} + \dfrac{3}{a^2 - a - 2}$

40. $\dfrac{-3}{a^2 + a - 2} + \dfrac{5}{a^2 - a - 6}$

41. $\dfrac{1}{8x^3 - 1} - \dfrac{1}{4x^2 - 1}$

42. $\dfrac{1}{27x^3 - 1} - \dfrac{1}{9x^2 - 1}$

43. $\dfrac{4}{4x^2 - 9} - \dfrac{6}{8x^2 - 6x - 9}$

44. $\dfrac{9}{9x^2 + 6x - 8} - \dfrac{6}{9x^2 - 4}$

45. $\dfrac{4a}{a^2 + 6a + 5} - \dfrac{3a}{a^2 + 5a + 4}$

46. $\dfrac{3a}{a^2 + 7a + 10} - \dfrac{2a}{a^2 + 6a + 8}$

47. $\dfrac{2x - 1}{x^2 + x - 6} - \dfrac{x + 2}{x^2 + 5x + 6}$

48. $\dfrac{4x + 1}{x^2 + 5x + 4} - \dfrac{x + 3}{x^2 + 4x + 3}$

49. $\dfrac{2x - 8}{3x^2 + 8x + 4} + \dfrac{x + 3}{3x^2 + 5x + 2}$

50. $\dfrac{5x + 3}{2x^2 + 5x + 3} - \dfrac{3x + 9}{2x^2 + 7x + 6}$

51. $\dfrac{2}{x^2 + 5x + 6} - \dfrac{4}{x^2 + 4x + 3} + \dfrac{3}{x^2 + 3x + 2}$

52. $\dfrac{-5}{x^2 + 3x - 4} + \dfrac{5}{x^2 + 2x - 3} + \dfrac{1}{x^2 + 7x + 12}$

53. $\dfrac{2x + 8}{x^2 + 5x + 6} - \dfrac{x + 5}{x^2 + 4x + 3} - \dfrac{x - 1}{x^2 + 3x + 2}$

54. $\dfrac{2x + 11}{x^2 + 9x + 20} - \dfrac{x + 1}{x^2 + 7x + 12} - \dfrac{x + 6}{x^2 + 8x + 15}$

55. $2 + \dfrac{3}{2x + 1}$

56. $3 - \dfrac{2}{2x + 3}$

57. $5 + \dfrac{2}{4 - t}$

58. $7 + \dfrac{3}{5 - t}$

59. $x - \dfrac{4}{2x + 3}$

60. $x - \dfrac{5}{3x + 4} + 1$

61. $\dfrac{x}{x + 2} + \dfrac{1}{2x + 4} - \dfrac{3}{x^2 + 2x}$

62. $\dfrac{x}{x + 3} + \dfrac{7}{3x + 9} - \dfrac{2}{x^2 + 3x}$

63. $\dfrac{1}{x} + \dfrac{x}{2x + 4} - \dfrac{2}{x^2 + 2x}$

64. $\dfrac{1}{x} + \dfrac{x}{3x + 9} - \dfrac{3}{x^2 + 3x}$

65. Let $f(x) = \dfrac{2x - 1}{4x - 16}$ and $g(x) = \dfrac{x - 3}{x - 4}$; find $f(x) - g(x)$.

66. Let $f(x) = \dfrac{2}{2x + 4}$ and $g(x) = \dfrac{x}{x - 4}$; find $f(x) + g(x)$.

67. Let $f(x) = \dfrac{2}{x + 4}$ and $g(x) = \dfrac{x - 1}{x^2 + 3x - 4}$; find $f(x) + g(x)$.

68. Let $f(x) = \dfrac{5}{3t - 2}$ and $g(x) = \dfrac{t - 3}{3t^2 + 7t - 6}$; find $f(x) - g(x)$.

69. Let $f(x) = \dfrac{2x}{x^2 - x - 2}$ and $g(x) = \dfrac{5}{x^2 + x - 6}$; find $f(x) + g(x)$.

70. Let $f(x) = \dfrac{7}{x^2 - x - 12}$ and $g(x) = \dfrac{5}{x^2 + x - 6}$; find $f(x) - g(x)$.

71. Let $f(x) = \dfrac{x}{9x^2 - 4}$ and $g(x) = \dfrac{1}{3x^2 - 4x - 4}$; find $f(x) + g(x)$.

72. Let $f(x) = \dfrac{1}{16x^2 - 1}$ and $g(x) = \dfrac{1}{64x^3 - 1}$; find $f(x) - g(x)$.

73. Let $f(x) = \dfrac{3x}{2x^2 - x - 1}$ and $g(x) = \dfrac{6}{2x^2 - 5x - 3}$; find $f(x) + g(x)$.

74. Let $f(x) = \dfrac{5x}{x^2 - 8x - 9}$ and $g(x) = \dfrac{4x}{x^2 - 10x + 9}$; find $f(x) - g(x)$.

75. Let $f(x) = \dfrac{5x}{x^2 - 13x + 36}$ and $g(x) = \dfrac{3x}{x^2 - 11x + 28}$; find $f(x) - g(x)$.

76. Let $f(x) = \dfrac{x + 4}{x^2 - 6x - 16}$ and $g(x) = \dfrac{x - 6}{x^2 - 11x + 24}$; find $f(x) - g(x)$.

Applying the Concepts

77. Number Problem Write an expression for the sum of a number and 4 times its reciprocal. Then, simplify that expression.

78. Skateboarding Suppose you are taking the 5-mile ride described in the section opener. If you have an average speed of 25 miles per hour during the first 2.5 miles of your trip and an average speed of 10 miles per hour for the second 2.5 miles, what is your average speed for your entire trip?

79. Optometry The formula $P = \dfrac{1}{a} + \dfrac{1}{b}$ is used by optometrists to help determine how strong to make the lenses for a pair of eyeglasses. If a is 10 and b is 0.2, find the corresponding value of P.

80. Quadratic Formula Later in the book we will work with the quadratic formula. The derivation of the formula requires that you can add the fractions below. Add the fractions.

$$-\frac{c}{a} + \left(\frac{b}{2a}\right)^2$$

81. Elliptical Orbits Consider two objects, A and B, that move in the same direction along an elliptical path at constant but different velocities.

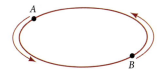

It can be shown that the time, T, it takes for the two objects to meet can be found from the formula

$$\frac{1}{T} = \frac{1}{t_A} - \frac{1}{t_B}$$

where t_A = time required for object A to orbit, and t_B = time required for object B to orbit.

a. If $t_A = 24$ months and $t_B = 30$ months, when will these two objects meet?

b. If $t_A = t_B$ what can one conclude?

82. Average Velocity If a car travels at a constant velocity, v_1, for 10 miles and then at a constant, but different velocity, v_2, for the next 10 miles, it can be shown that the car's average velocity, v_{avg}, over these 20 miles satisfies the equation

$$\frac{2}{v_{avg}} = \frac{1}{v_1} + \frac{1}{v_2}$$

Find the average velocity of a car that travels a constant 45 miles per hour for 10 miles and then increases to a constant 60 miles per hour for the next 10 miles. Round to the nearest tenth.

Maintaining Your Skills

Write each number in scientific notation.

83. 54,000

84. 768,000

85. 0.00034

86. 0.0359

Write each number in expanded form.

87. 6.44×10^3

88. 2.5×10^2

89. 6.44×10^{-3}

90. 2.5×10^{-2}

Simplify each expression as much as possible. Write all answers in scientific notation.

91. $(3 \times 10^8)(4 \times 10^{-5})$

92. $\dfrac{8 \times 10^{-3}}{4 \times 10^{-6}}$

Simplify.

93. $4(2x - 3) + 7$

94. $3^2 - 3(x - 5) + 9x$

95. $10 \div 5(x - 4) + 7$

96. $15 \div 3[(x - 3^2) + 2x] - (4x - 7)$

Solve.

97. $5 - 3(x - 4) = 4x + 3$

98. $7x - 4^2 + 16 \div 4 = 2(x + 4)$

Getting Ready for the Next Section

Divide.

99. $\dfrac{3}{4} \div \dfrac{5}{8}$

100. $\dfrac{2}{3} \div \dfrac{5}{6}$

101. $\dfrac{2}{3} \div \dfrac{8}{9}$

102. $\dfrac{5}{8} \div \dfrac{15}{16}$

Multiply.

103. $x\left(1 + \dfrac{2}{x}\right)$

104. $3\left(x + \dfrac{1}{3}\right)$

105. $3x\left(\dfrac{1}{x} - \dfrac{1}{3}\right)$

106. $3x\left(\dfrac{1}{x} + \dfrac{1}{3}\right)$

Factor.

107. $x^2 - 4$

108. $x^2 - x - 6$

Complex Fractions

6.5

OBJECTIVES

A Simplify complex fractions.

TICKET TO SUCCESS

Keep these questions in mind as you read through the section. Then respond in your own words and in complete sentences.

1. What is a complex fraction?
2. Explain how a least common denominator can be used to simplify a complex fraction.
3. When is it more efficient to convert a complex fraction to a problem involving division of rational expressions?
4. Which method of simplifying complex fractions do you prefer? Why?

gnohz/Shutterstock.com

Suppose a survey of college freshmen found that $\frac{1}{2}$ of the women and $\frac{2}{3}$ of the men live off campus. It was further found that $\frac{1}{4}$ of the women and $\frac{1}{6}$ of the men live in apartments rather than dorms or houses. If equal numbers of women and men were surveyed, what fraction of off-campus students live in apartments? To answer this question you will need to know how to work with complex fractions, which is our focus in this section.

The quotient of two fractions or two rational expressions is called a *complex fraction*. We begin this section by learning how to simplify complex fractions.

A Simplify Complex Fractions

EXAMPLE 1 Simplify $\dfrac{\frac{3}{4}}{\frac{5}{8}}$.

SOLUTION There are generally two methods that can be used to simplify complex fractions.

> *Method 1* We can multiply the numerator and denominator of the complex fraction by the LCD for both of the fractions, which in this case is 8.

$$\frac{\frac{3}{4}}{\frac{5}{8}} = \frac{\frac{3}{4} \cdot \mathbf{8}}{\frac{5}{8} \cdot \mathbf{8}} = \frac{6}{5}$$

PRACTICE PROBLEMS

1. Simplify $\dfrac{\frac{2}{3}}{\frac{5}{6}}$. (Try both methods.)

Answer

1. $\frac{4}{5}$

Method 2   To divide by $\dfrac{5}{8}$ we multiply by $\dfrac{8}{5}$.

$$\dfrac{\dfrac{3}{4}}{\dfrac{5}{8}} = \dfrac{3}{4} \cdot \dfrac{8}{5} = \dfrac{24}{20} = \dfrac{6}{5}$$ ∎

Here are some examples of complex fractions involving rational expressions. Most can be solved using either of the two methods shown in Example 1.

2. Simplify $\dfrac{\dfrac{1}{x} - \dfrac{1}{3}}{\dfrac{1}{x} + \dfrac{1}{3}}$.

EXAMPLE 2   Simplify $\dfrac{\dfrac{1}{x} + \dfrac{1}{y}}{\dfrac{1}{x} - \dfrac{1}{y}}$.

SOLUTION   This problem is most easily solved using Method 1. We begin by multiplying both the numerator and denominator by the quantity xy, which is the LCD for all the fractions.

$$\dfrac{\dfrac{1}{x} + \dfrac{1}{y}}{\dfrac{1}{x} - \dfrac{1}{y}} = \dfrac{\left(\dfrac{1}{x} + \dfrac{1}{y}\right) \cdot \boldsymbol{xy}}{\left(\dfrac{1}{x} - \dfrac{1}{y}\right) \cdot \boldsymbol{xy}}$$

$$= \dfrac{\dfrac{1}{x}\,(xy) + \dfrac{1}{y}\,(xy)}{\dfrac{1}{x}\,(xy) - \dfrac{1}{y}\,(xy)}$$

Apply the distributive property to distribute xy over both terms in the numerator and denominator.

$$= \dfrac{y + x}{y - x}$$ ∎

3. Simplify $\dfrac{\dfrac{x + 5}{x^2 - 16}}{\dfrac{x^2 - 25}{x - 4}}$.

EXAMPLE 3   Simplify $\dfrac{\dfrac{x - 2}{x^2 - 9}}{\dfrac{x^2 - 4}{x + 3}}$.

SOLUTION   Applying Method 2, we have

$$\dfrac{\dfrac{x - 2}{x^2 - 9}}{\dfrac{x^2 - 4}{x + 3}} = \dfrac{x - 2}{x^2 - 9} \cdot \dfrac{x + 3}{x^2 - 4}$$

$$= \dfrac{(x - 2)(x + 3)}{(x + 3)(x - 3)(x + 2)(x - 2)}$$

$$= \dfrac{1}{(x - 3)(x + 2)}$$ ∎

Answers

2. $\dfrac{3 - x}{3 + x}$

3. $\dfrac{1}{(x + 4)(x - 5)}$

EXAMPLE 4 Simplify $\dfrac{1 - \dfrac{4}{x^2}}{1 - \dfrac{1}{x} - \dfrac{6}{x^2}}$.

SOLUTION The easiest way to simplify this complex fraction is to multiply the numerator and denominator by the LCD, x^2.

$$\frac{1 - \dfrac{4}{x^2}}{1 - \dfrac{1}{x} - \dfrac{6}{x^2}} = \frac{\boldsymbol{x^2}\left(1 - \dfrac{4}{x^2}\right)}{\boldsymbol{x^2}\left(1 - \dfrac{1}{x} - \dfrac{6}{x^2}\right)} \qquad \text{Multiply numerator and denominator by } x^2.$$

$$= \frac{x^2 \cdot 1 - x^2 \cdot \dfrac{4}{x^2}}{x^2 \cdot 1 - x^2 \cdot \dfrac{1}{x} - x^2 \cdot \dfrac{6}{x^2}} \qquad \text{Distributive property}$$

$$= \frac{x^2 - 4}{x^2 - x - 6} \qquad \text{Simplify.}$$

$$= \frac{(x - 2)(x + 2)}{(x - 3)(x + 2)} \qquad \text{Factor.}$$

$$= \frac{x - 2}{x - 3} \qquad \text{Reduce.} \qquad \blacksquare$$

4. Simplify $\dfrac{1 - \dfrac{9}{x^2}}{1 - \dfrac{1}{x} - \dfrac{6}{x^2}}$.

EXAMPLE 5 Simplify $2 - \dfrac{3}{x + \dfrac{1}{3}}$.

SOLUTION First we simplify the expression that follows the subtraction sign.

$$2 - \frac{3}{x + \dfrac{1}{3}} = 2 - \frac{\boldsymbol{3} \cdot 3}{\boldsymbol{3}\left(x + \dfrac{1}{3}\right)} = 2 - \frac{9}{3x + 1}$$

Now we subtract by rewriting the first term, 2, with the LCD, $3x + 1$.

$$2 - \frac{9}{3x + 1} = \frac{2}{1} \cdot \frac{\boldsymbol{3x + 1}}{\boldsymbol{3x + 1}} - \frac{9}{3x + 1}$$

$$= \frac{6x + 2 - 9}{3x + 1}$$

$$= \frac{6x - 7}{3x + 1} \qquad \blacksquare$$

5. Simplify $2 + \dfrac{5}{x - \dfrac{1}{5}}$.

Answers

4. $\dfrac{x + 3}{x + 2}$

5. $\dfrac{10x + 23}{5x - 1}$

Problem Set 6.5

Moving Toward Success

"The sum of wisdom is that time is never lost that is devoted to work."

—Ralph Waldo Emerson, 1803–1882, American poet and essayist

1. Why is it important to hear not just the words your instructor uses to teach, but the meaning behind them as well?

2. How can being a good listener make your note taking more efficient?

A Simplify each of the following as much as possible. [Examples 1–5]

1. $\dfrac{\frac{3}{4}}{\frac{2}{3}}$

2. $\dfrac{\frac{5}{9}}{\frac{7}{12}}$

3. $\dfrac{\frac{1}{3} - \frac{1}{4}}{\frac{1}{2} + \frac{1}{8}}$

4. $\dfrac{\frac{1}{6} - \frac{1}{3}}{\frac{1}{4} - \frac{1}{8}}$

5. $\dfrac{3 + \frac{2}{5}}{1 - \frac{3}{7}}$

6. $\dfrac{2 + \frac{5}{6}}{1 - \frac{7}{8}}$

7. $\dfrac{\frac{1}{x}}{1 + \frac{1}{x}}$

8. $\dfrac{1 - \frac{1}{x}}{\frac{1}{x}}$

9. $\dfrac{1 + \frac{1}{a}}{1 - \frac{1}{a}}$

10. $\dfrac{1 - \frac{2}{a}}{1 - \frac{3}{a}}$

11. $\dfrac{\frac{1}{x} - \frac{1}{y}}{\frac{1}{x} + \frac{1}{y}}$

12. $\dfrac{\frac{1}{x} + \frac{2}{y}}{\frac{2}{x} + \frac{1}{y}}$

13. $\dfrac{\frac{x - 5}{x^2 - 4}}{\frac{x^2 - 25}{x + 2}}$

14. $\dfrac{\frac{3x + 1}{x^2 - 49}}{\frac{9x^2 - 1}{x - 7}}$

15. $\dfrac{\frac{4a}{2a^3 + 2}}{\frac{8a}{4a + 4}}$

16. $\dfrac{\frac{2a}{3a^3 - 3}}{\frac{4a}{6a - 6}}$

17. $\dfrac{1 - \dfrac{9}{x^2}}{1 - \dfrac{1}{x} - \dfrac{6}{x^2}}$

18. $\dfrac{4 - \dfrac{1}{x^2}}{4 + \dfrac{4}{x} + \dfrac{1}{x^2}}$

19. $\dfrac{2 + \dfrac{5}{a} - \dfrac{3}{a^2}}{2 - \dfrac{5}{a} + \dfrac{2}{a^2}}$

20. $\dfrac{3 + \dfrac{5}{a} - \dfrac{2}{a^2}}{3 - \dfrac{10}{a} + \dfrac{3}{a^2}}$

21. $\dfrac{27 - \dfrac{8}{x^3}}{3 + \dfrac{1}{x} - \dfrac{2}{x^2}}$

22. $\dfrac{64 + \dfrac{1}{x^3}}{4 - \dfrac{11}{x} - \dfrac{3}{x^2}}$

23. $\dfrac{1 + \dfrac{2}{x} + \dfrac{4}{x^2} + \dfrac{8}{x^3}}{1 - \dfrac{16}{x^4}}$

24. $\dfrac{27 + \dfrac{9}{x} + \dfrac{3}{x^2} + \dfrac{1}{x^3}}{81 - \dfrac{1}{x^4}}$

25. $\dfrac{2 + \dfrac{3}{x} - \dfrac{18}{x^2} - \dfrac{27}{x^3}}{2 + \dfrac{9}{x} + \dfrac{9}{x^2}}$

26. $\dfrac{3 + \dfrac{5}{x} - \dfrac{12}{x^2} - \dfrac{20}{x^3}}{3 + \dfrac{11}{x} + \dfrac{10}{x^2}}$

27. $\dfrac{1 + \dfrac{1}{x + 3}}{1 - \dfrac{1}{x + 3}}$

28. $\dfrac{1 + \dfrac{1}{x - 2}}{1 - \dfrac{1}{x - 2}}$

29. $\dfrac{1 - \dfrac{1}{a + 1}}{1 + \dfrac{1}{a - 1}}$

30. $\dfrac{\dfrac{1}{a - 1} + 1}{\dfrac{1}{a + 1} - 1}$

31. $\dfrac{\dfrac{1}{x + 3} + \dfrac{1}{x - 3}}{\dfrac{1}{x + 3} - \dfrac{1}{x - 3}}$

32. $\dfrac{\dfrac{1}{x + a} + \dfrac{1}{x - a}}{\dfrac{1}{x + a} - \dfrac{1}{x - a}}$

33. $\dfrac{\dfrac{y + 1}{y - 1} + \dfrac{y - 1}{y + 1}}{\dfrac{y + 1}{y - 1} - \dfrac{y - 1}{y + 1}}$

34. $\dfrac{\dfrac{y - 1}{y + 1} - \dfrac{y + 1}{y - 1}}{\dfrac{y - 1}{y + 1} + \dfrac{y + 1}{y - 1}}$

35. $1 - \dfrac{x}{1 - \dfrac{1}{x}}$

36. $x - \dfrac{1}{x - \dfrac{1}{2}}$

37. $1 + \dfrac{1}{1 + \dfrac{1}{1 + 1}}$

38. $1 - \dfrac{1}{1 - \dfrac{1}{1 - \dfrac{1}{2}}}$

39. $\dfrac{1 - \dfrac{1}{x + \dfrac{1}{2}}}{1 + \dfrac{1}{x + \dfrac{1}{2}}}$

40. $\dfrac{2 + \dfrac{1}{x - \dfrac{1}{3}}}{2 - \dfrac{1}{x - \dfrac{1}{3}}}$

41. $\dfrac{\dfrac{1}{x + h} - \dfrac{1}{x}}{h}$

42. $\dfrac{\dfrac{1}{(x + h)^2} - \dfrac{1}{x^2}}{h}$

43. $\dfrac{\dfrac{3}{ab} + \dfrac{4}{bc} - \dfrac{2}{ac}}{\dfrac{5}{abc}}$

44. $\dfrac{\dfrac{x}{yz} - \dfrac{y}{xz} + \dfrac{z}{xy}}{\dfrac{1}{x^2y^2} - \dfrac{1}{x^2z^2} + \dfrac{1}{y^2z^2}}$

45. $\dfrac{\dfrac{t^2 - 2t - 8}{t^2 + 7t + 6}}{\dfrac{t^2 - t - 6}{t^2 + 2t + 1}}$

46. $\dfrac{\dfrac{y^2 - 5y - 14}{y^2 + 3y - 10}}{\dfrac{y^2 - 8y + 7}{y^2 + 6y + 5}}$

47. $\dfrac{5 + \dfrac{4}{b - 1}}{\dfrac{7}{b + 5} - \dfrac{3}{b - 1}}$

48. $\dfrac{\dfrac{6}{x + 5} - 7}{\dfrac{8}{x + 5} - \dfrac{9}{x + 3}}$

49. $\dfrac{\dfrac{3}{x^2 - x - 6}}{\dfrac{2}{x + 2} - \dfrac{4}{x - 3}}$

50. $\dfrac{\dfrac{9}{a - 7} + \dfrac{8}{2a + 3}}{\dfrac{10}{2a^2 - 11a - 21}}$

51. $\dfrac{\dfrac{1}{m - 4} + \dfrac{1}{m - 5}}{\dfrac{1}{m^2 - 9m + 20}}$

52. $\dfrac{\dfrac{1}{k^2 - 7k + 12}}{\dfrac{1}{k - 3} + \dfrac{1}{k - 4}}$

53. Difference Quotient For each rational function below, find the difference quotient

$$\frac{f(x) - f(a)}{x - a}$$

a. $f(x) = \dfrac{4}{x}$

b. $f(x) = \dfrac{1}{x + 1}$

c. $f(x) = \dfrac{1}{x^2}$

54. Difference Quotient For each rational function below, find the difference quotient

$$\frac{f(x) - f(2)}{x - 2}$$

a. $f(x) = \dfrac{4}{x}$

b. $f(x) = \dfrac{1}{x + 1}$

c. $f(x) = \dfrac{1}{x^2}$

Applying the Concepts

55. Optics The formula $f = \dfrac{ab}{a + b}$ is used in optics to find the focal length of a lens. Show that the formula $f = (a^{-1} + b^{-1})^{-1}$ is equivalent to the preceding formula by rewriting it without the negative exponents and then simplifying the results.

56. Optics Show that the expression $(a^{-1} - b^{-1})^{-1}$ can be simplified to $\dfrac{ab}{b - a}$ by first writing it without the negative exponents and then simplifying the result.

57. Doppler Effect The change in the pitch of a sound (such as a train whistle) as an object passes is called the Doppler effect, named after C. J. Doppler (1803–1853). A person will *hear* a sound with a frequency, h, according to the formula

$$h = \frac{f}{1 + \dfrac{v}{s}}$$

where f is the actual frequency of the sound being produced, s is the speed of sound (about 740 miles per hour), and v is the velocity of the moving object.

a. Examine this fraction, and then explain why h and f approach the same value as v becomes smaller and smaller.

b. Solve this formula for v.

58. Work Problem A water storage tank has two drains. It can be shown that the time it takes to empty the tank if both drains are open is given by the formula

$$\frac{1}{\dfrac{1}{a} + \dfrac{1}{b}}$$

where a is the time it takes for the first drain to empty the tank, and b is the time for the second drain to empty the tank.

a. Simplify this complex fraction.

b. Find the amount of time needed to empty the tank using both drains if, used alone, the first drain empties the tank in 4 hours and the second drain can empty the tank in 3 hours.

Maintaining Your Skills

Solve each equation.

59. $3x + 60 = 15$

60. $3x - 18 = 4$

61. $3(y - 3) = 2(y - 2)$

62. $5(y + 2) = 4(y + 1)$

63. $10 - 2(x + 3) = x + 1$

64. $15 - 3(x - 1) = x - 2$

65. $x^2 - x - 12 = 0$

66. $3x^2 + x - 10 = 0$

67. $(x + 1)(x - 6) = -12$

68. $(x + 1)(x - 4) = -6$

Getting Ready for the Next Section

Multiply.

69. $x(y - 2)$

70. $x(y - 1)$

71. $6\left(\dfrac{x}{2} - 3\right)$

72. $6\left(\dfrac{x}{3} + 1\right)$

73. $xab \cdot \dfrac{1}{x}$

74. $xab\left(\dfrac{1}{b} + \dfrac{1}{a}\right)$

Factor.

75. $y^2 - 25$

76. $x^2 - 3x + 2$

77. $xa + xb$

78. $xy - y$

Solve.

79. $5x - 4 = 6$

80. $y^2 + y - 20 = 2y$

Extending the Concepts

Simplify each expression.

81. $\dfrac{\left(\frac{1}{3}\right) - \left(\frac{1}{3}\right)^2}{1 - \frac{1}{3}}$

82. $\dfrac{\left(\frac{1}{2}\right) - \left(\frac{1}{2}\right)^2}{1 - \frac{1}{2}}$

83. $\dfrac{\left(\frac{1}{9}\right) - \frac{1}{9}\left(\frac{1}{3}\right)^4}{1 - \frac{1}{3}}$

84. $\dfrac{\left(\frac{1}{6}\right) - \frac{1}{6}\left(\frac{1}{2}\right)^4}{1 - \frac{1}{2}}$

85. $\dfrac{1 + \dfrac{1}{1 - \frac{a}{b}}}{1 - \dfrac{3}{1 - \frac{a}{b}}}$

86. $\dfrac{1 - \dfrac{1}{\frac{a}{b} + 2}}{1 + \dfrac{3}{\frac{a}{2b} + 1}}$

87. $\dfrac{a^{-1} + b^{-1}}{2(ab)^{-1}}$

88. $\dfrac{(r^{-1} - s^{-1})^{-1}}{(rs)^{-2}}$

89. $\dfrac{(q^{-2} - t^{-2})^{-1}}{(t^{-1} - q^{-1})^{-1}}$

90. $\dfrac{(q^{-2} + t^{-2})^{-1}}{(t^{-1} + q^{-1})^{-1}}$

The only possible solution is $x = 2$. Checking this value back in the original equation gives

$$\frac{2}{2 - 2} + \frac{2}{3} \overset{?}{=} \frac{2}{2 - 2}$$

$$\frac{2}{0} + \frac{2}{3} \overset{?}{=} \frac{2}{0}$$

The first and last terms are undefined. The proposed solution, $x = 2$, does not check in the original equation. The solution set is the empty set. There is no solution to the original equation. ∎

When the proposed solution to an equation is not actually a solution, it is called an *extraneous solution*. In the last example, $x = 2$ is an extraneous solution.

EXAMPLE 4　Solve $\dfrac{5}{x^2 - 3x + 2} - \dfrac{1}{x - 2} = \dfrac{1}{3x - 3}$.

SOLUTION　Writing the equation again with the denominators in factored form, we have

$$\frac{5}{(x - 2)(x - 1)} - \frac{1}{x - 2} = \frac{1}{3(x - 1)}$$

The LCD is $3(x - 2)(x - 1)$. Multiplying through by the LCD, we have

$$\mathbf{3(x - 2)(x - 1)} \cdot \frac{5}{(x - 2)(x - 1)} - \mathbf{3(x - 2)(x - 1)} \cdot \frac{1}{(x - 2)}$$

$$= \mathbf{3(x - 2)(x - 1)} \cdot \frac{1}{3(x - 1)}$$

$$3 \cdot 5 - 3(x - 1) \cdot 1 = (x - 2) \cdot 1$$

$$15 - 3x + 3 = x - 2$$

$$-3x + 18 = x - 2$$

$$-4x + 18 = -2$$

$$-4x = -20$$

$$x = 5$$

Checking the proposed solution $x = 5$ in the original equation yields a true statement. Try it and see. ∎

EXAMPLE 5　Solve $3 + \dfrac{1}{x} = \dfrac{10}{x^2}$.

SOLUTION　To clear the equation of denominators, we multiply both sides by x^2.

$$\mathbf{x^2}\left(3 + \frac{1}{x}\right) = \mathbf{x^2}\left(\frac{10}{x^2}\right)$$

$$3(x^2) + \left(\frac{1}{x}\right)(x^2) = \left(\frac{10}{x^2}\right)(x^2)$$

$$3x^2 + x = 10$$

Rewrite in standard form and solve.

$$3x^2 + x - 10 = 0$$

$$(3x - 5)(x + 2) = 0$$

$$3x - 5 = 0 \quad \text{or} \quad x + 2 = 0$$

$$x = \frac{5}{3} \qquad\qquad x = -2$$

NOTE
In the process of solving the equation, we multiplied both sides by $3(x - 2)$, solved for x, and got $x = 2$ for our solution. But when x is 2, the quantity $3(x - 2) = 3(2 - 2) = 3(0) = 0$, which means we multiplied both sides of our equation by 0, which is not allowed under the multiplication property of equality.

4. Solve.

$$\frac{x}{x^2 - 9} - \frac{1}{x + 3} = \frac{1}{4x - 12}$$

NOTE
We can check the proposed solution in any of the equations obtained before multiplying through by the LCD. We cannot check the proposed solution in an equation obtained *after* multiplying through by the LCD since, if we have multiplied by 0, the resulting equations will not be equivalent to the original one.

5. Solve $1 - \dfrac{2}{x} = \dfrac{8}{x^2}$.

Answers
4. 9
5. −2, 4

The solution set is $\left\{-2, \frac{5}{3}\right\}$. Both solutions check in the original equation. Remember, we have to check *all solutions* any time we multiply both sides of the equation by an expression that contains the variable, just to be sure we haven't multiplied by 0.   ◼

6. Solve $\dfrac{y+1}{3(y+4)} = \dfrac{8}{(y+4)(y-4)}$.

EXAMPLE 6   Solve $\dfrac{y-4}{y^2-5y} = \dfrac{2}{y^2-25}$.

SOLUTION   Factoring each denominator, we find the LCD is $y(y-5)(y+5)$. Multiplying each side of the equation by the LCD clears the equation of denominators and leads us to our possible solutions.

$$\mathbf{y(y-5)(y+5)} \cdot \frac{y-4}{y(y-5)} = \frac{2}{(y-5)(y+5)} \cdot \mathbf{y(y-5)(y+5)}$$

$$(y+5)(y-4) = 2y$$

$$y^2 + y - 20 = 2y \qquad \text{Multiply out the left side.}$$

$$y^2 - y - 20 = 0 \qquad \text{Add } -2y \text{ to each side.}$$

$$(y-5)(y+4) = 0$$

$$y - 5 = 0 \quad \text{ or } \quad y + 4 = 0$$

$$y = 5 \qquad\qquad\qquad y = -4$$

The two possible solutions are 5 and -4. If we substitute -4 for y in the original equation, we find that it leads to a true statement. It is, therefore, a solution. However, if we substitute 5 for y in the original equation, we find that both sides of the equation are undefined. The only solution to our original equation is $y = -4$. The other possible solution $y = 5$ is extraneous.   ◼

B   Formulas

7. Solve for y: $x = \dfrac{y+2}{y-1}$.

EXAMPLE 7   Solve $x = \dfrac{y-4}{y-2}$ for y.

SOLUTION   To solve for y, we first multiply each side by $y-2$ to obtain

$$x(y-2) = y - 4$$

$$xy - 2x = y - 4 \qquad \text{Distributive property}$$

$$xy - y = 2x - 4 \qquad \text{Collect all terms containing } y \text{ on the left side.}$$

$$y(x-1) = 2x - 4 \qquad \text{Factor } y \text{ from each term on the left side.}$$

$$y = \frac{2x-4}{x-1} \qquad \text{Divide each side by } x-1.$$ ◼

8. Solve the formula $\dfrac{1}{a} = \dfrac{1}{x} + \dfrac{1}{b}$ for x.

EXAMPLE 8   Solve the formula $\dfrac{1}{x} = \dfrac{1}{b} + \dfrac{1}{a}$ for x.

SOLUTION   We begin by multiplying both sides by the least common denominator xab. As you can see from our previous examples, multiplying both sides of an equation by the LCD is equivalent to multiplying each term of both sides by the LCD.

Answers

6. 7

7. $y = \dfrac{x+2}{x-1}$

8. $x = \dfrac{ab}{b-a}$

$$xab \cdot \frac{1}{x} = \frac{1}{b} \cdot xab + \frac{1}{a} \cdot xab$$

$$ab = xa + xb$$

$$ab = (a + b)x \qquad \text{Factor } x \text{ from the right side.}$$

$$\frac{ab}{a + b} = x$$

We know we are finished because the variable we were solving for is alone on one side of the equation and does not appear on the other side. ■

C Graphing Rational Functions

In our next example, we investigate the graph of a rational function.

EXAMPLE 9 Graph the rational function $f(x) = \dfrac{6}{x - 2}$.

SOLUTION To find the y-intercept, we let x equal 0.

$$\text{When } x = 0: \qquad y = \frac{6}{0 - 2} = \frac{6}{-2} = -3 \qquad y\text{-intercept}$$

The graph will not cross the x-axis. If it did, we would have a solution to the equation

$$0 = \frac{6}{x - 2}$$

which has no solution because there is no number to divide 6 by to obtain 0.

The graph of our equation is shown in Figure 1 along with a table giving values of x and y that satisfy the equation. Notice that y is undefined when x is 2. This means that the graph will not cross the vertical line $x = 2$. (If it did, there would be a value of y for $x = 2$.) The line $x = 2$ is called a *vertical asymptote* of the graph. The graph will get very close to the vertical asymptote, but will never touch or cross it.

x	y
−4	−1
−1	−2
0	−3
1	−6
2	Undefined
3	6
4	3
5	2

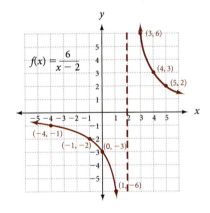

FIGURE 1

If you were to graph $y = \dfrac{6}{x}$ on the coordinate system in Figure 1, you would see that the graph of $y = \dfrac{6}{x - 2}$ is the graph of $y = \dfrac{6}{x}$ with all points shifted 2 units to the right. ■

9. Graph the rational function

$$f(x) = \frac{4}{x - 1}$$

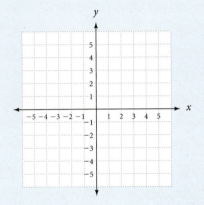

10. Graph $g(x) = \dfrac{4}{x + 1}$.

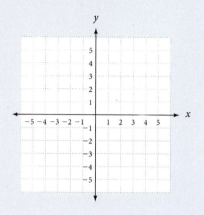

EXAMPLE 10 Graph $g(x) = \dfrac{6}{x + 2}$.

SOLUTION The only difference between this equation and the equation in Example 9 is in the denominator. This graph will have the same shape as the graph in Example 9, but the vertical asymptote will be $x = -2$ instead of $x = 2$. Figure 2 shows the graph.

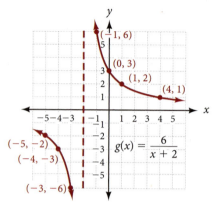

FIGURE 2

Notice that the graphs shown in Figures 1 and 2 are both graphs of functions because no vertical line will cross either graph in more than one place. Notice the similarities and differences in our two functions,

$$f(x) = \frac{6}{x - 2} \quad \text{and} \quad g(x) = \frac{6}{x + 2}$$

and their graphs. The vertical asymptotes shown in Figures 1 and 2 correspond to the fact that both $f(2)$ and $g(-2)$ are undefined. The domain for the function f is all real numbers except $x = 2$, while the domain for g is all real numbers except $x = -2$.

Problem Set 6.6

Moving Toward Success

"Laughter is the sun that drives winter from the human face."

—Victor Hugo, 1802–1885, French poet and dramatist

1. Why do you think eating healthy balanced meals leads to success in math?

2. Why do you think scheduling physical activity during your day can help you be successful in math?

A   Solve each of the following equations. [Examples 1–3]

1. $\dfrac{x}{5} + 4 = \dfrac{5}{3}$

2. $\dfrac{x}{5} = \dfrac{x}{2} - 9$

3. $\dfrac{a}{3} + 2 = \dfrac{4}{5}$

4. $\dfrac{a}{4} + \dfrac{1}{2} = \dfrac{2}{3}$

5. $\dfrac{y}{2} + \dfrac{y}{4} + \dfrac{y}{6} = 3$

6. $\dfrac{y}{3} - \dfrac{y}{6} + \dfrac{y}{2} = 1$

7. $\dfrac{5}{2x} = \dfrac{1}{x} + \dfrac{3}{4}$

8. $\dfrac{1}{2a} = \dfrac{2}{a} - \dfrac{3}{8}$

9. $\dfrac{1}{x} = \dfrac{1}{3} - \dfrac{2}{3x}$

10. $\dfrac{5}{2x} = \dfrac{2}{x} - \dfrac{1}{12}$

11. $\dfrac{2x}{x-3} + 2 = \dfrac{2}{x-3}$

12. $\dfrac{2}{x+5} = \dfrac{2}{5} - \dfrac{x}{x+5}$

13. $1 - \dfrac{1}{x} = \dfrac{12}{x^2}$

14. $2 + \dfrac{5}{x} = \dfrac{3}{x^2}$

15. $y - \dfrac{4}{3y} = -\dfrac{1}{3}$

16. $\dfrac{y}{2} - \dfrac{4}{y} = -\dfrac{7}{2}$

Let $f(x) = \dfrac{1}{x-3}$ and $g(x) = \dfrac{1}{x+3}$. Find x if

17. $f(x) + g(x) = \dfrac{5}{8}$

18. $f(x) - g(x) = \dfrac{2}{9}$

19. $\dfrac{f(x)}{g(x)} = 5$

20. $\dfrac{g(x)}{f(x)} = 5$

21. $f(x) = g(x)$

22. $f(x) = -g(x)$

Let $f(x) = \dfrac{4}{x+2}$ and $g(x) = \dfrac{4}{x-2}$. Find x if

23. $f(x) + g(x) = \dfrac{24}{5}$

24. $f(x) - g(x) = -\dfrac{4}{3}$

25. $\dfrac{f(x)}{g(x)} = -5$

26. $\dfrac{g(x)}{f(x)} = -7$

27. $f(x) = g(x)$

28. $f(x) = -g(x)$

29. Solve each equation.

   a. $6x - 2 = 0$

   b. $\dfrac{6}{x} - 2 = 0$

   c. $\dfrac{x}{6} - 2 = -\dfrac{1}{2}$

   d. $\dfrac{6}{x} - 2 = -\dfrac{1}{2}$

   e. $\dfrac{6}{x^2} + 6 = \dfrac{20}{x}$

30. Solve each equation.

   a. $5x - 2 = 0$

   b. $5 - \dfrac{2}{x} = 0$

   c. $\dfrac{x}{2} - 5 = -\dfrac{3}{4}$

   d. $\dfrac{2}{x} - 5 = -\dfrac{3}{4}$

   e. $-\dfrac{3}{x} + \dfrac{2}{x^2} = 5$

31. Work each problem according to the instructions given.

   a. Divide: $\dfrac{6}{x^2 - 2x - 8} \div \dfrac{x+3}{x+2}$

   b. Add: $\dfrac{6}{x^2 - 2x - 8} + \dfrac{x+3}{x+2}$

   c. Solve: $\dfrac{6}{x^2 - 2x - 8} + \dfrac{x+3}{x+2} = 2$

32. Work each problem according to the instructions given.

   a. Divide: $\dfrac{-10}{x^2 - 25} \div \dfrac{x-4}{x-5}$

   b. Add: $\dfrac{-10}{x^2 - 25} + \dfrac{x-4}{x-5}$

   c. Solve: $\dfrac{-10}{x^2 - 25} + \dfrac{x-4}{x-5} = \dfrac{4}{5}$

A Solve each equation. [Examples 2–6]

33. $\dfrac{x+2}{x+1} = \dfrac{1}{x+1} + 2$

34. $\dfrac{x+6}{x+3} = \dfrac{3}{x+3} + 2$

35. $\dfrac{3}{a-2} = \dfrac{2}{a-3}$

36. $\dfrac{5}{a+1} = \dfrac{4}{a+2}$

37. $6 - \dfrac{5}{x^2} = \dfrac{7}{x}$

38. $10 - \dfrac{3}{x^2} = -\dfrac{1}{x}$

39. $\dfrac{1}{x-1} - \dfrac{1}{x+1} = \dfrac{3x}{x^2-1}$

40. $\dfrac{5}{x-1} + \dfrac{2}{x-1} = \dfrac{4}{x+1}$

41. $\dfrac{2}{x-3} + \dfrac{x}{x^2-9} = \dfrac{4}{x+3}$

42. $\dfrac{2}{x+5} + \dfrac{3}{x+4} = \dfrac{2x}{x^2+9x+20}$

43. $\dfrac{3}{2} - \dfrac{1}{x-4} = \dfrac{-2}{2x-8}$

44. $\dfrac{2}{x} - \dfrac{1}{x+1} = \dfrac{-2}{5x+5}$

45. $\dfrac{t-4}{t^2-3t} = \dfrac{-2}{t^2-9}$

46. $\dfrac{t+3}{t^2-2t} = \dfrac{10}{t^2-4}$

47. $\dfrac{3}{y-4} - \dfrac{2}{y+1} = \dfrac{5}{y^2-3y-4}$

48. $\dfrac{1}{y+2} - \dfrac{2}{y-3} = \dfrac{-2y}{y^2-y-6}$

49. $\dfrac{2}{1+a} = \dfrac{3}{1-a} + \dfrac{5}{a}$

50. $\dfrac{1}{a+3} - \dfrac{a}{a^2-9} = \dfrac{2}{3-a}$

51. $\dfrac{3}{2x-6} - \dfrac{x+1}{4x-12} = 4$

52. $\dfrac{2x-3}{5x+10} + \dfrac{3x-2}{4x+8} = 1$

53. $\dfrac{y+2}{y^2-y} - \dfrac{6}{y^2-1} = 0$

54. $\dfrac{y+3}{y^2-y} - \dfrac{8}{y^2-1} = 0$

55. $\dfrac{4}{2x-6} - \dfrac{12}{4x+12} = \dfrac{12}{x^2-9}$

56. $\dfrac{1}{x+2} + \dfrac{1}{x-2} = \dfrac{4}{x^2-4}$

57. $\dfrac{2}{y^2-7y+12} - \dfrac{1}{y^2-9} = \dfrac{4}{y^2-y-12}$

58. $\dfrac{1}{y^2+5y+4} + \dfrac{3}{y^2-1} = \dfrac{-1}{y^2+3y-4}$

59. Solve the equation $6x^{-1} + 4 = 7$ by multiplying both sides by x. (Remember, $x^{-1} \cdot x = x^{-1} \cdot x^1 = x^0 = 1$.)

60. Solve the equation $3x^{-1} - 5 = 2x^{-1} - 3$ by multiplying both sides by x.

61. Solve the equation $1 + 5x^{-2} = 6x^{-1}$ by multiplying both sides by x^2.

62. Solve the equation $1 + 3x^{-2} = 4x^{-1}$ by multiplying both sides by x^2.

B [Examples 7, 8]

63. Solve the formula $\dfrac{1}{x} = \dfrac{1}{b} - \dfrac{1}{a}$ for x.

64. Solve $\dfrac{1}{x} = \dfrac{1}{a} - \dfrac{1}{b}$ for x.

65. Solve for R in the formula $\dfrac{1}{R} = \dfrac{1}{R_1} + \dfrac{1}{R_2}$.

66. Solve for R in the formula $\dfrac{1}{R} = \dfrac{1}{R_1} + \dfrac{1}{R_2} + \dfrac{1}{R_3}$.

Solve for y.

67. $x = \dfrac{y - 3}{y - 1}$

68. $x = \dfrac{y - 2}{y - 3}$

69. $x = \dfrac{2y + 1}{3y + 1}$

70. $x = \dfrac{3y + 2}{5y + 1}$

C Graph the following. [Examples 9, 10]

71. $f(x) = \dfrac{2}{x - 1}$

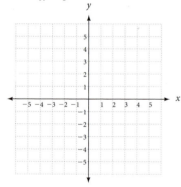

72. $f(x) = \dfrac{2}{x + 1}$

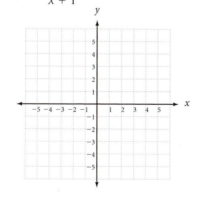

73. $f(x) = \dfrac{3}{x + 2}$

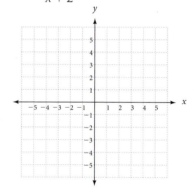

74. $f(x) = \dfrac{3}{x - 2}$

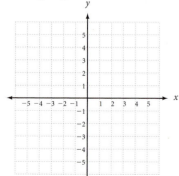

75. $f(x) = \dfrac{-2}{x - 3}$

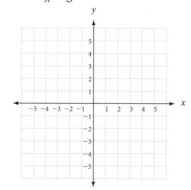

76. $f(x) = \dfrac{-3}{4x + 3}$

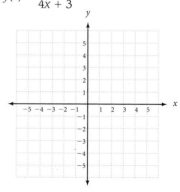

77. $f(x) = \dfrac{4}{3x - 2}$

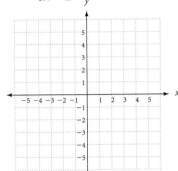

78. $f(x) = \dfrac{3}{2x - 1}$

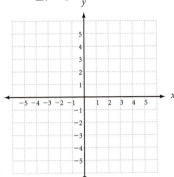

79. $f(x) = \dfrac{-5}{2x + 1}$

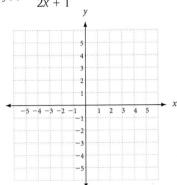

80. $f(x) = \dfrac{-3}{4x + 1}$

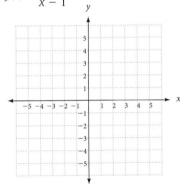

81. $f(x) = \dfrac{x - 1}{x + 1}$

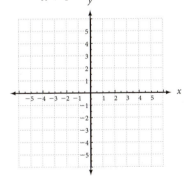

82. $f(x) = \dfrac{x + 1}{x - 1}$

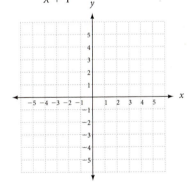

83. $f(x) = \dfrac{2x}{x - 1}$

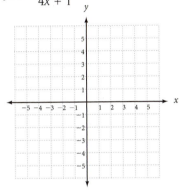

84. $f(x) = \dfrac{3x}{x + 1}$

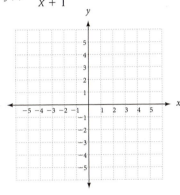

85. $f(x) = \dfrac{-2x}{x + 1}$

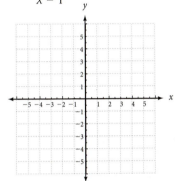

86. $f(x) = \dfrac{-3x}{x - 4}$

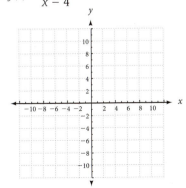

87. $f(x) = \dfrac{x - 3}{2x - 1}$

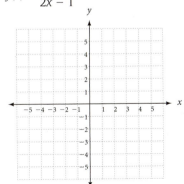

88. $f(x) = \dfrac{4x + 1}{2x - 3}$

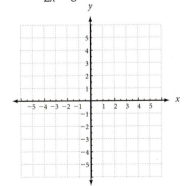

89. $f(x) = \dfrac{3x - 1}{2x + 3}$

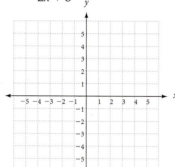

90. $f(x) = \dfrac{2x - 5}{3x + 4}$

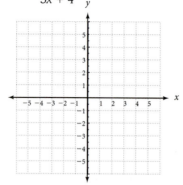

Applying the Concepts

91. An Identity An identity is an equation that is true for any value of the variable for which the expression is defined. Verify the following expression is an identity by simplifying the left side of the expression.

$$\frac{2}{x - y} - \frac{1}{y - x} = \frac{3}{x - y}$$

92. Harmonic Mean A number, h, is the harmonic mean of two numbers, n_1 and n_2, if $\frac{1}{h}$ is the mean (average) of $\frac{1}{n_1}$ and $\frac{1}{n_2}$.

  a. Write an equation relating the harmonic mean, h, to two numbers, n_1 and n_2 then solve the equation for h.

  b. Find the harmonic mean of 3 and 5.

93. Kayak Race In a kayak race, the participants must paddle a kayak 450 meters down a river and then return 450 meters up the river to the starting point (Figure 3). Susan has correctly deduced that the total time t (in seconds) depends on the speed c (in meters per second) of the water according to the following expression:

$$t = \frac{450}{v + c} + \frac{450}{v - c}$$

where v is the speed of the kayak in still water. Fill in the following table.

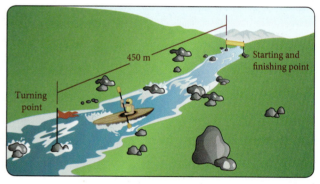

FIGURE 3

Time t (sec)	Speed of Kayak in Still Water v (m/sec)	Current of the River c (m/sec)
240		1
300		2
	4	3
	3	1
540	3	
	3	3

94. **Geometry** From plane geometry and the principle of similar triangles, the relationship between y_1, y_2, and h shown in Figure 4 can be expressed as

$$\frac{1}{h} = \frac{1}{y_1} + \frac{1}{y_2}$$

Two poles are 12 feet high and 8 feet high. If a wire is attached to the top of each one and stretched to the bottom of the other, what is the height above the ground at which the two wires will meet?

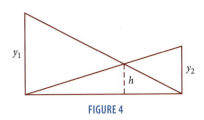

FIGURE 4

Maintaining Your Skills

95. **Number Problem** Twice the sum of a number and 3 is 16. Find the number.

96. **Number Problem** The sum of two consecutive odd integers is 48. Find the two integers.

97. **Geometry** The length of a rectangle is 3 meters less than twice the width. The perimeter is 42 meters. Find the length and width.

98. **Geometry** The smaller angle in a triangle is one fourth as large as the largest angle. The third angle is 9 degrees more than the smallest angle. Find the measure of all three angles.

99. **Consecutive Integers** The sum of the squares of two consecutive integers is 61. Find the integers.

100. **Consecutive Integers** The square of the sum of two consecutive integers is 121. Find the two integers.

101. **Geometry** The lengths of the sides of a right triangle are given by three consecutive integers. Find the lengths of the three sides.

102. **Geometry** The longest side of a right triangle is 8 inches more than the shortest side. The other side is 7 inches more than the shortest side. Find the lengths of the three sides.

Getting Ready for the Next Section

Multiply.

103. $39.3 \cdot 60$

104. $1{,}100 \cdot 60 \cdot 60$

Divide. Round to the nearest tenth, if necessary.

105. $65{,}000 \div 5{,}280$

106. $3{,}960{,}000 \div 5{,}280$

Multiply.

107. $2x\left(\dfrac{1}{x} + \dfrac{1}{2x}\right)$

108. $3x\left(\dfrac{1}{x} + \dfrac{1}{3x}\right)$

Solve.

109. $12(x + 3) + 12(x - 3) = 3(x^2 - 9)$

110. $40 + 2x = 60 - 3x$

111. $\dfrac{1}{10} - \dfrac{1}{12} = \dfrac{1}{x}$

112. $\dfrac{1}{x} + \dfrac{1}{2x} = 2$

Extending the Concepts

Solve each equation.

113. $\dfrac{12}{x} + \dfrac{8}{x^2} - \dfrac{75}{x^3} - \dfrac{50}{x^4} = 0$

114. $\dfrac{45}{x} + \dfrac{18}{x^2} - \dfrac{80}{x^3} - \dfrac{32}{x^4} = 0$

115. $\dfrac{1}{x^3} - \dfrac{1}{3x^2} - \dfrac{1}{4x} + \dfrac{1}{12} = 0$

116. $\dfrac{1}{x^3} - \dfrac{1}{2x^2} - \dfrac{1}{9x} + \dfrac{1}{18} = 0$

117. Solve for x. $\dfrac{2}{x} + \dfrac{4}{x + a} = \dfrac{-6}{a - x}$

118. Solve for x. $\dfrac{1}{b - x} - \dfrac{1}{x} = \dfrac{-2}{b + x}$

119. Solve for v. $\dfrac{s - vt}{t^2} = -16$

120. Solve for r. $A = P\left(1 + \dfrac{r}{n}\right)$

121. Solve for f. $\dfrac{1}{p} = \dfrac{1}{f} + \dfrac{1}{g}$

122. Solve for p. $h = \dfrac{v^2}{2g} + \dfrac{p}{c}$

Applications

TICKET TO SUCCESS

Keep these questions in mind as you read through the section. Then respond in your own words and in complete sentences.

1. Briefly list the steps in the Blueprint for Problem Solving that you have used previously to solve application problems.

2. Write an application problem for which the solution depends on solving the equation $\frac{1}{2} + \frac{1}{3} = \frac{1}{x}$.

3. What is a conversion factor in unit analysis?

4. What conversion factors would you use to convert feet per second to miles per hour?

Vladimir Mucibabic/Shutterstock.com

You and your roommate need to paint your house as part of an agreement with your landlord. You can complete the job by yourself in 12 hours and your roommate, who has some experience, can do the job in 9 hours. How long will it take you to do the job together?

In this section, we will work with some application problems like this, the solutions to which involve equations that contain rational expressions. As you will see, the solutions to the examples show only the essential steps from our Blueprint for Problem Solving. Recall that step 1 was done mentally; we read the problem and mentally list the items that are known and the items that are unknown. This is an essential part of problem solving. Now that you have had experience with application problems, you are doing step 1 automatically.

Also, we will look at a method of solving conversion problems that is called *unit analysis*. With unit analysis, we can convert expressions with units of feet per minute to equivalent expressions in miles per hour. This method of converting between different units of measure is used often in chemistry, physics, and engineering classes.

A Applications

EXAMPLE 1   One number is twice another. The sum of their reciprocals is 2. Find the numbers.

SOLUTION   Let x = the smaller number. The larger number is $2x$. Their reciprocals are $\frac{1}{x}$ and $\frac{1}{2x}$. The equation that describes the situation is

$$\frac{1}{x} + \frac{1}{2x} = 2$$

Multiplying both sides by the LCD $2x$, we have

$$2x \cdot \frac{1}{x} + 2x \cdot \frac{1}{2x} = 2x(2)$$

$$2 + 1 = 4x$$

$$3 = 4x$$

$$x = \frac{3}{4}$$

The smaller number is $\frac{3}{4}$. The larger is $2\left(\frac{3}{4}\right) = \frac{6}{4} = \frac{3}{2}$. Adding their reciprocals, we have

$$\frac{4}{3} + \frac{2}{3} = \frac{6}{3} = 2$$

The sum of the reciprocals of $\frac{3}{4}$ and $\frac{3}{2}$ is 2.   ∎

EXAMPLE 2   Two families from the same neighborhood plan a ski trip together. The first family makes the 455-mile trip at a speed 5 miles per hour faster than the second family. The second family takes a half-hour longer to make the trip. What are the speeds of the two families?

SOLUTION   The following table will be helpful in finding the equation necessary to solve this problem.

	d (distance)	r (rate)	t (time)
First Family			
Second Family			

If we let x be the speed of the second family, then the speed of the first family will be $x + 5$. Both families travel the same distance of 455 miles. Putting this information into the table we have

	d (distance)	r (rate)	t (time)
First Family	455	$x + 5$	
Second Family	455	x	

To fill in the last two spaces in the table, we use the relationship $d = r \cdot t$. Since the last column of the table is the time, we solve the equation $d = r \cdot t$ for t and get

$$t = \frac{d}{r}$$

Taking the distance and dividing by the rate (speed) for each family, we complete the table.

	d (distance)	r (rate)	t (time)
First Family	455	$x + 5$	$\frac{455}{x + 5}$
Second Family	455	x	$\frac{455}{x}$

Reading the problem again, we find that the time for the second family is longer than the time for the first family by one-half hour. In other words, the time for the second family can be found by adding one-half hour to the time for the first family, or

$$\frac{455}{x + 5} + \frac{1}{2} = \frac{455}{x}$$

Multiplying both sides by the LCD of $2x(x + 5)$ gives

$$2x \cdot (455) + x(x + 5) \cdot 1 = 455 \cdot 2(x + 5)$$

$$910x + x^2 + 5x = 910x + 4550$$

$$x^2 + 5x - 4550 = 0$$

$$(x + 70)(x - 65) = 0$$

$$x = -70 \qquad \text{or} \qquad x = 65$$

Since we cannot have a negative speed, the only solution is $x = 65$. Then

$$x + 5 = 65 + 5 = 70$$

The speed of the first family is 70 miles per hour, and the speed of the second family is 65 miles per hour. ∎

EXAMPLE 3 The speed of a boat in still water is 20 miles per hour. It takes the same amount of time for the boat to travel 3 miles downstream (with the current) as it does to travel 2 miles upstream (against the current). Find the speed of the current.

SOLUTION The following table will be helpful in finding the equation necessary to solve this problem.

	d (distance)	r (rate)	t (time)
Upstream			
Downstream			

3. A boat can travel at 15 miles per hour in still water. If it takes the same amount of time for the boat to travel 2 miles downstream as it does to travel 1 mile upstream, find the speed of the current.

Current

Answer
3. 5 mph

If we let x = the speed of the current, the speed (rate) of the boat upstream is $(20 - x)$ since it is traveling against the current. The rate downstream is $(20 + x)$ since the boat is then traveling with the current. The distance traveled upstream is 2 miles, and the distance traveled downstream is 3 miles. Putting the information given here into the table, we have

	d (distance)	*r* (rate)	*t* (time)
Upstream	2	$20 - x$	
Downstream	3	$20 + x$	

To fill in the last two spaces in the table we must use the relationship $d = r \cdot t$. Since we know the spaces to be filled in are in the time column, we solve the equation $d = r \cdot t$ for t and get

$$t = \frac{d}{r}$$

The completed table then is

	d (distance)	*r* (rate)	*t* (time)
Upstream	2	$20 - x$	$\dfrac{2}{20 - x}$
Downstream	3	$20 + x$	$\dfrac{3}{20 + x}$

Reading the problem again, we find that the time moving upstream is equal to the time moving downstream, or

$$\frac{2}{20 - x} = \frac{3}{20 + x}$$

Multiplying both sides by the LCD $(20 - x)(20 + x)$ gives

$$(20 + x) \cdot 2 = 3(20 - x)$$

$$40 + 2x = 60 - 3x$$

$$5x = 20$$

$$x = 4$$

The speed of the current is 4 miles per hour. ∎

EXAMPLE 4   The current of a river is 3 miles per hour. It takes a motorboat a total of 3 hours to travel 12 miles upstream and return 12 miles downstream. What is the speed of the boat in still water?

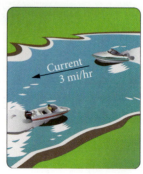

Current
3 mi/hr

SOLUTION   This time we let x = the speed of the boat in still water. Then, we fill in as much of the table as possible using the information given in the problem. For instance, since we let x = the speed of the boat in still water, the rate upstream (against the current) must be $x - 3$. The rate downstream (with the current) is $x + 3$.

	d	r	t
Upstream	12	$x - 3$	
Downstream	12	$x + 3$	

The last two boxes can be filled in using the relationship $t = \dfrac{d}{r}$

The completed table then is

	d (distance)	r (rate)	t (time)
Upstream	12	$x - 3$	$\dfrac{12}{x - 3}$
Downstream	12	$x - 3$	$\dfrac{12}{x - 3}$

The total time for the trip up and back is 3 hours.

$$\text{Time upstream} + \text{Time downstream} = \text{Total time}$$

$$\frac{12}{x - 3} + \frac{12}{x + 3} = 3$$

Multiplying both sides by $(x - 3)(x + 3)$, we have

$$12(x + 3) + 12(x - 3) = 3(x^2 - 9)$$

$$12x + 36 + 12x - 36 = 3x^2 - 27$$

$$3x^2 - 24x - 27 = 0$$

$$x^2 - 8x - 9 = 0 \qquad \text{Divide both sides by 3.}$$

$$(x - 9)(x + 1) = 0$$

$$x = 9 \qquad \text{or} \qquad x = -1$$

The speed of the motorboat in still water is 9 miles per hour. (We don't use $x = -1$ because the speed of the motorboat cannot be a negative number.) ■

EXAMPLE 5   An inlet pipe can fill a pool in 10 hours, and the drain can empty it in 12 hours. If the pool is empty and both the inlet pipe and drain are open, how long will it take to fill the pool?

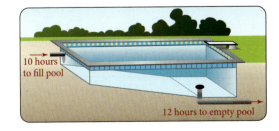

10 hours to fill pool

12 hours to empty pool

5. The hot-water faucet can fill a sink in 3 minutes. The drain will empty the sink in 4 minutes. If the hot-water faucet is on and the drain is open, how long will it take to fill the sink?

Answer
5. 12 minutes

SOLUTION   It is helpful to think in terms of how much work is done by each pipe in 1 hour.

Let x = the time it takes to fill the pool with both pipes open.

If the inlet pipe can fill the pool in 10 hours, then in 1 hour it is $\frac{1}{10}$ full. If the outlet pipe empties the pool in 12 hours, then in 1 hour it is $\frac{1}{12}$ empty. If the pool can be filled in x hours with both the inlet pipe and the drain open, then in 1 hour it is $\frac{1}{x}$ full when both pipes are open.

Here is the equation:

In 1 hour

$$\begin{bmatrix} \text{Amount filled} \\ \text{by inlet pipe} \end{bmatrix} - \begin{bmatrix} \text{Amount emptied} \\ \text{by the drain} \end{bmatrix} = \begin{bmatrix} \text{Fraction of pool filled} \\ \text{with both pipes open} \end{bmatrix}$$

$$\frac{1}{10} \qquad - \qquad \frac{1}{12} \qquad = \qquad \frac{1}{x}$$

Multiplying through by $60x$, we have

$$60x \cdot \frac{1}{10} - 60x \cdot \frac{1}{12} = 60x \cdot \frac{1}{x}$$

$$6x - 5x = 60$$

$$x = 60$$

It takes 60 hours to fill the pool if both the inlet pipe and the drain are open. ■

B Unit Analysis

In the 1950s, the United States had a spy plane, the U-2, that could fly at an altitude of 65,000 feet. Do you know how many miles are in 65,000 feet?

We can solve problems like this by using a method called *unit analysis*. With unit analysis, we analyze the units we are given and the units for which we are asked, and then multiply by the appropriate *conversion factor*. Since 1 mile is 5,280 feet, the conversion factor we use is

$$\frac{1 \text{ mile}}{5,280 \text{ feet}}$$

which is the number 1. Multiplying 65,000 feet by this conversion factor we have the following:

$$65,000 \text{ feet} = \frac{65,000 \text{ feet}}{1} \cdot \frac{1 \text{ mile}}{5,280 \text{ feet}}$$

65,000 ft

We treat the units common to the numerator and denominator in the same way we treat factors common to the numerator and denominator: we divide out common units, just as we divide out common factors. In the preceding expression, we have feet common to the numerator and denominator. Dividing them out leaves us with miles only. Here is the complete solution:

$$65,000 \text{ feet} = \frac{65,000 \text{ \cancel{feet}}}{1} \cdot \frac{1 \text{ mile}}{5,280 \text{ \cancel{feet}}}$$

$$= \frac{65,000}{5,280} \text{ mile}$$

$$= 12.3 \text{ miles, to the nearest tenth of a mile}$$

The key to solving a problem like this one lies in choosing the appropriate conversion factor. The fact that 1 mile = 5,280 feet yields two conversion factors, each of which is equal to the number 1. They are

$$\frac{1 \text{ mile}}{5{,}280 \text{ feet}} \quad \text{and} \quad \frac{5{,}280 \text{ feet}}{1 \text{ mile}}$$

The conversion factor we choose depends on the units we are given and the units with which we want to end up. Multiplying any expression by either of the two conversion factors leaves the value of the original expression unchanged because each of the conversion factors is simply the number 1.

EXAMPLE 6   In the first section of this chapter, we found a rider on the first Ferris wheel was traveling at approximately 39.3 feet per minute. Convert 39.3 feet per minute to miles per hour.

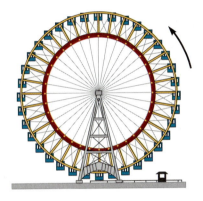

SOLUTION   We know that 5,280 feet = 1 mile and 60 minutes = 1 hour. Therefore, we have the following conversion factors, each of which is equal to 1.

$$\frac{5{,}280 \text{ feet}}{1 \text{ mile}} \quad \frac{1 \text{ mile}}{5{,}280 \text{ feet}} \quad \frac{60 \text{ minutes}}{1 \text{ hour}} \quad \frac{1 \text{ hour}}{60 \text{ minutes}}$$

The conversion factors we choose to multiply by are the ones that will allow us to divide out the units we are converting from and leave us with the units we are converting to. Specifically, we want to get rid of feet and be left with miles. Likewise, we want to get rid of minutes and be left with hours. Here is the conversion process that will accomplish these goals:

$$39.3 \text{ feet per minute} = \frac{39.3 \text{ feet}}{1 \text{ minute}} \cdot \frac{1 \text{ mile}}{5{,}280 \text{ feet}} \cdot \frac{60 \text{ minutes}}{1 \text{ hour}}$$

$$= \frac{39.3 \cdot 60 \text{ miles}}{5{,}280 \text{ hours}}$$

$$= 0.45 \text{ miles per hour, to the nearest hundredth} \qquad \blacksquare$$

6. In Practice Problem 7 of the first section of this chapter we found a rider on the Ferris wheel called *Colossus* was traveling at approximately 13.0 feet per second. Give the speed in miles per hour. Round to the nearest tenth.

Answer
6. 8.9 mph

7. Assume that the mistake in the advertisement is that feet per second should read feet per minute. Is 1,100 feet per minute a reasonable speed for a chair lift?

EXAMPLE 7   In 1993, a ski resort in Vermont advertised their new high-speed chair lift as "the world's fastest chair lift, with a speed of 1,100 feet per second." Show why the speed cannot be correct.

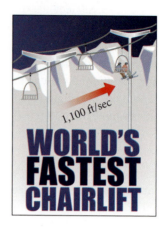

SOLUTION   To solve this problem, we can convert feet per second into miles per hour, a unit of measure we are more familiar with on an intuitive level.

$$1{,}100 \text{ feet per second} = \frac{1{,}100 \text{ feet}}{1 \text{ second}} \cdot \frac{1 \text{ mile}}{5{,}280 \text{ feet}} \cdot \frac{60 \text{ seconds}}{1 \text{ minute}} \cdot \frac{60 \text{ minutes}}{1 \text{ hour}}$$

$$= \frac{1{,}100 \cdot 60 \cdot 60 \text{ miles}}{5{,}280 \text{ hours}}$$

$$= 750 \text{ miles per hour}$$

Obviously, there was a mistake in the advertisement. ∎

Answer

7. 12.5 mph is a reasonable speed for a chair lift.

Problem Set 6.7

Moving Toward Success

"None are so old as those who have outlived enthusiasm."

—Henry David Thoreau, 1817–1862, American author and naturalist

1. What do you do if you have questions about the section or a homework problem outside of class?

2. How might you maintain or boost your enthusiasm toward this class?

Solve each of the following word problems. Be sure to show the equation in each case.

Number Problems

1. One number is 3 times another. The sum of their reciprocals is $\frac{20}{3}$. Find the numbers.

2. One number is 3 times another. The sum of their reciprocals is $\frac{4}{9}$. Find the numbers.

3. The sum of a number and its reciprocal is $\frac{10}{3}$. Find the number.

4. The sum of a number and twice its reciprocal is $\frac{27}{5}$. Find the number.

5. The sum of the reciprocals of two consecutive integers is $\frac{7}{12}$. Find the two integers.

6. Find two consecutive even integers, the sum of whose reciprocals is $\frac{3}{4}$.

7. If a certain number is added to the numerator and denominator of $\frac{7}{9}$, the result is $\frac{5}{6}$. Find the number.

8. Find the number you would add to both the numerator and denominator of $\frac{8}{11}$ so the result would be $\frac{6}{7}$.

9. Find the number you would add to both the numerator and the denominator of $\frac{9}{11}$ so the result would be $\frac{6}{7}$.

10. Find the number you would add to both the numerator and the denominator of $\frac{7}{15}$ so the result would be $\frac{5}{7}$.

Rate Problems

11. The speed of a boat in still water is 5 miles per hour. If the boat travels 3 miles downstream in the same amount of time it takes to travel 1.5 miles upstream, what is the speed of the current?

12. A boat, which moves at 18 miles per hour in still water, travels 14 miles downstream in the same amount of time it takes to travel 10 miles upstream. Find the speed of the current.

13. The current of a river is 2 miles per hour. A boat travels to a point 8 miles upstream and back again in 3 hours. What is the speed of the boat in still water?

14. A motorboat travels at 4 miles per hour in still water. It goes 12 miles upstream and 12 miles back again in a total of 8 hours. Find the speed of the current of the river.

15. Train A has a speed 15 miles per hour greater than that of train B. If train A travels 150 miles in the same time train B travels 120 miles, what are the speeds of the two trains?

16. A train travels 30 miles per hour faster than a car. If the train covers 120 miles in the same time the car covers 80 miles, what is the speed of each of them?

17. A small airplane flies 810 miles from Los Angeles to Portland, OR, with an average speed of 270 miles per hour. An hour and a half after the plane leaves, a Boeing 747 leaves Los Angeles for Portland. Both planes arrive in Portland at the same time. What was the average speed of the 747?

18. Lou leaves for a cross-country excursion traveling on a bicycle at 20 miles per hour. His friends are driving the trip and will meet him at several rest stops along the way. The first stop is scheduled 30 miles from the original starting point. If the people driving leave 15 minutes after Lou from the same place, how fast will they have to drive to reach the first rest stop at the same time as Lou?

19. A tour bus leaves Sacramento every Friday evening at 5:00 PM for a 270-mile trip to Las Vegas. This week, however, the bus leaves at 5:30 PM. To arrive in Las Vegas on time, the driver drives 6 miles per hour faster than usual. What is the bus's usual speed?

20. A bakery delivery truck leaves the bakery at 5:00 AM each morning on its 140-mile route. One day the driver gets a late start and does not leave the bakery until 5:30 AM. To finish her route on time the driver drives 5 miles per hour faster than usual. At what speed does she usually drive?

21. A 205-mile stretch of freeway has a speed limit of 70 miles per hour. Calculate how much faster, in minutes, a person driving at 75 miles per hour and a person driving at 80 miles per hour will finish the 205 miles than a person traveling the speed limit. Round to the nearest minute.

22. A 15-mile stretch of road has a speed limit of 55 miles per hour. Calculate how much faster, in seconds, a person driving at 60 miles per hour and a person driving at 65 miles per hour will finish the 15 miles than a person traveling the speed limit. Round to the nearest second.

Work Problems

23. A water tank can be filled by an inlet pipe in 8 hours. It takes twice that long for the outlet pipe to empty the tank. How long will it take to fill the tank if both pipes are open?

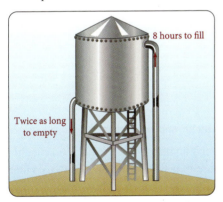

24. It takes 10 hours to fill a pool with the inlet pipe. It can be emptied in 15 hours with the outlet pipe. If the pool is half full to begin with, how long will it take to fill it from there if both pipes are open?

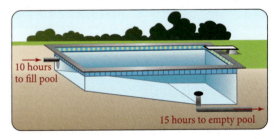

25. A sink can be filled from the faucet in 5 minutes. It takes only 3 minutes to empty the sink when the drain is open. If the sink is full and both the faucet and the drain are open, how long will it take to empty the sink?

26. A sink is one quarter full when both the faucet and the drain are opened. The faucet alone can fill the sink in 6 minutes, whereas it takes 8 minutes to empty it with the drain. How long will it take to fill the remaining three quarters of the sink?

27. A sink has two faucets: one for hot water and one for cold water. The sink can be filled by a cold-water faucet in 3.5 minutes. If both faucets are open, the sink is filled in 2.1 minutes. How long does it take to fill the sink with just the hot-water faucet open?

28. A water tank is being filled by two inlet pipes. Pipe A can fill the tank in $4\frac{1}{2}$ hours, but both pipes together can fill the tank in 2 hours. How long does it take to fill the tank using only pipe B?

29. A water tank is being filled by two inlet pipes. Pipe A can fill the tank three times as fast as pipe B. Both pipes can fill the tank in 4 hours. How long does it take to fill the tank using only pipe B?

30. A water tank is being filled by two inlet pipes. Pipe A can fill the tank in half the time it would take pipe B to fill the tank. Both pipes can fill the tank in 8 hours. How long does it take to fill the tank using only pipe A?

31. Painting If you can complete a job of painting a house in 12 hours and your friend can paint the house in 9 hours, how many hours would it take both of you to paint the house?

32. Painting You need to paint several rooms in your house. You could paint these rooms in 4 hours by yourself. Or your friend, who is not as experienced, could paint the rooms in 10 hours by himself. If you both work on painting these rooms, how long will it take?

Unit Analysis Problems

Give your answers to the following problems to the nearest tenth. Use 3.14 as an approximation for π.

33. The South Coast Shopping Mall in Costa Mesa, California, covers an area of 2,700,000 square feet. If 1 acre = 43,560 square feet, how many acres does the South Coast Shopping Mall cover?

34. The relationship between liters and cubic inches, both of which are measures of volume, is 0.0164 liters = 1 cubic inch. If a Ford Mustang has a motor with a displacement of 4.9 liters, what is the displacement in cubic inches?

35. The Forest chair lift at the Northstar ski resort in Lake Tahoe is 5,750 feet long. If a ride on this chair lift takes 11 minutes, what is the average speed of the lift in miles per hour?

36. The Bear Paw chair lift at the Northstar ski resort in Lake Tahoe is 790 feet long. If a ride on this chair lift takes 2.2 minutes, what is the average speed of the lift in miles per hour?

37. A sprinter runs 100 meters in 10.8 seconds. What is the sprinter's average speed in miles per hour? (1 meter = 3.28 feet)

38. A runner covers 400 meters in 49.8 seconds. What is the average speed of the runner in miles per hour?

39. A person riding a Ferris wheel with a diameter of 65 feet travels once around the wheel in 30 seconds. What is the average speed of the rider in miles per hour?

40. A person riding a Ferris wheel with a diameter of 102 feet travels once around the wheel in 3.5 minutes. What is the average speed of the rider in miles per hour?

41. A CD, when placed in a CD player, rotates at 300 rpm (meaning one revolution takes $\frac{1}{300}$ minute). Find the average speed of a point 2 inches from the center of the CD in miles per hour.

42. A house fan is running at 520 rpm. If the fan's blade tips are 8 inches from the center, what is the speed of the tips of the blades? What is the speed of a point 3 inches from the fan's center? Convert answers to miles per hour.

43. In order to produce the electricty we use in our homes, a utility scale wind turbine would need to rotate at 1000 rpm if a gear box was not used to convert the turbine's speed. What would be the tip speed of a turbine with a diameter of 232 feet if it was rotating at 1000 rpm? Convert your answer to miles per hour.

44. Golf The length of a golf course is the sum of the lengths of all 18 holes on the course. It is measured in yards. Round each of the lengths in the chart to the nearest hundred yards, then convert the result to miles, and round to the nearest tenth of a mile.

Longest Courses in U.S. Open History

Torrey Pines, South Course (2008)	7,643 yds
Bethpage State Park, Black Course (2009)	7,426 yds
Winged Foot Golf Course, West Course (2006)	7,264 yds
Oakmont Country Club (2007)	7,230 yds
Pinehurst, No. 2 Course (2005)	7,214 yds

Source: http://www.usopen.com

Miscellaneous Problems

45. Rhind Papyrus Nearly 4,000 years ago, Egyptians worked mathematical exercises involving reciprocals. The *Rhind Papyrus* we looked at in a previous chapter contains a wealth of such problems, and one of them is as follows:

Album/Oronoz/Newscom

> A quantity and its two thirds are added together, one third of this is added, then one third of the sum is taken, and the result is 10.

Write an equation and solve this exercise.

46. Photography For clear photographs, a camera must be focused properly. Professional photographers use a mathematical relationship relating the distance from the camera lens to the object being photographed, *a*; the distance from the lens to the film, *b*; and the focal length of the lens, *f*. These quantities, *a*, *b*, and *f*, are related by the equation

$$\frac{1}{a} + \frac{1}{b} = \frac{1}{f}$$

A camera has a focal length of 3 inches. If the lens is 5 inches from the film, how far should the lens be placed from the object being photographed for the camera to be perfectly focused?

Maintaining Your Skills

Perform the indicated operations.

47. $\dfrac{2a + 10}{a^3} \cdot \dfrac{a^2}{3a + 15}$

48. $\dfrac{4a + 8}{a^2 - a - 6} \div \dfrac{a^2 + 7a + 12}{a^2 - 9}$

49. $(x^2 - 9)\left(\dfrac{x + 2}{x + 3}\right)$

50. $\dfrac{1}{x + 4} + \dfrac{8}{x^2 - 16}$

51. $\dfrac{2x - 7}{x - 2} - \dfrac{x - 5}{x - 2}$

52. $2 + \dfrac{25}{5x - 1}$

Simplify each expression.

53. $\dfrac{\dfrac{1}{x} - \dfrac{1}{3}}{\dfrac{1}{x} + \dfrac{1}{3}}$

54. $\dfrac{1 - \dfrac{9}{x^2}}{1 - \dfrac{1}{x} - \dfrac{6}{x^2}}$

Solve each equation.

55. $\dfrac{x}{x - 3} + \dfrac{3}{2} = \dfrac{3}{x - 3}$

56. $1 - \dfrac{3}{x} = \dfrac{-2}{x^2}$

Chapter 6 Summary

■ Rational Numbers and Expressions [6.1]

1. $\frac{3}{4}$ is a rational number.
$\frac{x-3}{x^2-9}$ is a rational expression.

A *rational number* is any number that can be expressed as the ratio of two integers.

$$\text{Rational numbers} = \left\{ \frac{a}{b} \,\middle|\, a \text{ and } b \text{ are integers, } b \neq 0 \right\}$$

A *rational expression* is any quantity that can be expressed as the ratio of two polynomials:

$$\text{Rational expressions} = \left\{ \frac{P}{Q} \,\middle|\, P \text{ and } Q \text{ are polynomials, } Q \neq 0 \right\}$$

■ Properties of Rational Expressions [6.1]

If P, Q, and K are polynomials with $Q \neq 0$ and $K \neq 0$, then

$$\frac{P}{Q} = \frac{PK}{QK} \qquad \text{and} \qquad \frac{P}{Q} = \frac{\frac{P}{K}}{\frac{Q}{K}}$$

which is to say that multiplying or dividing the numerator and denominator of a rational expression by the same nonzero quantity always produces an equivalent rational expression.

■ Reducing to Lowest Terms [6.1]

2. $\dfrac{x-3}{x^2-9} = \dfrac{\cancel{x-3}}{(\cancel{x-3})(x+3)}$

$= \dfrac{1}{x+3}$

To reduce a rational expression to lowest terms, we first factor the numerator and denominator and then divide the numerator and denominator by any factors they have in common.

■ Dividing a Polynomial by a Monomial [6.2]

3. $\dfrac{15x^3 - 20x^2 + 10x}{5x}$

$= 3x^2 - 4x + 2$

To divide a polynomial by a monomial, divide each term of the polynomial by the monomial.

■ Long Division with Polynomials [6.2]

4.
$$\begin{array}{r} x - 2 \\ x-3{\overline{\smash{\big)}\,x^2 - 5x + 8}} \\ \underline{\cancel{+}\,x^2\,\cancel{-}\,3x}\,\downarrow \\ -2x + 8 \\ \underline{\cancel{+}\,2x\,\cancel{-}\,6} \\ 2 \end{array}$$

$= x - 2 + \dfrac{2}{x-3}$

If division with polynomials cannot be accomplished by dividing out factors common to the numerator and denominator, then we use a process similar to long division with whole numbers. The steps in the process are estimate, multiply, subtract, and bring down the next term.

Multiplication [6.3]

5. $\dfrac{x+1}{x^2-4} \cdot \dfrac{x+2}{3x+3}$

$= \dfrac{\cancel{(x+1)}\cancel{(x+2)}}{(x-2)\cancel{(x+2)}(3)\cancel{(x+1)}}$

$= \dfrac{1}{3(x-2)}$

To multiply two rational numbers or rational expressions, multiply numerators and multiply denominators. In symbols,

$$\frac{P}{Q} \cdot \frac{R}{S} = \frac{PR}{QS} \qquad (Q \neq 0 \text{ and } S \neq 0)$$

In practice, we don't really multiply, but rather, we factor and then divide out common factors.

Division [6.3]

6. $\dfrac{x^2-y^2}{x^3+y^3} \div \dfrac{x-y}{x^2-xy+y^2}$

$= \dfrac{x^2-y^2}{x^3+y^3} \cdot \dfrac{x^2-xy+y^2}{x-y}$

$= \dfrac{\cancel{(x+y)}\cancel{(x-y)}\cancel{(x^2-xy+y^2)}}{\cancel{(x+y)}\cancel{(x^2-xy+y^2)}\cancel{(x-y)}}$

$= 1$

To divide one rational expression by another, we use the definition of division to rewrite our division problem as an equivalent multiplication problem. To divide by a rational expression we multiply by its reciprocal. In symbols,

$$\frac{P}{Q} \div \frac{R}{S} = \frac{P}{Q} \cdot \frac{S}{R} = \frac{PS}{QR} \qquad (Q \neq 0, S \neq 0, R \neq 0)$$

Least Common Denominator [6.4]

7. The LCD for $\dfrac{2}{x-3}$ and $\dfrac{3}{5}$ is $5(x-3)$.

The *least common denominator*, LCD, for a set of denominators is the smallest quantity divisible by each of the denominators.

Addition and Subtraction [6.4]

8. $\dfrac{2}{x-3} + \dfrac{3}{5}$

$= \dfrac{2}{x-3} \cdot \dfrac{5}{5} + \dfrac{3}{5} \cdot \dfrac{x-3}{x-3}$

$= \dfrac{3x+1}{5(x-3)}$

If P, Q, and R represent polynomials, $R \neq 0$, then

$$\frac{P}{R} + \frac{Q}{R} = \frac{P+Q}{R} \quad \text{and} \quad \frac{P}{R} - \frac{Q}{R} = \frac{P-Q}{R}$$

When adding or subtracting rational expressions with different denominators, we must find the LCD for all denominators and change each rational expression to an equivalent expression that has the LCD.

Complex Fractions [6.5]

9. $\dfrac{\dfrac{1}{x} + \dfrac{1}{y}}{\dfrac{1}{x} - \dfrac{1}{y}} = \dfrac{xy\left(\dfrac{1}{x} + \dfrac{1}{y}\right)}{xy\left(\dfrac{1}{x} - \dfrac{1}{y}\right)}$

$= \dfrac{y+x}{y-x}$

A rational expression that contains, in its numerator or denominator, other rational expressions is called a *complex fraction*. One method of simplifying a complex fraction is to multiply the numerator and denominator by the LCD for all denominators.

Equations Involving Rational Expressions [6.6]

10. Solve $\dfrac{x}{2} + 3 = \dfrac{1}{3}$.

$6\left(\dfrac{x}{2}\right) + 6 \cdot 3 = 6 \cdot \dfrac{1}{3}$

$3x + 18 = 2$

$x = -\dfrac{16}{3}$

To solve an equation involving rational expressions, we first find the LCD for all denominators appearing on either side of the equation. We then multiply both sides by the LCD to clear the equation of all fractions and solve as usual.

Reduce to lowest terms. [6.1]

1. $\dfrac{125x^4yz^3}{35x^2y^4z^3}$

2. $\dfrac{a^3 - ab^2}{4a + 4b}$

3. $\dfrac{x^2 - 25}{x^2 + 10x + 25}$

4. $\dfrac{ax + x - 5a - 5}{ax - x - 5a + 5}$

Divide. If the denominator is a factor of the numerator, you may want to factor the numerator and divide out the common factor. [6.2]

5. $\dfrac{12x^3 + 8x^2 + 16x}{4x^2}$

6. $\dfrac{27a^2b^3 - 15a^3b^2 + 21a^4b^4}{-3a^2b^2}$

7. $\dfrac{x^{6n} - x^{5n}}{x^{3n}}$

8. $\dfrac{x^2 - x - 6}{x - 3}$

9. $\dfrac{5x^2 - 14xy - 24y^2}{x - 4y}$

10. $\dfrac{y^4 - 16}{y - 2}$

11. $\dfrac{8x^2 - 26x - 9}{2x - 7}$

12. $\dfrac{2y^3 - 9y^2 - 17y + 39}{2y - 3}$

Multiply and divide as indicated. [6.3]

13. $\dfrac{3}{4} \cdot \dfrac{12}{15} \div \dfrac{1}{3}$

14. $\dfrac{15x^2y}{8xy^2} \div \dfrac{10xy}{4x}$

15. $\dfrac{x^3 - 1}{x^4 - 1} \cdot \dfrac{x^2 - 1}{x^2 + x + 1}$

16. $\dfrac{a^2 + 5a + 6}{a + 1} \cdot \dfrac{a + 5}{a^2 + 2a - 3} \div \dfrac{a^2 + 7a + 10}{a^2 - 1}$

17. $\dfrac{ax + bx + 2a + 2b}{ax - 3a + bx - 3b} \div \dfrac{ax - bx - 2a + 2b}{ax - bx - 3a + 3b}$

18. $(4x^2 - 9) \cdot \dfrac{x + 3}{2x + 3}$

Add and subtract as indicated. [6.4]

19. $\dfrac{3}{5} - \dfrac{1}{10} + \dfrac{8}{15}$

20. $\dfrac{5}{x - 5} - \dfrac{x}{x - 5}$

21. $\dfrac{1}{x} + \dfrac{1}{x^2} + \dfrac{1}{x^3}$

22. $\dfrac{8}{y^2 - 16} - \dfrac{7}{y^2 - y - 12}$

23. $\dfrac{x - 2}{x^2 + 5x + 4} - \dfrac{x - 4}{2x^2 + 12x + 16}$

24. $3 + \dfrac{4}{5x - 2}$

Simplify each complex fraction. [6.5]

25. $\dfrac{1 + \dfrac{2}{3}}{1 - \dfrac{2}{3}}$

26. $\dfrac{\dfrac{4a}{2a^3 + 2}}{\dfrac{8a}{4a + 4}}$

27. $1 + \dfrac{1}{x + \dfrac{1}{x}}$

28. $\dfrac{1 - \dfrac{9}{x^2}}{1 - \dfrac{1}{x} - \dfrac{6}{x^2}}$

Solve each equation. [6.6]

29. $\dfrac{3}{x - 1} = \dfrac{3}{5}$

30. $\dfrac{x + 1}{3} + \dfrac{x - 3}{4} = \dfrac{1}{6}$

31. $\dfrac{5}{y + 1} = \dfrac{4}{y + 2}$

32. $\dfrac{x + 6}{x + 3} - 2 = \dfrac{3}{x + 3}$

33. $\dfrac{4}{x^2 - x - 12} + \dfrac{1}{x^2 - 9} = \dfrac{2}{x^2 - 7x + 12}$

34. $\dfrac{a + 4}{a^2 + 5a} = \dfrac{-2}{a^2 - 25}$

35. Distance, Rate, and Time A car makes a 120-mile trip 10 miles per hour faster than a truck. The truck takes 2 hours longer to make the trip. What are the speeds of the car and the truck? [6.7]

36. Average Speed A jogger covers 3.5 miles in 28 minutes. Find the average speed of the jogger in miles per hour. [6.7]

37. Unit Analysis The speed of sound is 1,088 feet per second. Convert the speed of sound to miles per hour. Round your answer to the nearest whole number. [6.7]

Chapter 6 Cumulative Review

Simplify.

1. $\left(-\dfrac{3}{2}\right)^3$

2. $\left(\dfrac{5}{6}\right)^{-2}$

3. $|-28 - 36| - 17$

4. $2^3 - 3(5^2 - 2^2)$

5. $64 \div 16 \cdot 4$

6. $125 \div 25 \cdot 5$

7. $\dfrac{3^4 - 6}{25(3^2 - 2^2)}$

8. $\dfrac{-2^3(19 - 5 \cdot 2)}{-7^2 + 13}$

9. Subtract $\dfrac{3}{4}$ from the product of -3 and $\dfrac{5}{12}$.

10. Write in symbols: The difference of $5a$ and $7b$ is greater than their sum.

Let $P(x) = 16x - 0.01x^2$ and find the following.

11. $P(10)$

12. $P(-10)$

Let $Q(x) = \dfrac{3}{5}x^2 + \dfrac{1}{2}x + 25$ and find the following.

13. $Q(10)$

14. $Q(-10)$

Simplify each of the following expressions.

15. $-3(5x + 4) + 12x$

16. $(x + 3)^2 - (x - 3)^2$

Subtract.

17. $\dfrac{6}{y^2 - 9} - \dfrac{5}{y^2 - y - 6}$

18. $\dfrac{y}{x^2 - y^2} - \dfrac{x}{x^2 - y^2}$

Multiply.

19. $\left(4t^2 + \dfrac{1}{3}\right)\left(3t^2 - \dfrac{1}{4}\right)$

20. $\dfrac{x^4 - 16}{x^3 - 8} \cdot \dfrac{x^2 + 2x + 4}{x^2 + 4}$

Divide.

21. $\dfrac{10x^{3n} - 15x^{4n}}{5x^n}$

22. $\dfrac{a^4 + a^3 - 1}{a + 2}$

Reduce to lowest terms. Write the answer with positive exponents only.

23. $\dfrac{x^{-5}}{x^{-8}}$

24. $\left(\dfrac{x^{-6}y^3}{x^{-3}y^{-4}}\right)^{-1}$

25. $\dfrac{x^3 + 2x^2 - 9x - 18}{x^2 - x - 6}$

26. $\dfrac{\dfrac{2a}{3a^3 - 3}}{\dfrac{4a}{6a - 6}}$

Find the equation of the line through the two points.

27. $(-6, -1)$ and $(-3, -5)$

28. $(-2, 6)$ and $(-2, 3)$

Factor each of the following expressions.

29. 168

30. $x^2 - 3x - 70$

31. $x^2 + 10x + 25 - y^2$

32. $y^3 + \dfrac{8}{27}x^3$

Solve the following equations.

33. $-\dfrac{3}{5}a + 3 = 15$

34. $7y - 6 = 2y + 9$

35. $\dfrac{2}{5}(15x - 2) - \dfrac{1}{5} = 5$

36. $\dfrac{2}{3}(9x - 2) + \dfrac{1}{3} = 4$

37. $\dfrac{3}{y - 2} = \dfrac{2}{y - 3}$

38. $2 - \dfrac{11}{x} = -\dfrac{12}{x^2}$

Solve each system.

39. $5x - 2y = -1$
$5x - 2y = 3x + 2$

40. $5x - 8y = -4$
$3x + 2y = -1$

Solve using any method.

41. $-3x + 2y = -1$
$4x - y = -16$

42. $x - 2y + z = -1$
$2x - 3z = -13$
$-2x + y - 2z = -1$

Solve each inequality, and graph the solution.

43. $-3(3x - 1) \le -2(3x - 3)$

44. $|2x + 3| - 4 < 1$

Graph on a rectangular coordinate system.

45. $6x - 5y = 30$

46. $3x - y < 4$

Use the graph to work Problems 47–50.

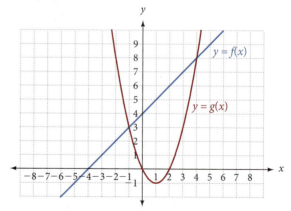

47. Find $(f - g)(-2)$.

48. Find $(f + g)(2)$.

49. Find $(f \circ g)(0)$.

50. Find $(g \circ f)(0)$.

Reduce to lowest terms. [6.1]

1. $\dfrac{x^2 - y^2}{x - y}$

2. $\dfrac{2x^2 - 5x + 3}{2x^2 - x - 3}$

3. Average Speed A person riding a Ferris wheel with a diameter of 75 feet travels once around the wheel in 40 seconds. What is the average speed of the rider in feet per second? Use 3.14 as an approximation for π and round your answer to the nearest tenth. [6.1]

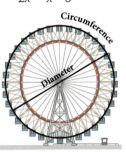

Circumference

Diameter

Divide. [6.2]

4. $\dfrac{24x^3y + 12x^2y^2 - 16xy^3}{4xy}$

5. $\dfrac{2x^3 - 9x^2 + 10}{2x - 1}$

Multiply and divide as indicated. [6.3]

6. $\dfrac{a^2 - 16}{5a - 15} \cdot \dfrac{10(a - 3)^2}{a^2 - 7a + 12}$

7. $\dfrac{a^4 - 81}{a^2 + 9} \div \dfrac{a^2 - 8a + 15}{4a - 20}$

8. $\dfrac{x^3 - 8}{2x^2 - 9x + 10} \div \dfrac{x^2 + 2x + 4}{2x^2 + x - 15}$

Add and subtract as indicated. [6.4]

9. $\dfrac{4}{21} + \dfrac{6}{35}$

10. $\dfrac{3}{4} - \dfrac{1}{2} + \dfrac{5}{8}$

11. $\dfrac{a}{a^2 - 9} + \dfrac{3}{a^2 - 9}$

12. $\dfrac{1}{x} + \dfrac{2}{x - 3}$

13. $\dfrac{4x}{x^2 + 6x + 5} - \dfrac{3x}{x^2 + 5x + 4}$

14. $\dfrac{2x + 8}{x^2 + 4x + 3} - \dfrac{x + 4}{x^2 + 5x + 6}$

Simplify each complex fraction. [6.5]

15. $\dfrac{3 - \dfrac{1}{a + 3}}{3 + \dfrac{1}{a + 3}}$

16. $\dfrac{1 - \dfrac{9}{x^2}}{1 + \dfrac{1}{x} - \dfrac{6}{x^2}}$

For the functions in Problems 17 and 18, evaluate the given difference quotients. [6.5]

a. $\dfrac{f(x + h) - f(x)}{h}$

b. $\dfrac{f(x) - f(a)}{x - a}$

17. $f(x) = 2x - 4$

18. $f(x) = x^2 + 5x - 1$

Solve each of the following equations. [6.6]

19. $\dfrac{1}{x} + 3 = \dfrac{4}{3}$

20. $\dfrac{x}{x - 3} + 3 = \dfrac{3}{x - 3}$

21. $\dfrac{y + 3}{2y} + \dfrac{5}{y - 1} = \dfrac{1}{2}$

22. $1 - \dfrac{1}{x} = \dfrac{6}{x^2}$

Solve the following applications. Be sure to show the equation in each case. [6.7]

23. Number Problem What number must be subtracted from the denominator of $\frac{10}{23}$ to make the result $\frac{1}{3}$?

24. Speed of a Boat The current of a river is 2 miles per hour. It takes a motorboat a total of 3 hours to travel 8 miles upstream and return 8 miles downstream. What is the speed of the boat in still water?

25. Filling a Pool An inlet pipe can fill a pool in 10 hours, and the drain can empty it in 15 hours. If the pool is half full and both the inlet pipe and the drain are left open, how long will it take to fill the pool the rest of the way?

26. Unit Analysis The top of Mount Whitney, the highest point in California, is 14,494 feet above sea level. Give this height in miles to the nearest tenth of a mile.

27. Unit Analysis The illustration shows the heights of the three tallest mountains in the world. Convert each height to miles and round to the nearest tenth.

The Greatest Heights

K2	28,238 ft
Mount Everest	29,035 ft
Kangchenjunga	28,208 ft

Source: Forrester Research, 2005

28. Unit Analysis A bullet fired from a gun travels a distance of 4,750 feet in 3.2 seconds. Find the average speed of the bullet in miles per hour. Round to the nearest whole number.

RATIONAL EXPRESSIONS AND RATIONAL FUNCTIONS

Group Project

Rational Expressions

Number of People	3
Time Needed	10–15 minutes
Equipment	Pencil and paper
Procedure	The four problems shown here all involve the same rational expressions. Often, students who have worked problems successfully on their homework have trouble when they take a test on rational expressions because the problems are mixed up and do not have similar instructions. Noticing similarities and differences between the types of problems involving rational expressions can help with this situation.

Problem 1: Add: $\dfrac{-2}{x^2 - 2x - 3} + \dfrac{3}{x^2 - 9}$.

Problem 2: Divide: $\dfrac{-2}{x^2 - 2x - 3} \div \dfrac{3}{x^2 - 9}$.

Problem 3: Solve: $\dfrac{-2}{x^2 - 2x - 3} + \dfrac{3}{x^2 - 9} = -1$.

Problem 4: Simplify: $\dfrac{\dfrac{-2}{x^2 - 2x - 3}}{\dfrac{3}{x^2 - 9}}$.

1. Which problems here do not require the use of a least common denominator?

2. Which two problems involve multiplying by the least common denominator?

3. Which of the problems will have an answer that is one or two numbers but no variables?

4. Work each of the four problems.

Ferris Wheel and The Third Man

Among the large Ferris wheels built around the turn of the twentieth century was one built in Vienna in 1897. It is the only one of those large wheels that is still in operation today. Known as the Riesenrad, it has a diameter of 197 feet and can carry a total of 800 people. A brochure that gives some statistics associated with the Riesenrad indicates that passengers riding it travel at 2 feet 6 inches per second. You can check the accuracy of this number by watching the movie *The Third Man.* In the movie, Orson Welles rides the Riesenrad through one complete revolution. Play *The Third Man* so you can view the Riesenrad in operation. Use the pause button and a timer to time how long it takes Orson Welles to ride once around the wheel. Then calculate his average speed during the ride. Use your results to either prove or disprove the claim that passengers travel at 2 feet 6 inches per second on the Riesenrad. When you have finished, write your procedures and results in essay form.

© Three Lions/Hulton Archive/Getty Images

LONDON FILMS / THE KOBAL COLLECTION

Rational Exponents and Roots

7

Sundial Bridge, Redding, CA

The Sundial Bridge in Redding, CA is a pedestrian bridge that spans the Sacramento River. Unique in its design, it boasts a glass floor that helps minimize any shadow on the water below. A prominent shadow on the water would disturb the abundant salmon population that has chosen this location to spawn. The bridge's tower stands at 217 feet and casts a shadow on a working sundial north of the bridge. It is here that visitors can see the earth's rotation as the shadow moves about one foot every minute.

The Sundial Bridge is called a cantilever span cable-stayed bridge, in which the bridge's platform is held up by cables attached to the tower. The distance d between the top of the bridge's tower to a cable's anchor point on the platform is given by $d = \sqrt{l^2 + h^2}$, where l is the length between the cable's anchor point and the tower base, and h is the height of the tower. This equation contains a radical expression, which we will work with in greater depth in this chapter.

Andy Z./Shutterstock.com

547

Key Words	Definition
Positive Square Root	The nonnegative number we square to get a quantity
Similar Radicals	Two radicals with the same index and same radicand
The Number i	The number such that $i = \sqrt{-1}$
Complex Number	Any number that can be put in the form $a + bi$

Chapter Outline

Rational Exponents

TICKET TO SUCCESS

Keep these questions in mind as you read through the section. Then respond in your own words and in complete sentences.

1. What is the definition for the positive square root?
2. What is the rational exponent theorem?
3. For the expression $a^{m/n}$, explain the significance of the numerator m and the significance of the denominator n in the exponent.
4. Briefly explain the concept behind the golden rectangle.

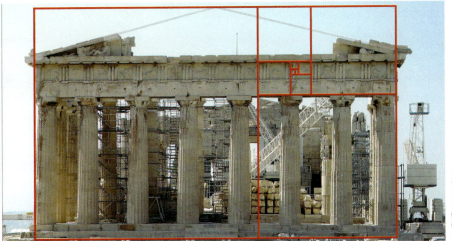

Jane Rix/Shutterstock.com

Figure 1 shows a square in which each of the four sides is 1 inch long. To find the square of the length of the diagonal c, we apply the Pythagorean theorem.

$$c^2 = 1^2 + 1^2$$
$$c^2 = 2$$

1 inch

c

1 inch

1 inch

FIGURE 1

Because we know that c is positive and that its square is 2, we call c the *positive square root* of 2, and we write $c = \sqrt{2}$. Associating numbers, such as $\sqrt{2}$, with the diagonal of a square or rectangle allows us to analyze some interesting items from geometry. One particularly interesting geometric object, the golden rectangle, that we will study in this section is shown in Figure 2 on the next page. You may have seen some real-life examples of this object such as the Parthenon, shown above. It is constructed from a right triangle, and the length of the diagonal is found from the Pythagorean theorem. We will come back to this figure at the end of this section.

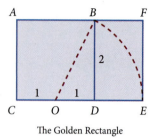

The Golden Rectangle

FIGURE 2

A Radical Expressions

Previously, we developed notation (exponents) to give us the square, cube, or any other power of a number. For instance, if we wanted the square of 3, we wrote $3^2 = 9$. If we wanted the cube of 3, we wrote $3^3 = 27$. In this section, we will develop notation that will take us in the reverse direction; that is, from the square of a number, say 25, back to the original number, 5.

> **Definition**
>
> If x is a nonnegative real number, then the expression $\sqrt{x}$ is called the **positive square root** of x and is the nonnegative number such that
>
> $$(\sqrt{x})^2 = x$$
>
> In words: $\sqrt{x}$ is the nonnegative number we square to get x.

The negative square root of x, $-\sqrt{x}$, is defined in a similar manner.

PRACTICE PROBLEMS

1. Give the two square roots of 36.

EXAMPLE 1   The positive square root of 64 is 8 because 8 is the positive number with the property $8^2 = 64$. The negative square root of 64 is -8 since -8 is the negative number whose square is 64. We can summarize both of these facts by saying

$$\sqrt{64} = 8 \quad \text{and} \quad -\sqrt{64} = -8 \qquad \blacksquare$$

The higher roots, cube roots, fourth roots, and so on have definitions similar to that of square roots.

> **Definition**
>
> If x is a real number and n is a positive integer, then
>
> Positive square root of x, $\sqrt{x}$, is such that $(\sqrt{x})^2 = x$ $x \geq 0$
>
> Cube root of x, $\sqrt[3]{x}$, is such that $(\sqrt[3]{x})^3 = x$
>
> Positive fourth root of x, $\sqrt[4]{x}$, is such that $(\sqrt[4]{x})^4 = x$ $x \geq 0$
>
> Fifth root of x, $\sqrt[5]{x}$, is such that $(\sqrt[5]{x})^5 = x$
>
> $$\vdots \qquad \vdots \qquad \vdots$$
>
> The **nth root of x**, $\sqrt[n]{x}$, is such that $(\sqrt[n]{x})^n = x$ $x \geq 0$ if n is even

Answers

1. 6, -6

The following is a table of the most common roots used in this book. Any of the roots that are unfamiliar should be memorized.

Square Roots		Cube Roots	Fourth Roots
$\sqrt{0} = 0$	$\sqrt{49} = 7$	$\sqrt[3]{0} = 0$	$\sqrt[4]{0} = 0$
$\sqrt{1} = 1$	$\sqrt{64} = 8$	$\sqrt[3]{1} = 1$	$\sqrt[4]{1} = 1$
$\sqrt{4} = 2$	$\sqrt{81} = 9$	$\sqrt[3]{8} = 2$	$\sqrt[4]{16} = 2$
$\sqrt{9} = 3$	$\sqrt{100} = 10$	$\sqrt[3]{27} = 3$	$\sqrt[4]{81} = 3$
$\sqrt{16} = 4$	$\sqrt{121} = 11$	$\sqrt[3]{64} = 4$	
$\sqrt{25} = 5$	$\sqrt{144} = 12$	$\sqrt[3]{125} = 5$	
$\sqrt{36} = 6$	$\sqrt{169} = 13$		

NOTATION An expression like $\sqrt[3]{8}$ that involves a root is called a *radical expression*. In the expression $\sqrt[3]{8}$, the 3 is called the *index*, the $\sqrt{}$ is the *radical sign,* and 8 is called the *radicand.* The index of a radical must be a positive integer greater than 1. If no index is written, it is assumed to be 2.

Roots and Negative Numbers

When dealing with negative numbers and radicals, the only restriction concerns negative numbers under even roots. We can have negative signs in front of radicals and negative numbers under odd roots and still obtain real numbers. Here are some examples to help clarify this. In the last section of this chapter, we will see how to deal with even roots of negative numbers.

EXAMPLE 2   Simplify each expression, if possible.

a. $\sqrt[3]{-8} = -2$ because $(-2)^3 = -8$.

b. $\sqrt{-4}$ is not a real number since there is no real number whose square is -4.

c. $-\sqrt{25} = -5$ is the negative square root of 25.

d. $\sqrt[5]{-32} = -2$ because $(-2)^5 = -32$.

e. $\sqrt[4]{-81}$ is not a real number since there is no real number we can raise to the fourth power and obtain -81. ■

Variables Under a Radical

From the preceding examples, it is clear that we must be careful that we do not try to take an even root of a negative number. For this reason, we will assume that all variables appearing under a radical sign represent nonnegative numbers.

EXAMPLE 3   Assume all variables represent nonnegative numbers and simplify each expression as much as possible.

a. $\sqrt{25a^4b^6} = 5a^2b^3$ because $(5a^2b^3)^2 = 25a^4b^6$.

b. $\sqrt[3]{x^6y^{12}} = x^2y^4$ because $(x^2y^4)^3 = x^6y^{12}$.

c. $\sqrt[4]{81r^8s^{20}} = 3r^2s^5$ because $(3r^2s^5)^4 = 81r^8s^{20}$. ■

2. Simplify, if possible.

a. $\sqrt[3]{-64}$

b. $\sqrt{-25}$

c. $-\sqrt{4}$

d. $\sqrt[5]{-1}$

e. $\sqrt[4]{-16}$

3. Simplify.

a. $\sqrt{81a^4b^8}$

b. $\sqrt[3]{8x^3y^9}$

c. $\sqrt[4]{81a^4b^8}$

Answers

2. a. -4 **b.** Not a real number

c. -2 **d.** -1

e. Not a real number

3. a. $9a^2b^4$ **b.** $2xy^3$ **c.** $3ab^2$

B Rational Numbers as Exponents

Next we develop a second kind of notation involving exponents that will allow us to designate square roots, cube roots, and so on in another way.

Consider the equation $x = 8^{1/3}$. Although we have not encountered fractional exponents before, let's assume that all the properties of exponents hold in this case. Cubing both sides of the equation, we have

$$x^3 = (8^{1/3})^3$$

$$x^3 = 8^{(1/3)(3)}$$

$$x^3 = 8^1$$

$$x^3 = 8$$

The last line tells us that x is the number whose cube is 8. It must be true, then, that x is the cube root of 8, $x = \sqrt[3]{8}$. Since we started with $x = 8^{1/3}$, it follows that

$$8^{1/3} = \sqrt[3]{8}$$

It seems reasonable, then, to define fractional exponents as indicating roots. Here is the formal definition.

Definition

For **fractional exponents**, if x is a real number and n is a positive integer greater than 1, then

$$x^{1/n} = \sqrt[n]{x} \qquad (x \geq 0 \text{ when } n \text{ is even})$$

In words: The quantity $x^{1/n}$ is the nth root of x.

With this definition we have a way of representing roots with exponents. Here are some examples.

EXAMPLE 4   Write each expression as a root and then simplify, if possible.

a. $8^{1/3} = \sqrt[3]{8} = 2$

b. $36^{1/2} = \sqrt{36} = 6$

c. $-25^{1/2} = -\sqrt{25} = -5$

d. $(-25)^{1/2} = \sqrt{-25}$, which is not a real number

e. $\left(\dfrac{4}{9}\right)^{1/2} = \sqrt{\dfrac{4}{9}} = \dfrac{2}{3}$ ∎

The properties of exponents developed in a previous chapter were applied to integer exponents only. We will now extend these properties to include rational exponents also. We do so without proof.

4. Simplify.

  a. $9^{1/2}$

  b. $27^{1/3}$

  c. $-49^{1/2}$

  d. $(-49)^{1/2}$

  e. $\left(\dfrac{16}{25}\right)^{1/2}$

Property of Exponents

If a and b are real numbers and r and s are rational numbers, and a and b are nonnegative whenever r and s indicate even roots, then

1. $a^r \cdot a^s = a^{r+s}$

4. $a^{-r} = \dfrac{1}{a^r}$ $\quad (a \neq 0)$

2. $(a^r)^s = a^{rs}$

5. $\left(\dfrac{a}{b}\right)^r = \dfrac{a^r}{b^r}$ $\quad (b \neq 0)$

3. $(ab)^r = a^r b^r$

6. $\dfrac{a^r}{a^s} = a^{r-s}$ $\quad (a \neq 0)$

There are times when rational exponents can simplify our work with radicals. Here are Examples 3b and 3c again, but this time we will work them using rational exponents.

EXAMPLE 5 Write the radical with a rational exponent then simplify.

SOLUTION $\sqrt[3]{x^6 y^{12}} = (x^6 y^{12})^{1/3}$

$= (x^6)^{1/3} (y^{12})^{1/3}$

$= x^2 y^4$ ∎

EXAMPLE 6 Write the radical with a rational exponent then simplify.

SOLUTION $\sqrt[4]{81 r^8 s^{20}} = (81 r^8 s^{20})^{1/4}$

$= 81^{1/4} (r^8)^{1/4} (s^{20})^{1/4}$

$= 3 r^2 s^5$ ∎

So far, the numerators of all the rational exponents we have encountered have been 1. The next theorem extends the work we can do with rational exponents to rational exponents with numerators other than 1.

The Rational Exponent Theorem

If a is a nonnegative real number, m is an integer, and n is a positive integer, then
$$a^{m/n} = (a^{1/n})^m = (a^m)^{1/n}$$

PROOF We can prove this theorem using the properties of exponents. Since $\dfrac{m}{n} = m\left(\dfrac{1}{n}\right)$ we have

$a^{m/n} = a^{m(1/n)}$ $\qquad$ $a^{m/n} = a^{(1/n)(m)}$

$= (a^m)^{1/n}$ $\qquad\quad$ $= (a^{1/n})^m$

Here are some examples that illustrate how we use this theorem.

Simplify.

5. $\sqrt[3]{8 x^3 y^9}$

6. $\sqrt[4]{81 a^4 b^8}$

NOTE
Unless we state otherwise, assume all variables throughout this section and problem set represent nonnegative numbers.

Answers
5. $2xy^3$
6. $3ab^2$

Simplify.

7. $9^{3/2}$

8. $16^{3/4}$

9. $8^{-2/3}$

10. $\left(\dfrac{16}{81}\right)^{-3/4}$

11. $x^{1/2} \cdot x^{1/4}$

12. $(y^{3/5})^{5/6}$

EXAMPLE 7   Simplify as much as possible.

SOLUTION $8^{2/3} = (8^{1/3})^2$ Rational exponent theorem

$\qquad\qquad\quad = 2^2$ Definition of fractional exponents

$\qquad\qquad\quad = 4$ The square of 2 is 4.   ■

EXAMPLE 8   Simplify as much as possible.

SOLUTION $25^{3/2} = (25^{1/2})^3$ Rational exponent theorem

$\qquad\qquad\quad = 5^3$ Definition of fractional exponents

$\qquad\qquad\quad = 125$ The cube of 5 is 125.   ■

EXAMPLE 9   Simplify as much as possible.

SOLUTION $9^{-3/2} = (9^{1/2})^{-3}$ Rational exponent theorem

$\qquad\qquad\quad = 3^{-3}$ Definition of fractional exponents

$\qquad\qquad\quad = \dfrac{1}{3^3}$ Negative exponent property

$\qquad\qquad\quad = \dfrac{1}{27}$ The cube of 3 is 27.   ■

EXAMPLE 10   Simplify as much as possible.

SOLUTION $\left(\dfrac{27}{8}\right)^{-4/3} = \left[\left(\dfrac{27}{8}\right)^{1/3}\right]^{-4}$ Rational exponent theorem

$\qquad\qquad\quad = \left(\dfrac{3}{2}\right)^{-4}$ Definition of fractional exponents

$\qquad\qquad\quad = \left(\dfrac{2}{3}\right)^{4}$ Negative exponent property

$\qquad\qquad\quad = \dfrac{16}{81}$ The fourth power of $\frac{2}{3}$ is $\frac{16}{81}$.   ■

EXAMPLE 11   Assume the variables represent positive quantities and simplify as much as possible.

SOLUTION $x^{1/3} \cdot x^{5/6} = x^{1/3+5/6}$ Product property for exponents

$\qquad\qquad\quad = x^{2/6+5/6}$ LCD is 6

$\qquad\qquad\quad = x^{7/6}$ Add fractions.   ■

EXAMPLE 12   Assume the variables represent positive quantities and simplify as much as possible.

SOLUTION $(y^{2/3})^{3/4} = y^{(2/3)(3/4)}$ Power property for exponents

$\qquad\qquad\quad = y^{1/2}$ Multiply fractions: $\frac{2}{3} \cdot \frac{3}{4} = \frac{6}{12} = \frac{1}{2}$.   ■

Answers

7. 27

8. 8

9. $\dfrac{1}{4}$

10. $\dfrac{27}{8}$

11. $x^{3/4}$

12. $y^{1/2}$

EXAMPLE 13   Assume the variables represent positive quantities and simplify as much as possible.

SOLUTION $\dfrac{z^{1/3}}{z^{1/4}} = z^{1/3-1/4}$ Quotient property for exponents

$= z^{4/12-3/12}$ LCD is 12.

$= z^{1/12}$ Subtract fractions.   ■

EXAMPLE 14   Assume the variables represent positive quantities and simplify as much as possible.

SOLUTION $\dfrac{(x^{-3}y^{1/2})^4}{x^{10}y^{3/2}} = \dfrac{(x^{-3})^4(y^{1/2})^4}{x^{10}y^{3/2}}$ Distributive property for exponents

$= \dfrac{x^{-12}y^2}{x^{10}y^{3/2}}$ Power property for exponents

$= x^{-22}y^{1/2}$ Quotient property for exponents

$= \dfrac{y^{1/2}}{x^{22}}$ Negative exponent property   ■

FACTS FROM GEOMETRY   The Pythagorean Theorem and Square Roots

Now that we have had some experience working with square roots, we can rewrite the Pythagorean theorem using a square root. If triangle ABC is a right triangle with $C = 90°$, then the length of the longest side is the *positive square root* of the sum of the squares of the other two sides (see Figure 3).

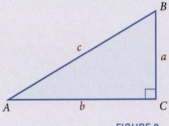

$c = \sqrt{a^2 + b^2}$

FIGURE 3

In the introduction to this section, we mentioned the golden rectangle. Its origins can be traced back more than 2,000 years to the Greek civilization that produced Pythagoras, Socrates, Plato, Aristotle, and Euclid. The most important mathematical work to come from that Greek civilization was Euclid's *Elements*, an elegantly written summary of all that was known about geometry at that time in history. Euclid's *Elements*, according to Howard Eves, an authority on the history of mathematics, exercised a greater influence on scientific thinking than any other work. Here is how we construct a golden rectangle from a square of side 2, using the same method that Euclid used in his *Elements*.

Assume the variables represent positive quantities and simplify as much as possible.

13. $\dfrac{z^{3/4}}{z^{2/3}}$

14. $\dfrac{(x^{1/3}y^{-3})^6}{x^4y^{10}}$

Answers

13. $z^{1/12}$

14. $\dfrac{1}{x^2y^{28}}$

FACTS FROM GEOMETRY   The Golden Rectangle

Step 1: Draw a square with a side of length two. Connect the midpoint of side
CD to corner B. (Note that we have labeled the midpoint of segment CD
with the letter O.)

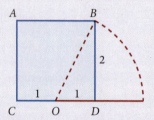

Step 2: Drop the diagonal OB from step 1 down so it aligns with side CD.

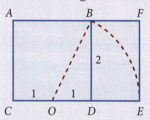

Step 3: Form rectangle ACEF. This is a golden rectangle.

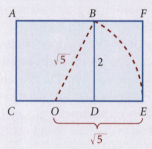

   All golden rectangles are constructed from squares. Every golden rectangle, no
matter how large or small it is, will have the same shape. To associate a number
with the shape of the golden rectangle, we use the ratio of its length to its width.
This ratio is called the *golden ratio*. To calculate the golden ratio, we must first find
the length of the diagonal we used to construct the golden rectangle. Figure 4
shows the golden rectangle that we constructed from a square of side 2. The
length of the diagonal OB is found by applying the Pythagorean theorem to
triangle OBD.

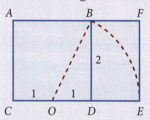

FIGURE 4

The length of segment OE is equal to the length of diagonal OB; both are $\sqrt{5}$.
Since the distance from C to O is 1, the length CE of the golden rectangle is
$1 + \sqrt{5}$. Now we can find the golden ratio:

$$\text{Golden ratio} = \frac{\text{length}}{\text{width}} = \frac{CE}{EF} = \frac{1 + \sqrt{5}}{2}$$

Problem Set 7.1

1. What is mathematical intuition and how is it important?
2. What does practice and repetition have to do with mathematics?

A Find each of the following roots, if possible. [Examples 1, 2]

1. $\sqrt{144}$

2. $-\sqrt{144}$

3. $\sqrt{-144}$

4. $\sqrt{-49}$

5. $-\sqrt{49}$

6. $\sqrt{49}$

7. $\sqrt[3]{-27}$

8. $-\sqrt[3]{27}$

9. $\sqrt[4]{16}$

10. $-\sqrt[4]{16}$

11. $\sqrt[4]{-16}$

12. $-\sqrt[4]{-16}$

13. $\sqrt{0.04}$

14. $\sqrt{0.81}$

15. $\sqrt[3]{0.008}$

16. $\sqrt[3]{0.125}$

17. $\sqrt[3]{125}$

18. $\sqrt[3]{-125}$

19. $-\sqrt[3]{216}$

20. $-\sqrt[3]{-216}$

21. $\sqrt{\dfrac{1}{36}}$

22. $\sqrt{\dfrac{9}{25}}$

23. $\sqrt[3]{\dfrac{8}{125}}$

24. $\sqrt[3]{-\dfrac{27}{216}}$

A Simplify each expression. Assume all variables represent nonnegative numbers. [Example 3]

25. $\sqrt{36a^8}$

26. $\sqrt{49a^{10}}$

27. $\sqrt[3]{27a^{12}}$

28. $\sqrt[3]{8a^{15}}$

29. $\sqrt[5]{32x^{10}y^5}$

30. $\sqrt[5]{32x^5y^{10}}$

31. $\sqrt[4]{16a^{12}b^{20}}$

32. $\sqrt[4]{81a^{24}b^8}$

B Simplify each expression. [Examples 4–6]

33. $36^{1/2}$

34. $49^{1/2}$

35. $-9^{1/2}$

36. $-16^{1/2}$

37. $8^{1/3}$

38. $-8^{1/3}$

39. $(-8)^{1/3}$

40. $-27^{1/3}$

41. $32^{1/5}$

42. $81^{1/4}$

43. $\left(\dfrac{81}{25}\right)^{1/2}$

44. $\left(\dfrac{64}{125}\right)^{1/3}$

B Use the rational exponent theorem to simplify each of the following as much as possible. [Examples 7, 8]

45. $27^{2/3}$

46. $8^{4/3}$

47. $25^{3/2}$

48. $81^{3/4}$

B Simplify each expression. Remember, negative exponents give reciprocals. [Examples 9, 10]

49. $27^{-1/3}$

50. $9^{-1/2}$

51. $81^{-3/4}$

52. $4^{-3/2}$

53. $\left(\dfrac{25}{36}\right)^{-1/2}$

54. $\left(\dfrac{16}{49}\right)^{-1/2}$

55. $\left(\dfrac{81}{16}\right)^{-3/4}$

56. $\left(\dfrac{27}{8}\right)^{-2/3}$

57. $16^{1/2} + 27^{1/3}$

58. $25^{1/2} + 100^{1/2}$

59. $8^{-2/3} + 4^{-1/2}$

60. $49^{-1/2} + 25^{-1/2}$

B Use the properties of exponents to simplify each of the following as much as possible. Assume all bases are positive. [Examples 11–14]

61. $x^{3/5} \cdot x^{1/5}$

62. $x^{3/4} \cdot x^{5/4}$

63. $(a^{3/4})^{4/3}$

64. $(a^{2/3})^{3/4}$

65. $\dfrac{x^{1/5}}{x^{3/5}}$

66. $\dfrac{x^{2/7}}{x^{5/7}}$

67. $\dfrac{x^{5/6}}{x^{2/3}}$

68. $\dfrac{x^{7/8}}{x^{8/7}}$

69. $(x^{3/5}y^{5/6}z^{1/3})^{3/5}$

70. $(x^{3/4}y^{1/8}z^{5/6})^{4/5}$

71. $\dfrac{a^{3/4}b^2}{a^{7/8}b^{1/4}}$

72. $\dfrac{a^{1/3}b^4}{a^{3/5}b^{1/3}}$

73. $\dfrac{(y^{2/3})^{3/4}}{(y^{1/3})^{3/5}}$

74. $\dfrac{(y^{5/4})^{2/5}}{(y^{1/4})^{4/3}}$

75. $\left(\dfrac{a^{-1/4}}{b^{1/2}}\right)^8$

76. $\left(\dfrac{a^{-1/5}}{b^{1/3}}\right)^{15}$

77. $\dfrac{(r^{-2}s^{1/3})^6}{r^8s^{3/2}}$

78. $\dfrac{(r^{-5}s^{1/2})^4}{r^{12}s^{5/2}}$

79. $\dfrac{(25a^6b^4)^{1/2}}{(8a^{-9}b^3)^{-1/3}}$

80. $\dfrac{(27a^3b^6)^{1/3}}{(81a^8b^{-4})^{1/4}}$

Applying the Concepts

81. Maximum Speed The maximum speed (v) that an automobile can travel around a curve of radius r without skidding is given by the equation

$$v = \left(\dfrac{5r}{2}\right)^{1/2}$$

where v is in miles per hour and r is measured in feet. What is the maximum speed a car can travel around a curve with a radius of 250 feet without skidding?

82. Golden Ratio The golden ratio is the ratio of the length to the width in any golden rectangle. The exact value of this number is $\dfrac{1+\sqrt{5}}{2}$. Use a calculator to find a decimal approximation to this number and round it to the nearest thousandth.

83. Rubix Cube Each square on the face of a rubix cube measures 0.75 inches.

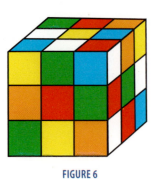

FIGURE 6

a. Find the length of the side of the face.

b. Find the length of the diagonal of the face. Write your answer in inches rounded to the nearest tenth.

c. If 1 inch is 2.54×10^{-5} km, give the length of the diagonal of the face in kilometers.

84. Geometry The length of each side of the cube shown in Figure 7 is 1 inch.

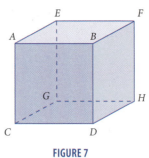

FIGURE 7

a. Find the length of the diagonal CH.

b. Find the length of the diagonal CF.

85. Comparing Graphs Identify the graph with the correct equation.

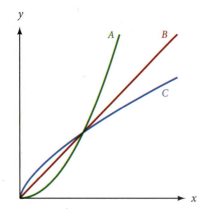

a. $y = x$

b. $y = x^2$

c. $y = x^{2/3}$

d. What are the two points of intersection of all three graphs?

86. Falling Objects The time t in seconds it takes an object to fall d feet is given by the equation

$$t = \frac{1}{4}\sqrt{d}$$

a. The Sears Tower in Chicago is 1,450 feet tall. How long would it take a penny to fall to the ground from the top of the Sears Tower? Round to the nearest hundreth of a second.

b. An object took 30 seconds to fall to the ground. From what distance must it have been dropped?

Maintaining Your Skills

87. $x^2(x^4 - x)$

88. $5x^2(2x^3 - x)$

89. $(x - 3)(x + 5)$

90. $(x - 2)(x + 2)$

91. $(x^2 - 5)^2$

92. $(x^2 + 5)^2$

93. $(x - 3)(x^2 + 3x + 9)$

94. $(x + 3)(x^2 - 3x + 9)$

Getting Ready for the Next Section

Simplify.

95. $x^2(x^4 - x^3)$

96. $(x^2 - 3)(x^2 + 5)$

97. $(3a - 2b)(4a - b)$

98. $(x^2 + 3)^2$

99. $(x^3 - 2)(x^3 + 2)$

100. $(a - b)(a^2 + ab + b^2)$

101. $\dfrac{15x^2y - 20x^4y^2}{5xy}$

102. $\dfrac{12a^3b^2 - 24a^2b^4}{3ab}$

Factor.

103. $x^2 - 3x - 10$

104. $x^2 + x - 12$

105. $6x^2 + 11x - 10$

106. $10x^2 - x - 3$

Use rules of exponents to simplify.

107. $x^{2/3} \cdot x^{4/3}$

108. $x^{1/4} \cdot x^{3/4}$

109. $(t^{1/2})^2$

110. $(x^{3/2})^2$

111. $\dfrac{x^{2/3}}{x^{1/3}}$

112. $\dfrac{x^{1/2}}{x^{1/2}}$

Extending the Concepts

113. Show that the expression $(a^{1/2} + b^{1/2})^2$ is not equal to $a + b$ by replacing a with 9 and b with 4 in both expressions and then simplifying each.

114. Show that the statement $(a^2 + b^2)^{1/2} = a + b$ is not, in general, true by replacing a with 3 and b with 4 and then simplifying both sides.

115. You may have noticed, if you have been using a calculator to find roots, that you can find the fourth root of a number by pressing the square root button twice. Written in symbols, this fact looks like this:

$$\sqrt{\sqrt{a}} = \sqrt[4]{a} \qquad (a \geq 0)$$

Show that this statement is true by rewriting each side with exponents instead of radical notation and then simplifying the left side.

116. Show that the following statement is true by rewriting each side with exponents instead of radical notation and then simplifying the left side.

$$\sqrt[3]{\sqrt{a}} = \sqrt[6]{a} \qquad (a \geq 0)$$

More Expressions Involving Rational Exponents

7.2

OBJECTIVES

A Multiply expressions with rational exponents.

B Divide expressions with rational exponents.

C Factor expressions with rational exponents.

D Add and subtract expressions with rational exponents.

TICKET TO SUCCESS

Keep these questions in mind as you read through the section. Then respond in your own words and in complete sentences.

1. When multiplying expressions with fractional exponents, when do we add the fractional exponents?
2. When can you use the FOIL method with expressions that contain rational exponents?
3. Explain how division with rational exponents is similar to division of a polynomial by a monomial.
4. What is the first thing you do when adding expressions that contain rational exponents but have different denominators?

© Lana Sundman / Alamy

Suppose you purchased 10 sets of baseball cards for $21 each, for a total investment of $210. Three years later, you find that each set is worth $30, which means that your 10 sets have a total value of $300.

You can calculate the annual rate of return on this investment using a formula that involves rational exponents. The annual rate of return will tell you at what interest rate you would have to invest your original $210 for it to be worth $300 three years later. As you will see at the end of this section, the annual rate of return on this investment is 12.6%, which is a good return on your money.

In this section, we will look at multiplication, division, factoring, and simplification of some expressions that resemble polynomials but contain rational exponents. The problems in this section will be of particular interest to you if you are planning to take either an engineering calculus class or a business calculus class. As was the case in the previous section, we will assume all variables represent nonnegative real numbers. That way, we will not have to worry about the possibility of introducing undefined terms—even roots of negative numbers—into any of our examples. Let's begin this section with a look at multiplication of expressions containing rational exponents.

A Multiplication with Rational Exponents

EXAMPLE 1 Multiply $x^{2/3}(x^{4/3} - x^{1/3})$.

SOLUTION Applying the distributive property and then simplifying the resulting terms, we have

$$x^{2/3}(x^{4/3} - x^{1/3}) = x^{2/3} \cdot x^{4/3} - x^{2/3} \cdot x^{1/3} \quad \text{\color{red}Distributive property}$$

$$= x^{6/3} - x^{3/3} \quad \text{\color{red}Add exponents.}$$

$$= x^2 - x \quad \text{\color{red}Simplify.} \qquad \blacksquare$$

EXAMPLE 2 Multiply $(x^{2/3} - 3)(x^{2/3} + 5)$.

SOLUTION Applying the FOIL method, we multiply as if we were multiplying two binomials.

$$(x^{2/3} - 3)(x^{2/3} + 5) = x^{2/3} \cdot x^{2/3} + 5x^{2/3} - 3x^{2/3} - 15$$

$$= x^{4/3} + 2x^{2/3} - 15 \qquad \blacksquare$$

EXAMPLE 3 Multiply $(3a^{1/3} - 2b^{1/3})(4a^{1/3} - b^{1/3})$.

SOLUTION Again, we use the FOIL method to multiply.

$$(3a^{1/3} - 2b^{1/3})(4a^{1/3} - b^{1/3})$$

$$= 3a^{1/3} \cdot 4a^{1/3} - 3a^{1/3} \cdot b^{1/3} - 2b^{1/3} \cdot 4a^{1/3} + 2b^{1/3} \cdot b^{1/3}$$

$$= 12a^{2/3} - 11a^{1/3}b^{1/3} + 2b^{2/3} \qquad \blacksquare$$

EXAMPLE 4 Expand $(t^{1/2} - 5)^2$.

SOLUTION We can use the definition of exponents and the FOIL method.

$$(t^{1/2} - 5)^2 = (t^{1/2} - 5)(t^{1/2} - 5)$$

$$= t^{1/2} \cdot t^{1/2} - 5t^{1/2} - 5t^{1/2} + 25$$

$$= t - 10t^{1/2} + 25$$

We can obtain the same result by using the formula for the square of a binomial, $(a - b)^2 = a^2 - 2ab + b^2$.

$$(t^{1/2} - 5)^2 = (t^{1/2})^2 - 2t^{1/2} \cdot 5 + 5^2$$

$$= t - 10t^{1/2} + 25 \qquad \blacksquare$$

EXAMPLE 5 Multiply $(x^{3/2} - 2^{3/2})(x^{3/2} + 2^{3/2})$.

SOLUTION This product has the form $(a - b)(a + b)$, which will result in the difference of two squares, $a^2 - b^2$.

$$(x^{3/2} - 2^{3/2})(x^{3/2} + 2^{3/2}) = (x^{3/2})^2 - (2^{3/2})^2$$

$$= x^3 - 2^3$$

$$= x^3 - 8 \qquad \blacksquare$$

EXAMPLE 6 Multiply $(a^{1/3} - b^{1/3})(a^{2/3} + a^{1/3}b^{1/3} + b^{2/3})$.

SOLUTION We can find this product by multiplying in columns.

$$
\begin{array}{r}
a^{2/3} + a^{1/3}b^{1/3} + b^{2/3} \\
a^{1/3} - b^{1/3} \\
\hline
a \quad + a^{2/3}b^{1/3} + a^{1/3}b^{2/3} \\
\quad - a^{2/3}b^{1/3} - a^{1/3}b^{2/3} - b \\
\hline
a \qquad\qquad\qquad - b
\end{array}
$$

The product is $a - b$. ∎

6. Multiply.
$(a^{1/3} + b^{1/3})(a^{2/3} - a^{1/3}b^{1/3} + b^{2/3})$

B Division with Rational Exponents

Our next example involves division with expressions that contain rational exponents. As you will see, this kind of division is very similar to division of a polynomial by a monomial.

NOTE
For Example 7 assume x and y are positive numbers.

EXAMPLE 7 Divide $\dfrac{15x^{2/3}y^{1/3} - 20x^{4/3}y^{2/3}}{5x^{1/3}y^{1/3}}$.

SOLUTION We can approach this problem in the same way we approached division by a monomial. We simply divide each term in the numerator by the term in the denominator.

$$
\frac{15x^{2/3}y^{1/3} - 20x^{4/3}y^{2/3}}{5x^{1/3}y^{1/3}} = \frac{15x^{2/3}y^{1/3}}{5x^{1/3}y^{1/3}} - \frac{20x^{4/3}y^{2/3}}{5x^{1/3}y^{1/3}}
$$
$$
= 3x^{1/3} - 4xy^{1/3} \quad ∎
$$

7. Divide $\dfrac{36x^{3/4}y^{1/4} - 18x^{5/4}y^{1/4}}{6x^{1/4}y^{1/4}}$.

C Factoring

The next three examples involve factoring. In the first example, we are told what to factor from each term of an expression.

EXAMPLE 8 Factor $3(x - 2)^{1/3}$ from $12(x - 2)^{4/3} - 9(x - 2)^{1/3}$, and then simplify, if possible.

SOLUTION This solution is similar to factoring out the greatest common factor.

$$
12(x - 2)^{4/3} - 9(x - 2)^{1/3} = 3(x - 2)^{1/3}[4(x - 2) - 3]
$$
$$
= 3(x - 2)^{1/3}(4x - 11) \quad ∎
$$

8. Factor $2(x - 3)^{1/2}$ from $8(x - 3)^{3/2} - 6(x - 3)^{1/2}$.

Although an expression containing rational exponents is not a polynomial—remember, a polynomial must have exponents that are whole numbers—we are going to treat the expressions that follow as if they were polynomials.

EXAMPLE 9 Factor $x^{2/3} - 3x^{1/3} - 10$ as if it were a trinomial.

SOLUTION We can think of $x^{2/3} - 3x^{1/3} - 10$ as if it is a trinomial in which the variable is $x^{1/3}$. To see this, replace $x^{1/3}$ with y to get

$$
y^2 - 3y - 10
$$

9. Factor $x^{2/3} - 4x^{1/3} - 21$.

Answers
6. $a + b$
7. $6x^{1/2} - 3x$
8. $2(x - 3)^{1/2}(4x - 15)$
9. $(x^{1/3} - 7)(x^{1/3} + 3)$

Since this trinomial in y factors as $(y - 5)(y + 2)$, we can factor our original expression similarly.

$$x^{2/3} - 3x^{1/3} - 10 = (x^{1/3} - 5)(x^{1/3} + 2)$$

Remember, with factoring, we can always multiply our factors to check that we have factored correctly. ∎

10. Factor $6x^{2/3} + 19x^{1/3} + 10$.

EXAMPLE 10 Factor $6x^{2/5} + 11x^{1/5} - 10$ as if it were a trinomial.

SOLUTION We can think of the expression in question as a trinomial in $x^{1/5}$.

$$6x^{2/5} + 11x^{1/5} - 10 = (3x^{1/5} - 2)(2x^{1/5} + 5)$$ ∎

D Addition and Subtraction

In our next example, we combine two expressions by applying the methods we used to add and subtract fractions or rational expressions.

11. Subtract.

$$(x^2 - 3)^{1/2} - \dfrac{x^2}{(x^2 - 3)^{1/2}}$$

EXAMPLE 11 Subtract $(x^2 + 4)^{1/2} - \dfrac{x^2}{(x^2 + 4)^{1/2}}$.

SOLUTION To combine these two expressions, we need to find a least common denominator, change to equivalent fractions, and subtract numerators. The least common denominator is $(x^2 + 4)^{1/2}$.

$$(x^2 + 4)^{1/2} - \frac{x^2}{(x^2 + 4)^{1/2}} = \frac{(x^2 + 4)^{1/2}}{1} \cdot \frac{(x^2 + 4)^{1/2}}{(x^2 + 4)^{1/2}} - \frac{x^2}{(x^2 + 4)^{1/2}}$$

$$= \frac{x^2 + 4 - x^2}{(x^2 + 4)^{1/2}}$$

$$= \frac{4}{(x^2 + 4)^{1/2}}$$ ∎

12. Find the annual rate of return on a coin collection that was purchased for $600 and sold 3 years later for $800. Round your answer to the nearest tenth of a percent.

EXAMPLE 12 If you purchase an investment for P dollars and t years later it is worth A dollars, then the annual rate of return r on that investment is given by the formula

$$r = \left(\frac{A}{P}\right)^{1/t} - 1$$

Find the annual rate of return on the baseball card collection in the section opener purchased for $210 and sold 3 years later for $300.

SOLUTION Using $A = 300$, $P = 210$, and $t = 3$ in the formula, we have

$$r = \left(\frac{300}{210}\right)^{1/3} - 1$$

The easiest way to simplify this expression is with a calculator.

$\boxed{(}\ 300\ \boxed{\div}\ 210\ \boxed{)}\ \boxed{\wedge}\ \boxed{(}\ 1\ \boxed{\div}\ 3\ \boxed{)}\ \boxed{-}\ 1\ \boxed{=}$

Allowing three decimal places, the result is 0.126. The annual return on the baseball cards is approximately 12.6%. To do as well with a savings account, we would have to invest the original $210 in an account that paid 12.6%, compounded annually. ∎

Answers

10. $(3x^{1/3} + 2)(2x^{1/3} + 5)$

11. $\dfrac{-3}{(x^2 - 3)^{1/2}}$

12. 10.1%

Problem Set 7.2

Moving Toward Success

"There is nothing in a caterpillar that tells you it's going to be a butterfly."

—R. Buckminster Fuller, 1895–1983, American architect and engineer

1. Go online and research learning styles (e.g., visual, auditory, kinesthetic) to determine how you learn best. You may prefer one style or a combination of styles. How do you prefer to learn?

2. How can you apply your preferred learning style to this class?

A Multiply. (Assume all variables in this problem set represent nonnegative real numbers.) [Examples 1–6]

1. $x^{2/3}(x^{1/3} + x^{4/3})$

2. $x^{2/5}(x^{3/5} - x^{8/5})$

3. $a^{1/2}(a^{3/2} - a^{1/2})$

4. $a^{1/4}(a^{3/4} + a^{7/4})$

5. $2x^{1/3}(3x^{8/3} - 4x^{5/3} + 5x^{2/3})$

6. $5x^{1/2}(4x^{5/2} + 3x^{3/2} + 2x^{1/2})$

7. $4x^{1/2}y^{3/5}(3x^{3/2}y^{-3/5} - 9x^{-1/2}y^{7/5})$

8. $3x^{4/5}y^{1/3}(4x^{6/5}y^{-1/3} - 12x^{-4/5}y^{5/3})$

9. $(x^{2/3} - 4)(x^{2/3} + 2)$

10. $(x^{2/3} - 5)(x^{2/3} + 2)$

11. $(a^{1/2} - 3)(a^{1/2} - 7)$

12. $(a^{1/2} - 6)(a^{1/2} - 2)$

13. $(4y^{1/3} - 3)(5y^{1/3} + 2)$

14. $(5y^{1/3} - 2)(4y^{1/3} + 3)$

15. $(5x^{2/3} + 3y^{1/2})(2x^{2/3} + 3y^{1/2})$

16. $(4x^{2/3} - 2y^{1/2})(5x^{2/3} - 3y^{1/2})$

17. $(t^{1/2} + 5)^2$

18. $(t^{1/2} - 3)^2$

19. $(x^{3/2} + 4)^2$

20. $(x^{3/2} - 6)^2$

21. $(a^{1/2} - b^{1/2})^2$

22. $(a^{1/2} + b^{1/2})^2$

23. $(2x^{1/2} - 3y^{1/2})^2$

24. $(5x^{1/2} + 4y^{1/2})^2$

25. $(a^{1/2} - 3^{1/2})(a^{1/2} + 3^{1/2})$

26. $(a^{1/2} - 5^{1/2})(a^{1/2} + 5^{1/2})$

27. $(x^{3/2} + y^{3/2})(x^{3/2} - y^{3/2})$

28. $(x^{5/2} + y^{5/2})(x^{5/2} - y^{5/2})$

29. $(t^{1/2} - 2^{3/2})(t^{1/2} + 2^{3/2})$

30. $(t^{1/2} - 5^{3/2})(t^{1/2} + 5^{3/2})$

31. $(2x^{3/2} + 3^{1/2})(2x^{3/2} - 3^{1/2})$

32. $(3x^{1/2} + 2^{3/2})(3x^{1/2} - 2^{3/2})$

33. $(x^{1/3} + y^{1/3})(x^{2/3} - x^{1/3}y^{1/3} + y^{2/3})$

34. $(x^{1/3} - y^{1/3})(x^{2/3} + x^{1/3}y^{1/3} + y^{2/3})$

35. $(a^{1/3} - 2)(a^{2/3} + 2a^{1/3} + 4)$

36. $(a^{1/3} + 3)(a^{2/3} - 3a^{1/3} + 9)$

37. $(2x^{1/3} + 1)(4x^{2/3} - 2x^{1/3} + 1)$

38. $(3x^{1/3} - 1)(9x^{2/3} + 3x^{1/3} + 1)$

39. $(t^{1/4} - 1)(t^{1/4} + 1)(t^{1/2} + 1)$

40. $(t^{1/4} - 2)(t^{1/4} + 2)(t^{1/2} + 4)$

B Divide. [Example 7]

41. $\dfrac{18x^{3/4} + 27x^{1/4}}{9x^{1/4}}$

42. $\dfrac{25x^{1/4} + 30x^{3/4}}{5x^{1/4}}$

43. $\dfrac{12x^{2/3}y^{1/3} - 16x^{1/3}y^{2/3}}{4x^{1/3}y^{1/3}}$

44. $\dfrac{12x^{4/3}y^{1/3} - 18x^{1/3}y^{4/3}}{6x^{1/3}y^{1/3}}$

45. $\dfrac{21a^{7/5}b^{3/5} - 14a^{2/5}b^{8/5}}{7a^{2/5}b^{3/5}}$

46. $\dfrac{24a^{9/5}b^{3/5} - 16a^{4/5}b^{8/5}}{8a^{4/5}b^{3/5}}$

C [Example 8]

47. Factor $3(x - 2)^{1/2}$ from $12(x - 2)^{3/2} - 9(x - 2)^{1/2}$.

48. Factor $4(x + 1)^{1/3}$ from $4(x + 1)^{4/3} + 8(x + 1)^{1/3}$.

49. Factor $5(x - 3)^{7/5}$ from $5(x - 3)^{12/5} - 15(x - 3)^{7/5}$.

50. Factor $6(x + 3)^{8/7}$ from $6(x + 3)^{15/7} - 12(x + 3)^{8/7}$.

51. Factor $3(x + 1)^{1/2}$ from $9x(x + 1)^{3/2} + 6(x + 1)^{1/2}$.

52. Factor $4x(x + 1)^{1/2}$ from $4x^2(x + 1)^{1/2} + 8x(x + 1)^{3/2}$.

C Factor each of the following as if it were a trinomial. [Examples 9–10]

53. $x^{2/3} - 5x^{1/3} + 6$ **54.** $x^{2/3} - x^{1/3} - 6$ **55.** $a^{2/5} - 2a^{1/5} - 8$ **56.** $a^{2/5} + 2a^{1/5} - 8$

57. $2y^{2/3} - 5y^{1/3} - 3$ **58.** $3y^{2/3} + 5y^{1/3} - 2$ **59.** $9t^{2/5} - 25$ **60.** $16t^{2/5} - 49$

61. $4x^{2/7} + 20x^{1/7} + 25$ **62.** $25x^{2/7} - 20x^{1/7} + 4$

Evaluate the following functions for the given value.

63. If $f(x) = x - 2\sqrt{x} - 8$ find $f(4)$.   **64.** If $g(x) = x - 2\sqrt{x} - 3$ find $g(9)$.

65. If $f(x) = 2x + 9\sqrt{x} - 5$ find $f(25)$.   **66.** If $f(t) = t - 2\sqrt{t} - 15$ find $f(9)$.

67. If $g(x) = 2x - \sqrt{x} - 6$ find $g\left(\dfrac{9}{4}\right)$.   **68.** If $f(x) = 2x + \sqrt{x} - 15$ find $f(9)$.

69. If $f(x) = x^{2/3} - 2x^{1/3} - 8$ find $f(-8)$.   **70.** If $g(x) = x^{2/3} + 4x^{1/3} - 12$ find $g(8)$.

D Simplify each of the following to a single fraction. [Example 11]

71. $\dfrac{3}{x^{1/2}} + x^{1/2}$ **72.** $\dfrac{2}{x^{1/2}} - x^{1/2}$ **73.** $x^{2/3} + \dfrac{5}{x^{1/3}}$ **74.** $x^{3/4} - \dfrac{7}{x^{1/4}}$

75. $\dfrac{3x^2}{(x^3 + 1)^{1/2}} + (x^3 + 1)^{1/2}$ **76.** $\dfrac{x^3}{(x^2 - 1)^{1/2}} + 2x(x^2 - 1)^{1/2}$ **77.** $\dfrac{x^2}{(x^2 + 4)^{1/2}} - (x^2 + 4)^{1/2}$ **78.** $\dfrac{x^5}{(x^2 - 2)^{1/2}} + 4x^3(x^2 - 2)^{1/2}$

Applying the Concepts

Round answers to the nearest tenth of a percent. [Example 12]

79. Investing A coin collection is purchased as an investment for $500 and sold 4 years later for $900. Find the annual rate of return on the investment.

80. Investing An investor buys stock in a company for $800. Five years later, the same stock is worth $1,600. Find the annual rate of return on the stocks.

81. Investing If you bought a set of trading cards for $50 and two years later the set is worth $65, what is the annual rate of return?

82. Investing If you bought a set of collectible baseball cards for $150 and five years later the set is worth $225, what is the annual rate of return?

Maintaining Your Skills

Reduce to lowest terms.

83. $\dfrac{x^2 - 9}{x^4 - 81}$

84. $\dfrac{6 - a - a^2}{3 - 2a - a^2}$

Divide.

85. $\dfrac{15x^2y - 20x^4y^2}{5xy}$

86. $\dfrac{12x^3y^2 - 24x^2y^3}{6xy}$

Divide using long division.

87. $\dfrac{10x^2 + 7x - 12}{2x + 3}$

88. $\dfrac{6x^2 - x - 35}{2x - 5}$

89. $\dfrac{x^3 - 125}{x - 5}$

90. $\dfrac{x^3 + 64}{x + 4}$

Getting Ready for the Next Section

91. $\sqrt{25}$

92. $\sqrt{4}$

93. $\sqrt{6^2}$

94. $\sqrt{3^2}$

95. $\sqrt{16x^4y^2}$

96. $\sqrt{4x^6y^8}$

97. $\sqrt{(5y)^2}$

98. $\sqrt{(8x^3)^2}$

99. $\sqrt[3]{27}$

100. $\sqrt[3]{-8}$

101. $\sqrt[3]{2^3}$

102. $\sqrt[3]{(-5)^3}$

103. $\sqrt[3]{8a^3b^3}$

104. $\sqrt[3]{64a^6b^3}$

Fill in the blank.

105. $50 = \underline{\hspace{1.5cm}} \cdot 2$

106. $12 = \underline{\hspace{1.5cm}} \cdot 3$

107. $48x^4y^3 = \underline{\hspace{1.5cm}} \cdot y$

108. $40a^5b^4 = \underline{\hspace{1.5cm}} \cdot 5a^2b$

109. $12x^7y^6 = \underline{\hspace{1.5cm}} \cdot 3x$

110. $54a^6b^2c^4 = \underline{\hspace{1.5cm}} \cdot 2b^2c$

Simplified Form for Radicals

TICKET TO SUCCESS

Keep these questions in mind as you read through the section. Then respond in your own words and in complete sentences.

1. Explain why the product property for radicals cannot be applied to the square root of a sum.
2. What is simplified form for an expression that contains a square root?
3. Why is it not necessarily true that $\sqrt{a^2} = a$?
4. What does it mean to rationalize the denominator in an expression?

The Burj Khalifa is a skyscraper in Dubai that contains the world's highest outdoor observation deck. The deck stands at an amazing 0.28 miles (1,483 feet) above the ground. Dubai is also known for its manmade palm-shaped islands that are the foundation for residential and commercial developments. Palm Jebel Ali, the largest of the islands, is approximately 22 miles from the Burj Khalifa.

Imagine visiting the observation deck of the Burj Khalifa. Would you be able to see Palm Jebel Ali from the observation deck? To answer this question, you must use the the distance formula

$$d = \sqrt{8000k + k^2}$$

where k is the number of miles above the earth's surface and d is the distance you can see from that point. In this section, we will work with expressions and formulas that contain radical notation, such as the distance formula above. Let's begin by stating the first two properties of radicals and provide a simplified form for radical expressions. Then, when you revisit this problem in the problem set, you will be able to solve it.

NOTE

There is no property for radicals that says the nth root of a sum is the sum of the nth roots; that is, in general,

$$\sqrt[n]{a + b} \neq \sqrt[n]{a} + \sqrt[n]{b}$$

A Simplified Form

We will now look at the first two properties of radicals. For these two properties, we will assume a and b are nonnegative real numbers whenever n is an even number.

Product Property for Radicals

$$\sqrt[n]{ab} = \sqrt[n]{a}\,\sqrt[n]{b}$$

In words: The nth root of a product is the product of the nth roots.

PROOF OF PRODUCT PROPERTY FOR RADICALS

$$\sqrt[n]{ab} = (ab)^{1/n} \qquad \text{Definition of fractional exponents}$$
$$= a^{1/n}b^{1/n} \qquad \text{Exponents distribute over products.}$$
$$= \sqrt[n]{a}\,\sqrt[n]{b} \qquad \text{Definition of fractional exponents}$$

Quotient Property for Radicals

$$\sqrt[n]{\frac{a}{b}} = \frac{\sqrt[n]{a}}{\sqrt[n]{b}} \qquad (b \neq 0)$$

In words: The nth root of a quotient is the quotient of the nth roots.

The proof of the quotient property is similar to the proof of the product property.

The two properties of radicals allow us to change the form of and simplify radical expressions without changing their value.

Simplified Form for Radical Expressions

A radical expression is in *simplified form* if

1. None of the factors of the radicand (the quantity under the radical sign) can be written as powers greater than or equal to the index—that is, no perfect squares can be factors of the quantity under a square root sign, no perfect cubes can be factors of what is under a cube root sign, and so forth;

2. There are no fractions under the radical sign; and

3. There are no radicals in the denominator.

Satisfying the first condition for simplified form actually amounts to taking as much out from under the radical sign as possible. The following examples illustrate the first condition for simplified form.

EXAMPLE 1 Write $\sqrt{50}$ in simplified form.

SOLUTION The largest perfect square that divides 50 is 25. We write 50 as $25 \cdot 2$ and apply the product property for radicals.

$$\sqrt{50} = \sqrt{25 \cdot 2} \qquad 50 = 25 \cdot 2$$
$$= \sqrt{25}\,\sqrt{2} \qquad \text{Product property}$$
$$= 5\sqrt{2} \qquad \sqrt{25} = 5$$

We have taken as much as possible out from under the radical sign—in this case, factoring 25 from 50 and then writing $\sqrt{25}$ as 5.

NOTE

Writing a radical expression in simplified form does not always result in a simpler-looking expression. Simplified form for radicals is a way of writing radicals so they are easiest to work with.

PRACTICE PROBLEMS

1. Write $\sqrt{18}$ in simplified form.

Answer

1. $3\sqrt{2}$

EXAMPLE 2 Write $\sqrt{48x^4y^3}$, where $x, y \geq 0$ in simplified form.

SOLUTION The largest perfect square that is a factor of the radicand is $16x^4y^2$. Applying the product property again, we have

$$\sqrt{48x^4y^3} = \sqrt{16x^4y^2 \cdot 3y}$$

$$= \sqrt{16x^4y^2} \sqrt{3y}$$

$$= 4x^2y\sqrt{3y} \qquad \blacksquare$$

EXAMPLE 3 Write $\sqrt[3]{40a^5b^4}$ in simplified form.

SOLUTION We now want to factor the largest perfect cube from the radicand. We write $40a^5b^4$ as $8a^3b^3 \cdot 5a^2b$ and proceed as we did in Examples 1 and 2.

$$\sqrt[3]{40a^5b^4} = \sqrt[3]{8a^3b^3 \cdot 5a^2b}$$

$$= \sqrt[3]{8a^3b^3} \sqrt[3]{5a^2b}$$

$$= 2ab\sqrt[3]{5a^2b} \qquad \blacksquare$$

Here are some further examples concerning the first condition for simplified form.

EXAMPLE 4 Write the expression in simplified form.

SOLUTION $\sqrt{12x^7y^6} = \sqrt{4x^6y^6 \cdot 3x}$

$\qquad\qquad = \sqrt{4x^6y^6} \sqrt{3x}$

$\qquad\qquad = 2x^3y^3\sqrt{3x}$

EXAMPLE 5 Write the expression in simplified form.

SOLUTION $\sqrt[3]{54a^6b^2c^4} = \sqrt[3]{27a^6c^3 \cdot 2b^2c}$

$\qquad\qquad = \sqrt[3]{27a^6c^3} \sqrt[3]{2b^2c}$

$\qquad\qquad = 3a^2c\sqrt[3]{2b^2c} \qquad \blacksquare$

B Rationalizing the Denominator

The quotient property of radicals is used to simplify a radical that contains a fraction.

EXAMPLE 6 Simplify $\sqrt{\dfrac{3}{4}}$.

SOLUTION Applying the quotient property for radicals, we have

$$\sqrt{\frac{3}{4}} = \frac{\sqrt{3}}{\sqrt{4}} \qquad \text{\color{orange}{Quotient property}}$$

$$= \frac{\sqrt{3}}{2} \qquad \text{\color{orange}{$\sqrt{4} = 2$}}$$

The last expression is in simplified form because it satisfies all three conditions for simplified form. $\qquad \blacksquare$

Write each expression in simplified form. Assume $x, y \geq 0$.

2. $\sqrt{50x^2y^3}$

> **NOTE**
> Unless we state otherwise, assume all variables throughout this section represent nonnegative numbers. Further, if a variable appears in a denominator, assume the variable cannot be 0.

3. $\sqrt[3]{54a^4b^3}$

4. $\sqrt{75x^5y^8}$

5. $\sqrt[4]{48a^8b^5c^4}$

6. Simplify $\sqrt{\dfrac{5}{9}}$.

Answers
2. $5xy\sqrt{2y}$
3. $3ab\sqrt[3]{2a}$
4. $5x^2y^4\sqrt{3x}$
5. $2a^2bc\sqrt[4]{3b}$
6. $\dfrac{\sqrt{5}}{3}$

7. Write $\sqrt{\dfrac{2}{3}}$ in simplified form.

NOTE
The idea behind rationalizing the denominator is to produce a perfect square under the square root sign in the denominator. This is accomplished by multiplying both the numerator and denominator by the appropriate radical.

Rationalize each denominator.

8. $\dfrac{5}{\sqrt{2}}$

9. $\dfrac{3\sqrt{5x}}{\sqrt{2y}}$

10. $\dfrac{5}{\sqrt[3]{9}}$

Answers

7. $\dfrac{\sqrt{6}}{3}$

8. $\dfrac{5\sqrt{2}}{2}$

9. $\dfrac{3\sqrt{10xy}}{2y}$

10. $\dfrac{5\sqrt[3]{3}}{3}$

EXAMPLE 7 Write $\sqrt{\dfrac{5}{6}}$ in simplified form.

SOLUTION Proceeding as in Example 6, we have

$$\sqrt{\frac{5}{6}} = \frac{\sqrt{5}}{\sqrt{6}}$$

The resulting expression satisfies the second condition for simplified form since neither radical contains a fraction. It does, however, violate condition 3 since it has a radical in the denominator. Getting rid of the radical in the denominator is called *rationalizing the denominator* and is accomplished, in this case, by multiplying the numerator and denominator by $\sqrt{6}$.

$$\frac{\sqrt{5}}{\sqrt{6}} = \frac{\sqrt{5}}{\sqrt{6}} \cdot \frac{\sqrt{6}}{\sqrt{6}}$$

$$= \frac{\sqrt{30}}{\sqrt{6^2}}$$

$$= \frac{\sqrt{30}}{6} \quad \blacksquare$$

EXAMPLE 8 Rationalize the denominator.

SOLUTION
$$\frac{4}{\sqrt{3}} = \frac{4}{\sqrt{3}} \cdot \frac{\sqrt{3}}{\sqrt{3}}$$

$$= \frac{4\sqrt{3}}{\sqrt{3^2}}$$

$$= \frac{4\sqrt{3}}{3} \quad \blacksquare$$

EXAMPLE 9 Rationalize the denominator.

SOLUTION
$$\frac{2\sqrt{3x}}{\sqrt{5y}} = \frac{2\sqrt{3x}}{\sqrt{5y}} \cdot \frac{\sqrt{5y}}{\sqrt{5y}}$$

$$= \frac{2\sqrt{15xy}}{\sqrt{(5y)^2}}$$

$$= \frac{2\sqrt{15xy}}{5y} \quad \blacksquare$$

When the denominator involves a cube root, we must multiply by a radical that will produce a perfect cube under the cube root sign in the denominator, as our next example illustrates.

EXAMPLE 10 Rationalize the denominator in $\dfrac{7}{\sqrt[3]{4}}$.

SOLUTION Since $4 = 2^2$, we can multiply both numerator and denominator by $\sqrt[3]{2}$ and obtain $\sqrt[3]{2^3}$ in the denominator.

$$\frac{7}{\sqrt[3]{4}} = \frac{7}{\sqrt[3]{2^2}}$$

$$= \frac{7}{\sqrt[3]{2^2}} \cdot \frac{\sqrt[3]{2}}{\sqrt[3]{2}}$$

$$= \frac{7\sqrt[3]{2}}{\sqrt[3]{2^3}}$$

$$= \frac{7\sqrt[3]{2}}{2} \quad \blacksquare$$

EXAMPLE 11 Simplify $\sqrt{\dfrac{12x^5y^3}{5z}}$.

SOLUTION We use the quotient property to write the numerator and denominator as two separate radicals.

$$\sqrt{\dfrac{12x^5y^3}{5z}} = \dfrac{\sqrt{12x^5y^3}}{\sqrt{5z}}$$

Simplifying the numerator, we have

$$\dfrac{\sqrt{12x^5y^3}}{\sqrt{5z}} = \dfrac{\sqrt{4x^4y^2}\sqrt{3xy}}{\sqrt{5z}}$$

$$= \dfrac{2x^2y\sqrt{3xy}}{\sqrt{5z}}$$

To rationalize the denominator, we multiply the numerator and denominator by $\sqrt{5z}$.

$$\dfrac{2x^2y\sqrt{3xy}}{\sqrt{5z}} \cdot \dfrac{\sqrt{5z}}{\sqrt{5z}} = \dfrac{2x^2y\sqrt{15xyz}}{\sqrt{(5z)^2}}$$

$$= \dfrac{2x^2y\sqrt{15xyz}}{5z} \quad \blacksquare$$

The Square Root of a Perfect Square

So far in this chapter we have assumed that all our variables are nonnegative when they appear under a square root symbol. There are times, however, when this is not the case.

Consider the following two statements:

$$\sqrt{3^2} = \sqrt{9} = 3 \quad \text{and} \quad \sqrt{(-3)^2} = \sqrt{9} = 3$$

Whether we operate on 3 or −3, the result is the same: both expressions simplify to 3. The other operation we have worked with in the past that produces the same result is absolute value; that is,

$$|3| = 3 \quad \text{and} \quad |-3| = 3$$

This leads us to the next property of radicals.

Absolute Value Property for Radicals

If a is a real number, then $\sqrt{a^2} = |a|$

The result of this discussion and the absolute value property is simply this:

If we know a is positive, then $\sqrt{a^2} = a$.

If we know a is negative, then $\sqrt{a^2} = |a|$.

If we don't know if a is positive or negative, then $\sqrt{a^2} = |a|$.

EXAMPLE 12 Simplify each expression. Do *not* assume the variables represent positive numbers.

a. $\sqrt{9x^2} = 3|x|$
b. $\sqrt{x^3} = |x|\sqrt{x}$
c. $\sqrt{x^2 - 6x + 9} = \sqrt{(x-3)^2} = |x-3|$
d. $\sqrt{x^3 - 5x^2} = \sqrt{x^2(x-5)} = |x|\sqrt{x-5}$ $\blacksquare$

11. Simplify $\sqrt{\dfrac{48x^3y^4}{7z}}$.

12. Simplify each expression. Do *not* assume the variables represent nonnegative numbers.

a. $\sqrt{16x^2}$

b. $\sqrt{25x^3}$

c. $\sqrt{x^2 + 10x + 25}$

d. $\sqrt{2x^3 + 7x^2}$

Answers

11. $\dfrac{4xy^2\sqrt{21xz}}{7z}$

12. a. $4|x|$ **b.** $5|x|\sqrt{x}$
c. $|x+5|$ **d.** $|x|\sqrt{2x+7}$

As you can see, we must use absolute value symbols when we take a square root of a perfect square, unless we know the base of the perfect square is a positive number. The same idea holds for higher even roots, but not for odd roots. With odd roots, no absolute value symbols are necessary.

13. Simplify each expression.

a. $\sqrt[3]{(-3)^3}$

b. $\sqrt[3]{(-1)^3}$

EXAMPLE 13 Simplify each expression.

a. $\sqrt[3]{(-2)^3} = \sqrt[3]{-8} = -2$

b. $\sqrt[3]{(-5)^3} = \sqrt[3]{-125} = -5$ ∎

We can extend this discussion to all roots as follows:

> **Extending the Absolute Value Property for Radicals**
>
> If a is a real number, then
> $$\sqrt[n]{a^n} = |a| \quad \text{if} \quad n \text{ is even}$$
> $$\sqrt[n]{a^n} = a \quad \text{if} \quad n \text{ is odd}$$

Answers

13. a. -3 **b.** -1

Problem Set 7.3

Moving Toward Success

"Judge a man by his questions rather than by his answers."

    —Voltaire, 1694–1778, French writer and philosopher

1. If your instructor writes something on the board in class, should you write it in your notes? Why or why not?

2. Should you ask questions in class? Why or why not?

A Use the product property for radicals to write each of the following expressions in simplified form. (Assume all variables are nonnegative through Problem 84.) [Examples 1–5]

1. $\sqrt{8}$

2. $\sqrt{32}$

3. $\sqrt{98}$

4. $\sqrt{75}$

5. $\sqrt{288}$

6. $\sqrt{128}$

7. $\sqrt{80}$

8. $\sqrt{200}$

9. $\sqrt{48}$

10. $\sqrt{27}$

11. $\sqrt{675}$

12. $\sqrt{972}$

13. $\sqrt[3]{54}$

14. $\sqrt[3]{24}$

15. $\sqrt[3]{128}$

16. $\sqrt[3]{162}$

17. $\sqrt[3]{432}$

18. $\sqrt[3]{1,536}$

19. $\sqrt[5]{64}$

20. $\sqrt[4]{48}$

21. $\sqrt{18x^3}$

22. $\sqrt{27x^5}$

23. $\sqrt[4]{32y^7}$

24. $\sqrt[5]{32y^7}$

25. $\sqrt[3]{40x^4y^7}$

26. $\sqrt[3]{128x^6y^2}$

27. $\sqrt{48a^2b^3c^4}$

28. $\sqrt{72a^4b^3c^2}$

29. $\sqrt[3]{48a^2b^3c^4}$

30. $\sqrt[3]{72a^4b^3c^2}$

31. $\sqrt[5]{64x^8y^{12}}$

32. $\sqrt[4]{32x^9y^{10}}$

33. $\sqrt[5]{243x^7y^{10}z^5}$

34. $\sqrt[5]{64x^8y^4z^{11}}$

Substitute the given numbers into the expression $\sqrt{b^2 - 4ac}$, and then simplify.

35. $a = 2, b = -6, c = 3$

36. $a = 6, b = 7, c = -5$

37. $a = 1, b = 2, c = 6$

38. $a = 2, b = 5, c = 3$

39. $a = \dfrac{1}{2}, b = -\dfrac{1}{2}, c = -\dfrac{5}{4}$

40. $a = \dfrac{7}{4}, b = -\dfrac{3}{4}, c = -2$

B   Rationalize the denominator in each of the following expressions. [Examples 6–11]

41. $\dfrac{2}{\sqrt{3}}$

42. $\dfrac{3}{\sqrt{2}}$

43. $\dfrac{5}{\sqrt{6}}$

44. $\dfrac{7}{\sqrt{5}}$

45. $\sqrt{\dfrac{1}{2}}$

46. $\sqrt{\dfrac{1}{3}}$

47. $\sqrt{\dfrac{1}{5}}$

48. $\sqrt{\dfrac{1}{6}}$

49. $\dfrac{4}{\sqrt[3]{2}}$

50. $\dfrac{5}{\sqrt[3]{3}}$

51. $\dfrac{2}{\sqrt[3]{9}}$

52. $\dfrac{3}{\sqrt[3]{4}}$

53. $\sqrt[4]{\dfrac{3}{2x^2}}$

54. $\sqrt[4]{\dfrac{5}{3x^2}}$

55. $\sqrt[4]{\dfrac{8}{y}}$

56. $\sqrt[4]{\dfrac{27}{y}}$

57. $\sqrt[3]{\dfrac{4x}{3y}}$

58. $\sqrt[3]{\dfrac{7x}{6y}}$

59. $\sqrt[3]{\dfrac{2x}{9y}}$

60. $\sqrt[3]{\dfrac{5x}{4y}}$

61. $\sqrt[4]{\dfrac{1}{8x^3}}$

62. $\sqrt[4]{\dfrac{8}{9x^3}}$

Write each of the following in simplified form.

63. $\sqrt{\dfrac{27x^3}{5y}}$

64. $\sqrt{\dfrac{12x^5}{7y}}$

65. $\sqrt{\dfrac{75x^3y^2}{2z}}$

66. $\sqrt[3]{\dfrac{50x^2y^3}{3z}}$

67. $\sqrt[3]{\dfrac{16a^4b^3}{9c}}$

68. $\sqrt[3]{\dfrac{54a^5b^4}{25c^2}}$

69. $\sqrt[3]{\dfrac{8x^3y^6}{9z}}$

70. $\sqrt[3]{\dfrac{27x^6y^3}{2z^2}}$

71. $\sqrt{\sqrt{x^2}}$

72. $\sqrt{\sqrt{2x^3}}$

73. $\sqrt[3]{\sqrt{xy}}$

74. $\sqrt{\sqrt{4x}}$

75. $\sqrt[3]{\sqrt[4]{a}}$

76. $\sqrt[6]{\sqrt[4]{x}}$

77. $\sqrt[3]{\sqrt[3]{6x^{10}}}$

78. $\sqrt[5]{\sqrt{x^{14}y^{11}z}}$

79. $\sqrt[4]{\sqrt[3]{a^{12}b^{24}c^{14}}}$

80. $\sqrt{\sqrt{4a^{17}}}$

81. $\sqrt[3]{\sqrt[5]{3a^{17}b^{16}c^{30}}}$

82. $\left(\sqrt[4]{\sqrt{x^4y^8z^9}}\right)^2$

83. $\left(\sqrt{\sqrt[3]{8ab^6}}\right)^2$

84. $\left(\sqrt[4]{\sqrt[3]{16x^8y^{12}z^3}}\right)^3$

Simplify each expression. Do *not* assume the variables represent positive numbers. [Examples 12, 13]

85. $\sqrt{25x^2}$

86. $\sqrt{49x^2}$

87. $\sqrt{27x^3y^2}$

88. $\sqrt{40x^3y^2}$

89. $\sqrt{x^2 - 10x + 25}$

90. $\sqrt{x^2 - 16x + 64}$

91. $\sqrt{4x^2 + 12x + 9}$

92. $\sqrt{16x^2 + 40x + 25}$

93. $\sqrt{4a^4 + 16a^3 + 16a^2}$

94. $\sqrt{9a^4 + 18a^3 + 9a^2}$

95. $\sqrt{4x^3 - 8x^2}$

96. $\sqrt{18x^3 - 9x^2}$

97. Show that the statement $\sqrt{a + b} = \sqrt{a} + \sqrt{b}$ is not true by replacing a with 9 and b with 16 and simplifying both sides.

98. Find a pair of values for a and b that will make the statement $\sqrt{a + b} = \sqrt{a} + \sqrt{b}$ true.

Applying the Concepts

99. Diagonal Distance The distance d between opposite corners of a rectangular room with length l and width w is given by

$$d = \sqrt{l^2 + w^2}$$

How far is it between opposite corners of a living room that measures 10 by 15 feet?

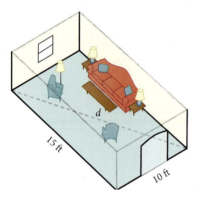

15 ft

10 ft

100. Radius of a Sphere The radius r of a sphere with volume V can be found by using the formula

$$r = \sqrt[3]{\frac{3V}{4\pi}}$$

Find the radius of a sphere with volume 9 cubic feet. Write your answer in simplified form. (Use $\frac{22}{7}$ for π.)

Volume

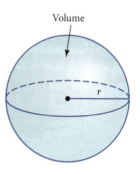

101. Radius of a Cylinder The radius r of a cylinder with volume V and height h can be found by using the formula

$$r = \sqrt{\frac{V}{h\pi}}$$

Find the radius of the cylinder below. Write your answer in simplified form. (Use $\frac{22}{7}$ for π.)

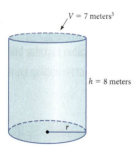

$V = 7$ meters^3

$h = 8$ meters

r

102. Radius of a Cone The radius r of a cone with a volume V and height h can be found by using the formula

$$r = \sqrt{\frac{3V}{h\pi}}$$

Find the radius of a cone with a volume of 6 cubic inches and a height of 7 inches. Write your asnwer in simplified form. (Use $\frac{22}{7}$ for π.)

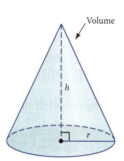

Volume

h

r

103. Diagonal of a Box The length of the diagonal of a rectangular box with length l, width w, and height h is given by $d = \sqrt{l^2 + w^2 + h^2}$.

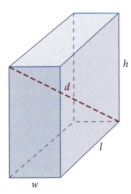

a. Find the length of the diagonal of a rectangular box that is 3 feet wide, 4 feet long, and 12 feet high.

b. Find the length of the diagonal of a rectangular box that is 2 feet wide, 4 feet high, and 6 feet long.

104. Distance to the Horizon Recall from the begining of this section that if you are at a point k miles above the surface of the Earth, the distance you can see, in miles, is approximated by the equation

$$d = \sqrt{8000k + k^2}.$$

a. How far can you see from a point that is 1 mile above the surface of the Earth?

b. How far can you see from a point that is 2 miles above the surface of the Earth?

c. How far can you see from a point that is 3 miles above the surface of the Earth?

d. The observation deck of the Burj Khalifa in Dubai stands at 0.28 miles above the ground. Dubai's manmade island Palm Jebel Ali is approximately 22 miles from the Burj Khalifa. Can you see Palm Jebel Ali from the observation deck?

Spiral of Roots The Pythagorean theorem can be used to construct the spiral shown here. This spiral is called the Spiral of Roots because each of the diagonals is the positive square root of one of the positive integers. Use this diagram to help you work Problems 105 and 106.

105. Construct your own spiral of roots by using a ruler. Draw the first triangle by using two 1-inch lines. The first diagonal will have a length of $\sqrt{2}$ inches. Each new triangle will be formed by drawing a 1-inch line segment at the end of the previous diagonal so that the angle formed is 90°.

106. Construct a spiral of roots by using line segments of length 2 inches. The length of the first diagonal will be $2\sqrt{2}$ inches. The length of the second diagonal will be $2\sqrt{3}$ inches.

The Spiral of Roots

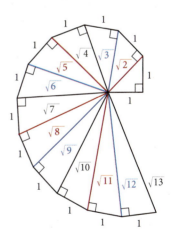

Getting Ready for the Next Section

Simplify the following.

107. $5x - 4x + 6x$

108. $12x + 8x - 7x$

109. $35xy^2 - 8xy^2$

110. $20a^2b + 33a^2b$

111. $\dfrac{1}{2}x + \dfrac{1}{3}x$

112. $\dfrac{2}{3}x + \dfrac{5}{8}x$

Write in simplified form for radicals.

113. $\sqrt{18}$

114. $\sqrt{8}$

115. $\sqrt{75xy^3}$

116. $\sqrt{12xy}$

117. $\sqrt[3]{8a^4b^2}$

118. $\sqrt[3]{27ab^2}$

Maintaining Your Skills

Perform the indicated operations.

119. $\dfrac{8xy^3}{9x^2y} \div \dfrac{16x^2y^2}{18xy^3}$

120. $\dfrac{25x^2}{5y^4} \cdot \dfrac{30y^3}{2x^5}$

121. $\dfrac{12a^2 - 4a - 5}{2a + 1} \cdot \dfrac{7a + 3}{42a^2 - 17a - 15}$

122. $\dfrac{20a^2 - 7a - 3}{4a + 1} \cdot \dfrac{25a^2 - 5a - 6}{5a + 2}$

123. $\dfrac{8x^3 + 27}{27x^3 + 1} \div \dfrac{6x^2 + 7x - 3}{9x^2 - 1}$

124. $\dfrac{27x^3 + 8}{8x^3 + 1} \div \dfrac{6x^2 + x - 2}{4x^2 - 1}$

Extending the Concepts

Factor each radicand into the product of prime factors. Then simplify each radical.

125. $\sqrt[3]{8,640}$

126. $\sqrt{8,640}$

127. $\sqrt[3]{10,584}$

128. $\sqrt{10,584}$

Assume a is a positive number, and rationalize each denominator.

129. $\dfrac{1}{\sqrt[10]{a^3}}$

130. $\dfrac{1}{\sqrt[12]{a^7}}$

131. $\dfrac{1}{\sqrt[20]{a^{11}}}$

132. $\dfrac{1}{\sqrt[15]{a^{13}}}$

133. Show that the two expressions $\sqrt{x^2 + 1}$ and $x + 1$ are not, in general, equal to each other by graphing $y = \sqrt{x^2 + 1}$ and $y = x + 1$ in the same viewing window.

134. Show that the two expressions $\sqrt{x^2 + 9}$ and $x + 3$ are not, in general, equal to each other by graphing $y = \sqrt{x^2 + 9}$ and $y = x + 3$ in the same viewing window.

135. Approximately how far apart are the graphs in Problem 133 when $x = 2$?

136. Approximately how far apart are the graphs in Problem 134 when $x = 2$?

137. For what value of x are the expressions $\sqrt{x^2 + 1}$ and $x + 1$ equal?

138. For what value of x are the expressions $\sqrt{x^2 + 9}$ and $x + 3$ equal?

Addition and Subtraction of Radical Expressions

OBJECTIVES

A Add and subtract radicals.

B Construct golden rectangles from squares.

TICKET TO SUCCESS

Keep these questions in mind as you read through the section. Then respond in your own words and in complete sentences.

1. What are similar radicals?

2. When can we add two radical expressions?

3. What is the first step when adding or subtracting expressions containing radicals?

4. How would you construct a golden rectangle from a square of side 6?

Carlos E. Santa Maria/Shutterstock.com

The yield sign often seen on roadways is an example of an equilateral triangle—a triangle with 3 sides of equal length. The ratio of the height of the sign to one of its sides can be found using addition and subtraction of radical expressions, which is the focus of this section.

We have been able to add and subtract polynomials by combining similar terms. The same idea applies to addition and subtraction of radical expressions.

A Combining Similar Terms

> **Definition**
>
> Two radicals are said to be **similar radicals** if they have the same index and the same radicand.

The expressions $5\sqrt[3]{7}$ and $-8\sqrt[3]{7}$ are similar since the index is 3 in both cases and the radicands are 7. The expressions $3\sqrt[4]{5}$ and $7\sqrt[3]{5}$ are not similar since they have different indices, and the expressions $2\sqrt[5]{8}$ and $3\sqrt[5]{9}$ are not similar because the radicands are not the same.

PRACTICE PROBLEMS

1. Combine $3\sqrt{5} - 2\sqrt{5} + 4\sqrt{5}$.

EXAMPLE 1   Combine $5\sqrt{3} - 4\sqrt{3} + 6\sqrt{3}$.

SOLUTION   All three radicals are similar. We apply the distributive property to get

$$5\sqrt{3} - 4\sqrt{3} + 6\sqrt{3} = (5 - 4 + 6)\sqrt{3}$$
$$= 7\sqrt{3} \qquad \blacksquare$$

2. Combine $4\sqrt{50} + 3\sqrt{8}$.

EXAMPLE 2   Combine $3\sqrt{8} + 5\sqrt{18}$.

SOLUTION   The two radicals do not seem to be similar. We must write each in simplified form before applying the distributive property.

$$3\sqrt{8} + 5\sqrt{18} = 3\sqrt{4 \cdot 2} + 5\sqrt{9 \cdot 2}$$
$$= 3\sqrt{4}\ \sqrt{2} + 5\sqrt{9}\ \sqrt{2}$$
$$= 3 \cdot 2\sqrt{2} + 5 \cdot 3\sqrt{2}$$
$$= 6\sqrt{2} + 15\sqrt{2}$$
$$= (6 + 15)\sqrt{2}$$
$$= 21\sqrt{2} \qquad \blacksquare$$

The result of Example 2 can be generalized to the following rule for sums and differences of radical expressions.

> **Rule Adding or Subtracting Radical Expressions**
>
> To add or subtract radical expressions, put each in simplified form and apply the distributive property if possible. We can add only similar radicals. We must write each expression in simplified form for radicals before we can tell if the radicals are similar.

3. Assume $x, y \geq 0$ and combine:
$\sqrt{18x^2y} - 3x\sqrt{50y}$

EXAMPLE 3   Combine $7\sqrt{75xy^3} - 4y\sqrt{12xy}$, where $x, y \geq 0$.

SOLUTION   We write each expression in simplified form and combine similar radicals.

$$7\sqrt{75xy^3} - 4y\sqrt{12xy} = 7\sqrt{25y^2}\ \sqrt{3xy} - 4y\sqrt{4}\ \sqrt{3xy}$$
$$= 35y\sqrt{3xy} - 8y\sqrt{3xy}$$
$$= (35y - 8y)\sqrt{3xy}$$
$$= 27y\sqrt{3xy} \qquad \blacksquare$$

4. Combine
$2\sqrt[3]{27a^2b^4} + 3b\sqrt[3]{125a^2b}$.

EXAMPLE 4   Combine $10\sqrt[3]{8a^4b^2} + 11a\sqrt[3]{27ab^2}$.

SOLUTION   Writing each radical in simplified form and combining similar terms, we have

$$10\sqrt[3]{8a^4b^2} + 11a\sqrt[3]{27ab^2} = 10\sqrt[3]{8a^3}\ \sqrt[3]{ab^2} + 11a\sqrt[3]{27}\ \sqrt[3]{ab^2}$$
$$= 20a\sqrt[3]{ab^2} + 33a\sqrt[3]{ab^2}$$
$$= 53a\sqrt[3]{ab^2} \qquad \blacksquare$$

Answers
1. $5\sqrt{5}$
2. $26\sqrt{2}$
3. $-12x\sqrt{2y}$
4. $21b\sqrt[3]{a^2b}$

EXAMPLE 5 Combine $\dfrac{\sqrt{3}}{2} + \dfrac{1}{\sqrt{3}}$.

SOLUTION We begin by writing the second term in simplified form.

$$\frac{\sqrt{3}}{2} + \frac{1}{\sqrt{3}} = \frac{\sqrt{3}}{2} + \frac{1}{\sqrt{3}} \cdot \frac{\sqrt{3}}{\sqrt{3}}$$

$$= \frac{\sqrt{3}}{2} + \frac{\sqrt{3}}{3}$$

$$= \frac{1}{2}\sqrt{3} + \frac{1}{3}\sqrt{3}$$

$$= \left(\frac{1}{2} + \frac{1}{3}\right)\sqrt{3}$$

$$= \frac{5}{6}\sqrt{3} = \frac{5\sqrt{3}}{6} \qquad \blacksquare$$

5. Combine $\dfrac{\sqrt{5}}{3} + \dfrac{1}{\sqrt{5}}$.

B The Golden Rectangle

Recall our discussion of the golden rectangle from the first section of this chapter. Now that we have practiced adding and subtracting radical expressions, let's put our new skills to use to find a golden rectangle of a different size.

EXAMPLE 6 Construct a golden rectangle from a square of side 4. Then show that the ratio of the length to the width is the golden ratio $\dfrac{1 + \sqrt{5}}{2}$.

SOLUTION Figure 1 shows the golden rectangle constructed from a square of side 4.

6. Construct a golden rectangle from a square of side 6. Then show that the ratio of the length to the width is the golden ratio.

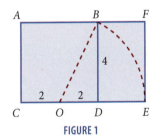

FIGURE 1

The length of the diagonal OB is found from the Pythagorean theorem.

$$OB = \sqrt{2^2 + 4^2} = \sqrt{4 + 16} = \sqrt{20} = 2\sqrt{5}$$

The ratio of the length to the width for the rectangle is the golden ratio.

$$\text{Golden ratio} = \frac{CE}{EF} = \frac{2 + 2\sqrt{5}}{4} = \frac{2(1 + \sqrt{5})}{2 \cdot 2} = \frac{1 + \sqrt{5}}{2} \qquad \blacksquare$$

As you can see, showing that the ratio of length to width in this rectangle is the golden ratio depends on our ability to write $\sqrt{20}$ as $2\sqrt{5}$ and our ability to reduce to lowest terms by factoring and then dividing out the common factor 2 from the numerator and denominator

Problem Set 7.4

Moving Toward Success

"We make a living by what we get, but we make a life by what we give."

—Winston Churchill, 1874–1965, English politician and Nobel Prize winner

1. Should you read the notes in the side columns of the section pages? Why or why not?
2. How does coming to class prepared help to reduce stress?

A Combine the following expressions. (Assume any variables under an even root are nonnegative.) [Examples 1–5]

1. $3\sqrt{5} + 4\sqrt{5}$

2. $6\sqrt{3} - 5\sqrt{3}$

3. $3x\sqrt{7} - 4x\sqrt{7}$

4. $6y\sqrt{a} + 7y\sqrt{a}$

5. $5\sqrt[3]{10} - 4\sqrt[3]{10}$

6. $6\sqrt[4]{2} + 9\sqrt[4]{2}$

7. $8\sqrt[5]{6} - 2\sqrt[5]{6} + 3\sqrt[5]{6}$

8. $7\sqrt[6]{7} - \sqrt[6]{7} + 4\sqrt[6]{7}$

9. $3x\sqrt{2} - 4x\sqrt{2} + x\sqrt{2}$

10. $5x\sqrt{6} - 3x\sqrt{6} - 2x\sqrt{6}$

11. $\sqrt{20} - \sqrt{80} + \sqrt{45}$

12. $\sqrt{8} - \sqrt{32} - \sqrt{18}$

13. $4\sqrt{8} - 2\sqrt{50} - 5\sqrt{72}$

14. $\sqrt{48} - 3\sqrt{27} + 2\sqrt{75}$

15. $5x\sqrt{8} + 3\sqrt{32x^2} - 5\sqrt{50x^2}$

16. $2\sqrt{50x^2} - 8x\sqrt{18} - 3\sqrt{72x^2}$

17. $5\sqrt[3]{16} - 4\sqrt[3]{54}$

18. $\sqrt[3]{81} + 3\sqrt[3]{24}$

19. $\sqrt[3]{x^4y^2} + 7x\sqrt[3]{xy^2}$

20. $2\sqrt[3]{x^8y^6} - 3y^2\sqrt[3]{8x^8}$

21. $5a^2\sqrt{27ab^3} - 6b\sqrt{12a^5b}$

22. $9a\sqrt{20a^3b^2} + 7b\sqrt{45a^5}$

23. $b\sqrt[3]{24a^5b} + 3a\sqrt[3]{81a^2b^4}$

24. $7\sqrt[3]{a^4b^3c^2} - 6ab\sqrt[3]{ac^2}$

25. $5x\sqrt[4]{3y^5} + y\sqrt[4]{243x^4y} + \sqrt[4]{48x^4y^5}$

26. $x\sqrt[4]{5xy^8} + y\sqrt[4]{405x^5y^4} + y^2\sqrt[4]{80x^5}$

27. $\dfrac{\sqrt{2}}{2} + \dfrac{1}{\sqrt{2}}$

28. $\dfrac{\sqrt{5}}{3} + \dfrac{1}{\sqrt{5}}$

29. $\dfrac{\sqrt{3}}{3} + \dfrac{1}{\sqrt{3}}$

30. $\dfrac{\sqrt{6}}{2} + \dfrac{1}{\sqrt{6}}$

31. $\sqrt{x} - \dfrac{1}{\sqrt{x}}$

32. $\sqrt{x} + \dfrac{1}{\sqrt{x}}$

33. $\dfrac{\sqrt{18}}{6} + \sqrt{\dfrac{1}{2}} + \dfrac{\sqrt{2}}{2}$

34. $\dfrac{\sqrt{12}}{6} + \sqrt{\dfrac{1}{3}} + \dfrac{\sqrt{3}}{3}$

35. $\sqrt{6} - \sqrt{\dfrac{2}{3}} + \sqrt{\dfrac{1}{6}}$

36. $\sqrt{15} - \sqrt{\dfrac{3}{5}} + \sqrt{\dfrac{5}{3}}$

37. $\sqrt[3]{25} + \dfrac{3}{\sqrt[3]{5}}$

38. $\sqrt[4]{8} + \dfrac{1}{\sqrt[4]{2}}$

The following problems apply to what you have learned about algebra with functions.

39. Let $f(x) = \sqrt{8x}$ and $g(x) = \sqrt{72x}$, then find

   a. $f(x) + g(x)$

   b. $f(x) - g(x)$

40. Let $f(x) = x + \sqrt{3}$ and $g(x) = x + 2\sqrt{3}$, then find

   a. $f(x) + g(x)$

   b. $f(x) - g(x)$

41. Let $f(x) = 3\sqrt{2x}$ and $g(x) = \sqrt{2x}$, then find

   a. $f(x) + g(x)$

   b. $f(x) - g(x)$

42. Let $f(x) = x\sqrt[3]{64}$ and $g(x) = x\sqrt{81}$, then find

   a. $f(x) + g(x)$

   b. $f(x) - g(x)$

43. Let $f(x) = x\sqrt{2}$ and $g(x) = 2x\sqrt{2}$, then find

   a. $f(x) + g(x)$

   b. $f(x) - g(x)$

44. Let $f(x) = 5 + 2\sqrt{5x}$ and $g(x) = 3\sqrt{5x}$, then find

   a. $f(x) + g(x)$

   b. $f(x) - g(x)$

45. Let $f(x) = \sqrt{2x} - 2$ and $g(x) = 2\sqrt{2x} + 5$ then find

   a. $f(x) + g(x)$

   b. $f(x) - g(x)$

46. Let $f(x) = 1 - 2\sqrt[3]{3x}$ and $g(x) = 2 + 3\sqrt[3]{3x}$, then find

   a. $f(x) + g(x)$

   b. $f(x) - g(x)$

47. Use a calculator to find a decimal approximation for $\sqrt{12}$ and for $2\sqrt{3}$.

48. Use a calculator to find decimal approximations for $\sqrt{50}$ and $5\sqrt{2}$.

49. Use a calculator to find a decimal approximation for $\sqrt{8} + \sqrt{18}$. Is it equal to the decimal approximation for $\sqrt{26}$ or $\sqrt{50}$?

50. Use a calculator to find a decimal approximation for $\sqrt{3} + \sqrt{12}$. Is it equal to the decimal approximation for $\sqrt{15}$ or $\sqrt{27}$?

Applying the Concepts

51. Golden Rectangle Construct a golden rectangle from a square of side 8. Then show that the ratio of the length to the width is the golden ratio $\frac{1 + \sqrt{5}}{2}$.

52. Golden Rectangle Construct a golden rectangle from a square of side 10. Then show that the ratio of the length to the width is the golden ratio $\frac{1 + \sqrt{5}}{2}$.

53. Golden Rectangle To show that all golden rectangles have the same ratio of length to width, construct a golden rectangle from a square of side $2x$. Then show that the ratio of the length to the width is the golden ratio.

54. Equilateral Triangle A yield sign, which is an equilateral triangle, has a side length of 30 inches. Find the ratio of the height to a side.

55. Equilateral Triangles A triangle is equilateral if it has three equal sides. The triangle in the figure is equilateral with each side of length $2x$. Find the ratio of the height to a side.

56. Equilateral Triangle The triangle in the figure is equilateral with each side of length $5x$. Find the ratio of the height to a side.

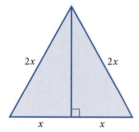

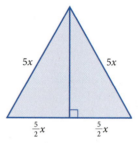

57. Pyramids Refer to the diagram of a square pyramid below. Find the ratio of the height h of the pyramid to the altitude a.

58. Pyramids Refer to the diagram of a square pyramid below. Find the ratio of the height h of the pyramid to the altitude a.

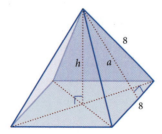

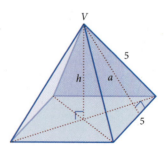

Maintaining Your Skills

Add or subtract as indicated.

59. $\dfrac{2a - 4}{a + 2} - \dfrac{a - 6}{a + 2}$

60. $\dfrac{2a - 3}{a - 2} - \dfrac{a - 1}{a - 2}$

61. $3 + \dfrac{4}{3 - t}$

62. $6 + \dfrac{2}{5 - t}$

63. $\dfrac{3}{2x - 5} - \dfrac{39}{8x^2 - 14x - 15}$

64. $\dfrac{2}{4x - 5} + \dfrac{9}{8x^2 - 38x + 35}$

65. $\dfrac{1}{x - y} - \dfrac{3xy}{x^3 - y^3}$

66. $\dfrac{1}{x + y} + \dfrac{3xy}{x^3 + y^3}$

Getting Ready for the Next Section

Simplify the following.

67. $3 \cdot 2$

68. $5 \cdot 7$

69. $(x + y)(4x - y)$

70. $(2x + y)(x - y)$

71. $(x + 3)^2$

72. $(3x - 2y)^2$

73. $(x - 2)(x + 2)$

74. $(2x + 5)(2x - 5)$

Simplify the following expressions.

75. $2\sqrt{18}$

76. $5\sqrt{36}$

77. $(\sqrt{6})^2$

78. $(\sqrt{2})^2$

79. $(3\sqrt{x})^2$

80. $(2\sqrt{y})^2$

Rationalize the denominator.

81. $\dfrac{\sqrt{3}}{\sqrt{2}}$

82. $\dfrac{\sqrt{5}}{\sqrt{6}}$

Extending the Concepts

Assume all variables represent positive numbers.

Simplify.

83. $\sqrt[5]{32x^5y^5} - y\sqrt[3]{27x^3}$

84. $\sqrt[6]{x^4} + 4\sqrt[3]{8x^2}$

85. $3\sqrt[9]{x^9y^{18}z^{27}} - 4\sqrt[6]{x^6y^{12}z^{18}}$

86. $4a\sqrt{b^4c^6} + 3b\sqrt[3]{a^3b^3c^9}$

87. $3c\sqrt[8]{4a^6b^{18}} + b\sqrt[4]{32a^3b^5c^4}$

88. $4x\sqrt[6]{16y^6z^8} - y\sqrt[3]{32x^3z^4}$

89. $3\sqrt[9]{8a^{12}b^9} + b\sqrt[3]{16a^4} - 8\sqrt[6]{4a^8b^6}$

90. $-ac\sqrt{108bc^3} - 4\sqrt[6]{27a^6b^3c^{15}} + 3\sqrt[4]{9a^4b^2c^{10}}$

Multiplication and Division of Radical Expressions

TICKET TO SUCCESS

Keep these questions in mind as you read through the section. Then respond in your own words and in complete sentences.

1. Explain why $(\sqrt{5} + \sqrt{2})^2 \neq 5 + 2$.
2. Explain in words how you would rationalize the denominator in the expression $\frac{\sqrt{3}}{\sqrt{5} - \sqrt{2}}$.
3. What are conjugates?
4. What result is guaranteed when multiplying radical expressions that are conjugates?

rook76/Shutterstock.com

Leonardo da Vinci, as a mathematician and an artist, used the golden ratio in several of his works. He called it the "divine proportion" and considered it as a basis for perfection in the human body. The Mona Lisa, for example, includes the golden ratio in several rectangles. In the above figure, we can see how several of these rectangles can be overlayed on the painting and how the edges of these new figures come to all the important focal points of the woman: her chin, her eye, her nose, and corner of her famously mysterious mouth. This painting seems to be made purposefully to line up with the golden rectangle. In this section we will expand our mathematical work with the golden ratio to include multiplication and division of radical expressions.

We have worked with the golden rectangle more than once in this chapter. Remember the following diagram, the general representation of a golden rectangle:

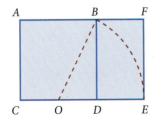

By now you know that in any golden rectangle constructed from a square (of any size) the ratio of the length to the width will be

$$\frac{1 + \sqrt{5}}{2}$$

which we call the golden ratio. What is interesting is that the smaller rectangle on the right of the figure, *BFED*, is also a golden rectangle. We will use the mathematics developed in this section to confirm this fact.

In this section we will look at multiplication and division of expressions that contain radicals. As you will see, multiplication of expressions that contain radicals is very similar to multiplication of polynomials. The division problems in this section are just an extension of the work we did previously when we rationalized denominators.

A Multiplication of Radical Expressions

EXAMPLE 1   Multiply $(3\sqrt{5})(2\sqrt{7})$.

SOLUTION   We can rearrange the order and grouping of the numbers in this product by applying the commutative and associative properties. Then, we apply the product property for radicals and multiply.

$$(3\sqrt{5})(2\sqrt{7}) = (3 \cdot 2)(\sqrt{5} \cdot \sqrt{7}) \quad \text{Commutative and associative properties}$$
$$= (3 \cdot 2)(\sqrt{5} \cdot \sqrt{7}) \quad \text{Product property for radicals}$$
$$= 6\sqrt{35} \qquad\qquad \text{Multiply.} \qquad\blacksquare$$

In practice, it is not necessary to show the first two steps.

EXAMPLE 2   Multiply $\sqrt{3}(2\sqrt{6} - 5\sqrt{12})$.

SOLUTION   Applying the distributive property, we have

$$\sqrt{3}(2\sqrt{6} - 5\sqrt{12}) = \sqrt{3} \cdot 2\sqrt{6} - \sqrt{3} \cdot 5\sqrt{12}$$
$$= 2\sqrt{18} - 5\sqrt{36}$$

Writing each radical in simplified form gives

$$2\sqrt{18} - 5\sqrt{36} = 2\sqrt{9} \cdot \sqrt{2} - 5\sqrt{36}$$
$$= 6\sqrt{2} - 30 \qquad\blacksquare$$

EXAMPLE 3   Multiply $(\sqrt{3} + \sqrt{5})(4\sqrt{3} - \sqrt{5})$.

SOLUTION   The same principle that applies when multiplying two binomials applies to this product. We must multiply each term in the first expression by each term in the second one. Any convenient method can be used. Let's use the FOIL method.

$$(\sqrt{3} + \sqrt{5})(4\sqrt{3} - \sqrt{5}) = \sqrt{3} \cdot 4\sqrt{3} - \sqrt{3} \cdot \sqrt{5} + \sqrt{5} \cdot 4\sqrt{3} - \sqrt{5} \cdot \sqrt{5}$$
$$\qquad\qquad\qquad\qquad\quad \text{F} \qquad\qquad \text{O} \qquad\qquad \text{I} \qquad\qquad \text{L}$$
$$= 4 \cdot 3 - \sqrt{15} + 4\sqrt{15} - 5$$
$$= 12 + 3\sqrt{15} - 5$$
$$= 7 + 3\sqrt{15} \qquad\blacksquare$$

EXAMPLE 4   Expand and simplify $(\sqrt{x} + 3)^2$.

SOLUTION 1   We can write this problem as a multiplication problem and proceed as we did in Example 3:

$$(\sqrt{x} + 3)^2 = (\sqrt{x} + 3)(\sqrt{x} + 3)$$

$$= \underset{F}{\sqrt{x} \cdot \sqrt{x}} + \underset{O}{3\sqrt{x}} + \underset{I}{3\sqrt{x}} + \underset{L}{3 \cdot 3}$$

$$= x + 3\sqrt{x} + 3\sqrt{x} + 9$$

$$= x + 6\sqrt{x} + 9$$

SOLUTION 2   We can obtain the same result by applying the formula for the square of a sum: $(a + b)^2 = a^2 + 2ab + b^2$.

$$(\sqrt{x} + 3)^2 = (\sqrt{x})^2 + 2(\sqrt{x})(3) + 3^2$$

$$= x + 6\sqrt{x} + 9 \qquad \blacksquare$$

EXAMPLE 5   Expand $(3\sqrt{x} - 2\sqrt{y})^2$ and simplify the result.

SOLUTION   Let's apply the formula for the square of a difference, $(a - b)^2 = a^2 - 2ab + b^2$.

$$(3\sqrt{x} - 2\sqrt{y})^2 = (3\sqrt{x})^2 - 2(3\sqrt{x})(2\sqrt{y}) + (2\sqrt{y})^2$$

$$= 9x - 12\sqrt{xy} + 4y \qquad \blacksquare$$

EXAMPLE 6   Expand and simplify $(\sqrt{x + 2} - 1)^2$.

SOLUTION   Applying the formula $(a - b)^2 = a^2 - 2ab + b^2$, we have

$$(\sqrt{x + 2} - 1)^2 = (\sqrt{x + 2})^2 - 2\sqrt{x + 2}(1) + 1^2$$

$$= x + 2 - 2\sqrt{x + 2} + 1$$

$$= x + 3 - 2\sqrt{x + 2} \qquad \blacksquare$$

EXAMPLE 7   Multiply $(\sqrt{6} + \sqrt{2})(\sqrt{6} - \sqrt{2})$.

SOLUTION   We notice the product is of the form $(a + b)(a - b)$, which always gives the difference of two squares, $a^2 - b^2$.

$$(\sqrt{6} + \sqrt{2})(\sqrt{6} - \sqrt{2}) = (\sqrt{6})^2 - (\sqrt{2})^2$$

$$= 6 - 2$$

$$= 4 \qquad \blacksquare$$

The two expressions $(\sqrt{6} + \sqrt{2})$ and $(\sqrt{6} - \sqrt{2})$ are called *conjugates*. In general, the conjugate of $\sqrt{a} + \sqrt{b}$ is $\sqrt{a} - \sqrt{b}$. If a and b are integers, multiplying conjugates of this form always produces a rational number.

---

4. Expand and simplify.
$(\sqrt{x} + 5)^2$

5. Expand $(5\sqrt{a} - 3\sqrt{b})^2$ and simplify the result.

6. Expand and simplify.
$(\sqrt{x + 3} - 1)^2$

7. Multiply.
$(\sqrt{5} + \sqrt{3})(\sqrt{5} - \sqrt{3})$

NOTE

We can prove that conjugates always multiply to yield a rational number as follows: If a and b are positive integers, then

$(\sqrt{a} + \sqrt{b})(\sqrt{a} - \sqrt{b})$
$= \sqrt{a} \cdot \sqrt{a} - \sqrt{a} \cdot \sqrt{b} + \sqrt{a} \cdot \sqrt{b} - \sqrt{b} \cdot \sqrt{b}$
$= a - \sqrt{ab} + \sqrt{ab} - b$
$= a - b$

which is rational if a and b are rational.

Answers

4. $x + 10\sqrt{x} + 25$
5. $25a - 30\sqrt{ab} + 9b$
6. $x + 4 - 2\sqrt{x + 3}$
7. 2

B Division of Radical Expressions

Division with radical expressions is the same as rationalizing the denominator. In a previous section we were able to divide $\sqrt{3}$ by $\sqrt{2}$ by rationalizing the denominator.

$$\frac{\sqrt{3}}{\sqrt{2}} = \frac{\sqrt{3}}{\sqrt{2}} \cdot \frac{\sqrt{2}}{\sqrt{2}} = \frac{\sqrt{6}}{2}$$

We can accomplish the same result with expressions such as

$$\frac{6}{\sqrt{5} - \sqrt{3}}$$

by multiplying the numerator and denominator by the conjugate of the denominator.

8. Divide $\dfrac{3}{\sqrt{7} - \sqrt{3}}$.

EXAMPLE 8 Divide $\dfrac{6}{\sqrt{5} - \sqrt{3}}$. (Rationalize the denominator.)

SOLUTION Since the product of two conjugates is a rational number, we multiply the numerator and denominator by the conjugate of the denominator.

$$\frac{6}{\sqrt{5} - \sqrt{3}} = \frac{6}{\sqrt{5} - \sqrt{3}} \cdot \frac{(\sqrt{5} + \sqrt{3})}{(\sqrt{5} + \sqrt{3})}$$

$$= \frac{6\sqrt{5} + 6\sqrt{3}}{(\sqrt{5})^2 - (\sqrt{3})^2}$$

$$= \frac{6\sqrt{5} + 6\sqrt{3}}{5 - 3}$$

$$= \frac{6\sqrt{5} + 6\sqrt{3}}{2}$$

The numerator and denominator of this last expression have a factor of 2 in common. We can reduce to lowest terms by factoring 2 from the numerator and then dividing both the numerator and denominator by 2.

$$= \frac{\cancel{2}(3\sqrt{5} + 3\sqrt{3})}{\cancel{2}}$$

$$= 3\sqrt{5} + 3\sqrt{3}$$ ∎

9. Rationalize the denominator.

$$\frac{\sqrt{10} - 3}{\sqrt{10} + 3}$$

EXAMPLE 9 Rationalize the denominator $\dfrac{\sqrt{5} - 2}{\sqrt{5} + 2}$.

SOLUTION To rationalize the denominator, we multiply the numerator and denominator by the conjugate of the denominator.

$$\frac{\sqrt{5} - 2}{\sqrt{5} + 2} = \frac{\sqrt{5} - 2}{\sqrt{5} + 2} \cdot \frac{(\sqrt{5} - 2)}{(\sqrt{5} - 2)}$$

$$= \frac{5 - 2\sqrt{5} - 2\sqrt{5} + 4}{(\sqrt{5})^2 - 2^2}$$

$$= \frac{9 - 4\sqrt{5}}{5 - 4}$$

$$= \frac{9 - 4\sqrt{5}}{1}$$

$$= 9 - 4\sqrt{5}$$ ∎

Answers

8. $\dfrac{3\sqrt{7} + 3\sqrt{3}}{4}$

9. $19 - 6\sqrt{10}$

Problem Set 7.5

Moving Toward Success

"Do not worry if you have built your castles in the air. They are where they should be. Now put the foundations under them."

—Henry David Thoreau, 1817–1862, American author and naturalist

1. Why should you expect to spend more time on challenging material?

2. Each chapter is a building block for the next. How can this fact help you and potentially hurt you if you're not careful?

A Multiply. (Assume all expressions appearing under a square root symbol represent nonnegative numbers throughout this problem set.) [Examples 1–7]

1. $\sqrt{6}\ \sqrt{3}$

2. $\sqrt{6}\ \sqrt{2}$

3. $(2\sqrt{3})(5\sqrt{7})$

4. $(3\sqrt{5})(2\sqrt{7})$

5. $(4\sqrt{6})(2\sqrt{15})(3\sqrt{10})$

6. $(4\sqrt{35})(2\sqrt{21})(5\sqrt{15})$

7. $(3\sqrt[3]{3})(6\sqrt[3]{9})$

8. $(2\sqrt[3]{2})(6\sqrt[3]{4})$

9. $\sqrt{3}(\sqrt{2} - 3\sqrt{3})$

10. $\sqrt{2}(5\sqrt{3} + 4\sqrt{2})$

11. $6\sqrt[3]{4}(2\sqrt[3]{2} + 1)$

12. $7\sqrt[3]{5}(3\sqrt[3]{25} - 2)$

13. $\sqrt[3]{4}(\sqrt[3]{2} + \sqrt[3]{6})$

14. $\sqrt[3]{5}(\sqrt[3]{8} - \sqrt[3]{25})$

15. $\sqrt[3]{x}(\sqrt[3]{x^2y^4} + \sqrt[3]{x^5y})$

16. $\sqrt[3]{x^2y}(2\sqrt[3]{x^4y^2} - \sqrt[3]{x^2y^4})$

17. $\sqrt[4]{2x^3}(\sqrt[4]{8x^6} + \sqrt[4]{16x^9})$

18. $\sqrt[4]{8y^2}(\sqrt[4]{6y^6} - \sqrt[4]{8y^9})$

19. $(\sqrt{3} + \sqrt{2})(3\sqrt{3} - \sqrt{2})$

20. $(\sqrt{5} - \sqrt{2})(3\sqrt{5} + 2\sqrt{2})$

21. $(\sqrt{x} + 5)(\sqrt{x} - 3)$

22. $(\sqrt{x} + 4)(\sqrt{x} + 2)$

23. $(3\sqrt{6} + 4\sqrt{2})(\sqrt{6} + 2\sqrt{2})$

24. $(\sqrt{7} - 3\sqrt{3})(2\sqrt{7} - 4\sqrt{3})$

25. $(\sqrt{3} + 4)^2$

26. $(\sqrt{5} - 2)^2$

27. $(\sqrt{x} - 3)^2$

28. $(\sqrt{x} + 4)^2$

29. $(2\sqrt{a} - 3\sqrt{b})^2$

30. $(5\sqrt{a} - 2\sqrt{b})^2$

31. $(\sqrt{x-4} + 2)^2$

32. $(\sqrt{x-3} + 2)^2$

33. $(\sqrt{x-5} - 3)^2$

34. $(\sqrt{x-3} - 4)^2$

35. $(\sqrt{3} - \sqrt{2})(\sqrt{3} + \sqrt{2})$

36. $(\sqrt{5} - \sqrt{2})(\sqrt{5} + \sqrt{2})$

37. $(\sqrt{a} + 7)(\sqrt{a} - 7)$

38. $(\sqrt{a} + 5)(\sqrt{a} - 5)$

39. $(5 - \sqrt{x})(5 + \sqrt{x})$

40. $(3 - \sqrt{x})(3 + \sqrt{x})$

41. $(\sqrt{x-4} + 2)(\sqrt{x-4} - 2)$

42. $(\sqrt{x+3} + 5)(\sqrt{x+3} - 5)$

43. $(\sqrt{3} + 1)^3$

44. $(\sqrt{5} - 2)^3$

45. $(\sqrt[3]{3} + \sqrt[3]{2})(\sqrt[3]{9} + \sqrt[3]{4})$

46. $(\sqrt[3]{5} + \sqrt[3]{3})(\sqrt[3]{9} + \sqrt[3]{25})$

47. $(\sqrt[3]{x^5} + \sqrt[3]{y})(\sqrt[3]{x} + \sqrt[3]{y^2})$

48. $(\sqrt[3]{x^7} - \sqrt[3]{y^4})(\sqrt[3]{x^2} - \sqrt[3]{y^2})$

B Rationalize the denominator in each of the following. [Examples 8, 9]

49. $\dfrac{1}{\sqrt{2}}$

50. $\dfrac{x}{\sqrt{2}}$

51. $\dfrac{1}{\sqrt{x}}$

52. $\dfrac{3}{\sqrt{x}}$

53. $\dfrac{4}{\sqrt{3}}$

54. $\dfrac{4w}{\sqrt{2x}}$

55. $\dfrac{2x}{\sqrt{6}}$

56. $\dfrac{6x}{\sqrt{3}}$

57. $\dfrac{4}{\sqrt{10x}}$

58. $\sqrt{\dfrac{4x^2}{3x}}$

59. $\dfrac{2}{\sqrt{8}}$

60. $\sqrt{\dfrac{16xy}{4x^2y}}$

61. $\sqrt{\dfrac{32x^3y}{4xy^2}}$

62. $\dfrac{4}{\sqrt{6x^2y^5}}$

63. $\sqrt{\dfrac{12a^3b^3c}{8ab^5c^2}}$

64. $\sqrt{\dfrac{18x^3y^2z^4}{16xy^3z^2}}$

65. $\sqrt[3]{\dfrac{16a^4b^2c^4}{2ab^3c}}$

66. $\sqrt[3]{\dfrac{9x^5y^5z}{3x^3y^7z^2}}$

67. $\dfrac{\sqrt{2}}{\sqrt{6}-\sqrt{2}}$

68. $\dfrac{\sqrt{5}}{\sqrt{5}+\sqrt{3}}$

69. $\dfrac{\sqrt{5}}{\sqrt{5}+1}$

70. $\dfrac{\sqrt{7}}{\sqrt{7}-1}$

71. $\dfrac{\sqrt{x}}{\sqrt{x}-3}$

72. $\dfrac{\sqrt{x}}{\sqrt{x}+2}$

73. $\dfrac{\sqrt{5}}{2\sqrt{5}-3}$

74. $\dfrac{\sqrt{7}}{3\sqrt{7}-2}$

75. $\dfrac{3}{\sqrt{x}-\sqrt{y}}$

76. $\dfrac{2}{\sqrt{x}+\sqrt{y}}$

77. $\dfrac{\sqrt{6}+\sqrt{2}}{\sqrt{6}-\sqrt{2}}$

78. $\dfrac{\sqrt{5}-\sqrt{3}}{\sqrt{5}+\sqrt{3}}$

79. $\dfrac{\sqrt{7}-2}{\sqrt{7}+2}$

80. $\dfrac{\sqrt{11}+3}{\sqrt{11}-3}$

81. $\dfrac{\sqrt{a}+\sqrt{b}}{\sqrt{a}-\sqrt{b}}$

82. $\dfrac{\sqrt{a}-\sqrt{b}}{\sqrt{a}+\sqrt{b}}$

83. $\dfrac{\sqrt{x}+2}{\sqrt{x}-2}$

84. $\dfrac{\sqrt{x}-3}{\sqrt{x}+3}$

85. $\dfrac{2\sqrt{3}-\sqrt{7}}{3\sqrt{3}+\sqrt{7}}$

86. $\dfrac{5\sqrt{6}+2\sqrt{2}}{\sqrt{6}-\sqrt{2}}$

87. $\dfrac{3\sqrt{x}+2}{1+\sqrt{x}}$

88. $\dfrac{5\sqrt{x}-1}{2+\sqrt{x}}$

89. $\dfrac{2}{\sqrt{3}}+\sqrt{12}$

90. $\sqrt{2}+\dfrac{5}{\sqrt{8}}$

91. $\dfrac{1}{\sqrt{5}}+\sqrt{20}$

92. $\dfrac{4}{\sqrt{2}}+\sqrt{72}$

93. $\dfrac{6}{\sqrt{12}}-\sqrt{75}$

94. $\dfrac{5}{\sqrt{7}}-\sqrt{63}+\sqrt{28}$

95. $\dfrac{1}{\sqrt{3}}+\sqrt{48}+\dfrac{4}{\sqrt{12}}$

96. $\dfrac{4}{\sqrt{2}}+\sqrt{8}+\dfrac{10}{\sqrt{50}}$

97. Show that the product $(\sqrt[3]{2}+\sqrt[3]{3})(\sqrt[3]{4}-\sqrt[3]{6}+\sqrt[3]{9})$ is 5.

98. Show that the product $(\sqrt[3]{x}+2)(\sqrt[3]{x^2}-2\sqrt[3]{x}+4)$ is $x+8$.

Each statement below is false. Correct the right side of each one.

99. $5(2\sqrt{3}) = 10\sqrt{15}$

100. $3(2\sqrt{x}) = 6\sqrt{3x}$

101. $(\sqrt{x} + 3)^2 = x + 9$

102. $(\sqrt{x} - 7)^2 = x - 49$

103. $(5\sqrt{3})^2 = 15$

104. $(3\sqrt{5})^2 = 15$

Applying the Concepts

105. Gravity If an object is dropped from the top of a 100-foot building, the amount of time t (in seconds) that it takes for the object to be h feet from the ground is given by the formula

$$t = \frac{\sqrt{100 - h}}{4}$$

How long does it take before the object is 50 feet from the ground? How long does it take to reach the ground? (When it is on the ground, h is 0.)

106. Gravity Use the formula given in Problem 105 to determine h if t is 1.25 seconds.

107. Golden Rectangle Rectangle *ACEF* in Figure 2 is a golden rectangle. If side *AC* is 6 inches, show that the smaller rectangle *BDEF* is also a golden rectangle.

108. Golden Rectangle Rectangle *ACEF* in Figure 2 is a golden rectangle. If side *AC* is 1 inch, show that the smaller rectangle *BDEF* is also a golden rectangle.

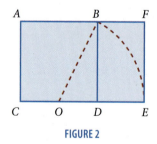

FIGURE 2

109. Golden Rectangle If side *AC* in Figure 2 is 2*x*, show the rectangle *BDEF* is a golden rectangle.

110. Golden Rectangle If side *AC* in Figure 2 is *x*, show that rectangle *BDEF* is a golden rectangle.

Maintaining Your Skills

Simplify each complex fraction.

111. $\dfrac{\frac{1}{4} - \frac{1}{3}}{\frac{1}{2} + \frac{1}{6}}$

112. $\dfrac{\frac{1}{8} - \frac{1}{3}}{\frac{1}{4} - \frac{1}{3}}$

113. $\dfrac{1 - \frac{2}{y}}{1 + \frac{2}{y}}$

114. $\dfrac{1 + \frac{3}{y}}{1 - \frac{3}{y}}$

115. $\dfrac{4 + \frac{4}{x} + \frac{1}{x^2}}{4 - \frac{1}{x^2}}$

116. $\dfrac{1 - \frac{1}{x} - \frac{6}{x^2}}{1 - \frac{9}{x^2}}$

Getting Ready for the Next Section

Simplify.

117. $(t + 5)^2$

118. $(x - 4)^2$

119. $\sqrt{x} \cdot \sqrt{x}$

120. $\sqrt{3x} \cdot \sqrt{3x}$

Solve.

121. $3x + 4 = 5^2$

122. $4x - 7 = 3^2$

123. $t^2 + 7t + 12 = 0$

124. $x^2 - 3x - 10 = 0$

125. $t^2 + 10t + 25 = t + 7$

126. $x^2 - 4x + 4 = x - 2$

127. $(x + 4)^2 = x + 6$

128. $(x - 6)^2 = x - 4$

Extending the Concepts

Simplify.

129. $\dfrac{x}{\sqrt{x-2}+4}$

130. $\dfrac{3}{\sqrt{x-4}-7}$

131. $\dfrac{x}{\sqrt{x+5}-5}$

132. $\dfrac{2}{\sqrt{2x-1}+3}$

133. $\dfrac{3x}{\sqrt{5x}+x}$

134. $\dfrac{4x}{\sqrt{2x^3}+2x}$

Equations with Radicals

OBJECTIVES

A Solve equations containing radicals.

B Graph simple square root and cube root equations in two variables.

TICKET TO SUCCESS

Keep these questions in mind as you read through the section. Then respond in your own words and in complete sentences.

1. What is the squaring property of equality?
2. Under what conditions do we obtain extraneous solutions to equations that contain radical expressions?
3. If we have raised both sides of an equation to a power, when is it not necessary to check for extraneous solutions?
4. Why does the graph of $y = \sqrt{x}$ appear in the first quadrant only?

George P. Chroma/Shutterstock.com

Imagine you are standing on the roof of a 100-foot-tall building. If you accidentally drop your car keys over the side of the building, how long will it take them to land in the first floor awning, 8 feet from the ground? How long before they hit the ground if they miss the awning? Both answers can be found by solving the following equation:

$$t = \frac{\sqrt{100 - h}}{4}$$

Solving equations like this that involve radicals is our focus in this section.

This section is concerned with solving equations that involve one or more radicals. The first step in solving an equation that contains a radical is to eliminate the radical from the equation. To do so, we need an additional property.

A Equations with Radicals

Squaring Property of Equality

If both sides of an equation are squared, the solutions to the original equation are solutions to the resulting equation.

We will never lose solutions to our equations by squaring both sides. We may, however, introduce *extraneous solutions*. Extraneous solutions satisfy the equation obtained by squaring both sides of the original equation, but they do not satisfy the original equation.

We know that if two real numbers a and b are equal, then so are their squares.

$$\text{If} \qquad a = b$$

$$\text{then} \qquad a^2 = b^2$$

However, extraneous solutions are introduced when we square opposites; that is, even though opposites are not equal, their squares are. For example,

$$5 = -5 \qquad \text{A false statement}$$

$$(5)^2 = (-5)^2 \qquad \text{Square both sides.}$$

$$25 = 25 \qquad \text{A true statement}$$

We are free to square both sides of an equation any time it is convenient. However, we must be aware that doing so may introduce extraneous solutions. We must, therefore, check all our solutions in the original equation if at any time we square both sides of the original equation.

PRACTICE PROBLEMS

1. Solve for x: $\sqrt{2x + 4} = 4$

EXAMPLE 1   Solve $\sqrt{3x + 4} = 5$ for x.

SOLUTION   We square both sides and proceed as usual.

$$\sqrt{3x + 4} = 5$$

$$(\sqrt{3x + 4})^2 = 5^2$$

$$3x + 4 = 25$$

$$3x = 21$$

$$x = 7$$

Checking $x = 7$ in the original equation, we have

$$\sqrt{3(7) + 4} \stackrel{?}{=} 5$$

$$\sqrt{21 + 4} \stackrel{?}{=} 5$$

$$\sqrt{25} \stackrel{?}{=} 5$$

$$5 = 5$$

The solution $x = 7$ satisfies the original equation. ∎

2. Solve $\sqrt{7x - 3} = -5$.

EXAMPLE 2   Solve $\sqrt{4x - 7} = -3$.

SOLUTION   Squaring both sides, we have

$$\sqrt{4x - 7} = -3$$

$$(\sqrt{4x - 7})^2 = (-3)^2$$

$$4x - 7 = 9$$

$$4x = 16$$

$$x = 4$$

Answers

1. 6

2. No solution

Checking $x = 4$ in the original equation gives

$$\sqrt{4(4) - 7} \overset{?}{=} -3$$

$$\sqrt{16 - 7} \overset{?}{=} -3$$

$$\sqrt{9} \overset{?}{=} -3$$

$$3 = -3 \qquad \text{A false statement}$$

The solution $x = 4$ produces a false statement when checked in the original equation. Since $x = 4$ was the only possible solution, there is no solution to the original equation. The possible solution $x = 4$ is an extraneous solution. It satisfies the equation obtained by squaring both sides of the original equation, but it does not satisfy the original equation. ∎

EXAMPLE 3 Solve $\sqrt{5x - 1} + 3 = 7$.

SOLUTION We must isolate the radical on the left side of the equation. If we attempt to square both sides without doing so, the resulting equation will also contain a radical. Adding -3 to both sides, we have

$$\sqrt{5x - 1} + 3 = 7$$

$$\sqrt{5x - 1} = 4$$

We can now square both sides and proceed as usual.

$$(\sqrt{5x - 1})^2 = 4^2$$

$$5x - 1 = 16$$

$$5x = 17$$

$$x = \frac{17}{5}$$

Checking $x = \frac{17}{5}$, we have

$$\sqrt{5\left(\frac{17}{5}\right) - 1} + 3 \overset{?}{=} 7$$

$$\sqrt{17 - 1} + 3 \overset{?}{=} 7$$

$$\sqrt{16} + 3 \overset{?}{=} 7$$

$$4 + 3 \overset{?}{=} 7$$

$$7 = 7$$

Our solution checks. ∎

EXAMPLE 4 Solve $t + 5 = \sqrt{t + 7}$.

SOLUTION This time, squaring both sides of the equation results in a quadratic equation.

$$(t + 5)^2 = (\sqrt{t + 7})^2 \qquad \text{Square both sides.}$$

$$t^2 + 10t + 25 = t + 7$$

$$t^2 + 9t + 18 = 0 \qquad \text{Standard form}$$

$$(t + 3)(t + 6) = 0 \qquad \text{Factor the left side.}$$

$$t + 3 = 0 \quad \text{or} \quad t + 6 = 0 \qquad \text{Set factors equal to 0.}$$

$$t = -3 \qquad\qquad t = -6$$

NOTE

The fact that there is no solution to the equation in Example 2 was obvious to begin with. Notice that the left side of the equation is the *positive* square root of $4x - 7$, which must be a nonnegative number. The right side of the equation is -3. Since we cannot have a number that is nonnegative equal to a negative number, there is no solution to the equation.

3. Solve $\sqrt{4x + 5} + 2 = 7$.

4. Solve $t - 6 = \sqrt{t - 4}$.

Answers
3. 5
4. 8

We must check each solution in the original equation.

Check $t = -3$ Check $t = -6$

$-3 + 5 \stackrel{?}{=} \sqrt{-3 + 7}$ $-6 + 5 \stackrel{?}{=} \sqrt{-6 + 7}$

$2 \stackrel{?}{=} \sqrt{4}$ $-1 \stackrel{?}{=} \sqrt{1}$

$2 = 2$ $-1 = 1$

A true statement   A false statement

Since $t = -6$ does not check, our only solution is $t = -3$. ∎

5. Solve $\sqrt{x - 9} = \sqrt{x} - 3$.

EXAMPLE 5   Solve $\sqrt{x - 3} = \sqrt{x} - 3$.

SOLUTION   We begin by squaring both sides. Note what happens when we square the right side of the equation, and compare the square of the right side with the square of the left side. You must convince yourself that these results are correct. (The note in the margin will help if you are having trouble convincing yourself that what is written below is true.)

$$(\sqrt{x - 3})^2 = (\sqrt{x} - 3)^2$$

$$x - 3 = x - 6\sqrt{x} + 9$$

Now we still have a radical in our equation, so we will have to square both sides again. Before we do, though, let's isolate the remaining radical.

$x - 3 = x - 6\sqrt{x} + 9$

$-3 = -6\sqrt{x} + 9$ Add $-x$ to each side.

$-12 = -6\sqrt{x}$ Add -9 to each side.

$2 = \sqrt{x}$ Divide each side by -6.

$4 = x$ Square each side.

Our only possible solution is $x = 4$, which we check in our original equation as follows:

$\sqrt{4 - 3} \stackrel{?}{=} \sqrt{4} - 3$

$\sqrt{1} \stackrel{?}{=} 2 - 3$

$1 = -1$ A false statement ∎

Substituting 4 for x in the original equation yields a false statement. Since 4 was our only possible solution, there is no solution to our equation.

Here is another example of an equation for which we must apply our squaring property twice before all radicals are eliminated.

6. Solve $\sqrt{x + 4} = 2 - \sqrt{3x}$.

EXAMPLE 6   Solve $\sqrt{x + 1} = 1 - \sqrt{2x}$.

SOLUTION   This equation has two separate terms involving radical signs. Squaring both sides gives

$x + 1 = 1 - 2\sqrt{2x} + 2x$

$-x = -2\sqrt{2x}$ Add $-2x$ and -1 to both sides.

$x^2 = 4(2x)$ Square both sides.

$x^2 - 8x = 0$ Standard form

NOTE

It is very important that you realize that the square of $(\sqrt{x} - 3)$ is not $x + 9$. Remember, when we square a difference with two terms, we use the formula

$(a - b)^2 = a^2 - 2ab + b^2$

Applying this formula to $(\sqrt{x} - 3)^2$, we have

$(\sqrt{x} - 3)^2$
$= (\sqrt{x})^2 - 2(\sqrt{x})(3) + 3^2$
$= x - 6\sqrt{x} + 9$

Answers

5. 9

6. 0

Our equation is a quadratic equation in standard form. To solve for x, we factor the left side and set each factor equal to 0.

$$x(x - 8) = 0 \qquad \text{Factor left side.}$$

$$x = 0 \quad \text{or} \quad x - 8 = 0 \qquad \text{Set factors equal to 0.}$$

$$x = 8$$

Since we squared both sides of our equation, we have the possibility that one or both of the solutions are extraneous. We must check each one in the original equation.

Check $x = 8$	Check $x = 0$
$\sqrt{8 + 1} \overset{?}{=} 1 - \sqrt{2 \cdot 8}$	$\sqrt{0 + 1} \overset{?}{=} 1 - \sqrt{2 \cdot 0}$
$\sqrt{9} \overset{?}{=} 1 - \sqrt{16}$	$\sqrt{1} \overset{?}{=} 1 - \sqrt{0}$
$3 \overset{?}{=} 1 - 4$	$1 \overset{?}{=} 1 - 0$
$3 = -3$	$1 = 1$
A false statement	A true statement

Since $x = 8$ does not check, it is an extraneous solution. Our only solution is $x = 0$. ∎

EXAMPLE 7   Solve $\sqrt{x + 1} = \sqrt{x + 2} - 1$.

SOLUTION   Squaring both sides we have

$$(\sqrt{x + 1})^2 = (\sqrt{x + 2} - 1)^2$$

$$x + 1 = x + 2 - 2\sqrt{x + 2} + 1$$

Once again we are left with a radical in our equation. Before we square each side again, we must isolate the radical on the right side of the equation.

$$x + 1 = x + 3 - 2\sqrt{x + 2} \qquad \text{Simplify the right side.}$$

$$1 = 3 - 2\sqrt{x + 2} \qquad \text{Add } -x \text{ to each side.}$$

$$-2 = -2\sqrt{x + 2} \qquad \text{Add } -3 \text{ to each side.}$$

$$1 = \sqrt{x + 2} \qquad \text{Divide each side by } -2.$$

$$1 = x + 2 \qquad \text{Square both sides.}$$

$$-1 = x \qquad \text{Add } -2 \text{ to each side.}$$

Checking our only possible solution, $x = -1$, in our original equation, we have

$$\sqrt{-1 + 1} \overset{?}{=} \sqrt{-1 + 2} - 1$$

$$\sqrt{0} \overset{?}{=} \sqrt{1} - 1$$

$$0 \overset{?}{=} 1 - 1$$

$$0 = 0 \qquad \text{A true statement.}$$

Our solution checks. ∎

It is also possible to raise both sides of an equation to powers greater than 2. We only need to check for extraneous solutions when we raise both sides of an equation to an even power. Raising both sides of an equation to an odd power will not produce extraneous solutions.

---

7. Solve $\sqrt{x + 2} = \sqrt{x + 3} - 1$.

Answer

7. -2

8. Solve $\sqrt[3]{3x - 7} = 2$.

EXAMPLE 8 Solve $\sqrt[3]{4x + 5} = 3$.

SOLUTION Cubing both sides we have

$$(\sqrt[3]{4x + 5})^3 = 3^3$$

$$4x + 5 = 27$$

$$4x = 22$$

$$x = \frac{22}{4}$$

$$x = \frac{11}{2}$$

We do not need to check $x = \frac{11}{2}$ since we raised both sides to an odd power. ∎

9. Graph each equation:
$$y = \sqrt{x} + 3$$

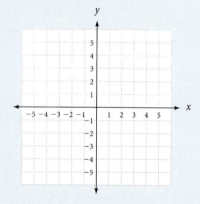

B Graphing Radicals

We end this section by looking at graphs of some equations that contain radicals.

EXAMPLE 9 Graph $y = \sqrt{x}$ and $y = \sqrt[3]{x}$.

SOLUTION The graphs are shown in Figures 1 and 2. Notice that the graph of $y = \sqrt{x}$ appears in the first quadrant only because in the equation $y = \sqrt{x}$, x and y cannot be negative.

The graph of $y = \sqrt[3]{x}$ appears in quadrants 1 and 3 since the cube root of a positive number is also a positive number and the cube root of a negative number is a negative number; that is, when x is positive, y will be positive and when x is negative, y will be negative.

The graphs of both equations will contain the origin since $y = 0$ when $x = 0$ in both equations.

$$y = \sqrt{x} + 3$$

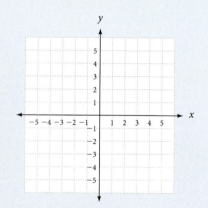

x	y
-4	undefined
-1	undefined
0	0
1	1
4	2
9	3
16	4

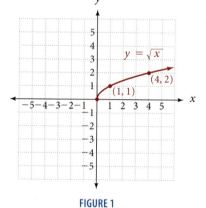

FIGURE 1

x	y
-27	-3
-8	-2
-1	-1
0	0
1	1
8	2
27	3

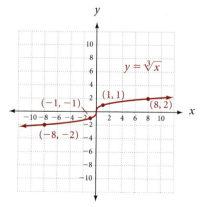

FIGURE 2

Answers

8. 5

9. See Solutions Section.

Problem Set 7.6

Moving Toward Success

"He that is good for making excuses is seldom good for anything else."

—Benjamin Franklin, 1706–1790, American scientist and diplomat

1. Should you go to your study group to just listen to the other members? Or should you take turns teaching the material? Explain.

2. Why is it important to take quick breaks during your study group sessions?

A Solve each of the following equations. [Examples 1–4, 8]

1. $\sqrt{2x + 1} = 3$

2. $\sqrt{3x + 1} = 4$

3. $\sqrt{4x + 1} = -5$

4. $\sqrt{6x + 1} = -5$

5. $\sqrt{2y - 1} = 3$

6. $\sqrt{3y - 1} = 2$

7. $\sqrt{5x - 7} = -1$

8. $\sqrt{8x + 3} = -6$

9. $\sqrt{2x - 3} - 2 = 4$

10. $\sqrt{3x + 1} - 4 = 1$

11. $\sqrt{4x + 1} + 3 = 2$

12. $\sqrt{5a - 3} + 6 = 2$

13. $\sqrt[3]{3x + 1} = 2$

14. $\sqrt[4]{4x + 1} = 3$

15. $\sqrt[3]{2x - 5} = 1$

16. $\sqrt[3]{5x + 7} = 2$

17. $\sqrt[3]{3a + 5} = -3$

18. $\sqrt[3]{2a + 7} = -2$

19. $\sqrt{y - 3} = y - 3$

20. $\sqrt{y + 3} = y - 3$

21. $\sqrt{a + 2} = a + 2$

22. $\sqrt{a + 10} = a - 2$

23. $\sqrt{2x + 3} = \dfrac{2x - 7}{3}$

24. $\sqrt{3x - 2} = \dfrac{2x - 3}{3}$

25. $\sqrt{4x - 3} = \dfrac{x + 3}{2}$

26. $\sqrt{4x + 5} = \dfrac{x + 3}{2}$

27. $\sqrt{7x + 2} = \dfrac{2x + 2}{3}$

28. $\sqrt{7x - 3} = \dfrac{4x - 1}{3}$

29. $\sqrt{2x + 4} = \sqrt{1 - x}$

30. $\sqrt{3x + 4} = -\sqrt{2x + 3}$

31. $\sqrt{4a + 7} = -\sqrt{a + 2}$

32. $\sqrt{7a - 1} = \sqrt{2a + 4}$

33. $\sqrt[4]{5x - 8} = \sqrt[4]{4x - 1}$

34. $\sqrt[4]{6x + 7} = \sqrt[4]{x + 2}$

35. $x + 1 = \sqrt{5x + 1}$

36. $x - 1 = \sqrt{6x + 1}$

37. $t + 5 = \sqrt{2t + 9}$

38. $t + 7 = \sqrt{2t + 13}$

39. $\sqrt{y - 8} = \sqrt{8 - y}$

40. $\sqrt{2y + 5} = \sqrt{5y + 2}$

41. $\sqrt[3]{3x + 5} = \sqrt[3]{5 - 2x}$

42. $\sqrt[3]{4x + 9} = \sqrt[3]{3 - 2x}$

A The following equations will require that you square both sides twice before all the radicals are eliminated.
[Examples 5–7]

43. $\sqrt{x - 8} = \sqrt{x} - 2$

44. $\sqrt{x + 3} = \sqrt{x} - 3$

45. $\sqrt{x + 1} = \sqrt{x} + 1$

46. $\sqrt{x - 1} = \sqrt{x} - 1$

47. $\sqrt{x + 8} = \sqrt{x - 4} + 2$

48. $\sqrt{x + 5} = \sqrt{x - 3} + 2$

49. $\sqrt{x - 5} - 3 = \sqrt{x - 8}$

50. $\sqrt{x - 3} - 4 = \sqrt{x - 3}$

51. $\sqrt{x + 4} = 2 - \sqrt{2x}$

52. $\sqrt{5x + 1} = 1 + \sqrt{5x}$

53. $\sqrt{2x + 4} = \sqrt{x + 3} + 1$

54. $\sqrt{2x - 1} = \sqrt{x - 4} + 2$

Let $f(x) = \sqrt{2x - 1}$. Find all values for the variable x that produce the following values of $f(x)$.

55. $f(x) = 0$

56. $f(x) = -10$

57. $f(x) = 2x - 1$

58. $f(x) = \sqrt{5x - 10}$

59. $f(x) = \sqrt{x - 4} + 2$

60. $f(x) = 9$

Let $g(x) = \sqrt{2x + 3}$. Find all values for the variable x that produce the following values of $g(x)$.

61. $g(x) = 0$

62. $g(x) = x$

63. $g(x) = -\sqrt{5x}$

64. $g(x) = \sqrt{x^2 - 5}$

Let $f(x) = \sqrt{2x} - 1$. Find all values for the variable x for which $f(x) = g(x)$.

65. $g(x) = 0$

66. $g(x) = 5$

67. $g(x) = \sqrt{2x + 5}$

68. $g(x) = x - 1$

Let $h(x) = \sqrt[3]{3x + 5}$. Find all values for the variable x for which $h(x) = f(x)$.

69. $f(x) = 2$

70. $f(x) = -1$

Let $h(x) = \sqrt[3]{5 - 2x}$. Find all values for the variable x for which $h(x) = f(x)$.

71. $f(x) = 3$

72. $f(x) = -1$

B Graph each equation. [Example 9]

73. $y = 2\sqrt{x}$

74. $y = -2\sqrt{x}$

75. $y = \sqrt{x} - 2$

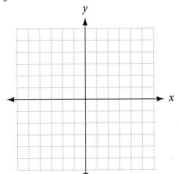

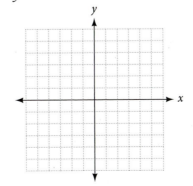

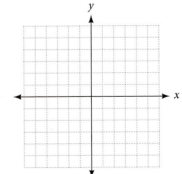

76. $y = \sqrt{x} + 2$

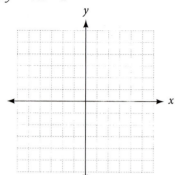

77. $y = \sqrt{x - 2}$

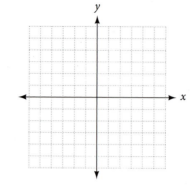

78. $y = \sqrt{x + 2}$

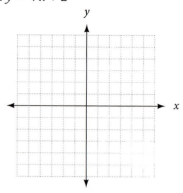

79. $y = 3\sqrt[3]{x}$

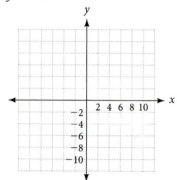

80. $y = -3\sqrt[3]{x}$

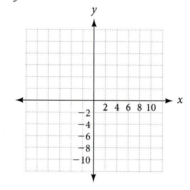

81. $y = \sqrt[3]{x} + 3$

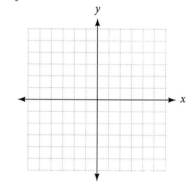

82. $y = \sqrt[3]{x} - 3$

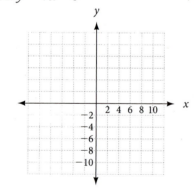

83. $y = \sqrt[3]{x + 3}$

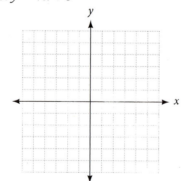

84. $y = \sqrt[3]{x - 3}$

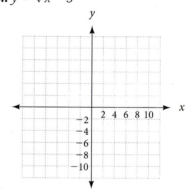

Applying the Concepts

85. Solve the following formula for h:

$$t = \frac{\sqrt{100 - h}}{4}$$

86. Solve the following formula for h:

$$t = \sqrt{\frac{2h - 40t}{g}}$$

87. Pendulum clock The length of time, T, in seconds it takes the pendulum of a grandfather clock to swing through one complete cycle is given by the formula

$$T = 2\pi \sqrt{\frac{L}{32}}$$

where L is the length, in feet, of the pendulum, and π is approximately $\frac{22}{7}$. How long must the pendulum be if one complete cycle takes 2 seconds?

1 sec

88. Falling Objects If you dropped your keys from the top of a 100-foot building, the equation to describe their height from the ground could be given by the equation,

$$t = \frac{\sqrt{100 - h}}{4}$$

How long until the keys are 8 feet from the ground? How long until they hit the ground?

Pollution A long straight river, 100 meters wide, is flowing at 1 meter per second. A pollutant is entering the river at a constant rate from one of its banks. As the pollutant disperses in the water, it forms a plume that is modeled by the equation $y = \sqrt{x}$. Use this information to answer the following questions.

89. How wide is the plume 25 meters down river from the source of the pollution?

90. How wide is the plume 100 meters down river from the source of the pollution?

91. How far down river from the source of the pollution does the plume reach halfway across the river?

92. How far down river from the source of the pollution does the plume reach the other side of the river?

Maintaining Your Skills

Multiply.

93. $\sqrt{2}(\sqrt{3} - \sqrt{2})$

94. $(\sqrt{x} - 4)(\sqrt{x} + 5)$

95. $(\sqrt{x} + 5)^2$

96. $(\sqrt{5} + \sqrt{3})(\sqrt{5} - \sqrt{3})$

Rationalize the denominator.

97. $\dfrac{\sqrt{x}}{\sqrt{x} + 3}$

98. $\dfrac{\sqrt{5} - \sqrt{3}}{\sqrt{5} + \sqrt{3}}$

Getting Ready for the Next Section

Simplify.

99. $\sqrt{25}$

100. $\sqrt{49}$

101. $\sqrt{12}$

102. $\sqrt{50}$

103. $(-1)^{15}$

104. $(-1)^{20}$

105. $(-1)^{50}$

106. $(-1)^5$

Solve.

107. $3x = 12$

108. $4 = 8y$

109. $4x - 3 = 5$

110. $7 = 2y - 1$

Perform the indicated operation.

111. $(3 + 4x) + (7 - 6x)$

112. $(2 - 5x) + (-1 + 7x)$

113. $(7 + 3x) - (5 + 6x)$

114. $(5 - 2x) - (9 - 4x)$

115. $(3 - 4x)(2 + 5x)$

116. $(8 + x)(7 - 3x)$

117. $2x(4 - 6x)$

118. $3x(7 + 2x)$

119. $(2 + 3x)^2$

120. $(3 + 5x)^2$

121. $(2 - 3x)(2 + 3x)$

122. $(4 - 5x)(4 + 5x)$

Complex Numbers

OBJECTIVES

A	Simplify square roots of negative numbers.
B	Simplify powers of i.
C	Solve for unknown variables by equating real parts and by equating imaginary parts of two complex numbers.
D	Add and subtract complex numbers.
E	Multiply complex numbers.
F	Divide complex numbers.

TICKET TO SUCCESS

Keep these questions in mind as you read through the section. Then respond in your own words and in complete sentences.

1. What is the number i?
2. Explain why every real number is a complex number.
3. What kind of number results when we multiply complex conjugates?
4. Explain how to divide complex numbers.

©NTERFOTO/Alamy

Known by some as the Prince of Mathematicians, Carl Friedrich Gauss made profound contributions to many fields of study. In his early twenties, he provided significant proof of the theorem behind complex numbers, greatly clarifying the concept for the academic world. The stamp shown here was issued by Germany in 1977 to commemorate the 200th anniversary of the birth of Carl Gauss. In this section, we will learn about complex numbers and how they are used in algebra.

Working with complex numbers gives us a way to solve a wider variety of equations. For example, the equation $x^2 = -9$ has no real number solutions since the square of a real number is always positive. We have been unable to work with square roots of negative numbers like $\sqrt{-25}$ and $\sqrt{-16}$ for the same reason.

Complex numbers allow us to expand our work with radicals to include square roots of negative numbers and solve equations like $x^2 = -9$ and $x^2 = -64$. Our work with complex numbers is based on the following definition.

rook76/Shutterstock.com

A Square Roots of Negative Numbers

> **Definition**
>
> The **number i** is such that $i = \sqrt{-1}$, which is the same as saying $i^2 = -1$.

The number i, as we have defined it here, is not a real number. Because of the way we have defined i, we can use it to simplify square roots of negative numbers.

> **Rule**
>
> If a is a positive number, then $\sqrt{-a}$ can always be written as $i\sqrt{a}$; that is,
>
> $$\sqrt{-a} = i\sqrt{a} \text{ if } a \text{ is a positive number.}$$

To justify our rule, we simply square the quantity $i\sqrt{a}$ to obtain $-a$. Here is what it looks like when we do so:

$$(i\sqrt{a})^2 = i^2 \cdot (\sqrt{a})^2$$
$$= -1 \cdot a$$
$$= -a$$

Here are some examples that illustrate the use of our new rule.

PRACTICE PROBLEMS

1. a. $\sqrt{-36}$

 b. $-\sqrt{-64}$

 c. $\sqrt{-18}$

 d. $-\sqrt{-19}$

EXAMPLE 1 Write each square root in terms of the number i.

 a. $\sqrt{-25} = i\sqrt{25} = i \cdot 5 = 5i$

 b. $\sqrt{-49} = i\sqrt{49} = i \cdot 7 = 7i$

 c. $\sqrt{-12} = i\sqrt{12} = i \cdot 2\sqrt{3} = 2i\sqrt{3}$

 d. $\sqrt{-17} = i\sqrt{17}$ ∎

B Powers of i

If we assume all the properties of exponents hold when the base is i, we can write any power of i as i, -1, $-i$, or 1. Using the fact that $i^2 = -1$, we have

$$i^1 = i$$

$$i^2 = -1$$

$$i^3 = i^2 \cdot i = -1(i) = -i$$

$$i^4 = i^2 \cdot i^2 = -1(-1) = 1$$

NOTE
In Examples 1c and 1d, we wrote i before the radical simply to avoid confusion. If we were to write the answer to 1c as $2\sqrt{3}i$, some people would think the i was under the radical sign and it is not.

Since $i^4 = 1$, i^5 will simplify to i, and we will begin repeating the sequence i, -1, $-i$, 1 as we simplify higher powers of i: any power of i simplifies to i, -1, $-i$, or 1. The easiest way to simplify higher powers of i is to write them in terms of i^2. For instance, to simplify i^{21}, we would write it as

$$(i^2)^{10} \cdot i \quad \text{because } 2 \cdot 10 + 1 = 21$$

Then, since $i^2 = -1$, we have

$$(-1)^{10} \cdot i = 1 \cdot i = i$$

Answers

1. a. $6i$ **b.** $-8i$

 c. $3i\sqrt{2}$ **d.** $-i\sqrt{19}$

EXAMPLE 2 Simplify as much as possible.

a. $i^{30} = (i^2)^{15} = (-1)^{15} = -1$

b. $i^{11} = (i^2)^5 \cdot i = (-1)^5 \cdot i = (-1)i = -i$

c. $i^{40} = (i^2)^{20} = (-1)^{20} = 1$ ∎

Simplify.

2. a. i^{20}

b. i^{23}

c. i^{50}

> **Definition**
>
> A **complex number** is any number that can be put in the form
>
> $$a + bi$$
>
> where a and b are real numbers and $i = \sqrt{-1}$. The form $a + bi$ is called **standard form** for complex numbers. The number a is called the **real part** of the complex number. The number b is called the **imaginary part** of the complex number.

Every real number is a complex number. For example, 8 can be written as $8 + 0i$. Likewise, $-\frac{1}{2}, \pi, \sqrt{3}$, and -9 are complex numbers because they can all be written in the form $a + bi$.

$$-\frac{1}{2} = -\frac{1}{2} + 0i \qquad \pi = \pi + 0i$$

$$\sqrt{3} = \sqrt{3} + 0i \qquad -9 = -9 + 0i$$

The real numbers occur when $b = 0$. When $b \neq 0$, we have complex numbers that contain i, such as $2 + 5i$, $6 - i$, $4i$, and $\frac{1}{2}i$. These numbers are called *imaginary numbers*. The diagram explains this further.

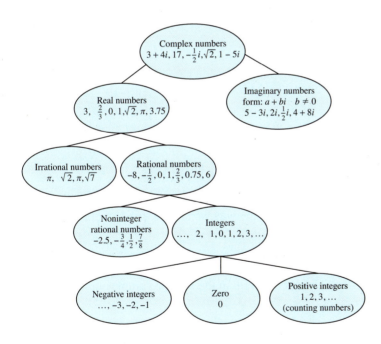

C Equality for Complex Numbers

Two complex numbers are equal if and only if their real parts are equal and their imaginary parts are equal; that is, for real numbers a, b, c, and d,

$$a + bi = c + di \quad \text{if and only if} \quad a = c \quad \text{and} \quad b = d$$

Answers

2. a. 1 **b.** $-i$ **c.** -1

3. Find x and y if
$$4x + 7i = 8 - 14yi.$$

EXAMPLE 3 Find x and y if $3x + 4i = 12 - 8yi$.

SOLUTION Since the two complex numbers are equal, their real parts are equal and their imaginary parts are equal.

$$3x = 12 \quad \text{and} \quad 4 = -8y$$
$$x = 4 \qquad\qquad y = -\frac{1}{2}$$ ∎

4. Find x and y if
$$(2x - 1) + 9i = 5 + (4y + 1)i.$$

EXAMPLE 4 Find x and y if $(4x - 3) + 7i = 5 + (2y - 1)i$.

SOLUTION The real parts are $4x - 3$ and 5. The imaginary parts are 7 and $2y - 1$.

$$4x - 3 = 5 \quad \text{and} \quad 7 = 2y - 1$$
$$4x = 8 \qquad\qquad 8 = 2y$$
$$x = 2 \qquad\qquad y = 4$$ ∎

D Addition and Subtraction of Complex Numbers

To add two complex numbers, add their real parts and add their imaginary parts; that is, if a, b, c, and d are real numbers, then

$$(a + bi) + (c + di) = (a + c) + (b + d)i$$

If we assume that the commutative, associative, and distributive properties hold for the number i, then the definition of addition is simply an extension of these properties.

We define subtraction in a similar manner. If a, b, c, and d are real numbers, then

$$(a + bi) - (c + di) = (a - c) + (b - d)i$$

5. Add or subtract as indicated.

a. $(2 + 6i) + (3 - 4i)$

b. $(6 + 5i) - (4 + 3i)$

c. $(7 - i) - (8 - 2i)$

EXAMPLE 5 Add or subtract as indicated.

a. $(3 + 4i) + (7 - 6i) = (3 + 7) + (4 - 6)i = 10 - 2i$

b. $(7 + 3i) - (5 + 6i) = (7 - 5) + (3 - 6)i = 2 - 3i$

c. $(5 - 2i) - (9 - 4i) = (5 - 9) + (-2 + 4)i = -4 + 2i$ ∎

E Multiplication of Complex Numbers

Since complex numbers have the same form as binomials, we find the product of two complex numbers the same way we find the product of two binomials.

6. Multiply $(2 + 3i)(1 - 4i)$.

EXAMPLE 6 Multiply $(3 - 4i)(2 + 5i)$.

SOLUTION Multiplying each term in the second complex number by each term in the first, we have

$$(3 - 4i)(2 + 5i) = 3 \cdot 2 + 3 \cdot 5i - 2 \cdot 4i - 5i(4i)$$

$$= 6 + 15i - 8i - 20i^2$$

Combining similar terms and using the fact that $i^2 = -1$, we can simplify as follows:

$$6 + 15i - 8i - 20i^2 = 6 + 7i - 20(-1)$$

$$= 6 + 7i + 20$$

$$= 26 + 7i$$

The product of the complex numbers $3 - 4i$ and $2 + 5i$ is the complex number $26 + 7i$. ∎

Answers

3. $x = 2, y = -\dfrac{1}{2}$

4. $x = 3, y = 2$

5. a. $5 + 2i$ **b.** $2 + 2i$ **c.** $-1 + i$

6. $14 - 5i$

EXAMPLE 7 Multiply $2i(4 - 6i)$.

SOLUTION Applying the distributive property gives us

$$2i(4 - 6i) = 2i \cdot 4 - 2i \cdot 6i$$

$$= 8i - 12i^2$$

$$= 12 + 8i$$ ∎

EXAMPLE 8 Expand $(3 + 5i)^2$.

SOLUTION We treat this like the square of a binomial. Remember, $(a + b)^2 = a^2 + 2ab + b^2$.

$$(3 + 5i)^2 = 3^2 + 2(3)(5i) + (5i)^2$$

$$= 9 + 30i + 25i^2$$

$$= 9 + 30i - 25$$

$$= -16 + 30i$$ ∎

EXAMPLE 9 Multiply $(2 - 3i)(2 + 3i)$.

SOLUTION This product has the form $(a - b)(a + b)$, which we know results in the difference of two squares, $a^2 - b^2$.

$$(2 - 3i)(2 + 3i) = 2^2 - (3i)^2$$

$$= 4 - 9i^2$$

$$= 4 + 9$$

$$= 13$$ ∎

The product of the two complex numbers $2 - 3i$ and $2 + 3i$ is the real number 13. The two complex numbers $2 - 3i$ and $2 + 3i$ are called *complex conjugates*. The fact that their product is a real number is very useful.

> **Definition**
>
> The complex numbers $a + bi$ and $a - bi$ are called **complex conjugates.** One important property they have is that their product is the real number $a^2 + b^2$. Here's why:
>
> $$(a + bi)(a - bi) = a^2 - (bi)^2$$
>
> $$= a^2 - b^2i^2$$
>
> $$= a^2 - b^2(-1)$$
>
> $$= a^2 + b^2$$

7. Multiply $-3i(2 + 3i)$.

8. Expand $(2 + 4i)^2$.

> **NOTE**
> We can obtain the same result by writing $(3 + 5i)^2$ as $(3 + 5i)$ times $(3 + 5i)$ and applying the FOIL method as we did with the problem in Example 6.

9. Multiply $(3 - 5i)(3 + 5i)$.

Answers
7. $9 - 6i$
8. $-12 + 16i$
9. 34

10. Divide $\dfrac{3 + 2i}{2 - 5i}$.

F Division with Complex Numbers

The fact that the product of two complex conjugates is a real number is the key to division with complex numbers.

EXAMPLE 10 Divide $\dfrac{2 + i}{3 - 2i}$.

SOLUTION We want a complex number in standard form that is equivalent to the quotient $\dfrac{2 + i}{3 - 2i}$. We need to eliminate i from the denominator. Multiplying the numerator and denominator by $3 + 2i$ will give us what we want.

$$\frac{2 + i}{3 - 2i} = \frac{2 + i}{3 - 2i} \cdot \frac{(3 + 2i)}{(3 + 2i)}$$

$$= \frac{6 + 4i + 3i + 2i^2}{9 - 4i^2}$$

$$= \frac{6 + 7i - 2}{9 + 4}$$

$$= \frac{4 + 7i}{13}$$

$$= \frac{4}{13} + \frac{7}{13}i$$

Dividing the complex number $2 + i$ by $3 - 2i$ gives the complex number $\dfrac{4}{13} + \dfrac{7}{13}i$. ■

11. Divide $\dfrac{3 + 2i}{i}$.

EXAMPLE 11 Divide $\dfrac{7 - 4i}{i}$.

SOLUTION The conjugate of the denominator is $-i$. Multiplying the numerator and denominator by this number, we have

$$\frac{7 - 4i}{i} = \frac{7 - 4i}{i} \cdot \frac{-i}{-i}$$

$$= \frac{-7i + 4i^2}{-i^2}$$

$$= \frac{-7i + 4(-1)}{-(-1)}$$

$$= -4 - 7i$$ ■

Answers
10. $-\dfrac{4}{29} + \dfrac{19}{29}i$
11. $2 - 3i$

Problem Set 7.7

Moving Toward Success

*"The most beautiful thing we can experience is the mysterious.
It is the source of all true art and science."*

—Albert Einstein, 1879–1955, German-born American physicist

1. Why is it important to still get good restful sleep between studying and going to class?
2. How would treating this class as a full-time job help you focus your time?

A Write the following in terms of i, and simplify as much as possible. [Example 4]

1. $\sqrt{-36}$

2. $\sqrt{-49}$

3. $-\sqrt{-25}$

4. $-\sqrt{-81}$

5. $\sqrt{-72}$

6. $\sqrt{-48}$

7. $-\sqrt{-12}$

8. $-\sqrt{-75}$

B Write each of the following as i, -1, $-i$, or 1. [Example 2]

9. i^{28}

10. i^{31}

11. i^{26}

12. i^{37}

13. i^{75}

14. i^{42}

C Find x and y so each of the following equations is true. [Examples 3, 4]

15. $2x + 3yi = 6 - 3i$

16. $4x - 2yi = 4 + 8i$

17. $2 - 5i = -x + 10yi$

18. $4 + 7i = 6x - 14yi$

19. $2x + 10i = -16 - 2yi$

20. $4x - 5i = -2 + 3yi$

21. $(2x - 4) - 3i = 10 - 6yi$

22. $(4x - 3) - 2i = 8 + yi$

23. $(7x - 1) + 4i = 2 + (5y + 2)i$

24. $(5x + 2) - 7i = 4 + (2y + 1)i$

D Combine the following complex numbers. [Example 5]

25. $(2 + 3i) + (3 + 6i)$

26. $(4 + i) + (3 + 2i)$

27. $(3 - 5i) + (2 + 4i)$

28. $(7 + 2i) + (3 - 4i)$

29. $(5 + 2i) - (3 + 6i)$

30. $(6 + 7i) - (4 + i)$

31. $(3 - 5i) - (2 + i)$

32. $(7 - 3i) - (4 + 10i)$

33. $[(3 + 2i) - (6 + i)] + (5 + i)$

34. $[(4 - 5i) - (2 + i)] + (2 + 5i)$

35. $[(7 - i) - (2 + 4i)] - (6 + 2i)$

36. $[(3 - i) - (4 + 7i)] - (3 - 4i)$

37. $(3 + 2i) - [(3 - 4i) - (6 + 2i)]$

38. $(7 - 4i) - [(-2 + i) - (3 + 7i)]$

39. $(4 - 9i) + [(2 - 7i) - (4 + 8i)]$

40. $(10 - 2i) - [(2 + i) - (3 - i)]$

E Find the following products. [Examples 6–9]

41. $3i(4 + 5i)$

42. $2i(3 + 4i)$

43. $6i(4 - 3i)$

44. $11i(2 - i)$

45. $(3 + 2i)(4 + i)$

46. $(2 - 4i)(3 + i)$

47. $(4 + 9i)(3 - i)$

48. $(5 - 2i)(1 + i)$

49. $(1 + i)^3$

50. $(1 - i)^3$

51. $(2 - i)^3$

52. $(2 + i)^3$

53. $(2 + 5i)^2$

54. $(3 + 2i)^2$

55. $(1 - i)^2$

56. $(1 + i)^2$

57. $(3 - 4i)^2$

58. $(6 - 5i)^2$

59. $(2 + i)(2 - i)$

60. $(3 + i)(3 - i)$

61. $(6 - 2i)(6 + 2i)$ **62.** $(5 + 4i)(5 - 4i)$ **63.** $(2 + 3i)(2 - 3i)$ **64.** $(2 - 7i)(2 + 7i)$

65. $(10 + 8i)(10 - 8i)$ **66.** $(11 - 7i)(11 + 7i)$

F Find the following quotients. Write all answers in standard form for complex numbers. [Examples 10, 11]

67. $\dfrac{2 - 3i}{i}$ **68.** $\dfrac{3 + 4i}{i}$ **69.** $\dfrac{5 + 2i}{-i}$ **70.** $\dfrac{4 - 3i}{-i}$

71. $\dfrac{4}{2 - 3i}$ **72.** $\dfrac{3}{4 - 5i}$ **73.** $\dfrac{6}{-3 + 2i}$ **74.** $\dfrac{-1}{-2 - 5i}$

75. $\dfrac{2 + 3i}{2 - 3i}$ **76.** $\dfrac{4 - 7i}{4 + 7i}$ **77.** $\dfrac{5 + 4i}{3 + 6i}$ **78.** $\dfrac{2 + i}{5 - 6i}$

Applying the Concepts

79. Electrical Circuits Complex numbers may be applied to electrical circuits. Electrical engineers use the fact that resistance R to electrical flow of the electrical current I and the voltage V are related by the formula $V = RI$. (Voltage is measured in volts, resistance in ohms, and current in amperes.) Find the resistance to electrical flow in a circuit that has a voltage $V = (80 + 20i)$ volts and current $I = (-6 + 2i)$ amps.

80. Electrical Circuits Refer to the information about electrical circuits in Problem 79, and find the current in a circuit that has a resistance of $(4 + 10i)$ ohms and a voltage of $(5 - 7i)$ volts.

Maintaining Your Skills

Solve each equation.

81. $\dfrac{t}{3} - \dfrac{1}{2} = -1$

82. $\dfrac{x}{x-2} + \dfrac{2}{3} = \dfrac{2}{x-2}$

83. $2 + \dfrac{5}{y} = \dfrac{3}{y^2}$

84. $1 - \dfrac{1}{y} = \dfrac{12}{y^2}$

Solve each application problem.

85. The sum of a number and its reciprocal is $\dfrac{41}{20}$. Find the number.

86. It takes an inlet pipe 8 hours to fill a tank. The drain can empty the tank in 6 hours. If the tank is full and both the inlet pipe and drain are open, how long will it take to drain the tank?

Extending the Concepts

87. Show that $-i$ and $\dfrac{1}{i}$ (the opposite and the reciprocal of i) are the same number.

88. Show that i^{2n+1} is the same as i for all positive even integers n.

89. Show that $x = 1 + i$ is a solution to the equation $x^2 - 2x + 2 = 0$.

90. Show that $x = 1 - i$ is a solution to the equation $x^2 - 2x + 2 = 0$.

91. Show that $x = 2 + i$ is a solution to the equation $x^3 - 11x + 20 = 0$.

92. Show that $x = 2 - i$ is a solution to the equation $x^3 - 11x + 20 = 0$.

■ Square Roots [7.1]

EXAMPLES

1. The number 49 has two square roots, 7 and −7. They are written like this:

$$\sqrt{49} = 7 \qquad -\sqrt{49} = -7$$

Every positive real number x has two square roots. The *positive square root* of x is written $\sqrt{x}$, and the *negative square root* of x is written $-\sqrt{x}$. Both the positive and the negative square roots of x are numbers we square to get x; that is,

$$\left. \begin{array}{l} (\sqrt{x})^2 = x \\ \text{and} \qquad (-\sqrt{x})^2 = x \end{array} \right\} \quad \text{for } x \geq 0$$

■ Higher Roots [7.1]

2. $\sqrt[3]{8} = 2$
$\sqrt[3]{-27} = -3$

In the expression $\sqrt[n]{a}$, n is the *index*, a is the *radicand*, and $\sqrt{}$ is the *radical sign*. The expression $\sqrt[n]{a}$ is such that

$$(\sqrt[n]{a})^n = a \qquad a \geq 0 \text{ when } n \text{ is even}$$

■ Rational Exponents [7.1, 7.2]

3. $25^{1/2} = \sqrt{25} = 5$
$8^{2/3} = (\sqrt[3]{8})^2 = 2^2 = 4$
$9^{3/2} = (\sqrt{9})^3 = 3^3 = 27$

Rational exponents are used to indicate roots. The relationship between rational exponents and roots is as follows:

$$a^{1/n} = \sqrt[n]{a} \qquad \text{and} \qquad a^{m/n} = (a^{1/n})^m = (a^m)^{1/n}$$

$$a \geq 0 \text{ when } n \text{ is even}$$

■ Properties of Radicals [7.3]

4. $\sqrt{4 \cdot 5} = \sqrt{4}\,\sqrt{5} = 2\sqrt{5}$
$\sqrt{\dfrac{7}{9}} = \dfrac{\sqrt{7}}{\sqrt{9}} = \dfrac{\sqrt{7}}{3}$

If a and b are nonnegative real numbers whenever n is even, then

1. $\sqrt[n]{ab} = \sqrt[n]{a}\,\sqrt[n]{b}$ Product property

2. $\sqrt[n]{\dfrac{a}{b}} = \dfrac{\sqrt[n]{a}}{\sqrt[n]{b}}$ $(b \neq 0)$ Quotient property

■ Simplified Form for Radicals [7.3]

5. $\sqrt{\dfrac{4}{5}} = \dfrac{\sqrt{4}}{\sqrt{5}}$

$= \dfrac{2}{\sqrt{5}} \cdot \dfrac{\sqrt{5}}{\sqrt{5}}$

$= \dfrac{2\sqrt{5}}{5}$

A radical expression is said to be in *simplified form*

1. If there is no factor of the radicand that can be written as a power greater than or equal to the index;

2. If there are no fractions under the radical sign; and

3. If there are no radicals in the denominator.

■ Addition and Subtraction of Radical Expressions [7.4]

6. $5\sqrt{3} - 7\sqrt{3} = (5 - 7)\sqrt{3}$
$= -2\sqrt{3}$
$\sqrt{20} + \sqrt{45} = 2\sqrt{5} + 3\sqrt{5}$
$= (2 + 3)\sqrt{5}$
$= 5\sqrt{5}$

We add and subtract radical expressions by using the distributive property to combine similar radicals. Similar radicals are radicals with the same index and the same radicand.

■ Multiplication of Radical Expressions [7.5]

7. $(\sqrt{x} + 2)(\sqrt{x} + 3)$
$= \sqrt{x}\,\sqrt{x} + 3\sqrt{x} + 2\sqrt{x} + 2 \cdot 3$
$= x + 5\sqrt{x} + 6$

We multiply radical expressions in the same way that we multiply polynomials. We can use the distributive property and the FOIL method.

■ Rationalizing the Denominator [7.3, 7.5]

8. $\dfrac{3}{\sqrt{2}} = \dfrac{3}{\sqrt{2}} \cdot \dfrac{\sqrt{2}}{\sqrt{2}}$

$\qquad = \dfrac{3\sqrt{2}}{2}$

$\dfrac{3}{\sqrt{5} - \sqrt{3}}$

$\qquad = \dfrac{3}{\sqrt{5} - \sqrt{3}} \cdot \dfrac{\sqrt{5} + \sqrt{3}}{\sqrt{5} + \sqrt{3}}$

$\qquad = \dfrac{3\sqrt{5} + 3\sqrt{3}}{5 - 3}$

$\qquad = \dfrac{3\sqrt{5} + 3\sqrt{3}}{2}$

When a fraction contains a square root in the denominator, we rationalize the denominator by multiplying numerator and denominator by

1. The square root itself if there is only one term in the denominator, or
2. The conjugate of the denominator if there are two terms in the denominator.

Rationalizing the denominator is also called division of radical expressions.

■ Squaring Property of Equality [7.6]

9. $\sqrt{2x + 1} = 3$
$(\sqrt{2x + 1})^2 = 3^2$
$\qquad 2x + 1 = 9$
$\qquad\qquad x = 4$

We may square both sides of an equation any time it is convenient to do so, as long as we check all resulting solutions in the original equation.

■ Complex Numbers [7.7]

10. $3 + 4i$ is a complex number.
Addition
$(3 + 4i) + (2 - 5i) = 5 - i$
Multiplication
$(3 + 4i)(2 - 5i)$
$\qquad = 6 - 15i + 8i - 20i^2$
$\qquad = 6 - 7i + 20$
$\qquad = 26 - 7i$
Division
$\dfrac{2}{3 + 4i} = \dfrac{2}{3 + 4i} \cdot \dfrac{3 - 4i}{3 - 4i}$

$\qquad = \dfrac{6 - 8i}{9 + 16}$

$\qquad = \dfrac{6}{25} - \dfrac{8}{25}i$

A *complex number* is any number that can be put in the form

$$a + bi$$

where a and b are real numbers and $i = \sqrt{-1}$. The *real part* of the complex number is a, and bi is the *imaginary part*.

If a, b, c, and d are real numbers, then we have the following definitions associated with complex numbers:

1. Equality

$$a + bi = c + di \quad \text{if and only if} \quad a = c \text{ and } b = d$$

2. Addition and subtraction

$$(a + bi) + (c + di) = (a + c) + (b + d)i$$

$$(a + bi) - (c + di) = (a - c) + (b - d)i$$

3. Multiplication

$$(a + bi)(c + di) = (ac - bd) + (ad + bc)i$$

4. Division is similar to rationalizing the denominator.

Simplify each expression as much as possible. [7.1]

1. $\sqrt{49}$

2. $(-27)^{1/3}$

3. $16^{1/4}$

4. $9^{3/2}$

5. $\sqrt[5]{32x^{15}y^{10}}$

6. $8^{-4/3}$

Use the properties of exponents to simplify each expression. Assume all bases represent positive numbers. [7.1]

7. $x^{2/3} \cdot x^{4/3}$

8. $(a^{2/3}b^{4/3})^3$

9. $\dfrac{a^{3/5}}{a^{1/4}}$

10. $\dfrac{a^{2/3}b^3}{a^{1/4}b^{1/3}}$

Multiply. [7.2]

11. $(3x^{1/2} + 5y^{1/2})(4x^{1/2} - 3y^{1/2})$

12. $(a^{1/3} - 5)^2$

13. Divide: $\dfrac{28x^{5/6} + 14x^{7/6}}{7x^{1/3}}$. (Assume $x > 0$.) [7.2]

14. Factor $2(x - 3)^{1/4}$ from $8(x - 3)^{5/4} - 2(x - 3)^{1/4}$. [7.2]

15. Simplify $x^{3/4} + \dfrac{5}{x^{1/4}}$ into a single fraction. (Assume $x > 0$.) [7.2]

Write each expression in simplified form for radicals. (Assume all variables represent nonnegative numbers.) [7.3]

16. $\sqrt{12}$

17. $\sqrt{50}$

18. $\sqrt[3]{16}$

19. $\sqrt{18x^2}$

20. $\sqrt{80a^3b^4c^2}$

21. $\sqrt[4]{32a^4b^5c^6}$

Rationalize the denominator in each expression. [7.3]

22. $\dfrac{3}{\sqrt{2}}$

23. $\dfrac{6}{\sqrt[3]{2}}$

Write each expression in simplified form. (Assume all variables represent positive numbers.) [7.3]

24. $\sqrt{\dfrac{48x^3}{7y}}$

25. $\sqrt[3]{\dfrac{40x^2y^3}{3z}}$

Combine the following expressions. (Assume all variables represent positive numbers.) [7.4]

26. $5x\sqrt{6} + 2x\sqrt{6} - 9x\sqrt{6}$

27. $\sqrt{12} + \sqrt{3}$

28. $\dfrac{3}{\sqrt{5}} + \sqrt{5}$

29. $3\sqrt{8} - 4\sqrt{72} + 5\sqrt{50}$

30. $3b\sqrt{27a^5b} + 2a\sqrt{3a^3b^3}$

31. $2x\sqrt[3]{xy^3z^2} - 6y\sqrt[3]{x^4z^2}$

Multiply. [7.5]

32. $\sqrt{2}(\sqrt{3} - 2\sqrt{2})$

33. $(\sqrt{x} - 2)(\sqrt{x} - 3)$

Rationalize the denominator. [7.5]

34. $\dfrac{3}{\sqrt{5} - 2}$

35. $\dfrac{\sqrt{7} + \sqrt{5}}{\sqrt{7} - \sqrt{5}}$

36. $\dfrac{3\sqrt{7}}{3\sqrt{7} - 4}$

Solve each equation. [7.6]

37. $\sqrt{4a + 1} = 1$

38. $\sqrt[3]{3x - 8} = 1$

39. $\sqrt{3x + 1} - 3 = 1$

40. $\sqrt{x + 4} = \sqrt{x} - 2$

Graph each equation. [7.6]

41. $y = 3\sqrt{x}$

42. $y = \sqrt[3]{x} + 2$

Write each of the following as i, -1, $-i$, or 1. [7.7]

43. i^{24}

44. i^{27}

Find x and y so that each of the following equations is true. [7.7]

45. $3 - 4i = -2x + 8yi$

46. $(3x + 2) - 8i = -4 + 2yi$

Combine the following complex numbers. [7.7]

47. $(3 + 5i) + (6 - 2i)$

48. $(2 + 5i) - [(3 + 2i) + (6 - i)]$

Multiply. [7.7]

49. $3i(4 + 2i)$

50. $(2 + 3i)(4 + i)$

51. $(4 + 2i)^2$

52. $(4 + 3i)(4 - 3i)$

Divide. Write all answers in standard form for complex numbers. [7.7]

53. $\dfrac{3 + i}{i}$

54. $\dfrac{-3}{2 + i}$

Simplify.

1. $6 + [2 - 5(7 - 2)]$ **2.** $3 - 15 \div 5 + 6 \cdot 2$

3. $2 + (5 - 7)^2 - (6 - 5)^3$ **4.** $\sqrt[4]{48}$

5. $8^{2/3} + 16^{1/4}$ **6.** $\dfrac{1 - \frac{2}{3}}{1 + \frac{1}{3}}$

7. $\left(-\dfrac{27}{8}\right)^{-1/3}$ **8.** $12 - [3 - 4 \div 2(7 - 5)]$

Reduce.

9. $\dfrac{105}{189}$ **10.** $\dfrac{6x^2 - 4xy - 16y^2}{2x - 4y}$

11. $\dfrac{x^2 - x - 6}{x^2 - 2x - 3}$ **12.** $\dfrac{6a^7b^{-3}}{36a^5b^{-2}}$

Multiply or divide.

13. $(\sqrt{x} - 3)(2\sqrt{x} - 5)$ **14.** $(7 - i)(3 + i)$

15. $\dfrac{4 + i}{4 + 3i}$ **16.** $\dfrac{\sqrt{3}}{6 - \sqrt{3}}$

17. $(2x - 5y)(3x - 2y)$ **18.** $(9x^2 - 25) \cdot \dfrac{x + 5}{3x - 5}$

19. $\dfrac{27x^3y^2}{13\,x^2y^4} \div \dfrac{9xy}{26y}$ **20.** $\dfrac{\sqrt{x} + \sqrt{y}}{\sqrt{x} - \sqrt{y}}$

Solve.

21. $\dfrac{3}{2} + \dfrac{1}{a - 2} = -\dfrac{1}{a - 2}$ **22.** $\sqrt{x - 2} = x - 2$

23. $(2x - 5)^2 = 12$ **24.** $0.03x^2 - 0.02x = 0.08$

25. $\sqrt{x - 3} = 3 - \sqrt{x}$ **26.** $3x^3 - 28x = -5x^2$

27. $\dfrac{1}{2}x^2 + \dfrac{3x}{8} = \dfrac{5}{16}$ **28.** $\dfrac{2}{3}(9x - 2) + \dfrac{1}{3} = 4$

29. $|3x - 1| - 2 = 6$ **30.** $\sqrt[3]{8 - 3x} = -1$

Solve each inequality, and graph the solution.

31. $-3 \le \dfrac{1}{4}x - 5 \le -1$ **32.** $|2x - 3| > 7$

Solve each system.

33. $2x - 3y = -16$
$\ 3x + y = -2$

34. $\dfrac{3}{4}x + \dfrac{1}{2}y = -9$
$\ \dfrac{1}{2}x = \dfrac{2}{3}y$

35. $2x + 4y + z = 6$
$\ 3x - y - 2z = 4$
$\ x + 2y + 3z = -10$

36. $6x + 8y = -9$
$\ -4x + 6y = -11$

Graph on a rectangular coordinate system.

37. $-2x + 3y = -6$ **38.** $x + 2y > 2$

Find the slope of the line passing through the points.

39. $(-3, 4)$ and $(-2, -3)$ **40.** $(-2, -4)$ and $(-3, 6)$

Rationalize the denominator.

41. $\dfrac{6}{\sqrt[4]{8}}$ **42.** $\sqrt[3]{\dfrac{8y}{4x}}$

Factor completely.

43. $625a^4 - 16b^4$

44. $24a^4 + 10a^2 - 6$

45. Inverse Variation y varies inversely with the square root of x. If $y = \dfrac{1}{2}$ when $x = 16$, find y when $x = 4$.

46. Joint Variation z varies jointly with x and the cube of y. If $z = 80$ when $x = 5$, and $y = 2$, find z when $x = 3$ and $y = -1$.

Let $f(x) = x^2 - 3x$ and $g(x) = x - 1$. Find the following.

47. $f(b)$ **48.** $(g - f)(3)$

49. $(g \circ f)(3)$ **50.** $(f \circ g)(-3)$

Use the illustration to answer Questions 51 and 52.

Most Visited Countries

France	79,300,000
United States	58,000,000
Spain	57,300,000
China	53,000,000
Italy	42,700,000

*Annual number of arrivals

Source: TenMojo.com

51. How many more arrivals were there in France than in Spain?

52. Find the difference between the number of people visting the United States and those visiting China.

Chapter 7 Test

Simplify each of the following. (Assume all variable bases are positive integers and all variable exponents are positive real numbers throughout this test.) [7.1]

1. $27^{-2/3}$

2. $\left(\dfrac{25}{49}\right)^{-1/2}$

3. $a^{3/4} \cdot a^{-1/3}$

4. $\dfrac{(x^{2/3}y^{-3})^{1/2}}{(x^{3/4}y^{1/2})^{-1}}$

5. $\sqrt{49x^8y^{10}}$

6. $\sqrt[5]{32x^{10}y^{20}}$

7. $\dfrac{(36a^8b^4)^{1/2}}{(27a^9b^6)^{1/3}}$

8. $\dfrac{(x^ny^{1/n})^n}{(x^{1/n}y^n)^{n^2}}$

Multiply. [7.2]

9. $2a^{1/2}(3a^{3/2} - 5a^{1/2})$

10. $(4a^{3/2} - 5)^2$

Factor. [7.2]

11. $3x^{2/3} + 5x^{1/3} - 2$

12. $9x^{2/3} - 49$

Write in simplified form. [7.3]

13. $\sqrt{125x^3y^5}$

14. $\sqrt[3]{40x^7y^8}$

15. $\sqrt{\dfrac{2}{3}}$

16. $\sqrt{\dfrac{12a^4b^3}{5c}}$

Combine. [7.4]

17. $\dfrac{4}{x^{1/2}} + x^{1/2}$

18. $\dfrac{x^2}{(x^2 - 3)^{1/2}} - (x^2 - 3)^{1/2}$

Combine. [7.4]

19. $3\sqrt{12} - 4\sqrt{27}$

20. $2\sqrt[3]{24a^3b^3} - 5a\sqrt[3]{3b^3}$

Multiply. [7.5]

21. $(\sqrt{x} + 7)(\sqrt{x} - 4)$

22. $(3\sqrt{2} - \sqrt{3})^2$

Rationalize the denominator. [7.5]

23. $\dfrac{5}{\sqrt{3} - 1}$

24. $\dfrac{\sqrt{x} - \sqrt{2}}{\sqrt{x} + \sqrt{2}}$

Solve for x. [7.6]

25. $\sqrt{3x + 1} = x - 3$

26. $\sqrt[3]{2x + 7} = -1$

27. $\sqrt{x + 3} = \sqrt{x + 4} - 1$

Graph. [7.6]

28. $y = \sqrt{x - 2}$

29. $y = \sqrt[3]{x} + 3$

30. Find x and y so that the following equation is true. [7.7]

$$(2x + 5) - 4i = 6 - (y - 3)i$$

Perform the indicated operations. [7.7]

31. $(3 + 2i) - [(7 - i) - (4 + 3i)]$

32. $(2 - 3i)(4 + 3i)$

33. $(5 - 4i)^2$

34. $\dfrac{2 - 3i}{2 + 3i}$

35. Show that i^{38} can be written as -1. [7.7]

Match each equation with its graph. [7.6]

36. $y = -\sqrt{x}$

37. $y = \sqrt{x} - 1$

38. $y = \sqrt{x + 1}$

39. $y = \sqrt[3]{x} - 1$

40. $y = \sqrt[3]{x - 1}$

41. $y = \sqrt[3]{x + 1}$

A.
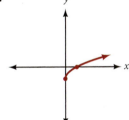

B.
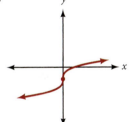

C.
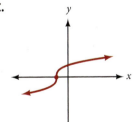

D.
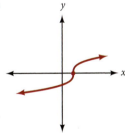

E.
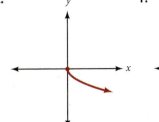

F.
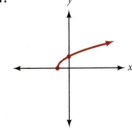

RATIONAL EXPONENTS AND ROOTS

Group Project

Constructing the Spiral of Roots

Number of People 3

Time Needed 20 minutes

Equipment Two sheets of graph paper (4 or 5 squares per inch) and pencils.

Background The spiral of roots gives us a way to visualize the positive square roots of the counting numbers, and in so doing, we see many line segments whose lengths are irrational numbers.

Procedure You are to construct a spiral of roots from a line segment 1 inch long. The graph paper you have contains either 4 or 5 squares per inch, allowing you to accurately draw 1-inch line segments. Because the lines on the graph paper are perpendicular to one another, if you are careful, you can use the graph paper to connect one line segment to another so that they form a right angle.

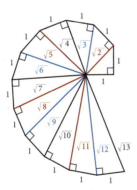

1. Fold one of the pieces of graph paper so it can be used as a ruler.

2. Use the folded paper to draw a line segment 1-inch long, just to the right of the middle of the unfolded paper. On the end of this segment, attach another segment of 1-inch length at a right angle to the first one.

Connect the end points of the segments to form a right triangle. Label each side of this triangle. When you are finished, your work should resemble Figure 1.

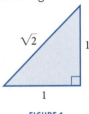

FIGURE 1

3. On the end of the hypotenuse of the triangle, attach a 1-inch line segment so that the two segments form a right angle. (Use the folded paper to do this.) Draw the hypotenuse of this triangle. Label all the sides of this second triangle. Your work should resemble Figure 2.

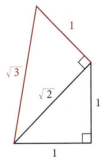

FIGURE 2

4. Continue to draw a new right triangle by attaching 1-inch line segments at right angles to the previous hypotenuse. Label all the sides of each triangle.

5. Stop when you have drawn a hypotenuse $\sqrt{8}$ inches long.

Maria Gaetana Agnesi (1718–1799)

January 9, 1999, was the 200th anniversary of the death of Maria Agnesi, the author of *Instituzioni analitiche ad uso della gioventu italiana* (1748), a calculus textbook considered to be the best book of its time and the first surviving mathematical work written by a woman. Maria Agnesi is also famous for a curve that is named for her. The curve is called the Witch of Agnesi in English, but its actual translation is the Locus of Agnesi. The foundation of the curve is shown in the figure. Research the Witch of Agnesi and then explain, in essay form, how the diagram in the figure is used to produce the Witch of Agnesi. Include a rough sketch of the curve, starting with a circle of diameter *a* as shown in the figure. Then, for comparison, sketch the curve again, starting with a circle of diameter 2*a*.

© Bettmann/CORBIS

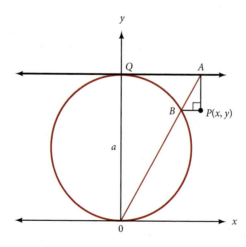

Quadratic Functions

The Trift footbridge is considered one of Europe's longest and highest rope suspension bridges for pedestrians. Nestled high in the Swiss Alps, the bridge spans 170 meters above the Trift Glacier. Up until a few years ago, mountaineers could reach one of the mountain huts of the Swiss Alpine Club by traversing the glacier. But as the glacier melted and receded, it was no longer high enough for climbers to reach the restaurant. In 2004, the bridge was built and now hangs 100 meters above the melting glacier. Suppose you drop a rock from the middle of the bridge and time how long before it strikes the glacier. The path, location, and velocity of a falling object, such as the rock, can be represented by a quadratic equation.

Tomas Sereda/Shutterstock.com

Up until now, we have worked with quadratic equations on a limited scale. In this chapter, we will explore these special equations, examine new methods to solve them, and plot their graphs on a coordinate system.

Key Words	Definition
The Quadratic Formula	$x = \dfrac{-b \pm \sqrt{b^2 - 4ac}}{2a}$
The Discriminant	The expression under the radical in the quadratic formula
Parabola	The graph of an equation in the form $y = ax^2 + bx + c$

Chapter Outline

8.1 Completing the Square

A   Solve quadratic equations by taking the square root of both sides.

B   Solve quadratic equations by completing the square.

C   Use quadratic equations to solve for missing parts of right triangles.

8.2 The Quadratic Formula

A   Solve quadratic equations by the quadratic formula.

B   Solve application problems using quadratic equations.

8.3 Additional Items Involving Solutions to Equations

A   Find the number and kinds of solutions to a quadratic equation by using the discriminant.

B   Find an unknown constant in a quadratic equation so that there is exactly one solution.

C   Find an equation from its solutions.

8.4 Equations Quadratic in Form

A   Solve equations that are reducible to a quadratic equation.

B   Solve application problems using quadratic equations.

8.5 Graphing Parabolas

A   Graph a parabola.

B   Solve application problems using information from a graph.

C   Find an equation from its graph.

8.6 Quadratic Inequalities

A   Solve quadratic inequalities and graph the solution sets.

B   Solve inequalities involving rational expressions.

Completing the Square

TICKET TO SUCCESS

Keep these questions in mind as you read through the section. Then respond in your own words and in complete sentences.

1. Explain the square root theorem.
2. What kind of equation do we solve using the method of completing the square?
3. Explain the strategy for completing the square on $x^2 - 16x = 4$.
4. What is the relationship between the shortest side and the longest side in a $30°-60°-90°$ triangle?

OBJECTIVES

A Solve quadratic equations by taking the square root of both sides.

B Solve quadratic equations by completing the square.

C Use quadratic equations to solve for missing parts of right triangles.

Terrie Delayn/Shutterstock.com

Suppose you are building a square koi pond in your backyard. Once complete, the pond covers 16 square feet. The following equation represents the dimensions of your pond:

$$x^2 = 16$$

where x represents the side length. What are the dimensions of your pond? In this section, we will learn how to solve equations, such as this one, by taking square roots of both sides. Let's begin.

A The Square Root Method

Consider the equation $x^2 = 16$ that we used to represent the koi pond. We could solve it by writing it in standard form, factoring the left side, and proceeding as we have done previously. We can shorten our work considerably, however, if we simply notice that x must be either the positive square root of 16 or the negative square root of 16; that is,

$$\text{If} \rightarrow \quad x^2 = 16$$

$$\text{then} \rightarrow \quad x = \sqrt{16} \quad \text{or} \quad x = -\sqrt{16}$$

$$x = 4 \qquad\qquad x = -4$$

We can generalize this result into the following theorem.

> **Square Root Theorem**
>
> If $a^2 = b$ where b is a real number, then $a = \sqrt{b}$ or $a = -\sqrt{b}$.

NOTATION The expression $a = \sqrt{b}$ or $a = -\sqrt{b}$ can be written in shorthand form as $a = \pm\sqrt{b}$. The symbol $\pm$ is read "plus or minus."

We can apply the square root theorem to some fairly complicated quadratic equations.

PRACTICE PROBLEMS

1. Solve $(3x + 2)^2 = 16$.

EXAMPLE 1 Solve $(2x - 3)^2 = 25$.

SOLUTION

$$(2x - 3)^2 = 25$$

$$2x - 3 = \pm\sqrt{25} \qquad \text{Square root theorem}$$

$$2x - 3 = \pm 5 \qquad \sqrt{25} = \pm 5$$

$$2x = 3 \pm 5 \qquad \text{Add 3 to both sides.}$$

$$x = \frac{3 \pm 5}{2} \qquad \text{Divide both sides by 2.}$$

The last equation can be written as two separate statements:

$$x = \frac{3 + 5}{2} \qquad \text{or} \qquad x = \frac{3 - 5}{2}$$

$$= \frac{8}{2} \qquad\qquad\qquad = \frac{-2}{2}$$

$$= 4 \qquad\qquad\qquad = -1$$

The solution set is $\{4, -1\}$. ∎

Notice that we could have solved the equation in Example 1 by expanding the left side, writing the resulting equation in standard form, and then factoring. The problem would look like this:

$$(2x - 3)^2 = 25 \qquad \text{Original equation}$$

$$4x^2 - 12x + 9 = 25 \qquad \text{Expand the left side.}$$

$$4x^2 - 12x - 16 = 0 \qquad \text{Add } -25 \text{ to each side.}$$

$$4(x^2 - 3x - 4) = 0 \qquad \text{Begin factoring.}$$

$$4(x - 4)(x + 1) = 0 \qquad \text{Factor completely.}$$

$$x - 4 = 0 \quad \text{or} \quad x + 1 = 0 \qquad \text{Set variable factors equal to 0.}$$

$$x = 4 \qquad\qquad x = -1$$

As you can see, solving the equation by factoring leads to the same two solutions.

> **NOTE**
>
> We cannot solve the equation in Example 2 by factoring. If we expand the left side and write the resulting equation in standard form, we are left with a quadratic equation that does not factor:
> Equation from Example 2
> $$(3x - 1)^2 = -12$$
> Expand the left side
> $$9x^2 - 6x + 1 = -12$$
> Standard form, but not factorable
> $$9x^2 - 6x + 13 = 0$$

2. Solve $(4x - 3)^2 = -50$.

EXAMPLE 2 Solve $(3x - 1)^2 = -12$ for x.

SOLUTION $(3x - 1)^2 = -12$

$$3x - 1 = \pm\sqrt{-12} \qquad \text{Square root theorem}$$

$$3x - 1 = \pm 2i\sqrt{3} \qquad \sqrt{-12} = 2i\sqrt{3}$$

Answers

1. $\frac{2}{3}, -2$

2. $\frac{3 \pm 5i\sqrt{2}}{4}$

$$3x = 1 \pm 2i\sqrt{3} \qquad \text{Add 1 to both sides.}$$

$$x = \frac{1 \pm 2i\sqrt{3}}{3} \qquad \text{Divide both sides by 3.}$$

The solution set is $\left\{ \dfrac{1 + 2i\sqrt{3}}{3}, \dfrac{1 - 2i\sqrt{3}}{3} \right\}$.

Both solutions are complex. Here is a check of the first solution:

When → $\qquad\qquad\qquad x = \dfrac{1 + 2i\sqrt{3}}{3}$

the equation → $\qquad\qquad (3x - 1)^2 = -12$

becomes → $\qquad \left(3 \cdot \dfrac{1 + 2i\sqrt{3}}{3} - 1 \right)^2 \stackrel{?}{=} -12$

$$(1 + 2i\sqrt{3} - 1)^2 \stackrel{?}{=} -12$$

$$(2i\sqrt{3})^2 \stackrel{?}{=} -12$$

$$4 \cdot i^2 \cdot 3 \stackrel{?}{=} -12$$

$$12(-1) \stackrel{?}{=} -12$$

$$-12 = -12 \qquad \blacksquare$$

EXAMPLE 3 Solve $x^2 + 6x + 9 = 12$.

SOLUTION We can solve this equation as we have the equations in Examples 1 and 2 if we first write the left side as $(x + 3)^2$.

$$x^2 + 6x + 9 = 12 \qquad \text{Original equation}$$

$$(x + 3)^2 = 12 \qquad \text{Write } x^2 + 6x + 9 \text{ as } (x + 3)^2.$$

$$x + 3 = \pm 2\sqrt{3} \qquad \text{Square root theorem}$$

$$x = -3 \pm 2\sqrt{3} \qquad \text{Add } -3 \text{ to each side.}$$

We have two irrational solutions: $-3 + 2\sqrt{3}$ and $-3 - 2\sqrt{3}$. What is important about this problem, however, is the fact that the equation was easy to solve because the left side was a perfect square trinomial. $\qquad \blacksquare$

B Completing the Square

Now we will introduce a new method of solving quadratic equations, called *completing the square*. The method of completing the square is simply a way of transforming any quadratic equation into an equation of the form found in the preceding three examples. By doing so, we will be able to obtain solutions, regardless of whether the equation can be factored.

The key to understanding the method of completing the square lies in recognizing the relationship between the last two terms of any perfect square trinomial whose leading coefficient is 1.

Consider the following list of perfect square trinomials and their corresponding binomial squares:

$$x^2 - 6x + 9 = (x - 3)^3$$

$$x^2 + 8x + 16 = (x + 4)^2$$

$$x^2 - 10x + 25 = (x - 5)^2$$

$$x^2 + 12x + 36 = (x + 6)^2$$

3. Solve $x^2 + 10x + 25 = 20$.

Answer

3. $-5 \pm 2\sqrt{5}$

In each case, the leading coefficient is 1. A more important observation comes from noticing the relationship between the linear and constant terms (middle and last terms) in each trinomial. Observe that the constant term in each case is the square of half the coefficient of x in the middle term. For example, in the last expression, the constant term 36 is the square of half of 12, where 12 is the coefficient of x in the middle term. (Notice also that the second terms in all the binomials on the right side are half the coefficients of the middle terms of the trinomials on the left side.) We can use these observations to build our own perfect square trinomials and, in doing so, solve some quadratic equations.

Consider the following equation:

$$x^2 + 6x = 3$$

We can think of the left side as having the first two terms of a perfect square trinomial. We need only add the correct constant term. If we take half the coefficient of x, we get 3. If we then square this quantity, we have 9. Adding the 9 to both sides, the equation becomes

$$x^2 + 6x + \mathbf{9} = 3 + \mathbf{9}$$

> **NOTE**
> This is the step in which we actually complete the square.

The left side is the perfect square $(x + 3)^2$; the right side is 12.

$$(x + 3)^2 = 12$$

The equation is now in the correct form. We can apply the square root theorem and finish the solution.

$$(x + 3)^2 = 12$$

$$x + 3 = \pm\sqrt{12} \qquad \text{Square root theorem}$$

$$x + 3 = \pm 2\sqrt{3}$$

$$x = -3 \pm 2\sqrt{3}$$

The solution set is $\{-3 + 2\sqrt{3}, -3 - 2\sqrt{3}\}$. The method just used is called *completing the square* since we complete the square on the left side of the original equation by adding the appropriate constant term.

---

4. Solve $x^2 + 3x - 4 = 0$ by completing the square.

EXAMPLE 4 Solve $x^2 + 5x - 2 = 0$ by completing the square.

SOLUTION We must begin by adding 2 to both sides. The left side of the equation, as it is, is not a perfect square because it does not have the correct constant term. We will simply "move" that term to the other side and use our own constant term.

$$x^2 + 5x = 2 \qquad \text{Add 2 to each side.}$$

We complete the square by adding the square of half the coefficient of the linear term to both sides.

$$x^2 + 5x + \frac{\mathbf{25}}{\mathbf{4}} = 2 + \frac{\mathbf{25}}{\mathbf{4}} \qquad \text{Half of 5 is } \frac{5}{2}, \text{ the square of which is } \frac{25}{4}$$

$$\left(x + \frac{5}{2}\right)^2 = \frac{33}{4} \qquad 2 + \frac{25}{4} = \frac{8}{4} + \frac{25}{4} = \frac{33}{4}$$

$$x + \frac{5}{2} = \pm\sqrt{\frac{33}{4}} \qquad \text{Square root theorem}$$

$$= \pm\frac{\sqrt{33}}{2} \qquad \text{Simplify the radical.}$$

$$x = -\frac{5}{2} \pm \frac{\sqrt{33}}{2} \qquad \text{Add } -\frac{5}{2} \text{ to both sides.}$$

Answers

4. $1, -4$

$$= \frac{-5 \pm \sqrt{33}}{2}$$

The solution set is $\left\{ \dfrac{-5 + \sqrt{33}}{2}, \dfrac{-5 - \sqrt{33}}{2} \right\}$.   ∎

NOTE
We can use a calculator to get decimal approximations to these solutions. If $\sqrt{33} \approx 5.74$, then

$$\frac{-5 + 5.74}{2} = 0.37$$

$$\frac{-5 - 5.74}{2} = -5.37$$

EXAMPLE 5   Solve $3x^2 - 8x + 7 = 0$ for x.

SOLUTION   We begin by adding -7 to both sides.

$$3x^2 - 8x + 7 = 0$$

$$3x^2 - 8x = -7 \qquad \text{Add } -7 \text{ to both sides.}$$

We cannot complete the square on the left side because the leading coefficient is not 1. We take an extra step and divide both sides by 3.

$$\frac{3x^2}{3} - \frac{8x}{3} = -\frac{7}{3}$$

$$x^2 - \frac{8}{3}x = -\frac{7}{3}$$

Half of $\dfrac{8}{3}$ is $\dfrac{4}{3}$, the square of which is $\dfrac{16}{9}$.

$$x^2 - \frac{8}{3}x + \mathbf{\frac{16}{9}} = -\frac{7}{3} + \mathbf{\frac{16}{9}} \qquad \text{Add } \tfrac{16}{9} \text{ to both sides.}$$

$$\left(x - \frac{4}{3} \right)^2 = -\frac{5}{9} \qquad \text{Simplify right side.}$$

$$x - \frac{4}{3} = \pm\sqrt{-\frac{5}{9}} \qquad \text{Square root theorem}$$

$$x - \frac{4}{3} = \pm\frac{i\sqrt{5}}{3} \qquad \sqrt{-\tfrac{5}{9}} = \tfrac{\sqrt{-5}}{3} = \tfrac{i\sqrt{5}}{3}$$

$$x = \frac{4}{3} \pm \frac{i\sqrt{5}}{3} \qquad \text{Add } \tfrac{4}{3} \text{ to both sides.}$$

$$x = \frac{4 \pm i\sqrt{5}}{3}$$

The solution set is $\left\{ \dfrac{4 + i\sqrt{5}}{3}, \dfrac{4 - i\sqrt{5}}{3} \right\}$.   ∎

5. Solve for x: $5x^2 - 3x + 2 = 0$.

Strategy   **To Solve a Quadratic Equation by Completing the Square**

To summarize the method used in the preceding two examples, we list the following steps:

Step 1   Write the equation in the form $ax^2 + bx = c$.

Step 2   If the leading coefficient is not 1, divide both sides by the coefficient so that the resulting equation has a leading coefficient of 1; that is, if $a \neq 1$, then divide both sides by a.

Step 3   Add the square of half the coefficient of the linear term to both sides of the equation.

Step 4   Write the left side of the equation as the square of a binomial, and simplify the right side if possible.

Step 5   Apply the square root theorem, and solve as usual.

Answer

5. $\dfrac{3 \pm i\sqrt{31}}{10}$

C   Quadratic Equations and Right Triangles

FACTS FROM GEOMETRY   **More Special Triangles**

The triangles shown in Figures 1 and 2 occur frequently in mathematics.

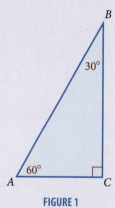
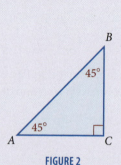

| FIGURE 1 | FIGURE 2 |

Note that both of the triangles are right triangles. We refer to the triangle in Figure 1 as a 30°–60°–90° triangle and the triangle in Figure 2 as a 45°–45°–90° triangle.

6. If the shortest side in a 30° – 60° – 90° triangle is 2 inches long, find the lengths of the other two sides.

EXAMPLE 6   If the shortest side in a 30°–60°–90° triangle is 1 inch, find the lengths of the other two sides.

SOLUTION   In Figure 3, triangle ABC is a 30°–60°–90° triangle in which the shortest side AC is 1 inch long. Triangle DBC is also a 30°–60°–90° triangle in which the shortest side DC is 1 inch long.

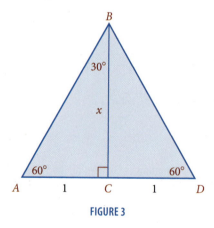

FIGURE 3

Notice that the large triangle ABD is an equilateral triangle because each of its interior angles is 60°. Each side of triangle ABD is 2 inches long. Side AB in triangle ABC is therefore 2 inches. To find the length of side BC, we use the Pythagorean theorem.

$$BC^2 + AC^2 = AB^2$$

$$x^2 + 1^2 = 2^2$$

$$x^2 + 1 = 4$$

$$x^2 = 3$$

$$x = \sqrt{3} \text{ inches}$$

Note that we write only the positive square root because x is the length of a side in a triangle and is therefore a positive number.   ■

Answers

6. 4 inches and $2\sqrt{3}$ inches

EXAMPLE 7   Table 1 gives the length of each chairlift at the Northstar at Tahoe Ski Resort. The table also shows the change in elevation from the beginning of each lift to the end. As you can see, the vertical rise of the Forest Double chair lift is 1,170 feet and the length of the chair lift is 5,750 feet. To the nearest foot, find the horizontal distance covered by a person riding this lift.

7. Table 1 gives the vertical rise of the Lookout Double chair lift as 960 feet and the length of the chair lift as 4,330 feet. To the nearest foot, find the horizontal distance covered by a person riding this lift.

TABLE 1		
FROM THE TRAIL MAP FOR THE NORTHSTAR AT TAHOE SKI RESORT		
Lift Information		
Lift	**Vertical Rise (feet)**	**Length (feet)**
Big Springs Gondola	480	4,100
Bear Paw Double	120	790
Echo Triple	710	4,890
Aspen Express Quad	900	5,100
Forest Double	1,170	5,750
Lookout Double	960	4,330
Comstock Express Quad	1,250	5,900
Rendezvous Triple	650	2,900
Schaffer Camp Triple	1,860	6,150
Chipmunk Tow Lift	28	280
Bear Cub Tow Lift	120	750

Source:skilifts.org

SOLUTION   Figure 4 is a model of the Forest Double chair lift. A rider gets on the lift at point A and exits at point B. The length of the lift is AB.
To find the horizontal distance covered by a person riding the chair lift, we use the Pythagorean theorem.

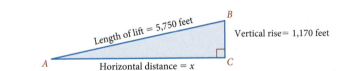

Length of lift = 5,750 feet B Vertical rise = 1,170 feet

A Horizontal distance = x C

FIGURE 4

$$5,750^2 = x^2 + 1,170^2 \qquad \text{Pythagorean theorem}$$

$$33,062,500 = x^2 + 1,368,900 \qquad \text{Simplify squares.}$$

$$x^2 = 33,062,500 - 1,368,900 \qquad \text{Solve for } x^2.$$

$$x^2 = 31,693,600 \qquad \text{Simplify the right side.}$$

$$x = \sqrt{31,693,600} \qquad \text{Square root theorem}$$

$$= 5,630 \text{ feet} \qquad \text{Rounded to the nearest foot}$$

A rider getting on the lift at point A and riding to point B will cover a horizontal distance of approximately 5,630 feet. ■

Answer

7. 4,222 feet

Problem Set 8.1

Moving Toward Success

"If you are planning for a year, sow rice; if you are planning for a decade, plant trees; if you are planning for a lifetime, educate people."

—Chinese proverb

1. If you focus solely on facts, you may not retain the information as well as if you focus on main ideas and concepts. Why is this true?

2. Make a conscious decision to be interested in what the instructor is teaching or what you're studying. How will this increase your success in this class?

A Solve the following equations. [Examples 1–3]

1. $x^2 = 25$

2. $x^2 = 16$

3. $y^2 = \dfrac{3}{4}$

4. $y^2 = \dfrac{5}{9}$

5. $x^2 + 12 = 0$

6. $x^2 + 8 = 0$

7. $4a^2 - 45 = 0$

8. $9a^2 - 20 = 0$

9. $(2y - 1)^2 = 25$

10. $(3y + 7)^2 = 1$

11. $(2a + 3)^2 = -9$

12. $(3a - 5)^2 = -49$

13. $x^2 + 8x + 16 = -27$

14. $x^2 - 12x + 36 = -8$

15. $4a^2 - 12a + 9 = -4$

16. $9a^2 - 12a + 4 = -9$

B Copy each of the following, and fill in the blanks so that the left side of each is a perfect square trinomial; that is, complete the square.

17. $x^2 + 12x +$ _____ $= (x +$ _____ $)^2$

18. $x^2 + 6x +$ _____ $= (x +$ _____ $)^2$

19. $x^2 - 4x +$ _____ $= (x -$ _____ $)^2$

20. $x^2 - 2x +$ _____ $= (x -$ _____ $)^2$

21. $a^2 - 10a +$ _____ $= (a -$ _____ $)^2$

22. $a^2 - 8a +$ _____ $= (a -$ _____ $)^2$

23. $x^2 + 5x +$ _____ $= (x +$ _____ $)^2$

24. $x^2 + 3x +$ _____ $= (x +$ _____ $)^2$

25. $y^2 - 7y +$ _____ $= (y -$ _____ $)^2$

26. $y^2 - y +$ _____ $= (y -$ _____ $)^2$

27. $x^2 + \dfrac{1}{2}x +$ _____ $= (x +$ _____ $)^2$

28. $x^2 - \dfrac{3}{4}x +$ _____ $= (x -$ _____ $)^2$

29. $x^2 + \dfrac{2}{3}x +$ _____ $= (x +$ _____ $)^2$

30. $x^2 - \dfrac{4}{5}x +$ _____ $= (x -$ _____ $)^2$

B Solve each of the following quadratic equations by completing the square. [Examples 4, 5]

31. $x^2 + 12x = -27$ **32.** $x^2 - 6x = 16$ **33.** $a^2 - 2a + 5 = 0$ **34.** $a^2 + 10a + 22 = 0$

35. $y^2 - 8y + 1 = 0$ **36.** $y^2 + 6y - 1 = 0$ **37.** $x^2 - 5x - 3 = 0$ **38.** $x^2 - 5x - 2 = 0$

39. $2x^2 - 4x - 8 = 0$ **40.** $3x^2 - 9x - 12 = 0$ **41.** $3t^2 - 8t + 1 = 0$ **42.** $5t^2 + 12t - 1 = 0$

43. $4x^2 - 3x + 5 = 0$ **44.** $7x^2 - 5x + 2 = 0$

45. For the equation $x^2 = -9$:

    a. Can it be solved by factoring?

    b. Solve it.

46. For the equation $x^2 - 10x + 18 = 0$:

    a. Can it be solved by factoring?

    b. Solve it.

47. Solve each equation below by the indicated method.

    a. Factor: $x^2 - 6x = 0$.

    b. Complete the square: $x^2 - 6x = 0$.

48. Solve each equation below by the indicated method.

    a. Factor: $x^2 + ax = 0$.

    b. Complete the square: $x^2 + ax = 0$.

49. Solve the equation $x^2 + 2x = 35$

    a. by factoring.

    b. by completing the square.

50. Solve the equation $8x^2 - 10x - 25 = 0$

    a. by factoring.

    b. by completing the square.

51. Is $x = -3 + \sqrt{2}$ a solution to $x^2 - 6x = 7$?

52. Is $x = 2 - \sqrt{5}$ a solution to $x^2 - 4x = 1$?

The next two problems will give you practice solving a variety of equations.

53. Solve each equation.

a. $5x - 7 = 0$

b. $5x - 7 = 8$

c. $(5x - 7)^2 = 8$

d. $\sqrt{5x - 7} = 0$

e. $\dfrac{5}{2} - \dfrac{7}{2x} = \dfrac{4}{x}$

54. Solve each equation.

a. $5x + 11 = 0$

b. $5x + 11 = 9$

c. $(5x + 11)^2 = 9$

d. $\sqrt{5x + 11} = 9$

e. $\dfrac{5}{3} - \dfrac{11}{3x} = \dfrac{3}{x}$

Simplify the left side of each equation, and then solve for x.

55. $(x + 5)^2 + (x - 5)^2 = 52$

56. $(2x + 1)^2 + (2x - 1)^2 = 10$

57. $(2x + 3)^2 + (2x - 3)^2 = 26$

58. $(3x + 2)^2 + (3x - 2)^2 = 26$

59. $(3x + 4)(3x - 4) - (x + 2)(x - 2) = -4$

60. $(5x + 2)(5x - 2) - (x + 3)(x - 3) = 29$

61. Fill in the table below given the following functions.
$f(x) = (2x - 3)^2, g(x) = 4x^2 - 12x + 9, h(x) = 4x^2 + 9$

x	$f(x)$	$g(x)$	$h(x)$
-2			
-1			
0			
1			
2			

62. Fill in the table below given the following functions.
$f(x) = \left(x + \dfrac{1}{2}\right)^2, g(x) = x^2 + x + \dfrac{1}{4}, h(x) = x^2 + \dfrac{1}{4}$

x	$f(x)$	$g(x)$	$h(x)$
-2			
-1			
0			
1			
2			

63. If $f(x) = (x - 3)^2$ find x if $f(x) = 0$.

64. If $f(x) = (3x - 5)^2$ find x if $f(x) = 0$.

65. If $f(x) = x^2 - 5x - 6$ find

  a. the x-intercepts.

  b. the value of x for which $f(x) = 0$.

  c. $f(0)$

  d. $f(1)$

66. If $f(x) = 9x^2 - 12x + 4$ find

  a. the x-intercepts.

  b. the value of x for which $f(x) = 0$.

  c. $f(0)$

  d. $f(1)$

Applying the Concepts

67. Geometry If the shortest side in a 30°–60°–90° triangle is $\frac{1}{2}$ inch long, find the lengths of the other two sides.

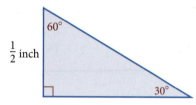

68. Geometry If the length of the longest side of a 30°–60°–90° triangle is x, find the length of the other two sides in terms of x

69. Geometry If the length of the shorter sides of a 45°–45°–90° triangle is 1 inch, find the length of the hypotenuse.

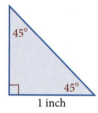

70. Geometry If the length of the shorter sides of a 45°–45°–90° triangle is x, find the length of the hypotenuse, in terms of x.

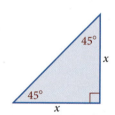

71. Chair Lift Use Table 1 from Example 7 to find the horizontal distance covered by a person riding the Bear Paw Double chair lift. Round your answer to the nearest foot.

72. Chair Lift Use Table 1 from Example 7 to find the horizontal distance covered by a person riding the Big Springs Gondola lift. Round your answer to the nearest foot.

TABLE 1		
FROM THE TRAIL MAP FOR THE NORTHSTAR AT TAHOE SKI RESORT		
Lift Information		
Lift	**Vertical Rise (feet)**	**Length (feet)**
Big Springs Gondola	480	4,100
Bear Paw Double	120	790
Echo Triple	710	4,890
Aspen Express Quad	900	5,100
Forest Double	1,170	5,750
Lookout Double	960	4,330
Comstock Express Quad	1,250	5,900
Rendezvous Triple	650	2,900
Schaffer Camp Triple	1,860	6,150
Chipmunk Tow Lift	28	280
Bear Cub Tow Lift	120	750
Source: skilifts.org		

73. Interest Rate Suppose a deposit of $3,000 in a savings account that paid an annual interest rate r (compounded yearly) is worth $3,456 after 2 years. Using the formula $A = P(1 + r)^t$, we have

$$3{,}456 = 3{,}000(1 + r)^2$$

Solve for r to find the annual interest rate.

74. Special Triangles In Figure 5, triangle ABC has angles 45° and 30° and height x. Find the lengths of sides AB, BC, and AC, in terms of x.

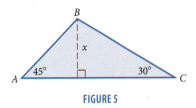

FIGURE 5

75. Length of an Escalator An escalator in a department store is to carry people a vertical distance of 20 feet between floors. How long is the escalator if it makes an angle of 45° with the ground? (See Figure 6.)

FIGURE 6

76. Dimensions of a Tent A two-person tent is to be made so the height at the center is 4 feet. If the sides of the tent are to meet the ground at an angle of 60° and the tent is to be 6 feet in length, how many square feet of material will be needed to make the tent? (Figure 7, assume that the tent has a floor and is closed at both ends.) Give your answer to the nearest tenth of a square foot.

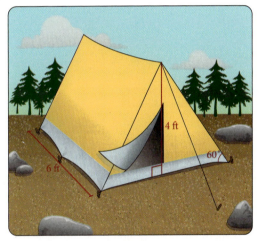

FIGURE 7

Maintaining Your Skills

Write each of the following in simplified form for radicals.

77. $\sqrt{45}$

78. $\sqrt{24}$

79. $\sqrt{27y^5}$

80. $\sqrt{8y^3}$

81. $\sqrt[3]{54x^6y^5}$

82. $\sqrt[3]{16x^9y^7}$

Rationalize the denominator.

83. $\dfrac{3}{\sqrt{2}}$

84. $\dfrac{5}{\sqrt{3}}$

85. $\dfrac{2}{\sqrt[3]{4}}$

86. $\dfrac{3}{\sqrt[3]{2}}$

Getting Ready for the Next Section

Simplify.

87. $\sqrt{49 - 4(6)(-5)}$

88. $\sqrt{49 - 4(6)2}$

89. $\sqrt{(-27)^2 - 4(0.1)(1,700)}$

90. $\sqrt{25 - 4(4)(-10)}$

91. $\dfrac{-7 + \sqrt{169}}{12}$

92. $\dfrac{-7 - \sqrt{169}}{12}$

93. Simplify $\sqrt{b^2 - 4ac}$ when $a = 6$, $b = 7$, and $c = -5$.

94. Simplify $\sqrt{b^2 - 4ac}$ when $a = 2$, $b = -6$, and $c = 3$.

Factor.

95. $27t^3 - 8$

96. $125t^3 + 1$

Extending the Concepts

Solve for x.

97. $(x + a)^2 + (x - a)^2 = 10a^2$

98. $(ax + 1)^2 + (ax - 1)^2 = 10$

Assume p and q are positive numbers and solve for x by completing the square on x.

99. $x^2 + px + q = 0$

100. $x^2 - px + q = 0$

101. $3x^2 + px + q = 0$

102. $3x^2 + 2px + q = 0$

Complete the square on x and also on y so that each equation below is written in the form

$$(x - a)^2 + (y - b)^2 = r^2$$

which you will see later in the book as the equation of a circle with center (a, b) and radius r.

103. $x^2 - 10x + y^2 - 6y = -30$

104. $x^2 - 2x + y^2 - 4y = 20$

The Quadratic Formula

TICKET TO SUCCESS

Keep these questions in mind as you read through the section. Then respond in your own words and in complete sentences.

1. What is the quadratic formula?
2. Explain how you can use the quadratic formula to solve an equation that is not quadratic.
3. When would factoring be a faster method than using the quadratic formula to solve a quadratic equation? When would it not?
4. When will the quadratic formula result in only one solution?

Voyagerix/Shutterstock.com

Imagine you and a friend are throwing rocks 30 feet down a well to the water below. You throw the rock with an initial velocity of 10 feet per second and your friend throws with an initial velocity of 7 feet per second. The relationships between distance the rocks travel s and the time t is given by

$$s = 10t + 16t^2 \quad \text{Your rock}$$

$$s = 7t + 16t^2 \quad \text{Your friend's rock}$$

How much sooner will your rock splash than your friend's? Equations such as these require the use of the quadratic formula. In this section, we will use the method of completing the square from the preceding section to derive the quadratic formula. The *quadratic formula* is a very useful tool in mathematics. It allows us to solve all types of quadratic equations.

A The Quadratic Theorem

The Quadratic Theorem

For any quadratic equation in the form $ax^2 + bx + c = 0$, where $a \neq 0$, the two solutions are

$$x = \frac{-b + \sqrt{b^2 - 4ac}}{2a} \quad \text{and} \quad x = \frac{-b - \sqrt{b^2 - 4ac}}{2a}$$

PROOF We will prove the quadratic theorem by completing the square on $ax^2 + bx + c = 0$.

$$ax^2 + bx + c = 0$$

$$ax^2 + bx = -c \qquad \text{Add } -c \text{ to both sides.}$$

$$x^2 + \frac{b}{a}x = -\frac{c}{a} \qquad \text{Divide both sides by } a.$$

To complete the square on the left side, we add the square of $\frac{1}{2}$ of $\frac{b}{a}$ to both sides $\left(\frac{1}{2} \text{ of } \frac{b}{a} \text{ is } \frac{b}{2a}\right)$.

$$x^2 + \frac{b}{a}x + \left(\frac{b}{2a}\right)^2 = -\frac{c}{a} + \left(\frac{b}{2a}\right)^2$$

We now simplify the right side as a separate step. We square the second term and combine the two terms by writing each with the least common denominator $4a^2$.

$$-\frac{c}{a} + \left(\frac{b}{2a}\right)^2 = -\frac{c}{a} + \frac{b^2}{4a^2} = \frac{4a}{4a}\left(\frac{-c}{a}\right) + \frac{b^2}{4a^2} = \frac{-4ac + b^2}{4a^2}$$

It is convenient to write this last expression as

$$\frac{b^2 - 4ac}{4a^2}$$

Continuing with the proof, we have

$$x^2 + \frac{b}{a}x + \left(\frac{b}{2a}\right)^2 = \frac{b^2 - 4ac}{4a^2}$$

$$\left(x + \frac{b}{2a}\right)^2 = \frac{b^2 - 4ac}{4a^2} \qquad \text{Write left side as a binomial square.}$$

$$x + \frac{b}{2a} = \pm\frac{\sqrt{b^2 - 4ac}}{2a} \qquad \text{Square root theorem}$$

$$x = -\frac{b}{2a} \pm \frac{\sqrt{b^2 - 4ac}}{2a} \qquad \text{Add } -\frac{b}{2a} \text{ to both sides.}$$

$$= \frac{-b \pm \sqrt{b^2 - 4ac}}{2a}$$

Our proof is now complete. What we have is this:

Definition

If our equation is in the form $ax^2 + bx + c = 0$ (standard form), where $a \neq 0$, the two solutions are always given by the formula

$$x = \frac{-b \pm \sqrt{b^2 - 4ac}}{2a}$$

This formula is known as the **quadratic formula**.

If we substitute the coefficients a, b, and c of any quadratic equation in standard form into the formula, we need only perform some basic arithmetic to arrive at the solution set.

PRACTICE PROBLEMS

1. Solve $6x^2 + 7x + 2 = 0$ using the quadratic formula.

EXAMPLE 1   Use the quadratic formula to solve $6x^2 + 7x - 5 = 0$.

SOLUTION   Using $a = 6$, $b = 7$, and $c = -5$ in the quadratic formula we have

$$x = \frac{-b \pm \sqrt{b^2 - 4ac}}{2a}$$

$$= \frac{-7 \pm \sqrt{49 - 4(6)(-5)}}{2(6)}$$

$$= \frac{-7 \pm \sqrt{49 + 120}}{12}$$

$$= \frac{-7 \pm \sqrt{169}}{12}$$

$$= \frac{-7 \pm 13}{12}$$

We separate the last equation into the two statements

$$x = \frac{-7 + 13}{12} \quad \text{or} \quad x = \frac{-7 - 13}{12}$$

$$x = \frac{1}{2} \qquad\qquad x = -\frac{5}{3}$$

The solution set is $\left\{\frac{1}{2}, -\frac{5}{3}\right\}$.   ■

Whenever the solutions to a quadratic equation are rational numbers, as they are in Example 1, it means that the original equation was solvable by factoring. To illustrate, let's solve the equation from Example 1 again, but this time by factoring.

$$6x^2 + 7x - 5 = 0 \qquad \text{Equation in standard form}$$

$$(3x + 5)(2x - 1) = 0 \qquad \text{Factor the left side.}$$

$$3x + 5 = 0 \quad \text{or} \quad 2x - 1 = 0 \qquad \text{Set factors equal to 0.}$$

$$x = -\frac{5}{3} \qquad\qquad x = \frac{1}{2}$$

When an equation can be solved by factoring, then factoring is usually the faster method of solution. It is best to try to factor first, and then if you have trouble factoring, go to the quadratic formula. It always works.

EXAMPLE 2   Solve $\dfrac{x^2}{3} - x = -\dfrac{1}{2}$.

SOLUTION   Multiplying through by 6 and writing the result in standard form, we have

$$2x^2 - 6x + 3 = 0$$

the left side of which is not factorable. Therefore, we use the quadratic formula with $a = 2$, $b = -6$, and $c = 3$. The two solutions are given by

$$x = \frac{-(-6) \pm \sqrt{36 - 4(2)(3)}}{2(2)}$$

$$= \frac{6 \pm \sqrt{12}}{4}$$

$$= \frac{6 \pm 2\sqrt{3}}{4} \qquad \sqrt{12} = \sqrt{4 \cdot 3} = \sqrt{4} \cdot \sqrt{3} = 2\sqrt{3}$$

We can reduce this last expression to lowest terms by factoring 2 from the numerator and denominator and then dividing the numerator and denominator by 2.

$$x = \frac{2(3 \pm \sqrt{3})}{2 \cdot 2} = \frac{3 \pm \sqrt{3}}{2}$$ ■

3. Solve $\dfrac{1}{x+4} - \dfrac{1}{x} = \dfrac{1}{2}$.

EXAMPLE 3   Solve $\dfrac{1}{x+2} - \dfrac{1}{x} = \dfrac{1}{3}$.

SOLUTION   To solve this equation, we must first put it in standard form. To do so, we must clear the equation of fractions by multiplying each side by the LCD for all the denominators, which is $3x(x+2)$. Multiplying both sides by the LCD, we have

$$3x(x+2)\left(\frac{1}{x+2} - \frac{1}{x}\right) = \frac{1}{3} \cdot 3x(x+2) \qquad \text{Multiply each by the LCD.}$$

$$3x(x+2)\cdot\frac{1}{x+2} - 3x(x+2)\cdot\frac{1}{x} = \frac{1}{3}\cdot 3x(x+2)$$

$$3x - 3(x+2) = x(x+2)$$

$$3x - 3x - 6 = x^2 + 2x \qquad \text{Multiply.}$$

$$-6 = x^2 + 2x \qquad \text{Simplify left side.}$$

$$0 = x^2 + 2x + 6 \qquad \text{Add 6 to each side.}$$

Since the right side of our last equation is not factorable, we use the quadratic formula. From our last equation, we have $a=1$, $b=2$, and $c=6$. Using these numbers for a, b, and c in the quadratic formula gives us

$$x = \frac{-2 \pm \sqrt{4-4(1)(6)}}{2(1)}$$

$$= \frac{-2 \pm \sqrt{4-24}}{2} \qquad \text{Simplify inside the radical.}$$

$$= \frac{-2 \pm \sqrt{-20}}{2} \qquad 4-24=-20$$

$$= \frac{-2 \pm 2i\sqrt{5}}{2} \qquad \sqrt{-20} = i\sqrt{20} = i\sqrt{4}\sqrt{5} = 2i\sqrt{5}$$

$$= \frac{2(-1 \pm i\sqrt{5})}{2} \qquad \text{Factor 2 from the numerator.}$$

$$= -1 \pm i\sqrt{5} \qquad \text{Divide numerator and denominator by 2.}$$

Since neither of the two solutions, $-1 + i\sqrt{5}$ nor $-1 - i\sqrt{5}$, will make any of the denominators in our original equation 0, they are both solutions.   ∎

Although the equation in our next example is not a quadratic equation, we solve it by using both factoring and the quadratic formula.

4. Solve $8t^3 - 27 = 0$.

EXAMPLE 4   Solve $27t^3 - 8 = 0$.

SOLUTION   It would be a mistake to add 8 to each side of this equation and then take the cube root of each side because we would lose two of our solutions. Instead, we factor the left side, and then set the factors equal to 0.

$$27t^3 - 8 = 0 \qquad \text{Equation in standard form}$$

$$(3t-2)(9t^2 + 6t + 4) = 0 \qquad \text{Factor as the difference of two cubes.}$$

$$3t - 2 = 0 \quad \text{or} \quad 9t^2 + 6t + 4 = 0 \qquad \text{Set each factor equal to 0.}$$

The first equation leads to a solution of $t = \dfrac{2}{3}$. The second equation does not factor, so we use the quadratic formula with $a=9$, $b=6$, and $c=4$:

Answers

3. $-2 \pm 2i$

4. $\dfrac{-3 \pm 3i\sqrt{3}}{4}, \dfrac{3}{2}$

$$t = \frac{-6 \pm \sqrt{36 - 4(9)(4)}}{2(9)}$$

$$= \frac{-6 \pm \sqrt{36 - 144}}{18}$$

$$= \frac{-6 \pm \sqrt{-108}}{18}$$

$$= \frac{-6 \pm 6i\sqrt{3}}{18} \qquad \sqrt{-108} = i\sqrt{36 \cdot 3} = 6i\sqrt{3}$$

$$= \frac{6(-1 \pm i\sqrt{3})}{6 \cdot 3} \qquad \text{Factor 6 from the numerator and denominator.}$$

$$= \frac{-1 \pm i\sqrt{3}}{3} \qquad \text{Divide out common factor 6.}$$

The three solutions to our original equation are

$$\frac{2}{3}, \qquad \frac{-1 + i\sqrt{3}}{3}, \qquad \text{and} \qquad \frac{-1 - i\sqrt{3}}{3} \qquad ■$$

B Applications

EXAMPLE 5   If an object is thrown downward with an initial velocity of 20 feet per second, the distance $s(t)$, in feet, it travels in t seconds is given by the function $s(t) = 20t + 16t^2$. How long does it take the object to fall 40 feet?

SOLUTION   We let $s(t) = 40$, and solve for t.

When → $\qquad s(t) = 40$

the function → $\quad s(t) = 20t + 16t^2$

becomes → $\qquad 40 = 20t + 16t^2$

$$16t^2 + 20t - 40 = 0$$

$$4t^2 + 5t - 10 = 0 \qquad \text{Divide by 4.}$$

Using the quadratic formula, we have

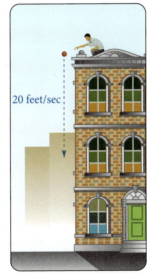

20 feet/sec

$$t = \frac{-5 \pm \sqrt{25 - 4(4)(-10)}}{2(4)}$$

$$= \frac{-5 \pm \sqrt{185}}{8}$$

$$= \frac{-5 + \sqrt{185}}{8} \qquad \text{or} \qquad \frac{-5 - \sqrt{185}}{8}$$

The second solution is impossible since it is a negative number and time t must be positive. It takes

$$t = \frac{-5 + \sqrt{185}}{8} \qquad \text{or approximately} \qquad \frac{-5 + 13.60}{8} \approx 1.08 \text{ seconds}$$

for the object to fall 40 feet.   ■

Recall that the relationship between profit, revenue, and cost is given by the formula

$$P(x) = R(x) - C(x)$$

where $P(x)$ is the profit, $R(x)$ is the total revenue, and $C(x)$ is the total cost of producing and selling x items.

5. An object thrown upward with an initial velocity of 32 feet per second rises and falls according to the equation $s = 32t - 16t^2$ where s is the height of the object above the ground at any time t. At what times t will the object be 12 feet above the ground?

Answer

5. $\frac{3}{2}$ seconds, $\frac{1}{2}$ second

EXAMPLE 6   A company produces and sells copies of an accounting program for home computers. The total weekly cost (in dollars) to produce x copies of the program is $C(x) = 8x + 500$, and the weekly revenue for selling all x copies of the program is $R(x) = 35x - 0.1x^2$. How many programs must be sold each week for the weekly profit to be $1,200?

SOLUTION   Substituting the given expressions for $R(x)$ and $C(x)$ in the equation $P(x) = R(x) - C(x)$, we have a polynomial in x that represents the weekly profit $P(x)$.

$$P(x) = R(x) - C(x)$$
$$= 35x - 0.1x^2 - (8x + 500)$$
$$= 35x - 0.1x^2 - 8x - 500$$
$$= -500 + 27x - 0.1x^2$$

By setting this expression equal to 1,200, we have a quadratic equation to solve that gives us the number of programs x that need to be sold each week to bring in a profit of $1,200.

$$1,200 = -500 + 27x - 0.1x^2$$

We can write this equation in standard form by adding the opposite of each term on the right side of the equation to both sides of the equation. Doing so produces the following equation:

$$0.1x^2 - 27x + 1,700 = 0$$

Applying the quadratic formula to this equation with $a = 0.1$, $b = -27$, and $c = 1,700$, we have

$$x = \frac{27 \pm \sqrt{(-27)^2 - 4(0.1)(1,700)}}{2(0.1)}$$

$$= \frac{27 \pm \sqrt{729 - 680}}{0.2}$$

$$= \frac{27 \pm \sqrt{49}}{0.2}$$

$$= \frac{27 \pm 7}{0.2}$$

Writing this last expression as two separate expressions, we have our two solutions:

$$x = \frac{27 + 7}{0.2} \quad \text{or} \quad x = \frac{27 - 7}{0.2}$$

$$= \frac{34}{0.2} \qquad\qquad = \frac{20}{0.2}$$

$$= 170 \qquad\qquad\quad = 100$$

The weekly profit will be $1,200 if the company produces and sells 100 programs or 170 programs.   ∎

What is interesting about the equation we solved in Example 6 is that it has rational solutions, meaning it could have been solved by factoring. But looking back at the equation, factoring does not seem like a reasonable method of solution because the coefficients are either very large or very small. So, there are times when using the quadratic formula is a faster method of solution, even though the equation you are solving is factorable.

Problem Set 8.2

Moving Toward Success

"A moment's insight is sometimes worth a lifetime's experience."

—Oliver Wendell Holmes, Jr., 1841–1935, American jurist

1. Why would beginning your study session with the most difficult topics make your session more efficient?

2. When studying, should you take breaks? Why or why not?

A Solve each equation in each problem using the quadratic formula. [Example 1]

1. a. $3x^2 + 4x - 2 = 0$

b. $3x^2 - 4x - 2 = 0$

c. $3x^2 + 4x + 2 = 0$

d. $2x^2 + 4x - 3 = 0$

e. $2x^2 - 4x + 3 = 0$

2. a. $3x^2 + 6x - 2 = 0$

b. $3x^2 - 6x - 2 = 0$

c. $3x^2 + 6x + 2 = 0$

d. $2x^2 + 6x + 3 = 0$

e. $2x^2 + 6x - 3 = 0$

3. a. $x^2 - 2x + 2 = 0$

b. $x^2 - 2x + 5 = 0$

c. $x^2 + 2x + 2 = 0$

4. a. $x^2 - 4x + 5 = 0$

b. $x^2 + 4x + 5 = 0$

c. $a^2 + 4a + 1 = 0$

A Solve each equation. Use factoring or the quadratic formula, whichever is appropriate. (Try factoring first. If you have any difficulty factoring, then go right to the quadratic formula.) [Examples 1–2]

5. $\frac{1}{6}x^2 - \frac{1}{2}x + \frac{1}{3} = 0$

6. $\frac{1}{4}x^2 + \frac{1}{4}x - \frac{1}{2} = 0$

7. $\frac{x^2}{2} + 1 = \frac{2x}{3}$

8. $\frac{x^2}{2} + \frac{2}{3} = -\frac{2x}{3}$

9. $y^2 - 5y = 0$

10. $2y^2 + 10y = 0$

11. $30x^2 + 40x = 0$

12. $50x^2 - 20x = 0$

13. $\frac{2t^2}{3} - t = -\frac{1}{6}$

14. $\dfrac{t^2}{3} - \dfrac{t}{2} = -\dfrac{3}{2}$

15. $0.01x^2 + 0.06x - 0.08 = 0$

16. $0.02x^2 - 0.03x + 0.05 = 0$

17. $2x + 3 = -2x^2$

18. $2x - 3 = 3x^2$

19. $100x^2 - 200x + 100 = 0$

20. $100x^2 - 600x + 900 = 0$

21. $\dfrac{1}{2}r^2 = \dfrac{1}{6}r - \dfrac{2}{3}$

22. $\dfrac{1}{4}r^2 = \dfrac{2}{5}r + \dfrac{1}{10}$

23. $(x - 3)(x - 5) = 1$

24. $(x - 3)(x + 1) = -6$

25. $(x + 3)^2 + (x - 8)(x - 1) = 16$

26. $(x - 4)^2 + (x + 2)(x + 1) = 9$

27. $\dfrac{x^2}{3} - \dfrac{5x}{6} = \dfrac{1}{2}$

28. $\dfrac{x^2}{6} + \dfrac{5}{6} = -\dfrac{x}{3}$

A Multiply both sides of each equation by its LCD. Then solve the resulting equation. [Example 3]

29. $\dfrac{1}{x + 1} - \dfrac{1}{x} = \dfrac{1}{2}$

30. $\dfrac{1}{x + 1} + \dfrac{1}{x} = \dfrac{1}{3}$

31. $\dfrac{1}{y - 1} + \dfrac{1}{y + 1} = 1$

32. $\dfrac{2}{y + 2} + \dfrac{3}{y - 2} = 1$

33. $\dfrac{1}{x + 2} + \dfrac{1}{x + 3} = 1$

34. $\dfrac{1}{x + 3} + \dfrac{1}{x + 4} = 1$

35. $\dfrac{6}{r^2 - 1} - \dfrac{1}{2} = \dfrac{1}{r + 1}$

36. $2 + \dfrac{5}{r - 1} = \dfrac{12}{(r - 1)^2}$

A Solve each equation. In each case you will have three solutions. [Example 4]

37. $x^3 - 8 = 0$

38. $x^3 - 27 = 0$

39. $8a^3 + 27 = 0$

40. $27a^3 + 8 = 0$

41. $125t^3 - 1 = 0$

42. $64t^3 + 1 = 0$

A Each of the following equations has three solutions. Look for the greatest common factor, then use the quadratic formula to find all solutions.

43. $2x^3 + 2x^2 + 3x = 0$

44. $6x^3 - 4x^2 + 6x = 0$

45. $3y^4 = 6y^3 - 6y^2$

46. $4y^4 = 16y^3 - 20y^2$ **47.** $6t^5 + 4t^4 = -2t^3$ **48.** $8t^5 + 2t^4 = -10t^3$

49. Which two of the expressions below are equivalent?

   a. $\dfrac{6 + 2\sqrt{3}}{4}$

   b. $\dfrac{3 + \sqrt{3}}{2}$

   c. $6 + \dfrac{\sqrt{3}}{2}$

50. Which two of the expressions below are equivalent?

   a. $\dfrac{8 - 4\sqrt{2}}{4}$

   b. $2 - 4\sqrt{2}$

   c. $2 - \sqrt{2}$

51. Solve $3x^2 - 5x = 0$

   a. by factoring.

   b. by the quadratic formula.

52. Solve $3x^2 + 23x - 70 = 0$

   a. by factoring.

   b. by the quadratic formula.

53. Can the equation $x^2 - 4x + 7 = 0$ be solved by factoring? Solve the equation.

54. Can the equation $x^2 = 5$ be solved by factoring? Solve the equation.

55. Is $x = -1 + i$ a solution to $x^2 + 2x = -2$?

56. Is $x = 2 + 2i$ a solution to $(x - 2)^2 = -4$?

57. Let $f(x) = x^2 - 2x - 3$. Find all values for the variable x, which produce the following values of $f(x)$.

   a. $f(x) = 0$

   b. $f(x) = -11$

   c. $f(x) = -2x + 1$

   d. $f(x) = 2x + 1$

58. Let $g(x) = x^2 + 16$. Find all values for the variable x, which produce the following values of $g(x)$.

   a. $g(x) = 0$

   b. $g(x) = 20$

   c. $g(x) = 8x$

   d. $g(x) = -8x$

59. Let $f(x) = \dfrac{10}{x^2}$. Find all values for the variable x, for which $f(x) = g(x)$.

   a. $g(x) = 3 + \dfrac{1}{x}$

   b. $g(x) = 8x - \dfrac{17}{x^2}$

   c. $g(x) = 0$

   d. $g(x) = 10$

60. Let $h(x) = \dfrac{x + 2}{x}$. Find all values for the variable x, for which $h(x) = f(x)$.

   a. $f(x) = 2$

   b. $f(x) = x + 2$

   c. $f(x) = x - 2$

   d. $f(x) = \dfrac{x - 2}{4}$

Applying the Concepts

61. Falling Object If you throw a rock down a 30-foot well with a downward velocity of 10 feet per second, the relationship between the distance the rock travels s and the time t is $s = 10t + 16t^2$. How long will it take for your rock to hit the water?

62. Coin Toss A coin is tossed upward with an initial velocity of 32 feet per second from a height of 16 feet above the ground. The equation giving the object's height h at any time t is $h = 16 + 32t - 16t^2$. Does the object ever reach a height of 32 feet?

63. Profit The total cost (in dollars) for a company to manufacture and sell x items per week is $C = 60x + 300$, whereas the revenue brought in by selling all x items is $R = 100x - 0.5x^2$. How many items must be sold to obtain a weekly profit of $300?

64. Profit Suppose a company manufactures and sells x picture frames each month with a total cost of $C = 1,200 + 3.5x$ dollars. If the revenue obtained by selling x frames is $R = 9x - 0.002x^2$, find the number of frames it must sell each month if its monthly profit is to be $2,300.

65. Photograph Cropping The following figure shows a photographic image on a 10.5-centimeter by 8.2-centimeter background. The overall area of the background is to be reduced to 80% of its original area by cutting off (cropping) equal strips on all four sides. What is the width of the strip that is cut from each side?

66. Area of a Garden A garden measures 20.3 meters by 16.4 meters. To double the area of the garden, strips of equal width are added to all four sides.

a. Draw a diagram that illustrates these conditions.

b. What are the new overall dimensions of the garden?

67. Area and Perimeter A rectangle has a perimeter of 20 yards and an area of 15 square yards.

   a. Write two equations that state these facts in terms of the rectangle's length, l, and its width, w.

   b. Solve the two equations from part a to determine the actual length and width of the rectangle.

   c. Explain why two answers are possible to part b.

68. Population Size Writing in 1829, former President James Madison made some predictions about the growth of the population of the United States. The populations he predicted fit the equation

$$y = 0.029x^2 - 1.39x + 42$$

where y is the population in millions of people x years from 1829.

   a. Use the equation to determine the approximate year President Madison would have predicted that the U.S. population would reach 100,000,000.

   b. If the U.S. population in 2010 was approximately 300 million, were President Madison's predictions accurate in the long term? Explain why or why not.

Maintaining Your Skills

Divide, using long division.

69. $\dfrac{8y^2 - 26y - 9}{2y - 7}$

70. $\dfrac{6y^2 + 7y - 18}{3y - 4}$

71. $\dfrac{x^3 + 9x^2 + 26x + 24}{x + 2}$

72. $\dfrac{x^3 + 6x^2 + 11x + 6}{x + 3}$

Simplify each expression. (Assume $x, y > 0$.)

73. $25^{1/2}$

74. $8^{1/3}$

75. $\left(\dfrac{9}{25}\right)^{3/2}$

76. $\left(\dfrac{16}{81}\right)^{3/4}$

77. $8^{-2/3}$

78. $4^{-3/2}$

79. $\dfrac{(49x^8y^{-4})^{1/2}}{(27x^{-3}y^9)^{-1/3}}$

80. $\dfrac{(x^{-2}y^{1/3})^6}{x^{-10}y^{3/2}}$

Getting Ready for the Next Section

Find the value of $b^2 - 4ac$ when

81. $a = 1, b = -3, c = -40$ **82.** $a = 2, b = 3, c = 4$ **83.** $a = 4, b = 12, c = 9$ **84.** $a = -3, b = 8, c = -1$

Solve.

85. $k^2 - 144 = 0$ **86.** $36 - 20k = 0$

Multiply.

87. $(x - 3)(x + 2)$ **88.** $(t - 5)(t + 5)$ **89.** $(x - 3)(x - 3)(x + 2)$ **90.** $(t - 5)(t + 5)(t - 3)$

Extending the Concepts

So far, all the equations we have solved have had coefficients that were rational numbers. Here are some equations that have irrational coefficients and some that have complex coefficients. Solve each equation. (Remember, $i^2 = -1$.)

91. $x^2 + \sqrt{3}x - 6 = 0$ **92.** $x^2 - \sqrt{5}x - 5 = 0$ **93.** $\sqrt{2}x^2 + 2x - \sqrt{2} = 0$

94. $\sqrt{7}x^2 + 2\sqrt{2}x - \sqrt{7} = 0$ **95.** $x^2 + ix + 2 = 0$ **96.** $x^2 + 3ix - 2 = 0$

Additional Items Involving Solutions to Equations

OBJECTIVES

A Find the number and kinds of solutions to a quadratic equation by using the discriminant.

B Find an unknown constant in a quadratic equation so that there is exactly one solution.

C Find an equation from its solutions.

TICKET TO SUCCESS

Keep these questions in mind as you read through the section. Then respond in your own words and in complete sentences.

1. What is the discriminant?
2. What kinds of solutions do we get when the discriminant of a quadratic equation is negative?
3. Explain why you would want find an unknown constant of a quadratic equation.
4. What does it mean for a solution to have multiplicity 3?

J. McPhail/Shutterstock.com

A shortstop can jump straight up with an initial velocity of 2 feet per second. If he wants to catch a line drive hit over his head, he needs to be able to jump 5 feet off the ground. The equation that models the jump is $h = -16t^2 + 2t$. Can he make the out? Problems like this can be answered without having to name the solutions. We can use an expression called the discriminant to determine if there are any real solutions to this equation. In this section, we will define the discriminant and use it to find the number and types of solutions an equation may have. We will also use the zero-factor property to build equations from their solutions, which may help us decide the initial velocity the shortstop will need to catch the line drive.

A The Discriminant

The quadratic formula

$$X = \frac{-b \pm \sqrt{b^2 - 4ac}}{2a}$$

gives the solutions to any quadratic equation in standard form. When working with quadratic equations, there are times when it is important only to know what kinds of solutions the equation has.

> **Definition**
>
> The expression under the radical in the quadratic formula is called the **discriminant:**
>
> $$\text{Discriminant} = D = b^2 - 4ac$$

The discriminant indicates the number and types of solutions to a quadratic equation, when the original equation has integer coefficients. For example, if we were to use the quadratic formula to solve the equation $2x^2 + 2x + 3 = 0$, we would find the discriminant to be

$$b^2 - 4ac = 2^2 - 4(2)(3) = -20$$

Since the discriminant appears under a square root symbol, we have the square root of a negative number in the quadratic formula. Our solutions therefore would be complex numbers. Similarly, if the discriminant were 0, the quadratic formula would yield

$$x = \frac{-b \pm \sqrt{0}}{2a} = \frac{-b \pm 0}{2a} = \frac{-b}{2a}$$

and the equation would have one rational solution, the number $\frac{-b}{2a}$.

The following table gives the relationship between the discriminant and the types of solutions to the equation.

For the equation $ax^2 + bx + c = 0$ where a, b, and c are integers and $a \neq 0$,

If the discriminant $b^2 - 4ac$ is	Then the equation will have
Negative	Two complex solutions containing i
Zero	One rational solution
A positive number that is also a perfect square	Two rational solutions
A positive number that is not a perfect square	Two irrational solutions

In the second and third cases, when the discriminant is 0 or a positive perfect square, the solutions are rational numbers. The quadratic equations in these two cases are the ones that can be factored.

EXAMPLE 1 For $x^2 - 3x - 40 = 0$, give the number and kinds of solutions.

SOLUTION Using $a = 1$, $b = -3$, and $c = -40$ in $b^2 - 4ac$, we have

$$(-3)^2 - 4(1)(-40) = 9 + 160 = 169.$$

The discriminant is a perfect square. The equation therefore has two rational solutions. ∎

EXAMPLE 2 For $2x^2 - 3x + 4 = 0$, give the number and kinds of solutions.

SOLUTION Using $a = 2$, $b = -3$, and $c = 4$, we have

$$b^2 - 4ac = (-3)^2 - 4(2)(4) = 9 - 32 = -23$$

The discriminant is negative, implying the equation has two complex solutions that contain i. ∎

PRACTICE PROBLEMS

For each equation, give the number and kinds of solutions.

1. $x^2 - 3x - 28 = 0$

2. $x^2 - 6x + 9 = 0$

Answers
1. Two rational
2. One rational

EXAMPLE 3 For $4x^2 - 12x + 9 = 0$, give the number and kinds of solutions.

SOLUTION Using $a = 4$, $b = -12$, and $c = 9$, the discriminant is

$$b^2 - 4ac = (-12)^2 - 4(4)(9) = 144 - 144 = 0$$

Since the discriminant is 0, the equation will have one rational solution. ■

EXAMPLE 4 For $x^2 + 6x = 8$, give the number and kinds of solutions.

SOLUTION First, we must put the equation in standard form by adding -8 to each side. If we do so, the resulting equation is

$$x^2 + 6x - 8 = 0$$

Now we identify a, b, and c as 1, 6, and -8, respectively.

$$b^2 - 4ac = 6^2 - 4(1)(-8) = 36 + 32 = 68$$

The discriminant is a positive number but not a perfect square. Therefore, the equation will have two irrational solutions. ■

B Finding an Unknown Constant

EXAMPLE 5 Find an appropriate k so that the equation $4x^2 - kx = -9$ has exactly one rational solution.

SOLUTION We begin by writing the equation in standard form.

$$4x^2 - kx + 9 = 0$$

Using $a = 4$, $b = -k$, and $c = 9$, we have

$$b^2 - 4ac = (-k)^2 - 4(4)(9)$$

$$= k^2 - 144$$

An equation has exactly one rational solution when the discriminant is 0. We set the discriminant equal to 0 and solve.

$$k^2 - 144 = 0$$

$$k^2 = 144$$

$$k = \pm 12$$

Choosing k to be 12 or -12 will result in an equation with one rational solution ■

C Building Equations from Their Solutions

Suppose we know that the solutions to an equation are $x = 3$ and $x = -2$. We can find equations with these solutions by using the zero-factor property. First, let's write our solutions as equations with 0 on the right side.

If →	$x = 3$	First solution
then →	$x - 3 = 0$	Add -3 to each side.
and if →	$x = -2$	Second solution
then →	$x + 2 = 0$	Add 2 to each side.

Give the number and kinds of solutions.

3. $3x^2 - 2x + 4 = 0$

4. $x^2 + 1 = 4x$

5. Find k so that the equation $9x^2 + kx = -4$ has exactly one rational solution.

Answers

3. Two complex

4. Two irrational

5. ± 12

Now, since both $x - 3$ and $x + 2$ are 0, their product must be 0 also. Therefore, we can write

$$(x - 3)(x + 2) = 0 \qquad \text{Zero-factor property}$$

$$x^2 - x - 6 = 0 \qquad \text{Multiply out the left side.}$$

Many other equations have 3 and -2 as solutions. For example, any constant multiple of $x^2 - x - 6 = 0$, such as $5x^2 - 5x - 30 = 0$, also has 3 and -2 as solutions. Similarly, any equation built from positive integer powers of the factors $x - 3$ and $x + 2$ will also have 3 and -2 as solutions. One such equation is

$$(x - 3)^2(x + 2) = 0$$

$$(x^2 - 6x + 9)(x + 2) = 0$$

$$x^3 - 4x^2 - 3x + 18 = 0$$

In mathematics, we distinguish between the solutions to this last equation and those to the equation $x^2 - x - 6 = 0$ by saying $x = 3$ is a solution of *multiplicity 2* in the equation $x^3 - 4x^2 - 3x + 18 = 0$ and a solution of *multiplicity 1* in the equation $x^2 - x - 6 = 0$.

6. Find an equation that has solutions $t = -2$, $t = 2$, and $t = 3$.

EXAMPLE 6 Find an equation that has solutions $t = 5$, $t = -5$, and $t = 3$.

SOLUTION First, we use the given solutions to write equations that have 0 on their right sides:

$$\text{If} \rightarrow \qquad t = 5 \qquad t = -5 \qquad t = 3$$

$$\text{then} \rightarrow \quad t - 5 = 0 \qquad t + 5 = 0 \qquad t - 3 = 0$$

Since $t - 5$, $t + 5$, and $t - 3$ are all 0, their product is also 0 by the zero-factor property. An equation with solutions of 5, -5, and 3 is

$$(t - 5)(t + 5)(t - 3) = 0 \qquad \text{Zero-factor property}$$

$$(t^2 - 25)(t - 3) = 0 \qquad \text{Multiply first two binomials.}$$

$$t^3 - 3t^2 - 25t + 75 = 0 \qquad \text{Complete the multiplication.}$$

The last line gives us an equation with solutions of 5, -5, and 3. Remember, many other equations have these same solutions. ■

7. Find an equation with solutions $x = -\frac{3}{4}$ and $x = \frac{1}{5}$.

EXAMPLE 7 Find an equation with solutions $x = -\frac{2}{3}$ and $x = \frac{4}{5}$.

SOLUTION The solution $x = -\frac{2}{3}$ can be rewritten as $3x + 2 = 0$ as follows:

$$x = -\frac{2}{3} \qquad \text{The first solution}$$

$$3x = -2 \qquad \text{Multiply each side by 3.}$$

$$3x + 2 = 0 \qquad \text{Add 2 to each side.}$$

Similarly, the solution $x = \frac{4}{5}$ can be rewritten as $5x - 4 = 0$.

$$x = \frac{4}{5} \qquad \text{The second solution}$$

$$5x = 4 \qquad \text{Multiply each side by 5.}$$

$$5x - 4 = 0 \qquad \text{Add } -4 \text{ to each side.}$$

Answers

6. $t^3 - 3t^2 - 4t + 12 = 0$

7. $20x^2 + 11x - 3 = 0$

Since both $3x + 2$ and $5x - 4$ are 0, their product is 0 also, giving us the equation we are looking for

$$(3x + 2)(5x - 4) = 0 \qquad \text{Zero-factor property}$$

$$15x^2 - 2x - 8 = 0 \qquad \text{Multiply.} \qquad \blacksquare$$

USING TECHNOLOGY

Graphing Calculators

Solving Equations

Now that we have explored the relationship between equations and their solutions, we can look at how a graphing calculator can be used in the solution process. To begin, let's solve the equation $x^2 = x + 2$ using techniques from algebra: writing it in standard form, factoring, and then setting each factor equal to 0.

$$x^2 - x - 2 = 0 \qquad \text{Standard form}$$

$$(x - 2)(x + 1) = 0 \qquad \text{Factor.}$$

$$x - 2 = 0 \quad \text{or} \quad x + 1 = 0 \qquad \text{Set each factor equal to 0.}$$

$$x = 2 \qquad\qquad x = -1 \qquad \text{Solve.}$$

Our original equation, $x^2 = x + 2$, has two solutions: $x = 2$ and $x = -1$. To solve the equation using a graphing calculator, we need to associate it with an equation (or equations) in two variables. One way to do this is to associate the left side with the equation $y = x^2$ and the right side of the equation with $y = x + 2$. To do so, we set up the functions list in our calculator this way:

$$Y_1 = X^2$$

$$Y_2 = X + 2$$

Window: X from -5 to 5, Y from -5 to 5

Graphing these functions in this window will produce a graph similar to the one shown in Figure 1.

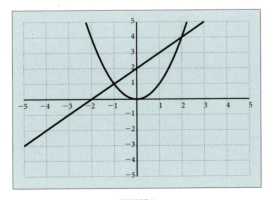

FIGURE 1

If we use the Trace feature to find the coordinates of the points of intersection, we find that the two curves intersect at $(-1, 1)$ and $(2, 4)$. We note that the x-coordinates of these two points match the solutions to the equation $x^2 = x + 2$, which we found using algebraic techniques. This makes sense because if two graphs intersect at a point (x, y), then the coordinates of that point satisfy both equations. If a point (x, y) satisfies both $y = x^2$ and $y = x + 2$, then, for that particular point, $x^2 = x + 2$. From this we conclude that the x-coordinates of the points of intersection are solutions to our original equation. Here is a summary of what we have discovered:

Conclusion I   If the graphs of two functions $y = f(x)$ and $y = g(x)$ intersect in the coordinate plane, then the x-coordinates of the points of intersection are solutions to the equation $f(x) = g(x)$.

A second method of solving our original equation $x^2 = x + 2$ graphically requires the use of one function instead of two. To begin, we write the equation in standard form as $x^2 - x - 2 = 0$. Next, we graph the function $y = x^2 - x - 2$. The x-intercepts of the graph are the points with y-coordinates of 0. Therefore, they satisfy the equation $0 = x^2 - x - 2$, which is equivalent to our original equation. The graph in Figure 2 shows $Y_1 = X^2 - X - 2$ in a window with X from -5 to 5 and Y from -5 to 5.

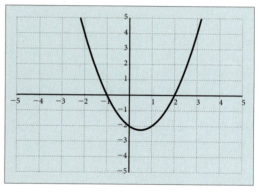

FIGURE 2

Using the Trace feature, we find that the x-intercepts of the graph are $x = -1$ and $x = 2$, which match the solutions to our original equation $x^2 = x + 2$. We can summarize the relationship between solutions to an equation and the intercepts of its associated graph this way:

Conclusion 2   If $y = f(x)$ is a function, then any x-intercept on the graph of $y = f(x)$ is a solution to the equation $f(x) = 0$.

Problem Set 8.3

Moving Toward Success

"The wisest mind has something yet to learn."

—George Santayana, 1863–1952, Spanish-born American philosopher

1. Should you still read the chapter even if you understand the concepts being taught in class? Why or why not?

2. If you receive a poor score on a quiz or test, do your study habits need to be reevaluated? Why or why not?

A Use the discriminant to find the number and kinds of solutions for each of the following equations. [Examples 1–4]

1. $x^2 - 6x + 5 = 0$

2. $x^2 - x - 12 = 0$

3. $4x^2 - 4x = -1$

4. $9x^2 + 12x = -4$

5. $x^2 + x - 1 = 0$

6. $x^2 - 2x + 3 = 0$

7. $2y^2 = 3y + 1$

8. $3y^2 = 4y - 2$

9. $x^2 - 9 = 0$

10. $4x^2 - 81 = 0$

11. $5a^2 - 4a = 5$

12. $3a = 4a^2 - 5$

B Determine k so that each of the following has exactly one real solution. [Example 5]

13. $x^2 - kx + 25 = 0$

14. $x^2 + kx + 25 = 0$

15. $x^2 = kx - 36$

16. $x^2 = kx - 49$

17. $4x^2 - 12x + k = 0$

18. $9x^2 + 30x + k = 0$

19. $kx^2 - 40x = 25$

20. $kx^2 - 2x = -1$

21. $3x^2 - kx + 2 = 0$

22. $5x^2 + kx + 1 = 0$

C For each of the following problems, find an equation that has the given solutions. [Examples 6, 7]

23. $x = 5, x = 2$

24. $x = -5, x = -2$

25. $t = -3, t = 6$

26. $t = -4, t = 2$

27. $y = 2, y = -2, y = 4$

28. $y = 1, y = -1, y = 3$

29. $x = \frac{1}{2}, x = 3$

30. $x = \frac{1}{3}, x = 5$

31. $t = -\dfrac{3}{4}, t = 3$ **32.** $t = -\dfrac{4}{5}, t = 2$ **33.** $x = 3, x = -3, x = \dfrac{5}{6}$ **34.** $x = 5, x = -5, x = \dfrac{2}{3}$

35. $a = -\dfrac{1}{2}, a = \dfrac{3}{5}$ **36.** $a = -\dfrac{1}{3}, a = \dfrac{4}{7}$ **37.** $x = -\dfrac{2}{3}, x = \dfrac{2}{3}, x = 1$ **38.** $x = -\dfrac{4}{5}, x = \dfrac{4}{5}, x = -1$

39. $x = 2, x = -2, x = 3, x = -3$ **40.** $x = 1, x = -1, x = 5, x = -5$

41. Find a possible $f(x)$ if $f(1) = 0$ and $f(-2) = 0$. **42.** Find a possible $f(x)$ if $f(4) = 0$ and $f(-3) = 0$.

43. Find a possible $f(x)$ if $f(3 + \sqrt{2}) = 0$ and $f(3 - \sqrt{2}) = 0$. **44.** Find a possible $f(x)$ if $f(-3 + \sqrt{15}) = 0$ and $f(-3 - \sqrt{15}) = 0$.

45. Find f a possible (x) if $f(2 + i) = 0$ and $f(2 - i) = 0$. **46.** Find a possible $f(x)$ if $f(3 + i\sqrt{5}) = 0$ and $f(3 - i\sqrt{5}) = 0$.

47. Find a possible $f(x)$ if $f\left(\dfrac{5 + \sqrt{7}}{2}\right) = 0$ and $f\left(\dfrac{5 - \sqrt{7}}{2}\right) = 0$. **48.** Find a possible $f(x)$ if $f\left(\dfrac{-4 + \sqrt{11}}{3}\right) = 0$ and $f\left(\dfrac{-4 - \sqrt{11}}{3}\right) = 0$.

49. Find a possible $f(x)$ if $f\left(\dfrac{3 + i\sqrt{5}}{2}\right) = 0$ and $f\left(\dfrac{3 - i\sqrt{5}}{2}\right) = 0$. **50.** Find a possible $f(x)$ if $f\left(\dfrac{-7 + i\sqrt{14}}{3}\right) = 0$ and $f\left(\dfrac{-7 - i\sqrt{14}}{3}\right) = 0$.

51. Find a possible $f(x)$ if $f(2 + i\sqrt{3}) = 0$ and $f(2 - i\sqrt{3}) = 0$ and $f(0) = 0$. **52.** Find a possible $f(x)$ if $f(5 + i\sqrt{7}) = 0$ and $f(5 - i\sqrt{7}) = 0$ and $f(0) = 0$.

53. Indicate which of the graphs represent functions with the following types of solutions.

  a. One real solution

  b. No real solution

  c. Two real solutions

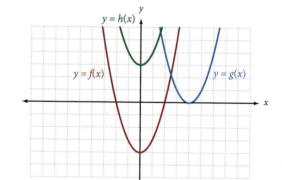

54. Find k, so that the function $9x^2 + kx + 4$ has

  a. exactly one real solution.

  b. two distinct real solutions.

  c. no real solution.

Maintaining Your Skills

Multiply. (Assume all variables represent positive numbers for the rest of the problems in this section.)

55. $a^4(a^{3/2} - a^{1/2})$ **56.** $(a^{1/2} - 5)(a^{1/2} + 3)$ **57.** $(x^{3/2} - 3)^2$ **58.** $(x^{1/2} - 8)(x^{1/2} + 8)$

Divide.

59. $\dfrac{30x^{3/4} - 25x^{5/4}}{5x^{1/4}}$

60. $\dfrac{45x^{5/3}y^{7/3} - 36x^{8/3}y^{4/3}}{9x^{2/3}y^{1/3}}$

61. Factor $5(x - 3)^{1/2}$ from $10(x - 3)^{3/2} - 15(x - 3)^{1/2}$.

62. Factor $2(x + 1)^{1/3}$ from $8(x + 1)^{4/3} - 2(x + 1)^{1/3}$.

Factor each of the following.

63. $2x^{2/3} - 11x^{1/3} + 12$

64. $9x^{2/3} + 12x^{1/3} + 4$

Getting Ready for the Next Section

Simplify.

65. $(x + 3)^2 - 2(x + 3) - 8$

66. $(x - 2)^2 - 3(x - 2) - 10$

67. $(2a - 3)^2 - 9(2a - 3) + 20$

68. $(3a - 2)^2 + 2(3a - 2) - 3$

69. $2(4a + 2)^2 - 3(4a + 2) - 20$

70. $6(2a + 4)^2 - (2a + 4) - 2$

Solve.

71. $x^2 = \dfrac{1}{4}$

72. $x^2 = -2$

73. $\sqrt{x} = -3$

74. $\sqrt{x} = 2$

75. $x + 3 = 4$

76. $x + 3 = -2$

77. $y^2 - 2y - 8 = 0$

78. $y^2 + y - 6 = 0$

79. $4y^2 + 7y - 2 = 0$

80. $6x^2 - 13x - 5 = 0$

Extending the Concepts

Find all solutions to the following equations. If rounding is necessary, round to the nearest hundredth. A calculator can be used in these problems.

81. $x^2 = 4x + 5$

82. $4x^2 = 8x + 5$

83. $x^2 - 1 = 2x$

84. $4x^2 - 1 = 4x$

85. $2x^3 - x^2 - 2x + 1 = 0$

86. $3x^3 - 2x^2 - 3x + 2 = 0$

87. $2x^3 + 2 = x^2 + 4x$

88. $3x^3 - 9x = 2x^2 - 6$

89. Baseball The section opener says that a shortstop has to jump five feet in the air to catch a line drive. If his jump is modeled by the equation $h = -16t^2 + 2t$, is it possible for him to catch the ball? Use the discriminant D to determine if there are any real solutions to this problem.

Equations Quadratic in Form

OBJECTIVES

A Solve equations that are reducible to a quadratic equation.

B Solve application problems using quadratic equations.

TICKET TO SUCCESS

Keep these questions in mind as you read through the section. Then respond in your own words and in complete sentences.

1. What does it mean for an equation to be quadratic in form?
2. What are all the circumstances in solving equations (that we have studied) in which it is necessary to check for extraneous solutions?
3. How would you start to solve the equation $x + \sqrt{x} - 6 = 0$?
4. What does it mean for a line segment to be divided in "extreme and mean ratio"?

Four Oaks/Shutterstock.com

An Olympic platform diver jumps off a platform of height h with an upward velocity of about 8 feet per second. The time it takes her to reach the water is modeled by the equation $16t^2 - 8t - h = 0$. We can use this equation to calculate how long before she passes the platform. We can also use it to determine how long until she reaches the water. Let's put our knowledge of quadratic equations to work to solve equations like this one.

A Equations Quadratic in Form

EXAMPLE 1 Solve $(x + 3)^2 - 2(x + 3) - 8 = 0$.

SOLUTION We can see that this equation is quadratic in form by replacing $x + 3$ with another variable, say y. Replacing $x + 3$ with y we have

$$y^2 - 2y - 8 = 0$$

We can solve this equation by factoring the left side and then setting each factor equal to 0.

$$y^2 - 2y - 8 = 0$$
$$(y - 4)(y + 2) = 0 \qquad \text{Factor.}$$
$$y - 4 = 0 \quad \text{or} \quad y + 2 = 0 \qquad \text{Set factors equal to 0.}$$
$$y = 4 \qquad\qquad y = -2$$

PRACTICE PROBLEMS

1. Solve
$(x - 2)^2 - 3(x - 2) - 10 = 0.$

Answers
1. 7, 0

Since our original equation was written in terms of the variable x, we would like our solutions in terms of x also. Replacing y with $x + 3$ and then solving for x we have

$$x + 3 = 4 \quad \text{or} \quad x + 3 = -2$$
$$x = 1 \qquad\qquad x = -5$$

The solutions to our original equation are 1 and -5.

   The method we have just shown lends itself well to other types of equations that are quadratic in form, as we will see. In this example, however, there is another method that works just as well. Let's solve our original equation again, but this time, let's begin by expanding $(x + 3)^2$ and $2(x + 3)$.

$$(x + 3)^2 - 2(x + 3) - 8 = 0$$
$$x^2 + 6x + 9 - 2x - 6 - 8 = 0 \qquad \text{Multiply.}$$
$$x^2 + 4x - 5 = 0 \qquad \text{Combine similar terms.}$$
$$(x - 1)(x + 5) = 0 \qquad \text{Factor.}$$
$$x - 1 = 0 \quad \text{or} \quad x + 5 = 0 \qquad \text{Set factors equal to 0.}$$
$$x = 1 \qquad\qquad x = -5$$

As you can see, either method produces the same result. ■

2. Solve $6x^4 - 13x^2 = 5$.

EXAMPLE 2   Solve $4x^4 + 7x^2 = 2$.

SOLUTION   This equation is quadratic in x^2. We can make it easier to look at by using the substitution $y = x^2$. (The choice of the letter y is arbitrary. We could just as easily use the substitution $m = x^2$.) Making the substitution $y = x^2$ and then solving the resulting equation we have

$$4y^2 + 7y = 2$$
$$4y^2 + 7y - 2 = 0 \qquad \text{Standard form}$$
$$(4y - 1)(y + 2) = 0 \qquad \text{Factor.}$$
$$4y - 1 = 0 \quad \text{or} \quad y + 2 = 0 \qquad \text{Set factors equal to 0.}$$
$$y = \frac{1}{4} \qquad\qquad y = -2$$

Now we replace y with x^2 to solve for x:

$$x^2 = \frac{1}{4} \qquad \text{or} \qquad x^2 = -2$$
$$x = \pm\sqrt{\frac{1}{4}} \qquad\qquad x = \pm\sqrt{-2} \qquad \text{Square root theorem}$$
$$x = \pm\frac{1}{2} \qquad\qquad x = \pm i\sqrt{2}$$

The solution set is $\left\{ \frac{1}{2}, -\frac{1}{2}, i\sqrt{2}, -i\sqrt{2} \right\}$. ■

3. Solve for x: $x - \sqrt{x} - 12 = 0$.

EXAMPLE 3   Solve $x + \sqrt{x} - 6 = 0$ for x.

SOLUTION   To see that this equation is quadratic in form, we have to notice that $(\sqrt{x})^2 = x$; that is, the equation can be rewritten as

$$(\sqrt{x})^2 + \sqrt{x} - 6 = 0$$

Answers

2. $\pm\dfrac{i\sqrt{3}}{3}, \pm\dfrac{\sqrt{10}}{2}$

3. 16

Replacing $\sqrt{x}$ with y and solving as usual, we have

$$y^2 + y - 6 = 0$$

$$(y + 3)(y - 2) = 0$$

$$y + 3 = 0 \quad \text{or} \quad y - 2 = 0$$

$$y = -3 \qquad\qquad y = 2$$

Again, to find x, we replace y with $\sqrt{x}$ and solve:

$$\sqrt{x} = -3 \quad \text{or} \quad \sqrt{x} = 2$$

$$x = 9 \qquad\qquad x = 4 \qquad \text{Square both sides of each equation.}$$

Since we squared both sides of each equation, we have the possibility of obtaining extraneous solutions. We have to check both solutions in our original equation.

When →	$x = 9$	When →	$x = 4$
the equation → $x + \sqrt{x} - 6 = 0$		the equation → $x + \sqrt{x} - 6 = 0$	
becomes →	$9 + \sqrt{9} - 6 \overset{?}{=} 0$	becomes →	$4 + \sqrt{4} - 6 \overset{?}{=} 0$
	$9 + 3 - 6 \overset{?}{=} 0$		$4 + 2 - 6 \overset{?}{=} 0$
	$6 \neq 0$		$0 = 0$

This means 9 is extraneous.

This means 4 is a solution.

The only solution to the equation $x + \sqrt{x} - 6 = 0$ is $x = 4$. ∎

We should note here that the two possible solutions, 9 and 4, to the equation in Example 3 can be obtained by another method. Instead of substituting for $\sqrt{x}$, we can isolate it on one side of the equation and then square both sides to clear the equation of radicals.

$$x + \sqrt{x} - 6 = 0$$

$$\sqrt{x} = -x + 6 \qquad \text{Isolate } \sqrt{x}.$$

$$x = x^2 - 12x + 36 \qquad \text{Square both sides.}$$

$$0 = x^2 - 13x + 36 \qquad \text{Add } -x \text{ to both sides.}$$

$$0 = (x - 4)(x - 9) \qquad \text{Factor.}$$

$$x - 4 = 0 \quad \text{or} \quad x - 9 = 0$$

$$x = 4 \qquad\qquad x = 9$$

We obtain the same two possible solutions. Since we squared both sides of the equation to find them, we would have to check each one in the original equation. As was the case in Example 3, only $x = 4$ is a solution; $x = 9$ is extraneous.

B Applications

EXAMPLE 4 If an object is tossed into the air with an upward velocity of 12 feet per second from the top of a building h feet high, the time it takes for the object to hit the ground below is given by the formula

$$16t^2 - 12t - h = 0$$

Solve this formula for t.

4. Solve $16t^2 - 10t - h = 0$ for t.

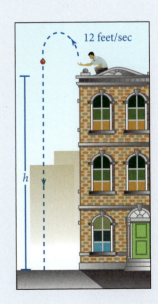

12 feet/sec

h

Answer

4. $t = \dfrac{5 \pm \sqrt{25 + 16h}}{16}$

SOLUTION   The formula is in standard form and is quadratic in t. The coefficients a, b, and c that we need to apply to the quadratic formula are $a = 16$, $b = -12$, and $c = -h$. Substituting these quantities into the quadratic formula, we have

$$t = \frac{12 \pm \sqrt{144 - 4(16)(-h)}}{2(16)}$$

$$= \frac{12 \pm \sqrt{144 + 64h}}{32}$$

We can factor the perfect square 16 from the two terms under the radical and simplify our radical somewhat.

$$t = \frac{12 \pm \sqrt{16(9 + 4h)}}{32}$$

$$= \frac{12 \pm 4\sqrt{9 + 4h}}{32}$$

Now we can reduce to lowest terms by factoring a 4 from the numerator and denominator.

$$t = \frac{\cancel{4}(3 \pm \sqrt{9 + 4h})}{\cancel{4} \cdot 8}$$

$$= \frac{3 \pm \sqrt{9 + 4h}}{8}$$

If we were given a value of h, we would find that one of the solutions to this last formula would be a negative number. Since time is always measured in positive units, we wouldn't use that solution.   ∎

More About the Golden Ratio

Previously, we derived the golden ratio $\frac{1 + \sqrt{5}}{2}$ by finding the ratio of length to width for a golden rectangle. The golden ratio was actually discovered before the golden rectangle by the Greeks who lived before Euclid. The early Greeks found the golden ratio by dividing a line segment into two parts so that the ratio of the shorter part to the longer part was the same as the ratio of the longer part to the whole segment. When they divided a line segment in this manner, they said it was divided in "extreme and mean ratio." Figure 1 illustrates a line segment divided this way.

$$\begin{array}{ccc} A & B & C \end{array}$$

FIGURE 1

If point B divides segment AC in "extreme and mean ratio," then

$$\frac{\text{Length of shorter segment}}{\text{Length of longer segment}} = \frac{\text{Length of longer segment}}{\text{Length of whole segment}}$$

$$\frac{AB}{BC} = \frac{BC}{AC}$$

EXAMPLE 5 If the length of segment *AB* in Figure 1 is 1 inch, find the length of *BC* so that the whole segment *AC* is divided in "extreme and mean ratio."

SOLUTION Using Figure 1 as a guide, if we let $x =$ the length of segment *BC*, then the length of *AC* is $x + 1$. If *B* divides *AC* into "extreme and mean ratio," then the ratio of *AB* to *BC* must equal the ratio of *BC* to *AC*. Writing this relationship using the variable x, we have

$$\frac{1}{x} = \frac{x}{x + 1}$$

If we multiply both sides of this equation by the LCD $x(x + 1)$ we have

$$x + 1 = x^2$$
$$0 = x^2 - x - 1 \quad \text{Write the equation in standard form.}$$

Since this last equation is not factorable, we apply the quadratic formula.

$$x = \frac{1 \pm \sqrt{(-1)^2 - 4(1)(-1)}}{2}$$

$$= \frac{1 \pm \sqrt{5}}{2}$$

Our equation has two solutions, which we approximate using decimals.

$$\frac{1 + \sqrt{5}}{2} \approx 1.618 \qquad \frac{1 - \sqrt{5}}{2} \approx -0.618$$

Since we originally let x equal the length of segment *BC*, we use only the positive solution to our equation. As you can see, the positive solution is the golden ratio. ∎

USING TECHNOLOGY

Graphing Calculators

More About Example 1

As we mentioned earlier, algebraic expressions entered into a graphing calculator do not have to be simplified to be evaluated. This fact applies to equations as well. We can graph the equation $y = (x + 3)^2 - 2(x + 3) - 8$ to assist us in solving the equation in Example 1. The graph is shown in Figure 2. Using the Zoom and Trace features at the x-intercepts gives us $x = 1$ and $x = -5$ as the solutions to the equation $0 = (x + 3)^2 - 2(x + 3) - 8$.

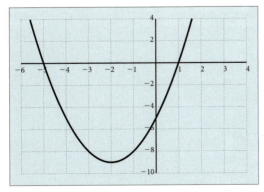

FIGURE 2

5. If the length of segment *AB* in Figure 1 is 2 inches, and *B* divides *AC* in "extreme and mean ratio," find *BC*, and then show that *BC* is twice the golden ratio.

Answer
5. $BC = 1 + \sqrt{5}$, which is twice the golden ratio

More About Example 2

Figure 3 shows the graph of $y = 4x^4 + 7x^2 - 2$. As we expect, the x-intercepts give the real number solutions to the equation $0 = 4x^4 + 7x^2 - 2$. The complex solutions do not appear on the graph.

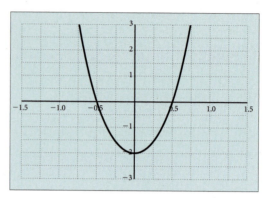

FIGURE 3

More About Example 3

In solving the equation in Example 3, we found that one of the possible solutions was an extraneous solution. If we solve the equation $x + \sqrt{x} - 6 = 0$ by graphing the function $y = x + \sqrt{x} - 6$, we find that the extraneous solution, 9, is not an x-intercept. Figure 4 shows that the only solution to the equation occurs at the x-intercept, 4.

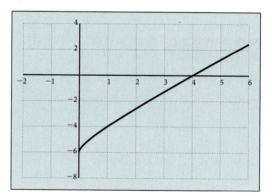

FIGURE 4

Problem Set 8.4

Moving Toward Success

"You've got to do your own growing, no matter how tall your grandfather was."

—Irish proverb

1. Why should you use different methods to study, such as reviewing class notes, working problems, making flashcards, outlining chapters, and studying with a group?

2. Would studying by mixing up the order of topics presented in the chapter better prepare you for an exam? Why or why not?

A Solve each equation. [Examples 1–2]

1. $(x - 3)^2 + 3(x - 3) + 2 = 0$

2. $(x + 4)^2 - (x + 4) - 6 = 0$

3. $2(x + 4)^2 + 5(x + 4) - 12 = 0$

4. $3(x - 5)^2 + 14(x - 5) - 5 = 0$

5. $x^4 - 10x^2 + 9 = 0$

6. $x^4 - 29x^2 + 100 = 0$

7. $x^4 - 7x^2 + 12 = 0$

8. $x^4 - 14x^2 + 45 = 0$

9. $x^4 - 6x^2 - 27 = 0$

10. $x^4 + 2x^2 - 8 = 0$

11. $x^4 + 9x^2 = -20$

12. $x^4 - 11x^2 = -30$

13. $(2a - 3)^2 - 9(2a - 3) = -20$

14. $(3a - 2)^2 + 2(3a - 2) = 3$

15. $2(4a + 2)^2 = 3(4a + 2) + 20$

16. $6(2a + 4)^2 = (2a + 4) + 2$

17. $6t^4 = -t^2 + 5$

18. $3t^4 = -2t^2 + 8$

19. $9x^4 - 49 = 0$

20. $25x^4 - 9 = 0$

A Solve each of the following equations. Remember, if you square both sides of an equation in the process of solving it, you have to check all solutions in the original equation. [Examples 1–3]

21. $x - 7\sqrt{x} + 10 = 0$

22. $x - 6\sqrt{x} + 8 = 0$

23. $t - 2\sqrt{t} - 15 = 0$

24. $t - 3\sqrt{t} - 10 = 0$

25. $6x + 11\sqrt{x} = 35$

26. $2x + \sqrt{x} = 15$

27. $x - 2\sqrt{x} - 8 = 0$

28. $x + 2\sqrt{x} - 3 = 0$

29. $x + 3\sqrt{x} - 18 = 0$

30. $x + 5\sqrt{x} - 14 = 0$

31. $2x + 9\sqrt{x} - 5 = 0$

32. $2x - \sqrt{x} - 6 = 0$

33. $(a - 2) - 11\sqrt{a - 2} + 30 = 0$

34. $(a - 3) - 9\sqrt{a - 3} + 20 = 0$

35. $(2x + 1) - 8\sqrt{2x + 1} + 15 = 0$

36. $(2x - 3) - 7\sqrt{2x - 3} + 12 = 0$

37. $(x^2 + 1)^2 - 2(x^2 + 1) - 15 = 0$

38. $(x^2 - 3)^2 - 4(x^2 - 3) - 12 = 0$

39. $(x^2 + 5)^2 - 6(x^2 + 5) - 27 = 0$

40. $(x^2 - 2)^2 - 6(x^2 - 2) - 7 = 0$

41. $x^{-2} - 3x^{-1} + 2 = 0$

42. $x^{-2} + x^{-1} - 2 = 0$

43. $y^{-4} - 5y^{-2} = 0$

44. $2y^{-4} + 10y^{-2} = 0$

45. $x^{2/3} + 4x^{1/3} - 12 = 0$

46. $x^{2/3} - 2x^{1/3} - 8 = 0$

47. $x^{4/3} + 6x^{2/3} + 9 = 0$

48. $x^{4/3} - 10x^{2/3} + 25 = 0$

49. Solve the formula $16t^2 - vt - h = 0$ for t.

50. Solve the formula $16t^2 + vt + h = 0$ for t.

51. Solve the formula $kx^2 + 8x + 4 = 0$ for x.

52. Solve the formula $k^2x^2 + kx + 4 = 0$ for x.

53. Solve $x^2 + 2xy + y^2 = 0$ for x by using the quadratic formula with $a = 1$, $b = 2y$, and $c = y^2$.

54. Solve $x^2 - 2xy + y^2 = 0$ for x by using the quadratic formula, with $a = 1$, $b = -2y$, and $c = y^2$.

Applying the Concepts

For Problems 55–56, t is in seconds.

55. Falling Object An object is tossed into the air with an upward velocity of 8 feet per second from the top of a building h feet high. The time it takes for the object to hit the ground below is given by the formula $16t^2 - 8t - h = 0$. Solve this formula for t.

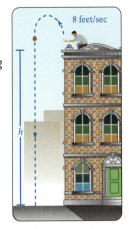

8 feet/sec

h

56. Diving Olympic diving platforms come in three different heights: 33 feet, 25 feet, and 17 feet. If a diver jumps off a platform with an upward velocity of 8 feet per second, the time it takes her to hit the water is modeled by the equation $16t^2 - 8t - h = 0$ where t is time and h is platform height. Find the time it would take the diver to reach the water if she jumped off of each of the platforms.

57. Saint Louis Arch The shape of the famous "Gateway to the West" arch in Saint Louis can be modeled by a parabola. The equation for one such parabola is

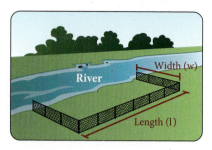
©Mark Karrass/Corbis

$$y = -\frac{1}{150}x^2 + \frac{21}{5}x$$

where x and y are in feet.

a. Sketch the graph of the arch's equation on a coordinate axis.

b. Approximately how far do you have to walk to get from one side of the arch to the other?

58. Area and Perimeter A total of 160 yards of fencing is to be used to enclose part of a lot that borders on a river. This situation is shown in the following diagram.

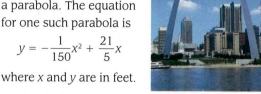

a. Write an equation that gives the relationship between the length and width and the 160 yards of fencing.

b. The formula for the area that is enclosed by the fencing and the river is $A = lw$. Solve the equation in part a for l, and then use the result to write the area in terms of w only.

c. Make a table that gives at least five possible values of w and associated area A.

d. From the pattern in your table shown in part c, what is the largest area that can be enclosed by the 160 yards of fencing? (Try some other table values if necessary.)

Golden Ratio Recall Figure 1 from this section and use it as a guide to working Problems 59–60. [Examples 5]

59. If AB is 4 inches, and B divides AC in "extreme and mean ratio," find BC, and then show that BC is 4 times the golden ratio.

60. If AB is $\frac{1}{2}$ inch, and B divides AC in "extreme and mean ratio," find BC, and then show that the ratio of BC to AB is the golden ratio.

Maintaining Your Skills

Combine, if possible.

61. $5\sqrt{7} - 2\sqrt{7}$

62. $6\sqrt{2} - 9\sqrt{2}$

63. $\sqrt{18} - \sqrt{8} + \sqrt{32}$

64. $\sqrt{50} + \sqrt{72} - \sqrt{8}$

65. $9x\sqrt{20x^3y^2} + 7y\sqrt{45x^5}$

66. $5x^2\sqrt{27xy^3} - 6y\sqrt{12x^5y}$

Multiply.

67. $(\sqrt{5} - 2)(\sqrt{5} + 8)$

68. $(2\sqrt{3} - 7)(2\sqrt{3} + 7)$

69. $(\sqrt{x} + 2)^2$

70. $(3 - \sqrt{x})(3 + \sqrt{x})$

Rationalize the denominator.

71. $\dfrac{\sqrt{7}}{\sqrt{7} - 2}$

72. $\dfrac{\sqrt{5} - \sqrt{2}}{\sqrt{5} + \sqrt{2}}$

Getting Ready for the Next Section

73. Evaluate $y = 3x^2 - 6x + 1$ for $x = 1$.

74. Evaluate $y = -2x^2 + 6x - 5$ for $x = \dfrac{3}{2}$.

75. Let $P(x) = -0.1x^2 + 27x - 500$ and find $P(135)$.

76. Let $P(x) = -0.1x^2 + 12x - 400$ and find $P(600)$.

Solve.

77. $0 = a(80)^2 + 70$

78. $0 = a(80)^2 + 90$

79. $x^2 - 6x + 5 = 0$

80. $x^2 - 3x - 4 = 0$

81. $-x^2 - 2x + 3 = 0$

82. $-x^2 + 4x + 12 = 0$

83. $2x^2 - 6x + 5 = 0$

84. $x^2 - 4x + 5 = 0$

Fill in the blanks to complete the square.

85. $x^2 - 6x + \square = (x - \square)^2$

86. $x^2 + 10x + \square = (x + \square)^2$

87. $y^2 + 2y + \square = (y + \square)^2$

88. $y^2 - 12y + \square = (y - \square)^2$

Find the x- and y-intercepts.

89. $y = x^3 - 4x$

90. $y = x^4 - 10x^2 + 9$

91. $y = 3x^3 + x^2 - 27x - 9$

92. $y = 2x^3 + x^2 - 8x - 4$

93. The graph of $y = 2x^3 - 7x^2 - 5x + 4$ crosses the x-axis at $x = 4$. Where else does it cross the x-axis?

94. The graph of $y = 6x^3 + x^2 - 12x + 5$ crosses the x-axis at $x = 1$. Where else does it cross the x-axis?

Graphing Parabolas

TICKET TO SUCCESS

Keep these questions in mind as you read through the section. Then respond in your own words and in complete sentences.

1. What is a parabola?
2. What part of the equation of a parabola determines whether the graph is concave up or concave down?
3. Suppose $f(x) = ax^2 + bx + c$ is the equation of a parabola. Explain how $f(4) = 1$ relates to the graph of the parabola.
4. Describe the ways you would find the vertex of a parabola.

FloridaStock/Shutterstock.com

A company selling beach umbrellas finds it can make a profit of P dollars each week by selling x umbrellas according to the formula $P(x) = -0.06x^2 + 4.5x - 1200$. How many umbrellas must it sell each week to make the maximum profit? What is that profit? Business problems like this one can be found by graphing a geometric figure known as a parabola. As we have seen in our previous work with linear equations, it is always possible to graph figures by making a table of values for P and x that satisfy the equation. As we have also seen, there are other methods that are faster and more accurate. In this section, we will begin our work with parabolas and look at several ways to graph them.

The solution set to the equation

$$y = x^2 - 3$$

consists of ordered pairs. One method of graphing the solution set is to find a number of ordered pairs that satisfy the equation and to graph them. We can obtain some ordered pairs that are solutions to $y = x^2 - 3$ by use of a table as follows:

x	$y = x^2 - 3$	y	Solutions
−3	$y = (-3)^2 - 3 = 9 - 3 = 6$	6	(−3, 6)
−2	$y = (-2)^2 - 3 = 4 - 3 = 1$	1	(−2, 1)
−1	$y = (-1)^2 - 3 = 1 - 3 = -2$	−2	(−1, −2)
0	$y = 0^2 - 3 = 0 - 3 = -3$	−3	(0, −3)
1	$y = 1^2 - 3 = 1 - 3 = -2$	−2	(1, −2)
2	$y = 2^2 - 3 = 4 - 3 = 1$	1	(2, 1)
3	$y = 3^2 - 3 = 9 - 3 = 6$	6	(3, 6)

Graphing these solutions and then connecting them with a smooth curve, we have the graph of $y = x^2 - 3$ (Figure 1).

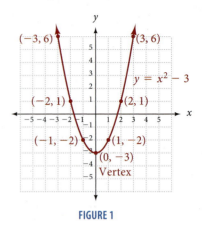

FIGURE 1

This graph is an example of a *parabola.* All equations of the form $y = ax^2 + bx + c$, $a \neq 0$, have parabolas for graphs.

Although it is always possible to graph parabolas by making a table of values of x and y that satisfy the equation, there are other methods that are faster and, in some cases, more accurate.

The important points associated with the graph of a parabola are the highest (or lowest) point on the graph and the x-intercepts. The y-intercepts can also be useful.

A   Intercepts for Parabolas

The graph of the equation $y = ax^2 + bx + c$ crosses the y-axis at $y = c$ since substituting $x = 0$ into $y = ax^2 + bx + c$ yields $y = c$.

Since the graph crosses the x-axis when $y = 0$, the x-intercepts are those values of x that are solutions to the quadratic equation $0 = ax^2 + bx + c$.

The Vertex of a Parabola

The highest or lowest point on a parabola is called the *vertex.* The vertex for the graph of $y = ax^2 + bx + c$ will occur when

$$x = \frac{-b}{2a}$$

To see this, we must transform the right side of $y = ax^2 + bx + c$ into an expression that contains x in just one of its terms. This is accomplished by completing the square on the first two terms. Here is what it looks like:

$$y = ax^2 + bx + c$$
$$y = a\left(x^2 + \frac{b}{a}x\right) + c$$
$$y = a\left[x^2 + \frac{b}{a}x + \left(\frac{b}{2a}\right)^2\right] + c - a\left(\frac{b}{2a}\right)^2$$
$$y = a\left(x + \frac{b}{2a}\right)^2 + \frac{4ac - b^2}{4a}$$

It may not look like it, but this last line indicates that the vertex of the graph of $y = ax^2 + bx + c$ has an x-coordinate of $\frac{-b}{2a}$. Since a, b, and c are constants, the only quantity that is varying in the last expression is the x in $\left(x + \frac{b}{2a}\right)^2$. Since the quantity $\left(x + \frac{b}{2a}\right)^2$ is the square of $x + \frac{b}{2a}$, the smallest it will ever be is 0, and that will happen when $x = \frac{-b}{2a}$.

NOTE
What we are doing here is attempting to explain why the vertex of a parabola always has an x-coordinate of $\frac{-b}{2a}$. But the explanation may not be easy to understand the first time you see it. It may be helpful to look over the examples in this section and the notes that accompany these examples and then come back and read over this discussion again.

We can use the vertex point along with the x- and y-intercepts to sketch the graph of any equation of the form $y = ax^2 + bx + c$. Here is a summary of the preceding information.

Graphing Parabolas

The graph of $y = ax^2 + bx + c$, $a \neq 0$, will have the following:

1. a y-intercept at $y = c$

2. x-intercepts (if they exist) at

$$x = \frac{-b \pm \sqrt{b^2 - 4ac}}{2a}$$

3. a vertex when $x = \dfrac{-b}{2a}$

EXAMPLE 1   Sketch the graph of $y = x^2 - 6x + 5$.

SOLUTION   To find the x-intercepts, we let $y = 0$ and solve for x.

$$0 = x^2 - 6x + 5$$

$$0 = (x - 5)(x - 1)$$

$$x = 5 \quad \text{or} \quad x = 1$$

To find the coordinates of the vertex, we first find

$$x = \frac{-b}{2a} = \frac{-(-6)}{2(1)} = 3$$

The x-coordinate of the vertex is 3. To find the y-coordinate, we substitute 3 for x in our original equation.

$$y = 3^2 - 6(3) + 5 = 9 - 18 + 5 = -4$$

The graph crosses the x-axis at 1 and 5 and has its vertex at $(3, -4)$. Plotting these points and connecting them with a smooth curve, we have the graph shown in Figure 2. The graph is a parabola that opens up, so we say the graph is *concave up*. The vertex is the lowest point on the graph. (Note that the graph crosses the y-axis at 5, which is the value of y we obtain when we let $x = 0$.)

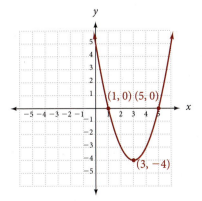

FIGURE 2

PRACTICE PROBLEMS

1. Graph $y = x^2 - 2x - 3$.

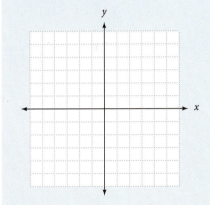

Answers

1–7. See Solutions Section.

Finding the Vertex by Completing the Square

Another way to locate the vertex of the parabola in Example 1 is by completing the square on the first two terms on the right side of the equation $y = x^2 - 6x + 5$. In this case, we would do so by adding 9 to and subtracting 9 from the right side of the equation. This amounts to adding 0 to the equation, so we know we haven't changed its solutions. This is what it looks like:

$$y = (x^2 - 6x) + 5$$
$$y = (x^2 - 6x + \mathbf{9}) + 5 - \mathbf{9}$$
$$y = (x - 3)^2 - 4$$

You may have to look at this last equation a while to see this, but when $x = 3$, then $y = (x - 3)^2 - 4 = 0^2 - 4 = -4$ is the smallest y will ever be. And that is why the vertex is at $(3, -4)$. As a matter of fact, this is the same kind of reasoning we used when we derived the formula $x = \frac{-b}{2a}$ for the x-coordinate of the vertex.

2. Graph $y = -x^2 + 2x + 8$.

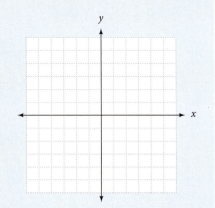

EXAMPLE 2   Graph $y = -x^2 - 2x + 3$.

SOLUTION   To find the x-intercepts, we let $y = 0$.

$$0 = -x^2 - 2x + 3$$
$$0 = x^2 + 2x - 3 \qquad \text{Multiply each side by } -1.$$
$$0 = (x + 3)(x - 1)$$
$$x = -3 \qquad \text{or} \qquad x = 1$$

The x-coordinate of the vertex is given by

$$x = \frac{-b}{2a} = \frac{-(-2)}{2(-1)} = \frac{2}{-2} = -1$$

To find the y-coordinate of the vertex, we substitute -1 for x in our original equation to get

$$y = -(-1)^2 - 2(-1) + 3 = -1 + 2 + 3 = 4$$

Our parabola has x-intercepts at -3 and 1 and a vertex at $(-1, 4)$. Figure 3 shows the graph. We say the graph is *concave down* since it opens downward.

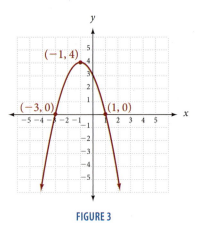

FIGURE 3

Again, we could have obtained the coordinates of the vertex by completing the square on the first two terms on the right side of our equation. To do so, we must first factor -1 from the first two terms. (Remember, the leading coefficient must be 1 to complete the square.) When we complete the square, we add 1 inside the parentheses, which actually decreases the right side of the equation by -1 since everything in the parentheses is multiplied by -1. To make up for it, we add 1 outside the parentheses.

$$y = -1(x^2 + 2x) + 3$$
$$y = -1(x^2 + 2x + \mathbf{1}) + 3 + \mathbf{1}$$
$$y = -1(x + 1)^2 + 4$$

The last line tells us that the *largest* value of y will be 4, and that will occur when $x = -1$. ∎

EXAMPLE 3   Graph $y = 3x^2 - 6x + 1$.

3. Graph $y = 2x^2 - 4x + 1$.

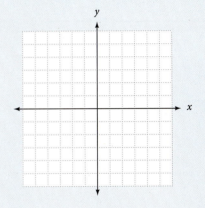

SOLUTION   To find the x-intercepts, we let $y = 0$ and solve for x.

$$0 = 3x^2 - 6x + 1$$

Since the right side of this equation does not factor, we can look at the discriminant to see what kind of solutions are possible. The discriminant for this equation is

$$b^2 - 4ac = 36 - 4(3)(1) = 24$$

Since the discriminant is a positive number but not a perfect square, the equation will have irrational solutions. This means that the x-intercepts are irrational numbers and will have to be approximated with decimals using the quadratic formula. Rather than use the quadratic formula, we will find some other points on the graph, but first let's find the vertex. Here are both methods of finding the vertex:

Method 1 Using the formula that gives us the x-coordinate of the vertex, we have

$$x = \frac{-b}{2a} = \frac{-(-6)}{2(3)} = 1$$

Substituting 1 for x in the equation gives us the y-coordinate of the vertex.

$$y = 3 \cdot 1^2 - 6 \cdot 1 + 1 = -2$$

Method 2 To complete the square on the right side of the equation, we factor 3 from the first two terms, add 1 inside the parentheses, and add -3 outside the parentheses (this amounts to adding 0 to the right side).

$$y = 3(x^2 - 2x \quad) + 1$$
$$y = 3(x^2 - 2x + \mathbf{1}) + 1 - \mathbf{3}$$
$$y = 3(x - 1)^2 - 2$$

In either case, the vertex is $(1, -2)$.

If we can find two points, one on each side of the vertex, we can sketch the graph. Let's let $x = 0$ and $x = 2$ since each of these numbers is the same distance from $x = 1$, and $x = 0$ will give us the y-intercept.

When $x = 0$	When $x = 2$
$y = 3(0)^2 - 6(0) + 1$	$y = 3(2)^2 - 6(2) + 1$
$= 0 - 0 + 1$	$= 12 - 12 + 1$
$= 1$	$= 1$

The two points just found are $(0, 1)$ and $(2, 1)$. Plotting these two points along with the vertex $(1, -2)$, we have the graph shown in Figure 4.

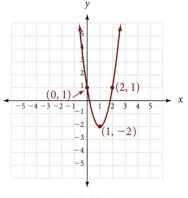

FIGURE 4

■

4. Graph $y = -x^2 + 4x - 5$.

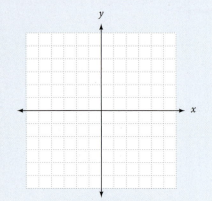

EXAMPLE 4   Graph $y = -2x^2 + 6x - 5$.

SOLUTION   Letting $y = 0$, we have

$$0 = -2x^2 + 6x - 5$$

Again, the right side of this equation does not factor. The discriminant is $b^2 - 4ac = 36 - 4(-2)(-5) = -4$, which indicates that the solutions are complex numbers. This means that our original equation does not have x-intercepts. The graph does not cross the x-axis. Let's find the vertex.

Method 1 Using our formula for the x-coordinate of the vertex, we have

$$x = \frac{-b}{2a} = \frac{-6}{2(-2)} = \frac{6}{4} = \frac{3}{2}$$

To find the y-coordinate, we let $x = \frac{3}{2}$.

$$y = -2\left(\frac{3}{2}\right)^2 + 6\left(\frac{3}{2}\right) - 5$$

$$= \frac{-18}{4} + \frac{18}{2} - 5$$

$$= \frac{-18 + 36 - 20}{4}$$

$$= -\frac{1}{2}$$

Method 2 Finding the vertex by completing the square is a more complicated matter. To make the coefficient of x^2 a 1, we must factor -2 from the first two terms. To complete the square inside the parentheses, we add $\frac{9}{4}$. Since each term inside the parentheses is multiplied by -2, we add $\frac{9}{2}$ outside the parentheses so that the net result is the same as adding 0 to the right side:

$$y = -\left(2x^2 - 3x\right) - 5$$

$$y = -2\left(x^2 - 3x + \frac{9}{4}\right) - 5 + \frac{9}{2}$$

$$y = -2\left(x - \frac{3}{2}\right)^2 - \frac{1}{2}$$

The vertex is $\left(\frac{3}{2}, -\frac{1}{2}\right)$. Since this is the only point we have so far, we must find two others. Let's let $x = 3$ and $x = 0$ since each point is the same distance from $x = \frac{3}{2}$ and on either side.

When $x = 3$ When $x = 0$

$$y = -2(3)^2 + 6(3) - 5 \qquad\qquad y = -2(0)^2 + 6(0) - 5$$

$$= -18 + 18 - 5 \qquad\qquad\qquad = 0 + 0 - 5$$

$$= -5 \qquad\qquad\qquad\qquad\qquad = -5$$

The two additional points on the graph are (3, −5) and (0, −5). Figure 5 shows the graph.

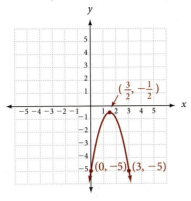

FIGURE 5

The graph is concave down. The vertex is the highest point on the graph. ■

By looking at the equations and graphs in Examples 1 through 4, we can conclude that the graph of $y = ax^2 + bx + c$ will be concave up when a is positive and concave down when a is negative. Taking this even further, if $a > 0$, then the vertex is the lowest point on the graph, and if $a < 0$, the vertex is the highest point on the graph. Finally, if we complete the square on x in the equation $y = ax^2 + bx + c$, $a \neq 0$, we can rewrite the equation of our parabola as $y = a(x - h)^2 + k$. When the equation is in this form, the vertex is at the point (h, k).

Here is a summary:

Graphing Parabolas II

The graph of

$$y = a(x - h)^2 + k, a \neq 0$$

will be a parabola with a vertex at (h, k). The vertex will be the highest point on the graph when $a < 0$, and it will be the lowest point on the graph when $a > 0$.

More About Parabolas

Now, we extend the work we did previously by considering parabolas that open left and right instead of up and down.

The graph of the equation $y = x^2 - 6x + 5$ is shown in Figure 6.

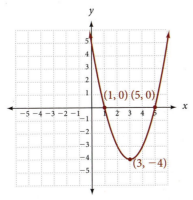

FIGURE 6

5. Graph $x = y^2 + 6y + 5$.

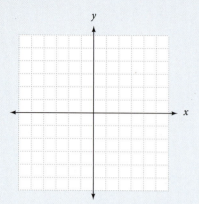

Note that the x-intercepts are 1 and 5, and the vertex is at $(3, -4)$.

If we interchange x and y in the equation $y = x^2 - 6x + 5$, we obtain the equation

$$x = y^2 - 6y + 5$$

We can expect the graph of this equation to be similar to the graph shown in Figure 6, except that the role of x and y will be interchanged. Example 5 gives more details.

EXAMPLE 5 Graph $x = y^2 - 6y + 5$.

SOLUTION Since the x-intercepts for $y = x^2 - 6x + 5$ are $x = 1$ and $x = 5$, we expect the y-intercepts for $x = y^2 - 6y + 5$ to be $y = 1$ and $y = 5$. To see that this is true, we let $x = 0$ in $x = y^2 - 6y + 5$, and solve for y.

$$0 = y^2 - 6y + 5$$
$$0 = (y - 1)(y - 5)$$

$$y - 1 = 0 \quad \text{or} \quad y - 5 = 0$$
$$y = 1 \qquad\qquad y = 5$$

The graph of $x = y^2 - 6y + 5$ will cross the y-axis at 1 and 5.

Since the graph in Figure 6 has its vertex at $x = 3, y = -4$, we expect the graph of $x = y^2 - 6y + 5$ to have its vertex at $y = 3, x = -4$. Completing the square on y shows that this is true.

$$x = (y^2 - 6y + 9) + 5 - 9$$
$$x = (y - 3)^2 - 4$$

This last line tells us the smallest value of x is -4, and it occurs when y is 3. Therefore, the vertex must be at the point $(-4, 3)$. Here is that graph:

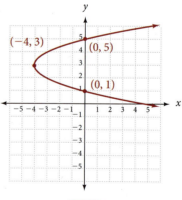

FIGURE 7

A summary of the discussion above is given below.

Parabolas That Open Left and Right

The graph of $x = ay^2 + by + c$ will have

1. an x-intercept at $x = c$

2. y-intercepts (if they exist) at

$$y = \frac{-b \pm \sqrt{b^2 - 4ac}}{2a}$$

3. a vertex when $y = \dfrac{-b}{2a}$

EXAMPLE 6 Graph $x = -y^2 - 2y + 3$.

SOLUTION The intercepts and vertex are found as shown below. The results are used to sketch the graph in Figure 8.

x-intercept: When $y = 0$, $x = 3$.

y-intercepts: When $x = 0$, we have $-y^2 - 2y + 3 = 0$, which yields solutions $y = -3$ and $y = 1$.

Vertex: $y = \dfrac{-b}{2a} = \dfrac{-(-2)}{2(-1)} = -1$

Substituting -1 for y in the equation gives us $x = 4$.
The vertex is at $(4, -1)$.

The graph of our equation is shown on the right side of Figure 8. Note the similarities to the graph of $y = -x^2 - 2x + 3$ shown on the left side.

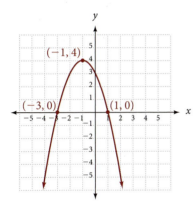
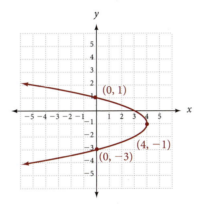

FIGURE 8

EXAMPLE 7 Graph $x = 3y^2 - 6y + 1$.

SOLUTION For this last example, we simply show the graph of our equation alongside the graph of the equation $y = 3x^2 - 6x + 1$. Both are shown in Figure 9.

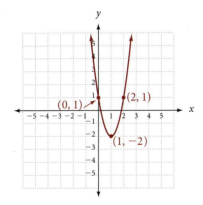
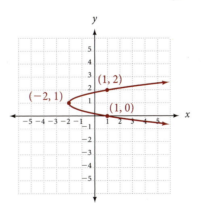

FIGURE 9

6. Graph $x = -y^2 + 2y + 3$.

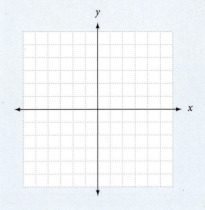

7. Graph $x = 2y^2 - 4y + 3$.

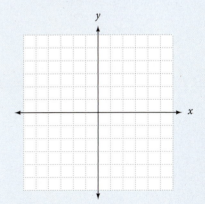

B Applications

EXAMPLE 8   A company selling copies of an accounting program for home computers finds that it will make a weekly profit of P dollars from selling x copies of the program, according to the equation

$$P(x) = -0.1x^2 + 27x - 500$$

How many copies of the program should it sell to make the largest possible profit, and what is the largest possible profit?

SOLUTION   Since the coefficient of x^2 is negative, we know the graph of this parabola will be concave down, meaning that the vertex is the highest point of the curve. We find the vertex by first finding its x-coordinate.

$$x = \frac{-b}{2a} = \frac{-27}{2(-0.1)} = \frac{27}{0.2} = 135$$

This represents the number of programs the company needs to sell each week to make a maximum profit. To find the maximum profit, we substitute 135 for x in the original equation. (A calculator is helpful for these kinds of calculations.)

$$P(135) = -0.1(135)^2 + 27(135) - 500$$

$$= -0.1(18{,}225) + 3{,}645 - 500$$

$$= -1{,}822.5 + 3{,}645 - 500$$

$$= 1{,}322.5$$

The maximum weekly profit is \$1,322.50 and is obtained by selling 135 programs a week.

∎

EXAMPLE 9   An art supply store finds that they can sell x sketch pads each week at p dollars each, according to the equation $x = 900 - 300p$. Graph the revenue equation $R = xp$. Then use the graph to find the price p that will bring in the maximum revenue. Finally, find the maximum revenue.

SOLUTION   As it stands, the revenue equation contains three variables. Since we are asked to find the value of p that gives us the maximum value of R, we rewrite the equation using just the variables R and p. Since $x = 900 - 300p$, we have

$$R = xp = (900 - 300p)p$$

The graph of this equation is shown in Figure 10. The graph appears in the first quadrant only since R and p are both positive quantities.

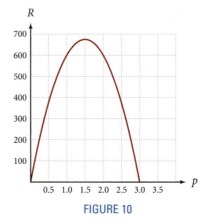

FIGURE 10

8. Find the largest value of y if $p(x) = -0.01x^2 + 25x - 400$.

9. Repeat Example 9 if the number of sketch pads they can sell each week is related to the price by the equation
$x = 800 - 200p$

Answer
8. 1,250 programs, \$15,225 weekly profit
9. \$800.00 at $p = \$2.00$

From the graph, we see that the maximum value of R occurs when $p = \$1.50$. We can calculate the maximum value of R from the equation.

When $\rightarrow$ $\qquad$ $p = 1.5$

the equation $\rightarrow$ $\qquad$ $R = (900 - 300p)p$

becomes $\rightarrow$ $\qquad$ $R = (900 - 300 \cdot 1.5)1.5$

$\qquad\qquad\qquad\quad = (900 - 450)1.5$

$\qquad\qquad\qquad\quad = 450 \cdot 1.5$

$\qquad\qquad\qquad\quad = 675$

The maximum revenue is \$675. It is obtained by setting the price of each sketch pad at $p = \$1.50$. $\qquad\qquad\blacksquare$

USING TECHNOLOGY

Graphing Calculators

If you have been using a graphing calculator for some of the material in this course, you are well aware that your calculator can draw all the graphs in this section very easily. It is important, however, that you be able to recognize and sketch the graph of any parabola by hand. It is a skill that all successful intermediate algebra students should possess, even if they are proficient in the use of a graphing calculator. My suggestion is that you work the problems in this section and problem set without your calculator. Then use your calculator to check your results.

C Finding the Equation from the Graph

EXAMPLE 10 As a human cannonball, David Smith Jr. is routinely shot out of a cannon. His flights reach heights of 80 feet before landing in a net 180 feet from the cannon. Sketch the graph of his path, and then find the equation of the graph.

SOLUTION We assume that the path taken by the human cannonball is a parabola. If the origin of the coordinate system is at the opening of the cannon, then the net that catches him will be at 180 on the x-axis. Figure 11 shows the graph:

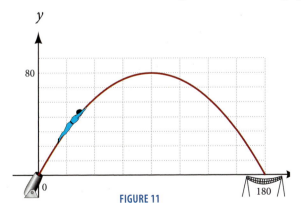

FIGURE 11

Since the curve is a parabola, we know the equation will have the form

$$y = a(x - h)^2 + k$$

10. How would the equation in Example 10 look if the maximum height were 100 feet? (Assume the maximum distance is still 180 feet.)

Answer
10. $y = -\frac{1}{81}(x - 90)^2 + 100$

Since the vertex of the parabola is at (90, 80), we can fill in two of the three constants in our equation, giving us

$$y = a(x - 90)^2 + 80$$

To find a, we note that the landing point will be (180, 0). Substituting the coordinates of this point into the equation, we solve for a

$$0 = a(180 - 90)^2 + 80$$

$$0 = a(90)^2 + 80$$

$$0 = 8{,}100a + 80$$

$$a = -\frac{80}{8{,}100} = -\frac{4}{405}$$

The equation that describes the path of the human cannonball is
$$y = -\frac{4}{405}(x - 90)^2 + 80 \text{ for } 0 \le x \le 180 \qquad \blacksquare$$

USING TECHNOLOGY

Graphing Calculators

Graph the equation found in Example 10 on a graphing calculator using the window shown below. (We will use this graph later in the book to find the angle between the cannon and the horizontal.)

Window: X from 0 to 190, increment 20

Y from 0 to 90, increment 10

On the TI-83/84, an increment of 20 for X means Xscl=20.

Problem Set 8.5

Moving Toward Success

"Write it on your heart that every day is the best day in the year."

—Ralph Waldo Emerson, 1803–1882, American poet and essayist

1. Why should you use all the exam time provided for you, even if you finish early?

2. What are some things you can do with any extra time to ensure the best test grade possible?

A For each of the following equations, give the *x*-intercepts and the coordinates of the vertex, and sketch the graph. [Examples 1, 2]

1. $y = x^2 + 2x - 3$

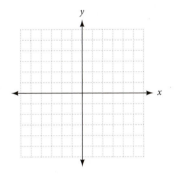

2. $y = x^2 - 2x - 3$

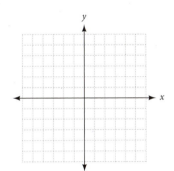

3. $y = -x^2 - 4x + 5$

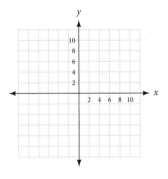

4. $y = x^2 + 4x - 5$

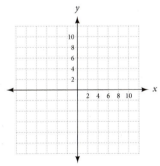

5. $y = x^2 - 1$

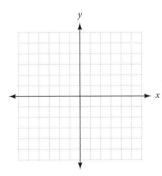

6. $y = x^2 - 4$

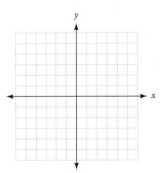

7. $y = -x^2 + 9$

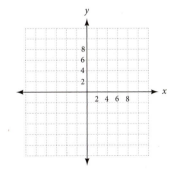

8. $y = -x^2 + 1$

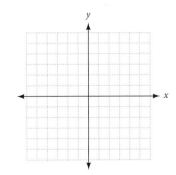

9. $y = 2x^2 - 4x - 6$

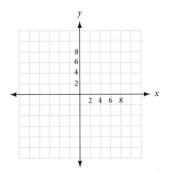

10. $y = 2x^2 + 4x - 6$

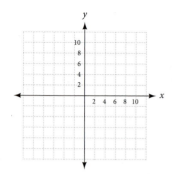

11. $y = x^2 - 2x - 4$

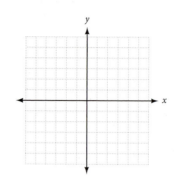

12. $y = x^2 - 2x - 2$

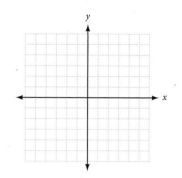

A Find the vertex and any two convenient points to sketch the graphs of the following. [Examples 3, 4]

13. $y = x^2 - 4x - 4$

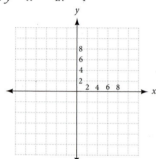

14. $y = x^2 - 2x + 3$

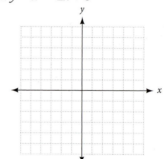

15. $y = -x^2 + 2x - 5$

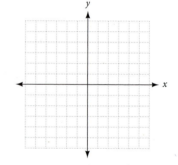

16. $y = -x^2 + 4x - 2$

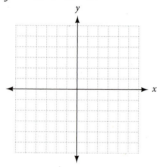

17. $y = x^2 + 1$

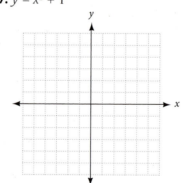

18. $y = x^2 + 4$

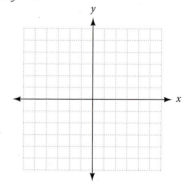

19. $y = -x^2 - 3$

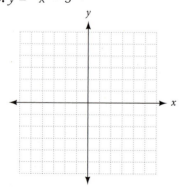

20. $y = -x^2 - 2$

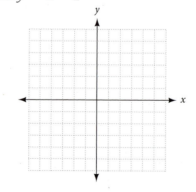

A For each of the following equations, find the coordinates of the vertex, and indicate whether the vertex is the highest point on the graph or the lowest point on the graph. (Do not graph.) [Examples 1–4]

21. $y = x^2 - 6x + 5$

22. $y = -x^2 + 6x - 5$

23. $y = -x^2 + 2x + 8$

24. $y = x^2 - 2x - 8$

25. $y = 12 + 4x - x^2$

26. $y = -12 - 4x + x^2$

27. $y = -x^2 - 8x$

28. $y = x^2 + 8x$

Graph each equation. [Examples 5–7]

29. $x = y^2 + 2y - 3$

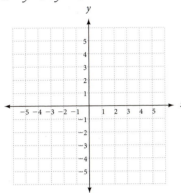

30. $x = y^2 - 2y - 3$

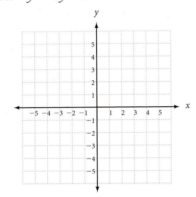

31. $x = -y^2 - 4y + 5$

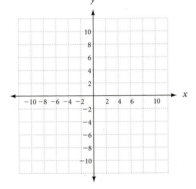

32. $x = y^2 + 4y - 5$

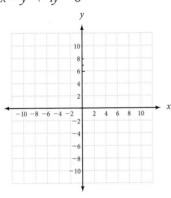

33. $x = y^2 - 1$

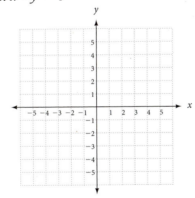

34. $x = y^2 - 4$

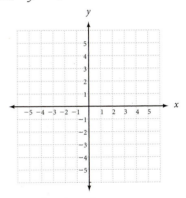

35. $x = -y^2 + 9$

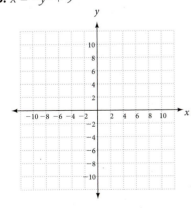

36. $x = -y^2 + 1$

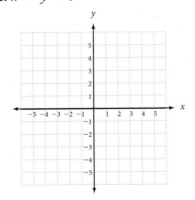

37. $x = 2y^2 - 4y - 6$

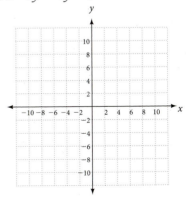

38. $x = 2y^2 + 4y - 6$

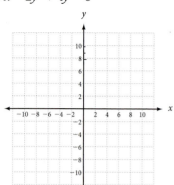

39. $x = y^2 - 2y - 4$

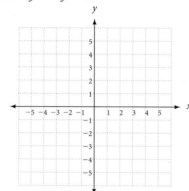

40. $x = y^2 - 2y + 3$

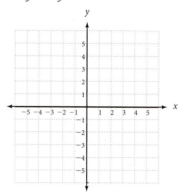

Applying the Concepts

41. Maximum Profit A company finds that it can make a profit of P dollars each month by selling x patterns, according to the formula $P(x) = -0.002x^2 + 3.5x - 800$. How many patterns must it sell each month to have a maximum profit? What is the maximum profit?

42. Maximum Profit

A company selling picture frames finds that it can make a profit of P dollars each month by selling x frames, according to the formula $P(x) = -0.002x^2 + 5.5x - 1,200$. How many frames must it sell each month to have a maximum profit? What is the maximum profit?

43. Maximum Height Chaudra is tossing a softball into the air with an underhand motion. The distance of the ball above her hand at any time is given by the function

$$h(t) = 32t - 16t^2$$

for $0 \le t \le 2$

where $h(t)$ is the height of the ball (in feet) and t is the time (in seconds). Find the times at which the ball is in her hand and the maximum height of the ball.

44. Maximum Area Justin wants to fence 3 sides of a rectangular exercise yard for his dog. The fourth side of the exercise yard will be a side of the house. He has 80 feet of fencing available. Find the dimensions of the exercise yard that will enclose the maximum area.

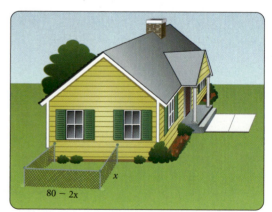

x

$80 - 2x$

45. Maximum Revenue A company that manufactures key chains knows that the number of key chains x it can sell each week is related to the price p of each key chain by the equation $x = 1{,}200 - 100p$. Graph the revenue equation $R = xp$. Then use the graph to find the price p that will bring in the maximum revenue. Finally, find the maximum revenue.

46. Maximum Revenue A company that manufactures CDs for home computers finds that it can sell x CDs each day at p dollars per CD, according to the equation $x = 800 - 100p$. Graph the revenue equation $R = xp$. Then use the graph to find the price p that will bring in the maximum revenue. Finally, find the maximum revenue.

47. Maximum Revenue The relationship between the number of calculators x a company sells each day and the price p of each calculator is given by the equation $x = 1{,}700 - 100p$. Graph the revenue equation $R = xp$, and use the graph to find the price p that will bring in the maximum revenue. Then find the maximum revenue.

48. Maximum Revenue The relationship between the number of pencil sharpeners x a company sells each week and the price p of each sharpener is given by the equation $x = 1{,}800 - 100p$. Graph the revenue equation $R = xp$, and use the graph to find the price p that will bring in the maximum revenue. Then find the maximum revenue.

C [Example 10]

49. Human Cannonball A human cannonball is shot from a cannon at the county fair. He reaches a height of 60 feet before landing in a net 180 feet from the cannon. Sketch the graph of his path, and then find the equation of the graph.

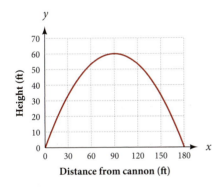

50. Human Cannonball Suppose the equation below gives the height $h(x)$ of the human cannonball after he has traveled x feet horizontally. What is $h(30)$, and what does it represent?

$$h(x) = -\frac{7}{640}(x - 80)^2 + 70$$

Getting Ready for the Next Section

Solve.

51. $x^2 - 2x - 8 = 0$

52. $x^2 - x - 12 = 0$

53. $6x^2 - x = 2$

54. $3x^2 - 5x = 2$

55. $x^2 - 6x + 9 = 0$

56. $x^2 + 8x + 16 = 0$

Maintaining Your Skills

Perform the indicated operations.

57. $(3 - 5i) - (2 - 4i)$

58. $2i(5 - 6i)$

59. $(3 + 2i)(7 - 3i)$

60. $(4 + 5i)^2$

61. $\dfrac{i}{3 + i}$

62. $\dfrac{2 + 3i}{2 - 3i}$

Extending the Concepts

Finding the equation from the graph For each of the next two problems, the graph is a parabola. In each case, find an equation in the form $y = a(x - h)^2 + k$ that describes the graph.

63.

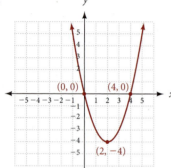

64.

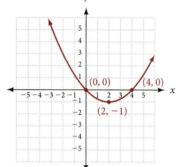

65. Interpreting Graphs The graph below shows the different paths taken by the human cannonball when his velocity out of the cannon is 50 mph and his cannon is inclined at varying angles.

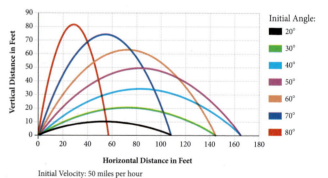

a. If his landing net is placed 108 feet from the cannon, at what angle should the cannon be inclined so that he lands in the net?

b. Approximately where do you think he would land if the cannon was inclined at 45°?

c. If the cannon was inclined at 45°, approximately what height do you think he would attain?

d. Do you think there is another angle for which he would travel the same distance he travels at 80°? Give an estimate of that angle.

e. The fact that every landing point can come from two different paths makes us think that the equations that give us the landing points must be what type of equations?

Quadratic Inequalities

TICKET TO SUCCESS

Keep these questions in mind as you read through the section. Then respond in your own words and in complete sentences.

1. What is the first step in solving a quadratic inequality?
2. Why would you draw the number lines of each factor from a quadratic inequality together?
3. How do you show that the endpoint of a line segment is not part of the graph of a quadratic inequality?
4. How would you use the graph of $y = ax^2 + bx + c$ to help you find the graph of $ax^2 + bx + c < 0$?

WilleeCole/Shutterstock.com

Suppose you needed to build a dog run with an area of at least 40 square feet. The length of the run has to be 3 times the width. What are the possibilities for the length? In order to answer this question, you need to know how to work with quadratic inequalities, the focus of this section.

A Quadratic Inequalities

Quadratic inequalities in one variable are inequalities of the form

$$ax^2 + bx + c < 0 \qquad ax^2 + bx + c > 0$$

$$ax^2 + bx + c \leq 0 \qquad ax^2 + bx + c \geq 0$$

where a, b, and c are constants, with $a \neq 0$. The technique we will use to solve inequalities of this type involves graphing. Suppose, for example, we wish to find the solution set for the inequality $x^2 - x - 6 > 0$. We begin by factoring the left side to obtain

$$(x - 3)(x + 2) > 0$$

We have two real numbers $x - 3$ and $x + 2$ whose product $(x - 3)(x + 2)$ is greater than zero; that is, their product is positive. The only way the product can be positive is either if both factors, $(x - 3)$ and $(x + 2)$, are positive or if they are both negative. To help visualize where $x - 3$ is positive and where it is negative, we draw a real number line and label it accordingly:

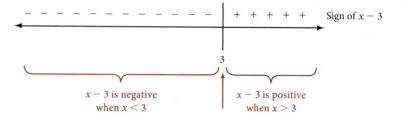

Here is a similar diagram showing where the factor $x + 2$ is positive and where it is negative:

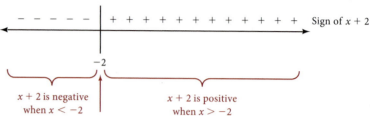

Drawing the two number lines together and eliminating the unnecessary numbers, we have

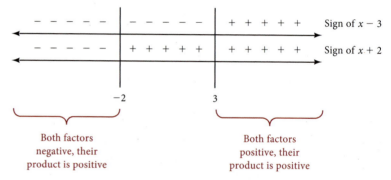

We can see from the preceding diagram that the graph of the solution to $x^2 - x - 6 > 0$ is

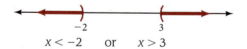

USING TECHNOLOGY

Graphical Solutions to Quadratic Inequalties

We can solve the preceding problem by using a graphing calculator to visualize where the product $(x - 3)(x + 2)$ is positive. First, we graph the function $y = (x - 3)(x + 2)$ as shown in Figure 1.

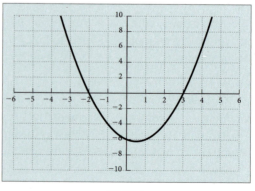

FIGURE 1

Next, we observe where the graph is above the x-axis. As you can see, the graph is above the x-axis to the right of 3 and to the left of -2, as shown in Figure 2.

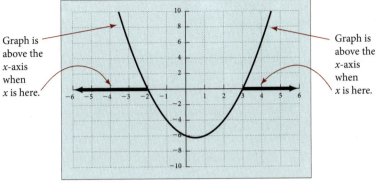

Graph is above the x-axis when x is here.

Graph is above the x-axis when x is here.

FIGURE 2

When the graph is above the x-axis, we have points whose y-coordinates are positive. Since these y-coordinates are the same as the expression $(x - 3)(x + 2)$, the values of x for which the graph of $y = (x - 3)(x + 2)$ is above the x-axis are the values of x for which the inequality $(x - 3)(x + 2) > 0$ is true. Therefore, our solution set is

EXAMPLE 1   Solve $x^2 - 2x - 8 \leq 0$ for x.

ALGEBRAIC SOLUTION   We begin by factoring.

$$x^2 - 2x - 8 \leq 0$$

$$(x - 4)(x + 2) \leq 0$$

The product $(x - 4)(x + 2)$ is negative or zero. The factors must have opposite signs. We draw a diagram showing where each factor is positive and where each factor is negative.

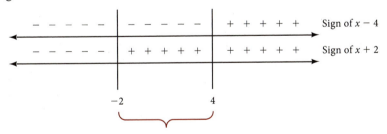

From the diagram, we have the graph of the solution set.

$$-2 \leq x \leq 4$$

When x is here, the graph is on or below the x-axis.

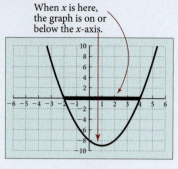

FIGURE 3

2. Solve for x.
$$2x^2 + 3x > 2$$

GRAPHICAL SOLUTION To solve this inequality with a graphing calculator, we graph the function $y = (x - 4)(x + 2)$ and observe where the graph is below the x-axis. These points have negative y-coordinates, which means that the product $(x - 4)(x + 2)$ is negative for these points. Figure 3 shows the graph of $y = (x - 4)(x + 2)$, along with the region on the x-axis where the graph contains points with negative y-coordinates.

As you can see, the graph is below the x-axis when x is between -2 and 4. Since our original inequality includes the possibility that $(x - 4)(x + 2)$ is 0, we include the endpoints, -2 and 4, with our solution set.

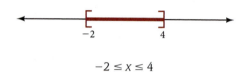

$$-2 \leq x \leq 4$$

EXAMPLE 2 Solve $6x^2 - x \geq 2$ for x.

ALGEBRAIC SOLUTION

$$6x^2 - x \geq 2$$
$$6x^2 - x - 2 \geq 0 \qquad \text{Standard form}$$
$$(3x - 2)(2x + 1) \geq 0$$

The product is positive, so the factors must agree in sign. Here is the diagram showing where that occurs:

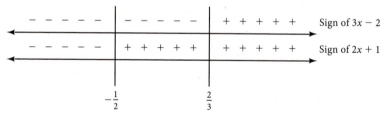

Since the factors agree in sign below $-\frac{1}{2}$ and above $\frac{2}{3}$, the graph of the solution set is

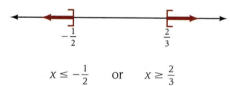

$$x \leq -\frac{1}{2} \qquad \text{or} \qquad x \geq \frac{2}{3}$$

GRAPHICAL SOLUTION To solve this inequality with a graphing calculator, we graph the function $y = (3x - 2)(2x + 1)$ and observe where the graph is above the x-axis. These are the points that have positive y-coordinates, which means that the product $(3x - 2)(2x + 1)$ is positive for these points. Figure 4 shows the graph of $y = (3x - 2)(2x + 1)$, along with the regions on the x-axis where the graph is on or above the x-axis.

To find the points where the graph crosses the x-axis, we need to use either the Trace and Zoom features to zoom in on each point, or the calculator function that finds the intercepts automatically (on the TI-83/84 this is the root/zero function under the CALC key). Whichever method we use, we will obtain the following result:

Graph is on or above the x-axis when x is here.

Graph is on or above the x-axis when x is here.

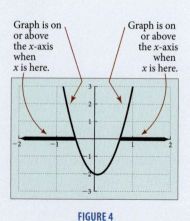

FIGURE 4

Answers

2. $x < -2$ or $x > \frac{1}{2}$

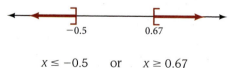

$$x \leq -0.5 \qquad \text{or} \qquad x \geq 0.67$$

EXAMPLE 3   Solve $x^2 - 6x + 9 \geq 0$.

ALGEBRAIC SOLUTION

$$x^2 - 6x + 9 \geq 0$$

$$(x - 3)^2 \geq 0$$

This is a special case in which both factors are the same. Since $(x - 3)^2$ is always positive or zero, the solution set is all real numbers; that is, any real number that is used in place of x in the original inequality will produce a true statement.

GRAPHICAL SOLUTION   The graph of $y = (x - 3)^2$ is shown in Figure 5. Notice that it touches the x-axis at 3 and is above the x-axis everywhere else. This means that every point on the graph has a y-coordinate greater than or equal to 0, no matter what the value of x. The conclusion that we draw from the graph is that the inequality $(x - 3)^2 \geq 0$ is true for all values of x. ■

B Inequalities Involving Rational Expressions

EXAMPLE 4   Solve $\dfrac{x - 4}{x + 1} \leq 0$.

SOLUTION   The inequality indicates that the quotient of $(x - 4)$ and $(x + 1)$ is negative or 0 (less than or equal to 0). We can use the same reasoning we used to solve the first three examples, because quotients are positive or negative under the same conditions that products are positive or negative. Here is the diagram that shows where each factor is positive and where each factor is negative:

Between -1 and 4 the factors have opposite signs, making the quotient negative. Thus, the region between -1 and 4 is where the solutions lie, because the original inequality indicates the quotient $\dfrac{x - 4}{x + 1}$ is negative. The solution set and its graph are shown here:

$$-1 < x \leq 4$$

Notice that the left endpoint is open; that is, it is not included in the solution set because $x = -1$ would make the denominator in the original inequality 0. It is important to check all endpoints of solution sets to inequalities that involve rational expressions. ■

EXAMPLE 5   Solve $\dfrac{3}{x - 2} - \dfrac{2}{x - 3} > 0$.

SOLUTION   We begin by adding the two rational expressions on the left side. The common denominator is $(x - 2)(x - 3)$.

$$\frac{3}{x - 2} \cdot \frac{(x - 3)}{(x - 3)} - \frac{2}{x - 3} \cdot \frac{(x - 2)}{(x - 2)} > 0$$

---

3. Solve $x^2 + 10x + 25 < 0$.

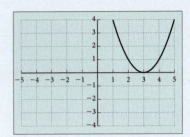

FIGURE 5

4. Solve $\dfrac{x + 3}{x - 2} \geq 0$.

5. Solve $\dfrac{1}{x - 4} - \dfrac{2}{x - 3} > 0$.

Answers
3. No solution
4. $x \leq -3$ or $x > 2$
5. $x < 3$ or $4 < x < 5$

$$\frac{3x - 9 - 2x + 4}{(x - 2)(x - 3)} > 0$$

$$\frac{x - 5}{(x - 2)(x - 3)} > 0$$

This time the quotient involves three factors. Here is the diagram that shows the signs of the three factors:

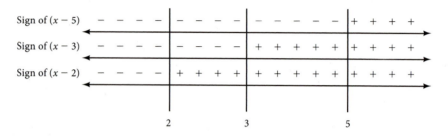

The original inequality indicates that the quotient is positive. For this to happen, either all three factors must be positive, or exactly two factors must be negative. Looking back at the diagram, we see the regions that satisfy these conditions are between 2 and 3 or above 5. Here is our solution set:

$$2 < x < 3 \text{ or } x > 5$$

∎

Problem Set 8.6

Moving Toward Success

"A life spent making mistakes is not only more honorable, but more useful than a life spent doing nothing."

—George Bernard Shaw, 1856–1950, Irish playwright and Nobel Prize winner

1. Why is it important to review your graded test when you get it back?

2. How will you address any mistakes you made on the test to ensure you don't continue to make the same ones on a future test?

A Solve each of the following inequalities and graph the solution set. [Examples 1–3]

1. $x^2 + x - 6 > 0$

2. $x^2 + x - 6 < 0$

3. $x^2 - x - 12 \leq 0$

4. $x^2 - x - 12 \geq 0$

5. $x^2 + 5x \geq -6$

6. $x^2 - 5x > 6$

7. $6x^2 < 5x - 1$

8. $4x^2 \geq -5x + 6$

9. $x^2 - 9 < 0$

10. $x^2 - 16 \geq 0$

11. $4x^2 - 9 \geq 0$

12. $9x^2 - 4 < 0$

13. $2x^2 - x - 3 < 0$

14. $3x^2 + x - 10 \geq 0$

15. $x^2 - 4x + 4 \geq 0$

16. $x^2 - 4x + 4 < 0$

17. $x^2 - 10x + 25 < 0$

18. $x^2 - 10x + 25 > 0$

19. $(x - 2)(x - 3)(x - 4) > 0$

20. $(x - 2)(x - 3)(x - 4) < 0$

21. $(x + 1)(x + 2)(x + 3) \leq 0$

22. $(x + 1)(x + 2)(x + 3) \geq 0$

A Solve each inequality by inspection, without showing any work. [Example 3]

23. $(x - 1)^2 < 0$

24. $(x + 2)^2 < 0$

25. $(x - 1)^2 \leq 0$

26. $(x + 2)^2 \leq 0$

27. $(x - 1)^2 \geq 0$

28. $(x + 2)^2 \geq 0$

29. $\dfrac{1}{(x - 1)^2} \geq 0$

30. $\dfrac{1}{(x + 2)^2} > 0$

31. $x^2 - 6x + 9 < 0$

32. $x^2 - 6x + 9 \leq 0$

33. $x^2 - 6x + 9 > 0$

34. $\dfrac{1}{x^2 - 6x + 9} > 0$

B [Examples 4–5]

35. $\dfrac{x - 1}{x + 4} \leq 0$

36. $\dfrac{x + 4}{x - 1} \leq 0$

37. $\dfrac{3x - 8}{x + 6} > 0$

38. $\dfrac{5x - 3}{x + 1} < 0$

39. $\dfrac{x - 2}{x - 6} > 0$

40. $\dfrac{x - 1}{x - 3} \geq 0$

41. $\dfrac{x - 2}{(x + 3)(x - 4)} < 0$

42. $\dfrac{x - 1}{(x + 2)(x - 5)} < 0$

43. $\dfrac{2}{x - 4} - \dfrac{1}{x - 3} > 0$

44. $\dfrac{4}{x + 3} - \dfrac{3}{x + 2} > 0$

45. Write each statement using inequality notation.

    a. $x - 1$ is always positive.

    b. $x - 1$ is never negative.

    c. $x - 1$ is greater than or equal to 0.

46. Match each expression on the left with a phrase on the right.

    a. $(x - 1)^2 \geq 0$ **i.** Never true

    b. $(x - 1)^2 < 0$ **ii.** Sometimes true

    c. $(x - 1)^2 \leq 0$ **iii.** Always true

47. The graph of $y = x^2 - 4$ is shown in Figure 6. Use the graph to write the solution set for each of the following:

a. $x^2 - 4 < 0$

b. $x^2 - 4 > 0$

c. $x^2 - 4 = 0$

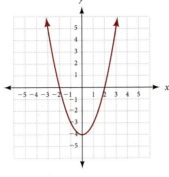

FIGURE 6

48. The graph of $y = 4 - x^2$ is shown in Figure 7. Use the graph to write the solution set for each of the following:

a. $4 - x^2 < 0$

b. $4 - x^2 > 0$

c. $4 - x^2 = 0$

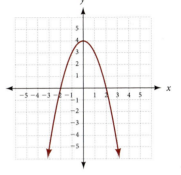

FIGURE 7

49. The graph of $y = x^2 - 3x - 10$ is shown in Figure 8. Use the graph to write the solution set for each of the following:

a. $x^2 - 3x - 10 < 0$

b. $x^2 - 3x - 10 > 0$

c. $x^2 - 3x - 10 = 0$

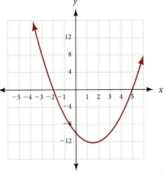

FIGURE 8

50. The graph of $y = x^2 + x - 12$ is shown in Figure 9. Use the graph to write the solution set for each of the following:

a. $x^2 + x - 12 < 0$

b. $x^2 - x - 12 > 0$

c. $x^2 + x - 12 = 0$

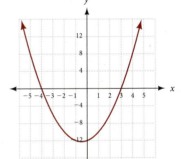

FIGURE 9

51. The graph of $y = x^3 - 3x^2 - x + 3$ is shown in Figure 10. Use the graph to write the solution set for each of the following:

a. $x^3 - 3x^2 - x + 3 < 0$

b. $x^3 - 3x^2 - x + 3 > 0$

c. $x^3 - 3x^2 - x + 3 = 0$

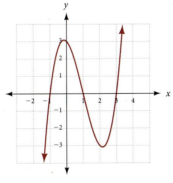

FIGURE 10

52. The graph of $y = x^3 + 4x^2 - 4x - 16$ is shown in Figure 11. Use the graph to write the solution set for each of the following:

a. $x^3 + 4x^2 - 4x - 16 < 0$

b. $x^3 + 4x^2 - 4x - 16 > 0$

c. $x^3 + 4x^2 - 4x - 16 = 0$

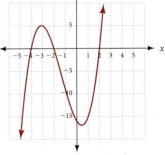

FIGURE 11

Applying the Concepts

53. Dimensions of a Rectangle The length of a rectangle is 3 inches more than twice the width. If the area is to be at least 44 square inches, what are the possibilities for the width?

54. Area A dog run has to have an area of at least 21 square feet. If the length of the run is to be one foot less than twice the width, what are the possibilities for the width?

55. Revenue A manufacturer of headphones knows that the weekly revenue produced by selling x headphones is given by the equation $R = 1{,}300p - 100p^2$, where p is the price of each set of headphones. What price should she charge for each set of headphones if she wants her weekly revenue to be at least $4,000?

56. Revenue A manufacturer of small calculators knows that the weekly revenue produced by selling x calculators is given by the equation $R = 1{,}700p - 100p^2$, where p is the price of each calculator. What price should be charged for each calculator if the revenue is to be at least $7,000 each week?

Maintaining Your Skills

Use a calculator to evaluate. Give answers to 4 decimal places.

57. $\dfrac{50{,}000}{32{,}000}$

58. $\dfrac{2.4362}{1.9758} - 1$

59. $\dfrac{1}{2}\left(\dfrac{4.5926}{1.3876} - 2\right)$

60. $1 + \dfrac{0.06}{12}$

Solve each equation.

61. $\sqrt{3t - 1} = 2$

62. $\sqrt{4t + 5} + 7 = 3$

63. $\sqrt{x + 3} = x - 3$

64. $\sqrt{x + 3} = \sqrt{x} - 3$

Graph each equation.

65. $y = \sqrt[3]{x - 1}$

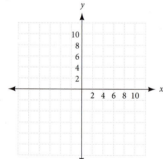

66. $y = \sqrt[3]{x} - 1$

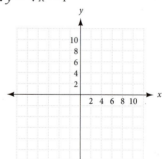

Extending the Concepts

Graph the solution set for each inequality.

67. $x^2 - 2x - 1 < 0$

68. $x^2 - 6x + 7 < 0$

69. $x^2 - 8x + 13 > 0$

70. $x^2 - 10x + 18 > 0$

Chapter 8 Summary

EXAMPLES

■ Square Root Theorem [8.1]

1. If $(x - 3)^2 = 25$
then $x - 3 = \pm 5$
$x = 3 \pm 5$
$x = 8$ or $x = -2$

If $a^2 = b$, where b is a real number, then

$$a = \sqrt{b} \quad \text{or} \quad a = -\sqrt{b}$$

which can be written as $a = \pm\sqrt{b}$.

■ To Solve a Quadratic Equation by Completing the Square [8.1]

2. Solve $x^2 - 6x - 6 = 0$
$x^2 - 6x = 6$
$x^2 - 6x + \mathbf{9} = 6 + \mathbf{9}$
$(x - 3)^2 = 15$
$x - 3 = \pm\sqrt{15}$
$x = 3 \pm \sqrt{15}$

Step 1: Write the equation in the form $ax^2 + bx = c$.

Step 2: If $a \neq 1$, divide through by the constant a so the coefficient of x^2 is 1.

Step 3: Complete the square on the left side by adding the square of $\frac{1}{2}$ the coefficient of x to both sides.

Step 4: Write the left side of the equation as the square of a binomial. Simplify the right side if possible.

Step 5: Apply the square root theorem, and solve as usual.

■ The Quadratic Theorem [8.2]

3. If $2x^2 + 3x - 4 = 0$, then

$x = \dfrac{-3 \pm \sqrt{9 - 4(2)(-4)}}{2(2)}$

$= \dfrac{-3 \pm \sqrt{41}}{4}$

For any quadratic equation in the form $ax^2 + bx + c = 0$, $a \neq 0$, the two solutions are

$$x = \frac{-b \pm \sqrt{b^2 - 4ac}}{2a}$$

This last equation is known as the *quadratic formula*.

■ The Discriminant [8.3]

4. The discriminant for
$x^2 + 6x + 9 = 0$
is $D = 36 - 4(1)(9) = 0$,
which means the equation
has one rational solution.

The expression $b^2 - 4ac$ that appears under the radical sign in the quadratic formula is known as the *discriminant*.

We can classify the solutions to $ax^2 + bx + c = 0$:

The solutions are	When the discriminant is
Two complex numbers containing i	Negative
One rational number	Zero
Two rational numbers	A positive perfect square
Two irrational numbers	A positive number, but not a perfect square

■ Equations Quadratic in Form [8.4]

5. The equation
$x^4 - x^2 - 12 = 0$ is
quadratic in x^2. Letting
$y = x^2$ we have

$$y^2 - y - 12 = 0$$
$$(y - 4)(y + 3) = 0$$
$$y = 4 \quad \text{or} \quad y = -3$$

Resubstituting x^2 for y, we
have

$$x^2 = 4 \quad \text{or} \quad x^2 = -3$$
$$x = \pm 2 \qquad x = \pm i\sqrt{3}$$

There are a variety of equations whose form is quadratic. We solve most of them by making a substitution so the equation becomes quadratic, and then solving the equation by factoring or the quadratic formula. For example,

The equation	*is quadratic in*
$(2x - 3)^2 + 5(2x - 3) - 6 = 0$	$2x - 3$
$4x^4 - 7x^2 - 2 = 0$	x^2
$2x - 7\sqrt{x} + 3 = 0$	$\sqrt{x}$

■ Graphing Parabolas [8.5]

6. The graph of $y = x^2 - 4$ will
be a parabola. It will cross
the x-axis at 2 and -2, and
the vertex will be $(0, -4)$.

The graph of any equation of the form

$$y = ax^2 + bx + c \qquad a \neq 0$$

is a *parabola*. The graph is *concave up* if $a > 0$ and *concave down* if $a < 0$. The highest or lowest point on the graph is called the *vertex* and always has an x-coordinate of $x = \frac{-b}{2a}$.

■ Quadratic Inequalities [8.6]

7. Solve $x^2 - 2x - 8 > 0$. We
factor and draw the sign
diagram:

$$(x - 4)(x + 2) > 0$$

The solution is $x < -2$ or
$x > 4$.

We solve quadratic inequalities by manipulating the inequality to get 0 on the right side and then factoring the left side. We then make a diagram that indicates where the factors are positive and where they are negative. From this sign diagram and the original inequality, we graph the appropriate solution set.

Solve each equation. [8.1]

1. $(2t - 5)^2 = 25$

2. $(3t - 2)^2 = 4$

3. $(3y - 4)^2 = -49$

4. $(2x + 6)^2 = 12$

Solve by completing the square. [8.1]

5. $2x^2 + 6x - 20 = 0$

6. $3x^2 + 15x = -18$

7. $a^2 + 9 = 6a$

8. $a^2 + 4 = 4a$

9. $2y^2 + 6y = -3$

10. $3y^2 + 3 = 9y$

Solve each equation. [8.2]

11. $\frac{1}{6}x^2 + \frac{1}{2}x - \frac{5}{3} = 0$

12. $8x^2 - 18x = 0$

13. $4t^2 - 8t + 19 = 0$

14. $100x^2 - 200x = 100$

15. $0.06a^2 + 0.05a = 0.04$

16. $9 - 6x = -x^2$

17. $(2x + 1)(x - 5) - (x + 3)(x - 2) = -17$

18. $2y^3 + 2y = 10y^2$

19. $5x^2 = -2x + 3$

20. $x^3 - 27 = 0$

21. $3 - \frac{2}{x} + \frac{1}{x^2} = 0$

22. $\frac{1}{x - 3} + \frac{1}{x + 2} = 1$

23. Profit The total cost (in dollars) for a company to produce x items per week is $C = 7x + 400$. The revenue for selling all x items is $R = 34x - 0.1x^2$. How many items must it produce and sell each week for its weekly profit to be $1,300? [8.2]

24. Profit The total cost (in dollars) for a company to produce x items per week is $C = 70x + 300$. The revenue for selling all x items is $R = 110x - 0.5x^2$. How many items must it produce and sell each week for its weekly profit to be $300? [8.2]

Use the discriminant to find the number and kind of solutions for each equation. [8.3]

25. $2x^2 - 8x = -8$

26. $4x^2 - 8x = -4$

27. $2x^2 + x - 3 = 0$

28. $5x^2 + 11x = 12$

29. $x^2 - x = 1$

30. $x^2 - 5x = -5$

31. $3x^2 + 5x = -4$

32. $4x^2 - 3x = -6$

Determine k so that each equation has exactly one real solution. [8.3]

33. $25x^2 - kx + 4 = 0$

34. $4x^2 + kx + 25 = 0$

35. $kx^2 + 12x + 9 = 0$

36. $kx^2 - 16x + 16 = 0$

37. $9x^2 + 30x + k = 0$

38. $4x^2 + 28x + k = 0$

For each of the following problems, find an equation that has the given solutions. [8.3]

39. $x = 3, x = 5$

40. $x = -2, x = 4$

41. $y = \frac{1}{2}, y = -4$

42. $t = 3, t = -3, t = 5$

Find all solutions. [8.4]

43. $(x - 2)^2 - 4(x - 2) - 60 = 0$

44. $6(2y + 1)^2 - (2y + 1) - 2 = 0$

45. $x^4 - x^2 = 12$

46. $x - \sqrt{x} - 2 = 0$

47. $2x - 11\sqrt{x} = -12$

48. $\sqrt{x + 5} = \sqrt{x} + 1$

49. $\sqrt{y + 21} + \sqrt{y} = 7$

50. $\sqrt{y + 9} - \sqrt{y - 6} = 3$

51. Projectile Motion An object is tossed into the air with an upward velocity of 10 feet per second from the top of a building h feet high. The time it takes for the object to hit the ground below is given by the formula $16t^2 - 10t - h = 0$. Solve this formula for t. [8.4]

52. Projectile Motion An object is tossed into the air with an upward velocity of v feet per second from the top of a 10-foot wall. The time it takes for the object to hit the ground below is given by the formula $16t^2 - vt - 10 = 0$. Solve this formula for t. [8.4]

Find the x-intercepts, if they exist, and the vertex for each parabola. Then use them to sketch the graph. [8.5]

53. $y = x^2 - 6x + 8$

54. $y = x^2 - 4$

Solve each inequality and graph the solution set. [8.6]

55. $x^2 - x - 2 < 0$

56. $3x^2 - 14x + 8 \leq 0$

57. $2x^2 + 5x - 12 \geq 0$

Simplify.

1. $11 + 20 \div 5 - 3 \cdot 5$

2. $4(15 - 19)^2 - 3(17 - 19)^3$

3. $\left(-\dfrac{2}{3}\right)^3$

4. $\left(-\dfrac{5}{4}\right)^2$

5. $4 + 8x - 3(5x - 2)$

6. $3 - 5[2x - 4(x - 2)]$

7. $\left(\dfrac{x^{-5}y^4}{x^{-2}y^{-3}}\right)^{-1}$

8. $\left(\dfrac{a^{-1}b^{-2}}{a^3b^{-3}}\right)^{-2}$

9. $\sqrt[3]{32}$

10. $\sqrt{50x^3}$

11. $8^{-2/3} + 25^{-1/2}$

12. $\left(\dfrac{8}{27}\right)^{-2/3}$

13. $\dfrac{1 - \frac{3}{4}}{1 + \frac{3}{4}}$

14. $\dfrac{2 + \frac{1}{5}}{1 - \frac{2}{3}}$

Reduce.

15. $\dfrac{5x^2 - 26xy - 24y^2}{5x + 4y}$

16. $\dfrac{x^2 - x - 6}{x + 2}$

Divide.

17. $\dfrac{3 + i}{i}$

18. $\dfrac{7 - i}{3 - 2i}$

Solve.

19. $\dfrac{7}{5}a - 6 = 15$

20. $\dfrac{3}{5}x - \dfrac{1}{2} = 3$

21. $|a| - 6 = 3$

22. $|a - 2| = -1$

23. $\dfrac{y}{3} + \dfrac{5}{y} = -\dfrac{8}{3}$

24. $\dfrac{a}{2} + \dfrac{3}{a - 3} = \dfrac{a}{a - 3}$

25. $(3x - 4)^2 = 18$

26. $3y^3 - y = 5y^2$

27. $\dfrac{2}{15}x^2 + \dfrac{1}{3}x + \dfrac{1}{5} = 0$

28. $0.06a^2 + 0.01a = -0.02$

29. $\sqrt{y + 3} = y + 3$

30. $\sqrt{x - 2} = 2 - \sqrt{x}$

31. $x - 3\sqrt{x} - 54 = 0$

32. $x^{2/3} - x^{1/3} = 6$

Solve each inequality and graph the solution.

33. $5 \le \dfrac{1}{4}x + 3 \le 8$

34. $|4x - 3| \ge 5$

35. $x^2 + 2x < 8$

36. $\dfrac{x - 1}{x + 3} \ge 0$

Solve each system.

37. $\begin{aligned} 3x - y &= 22 \\ -6x + 2y &= -4 \end{aligned}$

38. $\begin{aligned} 4x - 8y &= 6 \\ 6x - 12y &= 6 \end{aligned}$

Graph on a rectangular coordinate system.

39. $2x - 3y = 12$

40. $y = x^2 - x - 2$

41. Find the equation of the line passing through the points $\left(\frac{3}{2}, \frac{4}{3}\right)$ and $\left(\frac{1}{4}, -\frac{1}{3}\right)$.

42. Find the equation of the line perpendicular to $2x - 4y = 8$ and containing the point $(1, 5)$.

Factor completely.

43. $x^2 + 8x + 16 - y^2$

44. $(x - 1)^2 - 5(x - 1) + 4$

Rationalize the denominator.

45. $\dfrac{7}{\sqrt[3]{9}}$

46. $\dfrac{\sqrt{6}}{\sqrt{6} + \sqrt{2}}$

If $A = \{1, 2, 3, 4\}$ and $B = \{0, 2, 4, 6\}$, find the following.

47. $\{x \mid x \in B \text{ and } x \in A\}$

48. $\{x \mid x \notin A \text{ and } x \in B\}$

49. Geometry Find all three angles in a triangle if the smallest angle is one-fourth the largest angle and the remaining angle is 30° more than the smallest angle.

50. Investing A total of $8,000 is invested in two accounts. One account earns 5% per year, and the other earns 4% per year. If the total interest for the first year is $380; how much was invested in each account?

51. Inverse Variation y varies inversely with the square of x. If $y = 4$ when $x = \frac{5}{3}$, find y when $x = \frac{8}{3}$.

52. Direct Variation w varies directly with the square root of c. If w is 8 when c is 16, find w when c is 9.

The chart shows the increase in sales for different snack food. Use it to answer the following questions.

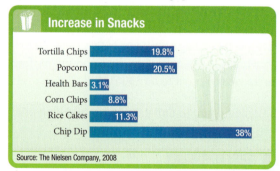

Increase in Snacks

Tortilla Chips	19.8%
Popcorn	20.5%
Health Bars	3.1%
Corn Chips	8.8%
Rice Cakes	11.3%
Chip Dip	38%

Source: The Nielsen Company, 2008

53. If a market sold $2,500 worth of popcorn last year, how much should it sell this year?

54. If a market sold $3,000 worth of tortilla chips last year, how much should it sell this year?

Chapter 8 Test

Solve each equation. [8.1, 8.2]

1. $(2x + 4)^2 = 25$

2. $(2x - 6)^2 = -8$

3. $y^2 - 10y + 25 = -4$

4. $(y + 1)(y - 3) = -6$

5. $8t^3 - 125 = 0$

6. $\dfrac{1}{a + 2} - \dfrac{1}{3} = \dfrac{1}{a}$

7. Solve the formula $64(1 + r)^2 = A$ for r. [8.1]

8. Solve $x^2 - 4x = -2$ by completing the square [8.1]

9. If the length of the longest side of 30°–60°–90° triangle is 4 inches, find the lengths of the other two sides. [8.1]

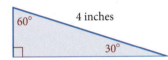

10. Projectile Motion An object thrown upward with an initial velocity of 32 feet per second will rise and fall according to the equation $s(t) = 32t - 16t^2$, where s is its distance above the ground at time t. At what times will the object be 12 feet above the ground? [8.2]

11. Revenue The total weekly cost for a company to make x ceramic coffee cups is given by the formula $C(x) = 2x + 100$. If the weekly revenue from selling all x cups is $R(x) = 25x - 0.2x^2$, how many cups must it sell a week to make a profit of $200 a week? [8.2]

12. Find k so that $kx^2 = 12x - 4$ has one rational solution. [8.3]

13. Use the discriminant to identify the number and kind of solutions to $2x^2 - 5x = 7$. [8.3]

Find equations that have the given solutions. [8.3]

14. $x = 5, x = -\dfrac{2}{3}$

15. $x = 2, x = -2, x = 7$

Solve each equation. [8.4]

16. $4x^4 - 7x^2 - 2 = 0$

17. $(2t + 1)^2 - 5(2t + 1) + 6 = 0$

18. $2t - 7\sqrt{t} + 3 = 0$

19. Projectile Motion An object is tossed into the air with an upward velocity of 14 feet per second from the top of a building h feet high. The time it takes for the object to hit the ground below is given by the formula $16t^2 - 14t - h = 0$. Solve this formula for t. [8.4]

Sketch the graph of each of the following. Give the coordinates of the vertex in each case. [8.5]

20. $y = x^2 - 2x - 3$

21. $y = -x^2 + 2x + 8$

22. Find an equation in the form $y = a(x - h)^2 + k$ that describes the graph of the parabola. [8.5]

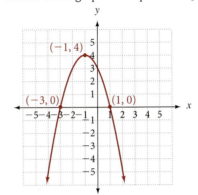

23. Profit Find the maximum weekly profit for a company with weekly costs of $C = 5x + 100$ and weekly revenue of $R = 25x - 0.1x^2$. [8.5]

Solve the following inequalities and graph the solution on a number line. [8.6]

24. $x^2 - x - 6 \le 0$

25. $2x^2 + 5x > 3$

26. The graph of $y = x^3 + 2x^2 - x - 2$ is shown below. Use the graph to write the solution set for each of the following. [8.6]

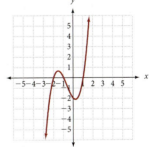

a. $x^3 + 2x^2 - x - 2 < 0$

b. $x^3 + 2x^2 - x - 2 > 0$

c. $x^3 + 2x^2 - x - 2 = 0$

QUADRATIC FUNCTIONS

Group Project

Maximum Volume of a Box

Number of People 5

Time Needed 30 minutes

Equipment Graphing calculator and five pieces of graph paper

Background For many people, having a concrete model to work with allows them to visualize situations that they would have difficulty with if they had only a written description to work with. The purpose of this project is to rework a problem we have worked previously but this time with a concrete model.

Procedure You are going to make boxes of varying dimensions from rectangles that are 11 centimeters wide and 17 centimeters long.

1. Cut a rectangle from your graph paper that is 11 squares for the width and 17 squares for the length. Pretend that each small square is 1 centimeter by 1 centimeter. Do this with five pieces of paper.

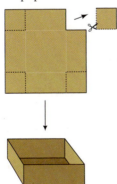

2. One person cuts off one square from each corner of their paper, then folds up the sides to form a box. Write down the length, width, and height of this box. Then calculate its volume.

3. The next person cuts a square that is two units on a side from each corner of their paper, then folds up the sides to form a box. Write down the length, width, and height of this box. Then calculate its volume.

4. The next person follows the same procedure, cutting a still larger square from each corner of their paper. This continues until the squares that are to be cut off are larger than the original piece of paper.

TABLE 1				
VOLUME OF A BOX				
Side of Square x (cm)	Length of Box L	Width of Box W	Height of Box H	Volume of Box V
1				
2				
3				
4				
5				

5. Enter the data from each box you have created into the table above. Then graph all the points (x, V) from the table.

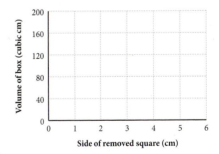

6. Using what you have learned from filling in the table, write a formula for the volume of the box that is created when a square of side x is cut from each corner of the original piece of paper.

7. Graph the equation you found in Step 6 on a graphing calculator. Use the result to connect the points you plotted in the graph.

8. Use the graphing calculator to find the value of x that will give the maximum volume of the box. Your answer should be accurate to the nearest hundredth.

Arlie O. Petters

It can seem at times as if all the mathematicians of note lived 100 or more years ago. However, that is not the case. There are mathematicians doing research today, who are discovering new mathematical ideas and extending what is known about mathematics. One of the current group of research mathematicians is Arlie O. Petters. Dr. Petters earned his Ph.D. from the Massachusetts Institute of Technology in 1991 and currently is working in the mathematics department at Duke University. Use the Internet to find out more about Dr. Petters' education, awards, and research interests. Then use your results to give a profile, in essay form, of a present-day working mathematician.

Arlie O. Petters, Massachusetts Institute of Technology

Exponential and Logarithmic Functions

The Metropolitan Museum of Art in New York City, often called "The Met," is one of the largest art galleries in the world. The museum curates more than two million pieces of art in their permanent collection. The collection includes pieces that span centuries and represent every corner of the earth, such as a set of flints from the Lower Paleolithic period (300,000 BC–75,000 BC), an ancient Egyptian temple made of sandstone from 15 BC (shown below), Chinese calligraphy and painting works from the 19th century, and even modern art paintings by Pablo Picasso and Jackson Pollock from the 20th century. The remarkable works of art draw more than 4 million visitors each year.

The art in the Metropolitan Museum of Art is considered priceless. But let's say you have acquired a painting that a family member had purchased fifty years ago for $200. If you know the painting's value has doubled every 5 years, how much would the painting be worth now? To answer this question, we would need to set up the following *exponential function*:

$$V(t) = 200 \cdot 2^{t/5} \text{ for } t > 0$$

where t is the number of years since the painting was purchased and $V(t)$ is its value (in dollars) at that time. Notice our unknown variable is an exponent, which is why this function is called an *exponential function*. We will begin our work in this chapter working with this type of function.

sepavo/Shutterstock.com

Preview

Key Words	Definition
Exponential Function	Any function that can be written in the form $f(x) = b^x$
One-to-One Function	Any function where every element in the range comes from exactly one element in the domain
Common Logarithm	A logarithm with a base of 10
Natural Logarithm	A logarithm with a base of e

Chapter Outline

Exponential Functions

TICKET TO SUCCESS

Keep these questions in mind as you read through the section. Then respond in your own words and in complete sentences.

1. What is an exponential function?
2. In an exponential function, explain why the base b cannot equal 1.
3. Explain continuously compounded interest.
4. What is the special number e in an exponential function?

Yehuda Boltshauser/Shutterstock.com

Imagine yourself visiting a miniature golf course with a friend. At one hole, you hit your golf ball up a ramp and through the front door of a castle. Your ball drops out of the castle's side from a 2-foot-high pipe, and bounces toward the hole. Your friend putts her ball, but it misses the ramp and falls into a gutter. The ball rolls out a pipe 1 foot above the ground and also bounces toward the hole, but not as far and with fewer bounces as yours.

Let's take a moment to picture the bounce height of each ball as it drops from its respective pipe. Suppose the height h of the first bounce reaches $\frac{1}{2}$ the pipe height. The second bounce reaches $\frac{1}{2}$ the height of the first bounce, and so on, as shown in Figure 1.

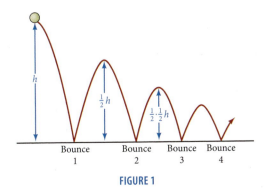

FIGURE 1

In other words, the maximum height of any bounce is $\frac{1}{2}$ the height of the previous bounce. The following table shows the bounce heights of the two golf balls.

TABLE 1		
	Your Ball	**Friend's Ball**
Initial height (feet)	$A = 2$	$A = 1$
Bounce 1	$h = \frac{1}{2}(2) = 1$	$h = \frac{1}{2}(1) = \frac{1}{2}$
Bounce 2	$h = \frac{1}{2}\left(\frac{1}{2}(2)\right) = 2\left(\frac{1}{2}\right)^2$	$h = \frac{1}{2}\left(\frac{1}{2}(1)\right) = \left(\frac{1}{2}\right)^2$
Bounce 3	$h = \frac{1}{2}\left(\frac{1}{2}\left(\frac{1}{2}(2)\right)\right) = 2\left(\frac{1}{2}\right)^3$	$h = \frac{1}{2}\left(\frac{1}{2}\left(\frac{1}{2}(1)\right)\right) = \left(\frac{1}{2}\right)^3$
Bounce 4	$h = \frac{1}{2}\left(\frac{1}{2}\left(\frac{1}{2}\left(\frac{1}{2}(2)\right)\right)\right) = 2\left(\frac{1}{2}\right)^4$	$h = \frac{1}{2}\left(\frac{1}{2}\left(\frac{1}{2}\left(\frac{1}{2}(1)\right)\right)\right) = \left(\frac{1}{2}\right)^4$
$\vdots$	$\vdots$	$\vdots$
Bounce n	$h = A\left(\frac{1}{2}\right)^n$	$h = A\left(\frac{1}{2}\right)^n$

The last equations are exponential in form. Functions in this form are the focus of this section. When considering a bouncing ball, we are considering the behavior of an exponential function in an intuitive way.

A Evaluating Exponential Functions

Let's begin with a formal definition that classifies all exponential functions.

> **Definition**
>
> An **exponential function** is any function that can be written in the form
> $$f(x) = b^x$$
> where b is a positive real number other than 1.

Each of the following is an exponential function:

$$f(x) = 2^x \qquad y = 3^x \qquad f(x) = \left(\frac{1}{4}\right)^x$$

The first step in becoming familiar with exponential functions is to find some values for specific exponential functions.

PRACTICE PROBLEMS

1. If $f(x) = 4^x$, find
  a. $f(0)$
  b. $f(1)$
  c. $f(2)$
  d. $f(3)$
  e. $f(-1)$
  f. $f(-2)$

EXAMPLE 1   If the exponential functions f and g are defined by

$$f(x) = 2^x \qquad \text{and} \qquad g(x) = 3^x \qquad \text{then}$$

$$f(0) = 2^0 = 1 \qquad\qquad g(0) = 3^0 = 1$$

$$f(1) = 2^1 = 2 \qquad\qquad g(1) = 3^1 = 3$$

$$f(2) = 2^2 = 4 \qquad\qquad g(2) = 3^2 = 9$$

$$f(3) = 2^3 = 8 \qquad\qquad g(3) = 3^3 = 27$$

$$f(-2) = 2^{-2} = \frac{1}{2^2} = \frac{1}{4} \qquad g(-2) = 3^{-2} = \frac{1}{3^2} = \frac{1}{9}$$

$$f(-3) = 2^{-3} = \frac{1}{2^3} = \frac{1}{8} \qquad g(-3) = 3^{-3} = \frac{1}{3^3} = \frac{1}{27}$$ ∎

Answers

1. a. 1 **b.** 4 **c.** 16 **d.** 64 **e.** $\frac{1}{4}$ **f.** $\frac{1}{16}$

Half-life is a term used in science and mathematics that quantifies the amount of decay or reduction in an element. The half-life of iodine-131 is 8 days, which means that every 8 days a sample of iodine-131 will decrease to half of its original amount. If we start with A_0 micrograms of iodine-131, then after t days the sample will contain

$$A(t) = A_0 \cdot 2^{-t/8}$$

micrograms of iodine-131.

EXAMPLE 2 A patient is administered a 1,200-microgram dose of iodine-131. How much iodine-131 will be in the patient's system after 10 days and after 16 days?

SOLUTION The initial amount of iodine-131 is $A_0 = 1,200$, so the function that gives the amount left in the patient's system after t days is

$$A(t) = 1,200 \cdot 2^{-t/8}$$

After 10 days, the amount left in the patient's system is

$$A(10) = 1,200 \cdot 2^{-10/8}$$

$$= 1,200 \cdot 2^{-1.25}$$

$$\approx 504.5 \text{ micrograms}$$

After 16 days, the amount left in the patient's system is

$$A(16) = 1,200 \cdot 2^{-16/8}$$

$$= 1,200 \cdot 2^{-2}$$

$$= 300 \text{ micrograms} \qquad \blacksquare$$

B Graphing Exponential Functions

We will now turn our attention to the graphs of exponential functions. Since the notation y is easier to use when graphing, and $y = f(x)$, for convenience we will write the exponential functions as

$$y = b^x$$

EXAMPLE 3 Sketch the graph of the exponential function $y = 2^x$.

SOLUTION Using the results of Example 1, we have the following table. Graphing the ordered pairs given in the table and connecting them with a smooth curve, we have the graph of $y = 2^x$ shown in Figure 2.

x	y
−3	$\frac{1}{8}$
−2	$\frac{1}{4}$
−1	$\frac{1}{2}$
0	1
1	2
2	4
3	8

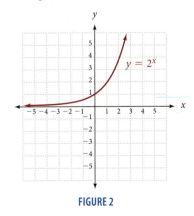

FIGURE 2

2. Referring to Example 2, how much iodine-131 will be in the patient's system after 12 days and after 24 days?

NOTE Recall that the symbol $\approx$ is read "is approximately equal to."

3. Graph $y = 3^x$.

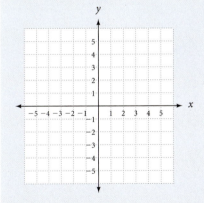

Answers
2. $A(12) \approx 424.3$ micrograms
$A(24) = 150$ micrograms
3. See Solutions Section.

Notice that the graph does not cross the *x*-axis. It *approaches* the *x*-axis—in fact, we can get it as close to the *x*-axis as we want without it actually intersecting the *x*-axis. For the graph of $y = 2^x$ to intersect the *x*-axis, we would have to find a value of *x* that would make $2^x = 0$. Because no such value of *x* exists, the graph of $y = 2^x$ cannot intersect the *x*-axis. ∎

4. Graph $y = \left(\frac{1}{2}\right)^x$.

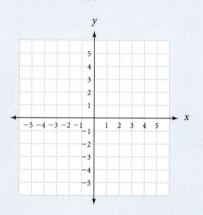

EXAMPLE 4 Sketch the graph of $y = \left(\frac{1}{3}\right)^x$.

SOLUTION The table shown here gives some ordered pairs that satisfy the equation. Using the ordered pairs from the table, we have the graph shown in Figure 3.

x	y
-3	27
-2	9
-1	3
0	1
1	$\frac{1}{3}$
2	$\frac{1}{9}$
3	$\frac{1}{27}$

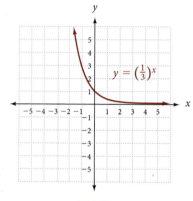

FIGURE 3 ∎

The graphs of all exponential functions have two things in common: each crosses the *y*-axis at (0, 1) since $b^0 = 1$; and none can cross the *x*-axis since $b^x = 0$ is impossible because of the restrictions on *b*.

Figures 4 and 5 show some families of exponential curves to help you become more familiar with them on an intuitive level.

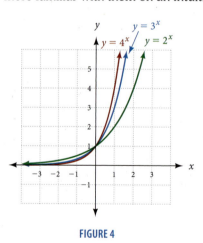

FIGURE 4

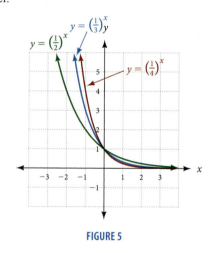

FIGURE 5

Among the many applications of exponential functions are the applications having to do with interest-bearing accounts. Here are the details.

C Applications

If *P* dollars are deposited in an account with annual interest rate *r*, compounded *n* times per year, then the amount of money in the account after *t* years is given by the formula

$$A(t) = P\left(1 + \frac{r}{n}\right)^{nt}$$

Answer

4. See Solutions Section.

5. Repeat Example 5 if $600 is deposited in an account that earns 6% annual interest, compounded monthly.

EXAMPLE 5   Suppose you deposit $500 in an account with an annual interest rate of 8% compounded quarterly. Find an equation that gives the amount of money in the account after t years. Then find

a. the amount of money in the account after 5 years.
b. the number of years it will take for the account to contain $1,000.

SOLUTION   First, we note that $P = 500$ and $r = 0.08$. Interest that is compounded quarterly is compounded 4 times a year, giving us $n = 4$. Substituting these numbers into the preceding formula, we have our function

$$A(t) = 500\left(1 + \frac{0.08}{4}\right)^{4t} = 500(1.02)^{4t}$$

a. To find the amount after 5 years, we let $t = 5$:

$$A(5) = 500(1.02)^{4 \cdot 5}$$

$$= 500(1.02)^{20}$$

$$\approx \$742.97$$

Our answer is found on a calculator and then rounded to the nearest cent.

b. To see how long it will take for this account to total $1,000, we graph the equation $Y_1 = 500(1.02)^{4X}$ on a graphing calculator and then look to see where it intersects the line $Y_2 = 1,000$. The two graphs are shown in Figure 6.

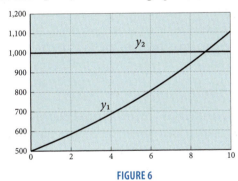

FIGURE 6

Using Zoom and Trace, or the Intersect function on the graphing calculator, we find that the two curves intersect at $X \approx 8.75$ and $Y = 1,000$. This means that our account will contain $1,000 after the money has been on deposit for 8.75 years. ∎

The Natural Exponential Function

A very commonly occurring exponential function is based on a special number we denote with the letter e. The number e is a number like π. It is irrational and occurs in many formulas that describe the world around us. Like π, it can be approximated with a decimal number. Whereas π is approximately 3.1416, e is approximately 2.7183. (If you have a calculator with a key labeled $\boxed{e^x}$, you can use it to find e^1 to find a more accurate approximation to e.) We cannot give a more precise definition of the number e without using some of the topics taught in calculus. For the work we are going to do with the number e, we only need to know that it is an irrational number that is approximately 2.7183.

Answers
5. a. $809.31   **b.** About 8.5 years

Here are a table and graph for the natural exponential function.

$$y = f(x) = e^x$$

x	$f(x) = e^x$
-2	$f(-2) = e^{-2} = \frac{1}{e^2} = 0.135$
-1	$f(-1) = e^{-1} = \frac{1}{e} \approx 0.368$
0	$f(0) = e^0 = 1$
1	$f(1) = e^1 = e \approx 2.72$
2	$f(2) = e^2 \approx 7.39$
3	$f(3) = e^3 \approx 20.09$

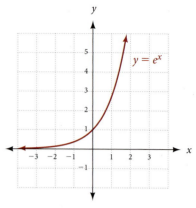

$y = e^x$

FIGURE 7

One common application of natural exponential functions is with interest-bearing accounts. In Example 5, we worked with the formula

$$A = P\left(1 + \frac{r}{n}\right)^{nt}$$

which gives the amount of money in an account if P dollars are deposited for t years at annual interest rate r, compounded n times per year. In Example 5, the number of compounding periods was 4. What would happen if we let the number of compounding periods become larger and larger, so that we compounded the interest every day, then every hour, then every second, and so on? If we take this as far as it can go, we end up compounding the interest every moment. When this happens, we have an account with interest that is compounded continuously, and the amount of money in such an account depends on the number e.

CONTINUOUSLY COMPOUNDED INTEREST If P dollars are deposited in an account with annual interest rate r, compounded continuously, then the amount of money in the account after t years is given by the formula

$$A(t) = Pe^{rt}$$

6. Suppose you deposit $600 in an account with an annual interest rate of 6% compounded continuously. How much money is in the account after 5 years?

EXAMPLE 6   Suppose you deposit $500 in an account with an annual interest rate of 8% compounded continuously. Find an equation that gives the amount of money in the account after t years. Then find the amount of money in the account after 5 years.

SOLUTION   Since the interest is compounded continuously, we use the formula $A(t) = Pe^{rt}$. Substituting $P = 500$ and $r = 0.08$ into this formula we have

$$A(t) = 500e^{0.08t}$$

After 5 years, this account will contain

$$A(5) = 500e^{0.08 \cdot 5}$$

$$= 500e^{0.4}$$

$$\approx \$745.91$$

to the nearest cent. Compare this result with the answer to Example 5a. ∎

Answer

6. $809.92

Problem Set 9.1

A Let $f(x) = 3^x$ and $g(x) = \left(\frac{1}{2}\right)^x$, and evaluate each of the following. [Examples 1, 2]

1. $g(0)$

2. $f(0)$

3. $g(-1)$

4. $g(-4)$

5. $f(-3)$

6. $f(-1)$

7. $f(2) + g(-2)$

8. $f(2) - g(-2)$

B Graph each of the following functions. [Examples 3, 4]

9. $y = 4^x$

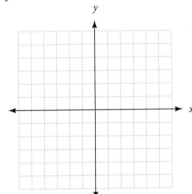

10. $y = 2^{-x}$

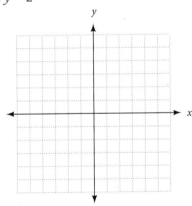

11. $y = 3^{-x}$

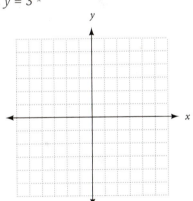

12. $y = \left(\frac{1}{3}\right)^{-x}$

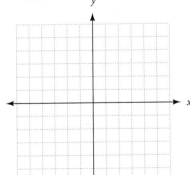

13. $y = 2^{x+1}$

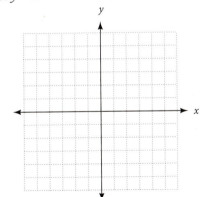

14. $y = 2^{x-3}$

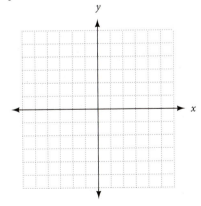

15. $y = e^x$

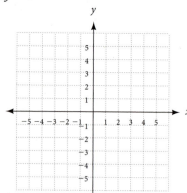

16. $y = e^{-x}$

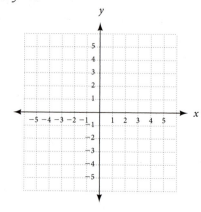

Graph each of the following functions on the same coordinate system for positive values of x only.

17. $y = 2x, y = x^2, y = 2^x$

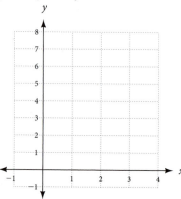

18. $y = 3x, y = x^3, y = 3^x$

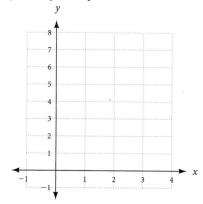

19. On a graphing calculator, graph the family of curves $y = b^x, b = 2, 4, 6, 8$.

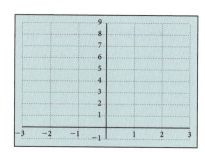

20. On a graphing calculator, graph the family of curves $y = b^x, b = \dfrac{1}{2}, \dfrac{1}{4}, \dfrac{1}{6}, \dfrac{1}{8}$.

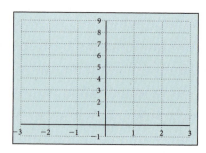

Applying the Concepts

21. Bouncing Ball A golf ball is manufactured so that if it is dropped from A feet above the ground onto a hard surface, the maximum height of each bounce will be $\frac{1}{2}$ of the height of the previous bounce. Find an exponential equation that gives the height h the ball will attain during the nth bounce. If the ball is dropped from 10 feet above the ground onto a hard surface, how high will it bounce on the 8th bounce?

22. Bouncing Ball Suppose a ball is dropped from a height of 6 feet above the ground. Each subsequent bounce is $\frac{2}{3}$ the height of the previous bounce, as shown in the illustration below. Find an exponential equation that gives the height h the ball will attain during the nth bounce. How high will it bounce on the fifth bounce?

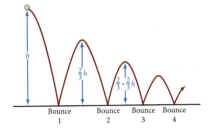

23. Exponential Decay The half-life of iodine-131 is 8 days. If a patient is administered a 1,400-microgram dose of iodine-131, how much iodine-131 will be in the patient's system after 8 days and after 11 days? (See Example 2.)

24. Exponential Decay The half-life of caffeine is about 6 hours. The amount of caffeine remaining in a person's system can be given by the formula, $C = C_0 \cdot 2^{-t/6}$, where C_0 is the amount of caffeine in the system initially and t is time in hours since ingestion. If a person drinks a cup of coffee containing 150 mg of caffeine at 8:00 AM, how much caffeine is in their body after 12 hours? At midnight?

25. Exponential Decay The half-life of ibuprofen is about 2 hours. The amount of ibuprofen remaining in a person's system can be given by the formula, $I = I_0 \cdot 2^{-t/2}$, where I_0 is the amount of ibuprofen in the system initially and t is time in hours since ingestion. If a person takes 600 mg of ibuprofen, how much ibuprofen is still in their body after 4 hours? After 8 hours?

26. Exponential Growth Automobiles built before 1993 use Freon in their air conditioners. The federal government now prohibits the manufacture of Freon. Because the supply of Freon is decreasing, the price per pound is increasing exponentially. Suppose estimates put the formula for the price per pound of Freon at $p(t) = 1.89(1.25)^t$, where t is the number of years since 1990. Find the price of Freon in 1995 and 1990. How much will Freon cost in the year 2014?

27. Compound Interest Suppose you deposit $1,200 in an account with an annual interest rate of 6% compounded quarterly. (See Example 5.)

  a. Find an equation that gives the amount of money in the account after t years.

  b. Find the amount of money in the account after 8 years.

  c. How many years will it take for the account to contain $2,400?

  d. If the interest were compounded continuously, how much money would the account contain after 8 years?

28. Compound Interest Suppose you deposit $500 in an account with an annual interest rate of 8% compounded monthly.

  a. Find an equation that gives the amount of money in the account after t years.

  b. Find the amount of money in the account after 5 years.

  c. How many years will it take for the account to contain $1,000?

  d. If the interest were compounded continuously, how much money would the account contain after 5 years?

29. Health Care In 1990, $699 billion were spent on health care expenditures. The amount of money, E, in billions spent on health care expenditures can be estimated using the function $E(t) = 78.16(1.11)^t$, where t is time in years since 1970 (U.S. Census Bureau).

  a. How close was the estimate determined by the function in estimating the actual amount of money spent on health care expenditures in 1990?

  b. What were the expected health care expenditures for 2008, 2009, and 2010? Round to the nearest billion.

30. Exponential Growth The cost of a can of Coca-Cola on January 1, 1960, was 10 cents. The function below gives the cost of a can of Coca-Cola t years after that.

$$C(t) = 0.10e^{0.0576t}$$

  a. Use the function to fill in the table below. (Round to the nearest cent.)

Years Since 1960 t	Cost $C(t)$
0	$0.10
15	
40	
50	
90	

  b. Use the table to find the cost of a can of Coca-Cola at the beginning of the year 2000.

  c. In what year will a can of Coca-Cola cost $17.84?

31. Value of a Painting A painting is purchased as an investment for $150. If the painting's value doubles every 3 years, then its value is given by the function

$$V(t) = 150 \cdot 2^{t/3} \quad \text{for } t \geq 0$$

where t is the number of years since it was purchased, and $V(t)$ is its value (in dollars) at that time. Graph this function.

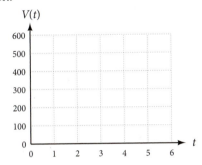

33. Value of a Painting When will the painting mentioned in Problem 31 be worth $600?

35. Value of a Crane The function

$$V(t) = 450{,}000(1 - 0.30)^t$$

where V is value and t is time in years, can be used to find the value of a crane for the first 6 years of use.

a. What is the value of the crane after 3 years and 6 months?

b. State the domain of this function.

c. Sketch the graph of this function.

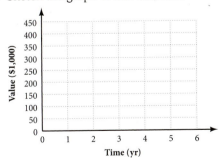

d. State the range of this function.

e. After how many years will the crane be worth only $85,000?

32. Value of a Painting A painting is purchased as an investment for $125. If the painting's value doubles every 5 years, then its value is given by the function

$$V(t) = 125 \cdot 2^{t/5} \quad \text{for } t \geq 0$$

where t is the number of years since it was purchased, and $V(t)$ is its value (in dollars) at that time. Graph this function.

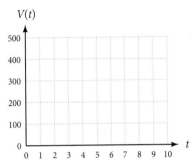

34. Value of a Painting When will the painting mentioned in Problem 32 be worth $250?

36. Value of a Printing Press The function $V(t) = 375{,}000(1 - 0.25)^t$, where V is value and t is time in years, can be used to find the value of a printing press during the first 7 years of use.

a. What is the value of the printing press after 4 years and 9 months?

b. State the domain of this function.

c. Sketch the graph of this function.

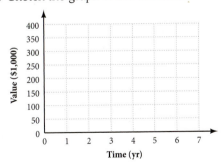

d. State the range of this function.

e. After how many years will the printing press be worth only $65,000?

Maintaining Your Skills

For each of the following relations, specify the domain and range, then indicate which are also functions.

37. $\{(1, 2), (3, 4), (4, 1)\}$

38. $\{(-2, 6), (-2, 8), (2, 3)\}$

State the domain for each of the following functions.

39. $y = \sqrt{3x + 1}$

40. $y = \dfrac{-4}{x^2 + 2x - 35}$

If $f(x) = 2x^2 - 18$ and $g(x) = 2x - 6$, find

41. $f(0)$

42. $g(f(0))$

43. $\dfrac{g(x + h) - g(x)}{h}$

44. $\dfrac{g(x)}{f(x)}$

Factor completely.

45. $2x^2 + 5x - 12$

46. $9x^2 - 33x - 12$

47. $5x^3 - 22x^2 + 8x$

48. $5x^3 - 3x^2 - 10x + 6$

49. $x^4 - 64x$

50. $5x^2 + 50x + 105$

Getting Ready for the Next Section

Solve each equation for y.

51. $x = 2y - 3$

52. $x = \dfrac{y + 7}{5}$

53. $x = y^2 - 2$

54. $x = (y + 4)^3$

55. $x = \dfrac{y - 4}{y - 2}$

56. $x = \dfrac{y + 5}{y - 3}$

57. $x = \sqrt{y - 3}$

58. $x = \sqrt{y} + 5$

The Inverse of a Function

TICKET TO SUCCESS

Keep these questions in mind as you read through the section. Then respond in your own words and in complete sentences.

1. What is the inverse of a function?
2. What is the relationship between the graph of a function and the graph of its inverse?
3. What is a one-to-one function?
4. In words, explain inverse function notation.

CREATISTA/Shutterstock.com

Suppose you are taking a train from Chicago, Illinois to Jacksonville, Florida. The route takes you from Chicago south to New Orleans, Louisiana, and then east to Jacksonville. On your return trip, the train heads west from Jacksonville toward New Orleans, and then north home to Chicago. The route you take to return home acts as the *inverse* of your original route because you reversed your direction. As you will see, the relationship between a function and its inverse function is similar to the relationship between your train route from Chicago to Jacksonville and your route home again.

A Finding the Inverse Relation

Suppose the function f is given by

$$f = \{(1, 4), (2, 5), (3, 6), (4, 7)\}$$

The inverse of f is obtained by reversing the order of the coordinates in each ordered pair in f. The inverse of f is the relation given by

$$g = \{(4, 1), (5, 2), (6, 3), (7, 4)\}$$

It is obvious that the domain of f is now the range of g, and the range of f is now the domain of g. Every function (or relation) has an inverse that is obtained from the original function by interchanging the components of each ordered pair.

PRACTICE PROBLEMS

1. If $f(x) = 4x + 1$, find the equation for the inverse of f.

Suppose a function f is defined with an equation instead of a list of ordered pairs. We can obtain the equation of the inverse of f by interchanging the role of x and y in the equation for f.

EXAMPLE 1 If the function f is defined by $f(x) = 2x - 3$, find the equation that represents the inverse of f.

SOLUTION Since the inverse of f is obtained by interchanging the components of all the ordered pairs belonging to f, and each ordered pair in f satisfies the equation $y = 2x - 3$, we simply exchange x and y in the equation $y = 2x - 3$ to get the formula for the inverse of f.

$$x = 2y - 3$$

We now solve this equation for y in terms of x.

$$x + 3 = 2y$$

$$\frac{x + 3}{2} = y$$

$$y = \frac{x + 3}{2}$$

The last line gives the equation that defines the inverse of f. Let's compare the graphs of f and its inverse as given above. (See Figure 1.)

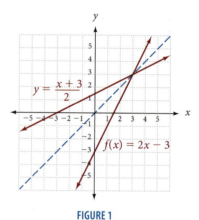

FIGURE 1

The graphs of f and its inverse have symmetry about the line $y = x$. This is a reasonable result since the one function was obtained from the other by interchanging x and y in the equation. The ordered pairs (a, b) and (b, a) always have symmetry about the line $y = x$. ∎

B Graph a Function with its Inverse

2. Graph $y = x^2 + 1$ and its inverse. Give the equation for the inverse.

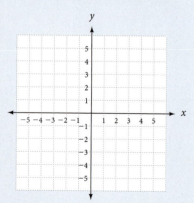

EXAMPLE 2 Graph the function $y = x^2 - 2$ and its inverse. Give the equation for the inverse.

SOLUTION We can obtain the graph of the inverse of $y = x^2 - 2$ by graphing $y = x^2 - 2$ by the usual methods and then reflecting the graph about the line $y = x$

Answers

1. $y = \frac{x - 1}{4}$

2. See Solutions Section.

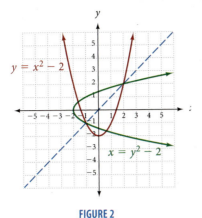

FIGURE 2

The equation that corresponds to the inverse of $y = x^2 - 2$ is obtained by interchanging x and y to get $x = y^2 - 2$.

We can solve the equation $x = y^2 - 2$ for y in terms of x as follows:

$$x = y^2 - 2$$

$$x + 2 = y^2$$

$$y = \pm\sqrt{x + 2}$$

∎

Comparing the graphs from Examples 1 and 2, we observe that the inverse of a function is not always a function. In Example 1, both f and its inverse have graphs that are nonvertical straight lines and therefore both represent functions. In Example 2, the inverse of function f is not a function since a vertical line crosses it in more than one place.

One-to-One Functions

We can distinguish between those functions with inverses that are also functions and those functions with inverses that are not functions with the following definition.

> **Definition**
>
> A function is a **one-to-one function** if every element in the range comes from exactly one element in the domain.

This definition indicates that a one-to-one function will yield a set of ordered pairs in which no two different ordered pairs have the same second coordinates. For example, the function

$$f = \{(2, 3), (-1, 3), (5, 8)\}$$

is not one-to-one because the element 3 in the range comes from both 2 and -1 in the domain. On the other hand, the function

$$g = \{(5, 7), (3, -1), (4, 2)\}$$

is a one-to-one function because every element in the range comes from only one element in the domain.

Horizontal Line Test

If we have the graph of a function, then we can determine if the function is one-to-one with the following test. If a horizontal line crosses the graph of a function in more than one place, then the function is not a one-to-one function because the points at which the horizontal line crosses the graph will be points with the same y-coordinates but different x-coordinates. Therefore, the function will have an element in the range (the y-coordinate) that comes from more than one element in the domain (the x-coordinates).

Of the functions we have covered previously, all the linear functions and exponential functions are one-to-one functions because no horizontal lines can be found that will cross their graphs in more than one place.

Functions Whose Inverses Are Also Functions

Because one-to-one functions do not repeat second coordinates, when we reverse the order of the ordered pairs in a one-to-one function, we obtain a relation in which no two ordered pairs have the same first coordinate—by definition, this relation must be a function. In other words, every one-to-one function has an inverse that is itself a function. Because of this, we can use function notation to represent that inverse.

> **Property**
>
> If $y = f(x)$ is a one-to-one function, then the inverse of f is also a function and can be denoted by $y = f^{-1}(x)$.

NOTE

The notation f^{-1} does not represent the reciprocal of f; that is, the -1 in this notation is not an exponent. The notation f^{-1} is defined as representing the inverse function for a one-to-one function.

To illustrate, in Example 1 we found the inverse of $f(x) = 2x - 3$ was the function $y = \frac{x + 3}{2}$. We can write this inverse function with inverse function notation as

$$f^{-1}(x) = \frac{x + 3}{2}$$

However, the inverse of the function in Example 2 is not itself a function, so we do not use the notation $f^{-1}(x)$ to represent it.

3. Find the inverse of
$$g(x) = \frac{x - 3}{x + 1}.$$

EXAMPLE 3 Find the inverse of $g(x) = \frac{x - 4}{x - 2}$.

SOLUTION To find the inverse for g, we begin by replacing $g(x)$ with y to obtain

$$y = \frac{x - 4}{x - 2} \qquad \text{Original function}$$

To find an equation for the inverse, we exchange x and y.

$$x = \frac{y - 4}{y - 2} \qquad \text{Inverse of original function}$$

To solve for y, we first multiply each side by $y - 2$ to obtain

$$x(y - 2) = y - 4$$

$$xy - 2x = y - 4 \qquad \text{Distributive property}$$

$$xy - y = 2x - 4 \qquad \text{Collect all terms containing } y \text{ on the left side.}$$

$$y(x - 1) = 2x - 4 \qquad \text{Factor } y \text{ from each term on the left side.}$$

$$y = \frac{2x - 4}{x - 1} \qquad \text{Divide each side by } x - 1.$$

Answer

3. $y = \frac{-x - 3}{x - 1} = \frac{3 + x}{1 - x}$

Because our original function is one-to-one, as verified by the graph in Figure 3, its inverse is also a function. Therefore, we can use inverse function notation to write

$$g^{-1}(x) = \frac{2x - 4}{x - 1}$$

NOTE
Asymptotes will be covered in depth more in the next chapter.

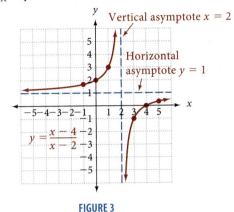

FIGURE 3

EXAMPLE 4   Graph the function $y = 2^x$ and its inverse $x = 2^y$.

SOLUTION   We graphed $y = 2^x$ in the preceding section. We simply reflect its graph about the line $y = x$ to obtain the graph of its inverse $x = 2^y$. (See Figure 4.)

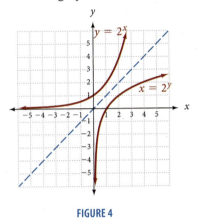

FIGURE 4

As you can see from the graph, $x = 2^y$ is a function. However, we do not have the mathematical tools to solve this equation for y. Therefore, we are unable to use the inverse function notation to represent this function. In the next section, we will give a definition that solves this problem. For now, we simply leave the equation as $x = 2^y$.

Functions, Relations, and Inverses—A Summary

Here is a summary of some of the things we know about functions, relations, and their inverses:

1. Every function is a relation, but not every relation is a function.

2. Every function has an inverse, but only one-to-one functions have inverses that are also functions.

3. The domain of a function is the range of its inverse, and the range of a function is the domain of its inverse.

4. If $y = f(x)$ is a one-to-one function, then we can use the notation $y = f^{-1}(x)$ to represent its inverse function.

5. The graph of a function and its inverse have symmetry about the line $y = x$.

6. If (a, b) belongs to the function f, then the point (b, a) belongs to its inverse.

4. Graph the function $y = 3^x$ and its inverse $x = 3^y$.

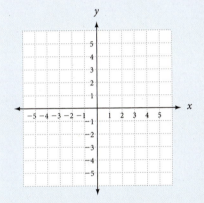

Answer
4. See Solutions Section.

Problem Set 9.2

Moving Toward Success

"The beautiful thing about learning is nobody can take it away from you."

   —B.B. King, 1925–present, American Blues guitarist and songwriter

1. Why will increasing the effectiveness of the time you spend learning help you?

2. How will keeping the same studying schedule you had at the beginning of this class benefit you?

A   For each of the following one-to-one functions, find the equation of the inverse. Write the inverse using the notation $f^{-1}(x)$.
[Examples 1, 3]

1. $f(x) = 3x - 1$

2. $f(x) = 2x - 5$

3. $f(x) = x^3$

4. $f(x) = x^3 - 2$

5. $f(x) = \dfrac{x - 3}{x - 1}$

6. $f(x) = \dfrac{x - 2}{x - 3}$

7. $f(x) = \dfrac{x - 3}{4}$

8. $f(x) = \dfrac{x + 7}{2}$

9. $f(x) = \dfrac{1}{2}x - 3$

10. $f(x) = \dfrac{1}{3}x + 1$

11. $f(x) = \dfrac{2x + 1}{3x + 1}$

12. $f(x) = \dfrac{3x + 2}{5x + 1}$

B   For each of the following relations, sketch the graph of the relation and its inverse, and write an equation for the inverse.
[Examples 2, 4]

13. $y = 2x - 1$

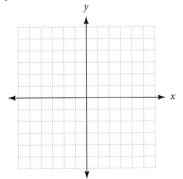

14. $y = 3x + 1$

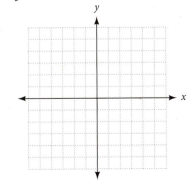

15. $y = x^2 - 3$

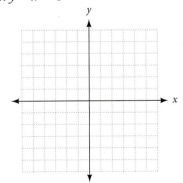

16. $y = x^2 + 1$

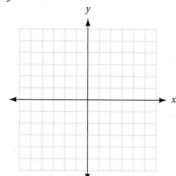

17. $y = x^2 - 2x - 3$

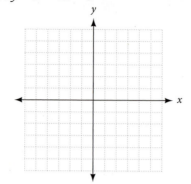

18. $y = x^2 + 2x - 3$

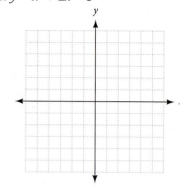

19. $y = 3^x$

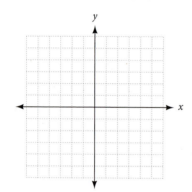

20. $y = \left(\dfrac{1}{2}\right)^x$

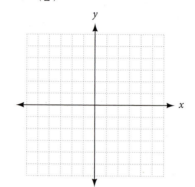

21. $y = 4$

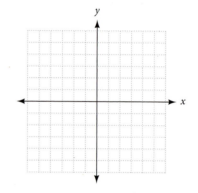

22. $y = -2$

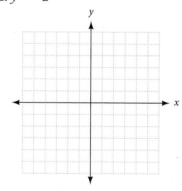

23. $y = \dfrac{1}{2}x^3$

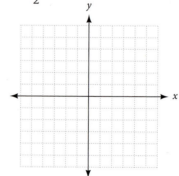

24. $y = x^3 - 2$

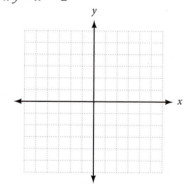

25. $y = \dfrac{1}{2}x + 2$

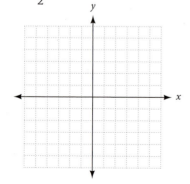

26. $y = \dfrac{1}{3}x - 1$

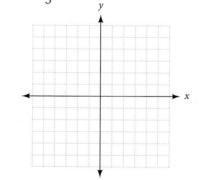

27. $y = \sqrt{x + 2}$

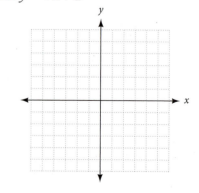

28. $y = \sqrt{x} + 2$

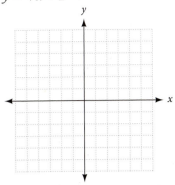

29. Determine if the following functions are one-to-one.

a.

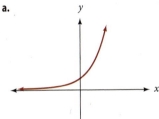

b.

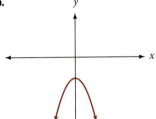

c.

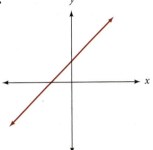

30. Could the following tables of values represent ordered pairs from one-to-one functions?

a.

x	y
−2	5
−1	4
0	3
1	4
2	5

b.

x	y
1.5	0.1
2.0	0.2
2.5	0.3
3.0	0.4
3.5	0.5

31. If $f(x) = 3x - 2$, then $f^{-1}(x) = \frac{x+2}{3}$. Use these two functions to find

a. $f(2)$

b. $f^{-1}(2)$

c. $f(f^{-1}(2))$

d. $f^{-1}(f(2))$

32. If $f(x) = \frac{1}{2}x + 5$, then $f^{-1}(x) = 2x - 10$. Use these two functions to find

a. $f(-4)$

b. $f^{-1}(-4)$

c. $f(f^{-1}(-4))$

d. $f^{-1}(f(-4))$

33. Let $f(x) = \frac{1}{x}$, and find $f^{-1}(x)$.

34. Let $f(x) = \frac{a}{x}$, and find $f^{-1}(x)$. (a is a real number constant.)

Maintaining Your Skills

Solve each equation.

35. $(2x - 1)^2 = 25$

36. $(3x + 5)^2 = -12$

37. What number would you add to $x^2 - 10x$ to make it a perfect square trinomial?

38. What number would you add to $x^2 - 5x$ to make it a perfect square trinomial?

Solve by completing the square.

39. $x^2 - 10x + 8 = 0$

40. $x^2 - 5x + 4 = 0$

41. $3x^2 - 6x + 6 = 0$

42. $4x^2 - 16x - 8 = 0$

Getting Ready for the Next Section

Simplify.

43. 3^{-2}

44. 2^3

Solve.

45. $2 = 3x$

46. $3 = 5x$

47. $4 = x^3$

48. $12 = x^2$

Fill in the blanks to make each statement true.

49. $8 = 2^{\square}$

50. $27 = 3^{\square}$

51. $10{,}000 = 10^{\square}$

52. $1{,}000 = 10^{\square}$

53. $81 = 3^{\square}$

54. $81 = 9^{\square}$

55. $6 = 6^{\square}$

56. $1 = 5^{\square}$

Extending the Concepts

For each of the following functions, find $f^{-1}(x)$. Then show that $f(f^{-1}(x)) = x$.

57. $f(x) = 3x + 5$

58. $f(x) = 6 - 8x$

59. $f(x) = x^3 + 1$

60. $f(x) = x^3 - 8$

61. $f(x) = \dfrac{x - 4}{x - 2}$

62. $f(x) = \dfrac{x - 3}{x - 1}$

Logarithms Are Exponents

OBJECTIVES

A Convert between logarithmic form and exponential form.

B Use the definition of logarithms to solve simple logarithmic equations.

C Sketch the graph of a logarithmic function.

D Simplify expressions involving logarithms.

TICKET TO SUCCESS

Keep these questions in mind as you read through the section. Then respond in your own words and in complete sentences.

1. What is a logarithm?

2. What is the relationship between $y = 2^x$ and $y = \log_2 x$? How are their graphs related?

3. Will the graph of $y = \log_b x$ ever appear in the second or third quadrants? Explain.

4. How would you use the identity $b^{\log_b x} = x$ to simplify an equation involving a logarithm?

JustASC/Shutterstock.com

In January 2010, news organizations reported that an earthquake had rocked the tiny island nation of Haiti and caused massive destruction. They reported the strength of the earthquake as having measured 7.0 on the Richter scale. For comparison, Table 1 gives the Richter magnitude of a number of other earthquakes.

TABLE 1		
EARTHQUAKES		
Year	**Earthquake**	**Richter Magnitude**
1971	Los Angeles	6.6
1985	Mexico City	8.1
1989	San Francisco	7.1
1992	Kobe, Japan	7.2
1994	Northridge	6.6
1999	Armenia, Colombia	6.0
2004	Indian Ocean	9.1
2010	Haiti	7.0
Source:earthquake.usgs.gov		

Although the sizes of the numbers in the table do not seem to be very different, the intensity of the earthquakes they measure can be very different. For example, the 1989 San Francisco earthquake was more than 10 times stronger than the 1999 earthquake in Colombia. The reason behind this is that the Richter scale is a *logarithmic scale*. In this section, we start our work with logarithms, which will give you an understanding of the Richter scale. Let's begin.

A   Logarithmic Functions

As you know from your work in the previous sections, equations of the form

$$y = b^x \qquad b > 0, b \neq 1$$

are called exponential functions. Because the equation of the inverse of a function can be obtained by exchanging x and y in the equation of the original function, the inverse of an exponential function must have the form

$$x = b^y \qquad b > 0, b \neq 1$$

Now, this last equation is actually the equation of a logarithmic function, as the following definition indicates:

Definition

The equation $y = \log_b x$ is read "y is the logarithm to the base b of x" and is equivalent to the equation

$$x = b^y \qquad b > 0, b \neq 1$$

In words: y is the number we raise b to in order to get x.

NOTATION   When an equation is in the form $x = b^y$, it is said to be in exponential form. On the other hand, if an equation is in the form $y = \log_b x$, it is said to be in logarithmic form.

    Here are some equivalent statements written in both forms.

Exponential Form		Logarithmic Form
$8 = 2^3$	⇔	$\log_2 8 = 3$
$25 = 5^2$	⇔	$\log_5 25 = 2$
$0.1 = 10^{-1}$	⇔	$\log_{10} 0.1 = -1$
$\frac{1}{8} = 2^{-3}$	⇔	$\log_2 \frac{1}{8} = -3$
$r = z^s$	⇔	$\log_z r = s$

B   Logarithmic Equations

EXAMPLE 1   Solve $\log_3 x = -2$ for x.

SOLUTION   In exponential form the equation looks like this:

$$x = 3^{-2}$$

$$x = \frac{1}{9}$$

The solution is $\frac{1}{9}$.   ■

NOTE
The ability to change from logarithmic form to exponential form is the first important thing to know about logarithms.

PRACTICE PROBLEMS

1. Solve for x: $\log_2 x = 3$.

Answer
1. 8

EXAMPLE 2 Solve $\log_x 4 = 3$.

SOLUTION Again, we use the definition of logarithms to write the equation in exponential form.

$$4 = x^3$$

Taking the cube root of both sides, we have

$$\sqrt[3]{4} = \sqrt[3]{x^3}$$

$$x = \sqrt[3]{4}$$

The solution set is $\{\sqrt[3]{4}\}$. ∎

EXAMPLE 3 Solve $\log_8 4 = x$.

SOLUTION We write the equation again in exponential form.

$$4 = 8^x$$

Since both 4 and 8 can be written as powers of 2, we write them in terms of powers of 2.

$$2^2 = (2^3)^x$$

$$2^2 = 2^{3x}$$

The only way the left and right sides of this last line can be equal is if the exponents are equal—that is, if

$$2 = 3x$$

$$\text{or}\quad x = \frac{2}{3}$$

The solution is $\frac{2}{3}$. We check as follows:

$$\log_8 4 = \frac{2}{3} \;\Leftrightarrow\; 4 = 8^{2/3}$$

$$4 = (\sqrt[3]{8})^2$$

$$4 = 2^2$$

$$4 = 4$$

The solution checks when used in the original equation. ∎

C Graphing Logarithmic Functions

Graphing logarithmic functions can be done using the graphs of exponential functions and the fact that the graphs of inverse functions have symmetry about the line $y = x$. Here's an example to illustrate.

EXAMPLE 4 Graph the equation $y = \log_2 x$.

SOLUTION The equation $y = \log_2 x$ is, by definition, equivalent to the exponential equation

$$x = 2^y$$

which is the equation of the inverse of the function

$$y = 2^x$$

2. Solve $\log_x 5 = 2$.

3. Solve $\log_9 27 = x$.

NOTE
The first step in each of these first three examples is the same. In each case the first step in solving the equation is to put the equation in exponential form.

4. Graph the equation
$$y = \log_3 x.$$

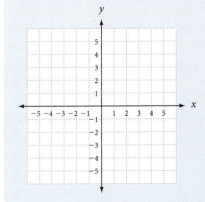

Answers
2. $\sqrt{5}$

3. $\frac{3}{2}$

4. See Solutions Section.

NOTE
From the graph of $y = \log_2 x$ it is apparent that $x > 0$ since the graph does not appear to the left of the y-axis. In general, the only variable in the expression $y = \log_b x$ that can be negative is y.

We simply reflect the graph of $y = 2^x$ about the line $y = x$ to get the graph of $x = 2^y$, which is also the graph of $y = \log_2 x$. (See Figure 1.)

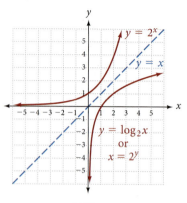

FIGURE 1

It is apparent from the graph that $y = \log_2 x$ is a function since no vertical line will cross its graph in more than one place. The same is true for all logarithmic equations of the form $y = \log_b x$, where b is a positive number other than 1. Note also that the graph of $y = \log_b x$ will always appear to the right of the y-axis, meaning that x will always be positive in the equation $y = \log_b x$. ■

USING TECHNOLOGY

Graphing Logarithmic Functions

As demonstrated in Example 4, we can graph the logarithmic function $y = \log_2 x$ as the inverse of the exponential function $y = 2^x$. Your graphing calculator most likely has a command to do this. First, define the exponential function as $Y_1 = 2^x$. To see the line of symmetry, define a second function $Y_2 = x$. Set the window variables so that

$$-6 \le x \le 6; \; -6 \le y \le 6$$

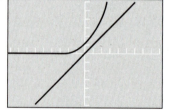

FIGURE 2

and use your zoom-square command to graph both functions. Your graph should look similar to the one shown in Figure 2. Now use the appropriate command to graph the inverse of the exponential function defined as Y_1 (Figure 3).

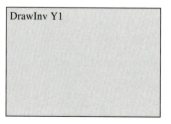

FIGURE 3

D Two Special Identities

If b is a positive real number other than 1, then each of the following is a consequence of the definition of a logarithm:

$$(1)\ b^{\log_b x} = x \quad \text{and} \quad (2)\ \log_b b^x = x$$

The justifications for these identities are similar. Let's consider only the first one. Consider the equation

$$y = \log_b x$$

By definition, it is equivalent to

$$x = b^y$$

Substituting $\log_b x$ for y in the last line gives us

$$x = b^{\log_b x}$$

The next examples in this section show how these two special properties can be used to simplify equations involving logarithms.

EXAMPLE 5   Simplify $\log_2 8$.

SOLUTION   Substitute 2^3 for 8.

$$\log_2 8 = \log_2 2^3$$
$$= 3 \qquad \blacksquare$$

EXAMPLE 6   Simplify $\log_{10} 10{,}000$.

SOLUTION   10,000 can be written as 10^4.

$$\log_{10} 10{,}000 = \log_{10} 10^4$$
$$= 4 \qquad \blacksquare$$

EXAMPLE 7   Simplify $\log_b b$ $(b > 0, b \ne 1)$.

SOLUTION   Since $b^1 = b$, we have

$$\log_b b = \log_b b^1$$
$$= 1 \qquad \blacksquare$$

EXAMPLE 8   Simplify $\log_b 1$ $(b > 0, b \ne 1)$.

SOLUTION   Since $1 = b^0$, we have

$$\log_b 1 = \log_b b^0$$
$$= 0 \qquad \blacksquare$$

EXAMPLE 9   Simplify $\log_4 (\log_5 5)$.

SOLUTION   Since $\log_5 5 = 1$,

$$\log_4 (\log_5 5) = \log_4 1$$
$$= 0 \qquad \blacksquare$$

5. Simplify $\log_3 27$.

6. Simplify $\log_{10} 1{,}000$.

7. Simplify $\log_6 6$.

8. Simplify $\log_3 1$.

9. Simplify $\log_2 (\log_8 8)$.

Answers
5. 3
6. 3
7. 1
8. 0
9. 0

Application

As we mentioned in the introduction to this section, one application of logarithms is in measuring the magnitude of an earthquake. If an earthquake has a shock wave T times greater than the smallest shock wave that can be measured on a seismograph, then the magnitude M of the earthquake, as measured on the Richter scale, is given by the formula

$$M = \log_{10} T$$

(When we talk about the size of a shock wave, we are talking about its amplitude. The amplitude of a wave is half the difference between its highest point and its lowest point, as shown in Figure 4.)

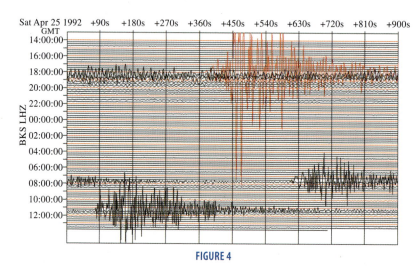

FIGURE 4

To illustrate the discussion, an earthquake that produces a shock wave that is 10,000 times greater than the smallest shock wave measurable on a seismograph will have a magnitude M on the Richter scale of

$$M = \log_{10} 10{,}000 = 4$$

10. Repeat Example 10 if $M = 6$.

EXAMPLE 10   If an earthquake has a magnitude of $M = 5$ on the Richter scale, what can you say about the size of its shock wave?

SOLUTION   To answer this question, we put $M = 5$ into the formula $M = \log_{10} T$ to obtain

$$5 = \log_{10} T$$

Writing this equation in exponential form, we have

$$T = 10^5 = 100{,}000$$

We can say that an earthquake that measures 5 on the Richter scale has a shock wave 100,000 times greater than the smallest shock wave measurable on a seismograph.

∎

From Example 10 and the discussion that preceded it, we find that an earthquake of magnitude 5 has a shock wave that is 10 times greater than an earthquake of magnitude 4 because 100,000 is 10 times 10,000.

Answer

10. $T = 1{,}000{,}000$

Problem Set 9.3

Moving Toward Success

"Just as iron rusts from disuse, even so does inaction spoil the intellect."

—Leonardo da Vinci, 1452–1519, Italian painter and sculptor

1. Why is it a poor choice to just come to class and not read the textbook?
2. Why is it a poor choice to rarely come to class and only read the textbook?

A Write each of the following equations in logarithmic form.

1. $2^4 = 16$

2. $3^2 = 9$

3. $125 = 5^3$

4. $16 = 4^2$

5. $0.01 = 10^{-2}$

6. $0.001 = 10^{-3}$

7. $2^{-5} = \dfrac{1}{32}$

8. $4^{-2} = \dfrac{1}{16}$

9. $\left(\dfrac{1}{2}\right)^{-3} = 8$

10. $\left(\dfrac{1}{3}\right)^{-2} = 9$

11. $27 = 3^3$

12. $81 = 3^4$

A Write each of the following equations in exponential form.

13. $\log_{10} 100 = 2$

14. $\log_2 8 = 3$

15. $\log_2 64 = 6$

16. $\log_2 32 = 5$

17. $\log_8 1 = 0$

18. $\log_9 9 = 1$

19. $\log_{10} 0.001 = -3$

20. $\log_{10} 0.0001 = -4$

21. $\log_6 36 = 2$

22. $\log_7 49 = 2$

23. $\log_5 \dfrac{1}{25} = -2$

24. $\log_3 \dfrac{1}{81} = -4$

B Solve each of the following equations for x. [Examples 1–3]

25. $\log_3 x = 2$

26. $\log_4 x = 3$

27. $\log_5 x = -3$

28. $\log_2 x = -4$

29. $\log_2 16 = x$

30. $\log_3 27 = x$

31. $\log_8 2 = x$

32. $\log_{25} 5 = x$

33. $\log_x 4 = 2$

34. $\log_x 16 = 4$

35. $\log_x 5 = 3$

36. $\log_x 8 = 2$

C Sketch the graph of each of the following logarithmic equations. [Example 4]

37. $y = \log_3 x$

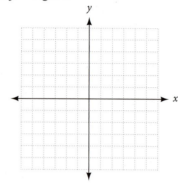

38. $y = \log_{1/2} x$

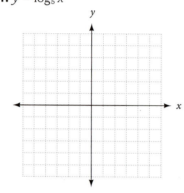

39. $y = \log_{1/3} x$

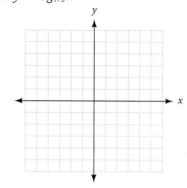

40. $y = \log_4 x$

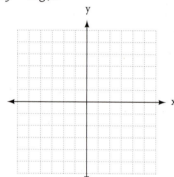

41. $y = \log_5 x$

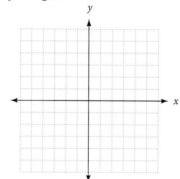

42. $y = \log_{1/5} x$

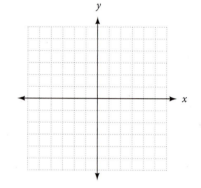

43. $y = \log_{10} x$

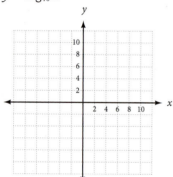

44. $y = \log_{1/4} x$

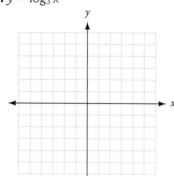

D Simplify each of the following. [Example 5–9]

45. $\log_2 16$

46. $\log_3 9$

47. $\log_{25} 125$

48. $\log_9 27$

49. $\log_{10} 1{,}000$

50. $\log_{10} 10{,}000$

51. $\log_3 3$

52. $\log_4 4$

53. $\log_5 1$

54. $\log_{10} 1$

55. $\log_3(\log_6 6)$

56. $\log_5(\log_3 3)$

57. $\log_4[\log_2(\log_2 16)]$

58. $\log_4[\log_3(\log_2 8)]$

Applying the Concepts

$pH = -\log_{10}[H+]$, where [H+] is the concentration of the hydrogen ions in solution. An acid solution has a pH below 7, and a basic solution has a pH higher than 7.

59. In distilled water, the concentration of hydrogen ions is $[H^+] = 10^{-7}$. What is the pH?

60. Find the pH of a bottle of vinegar if the concentration of hydrogen ions is $[H^+] = 10^{-3}$.

61. A hair conditioner has a pH of 6. Find the concentration of hydrogen ions, $[H^+]$, in the conditioner.

62. If a glass of orange juice has a pH of 4, what is the concentration of hydrogen ions, $[H^+]$, in the juice?

63. Magnitude of an Earthquake Find the magnitude M of an earthquake with a shock wave that measures $T = 100$ on a seismograph.

64. Magnitude of an Earthquake Find the magnitude M of an earthquake with a shock wave that measures $T = 100,000$ on a seismograph.

65. Shock Wave If an earthquake has a magnitude of 8 on the Richter scale, how many times greater is its shock wave than the smallest shock wave measurable on a seismograph?

66. Shock Wave Use Table 1 at the beginning of the section to find how many times greater the shock wave for the 2010 Haiti earthquake is than the smallest shock wave measurable.

Maintaining Your Skills

Fill in the blanks to complete the square.

67. $x^2 + 10x + \boxed{} = (x + \boxed{})^2$

68. $x^2 + 4x + \boxed{} = (x + \boxed{})^2$

69. $y^2 - 2y + \boxed{} = (y - \boxed{})^2$

70. $y^2 + 3y + \boxed{} = (y + \boxed{})^2$

Solve.

71. $-y^2 = 9$

72. $7 + y^2 = 11$

73. $-x^2 - 8 = -4$

74. $10x^2 = 100$

75. $2x^2 + 4x - 3 = 0$

76. $3x^2 + 4x - 2 = 0$

77. $(2y - 3)(2y - 1) = -4$

78. $(y - 1)(3y - 3) = 10$

79. $t^3 - 125 = 0$

80. $8t^3 + 1 = 0$

81. $4x^5 - 16x^4 = 20x^3$

82. $3x^4 + 6x^2 = 6x^3$

83. $\dfrac{1}{x - 3} + \dfrac{1}{x + 2} = 1$

84. $\dfrac{1}{x + 3} + \dfrac{1}{x - 2} = 1$

Getting Ready for the Next Section

Simplify.

85. $8^{2/3}$

86. $27^{2/3}$

Solve.

87. $(x + 2)(x) = 2^3$

88. $(x + 3)(x) = 2^2$

89. $\dfrac{x - 2}{x + 1} = 9$

90. $\dfrac{x + 1}{x - 4} = 25$

Write in exponential form.

91. $\log_2 [(x + 2)(x)] = 3$

92. $\log_4 [x(x - 6)] = 2$

93. $\log_3 \left(\dfrac{x - 2}{x + 1} \right) = 4$

94. $\log_5 \left(\dfrac{x - 1}{x - 4} \right) = 2$

Extending the Concepts

95. The graph of the exponential function $y = f(x) = b^x$ is shown here. Use the graph to complete parts a through d.

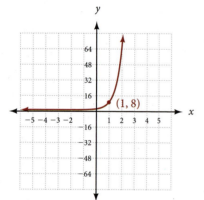

96. The graph of the exponential function $y = f(x) = b^x$ is shown here. Use the graph to complete parts a through d.

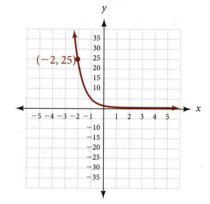

a. Fill in the table.

x	f(x)
−1	
0	
1	
2	

b. Fill in the table.

x	f⁻¹(x)
	−1
	0
	1
	2

c. Find the equation for $f(x)$.

d. Find the equation for $f^{-1}(x)$.

a. Fill in the table.

x	f(x)
−1	
0	
1	
2	

b. Fill in the table.

x	f⁻¹(x)
	−1
	0
	1
	2

c. Find the equation for $f(x)$.

d. Find the equation for $f^{-1}(x)$.

Properties of Logarithms

OBJECTIVES

A Use the properties of logarithms to convert between expanded form and single logarithms.

B Use the properties of logarithms to solve equations that contain logarithms.

TICKET TO SUCCESS

Keep these questions in mind as you read through the section. Then respond in your own words and in complete sentences.

1. Explain why the following statement is false: "The logarithm of a product is the product of the logarithms."

2. In words, explain the quotient property of logarithms.

3. Explain the difference between $\log_b m + \log_b n$ and $\log_b(m + n)$. Are they equivalent?

4. Explain the difference between $\log_b(mn)$ and $(\log_b m)(\log_b n)$. Are they equivalent?

Andy Dean Photography/Shutterstock.com

A *decibel* is a measurement of sound intensity, more specifically, the difference in intensity between two sounds. This difference is equal to ten times the logarithm of the ratio of the two sounds.

The following table and bar chart compare the intensity of sounds we are familar with.

Decibels	Comparable to
10	Calm breathing
20	Whisper
60	Normal conversation
80	City traffic
90	Lawn mower
110	Amplified music
120	Snowmobile
140	Fireworks
180	Aircraft at takeoff

Source: hearingaidknow.com

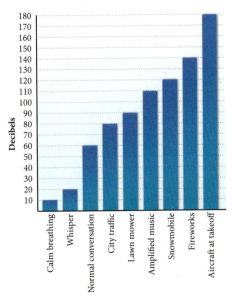

The precise definition for a *decibel* is

$$D = 10 \log_{10}\left(\frac{I}{I_0}\right)$$

where I is the intensity of the sound being measured, and I_0 is the intensity of the least audible sound. In this section, we will see that the preceding formula can also be written as

$$D = 10(\log_{10} I - \log_{10} I_0)$$

The rules we use to rewrite expressions containing logarithms are called the *properties of logarithms*. There are three of them.

A Properties of Logarithms

For the following three properties, x, y, and b are all positive real numbers, $b \neq 1$, and r is any real number.

Product Property for Logarithms

$$\log_b (xy) = \log_b x + \log_b y$$

In words: The logarithm of a *product* is the *sum* of the logarithms.

Quotient Property for Logarithms

$$\log_b\left(\frac{x}{y}\right) = \log_b x - \log_b y$$

In words: The logarithm of a *quotient* is the *difference* of the logarithms.

Power Property for Logarithms

$$\log_b x^r = r \log_b x$$

In words: The logarithm of a number raised to a *power* is the *product* of the power and the logarithm of the number.

PROOF OF THE PRODUCT PROPERTY FOR LOGARITHMS To prove the product property for logarithms, we simply apply the first identity for logarithms given in the preceding section.

$$b^{\log_b xy} = xy = (b^{\log_b x})(b^{\log_b y}) = b^{\log_b x + \log_b y}$$

Since the first and last expressions are equal and the bases are the same, the exponents $\log_b xy$ and $\log_b x + \log_b y$ must be equal. Therefore,

$$\log_b xy = \log_b x + \log_b y$$

The proofs of the quotient property and the power property for logartihms proceed in much the same manner, so we will omit them here. The examples that follow show how the three properties can be used.

PRACTICE PROBLEMS

1. Expand, using the properties of logarithms:

$$\log_3 \frac{5a}{b}$$

Answer

1. $\log_3 5 + \log_3 a - \log_3 b$

EXAMPLE 1   Expand, using the properties of logarithms: $\log_5 \dfrac{3xy}{z}$.

SOLUTION   Applying the quotient property for logarithms, we can write the quotient of $3xy$ and z in terms of a difference.

$$\log_5 \frac{3xy}{z} = \log_5 3xy - \log_5 z$$

Applying the product property to the product $3xy$, we write it in terms of addition.

$$\log_5 \frac{3xy}{z} = \log_5 3 + \log_5 x + \log_5 y - \log_5 z \qquad \blacksquare$$

EXAMPLE 2   Expand, using the properties of logarithms:

$$\log_2 \frac{x^4}{\sqrt{y} \cdot z^3}$$

SOLUTION   We write $\sqrt{y}$ as $y^{1/2}$ and apply the properties.

$$\log_2 \frac{x^4}{\sqrt{y} \cdot z^3} = \log_2 \frac{x^4}{y^{1/2}z^3} \qquad \sqrt{y} = y^{1/2}$$

$$= \log_2 x^4 - \log_2(y^{1/2} \cdot z^3) \qquad \begin{array}{l}\text{Quotient property}\\ \text{for logarithms}\end{array}$$

$$= \log_2 x^4 - (\log_2 y^{1/2} + \log_2 z^3) \qquad \begin{array}{l}\text{Product property}\\ \text{for logarithms}\end{array}$$

$$= \log_2 x^4 - \log_2 y^{1/2} - \log_2 z^3 \qquad \text{Remove parentheses.}$$

$$= 4 \log_2 x - \frac{1}{2} \log_2 y - 3 \log_2 z \qquad \begin{array}{l}\text{Power property}\\ \text{for logarithms}\end{array} \qquad \blacksquare$$

We can also use the three properties to write an expression in expanded form as just one logarithm.

EXAMPLE 3   Write the following as a single logarithm:

$$2 \log_{10} a + 3 \log_{10} b - \frac{1}{3} \log_{10} c$$

SOLUTION   We begin by applying the power property.

$$2 \log_{10} a + 3 \log_{10} b - \frac{1}{3} \log_{10} c$$

$$= \log_{10} a^2 + \log_{10} b^3 - \log_{10} c^{1/3} \qquad \begin{array}{l}\text{Power property}\\ \text{for logarithms}\end{array}$$

$$= \log_{10}(a^2 \cdot b^3) - \log_{10} c^{1/3} \qquad \begin{array}{l}\text{Product property}\\ \text{for logarithms}\end{array}$$

$$= \log_{10} \frac{a^2 b^3}{c^{1/3}} \qquad \begin{array}{l}\text{Quotient property}\\ \text{for logarithms}\end{array}$$

$$= \log_{10} \frac{a^2 b^3}{\sqrt[3]{c}} \qquad c^{1/3} = \sqrt[3]{c} \qquad \blacksquare$$

B Equations with Logarithms

The properties of logarithms along with the definition of logarithms are useful in solving equations that involve logarithms.

EXAMPLE 4   Solve $\log_2(x + 2) + \log_2 x = 3$ for x.

SOLUTION   Applying the product property to the left side of the equation allows us to write it as a single logarithm.

$$\log_2(x + 2) + \log_2 x = 3$$

$$\log_2[(x + 2)(x)] = 3$$

---

2. Expand $\log_{10} \frac{x^2}{\sqrt[3]{y}}$.

3. Write as a single logarithm.
$$3 \log_4 x + \log_4 y - 2 \log_4 z$$

4. Solve for x:
$$\log_2(x + 3) + \log_2 x = 2$$

Answers

2. $2 \log_{10} x - \frac{1}{3} \log_{10} y$

3. $\log_4 \frac{x^3 y}{z^2}$

4. 1

The last line can be written in exponential form using the definition of logarithms.

$$(x + 2)(x) = 2^3$$

Solve as usual.

$$x^2 + 2x = 8$$

$$x^2 + 2x - 8 = 0$$

$$(x + 4)(x - 2) = 0$$

$$x + 4 = 0 \quad \text{or} \quad x - 2 = 0$$

$$x = -4 \qquad\qquad x = 2$$

In the previous section, we noted the fact that x in the expression $y = \log_b x$ cannot be a negative number. Since substitution of $x = -4$ into the original equation gives

$$\log_2(-2) + \log_2(-4) = 3$$

which contains logarithms of negative numbers, we cannot use -4 as a solution. The solution set is {2}. ■

Problem Set 9.4

Moving Toward Success

"Our greatest weakness lies in giving up. The most certain way to succeed is always to try just one more time."

—Thomas Edison, 1847–1931, American inventor and entrepreneur

1. What can you do to stay focused in this course if things are not going well for you?
2. How is patience important while taking this class?

A Use the three properties of logarithms given in this section to expand each expression as much as possible. [Examples 1, 2]

1. $\log_3 4x$

2. $\log_2 5x$

3. $\log_6 \dfrac{5}{x}$

4. $\log_3 \dfrac{x}{5}$

5. $\log_2 y^5$

6. $\log_7 y^3$

7. $\log_9 \sqrt[3]{z}$

8. $\log_8 \sqrt{z}$

9. $\log_6 x^2 y^4$

10. $\log_{10} x^2 y^4$

11. $\log_5 \sqrt{x} \cdot y^4$

12. $\log_8 \sqrt[3]{x y^6}$

13. $\log_b \dfrac{xy}{z}$

14. $\log_b \dfrac{3x}{y}$

15. $\log_{10} \dfrac{4}{xy}$

16. $\log_{10} \dfrac{5}{4y}$

17. $\log_{10} \dfrac{x^2 y}{\sqrt{z}}$

18. $\log_{10} \dfrac{\sqrt{x} \cdot y}{z^3}$

19. $\log_{10} \dfrac{x^3 \sqrt{y}}{z^4}$

20. $\log_{10} \dfrac{x^4 \sqrt[3]{y}}{\sqrt{z}}$

21. $\log_b \sqrt[3]{\dfrac{x^2 y}{z^4}}$

22. $\log_b \sqrt[4]{\dfrac{x^4 y^3}{z^5}}$

A Write each expression as a single logarithm. [Example 3]

23. $\log_b x + \log_b z$

24. $\log_b x - \log_b z$

25. $2 \log_3 x - 3 \log_3 y$

26. $4 \log_2 x + 5 \log_2 y$

27. $\frac{1}{2} \log_{10} x + \frac{1}{3} \log_{10} y$

28. $\frac{1}{3} \log_{10} x - \frac{1}{4} \log_{10} y$

29. $3 \log_2 x + \frac{1}{2} \log_2 y - \log_2 z$

30. $2 \log_3 x + 3 \log_3 y - \log_3 z$

31. $\frac{1}{2} \log_2 x - 3 \log_2 y - 4 \log_2 z$

32. $3 \log_{10} x - \log_{10} y - \log_{10} z$

33. $\frac{3}{2} \log_{10} x - \frac{3}{4} \log_{10} y - \frac{4}{5} \log_{10} z$

34. $3 \log_{10} x - \frac{4}{3} \log_{10} y - 5 \log_{10} z$

B Solve each of the following equations. [Example 4]

35. $\log_2 x + \log_2 3 = 1$

36. $\log_3 x + \log_3 3 = 1$

37. $\log_3 x - \log_3 2 = 2$

38. $\log_3 x + \log_3 2 = 2$

39. $\log_3 x + \log_3(x - 2) = 1$

40. $\log_6 x + \log_6(x - 1) = 1$

41. $\log_3(x + 3) - \log_3(x - 1) = 1$

42. $\log_4(x - 2) - \log_4(x + 1) = 1$

43. $\log_2 x + \log_2(x - 2) = 3$

44. $\log_4 x + \log_4(x + 6) = 2$

45. $\log_8 x + \log_8(x - 3) = \frac{2}{3}$

46. $\log_{27} x + \log_{27}(x + 8) = \frac{2}{3}$

47. $\log_5 \sqrt{x} + \log_5 \sqrt{6x + 5} = 1$

48. $\log_2 \sqrt{x} + \log_2 \sqrt{6x + 5} = 1$

Applying the Concepts

49. Decibel Formula Use the properties of logarithms to rewrite the decibel formula $D = 10 \log_{10}\left(\frac{I}{I_0}\right)$ as

$$D = 10(\log_{10} I - \log_{10} I_0).$$

50. Decibel Formula In the decibel formula $D = 10 \log_{10}\left(\frac{I}{I_0}\right)$, the threshold of hearing, I_0, is

$$I_0 = 10^{-12} \text{ watts/meter}^2$$

Substitute 10^{-12} for I_0 in the decibel formula, and then show that it simplifies to

$$D = 10(\log_{10} I + 12)$$

51. Acoustic Powers The formula $N = \log_{10}\frac{P_1}{P_2}$ is used in radio electronics to find the ratio of the acoustic powers of two electric circuits in terms of their electric powers. Find N if P_1 is 100 and P_2 is 1. Then use the same two values of P_1 and P_2 to find N in the formula $N = \log_{10} P_1 - \log_{10} P_2$.

52. Henderson-Hasselbalch Formula Doctors use the Henderson-Hasselbalch formula to calculate the pH of a person's blood. This formula is represented as

$$\text{pH} = 6.1 + \log_{10}\left(\frac{x}{y}\right)$$

where x is the base concentration and y is the acidic concentration. Rewrite the Henderson–Hasselbalch formula so that the logarithm of a quotient is not involved.

Maintaining Your Skills

Divide.

53. $\dfrac{12x^2 + y^2}{36}$

54. $\dfrac{x^2 + 4y^2}{16}$

55. Divide $25x^2 + 4y^2$ by 100.

56. Divide $4x^2 + 9y^2$ by 36.

Use the discriminant to find the number and kind of solutions to the following equations.

57. $2x^2 - 5x + 4 = 0$

58. $4x^2 - 12x = -9$

For each of the following problems, find an equation with the given solutions.

59. $x = -3, x = 5$

60. $x = 2, x = -2, x = 1$

61. $y = \dfrac{2}{3}, y = 3$

62. $y = -\dfrac{3}{5}, y = 2$

Getting Ready for the Next Section

Simplify.

63. 5^0 **64.** 4^1 **65.** $\log_3 3$ **66.** $\log_5 5$ **67.** $\log_b b^4$ **68.** $\log_a a^k$

Use a calculator to find each of the following. Write your answer in scientific notation with the first number in each answer rounded to the nearest tenth.

69. $10^{-5.6}$ **70.** $10^{-4.1}$

Divide and round to the nearest whole number.

71. $\dfrac{2.00 \times 10^8}{3.96 \times 10^6}$ **72.** $\dfrac{3.25 \times 10^{12}}{1.72 \times 10^{10}}$

Common Logarithms and Natural Logarithms

OBJECTIVES

A Use a calculator to find common logarithms.

B Use a calculator to find a number given its common logarithm.

C Simplify expressions containing natural logarithms.

TICKET TO SUCCESS

Keep these questions in mind as you read through the section. Then respond in your own words and in complete sentences.

1. What is a common logarithm?
2. What is a natural logarithm?
3. What is an antilogarithm?
4. Find ln *e*, and explain how you arrived at your answer.

Limpopo/Shutterstock.com

Acid rain is precipitation with abnormally high levels of sulphur dioxide and nitrous dioxide, both caused by air pollution. The gas emissions react with water molecules in the air to create acid rain. In the United States, acid rain affects the East Coast more than any other area of the country. Harvard University in Cambridge, Massachusetts proudly displays many bronze and marble statues on their campus grounds. But every winter, the university must cover the statues to prevent erosion from acid rain.

The scale that measures acidity levels in aqueous solutions, such as acid rain, is called the pH scale. The range for pH is from 0 to 14, as shown in Figure 1. Pure water, a neutral solution, has a pH of 7. An acidic solution, such as lemon juice or acid rain, will have a pH less than 7, and an alkaline solution, such as baking soda or ammonia, has a pH above 7. The definition for pH involves the following *common logarithm*:

$$pH = -\log[H+]$$

where [H+] is the concentration of hydrogen ions in moles per liter.

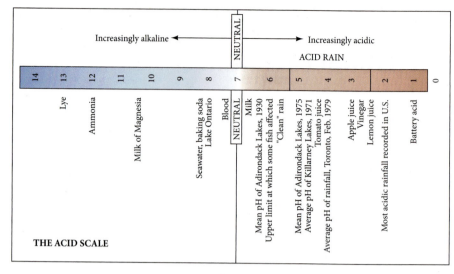

FIGURE 1

We will now continue our discussion of common logarithms.

A Common Logarithms

There are two kinds of logarithms that occur more frequently than other logarithms. They are logarithms with base 10 and natural logarithms, or logarithms with base e. Logarithms with a base of 10 are very common because our number system is a base-10 number system. For this reason, we call base-10 logarithms *common logarithms*.

> **Definition**
>
> A **common logarithm** is a logarithm with a base of 10. Because common logarithms are used so frequently, it is customary, in order to save time, to omit notating the base; that is,
>
> $$\log_{10} x = \log x$$
>
> When the base is not shown, it is assumed to be 10.

Common logarithms of powers of 10 are simple to evaluate. We need only recognize that $\log 10 = \log_{10} 10 = 1$ and apply the power property for logarithms: $\log_b x^r = r \log_b x$.

$$\log 1{,}000 = \log 10^3 \quad = 3 \log 10 \quad = 3(1) \quad = 3$$
$$\log 100 \quad = \log 10^2 \quad = 2 \log 10 \quad = 2(1) \quad = 2$$
$$\log 10 \quad = \log 10^1 \quad = 1 \log 10 \quad = 1(1) \quad = 1$$
$$\log 1 \quad = \log 10^0 \quad = 0 \log 10 \quad = 0(1) \quad = 0$$
$$\log 0.1 \quad = \log 10^{-1} = -1 \log 10 = -1(1) = -1$$
$$\log 0.01 \quad = \log 10^{-2} = -2 \log 10 = -2(1) = -2$$
$$\log 0.001 = \log 10^{-3} = -3 \log 10 = -3(1) = -3$$

To find common logarithms of numbers that are not powers of 10, we use a calculator with a $\boxed{\log}$ key.

NOTE
Remember, when the base is not written it is assumed to be 10.

Check the following logarithms to be sure you know how to use your calculator. (These answers have been rounded to the nearest ten-thousandth.)

$$\log 7.02 \approx 0.8463$$

$$\log 1.39 \approx 0.1430$$

$$\log 6.00 \approx 0.7782$$

$$\log 9.99 \approx 0.9996$$

PRACTICE PROBLEMS

EXAMPLE 1　　Use a calculator to find log 2,760.

1. Find log 27,600.

SOLUTION　　$\log 2{,}760 \approx 3.4409$

To work this problem on a scientific calculator, we simply enter the number 2,760 and press the key labeled $\boxed{\log}$. On a graphing calculator, we press the $\boxed{\log}$ key first, then 2,760.

　　The 3 in the answer is called the *characteristic,* and the decimal part of the logarithm is called the *mantissa.* ■

EXAMPLE 2　　Find log 0.0391.

2. Find log 0.00391.

SOLUTION　　$\log 0.0391 \approx -1.4078$　　■

EXAMPLE 3　　Find log 0.00523.

3. Find log 0.00952.

SOLUTION　　$\log 0.00523 \approx -2.2815$　　■

B　Antilogarithms

EXAMPLE 4　　Find x if $\log x = 3.8774$.

4. Find x if $\log x = 3.9786$.

SOLUTION　　We are looking for the number whose logarithm is 3.8774. On a scientific calculator, we enter 3.8774 and press the key labeled $\boxed{10^x}$. On a graphing calculator we press $\boxed{10^x}$ first, then 3.8774. The result is 7,540 to four significant digits. Here's why:

$$\text{If} \quad \log x = 3.8774$$

$$\text{then} \quad x = 10^{3.8774}$$

$$\approx 7{,}540$$

The number 7,540 is called the *antilogarithm* or just *antilog* of 3.8774; that is, 7,540 is the number whose logarithm is 3.8774.　　■

EXAMPLE 5　　Find x if $\log x = -2.4179$.

5. Find x if $\log x = -1.5901$.

SOLUTION　　Using the $\boxed{10^x}$ key, the result is 0.00382.

$$\text{If} \quad \log x = -2.4179$$

$$\text{then} \quad x = 10^{-2.4179}$$

$$\approx 0.00382$$

The antilog of −2.4179 is 0.00382; that is, the logarithm of 0.00382 is −2.4179.　　■

Answers
1. 4.4409
2. −2.4078
3. −2.0214
4. 9,519
5. 0.0257

Applications

Previously, we found that the magnitude M of an earthquake that produces a shock wave T times larger than the smallest shock wave that can be measured on a seismograph is given by the formula

$$M = \log_{10} T$$

We can rewrite this formula using our shorthand notation for common logarithms as

$$M = \log T$$

6. Find T if an earthquake measures 5.5 on the Richter scale.

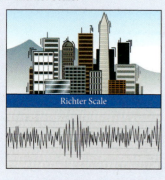

EXAMPLE 6   The San Francisco earthquake of 1906 is estimated to have measured 8.3 on the Richter scale. The San Fernando earthquake of 1971 measured 6.6 on the Richter scale. Find T for each earthquake, and then give some indication of how much stronger the 1906 earthquake was than the 1971 earthquake.

SOLUTION   For the 1906 earthquake:

$$\text{If } \log T = 8.3, \text{ then } T \approx 2.00 \times 10^8.$$

For the 1971 earthquake:

$$\text{If } \log T = 6.6, \text{ then } T \approx 3.98 \times 10^6.$$

Dividing the two values of T and rounding our answer to the nearest whole number, we have

$$\frac{2.00 \times 10^8}{3.98 \times 10^6} \approx 50$$

The shock wave for the 1906 earthquake was approximately 50 times stronger than the shock wave for the 1971 earthquake. ∎

Recall our discussion of the pH scale from the introduction to this section. Let's use the common logarithm $pH = -\log[H^+]$ used to define pH to work through the following examples.

7. Find the concentration of the hydrogen ion in a can of cola if the pH is 4.1.

EXAMPLE 7   Normal rainwater has a pH of 5.6. What is the concentration of hydrogen ions in normal rainwater?

SOLUTION   Substituting 5.6 for pH in the formula $pH = -\log[H^+]$, we have

$$5.6 = -\log[H^+] \qquad \text{Substitution}$$

$$\log[H^+] = -5.6 \qquad \text{Isolate the logarithm.}$$

$$[H^+] = 10^{-5.6} \qquad \text{Write in exponential form.}$$

$$\approx 2.5 \times 10^{-6} \text{ mole per liter} \qquad \text{Answer in scientific notation.}$$

∎

Answers

6. 3.16×10^5

7. 7.9×10^{-5}

EXAMPLE 8   The concentration of hydrogen ions in a sample of acid rain known to kill fish is 3.2×10^{-5} mole per liter. Find the pH of this acid rain to the nearest tenth.

SOLUTION   Substituting 3.2×10^{-5} for [H⁺] in the formula pH $= -\log[\text{H}^+]$, we have

$$\text{pH} = -\log[3.2 \times 10^{-5}] \qquad \text{Substitution}$$

$$\approx -(-4.5) \qquad \text{Evaluate the logarithm.}$$

$$= 4.5 \qquad \text{Simplify.} \qquad ■$$

C Natural Logarithms

Suppose you pour yourself a cup of hot coffee when you begin to make breakfast. Newton's Law of Cooling can be used to model the temperature change of the coffee while sitting at room temperature. The equation used to determine the change in heat is

$$T = e^{kt+C} + R$$

where T is the temperature of the coffee at time t, R is the temperature of the room and k and C are constants. The e is used to denote the base of a natural logarithm, which leads us to our next definition.

> **Definition**
>
> A **natural logarithm** is a logarithm with a base of e. The natural logarithm of x is denoted by ln x; that is,
>
> $$\ln x = \log_e x$$

We can assume that all our properties of exponents and logarithms hold for expressions with a base of e since e is a real number. Here are some examples intended to make you more familiar with the number e and natural logarithms.

EXAMPLE 9   Simplify each of the following expressions.

a. $e^0 = 1$

b. $e^1 = e$

c. $\ln e = 1$ In exponential form, $e^1 = e$.

d. $\ln 1 = 0$ In exponential form, $e^0 = 1$.

e. $\ln e^3 = 3$

f. $\ln e^{-4} = -4$

g. $\ln e^t = t$ ■

---

8. The concentration of hydrogen ions in a sample of acid rain is 1.8×10^{-5}. Find the pH. Round to the nearest tenth.

9. Simplify each expression.
   a. $\ln e^2$
   b. $\ln e^4$
   c. $\ln e^{-2}$
   d. $\ln e^x$

Answers

8. 4.7

9. a. 2   b. 4   c. −2   d. x

10. Expand ln Pe^{rt}.

EXAMPLE 10   Use the properties of logarithms to expand the expression ln Ae^{5t}.

SOLUTION   Since the properties of logarithms hold for natural logarithms, we have

$$\ln Ae^{5t} = \ln A + \ln e^{5t}$$

$$= \ln A + 5t \ln e$$

$$= \ln A + 5t \qquad \text{Because } \ln e = 1 \qquad ∎$$

11. If ln 5 = 1.6094 and ln 7 = 1.9459, find

a. ln 35
b. ln 0.2
c. ln 49

EXAMPLE 11   If ln 2 = 0.6931 and ln 3 = 1.0986, find

a. ln 6   **b.** ln 0.5   **c.** ln 8

SOLUTION   **a.** Since 6 = 2 · 3, we have

$$\ln 6 = \ln (2 \cdot 3)$$

$$= \ln 2 + \ln 3$$

$$= 0.6931 + 1.0986$$

$$= 1.7917$$

b. Writing 0.5 as $\frac{1}{2}$ and applying the quotient property for logarithms gives us

$$\ln 0.5 = \ln \frac{1}{2}$$

$$= \ln 1 - \ln 2$$

$$= 0 - 0.6931$$

$$= -0.6931$$

c. Writing 8 as 2^3 and applying the power property for logarithms, we have

$$\ln 8 = \ln 2^3$$

$$= 3 \ln 2$$

$$= 3(0.6931)$$

$$= 2.0793 \qquad ∎$$

Answers

10. ln $P + rt$

11. a. 3.5553   **b.** −1.6094
   c. 3.8918

Problem Set 9.5

Moving Toward Success

"We don't stop playing because we grow old; we grow old because we stop playing."

—George Bernard Shaw, 1856–1950, Irish playwright and Nobel Prize winner

1. Think of some fun things you may rather be doing than studying. How can you work those things into your day and still meet your study schedule?

2. How might you reward yourself for doing something well in this class?

A Find the following logarithms. Round to four decimal places. [Examples 1–3]

1. log 378

2. log 426

3. log 37.8

4. log 42,600

5. log 3,780

6. log 0.4260

7. log 0.0378

8. log 0.0426

9. log 37,800

10. log 4,900

11. log 600

12. log 900

13. log 2,010

14. log 10,200

15. log 0.00971

16. log 0.0312

17. log 0.0314

18. log 0.00052

19. log 0.399

20. log 0.111

B Find x in the following equations. [Examples 4, 5]

21. $\log x = 2.8802$

22. $\log x = 4.8802$

23. $\log x = -2.1198$

24. $\log x = -3.1198$

25. $\log x = 3.1553$

26. $\log x = 5.5911$

27. $\log x = -5.3497$

28. $\log x = -1.5670$

29. $\log x = -7.0372$

30. $\log x = -4.2000$

31. $\log x = 10$

32. $\log x = -1$

33. $\log x = -10$

34. $\log x = 1$

35. $\log x = 20$

36. $\log x = -20$

37. $\log x = -2$

38. $\log x = 4$

39. $\log x = \log_2 8$

40. $\log x = \log_3 9$

C Simplify each of the following expressions. [Example 9]

41. $\ln e$ **42.** $\ln 1$ **43.** $\ln e^5$ **44.** $\ln e^{-3}$ **45.** $\ln e^x$ **46.** $\ln e^y$

C Use the properties of logarithms to expand each of the following expressions. [Example 10]

47. $\ln 10e^{3t}$ **48.** $\ln 10e^{4t}$ **49.** $\ln Ae^{-2t}$ **50.** $\ln Ae^{-3t}$

C If $\ln 2 = 0.6931$, $\ln 3 = 1.0986$, and $\ln 5 = 1.6094$, find each of the following. [Example 11]

51. $\ln 15$ **52.** $\ln 10$ **53.** $\ln \dfrac{1}{3}$ **54.** $\ln \dfrac{1}{5}$

55. $\ln 9$ **56.** $\ln 25$ **57.** $\ln 16$ **58.** $\ln 81$

Use a calculator to evaluate each expression. Round your answers four decimal places, if necessary.

59. a. $\dfrac{\log 25}{\log 15}$ **b.** $\log \dfrac{25}{15}$ **60. a.** $\dfrac{\log 12}{\log 7}$ **b.** $\log \dfrac{12}{7}$

61. a. $\dfrac{\log 4}{\log 8}$ **b.** $\log \dfrac{4}{8}$ **62. a.** $\dfrac{\log 3}{\log 9}$ **b.** $\log \dfrac{3}{9}$

Applying the Concepts

Measuring Acidity Previously we indicated that the pH of a solution is defined in terms of logarithms as

$$pH = -\log[H^+]$$

where $[H^+]$ is the concentration of hydrogen ions in that solution. Round to the nearest hundredth.

63. Find the pH of orange juice if the concentration of hydrogen ions in the juice is $[H^+] = 6.50 \times 10^{-4}$.

64. Find the pH of milk if the concentration of hydrogen ions in milk is $[H^+] = 1.88 \times 10^{-6}$.

65. Find the concentration of hydrogen ions in a glass of wine if the pH is 4.75.

66. Find the concentration of hydrogen ions in a bottle of vinegar if the pH is 5.75.

The Richter Scale Find the relative size T of the shock wave of earthquakes with the following magnitudes, as measured on the Richter scale. Round to the nearest hundredth.

67. 5.5

68. 6.6

69. 8.3

70. 8.7

71. Shock Wave How much larger is the shock wave of an earthquake that measures 6.5 on the Richter scale than one that measures 5.5 on the same scale?

72. Shock Wave How much larger is the shock wave of an earthquake that measures 8.5 on the Richter scale than one that measures 5.5 on the same scale?

73. Earthquake The chart below is a partial listing of earthquakes that were recorded in Canada in 2000. Complete the chart by computing the magnitude on the Richter scale, M, or the number of times the associated shock wave is larger than the smallest measurable shock wave T.

Location	Date	Magnitude, M	Shock Wave, T
Moresby Island	January 23	4.0	
Vancouver Island	April 30		1.99×10^5
Quebec City	June 29	3.2	
Mould Bay	November 13	5.2	
St. Lawrence	December 14		5.01×10^3

Source: National Resources Canada, National Earthquake Hazards Program.

74. Earthquake On January 6, 2001, an earthquake with a magnitude of 7.7 on the Richter scale hit southern India (*National Earthquake Information Center*). By what factor was this earthquake's shock wave greater than the smallest measurable shock wave?

Depreciation The annual rate of depreciation r on a used car that is purchased for P dollars and is worth W dollars t years later can be found from the formula

$$\log(1 - r) = \frac{1}{t} \log \frac{W}{P}$$

75. Find the annual rate of depreciation on a car that is purchased for $9,000 and sold 5 years later for $4,500.

76. Find the annual rate of depreciation on a car that is purchased for $9,000 and sold 4 years later for $3,000.

Two cars depreciate in value according to the following depreciation tables. In each case, find the annual rate of depreciation.

77.

Age in Years	Value in Dollars
New	7,550
5	5,750

78.

Age in Years	Value in Dollars
New	7,550
3	5,750

79. Getting Close to e Use a calculator to complete the following table.

x	$(1 + x)^{1/x}$
1	
0.5	
0.1	
0.01	
0.001	
0.0001	
0.00001	

What number does the expression $(1 + x)^{1/x}$ seem to approach as x gets closer and closer to zero?

80. Getting Close to e Use a calculator to complete the following table.

x	$\left(1 + \dfrac{1}{x}\right)^x$
1	
10	
50	
100	
500	
1,000	
10,000	
1,000,000	

What number does the expression $\left(1 + \dfrac{1}{x}\right)^x$ seem to approach as x gets larger and larger?

Maintaining Your Skills

Solve each equation.

81. $(y + 3)^2 + y^2 = 9$

82. $(2y + 4)^2 + y^2 = 4$

83. $(x + 3)^2 + 1^2 = 2$

84. $(x - 3)^2 + (-1)^2 = 10$

Solve each equation.

85. $x^4 - 2x^2 - 8 = 0$

86. $x^{2/3} - 5x^{1/3} + 6 = 0$

87. $2x - 5\sqrt{x} + 3 = 0$

88. $(3x + 1) - 6\sqrt{3x + 1} + 8 = 0$

Getting Ready for the Next Section

Solve. Give answers to four decimal places if necessary.

89. $5(2x + 1) = 12$

90. $4(3x - 2) = 21$

Use a calculator to evaluate. Give answers to four decimal places.

91. $\dfrac{100,000}{32,000}$

92. $\dfrac{1.4982}{3.5681} + 3$

93. $\dfrac{1}{2}\left(\dfrac{-0.6931}{1.4289} + 3\right)$

94. $1 + \dfrac{0.04}{52}$

Use the power rule to rewrite the following logarithms.

95. $\log 1.05^t$

96. $\log 1.033^t$

Use identities to simplify.

97. $\ln e^{0.05t}$

98. $\ln e^{-0.000121t}$

Exponential Equations and Change of Base

TICKET TO SUCCESS

Keep these questions in mind as you read through the section. Then respond in your own words and in complete sentences.

1. What is doubling time?
2. What is an exponential equation and how would logarithms help you solve it?
3. What is the change-of-base property?
4. Write an application modeled by the equation $A = 10,000\left(1 + \frac{0.08}{2}\right)^{2(5)}$.

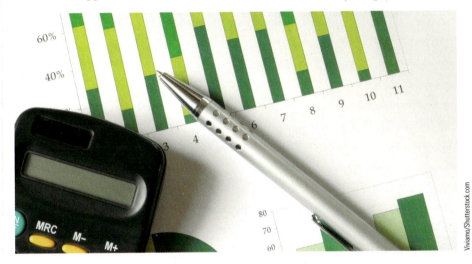

Viviamo/Shutterstock.com

In this section, we'll be working a lot with equations that involve exponential growth. Many real-life examples represent this type of growth, such as viral infections, nuclear chain reactions, internet traffic, and compound interest. As you proceed, you will encounter many problems that involve the last example of compound interest. We will use logarithms to help solve these problems.

The term *doubling time* is used when discussing compound interest to describe the time it takes for a quantity to double. For example, if you invest $5,000 in an account that pays 5% annual interest, compounded quarterly, you may want to know how long it will take for your money to double in value. You can find this doubling time if you can solve the equation

$$10,000 = 5,000 \, (1.0125)^{4t}$$

As you will see as you progress through this section, logarithms are the key to solving equations of this type.

A Exponential Equations

Logarithms are very important in solving equations in which the variable appears as an exponent. The equation

$$5^x = 12$$

is an example of one such equation. Equations of this form are called *exponential equations*. Since the quantities 5^x and 12 are equal, so are their common logarithms. We begin our solution by taking the logarithm of both sides.

$$\log 5^x = \log 12$$

We now apply the power property for logarithms, $\log x^r = r \log x$, to turn x from an exponent into a coefficient.

$$x \log 5 = \log 12$$

Dividing both sides by $\log 5$ gives us

$$x = \frac{\log 12}{\log 5}$$

If we want a decimal approximation of the solution, we can find $\log 12$ and $\log 5$ on a calculator and divide.

$$x \approx \frac{1.0792}{0.6990}$$

$$\approx 1.5439$$

The complete problem looks like this:

$$5^x = 12$$

$$\log 5^x = \log 12$$

$$x \log 5 = \log 12$$

$$x = \frac{\log 12}{\log 5}$$

$$\approx \frac{1.0792}{0.6990}$$

$$\approx 1.5439$$

Here is another example of solving an exponential equation using logarithms.

PRACTICE PROBLEMS

1. Solve for x: $12^{x+2} = 20$.

EXAMPLE 1   Solve $25^{2x+1} = 15$ for x.

SOLUTION   Taking the logarithm of both sides and then writing the exponent $(2x + 1)$ as a coefficient, we proceed as follows:

$$25^{2x+1} = 15$$

$$\log 25^{2x+1} = \log 15 \qquad \text{Take the log of both sides.}$$

$$(2x + 1) \log 25 = \log 15 \qquad \text{Power property of logarithms}$$

$$2x + 1 = \frac{\log 15}{\log 25} \qquad \text{Divide by } \log 25.$$

$$2x = \frac{\log 15}{\log 25} - 1 \qquad \text{Add } -1 \text{ to both sides.}$$

$$x = \frac{1}{2}\left(\frac{\log 15}{\log 25} - 1\right) \qquad \text{Multiply both sides by } \tfrac{1}{2}.$$

Using a calculator, we can write a decimal approximation of the answer.

$$x \approx \frac{1}{2}\left(\frac{1.1761}{1.3979} - 1\right)$$

$$\approx \frac{1}{2}(0.8413 - 1)$$

$$\approx \frac{1}{2}(-0.1587)$$

$$\approx -0.0793 \qquad \blacksquare$$

Answer

1. -0.7945 or -0.7944, depending on whether you round your intermediate answers

If you invest P dollars in an account with an annual interest rate r that is compounded n times a year, then t years later the amount of money in that account will be

$$A = P\left(1 + \frac{r}{n}\right)^{nt}$$

EXAMPLE 2 How long does it take for $5,000 to double if it is deposited in an account that yields 5% interest compounded once a year?

SOLUTION Substituting $P = 5,000$, $r = 0.05$, $n = 1$, and $A = 10,000$ into our formula, we have

$$10,000 = 5,000(1 + 0.05)^t$$

$$10,000 = 5,000(1.05)^t$$

$$2 = (1.05)^t \qquad \text{Divide by 5,000.}$$

This is an exponential equation. We solve by taking the logarithm of both sides.

$$\log 2 = \log(1.05)^t$$

$$\log 2 = t \log 1.05$$

Dividing both sides by $\log 1.05$, we have

$$t = \frac{\log 2}{\log 1.05}$$

$$\approx 14.2$$

It takes a little over 14 years for $5,000 to double if it earns 5% interest per year, compounded once a year. ∎

B Change of Base

There is a fourth property of logarithms we have not yet considered. This last property allows us to change from one base to another and is therefore called the *change-of-base property*.

> **Change-of-Base Property for Logarithms**
>
> If a and b are both positive numbers other than 1, and if $x > 0$, then
>
> $$\log_a x = \frac{\log_b x}{\log_b a}$$
>
> ↑   ↑
>
> Base a Base b

The logarithm on the left side has a base of a, and both logarithms on the right side have a base of b. This allows us to change from base a to any other base b that is a positive number other than 1. Here is a proof of the change-of-base property for logarithms.

PROOF We begin by writing the identity

$$a^{\log_a x} = x$$

Taking the logarithm base b of both sides and writing the exponent $\log_a x$ as a coefficient, we have

$$\log_b a^{\log_a x} = \log_b x$$

$$\log_a x \log_b a = \log_b x$$

2. How long does it take for $5,000 to double if it is invested in an account that pays 11% interest compounded once a year?

Answer
2. About 6.64 years

Dividing both sides by $\log_b a$, we have the desired result:

$$\frac{\log_a x \log_b a}{\log_b a} = \frac{\log_b x}{\log_b a}$$

$$\log_a x = \frac{\log_b x}{\log_b a}$$

We can use this property to find logarithms we could not otherwise compute on our calculators—that is, logarithms with bases other than 10 or e. The next example illustrates the use of this property.

3. Find $\log_6 14$.

EXAMPLE 3   Find $\log_8 24$.

SOLUTION   Since we do not have base-8 logarithms on our calculators, we can change this expression to an equivalent expression that contains only base-10 logarithms:

$$\log_8 24 = \frac{\log 24}{\log 8} \qquad \text{Change-of-base property for logarithms}$$

Don't be confused. We did not just drop the base, we changed to base 10. We could have written the last line like this:

$$\log_8 24 = \frac{\log_{10} 24}{\log_{10} 8}$$

From our calculators, we write

$$\log_8 24 \approx \frac{1.3802}{0.9031}$$

$$\approx 1.5283 \qquad \blacksquare$$

C   Applications

4. How long will it take the city in Example 4 to grow to 75,000 people from its current population of 32,000?

EXAMPLE 4   Suppose that the population of a small city is 32,000 in the beginning of 2010. The city council assumes that the population size t years later can be estimated by the equation

$$P = 32{,}000 e^{0.05t}$$

Approximately when will the city have a population of 50,000?

SOLUTION   We substitute 50,000 for P in the equation and solve for t.

$$50{,}000 = 32{,}000 e^{0.05t}$$

$$1.5625 = e^{0.05t} \qquad \frac{50{,}000}{32{,}000} = 1.5625$$

To solve this equation for t, we can take the natural logarithm of each side.

$$\ln 1.5625 = \ln e^{0.05t}$$

$$\ln 1.5625 = 0.05t \ln e \qquad \text{Power property for logarithms}$$

$$\ln 1.5625 = 0.05t \qquad \text{Because } \ln e = 1$$

$$t = \frac{\ln 1.5625}{0.05} \qquad \text{Divide each side by 0.05.}$$

$$\approx 8.93 \text{ years}$$

We can estimate that the population will reach 50,000 toward the end of 2018. $\blacksquare$

Answers
3. 1.4728 or 1.4729
4. Approximately 17.04 years

Problem Set 9.6

Moving Toward Success

"Wonder is the beginning of wisdom."

—Greek proverb

1. Can you study too much? Why or why not?
2. What do you plan to do to help yourself if you feel burned out during this class?

A Solve each exponential equation. Use a calculator to write the answer to four decimal places. [Example 1]

1. $3^x = 5$

2. $4^x = 3$

3. $5^x = 3$

4. $3^x = 4$

5. $5^{-x} = 12$

6. $7^{-x} = 8$

7. $12^{-x} = 5$

8. $8^{-x} = 7$

9. $8^{x+1} = 4$

10. $9^{x+1} = 3$

11. $4^{x-1} = 4$

12. $3^{x-1} = 9$

13. $3^{2x+1} = 2$

14. $2^{2x+1} = 3$

15. $3^{1-2x} = 2$

16. $2^{1-2x} = 3$

17. $15^{3x-4} = 10$

18. $10^{3x-4} = 15$

19. $6^{5-2x} = 4$

20. $9^{7-3x} = 5$

B Use the change-of-base property and a calculator to find a decimal approximation for each of the following logarithms. [Example 3]

21. $\log_8 16$

22. $\log_9 27$

23. $\log_{16} 8$

24. $\log_{27} 9$

25. $\log_7 15$

26. $\log_3 12$

27. $\log_{15} 7$

28. $\log_{12} 3$

29. $\log_8 240$

30. $\log_6 180$

31. $\log_4 321$

32. $\log_5 462$

Find a decimal approximation for each of the following natural logarithms.

33. $\ln 345$

34. $\ln 3{,}450$

35. $\ln 0.345$

36. $\ln 0.0345$

37. $\ln 10$

38. $\ln 100$

39. $\ln 45{,}000$

40. $\ln 450{,}000$

Applying the Concepts

Find estimates for the following.

41. Compound Interest How long will it take for $500 to double if it is invested at 6% annual interest compounded twice a year?

42. Compounded Interest How long will it take for $500 to double if it is invested at 6% annual interest compounded 12 times a year?

43. Compound Interest How long will it take for $1,000 to triple if it is invested at 12% annual interest compounded 6 times a year?

44. Compound Interest How long will it take for $1,000 to become $4,000 if it is invested at 12% annual interest compounded 6 times a year?

45. Doubling Time How long does it take for an amount of money P to double itself if it is invested at 8% interest compounded 4 times a year?

46. Tripling Time How long does it take for an amount of money P to triple itself if it is invested at 8% interest compounded 4 times a year?

47. Tripling Time If a $25 investment is worth $75 today, how long ago must that $25 have been invested at 6% interest compounded twice a year?

48. Doubling Time If a $25 investment is worth $50 today, how long ago must that $25 have been invested at 6% interest compounded twice a year?

Recall that if P dollars are invested in an account with annual interest rate r, compounded continuously, then the amount of money in the account after t years is given by the formula $A(t) = Pe^{rt}$.

49. Continuously Compounded Interest Repeat Problem 41 if the interest is compounded continuously.

50. Continuously Compounded Interest Repeat Problem 44 if the interest is compounded continuously.

51. Continuously Compounded Interest How long will it take $500 to triple if it is invested at 6% annual interest, compounded continuously?

52. Continuously Compounded Interest How long will it take $500 to triple if it is invested at 12% annual interest, compounded continuously?

Maintaining Your Skills

Find the vertex for each parabola. Is the vertex the lowest or the highest point of the parabola?

53. $y = 2x^2 + 8x - 15$

54. $y = 3x^2 - 9x - 10$

55. $y = 12x - 4x^2$

56. $y = 18x - 6x^2$

57. Maximum Height An object is projected into the air with an initial upward velocity of 64 feet per second. Its height h at any time t is given by the formula $h = 64t - 16t^2$. Find the time at which the object reaches its maximum height. Then, find the maximum height.

58. Maximum Height An object is projected into the air with an initial upward velocity of 64 feet per second from the top of a building 40 feet high. If the height h of the object t seconds after it is projected into the air is $h = 40 + 64t - 16t^2$, find the time at which the object reaches its maximum height. Then, find the maximum height it attains.

Extending the Concepts

59. Exponential Growth Suppose that the population in a small city is 32,000 at the beginning of 2010 and that the city council assumes that the population size t years later can be estimated by the equation

$$P(t) = 32,000e^{0.05t}$$

Approximately when will the city have a population of 64,000?

60. Exponential Growth Suppose the population of a city is given by the equation

$$P(t) = 100,000e^{0.05t}$$

where t is the number of years from the present time. How large is the population now? (Now corresponds to a certain value of t. Once you realize what that value of t is, the problem becomes very simple.)

61. Exponential Growth Suppose the population of a city is given by the equation

$$P(t) = 15,000e^{0.04t}$$

where t is the number of years from the present time. About how long will it take for the population to reach 45,000?

62. Exponential Growth Suppose the population of a city is given by the equation

$$P(t) = 15,000e^{0.08t}$$

where t is the number of years from the present time. About how long will it take for the population to reach 45,000?

63. Solve the formula $A = Pe^{rt}$ for t.

64. Solve the formula $A = Pe^{-rt}$ for t.

65. Solve the formula $A = P \cdot 2^{-kt}$ for t.

66. Solve the formula $A = P \cdot 2^{kt}$ for t.

67. Solve the formula $A = P(1 - r)^t$ for t.

68. Solve the formula $A = P(1 + r)^t$ for t.

69. The graphs of a function and its inverse are shown in the figure. Use the graphs to find the following:

a. $f(0)$

b. $f(1)$

c. $f(2)$

d. $f^{-1}(1)$

e. $f^{-1}(2)$

f. $f^{-1}(5)$

g. $f^{-1}(f(2))$

h. $f(f^{-1}(5))$

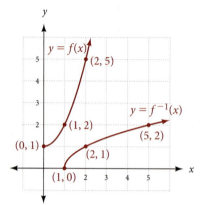

Chapter 9 Summary

EXAMPLES

■ Exponential Functions [9.1]

1. For the exponential function $f(x) = 2^x$,
$$f(0) = 2^0 = 1$$
$$f(1) = 2^1 = 2$$
$$f(2) = 2^2 = 4$$
$$f(3) = 2^3 = 8$$

Any function of the form
$$f(x) = b^x$$
where $b > 0$ and $b \neq 1$, is an *exponential function*.

■ One-to-One Functions [9.2]

2. The function $f(x) = x^2$ is not one-to-one because 9, which is in the range, comes from both 3 and -3 in the domain.

A function is a *one-to-one function* if every element in the range comes from exactly one element in the domain.

■ Inverse Functions [9.2]

3. The inverse of $f(x) = 2x - 3$ is
$$f^{-1}(x) = \frac{x + 3}{2}$$

The *inverse* of a function is obtained by reversing the order of the coordinates of the ordered pairs belonging to the function. Only one-to-one functions have inverses that are also functions.

■ Definition of Logarithms [9.3]

4. The definition allows us to write expressions like
$$y = \log_3 27$$
equivalently in exponential form as
$$3^y = 27$$
which makes it apparent that y is 3.

If b is a positive number not equal to 1, then the expression
$$y = \log_b x$$
is equivalent to $x = b^y$; that is, in the expression $y = \log_b x$, y is the number to which we raise b in order to get x. Expressions written in the form $y = \log_b x$ are said to be in *logarithmic form*. Expressions like $x = b^y$ are in *exponential form*.

■ Two Special Identities [9.3]

5. Examples of the two special identities are
$$5^{\log_5 12} = 12$$
and
$$\log_8 8^3 = 3$$

For $b > 0$, $b \neq 1$, the following two expressions hold for all positive real numbers x:

(1) $b^{\log_b x} = x$

(2) $\log_b b^x = x$

■ Properties of Logarithms [9.4]

6. We can rewrite the expression
$$\log_{10} \frac{45^6}{273}$$
using the properties of logarithms, as
$$6 \log_{10} 45 - \log_{10} 273$$

If x, y, and b are positive real numbers, $b \neq 1$, and r is any real number, then:

1. $\log_b (xy) = \log_b x + \log_b y$ Product property for logarithms

2. $\log_b \left(\dfrac{x}{y} \right) = \log_b x - \log_b y$ Quotient property for logarithms

3. $\log_b x^r = r \log_b x$ Power property for logarithms

■ Common Logarithms [9.5]

7. $\log_{10} 10{,}000 = \log 10{,}000$
$\qquad\qquad\quad = \log 10^4$
$\qquad\qquad\quad = 4$

Common logarithms are logarithms with a base of 10. To save time in writing, we omit the base when working with common logarithms; that is,

$$\log x = \log_{10} x$$

■ Natural Logarithms [9.5]

8. $\ln e = 1$
$\quad \ln 1 = 0$

Natural logarithms, written $\ln x$, are logarithms with a base of e, where the number e is an irrational number (like the number π). A decimal approximation for e is 2.7183. All the properties of exponents and logarithms hold when the base is e.

■ Change of Base [9.6]

9. $\log_6 475 = \dfrac{\log 475}{\log 6}$

$\qquad\qquad \approx \dfrac{2.6767}{0.7782}$

$\qquad\qquad \approx 3.44$

If x, a, and b are positive real numbers, $a \neq 1$ and $b \neq 1$, then

$$\log_a x = \frac{\log_b x}{\log_b a}$$

> ### 🚫 COMMON MISTAKES
>
> The most common mistakes that occur with logarithms come from trying to apply the three properties of logarithms to situations in which they don't apply. For example, a very common mistake looks like this:
>
> $$\frac{\log 3}{\log 2} = \log 3 - \log 2 \qquad \text{Mistake}$$
>
> This is not a property of logarithms. To write the equation $\log 3 - \log 2$, we would have to start with
>
> $$\log \frac{3}{2} \qquad NOT \qquad \frac{\log 3}{\log 2}$$
>
> There is a difference.

Let $f(x) = 2^x$ and $g(x) = \left(\frac{1}{3}\right)^x$, and find the following. [9.1]

1. $f(4)$ **2.** $f(-1)$

3. $g(2)$ **4.** $f(2) - g(-2)$

5. $f(-1) + g(1)$ **6.** $g(-1) + f(2)$

7. The graph of $y = f(x)$ **8.** The graph of $y = g(x)$

Sketch the graph of the given relation and its inverse. Write an equation for the inverse. [9.2]

9. $y = 2x + 1$ **10.** $y = x^2 - 4$

For each of the following functions, find the equation of the inverse. Write the inverse using the notation $f^{-1}(x)$ if the inverse is itself a function. [9.2]

11. $f(x) = 2x + 3$ **12.** $f(x) = x^2 - 1$

13. $f(x) = \frac{1}{2}x + 2$ **14.** $f(x) = 4 - 2x^2$

Write each equation in logarithmic form. [9.3]

15. $3^4 = 81$ **16.** $7^2 = 49$

17. $0.01 = 10^{-2}$ **18.** $2^{-3} = \frac{1}{8}$

Write each equation in exponential form. [9.3]

19. $\log_2 8 = 3$ **20.** $\log_3 9 = 2$

21. $\log_4 2 = \frac{1}{2}$ **22.** $\log_4 4 = 1$

Solve for x. [9.3]

23. $\log_5 x = 2$ **24.** $\log_{16} 8 = x$

25. $\log_x 0.01 = -2$

Graph each equation. [9.3]

26. $y = \log_2 x$ **27.** $y = \log_{1/2} x$

Simplify each expression. [9.3]

28. $\log_4 16$ **29.** $\log_{27} 9$

30. $\log_4(\log_3 3)$

Use the properties of logarithms to expand each expression. [9.4]

31. $\log_2 5x$ **32.** $\log_{10} \frac{2x}{y}$

33. $\log_a \frac{y^3 \sqrt{x}}{z}$ **34.** $\log_{10} \frac{x^2}{y^3 z^4}$

Write each expression as a single logarithm. [9.4]

35. $\log_2 x + \log_2 y$ **36.** $\log_3 x - \log_3 4$

37. $2 \log_a 5 - \frac{1}{2} \log_a 9$

38. $3 \log_2 x + 2 \log_2 y - 4 \log_2 z$

Solve each equation. [9.4]

39. $\log_2 x + \log_2 4 = 3$ **40.** $\log_2 x - \log_2 3 = 1$

41. $\log_3 x + \log_3(x - 2) = 1$

42. $\log_4(x + 1) - \log_4(x - 2) = 1$

43. $\log_6(x - 1) + \log_6 x = 1$

44. $\log_4(x - 3) + \log_4 x = 1$

Evaluate. Round to four decimal places. [9.5]

45. $\log 346$ **46.** $\log 0.713$

Find x. [9.5]

47. $\log x = 3.9652$ **48.** $\log x = -1.6003$

Simplify. [9.5]

49. $\ln e$ **50.** $\ln 1$

51. $\ln e^2$ **52.** $\ln e^{-4}$

Use the formula $pH = -\log[H^+]$ to find the pH of a solution with the given hydrogen ion concentration. Round to the nearest tenth. [9.5]

53. $[H^+] = 7.9 \times 10^{-3}$ **54.** $[H^+] = 8.1 \times 10^{-6}$

Find $[H^+]$ for a solution with the given pH. [9.5]

55. $pH = 2.7$ **56.** $pH = 7.5$

Solve each equation. [9.6]

57. $4^x = 8$ **58.** $4^{3x+2} = 5$

Evaluate using the change-of-base property and a calculator. Round answers to the nearest hundredth. [9.6]

59. $\log_{16} 8$ **60.** $\log_{12} 421$

Use the formula $A = P\left(1 + \frac{r}{n}\right)^{nt}$ to solve the following problem. [9.6]

61. Investing How long does it take $5,000 to double if it is deposited in an account that pays 16% annual interest compounded once a year?

Simplify.

1. $-8 + 2[5 - 3(-2 - 3)]$

2. $6(2x - 3) + 4(3x - 2)$

3. $\dfrac{3}{4} \div \dfrac{1}{8} \cdot \dfrac{3}{2}$

4. $\dfrac{3}{4} - \dfrac{1}{8} + \dfrac{3}{2}$

5. $3\sqrt{48} - 3\sqrt{75} + 2\sqrt{27}$

6. $(\sqrt{6} + 3\sqrt{2})(2\sqrt{6} + \sqrt{2})$

7. $[(6 + 2i) - (3 - 4i)] - (5 - i)$

8. $(1 + i)^2$

9. $\log_6(\log_5 5)$

10. $\log_5[\log_2(\log_3 9)]$

11. $\dfrac{1 - \dfrac{1}{x - 3}}{1 + \dfrac{1}{x - 3}}$

12. $1 + \dfrac{x}{1 + \dfrac{1}{x}}$

Reduce the following to lowest terms.

13. $\dfrac{452}{791}$

14. $\dfrac{468}{585}$

Multiply.

15. $\left(3t^2 + \dfrac{1}{4}\right)\left(4t^2 - \dfrac{1}{3}\right)$

16. $(3x - 2)(x^2 - 3x - 2)$

Divide.

17. $\dfrac{y^2 + 7y + 7}{y + 2}$

18. $\dfrac{9x^2 + 9x - 18}{3x - 4}$

Add or subtract.

19. $\dfrac{-3}{x^2 - 2x - 8} + \dfrac{4}{x^2 - 16}$

20. $\dfrac{7}{4x^2 - x - 3} - \dfrac{1}{4x^2 - 7x + 3}$

Solve.

21. $6 - 3(2x - 4) = 2$

22. $\dfrac{2}{3}(6x - 5) + \dfrac{1}{3} = 13$

23. $28x^2 = 3x + 1$

24. $2x - 1 = x^2$

25. $\dfrac{1}{x + 3} + \dfrac{1}{x - 2} = 1$

26. $\dfrac{1}{x^2 + 3x - 4} + \dfrac{3}{x^2 - 1} = \dfrac{-1}{x^2 + 5x + 4}$

27. $x - 3\sqrt{x} + 2 = 0$

28. $3(4y - 1)^2 + (4y - 1) - 10 = 0$

29. $\log_3 x = 3$

30. $\log_x 0.1 = -1$

31. $\log_3(x - 3) - \log_3(x + 2) = 1$

32. $\log_4(x - 5) + \log_4(x + 1) = 2$

Solve each system.

33. $\begin{aligned} -9x + 3y &= -1 \\ 5x - 2y &= -2 \end{aligned}$

34. $\begin{aligned} 4x + 7y &= -3 \\ x &= -2y - 2 \end{aligned}$

35. $\begin{aligned} x + 2y &= 4 \\ x + 2z &= 1 \\ y - 2z &= 5 \end{aligned}$

36. $\begin{aligned} 2x + 3y - 8z &= 2 \\ 3x - y + 2z &= 10 \\ 4x + y + 8z &= 16 \end{aligned}$

Graph each line.

37. $5x + 4y = 20$

38. $x = 3$

Write in symbols.

39. The difference of 7 and y is $-x$.

40. The difference of $9a$ and $4b$ is less than their sum.

Write in scientific notation.

41. 0.0000972

42. 47,000,000

Simplify.

43. $(2.6 \times 10^{-6})(1.4 \times 10^{19})$

44. $\dfrac{(5.6 \times 10^3)(1.5 \times 10^{-6})}{9.8 \times 10^{11}}$

Factor completely.

45. $8a^3 - 125$

46. $50a^4 + 10a^2 - 4$

Specify the domain and range for the relation and state whether the relation is a function.

47. $\{(2, -3), (2, -1), (-3, 3)\}$

48. $\{(3, 5), (4, 5), (6, 4)\}$

Use the illustration to work Problems 49 and 50.

49. Estimate the population of the United States in millions.

50. Estimate the world population in millions.

Graph each exponential function. [9.1]

1. $f(x) = 2^x$

2. $g(x) = 3^{-x}$

Sketch the graph of each function and its inverse. Find $f^{-1}(x)$ for Problem 3. [9.2]

3. $f(x) = 2x - 3$

4. $f(x) = x^2 - 4$

Solve for x. [9.3]

5. $\log_4 x = 3$

6. $\log_x 5 = 2$

Graph each of the following. [9.3]

7. $y = \log_2 x$

8. $y = \log_{\frac{1}{2}} x$

Evaluate each of the following. Round to the nearest hundredth if necessary. [9.3, 9.4, 9.5]

9. $\log_8 4$

10. $\log_7 21$

11. $\log 23{,}400$

12. $\log 0.0123$

13. $\ln 46.2$

14. $\ln 0.0462$

Use the properties of logarithms to expand each expression. [9.4]

15. $\log_2 \dfrac{6x^2}{y}$

16. $\log \dfrac{\sqrt{x}}{(y^4)\sqrt[5]{z}}$

Write each expression as a single logarithm. [9.4]

17. $2 \log_3 x - \dfrac{1}{2} \log_3 y$

18. $\dfrac{1}{3} \log x - \log y - 2 \log z$

Use a calculator to find x. Round to the nearest thousandth. [9.5]

19. $\log x = 4.8476$

20. $\log x = -2.6478$

Solve for x. Round to the nearest thousandth if necessary. [9.4, 9.6]

21. $5 = 3^x$

22. $4^{2x-1} = 8$

23. $\log_5 x - \log_5 3 = 1$

24. $\log_2 x + \log_2(x - 7) = 3$

25. Use the formula pH $= -\log[\text{H}^+]$ to find the pH of a solution in which $[\text{H}^+] = 6.6 \times 10^{-7}$. [9.5]

26. Compound Interest If \$400 is deposited in an account that earns 10% annual interest compounded twice a year, how much money will be in the account after 5 years? [9.1]

27. Compound Interest How long will it take \$600 to become \$1,800 if the \$600 is deposited in an account that earns 8% annual interest compounded four times a year? [9.6]

Determine if the function is one-to-one. [9.2]

28. $f = \{(1, 2), (2, 3), (3, 4), (4, 5)\}$

29. $g = \{(0, 1), (1, 0), (-1, 0), (-2, 1)\}$

30.

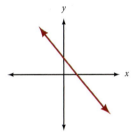

31.

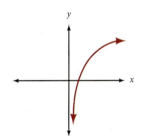

32.

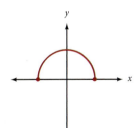

33.

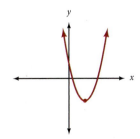

EXPONENTIAL AND LOGARITHMIC FUNCTIONS

Group Project

Two Depreciation Models

Number of People 3

Time Needed 20 minutes

Equipment Paper, pencil, and graphing calculator

Background Suppose a consumer magazine contained an article on leasing a computer. The original price of the computer was $2,579. The term of the lease was 24 months. At the end of the lease, the computer could be purchased for its residual value, $188. This is enough information to find an equation that will give us the value of the computer t months after it has been purchased.

Procedure We will find models for two types of depreciation: linear depreciation and exponential depreciation. Here are the general equations:

Linear Depreciation *Exponential Depreciation*

$$V = mt + b \qquad V = V_0 e^{-kt}$$

1. Let t represent time in months and V represent the value of the computer at time t; then find two ordered pairs (t, V) that correspond to the initial price of the computer and another ordered pair that corresponds to the residual value of $188 when the computer is 2 years old.

2. Use the two ordered pairs to find m and b in the linear depreciation model; then write the equation that gives us linear depreciation.

NEW APPLE MAC PRO
QUAD-CORE XEON PROCESSOR

- One 2.8GHz Quad-Core Intel Xeon "Nehalem"
- 3GB DDR3 ECC SDRAM
- 1TB 7200rpm Serial ATA 3Gb/s hard drive
- Apple LED Cinema Display (24" flat panel)
- ATI Radeon HD 5770 1GB PCI Express Graphics Card
- One 18x double-layer SuperDrive
- Apple Magic Mouse
- Apple Keyboard with Numeric Keypad
- Apple iWork
- Apple Aperture

$149 $\dfrac{\text{Mo. Business Lease}}{\textbf{24 Mos.}}$

or buy today for $2,579

---

3. Use the two ordered pairs to find V_0 and k in the exponential depreciation model; then write the equation that gives us exponential depreciation.

4. Graph each of the equations on your graphing calculator; then sketch the graphs on the following templates.

5. Find the value of the computer after 1 year, using both models.

6. Which of the two models do you think best describes the depreciation of a computer?

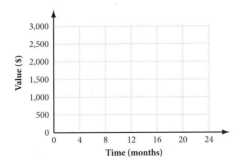

FIGURE 1

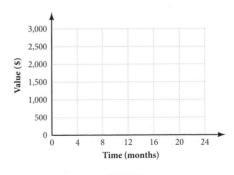

FIGURE 2

Drag Racing

The movie *Heart Like a Wheel* is based on the racing career of drag racer Shirley Muldowney. The movie includes a number of races. Choose four races as the basis for your report. For each race you choose, give a written description of the events leading to the race, along with the details of the race itself. Then draw a graph that shows the speed of the dragster as a function of time. Do this for

© picturesbyrob / Alamy

each race. One such graph from earlier in the chapter is shown here as an example. (For each graph, label the horizontal axis from 0 to 12 seconds, and label the vertical axis from 0 to 200 miles per hour.) Follow each graph with a description of any significant events that happen during the race (such as a motor malfunction or a crash) and their correlation to the graph.

Elapsed Time (sec)	Speed (mph)
0	0.0
1	72.7
2	129.9
3	162.8
4	192.2
5	212.4
6	228.1

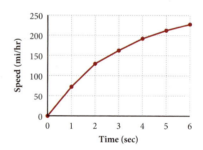

Conic Sections

An extinct marine organism, called an Archaeocyatha, lived 525 mllion years ago and is considered one of the first reef builders. The Archaeocyatha exhibited a sessile lifestyle and massed together to form reefs in warm tropical waters. Upon their extinction, the organism's shape, called a elliptic paraboloid (shown in the illustration), left behind detailed fossils on the sea floor.

The fossils of the Archaeocyatha, some found off the south coast of Australia, are considered *conic sections* of the organism's once three-dimensional body. This is because each fossil appears as a curve created by the intersection of a plane, the rock face, and the oval cone shape of the organism. The following illustration shows additional conic sections that we will encounter in this chapter.

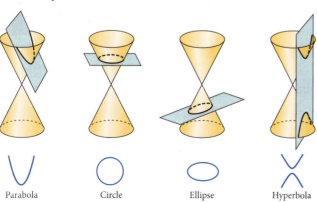

Parabola Circle Ellipse Hyperbola

Preview

Key Words	Definition
The Distance Formula	The distance between two points is given by the equation $d = \sqrt{(x_2 - x_1)^2 + (y_2 - y_1)^2}$
The Equation of a Circle	The equation $(x - a)^2 + (y - b)^2 = r^2$ gives a graph of a circle with center at (a, b) and radius r
The Equation of an Ellipse	The equation $\frac{(x - h)^2}{a^2} + \frac{(y - k)^2}{b^2} = 1$ gives a graph of an ellipse with center at (h, k)
The Equation of a Hyperbola	The equations $\frac{(x - h)^2}{a^2} - \frac{(y - k)^2}{b^2} = 1$ and $\frac{(y - k)^2}{b^2} - \frac{(x - h)^2}{a^2} = 1$ give graphs of hyperbolas each with a center at (h, k)

Chapter Outline

10.1 The Circle

A Use the distance formula.

B Write the equation of a circle given its center and radius.

C Find the center and radius of a circle from its equation, and then sketch the graph.

10.2 Ellipses and Hyperbolas

A Graph an ellipse.

B Graph a hyperbola.

10.3 Second-Degree Inequalities and Nonlinear Systems

A Graph second-degree inequalities.

B Solve systems of nonlinear equations.

C Graph the solution sets to systems of inequalities.

The Circle

TICKET TO SUCCESS

Keep these questions in mind as you read through the section. Then respond in your own words and in complete sentences.

1. What is the distance formula?
2. What is the mathematical definition of a circle?
3. How are the distance formula and the equation of a circle related?
4. How would you find the center and radius of a circle that can be identified by the equation $(x - a)^2 + (y - b)^2 = r^2$?

Matthew Jacques/Shutterstock.com

Because of their perfect symmetry, circles have been used for thousands of years in many disciplines, including art, science, and religion. For example, Stonehenge, a 4,500-year-old site in England, has a circular arrangement that is thought to have both religious and astronomical significance. We will work extensively with circles in this section. But before we do, we must first derive what is known as the *distance formula*.

A The Distance Formula

Suppose (x_1, y_1) and (x_2, y_2) are any two points in the first quadrant. (Actually, we could choose the two points to be anywhere on the coordinate plane. It is just more convenient to have them in the first quadrant.) We can name the points P_1 and P_2, respectively, and draw the diagram shown in Figure 1.

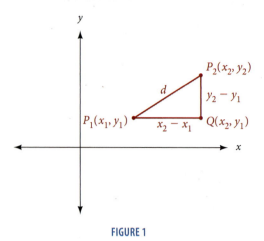

FIGURE 1

Notice the coordinates of point Q. The x-coordinate is x_2 since Q is directly below point P_2. The y-coordinate of Q is y_1 since Q is directly across from point P_1. It is evident from the diagram that the length from point P_2 to point Q, which is denoted as P_2Q, is $y_2 - y_1$ and the length of P_1Q is $x_2 - x_1$. Using the Pythagorean theorem, we have

$$(P_1P_2)^2 = (P_1Q)^2 + (P_2Q)^2$$

or

$$d^2 = (x_2 - x_1)^2 + (y_2 - y_1)^2$$

Taking the square root of both sides, we have

$$d = \sqrt{(x_2 - x_1)^2 + (y_2 - y_1)^2}$$

We know this is the positive square root since d is the distance from P_1 to P_2 and therefore must be positive. This formula is called the *distance formula*.

PRACTICE PROBLEMS

1. Find the distance between $(-4, 1)$ and $(2, 5)$.

EXAMPLE 1 Find the distance between $(3, 5)$ and $(2, -1)$.

SOLUTION If we let $(3, 5)$ be (x_1, y_1) and $(2, -1)$ be (x_2, y_2) and apply the distance formula, we have

$$d = \sqrt{(2 - 3)^2 + (-1 - 5)^2}$$

$$= \sqrt{(-1)^2 + (-6)^2}$$

$$= \sqrt{1 + 36}$$

$$= \sqrt{37} \qquad \blacksquare$$

2. Find x if the distance from $(x, 2)$ to $(3, -1)$ is $\sqrt{10}$.

EXAMPLE 2 Find x if the distance from $(x, 5)$ to $(3, 4)$ is $\sqrt{2}$.

SOLUTION Using the distance formula, we have

$$\sqrt{2} = \sqrt{(x - 3)^2 + (5 - 4)^2}$$

$$2 = (x - 3)^2 + 1^2$$

$$2 = x^2 - 6x + 9 + 1$$

$$0 = x^2 - 6x + 8 \qquad \text{\textcolor{orange}{Subtract 2 from both sides.}}$$

$$0 = (x - 4)(x - 2)$$

$$x = 4 \quad \text{or} \quad x = 2$$

Answers

1. $2\sqrt{13}$

2. 4 or 2

The two solutions are 4 and 2, which indicates there are two points, $(4, 5)$ and $(2, 5)$, which are $\sqrt{2}$ units from $(3, 4)$. $\qquad \blacksquare$

B Circles

We can model circles very easily in algebra by using equations that are based on the distance formula.

> **Circle Theorem**
>
> The equation of the circle with center at (a, b) and radius r is given by
>
> $$(x - a)^2 + (y - b)^2 = r^2$$

PROOF By definition, all points on the circle are a distance r from the center (a, b). If we let (x, y) represent any point on the circle, then (x, y) is r units from (a, b). Applying the distance formula, we have

$$r = \sqrt{(x - a)^2 + (y - b)^2}$$

Squaring both sides of this equation gives the equation of the circle.

$$(x - a)^2 + (y - b)^2 = r^2$$

We can use the circle theorem to find the equation of a circle, given its center and radius, or to find its center and radius, given the equation.

EXAMPLE 3 Find the equation of the circle with center at $(-3, 2)$ having a radius of 5.

SOLUTION We have $(a, b) = (-3, 2)$ and $r = 5$. Applying the circle theorem yields

$$[x - (-3)]^2 + (y - 2)^2 = 5^2$$

$$(x + 3)^2 + (y - 2)^2 = 25 \quad \blacksquare$$

EXAMPLE 4 Give the equation of the circle with radius 3 whose center is at the origin.

SOLUTION The coordinates of the center are $(0, 0)$, and the radius is 3. The equation must be

$$(x - 0)^2 + (y - 0)^2 = 3^2$$

$$x^2 + y^2 = 9 \quad \blacksquare$$

We can see from Example 4 that the equation of any circle with its center at the origin and radius r will be

$$x^2 + y^2 = r^2$$

3. Find the equation of the circle with center at $(4, -3)$ having a radius of 2.

4. Give the equation of the circle with radius 5 whose center is at the origin.

Answers
3. $(x - 4)^2 + (y + 3)^2 = 4$
4. $x^2 + y^2 = 25$

5. Find the center and radius of the circle whose equation is $(x - 3)^2 + (y - 4)^2 = 9$ and then sketch the graph.

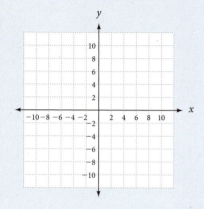

C Graphing Circles

EXAMPLE 5   Find the center and radius, and sketch the graph, of the circle with an equation of

$$(x - 1)^2 + (y + 3)^2 = 4$$

SOLUTION   Writing the equation in the form

$$(x - a)^2 + (y - b)^2 = r^2$$

we have

$$(x - 1)^2 + [y - (-3)]^2 = 2^2$$

The center is at $(1, -3)$ and the radius is 2. The graph is shown in Figure 2.

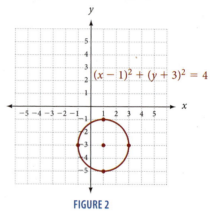

FIGURE 2   ∎

6. Graph.
$$x^2 + y^2 - 6x + 4y - 3 = 0$$

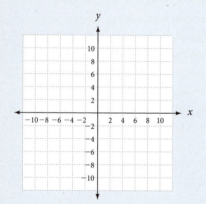

EXAMPLE 6   Sketch the graph of $x^2 + y^2 + 6x - 4y - 12 = 0$.

SOLUTION   To sketch the graph, we must find the center and radius. The center and radius can be identified if the equation has the form

$$(x - a)^2 + (y - b)^2 = r^2$$

The original equation can be written in this form by completing the squares on x and y.

$$x^2 + y^2 + 6x - 4y - 12 = 0$$
$$x^2 + 6x + y^2 - 4y = 12$$
$$x^2 + 6x + \mathbf{9} + y^2 - 4y + \mathbf{4} = 12 + \mathbf{9} + \mathbf{4}$$
$$(x + 3)^2 + (y - 2)^2 = 25$$
$$(x + 3)^2 + (y - 2)^2 = 5^2$$

From the last line, it is apparent that the center is at $(-3, 2)$ and the radius is 5. The graph is shown in Figure 3.

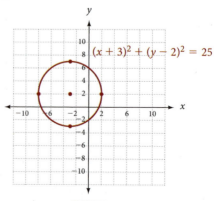

FIGURE 3   ∎

Answer
5. See Solutions Section.
6. See Solutions Section.

Problem Set 10.1

Moving Toward Success

"Spread love everywhere you go. Let no one ever come to you without leaving happier."

—Mother Teresa, 1910–1997, Albanian-born humanitarian and missionary

1. How will communicating your feelings about this class in an appropriate manner aid your success in it?
2. Why is maintaining motivation in this class more important now than ever?

A Find the distance between the following points. [Examples 1, 2]

1. (3, 7) and (6, 3)

2. (4, 7) and (8, 1)

3. (0, 9) and (5, 0)

4. (−3, 0) and (0, 4)

5. (3, −5) and (−2, 1)

6. (−8, 9) and (−3, −2)

7. (−1, −2) and (−10, 5)

8. (−3, −8) (−1, 6)

9. Find x so the distance between $(x, 2)$ and $(1, 5)$ is $\sqrt{13}$.

10. Find x so the distance between $(−2, 3)$ and $(x, 1)$ is 3.

11. Find y so the distance between $(7, y)$ and $(8, 3)$ is 1.

12. Find y so the distance between $(3, −5)$ and $(3, y)$ is 9.

B Write the equation of the circle with the given center and radius. [Examples 3, 4]

13. Center (2, 3); $r = 4$

14. Center (3, −1); $r = 5$

15. Center (3, −2); $r = 3$

16. Center (−2, 4); $r = 1$

17. Center (−5, −1); $r = \sqrt{5}$

18. Center (−7, −6); $r = \sqrt{3}$

19. Center (0, −5); $r = 1$

20. Center (0, −1); $r = 7$

21. Center (0, 0); $r = 2$

22. Center (0, 0); $r = 5$

C Give the center and radius, and sketch the graph of each of the following circles. [Examples 5, 6]

23. $x^2 + y^2 = 4$

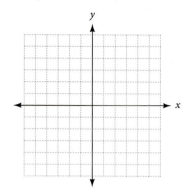

24. $x^2 + y^2 = 16$

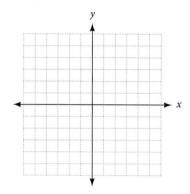

25. $(x - 1)^2 + (y - 3)^2 = 25$

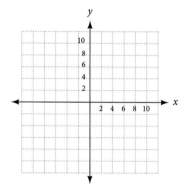

26. $(x - 4)^2 + (y - 1)^2 = 36$

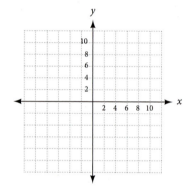

27. $(x + 2)^2 + (y - 4)^2 = 8$

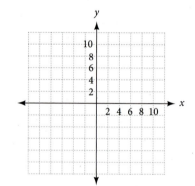

28. $(x - 3)^2 + (y + 1)^2 = 12$

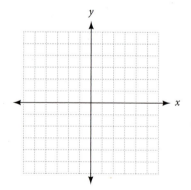

29. $(x + 1)^2 + (y + 1)^2 = 1$

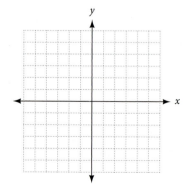

30. $(x + 3)^2 + (y + 2)^2 = 9$

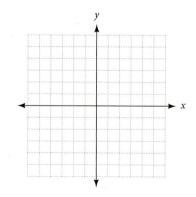

31. $x^2 + y^2 - 6y = 7$

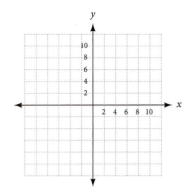

32. $x^2 + y^2 + 10x = 0$

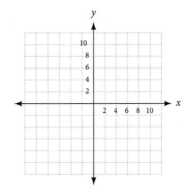

33. $x^2 + y^2 - 4x - 6y = -4$

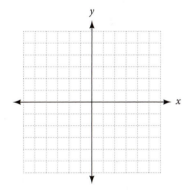

34. $x^2 + y^2 - 4x + 2y = 4$

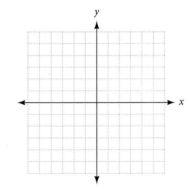

35. $x^2 + y^2 + 2x + y = \dfrac{11}{4}$

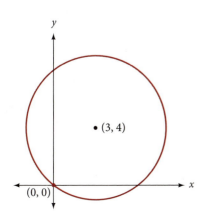

36. $x^2 + y^2 - 6x - y = -\dfrac{1}{4}$

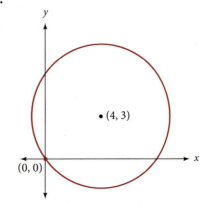

Both of the following circles pass through the origin. In each case, find the equation.

37.

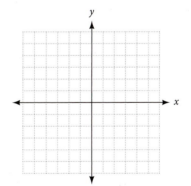

38.

39. Find the equations of circles *A*, *B*, and *C* in the following diagram. The three points are the centers of the three circles.

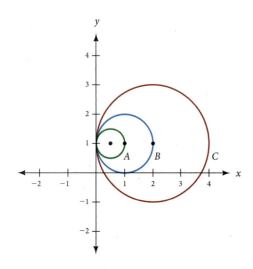

40. Each of the following circles passes through the origin. The centers are as shown. Find the equation of each circle.

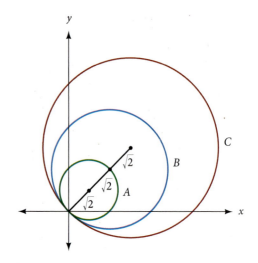

41. Find the equation of each of the three circles shown here.

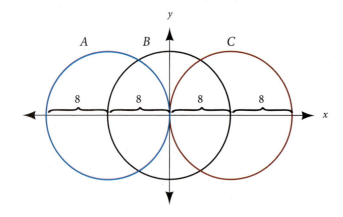

42. Find the equation of each of the three circles shown here.

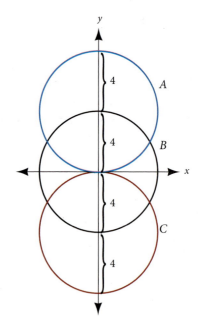

43. Find the equation of each of the three circles shown here.

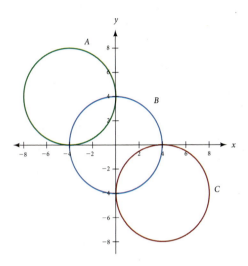

44. A parabola and a circle each contain the points $(-4, 0)$, $(0, 4)$, and $(4, 0)$. Sketch the graph of each curve on the same coordinate system, then write an equation for each curve.

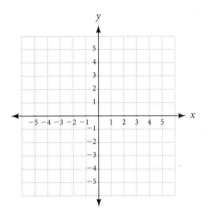

45. Ferris Wheel A giant Ferris wheel has a diameter of 240 feet and sits 12 feet above the ground. As shown in the diagram below, the wheel is 500 feet from the entrance to the park. The *xy*-coordinate system containing the wheel has its origin on the ground at the center of the entrance. Write an equation that models the shape of the wheel.

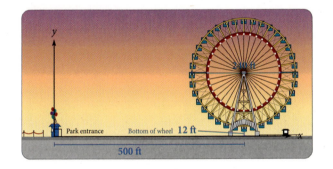

46. Magic Rings A magician is holding two rings that seem to lie in the same plane and intersect in two points. Each ring is 10 inches in diameter.

a. Find the equation of each ring if a coordinate system is placed with its origin at the center of the first ring and the *x*-axis contains the center of the second ring.

b. Find the equation of each ring if a coordinate system is placed with its origin at the center of the second ring and the *x*-axis contains the center of the first ring.

Find the equation of the inverse of each of the following functions. Write the inverse using the notation $f^{-1}(x)$, if the inverse is itself a function.

47. $f(x) = 3^x$

48. $f(x) = 5^x$

49. $f(x) = 2x + 3$

50. $f(x) = 3x - 2$

51. $f(x) = \dfrac{x + 3}{5}$

52. $f(x) = \dfrac{x - 2}{6}$

Getting Ready for the Next Section

Solve.

53. $y^2 = 9$

54. $x^2 = 25$

55. $-y^2 = 4$

56. $-x^2 = 16$

57. $\dfrac{-x^2}{9} = 1$

58. $\dfrac{y^2}{100} = 1$

59. Divide $4x^2 + 9y^2$ by 36.

60. Divide $25x^2 + 4y^2$ by 100.

Find the x-intercepts and the y-intercepts.

61. $3x - 4y = 12$

62. $y = 3x^2 + 5x - 2$

63. If $\dfrac{x^2}{25} + \dfrac{y^2}{9} = 1$, find y when x is 3.

64. If $\dfrac{x^2}{25} + \dfrac{y^2}{9} = 1$, find y when x is -4.

Extending the Concepts

A circle is *tangent to* a line if it touches, but does not cross, the line.

65. Find the equation of the circle with its center at $(2, 3)$ if the circle is tangent to the y-axis.

66. Find the equation of the circle with its center at $(3, 2)$ if the circle is tangent to the x-axis.

67. Find the equation of the circle with its center at $(2, 3)$ if the circle is tangent to the vertical line $x = 4$.

68. Find the equation of the circle with its center at $(3, 2)$ if the circle is tangent to the horizontal line $y = 6$.

Find the distance from the origin to the center of each of the following circles.

69. $x^2 + y^2 - 6x + 8y = 144$

70. $x^2 + y^2 - 8x + 6y = 144$

71. $x^2 + y^2 - 6x - 8y = 144$

72. $x^2 + y^2 + 8x + 6y = 144$

73. If we were to solve the equation $x^2 + y^2 = 9$ for y, we would obtain the equation $y = \pm\sqrt{9 - x^2}$. This last equation is equivalent to the two equations $y = \sqrt{9 - x^2}$, in which y is always positive, and $y = -\sqrt{9 - x^2}$, in which y is always negative. Look at the graph of $x^2 + y^2 = 9$ in Example 6 of this section and indicate what part of the graph each of the two equations corresponds to.

74. Solve the equation $x^2 + y^2 = 9$ for x, and then indicate what part of the graph in Example 6 each of the two resulting equations corresponds to.

Ellipses and Hyperbolas

TICKET TO SUCCESS

Keep these questions in mind as you read through the section. Then respond in your own words and in complete sentences.

1. What is an ellipse?
2. How would you find the x- and y-intercepts of an ellipse by looking at its equation in standard form?
3. Explain how to find the asymptotes of a graph of a hyperbola.
4. How would you find the vertices of a hyperbola with a center at (h, k)?

mv2/ZUMA Press/Newscom

The National Statuary Hall in the United States Capitol is known for its 101 statues of prominent Americans. Each day, thousands of tourists visit the two-story domed room to view the bronze and marble masterpieces. The hall is also known for its remarkable acoustics. If a visitor stands on the west side of the room, he can hear the whispered voice of another visitor standing in the far east side of the room. For this very reason, Statuary Hall is a famous example of a whispering gallery—a semi-elliptical domed structure that allows sound in one part of the room to be heard in another part of the room. In this section, we will introduce equations that produce *ellipses*, such as the one found in the dome of Statuary Hall.

A Ellipses

We begin this section with an introductory look at equations that produce ellipses. To simplify our work, we consider only ellipses that are centered about the origin.

Suppose we want to graph the equation

$$\frac{x^2}{25} + \frac{y^2}{9} = 1$$

We can find the y-intercepts by letting $x = 0$, and we can find the x-intercepts by letting $y = 0$.

NOTE
We can find other ordered pairs on the graph by substituting values for x (or y) and then solving for y (or x). For example, if $x = 3$, then

$$\frac{3^2}{25} + \frac{y^2}{9} = 1$$

$$\frac{9}{25} + \frac{y^2}{9} = 1$$

$$0.36 + \frac{y^2}{9} = 1$$

$$\frac{y^2}{9} = 0.64$$

$$y^2 = 5.76$$

$$y = \pm 2.4$$

This would give us the two ordered pairs (3, –2.4) and (3, 2.4).

When $x = 0$	When $y = 0$
$\dfrac{0^2}{25} + \dfrac{y^2}{9} = 1$	$\dfrac{x^2}{25} + \dfrac{0^2}{9} = 1$
$y^2 = 9$	$x^2 = 25$
$y = \pm 3$	$x = \pm 5$

The graph crosses the y-axis at (0, 3) and (0, −3) and the x-axis at (5, 0) and (−5, 0). Graphing these points and then connecting them with a smooth curve gives the graph shown in Figure 1.

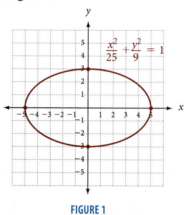

FIGURE 1

A graph of the type shown in Figure 1 is called an *ellipse*. If we were to find some other ordered pairs that satisfy our original equation, we would find that their graphs lie on the ellipse. Also, the coordinates of any point on the ellipse will satisfy the equation. We can generalize these results as follows.

> **The Ellipse**
>
> The graph of any equation of the form
>
> $$\frac{x^2}{a^2} + \frac{y^2}{b^2} = 1 \qquad \text{Standard form}$$
>
> is an **ellipse** centered at the origin. The ellipse will cross the x-axis at $(a, 0)$ and $(-a, 0)$. It will cross the y-axis at $(0, b)$ and $(0, -b)$. When a and b are equal, the ellipse will be a circle.

The most convenient way to graph an ellipse is to locate the intercepts.

EXAMPLE 1   Sketch the graph of $4x^2 + 9y^2 = 36$.

SOLUTION   To write the equation in the form

$$\frac{x^2}{a^2} + \frac{y^2}{b^2} = 1$$

we must divide both sides by 36.

$$\frac{4x^2}{36} + \frac{9y^2}{36} = \frac{36}{36}$$

$$\frac{x^2}{9} + \frac{y^2}{4} = 1$$

PRACTICE PROBLEMS

1. Graph $25x^2 + 4y^2 = 100$. (*Hint:* First, divide both sides by 100.)

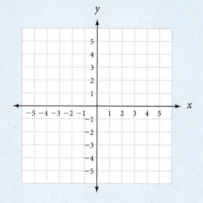

Answers
1. For all answers in this section see Solutions to Selected Practice Problems.

The graph crosses the x-axis at (3, 0), (−3, 0) and the y-axis at (0, 2), (0, −2), as shown in Figure 2.

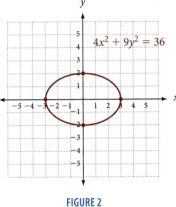

FIGURE 2   ■

B Hyperbolas

Figure 3 shows Europa, one of Jupiter's moons, as it was photographed by the Galileo space probe in the late 1990s. To speed up the trip from Earth to Jupiter—nearly a billion miles—Galileo made use of the *slingshot effect.* This involves flying a hyperbolic path very close to a planet, so that gravity can be used to gain velocity as the space probe hooks around the planet (Figure 4).

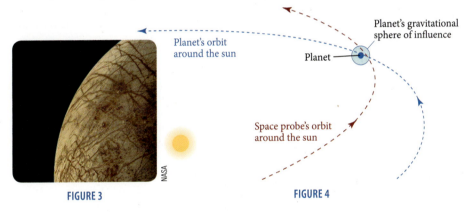

FIGURE 3   **FIGURE 4**

We use the next part of this section to consider equations that produce hyperbolas.

Consider the equation

$$\frac{x^2}{9} - \frac{y^2}{4} = 1$$

If we were to find a number of ordered pairs that are solutions to the equation and connect their graphs with a smooth curve, we would have Figure 5.

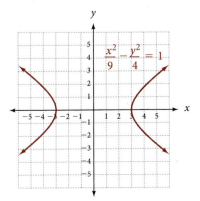

FIGURE 5

This graph is an example of a *hyperbola*. Notice that the graph has x-intercepts at $(3, 0)$ and $(-3, 0)$. The graph has no y-intercepts and hence does not cross the y-axis since substituting $x = 0$ into the equation yields

$$\frac{0^2}{9} - \frac{y^2}{4} = 1$$

$$-y^2 = 4$$

$$y^2 = -4$$

for which there is no real solution. We can, however, use the number below y^2 to help sketch the graph. If we draw a rectangle that has its sides parallel to the x- and y-axes and that passes through the x-intercepts and the points on the y-axis corresponding to the square roots of the number below y^2, 2 and -2, it looks like the rectangle in Figure 6. The lines that connect opposite corners of the rectangle are called *asymptotes*. The graph of the hyperbola

$$\frac{x^2}{9} - \frac{y^2}{4} = 1$$

will approach these lines. Figure 6 is the graph.

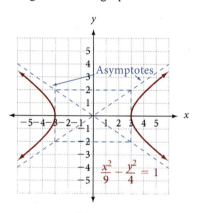

FIGURE 6

2. Graph the equation:
$$\frac{x^2}{25} - \frac{y^2}{9} = 1$$

EXAMPLE 2   Graph the equation $\dfrac{y^2}{9} - \dfrac{x^2}{16} = 1$.

SOLUTION   In this case, the y-intercepts are 3 and -3, and the x-intercepts do not exist. We can use the square roots of the number below x^2, however, to find the asymptotes associated with the graph. The sides of the rectangle used to draw the asymptotes must pass through 3 and -3 on the y-axis and 4 and -4 on the x-axis. (See Figure 7.)

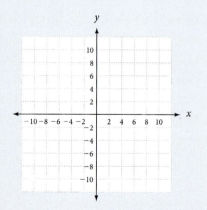

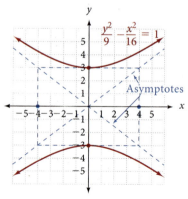

FIGURE 7

Here is a summary of what we have for hyperbolas.

The Hyperbola

The graph of the equation will be a hyperbola centered at the origin. The graph will have x-intercepts at $-a$ and a.

$$\frac{x^2}{a^2} - \frac{y^2}{b^2} = 1$$

The graph of the equation will be a hyperbola centered at the origin. The graph will have y-intercepts at $-a$ and a.

$$\frac{y^2}{a^2} - \frac{x^2}{b^2} = 1$$

As an aid in sketching either of the preceding equations, the asymptotes can be found by drawing lines through opposite corners of the rectangle whose sides pass through $-a, a, -b,$ and b on the axes.

Ellipses and Hyperbolas Not Centered at the Origin

The following equation is that of an ellipse with its center at the point (4, 1):

$$\frac{(x-4)^2}{9} + \frac{(y-1)^2}{4} = 1$$

To see why the center is at (4, 1) we substitute x' (read "x prime") for $x - 4$ and y' for $y - 1$ in the equation. That is,

$$\text{If} \rightarrow \qquad\qquad x' = x - 4$$

$$\text{and} \rightarrow \qquad\qquad y' = y - 1$$

$$\text{the equation} \rightarrow \qquad \frac{(x-4)^2}{9} + \frac{(y-1)^2}{4} = 1$$

$$\text{becomes} \rightarrow \qquad \frac{(x')^2}{9} + \frac{(y')^2}{4} = 1$$

This is the equation of an ellipse in a coordinate system with an x'-axis and a y'-axis. We call this new coordinate system the $x'y'$-*coordinate system*. The center of our ellipse is at the origin in the $x'y'$-coordinate system. The question is this: What are the coordinates of the center of this ellipse in the original xy-coordinate system? To answer this question, we go back to our original substitutions:

$$x' = x - 4$$
$$y' = y - 1$$

In the $x'y'$-coordinate system, the center of our ellipse is at $x' = 0, y' = 0$ (the origin of the $x'y'$ system). Substituting these numbers for x' and y', we have

$$0 = x - 4$$
$$0 = y - 1$$

Solving these equations for x and y will give us the coordinates of the center of our ellipse in the $x'y'$-coordinate system. As you can see, the solutions are $x = 4$ and $y = 1$. Therefore, in the $x'y'$-coordinate system, the center of our ellipse is at the point (4, 1). Figure 8 shows the graph.

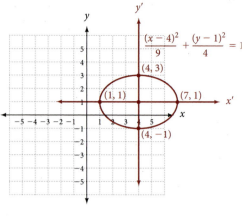

FIGURE 8

The coordinates of all points labeled in Figure 8 are given with respect to the *xy*-coordinate system. The *x'* and *y'* are shown simply for reference in our discussion. Note that the horizontal distance from the center to the vertices is 3—the square root of the denominator of the $(x - 4)^2$ term. Likewise, the vertical distance from the center to the other vertices is 2—the square root of the denominator of the $(y - 1)^2$ term.

We summarize the information above with the following:

An Ellipse with Center at (h, k)

The graph of the equation

$$\frac{(x - h)^2}{a^2} + \frac{(y - k)^2}{b^2} = 1$$

will be an ellipse with center at (h, k). The vertices of the ellipse will be at the points, $(h + a, k)$, $(h - a, k)$, $(h, k + b)$, and $(h, k - b)$.

---

3. Graph the ellipse:
$$4x^2 + 9y^2 - 24x + 36y + 36 = 0$$

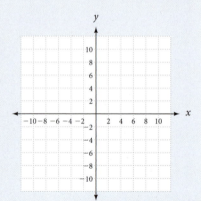

EXAMPLE 3   Graph the ellipse: $x^2 + 9y^2 + 4x - 54y + 76 = 0$.

SOLUTION   To identify the coordinates of the center, we must complete the square on *x* and also on *y*. To begin, we rearrange the terms so that those containing *x* are together, those containing *y* are together, and the constant term is on the other side of the equal sign. Doing so gives us the following equation:

$$x^2 + 4x \qquad + 9y^2 - 54y \qquad = -76$$

Before we can complete the square on *y*, we must factor 9 from each term containing *y*.

$$x^2 + 4x \qquad + 9(y^2 - 6y) \qquad = -76$$

To complete the square on *x*, we add 4 to each side of the equation. To complete the square on *y*, we add 9 inside the parentheses. This increases the left side of the equation by 81 since each term within the parentheses is multiplied by 9. Therefore, we must also add 81 to the right side of the equation.

$$x^2 + 4x + \mathbf{4} + 9(y^2 - 6y + \mathbf{9}) = -76 + \mathbf{4} + \mathbf{81}$$

$$(x + 2)^2 + 9(y - 3)^2 = 9$$

To identify the distances to the vertices, we divide each term on both sides by 9.

$$\frac{(x + 2)^2}{9} + \frac{9(y - 3)^2}{9} = \frac{9}{9}$$

$$\frac{(x + 2)^2}{9} + \frac{(y - 3)^2}{1} = 1$$

The graph is an ellipse with center at $(-2, 3)$, as shown in Figure 9.

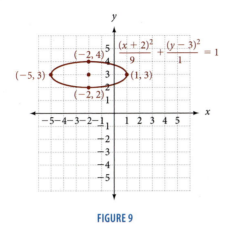

FIGURE 9 ■

The ideas associated with graphing hyperbolas whose centers are not at the origin parallel the ideas just presented about graphing ellipses whose centers have been moved off the origin. Without showing the justification for doing so, we state the following guidelines for graphing hyperbolas:

Hyperbolas with Center at (h, k)

The graphs of the equations

$$\frac{(x - h)^2}{a^2} - \frac{(y - k)^2}{b^2} = 1 \text{ and } \frac{(y - k)^2}{b^2} - \frac{(x - h)^2}{a^2} = 1$$

will be hyperbolas with their centers at (h, k). The vertices of the graph of the first equation will be at the points $(h + a, k)$, and $(h - a, k)$, and the vertices for the graph of the second equation will be at $(h, k + b)$, $(h, k - b)$. In either case, the asymptotes can be found by connecting opposite corners of the rectangle that contains the four points $(h + a, k)$, $(h - a, k)$, $(h, k + b)$, and $(h, k - b)$.

EXAMPLE 4   Graph the hyperbola $4x^2 - y^2 + 4y - 20 = 0$.

SOLUTION   To identify the coordinates of the center of the hyperbola, we need to complete the square on y. (Because there is no linear term in x, we do not need to complete the square on x. The x-coordinate of the center will be $x = 0$.)

$$4x^2 - y^2 + 4y - 20 = 0$$

$$4x^2 - y^2 + 4y = 20 \qquad \text{Add 20 to each side.}$$

$$4x^2 - 1(y^2 - 4y) = 20 \qquad \text{Factor } -1 \text{ from each term containing } y.$$

To complete the square on y, we add 4 to the terms inside the parentheses. Doing so adds -4 to the left side of the equation because everything inside the parentheses is multiplied by -1. To keep from changing the equation we must also add -4 to the right side.

4. Graph the hyperbola:

$$9y^2 - x^2 - 36y - 6x + 18 = 0$$

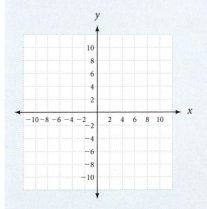

$$4x^2 - 1(y^2 - 4y + \mathbf{4}) = 20 - \mathbf{4} \qquad \text{Add } -4 \text{ to each side.}$$

$$4x^2 - 1(y - 2)^2 = 16 \qquad y^2 - 4y + 4 = (y - 2)^2$$

$$\frac{4x^2}{16} - \frac{(y - 2)^2}{16} = \frac{16}{16} \qquad \text{Divide each side by 16.}$$

$$\frac{x^2}{4} - \frac{(y - 2)^2}{16} = 1 \qquad \text{Simplify each term.}$$

This is the equation of a hyperbola with center at $(0, 2)$. The graph opens to the right and left as shown in Figure 10.

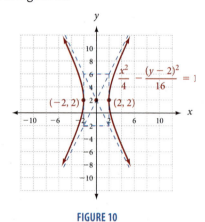

FIGURE 10

■

Problem Set 10.2

Moving Toward Success

"Our attitude toward life determines life's attitude towards us."

—Earl Nightingale, 1921–1989, American motivational
author and radio broadcaster

1. Explain why cramming for a test would increase
   your anxiety during the test.
2. Why will staying positive before and during a
   test reduce your stress level?

A Graph each of the following. Be sure to label both the *x*- and *y*-intercepts. [Example 1]

1. $\dfrac{x^2}{9} + \dfrac{y^2}{16} = 1$

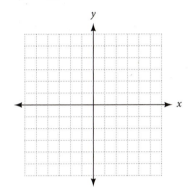

2. $\dfrac{x^2}{25} + \dfrac{y^2}{4} = 1$

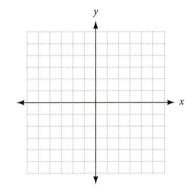

3. $\dfrac{x^2}{16} + \dfrac{y^2}{9} = 1$

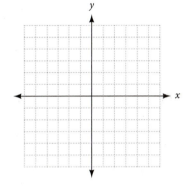

4. $\dfrac{x^2}{4} + \dfrac{y^2}{25} = 1$

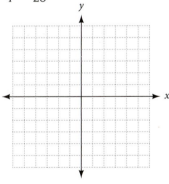

5. $\dfrac{x^2}{3} + \dfrac{y^2}{4} = 1$

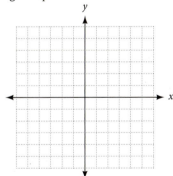

6. $\dfrac{x^2}{4} + \dfrac{y^2}{3} = 1$

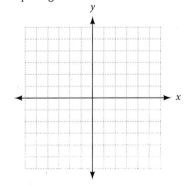

7. $4x^2 + 25y^2 = 100$

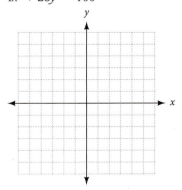

8. $4x^2 + 9y^2 = 36$

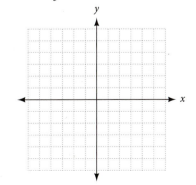

9. $x^2 + 8y^2 = 16$

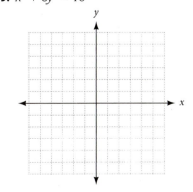

10. $12x^2 + y^2 = 36$

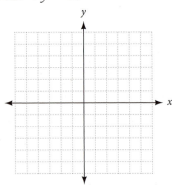

B Graph each of the following. Show the intercepts and the asymptotes in each case. [Example 2]

11. $\dfrac{x^2}{9} - \dfrac{y^2}{16} = 1$

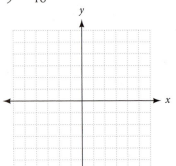

12. $\dfrac{x^2}{25} - \dfrac{y^2}{4} = 1$

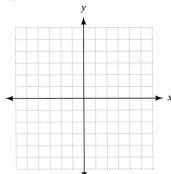

13. $\dfrac{x^2}{16} - \dfrac{y^2}{9} = 1$

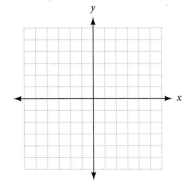

14. $\dfrac{x^2}{4} - \dfrac{y^2}{25} = 1$

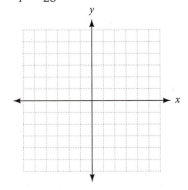

15. $\dfrac{y^2}{9} - \dfrac{x^2}{16} = 1$

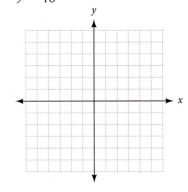

16. $\dfrac{y^2}{25} - \dfrac{x^2}{4} = 1$

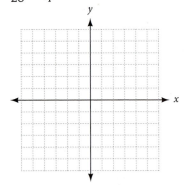

17. $\dfrac{y^2}{36} - \dfrac{x^2}{4} = 1$

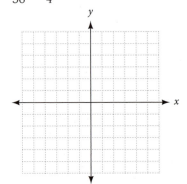

18. $\dfrac{y^2}{4} - \dfrac{x^2}{36} = 1$

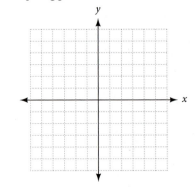

19. $x^2 - 4y^2 = 4$

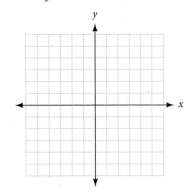

20. $y^2 - 4x^2 = 4$

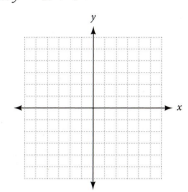

21. $16y^2 - 9x^2 = 144$

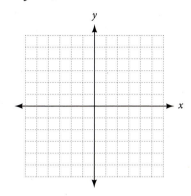

22. $4y^2 - 25x^2 = 100$

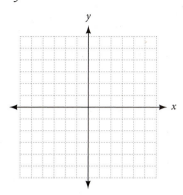

Find the x- and y-intercepts, if they exist, for each of the following. Do not graph.

23. $0.4x^2 + 0.9y^2 = 3.6$

24. $1.6x^2 + 0.9y^2 = 14.4$

25. $\dfrac{x^2}{0.04} - \dfrac{y^2}{0.09} = 1$

26. $\dfrac{y^2}{0.16} - \dfrac{x^2}{0.25} = 1$

27. $\dfrac{25x^2}{9} + \dfrac{25y^2}{4} = 1$

28. $\dfrac{16x^2}{9} + \dfrac{16y^2}{25} = 1$

B Graph each of the following ellipses. In each case, label the coordinates of the center and the vertices. [Example 3]

29. $\dfrac{(x - 4)^2}{4} + \dfrac{(y - 2)^2}{9} = 1$

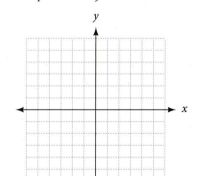

30. $\dfrac{(x - 2)^2}{4} + \dfrac{(y - 4)^2}{9} = 1$

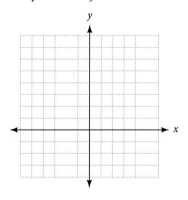

31. $4x^2 + y^2 - 4y - 12 = 0$

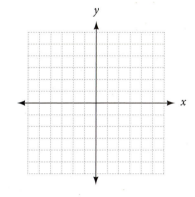

32. $4x^2 + y^2 - 24x - 4y + 36 = 0$

33. $9x^2 + y^2 + 4y - 54x + 76 = 0$

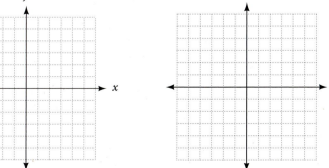

34. $4x^2 + y^2 - 16x + 2y + 13 = 0$

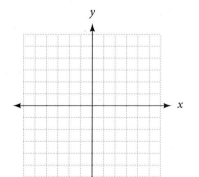

B Graph each of the following hyperbolas. In each case, label the coordinates of the center and the vertices and show the asymptotes. [Example 4]

35. $\dfrac{(x-2)^2}{16} - \dfrac{y^2}{4} = 1$

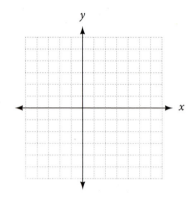

36. $\dfrac{(y-2)^2}{16} - \dfrac{x^2}{4} = 1$

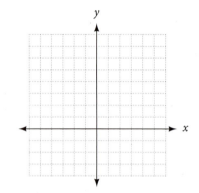

37. $9y^2 - x^2 - 4x + 54y + 68 = 0$

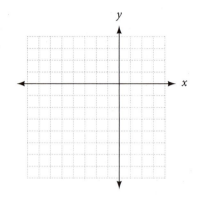

38. $4x^2 - y^2 - 24x + 4y + 28 = 0$

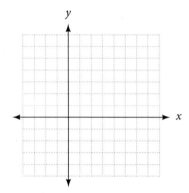

39. $4y^2 - 9x^2 - 16y + 72x - 164 = 0$

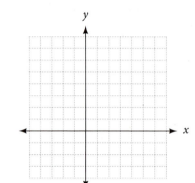

40. $4x^2 - y^2 - 16x - 2y + 11 = 0$

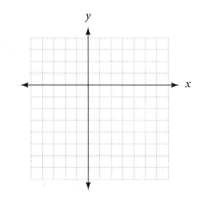

Solve.

41. Find y when x is 4 in the equation $\dfrac{x^2}{25} + \dfrac{y^2}{9} = 1$.

42. Find x when y is 3 in the equation $\dfrac{x^2}{4} + \dfrac{y^2}{25} = 1$.

43. Find y when x is 1.8 in $16x^2 + 9y^2 = 144$.

44. Find y when x is 1.6 in $49x^2 + 4y^2 = 196$.

45. Give the equations of the two asymptotes in the graph you found in Problem 15.

46. Give the equations of the two asymptotes in the graph you found in Problem 16.

Find an equation for each graph.

47.

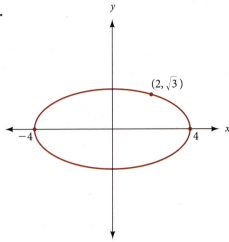

48.

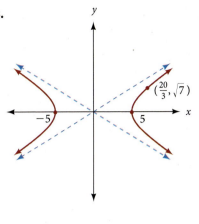

Applying the Concepts

The diagram shows the minor axis and the major axis for an ellipse.

In any ellipse, the length of the major axis is $2a$, and the length of the minor axis is $2b$ (these are the same a and b that appear in the general equation of an ellipse). Each of the two points shown on the major axis is a focus of the ellipse. If the distance from the center of the ellipse to each focus is c, then it is always true that $a^2 = b^2 + c^2$. You will need this information for some of the problems that follow.

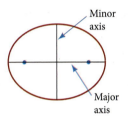

49. Archway A new theme park is planning an archway at its main entrance. The arch is to be in the form of a semi-ellipse with the major axis as the span. If the span is to be 40 feet and the height at the center is to be 10 feet, what is the equation of the ellipse? How far left and right of center could a 6-foot man walk upright under the arch?

50. Whispering Gallery An architecture student decides to build a model of the National Statuary Hall for his senior project. His model is the shape of a semi-ellipse as shown in the diagram. On the floor of his model, the student labels the focus c 12 inches from the center. The height h from the center of his model to its ceiling is 15 inches. Use this information to find the equation of the ellipse that represents the model's roof.

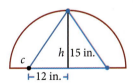

51. Elliptical Pool Table A children's science museum plans to build an elliptical pool table to demonstrate that a ball rolled from a particular point (focus) will always go into a hole located at another particular point (the other focus). The focus needs to be 1 foot from the vertex of the ellipse. If the table is to be 8 feet long, how wide should it be? *Hint:* The distance from the center to each focus point is represented by c and is found by using the equation $a^2 = b^2 + c^2$.

52. Lithotripter A lithotripter, used to break up kidney stones, is based on the ellipse $\frac{x^2}{36} + \frac{y^2}{25} = 1$. Determine how many units the kidney stone and the wave source (focus points) must be placed from the center of the ellipse. *Hint:* The distance from the center to each focus point is represented by c and is found by using the equation $a^2 = b^2 + c^2$.

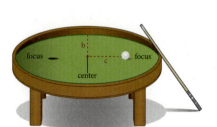

Maintaining Your Skills

Let $f(x) = \frac{2}{x-2}$ and $g(x) = \frac{2}{x+2}$. Find the following:

53. $f(4)$

54. $g(4)$

55. $f(g(0))$

56. $g(f(0))$

57. $f(x) + g(x)$

58. $f(x) - g(x)$

Getting Ready for the Next Section

59. Which of the following are solutions to $x^2 + y^2 < 16$?
(0, 0)   (4, 0)   (0, 5)

60. Which of the following are solutions to $y \le x^2 - 2$?
(0, 0)   (−2, 0)   (0, −2)

Expand and multiply.

61. $(2y + 4)^2$

62. $(y + 3)^2$

63. Solve $x - 2y = 4$ for x.

64. Solve $2x + 3y = 6$ for y.

Simplify.

65. $x^2 - 2(x^2 - 3)$

66. $x^2 + (x^2 - 4)$

Factor.

67. $5y^2 + 16y + 12$

68. $3x^2 + 17x - 28$

Solve.

69. $y^2 = 4$

70. $x^2 = 25$

71. $-x^2 + 6 = 2$

72. $5y^2 + 16y + 12 = 0$

Second-Degree Inequalities and Nonlinear Systems

TICKET TO SUCCESS

Keep these questions in mind as you read through the section. Then respond in your own words and in complete sentences.

1. What is a second-degree inequality and how might its graph differ from a linear inequality?
2. What is the significance of a broken line when graphing second-degree inequalities?
3. Is solving a system of nonlinear equations any different from solving a system of linear equations? Explain.
4. When solving a nonlinear system whose graphs are both circles, how many solutions sets can you expect?

A Graph second-degree inequalities.

B Solve systems of nonlinear equations.

C Graph the solution sets to systems of inequalities.

walex101/Shutterstock.com

Comets and planets both orbit the sun in an elliptical path. If the equation for the Earth and the comet are known, can you predict if the two objects will collide? In this section, we will work with systems of nonlinear equations like these to predict whether ellipses and other nonlinear pathways will intersect.

Previously, we graphed linear inequalities by first graphing the boundary and then choosing a test point not on the boundary to indicate the region used for the solution set. The problems in this section are very similar. We will use the same general methods for graphing second-degree inequalities in two variables that we used when we graphed linear inequalities in two variables.

A Second-Degree Inequalities

A *second-degree inequality* is an inequality that contains at least one squared variable.

EXAMPLE 1   Graph $x^2 + y^2 < 16$.

SOLUTION   The boundary is $x^2 + y^2 = 16$, which is a circle with center at the origin and a radius of 4. Since the inequality sign is $<$, the boundary is not included in the solution set and therefore must be represented with a broken line. The graph of the boundary is shown in Figure 1 on the next page.

PRACTICE PROBLEMS

1. Graph $x^2 + y^2 > 9$.

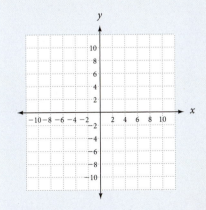

Answer
1–3. See Solutions Section.

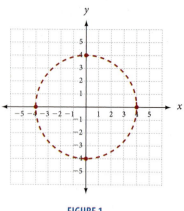

FIGURE 1

The solution set for $x^2 + y^2 < 16$ is either the region inside the circle or the region outside the circle. To see which region represents the solution set, we choose a convenient point not on the boundary and test it in the original inequality. The origin $(0, 0)$ is a convenient point. Since the origin satisfies the inequality $x^2 + y^2 < 16$, all points in the same region will also satisfy the inequality. The graph of the solution set is shown in Figure 2.

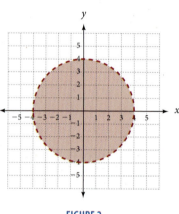

FIGURE 2

2. Graph $y \geq x^2 + 3$.

EXAMPLE 2   Graph the inequality $y \leq x^2 - 2$.

SOLUTION   The parabola $y = x^2 - 2$ is the boundary and is included in the solution set. Using $(0, 0)$ as the test point, we see that $0 \leq 0^2 - 2$ is a false statement, which means that the region containing $(0, 0)$ is not in the solution set. (See Figure 3.)

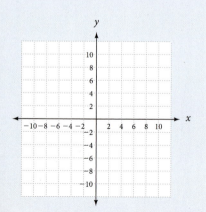

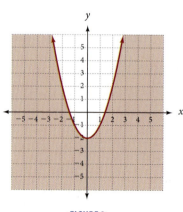

FIGURE 3

EXAMPLE 3   Graph $4y^2 - 9x^2 < 36$.

SOLUTION   The boundary is the hyperbola $4y^2 - 9x^2 = 36$ and is not included in the solution set. Testing $(0, 0)$ in the original inequality yields a true statement, which means that the region containing the origin is the solution set. (See Figure 4.)

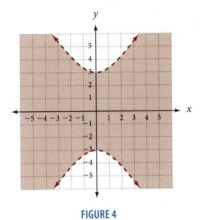

FIGURE 4

■

B   Nonlinear Systems

In an earlier chapter, we worked with equations of linear systems. Now we will practice solving nonlinear systems, which include at least one nonlinear equation. Remember, nonlinear equations can produce graphs in the shapes of circles, ellipses, hyperbolas, and parabolas.

EXAMPLE 4   Solve the system.

$$x^2 + y^2 = 4$$
$$x - 2y = 4$$

SOLUTION   In this case the substitution method is the most convenient. Solving the second equation for x in terms of y, we have

$$x - 2y = 4$$
$$x = 2y + 4$$

We now substitute $2y + 4$ for x in the first equation in our original system and proceed to solve for y.

$$(2y + 4)^2 + y^2 = 4$$

$4y^2 + 16y + 16 + y^2 = 4$ Expand $(2y + 4)^2$.

$5y^2 + 16y + 16 = 4$ Simplify left side.

$5y^2 + 16y + 12 = 0$ Add -4 to each side.

$(5y + 6)(y + 2) = 0$ Factor.

$5y + 6 = 0$ or $y + 2 = 0$ Set factors equal to 0.

$y = -\dfrac{6}{5}$ $y = -2$ Solve.

---

3. Graph $9x^2 - 4y^2 > 36$.

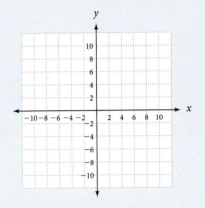

4. Solve the system.
$$x^2 + y^2 = 9$$
$$x - y = 3$$

Answer
4. $(3, 0)$, $(0, -3)$

These are the y-coordinates of the two solutions to the system. Substituting $y = -\frac{6}{5}$ into $x - 2y = 4$ and solving for x gives us $x = \frac{8}{5}$. Using $y = -2$ in the same equation yields $x = 0$. The two solutions to our system are $\left(\frac{8}{5}, -\frac{6}{5}\right)$ and $(0, -2)$. Although graphing the system is not necessary, it does help us visualize the situation (See Figure 5.)

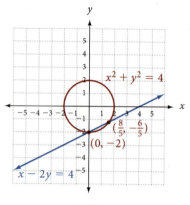

FIGURE 5

■

5. Solve the system.

$$16x^2 - 4y^2 = 64$$
$$x^2 + y^2 = 4$$

EXAMPLE 5 Solve the system.

$$16x^2 - 4y^2 = 64$$

$$x^2 + y^2 = 9$$

SOLUTION Since each equation is of the second degree in both x and y, it is easier to solve this system by eliminating one of the variables by addition. To eliminate y, we multiply the bottom equation by 4 and add the result to the top equation.

$$16x^2 - 4y^2 = 64$$
$$\underline{4x^2 + 4y^2 = 36}$$
$$20x^2 = 100$$
$$x^2 = 5$$
$$x = \pm\sqrt{5}$$

The x-coordinates of the points of intersection are $\sqrt{5}$ and $-\sqrt{5}$. We substitute each back into the second equation in the original system and solve for y.

When → $\qquad\qquad x = \sqrt{5}$

the equation → $\qquad x^2 + y^2 = 9$

becomes → $\qquad (\sqrt{5})^2 + y^2 = 9$

$$5 + y^2 = 9$$
$$y^2 = 4$$
$$y = \pm 2$$

When → $\qquad\qquad x = -\sqrt{5}$

the equation → $\qquad x^2 + y^2 = 9$

becomes → $\qquad (-\sqrt{5})^2 + y^2 = 9$

$$5 + y^2 = 9$$
$$y^2 = 4$$
$$y = \pm 2$$

Answer

5. $(2, 0), (-2, 0)$

The four points of intersection are $(\sqrt{5}, 2)$, $(\sqrt{5}, -2)$, $(-\sqrt{5}, 2)$, and $(-\sqrt{5}, -2)$. Graphically, the situation is as shown in Figure 6.

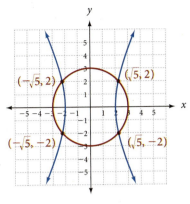

FIGURE 6

∎

EXAMPLE 6   Solve the system.

$$x^2 - 2y = 2$$

$$y = x^2 - 3$$

SOLUTION   We can solve this system using the substitution method. Replacing y in the first equation with $x^2 - 3$ from the second equation, we have

$$x^2 - 2(x^2 - 3) = 2$$

$$-x^2 + 6 = 2$$

$$x^2 = 4$$

$$x = \pm 2$$

Using either 2 or -2 in the equation $y = x^2 - 3$ gives us $y = 1$. The system has two solutions: $(2, 1)$ and $(-2, 1)$.   ∎

EXAMPLE 7   The sum of the squares of two numbers is 34. The difference of their squares is 16. Find the two numbers.

SOLUTION   Let x and y be the two numbers. The sum of their squares is $x^2 + y^2$, and the difference of their squares is $x^2 - y^2$. (We can assume here that x^2 is the larger number.) The system of equations that describes the situation is

$$x^2 + y^2 = 34$$

$$x^2 - y^2 = 16$$

We can eliminate y by simply adding the two equations. The result of doing so is

$$2x^2 = 50$$

$$x^2 = 25$$

$$x = \pm 5$$

Substituting $x = 5$ into either equation in the system gives $y = \pm 3$. Using $x = -5$ gives the same results, $y = \pm 3$. The four pairs of numbers that are solutions to the original problem are 5 and 3, -5 and 3, 5 and -3, -5 and -3.   ∎

6. Solve the system.
$$x^2 + y^2 = 4$$
$$y = x^2 - 4$$

7. One number is two less than the square of another number. The sum of the squares of the two numbers is 58. Find the two numbers.

Answers
6. $(2, 0)$, $(-2, 0)$, $(\sqrt{3}, -1)$, $(-\sqrt{3}, -1)$
7. 3, 7 and -3, 7

We now turn our attention to systems of inequalities. To solve a system of inequalities by graphing, we simply graph each inequality on the same set of axes. The solution set for the system is the region common to both graphs—the intersection of the individual solution sets.

C Systems of Inequalities

EXAMPLE 8   Graph the solution set for the system.

$$x^2 + y^2 \leq 9$$
$$\frac{x^2}{4} + \frac{y^2}{25} \geq 1$$

SOLUTION   The boundary for the top inequality is a circle with center at the origin and a radius of 3. The solution set lies inside the boundary. The boundary for the second inequality is an ellipse. In this case, the solution set lies outside the boundary. (See Figure 7.) The solution set for the system is shown in Figure 8.

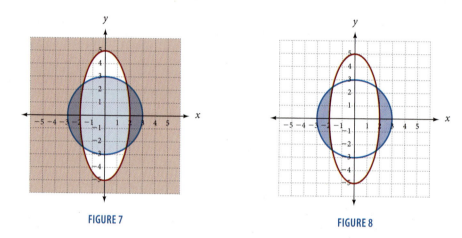

FIGURE 7

FIGURE 8

The solution set is the intersection of the two individual solution sets.   ∎

8. Graph the solution set for the system.

$$x^2 + y^2 \geq 9$$
$$\frac{x^2}{4} + \frac{y^2}{25} \leq 1$$

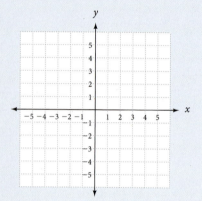

Answer

8. See Solutions Section.

Problem Set 10.3

Moving Toward Success

"He who would learn to fly one day must first learn to stand and walk and run and climb and dance; one cannot fly into flying."

—Friedrich Nietzsche, 1844–1900, German philosopher

1. Why is it important to go into an exam knowing you can work any problem on your list of difficult problems?

2. How would visualizing yourself working through possible problems increase your potential for success during a test?

A Graph each of the following inequalities. [Examples 1–3]

1. $x^2 + y^2 \leq 49$

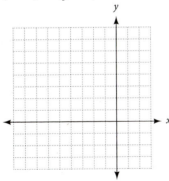

2. $x^2 + y^2 < 49$

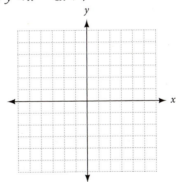

3. $(x - 2)^2 + (y + 3)^2 < 16$

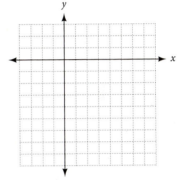

4. $(x + 3)^2 + (y - 2)^2 \geq 25$

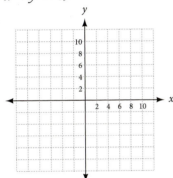

5. $y < x^2 - 6x + 7$

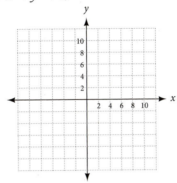

6. $y \geq x^2 + 2x - 8$

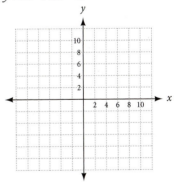

7. $4x^2 + 25y^2 \leq 100$

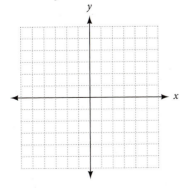

8. $25x^2 - 4y^2 > 100$

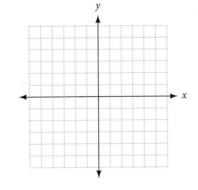

B Solve each of the following systems of equations. [Examples 4–6]

9. $x^2 + y^2 = 9$
  $2x + y = 3$

10. $x^2 + y^2 = 9$
  $x + 2y = 3$

11. $x^2 + y^2 = 16$
  $x + 2y = 8$

12. $x^2 + y^2 = 16$
  $x - 2y = 8$

13. $x^2 + y^2 = 25$
  $x^2 - y^2 = 25$

14. $x^2 + y^2 = 4$
  $2x^2 - y^2 = 5$

15. $x^2 + y^2 = 9$
  $y = x^2 - 3$

16. $x^2 + y^2 = 4$
  $y = x^2 - 2$

17. $x^2 + y^2 = 16$
  $y = x^2 - 4$

18. $x^2 + y^2 = 1$
  $y = x^2 - 1$

19. $3x + 2y = 10$
  $y = x^2 - 5$

20. $4x + 2y = 10$
  $y = x^2 - 10$

21. $y = x^2 + 2x - 3$
  $y = -x + 1$

22. $y = -x^2 - 2x + 3$
  $y = -x + 1$

23. $y = x^2 - 6x + 5$
  $y = x - 5$

24. $y = x^2 - 2x - 4$
  $y = x - 4$

25. $4x^2 - 9y^2 = 36$
  $4x^2 + 9y^2 = 36$

26. $4x^2 + 25y^2 = 100$
  $4x^2 - 25y^2 = 100$

27. $x - y = 4$
  $x^2 + y^2 = 16$

28. $x + y = 2$
  $x^2 - y^2 = 4$

29. In a previous problem set, you found the equations for the three circles below. The equations are

Circle A Circle B Circle C

$(x + 8)^2 + y^2 = 64$ $x^2 + y^2 = 64$ $(x - 8)^2 + y^2 = 64$

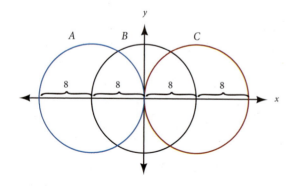

a. Find the points of intersection of circles A and B.

b. Find the points of intersection of circles B and C.

30. A magician is holding two rings that seem to lie in the same plane and intersect in two points. Each ring is 10 inches in diameter. If a coordinate system is placed with its origin at the center of the first ring and the x-axis contains the center of the second ring, then the equations are as follows:

First Ring   Second Ring

$x^2 + y^2 = 25$ $(x - 5)^2 + y^2 = 25$

Find the points of intersection of the two rings.

C Graph the solution sets to the following systems. [Example 8]

31. $x^2 + y^2 < 9$
$y \geq x^2 - 1$

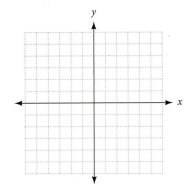

32. $x^2 + y^2 \leq 16$
$y < x^2 + 2$

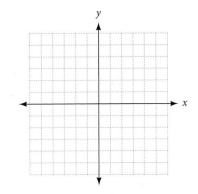

33. $\dfrac{x^2}{9} + \dfrac{y^2}{25} \leq 1$
$\dfrac{x^2}{4} - \dfrac{y^2}{9} > 1$

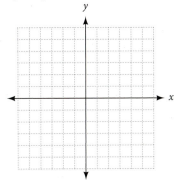

34. $\dfrac{x^2}{4} + \dfrac{y^2}{16} \geq 1$
$\dfrac{x^2}{9} - \dfrac{y^2}{25} < 1$

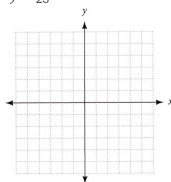

35. $4x^2 + 9y^2 \leq 36$
$y > x^2 + 2$

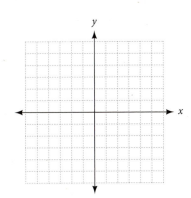

36. $9x^2 + 4y^2 \geq 36$
$y < x^2 + 1$

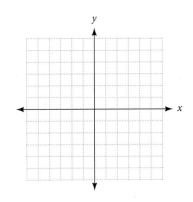

37. A parabola and a circle each contain the points $(-4, 0), (0, 4),$ and $(4, 0),$ as shown below. The equations for each of the curves are

Circle	Parabola
$x^2 + y^2 = 16$	$y = 4 - \dfrac{1}{4}x^2$

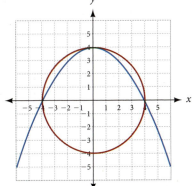

Write a system of inequalities that describes the regions in quadrants I and II that are between the two curves, if the boundaries are not included with the shaded region.

38. Find a system of inequalities that describes the shaded region in the figure below if the boundaries of the shaded region are included and equations of the circles are given by

Circle A	Circle B	Circle C
$(x + 8)^2 + y^2 = 64$	$x^2 + y^2 = 64$	$(x - 8)^2 + y^2 = 64$

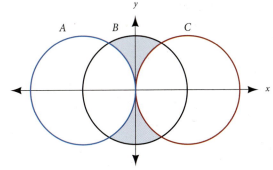

Applying the Concepts

39. Number Problems The sum of the squares of two numbers is 89. The difference of their squares is 39. Find the numbers.

40. Number Problems The difference of the squares of two numbers is 35. The sum of their squares is 37. Find the numbers.

41. Number Problems One number is 3 less than the square of another. Their sum is 9. Find the numbers.

42. Number Problems The square of one number is 2 less than twice the square of another. The sum of the squares of the two numbers is 25. Find the numbers.

Maintaining Your Skills

43. Let $g(x) = \left(x + \dfrac{2}{5}\right)^2$. Find all values for the variable x for which $g(x) = 0$.

44. Let $f(x) = (x + 2)^2 - 25$. Find all values for the variable x for which $f(x) = 0$.

For the problems below, let $f(x) = x^2 + 4x - 4$. Find all values for the variable x for which $f(x) = g(x)$.

45. $g(x) = 1$

46. $g(x) = -7$

47. $g(x) = x - 6$

48. $g(x) = x + 6$

Chapter 10 Summary

■ Distance Formula [10.1]

1. The distance between $(5, 2)$ and $(-1, 1)$ is
$$d = \sqrt{(5 + 1)^2 + (2 - 1)^2}$$
$$= \sqrt{37}$$

The distance between the two points (x_1, y_1) and (x_2, y_2) is given by the formula
$$d = \sqrt{(x_2 - x_1)^2 + (y_2 - y_1)^2}$$

■ The Circle [10.1]

2. The graph of the circle $(x - 3)^2 + (y + 2)^2 = 25$ will have its center at $(3, -2)$ and the radius will be 5.

The graph of any equation of the form
$$(x - a)^2 + (y - b)^2 = r^2$$
will be a circle having its center at (a, b) and a radius of r.

■ The Ellipse [10.2]

3. The ellipse $\frac{x^2}{9} + \frac{y^2}{4} = 1$ will cross the x-axis at 3 and -3 and will cross the y-axis at 2 and -2.

Any equation that can be put in the form
$$\frac{x^2}{a^2} + \frac{y^2}{b^2} = 1$$
will have an ellipse for its graph. The x-intercepts will be at a and $-a$, and the y-intercepts will be at b and $-b$.

■ The Hyperbola [10.2]

4. The hyperbola $\frac{x^2}{4} - \frac{y^2}{9} = 1$ will cross the x-axis at 2 and -2. It will not cross the y-axis.

The graph of an equation that can be put in either of the forms
$$\frac{x^2}{a^2} - \frac{y^2}{b^2} = 1 \quad \text{or} \quad \frac{y^2}{a^2} - \frac{x^2}{b^2} = 1$$
will be a hyperbola. The x-intercepts, for the first equation, will be at a and $-a$. The y-intercepts, for the second equation, will be at a and $-a$. Two straight lines, called *asymptotes,* are associated with the graph of every hyperbola. Although the asymptotes are not part of the hyperbola, they are useful in sketching the graph.

■ Second-Degree Inequalities [10.3]

5. The graph of the inequality
$$x^2 + y^2 < 9$$
is all points inside the circle with center at the origin and radius 3. The circle itself is not part of the solution and therefore is shown with a broken curve.

We graph second-degree inequalities in much the same way that we graphed linear inequalities; that is, we begin by graphing the boundary, using a solid curve if the boundary is included in the solution (this happens when the inequality symbol is $\geq$ or $\leq$) or a broken curve if the boundary is not included in the solution (when the inequality symbol is $>$ or $<$). After we have graphed the boundary, we choose a test point that is not on the boundary and try it in the original inequality. A true statement indicates we are in the region of the solution. A false statement indicates we are not in the region of the solution.

6. We can solve the system

$$x^2 + y^2 = 4$$
$$x = 2y + 4$$

by substituting $2y + 4$ from the second equation for x in the first equation:

$$(2y + 4)^2 + y^2 = 4$$
$$4y^2 + 16y + 16 + y^2 = 4$$
$$5y^2 + 16y + 12 = 0$$
$$(5y + 6)(y + 2) = 0$$
$$y = -\frac{6}{5} \quad \text{or} \quad y = -2$$

Substituting these values of y into the second equation in our system gives $x = \frac{8}{5}$ and $x = 0$. The solutions are $\left(\frac{8}{5}, -\frac{6}{5}\right)$ and $(0, -2)$.

■ Systems of Nonlinear Equations [10.3]

A system of nonlinear equations is two equations, at least one of which is not linear, considered at the same time. The solution set for the system consists of all ordered pairs that satisfy both equations. In most cases we use the substitution method to solve these systems; however, the addition method can be used if like variables are raised to the same power in both equations. It is sometimes helpful to graph each equation in the system on the same set of axes to anticipate the number and approximate positions of the solutions.

Find the distance between the following points. [10.1]

1. $(2, 6)$, $(-1, 5)$ **2.** $(3, -4)$, $(1, -1)$

3. $(0, 3)$, $(-4, 0)$ **4.** $(-3, 7)$, $(-3, -2)$

5. Find x so that the distance between $(x, -1)$ and $(2, -4)$ is 5. [10.1]

6. Find y so that the distance between $(3, -4)$ and $(-3, y)$ is 10. [10.1]

Find the equation of each circle. [10.1]

7. Center at the origin, x-intercepts ± 5

8. Center at the origin, y-intercepts ± 3

9. Center at $(-2, 3)$ and passing through the point $(2, 0)$

10. Center at $(-6, 8)$ and passing through the origin

Give the center and radius of each circle, and then sketch the graph. [10.1]

11. $x^2 + y^2 = 4$

12. $(x - 3)^2 + (y + 1)^2 = 16$

13. $x^2 + y^2 - 6x + 4y = -4$

14. $x^2 + y^2 + 4x - 2y = 4$

Graph each of the following. Label the x- and y-intercepts. [10.2]

15. $\dfrac{x^2}{4} + \dfrac{y^2}{9} = 1$ **16.** $4x^2 + y^2 = 16$

Graph the following. Show the asymptotes. [10.2]

17. $\dfrac{x^2}{4} - \dfrac{y^2}{9} = 1$ **18.** $4x^2 - y^2 = 16$

Graph each of the following inequalities. [10.3]

19. $x^2 + y^2 < 9$ **20.** $(x + 2)^2 + (y - 1)^2 \leq 4$

21. $y \geq x^2 - 1$ **22.** $9x^2 + 4y^2 \leq 36$

Graph the solution set for each system. [10.3]

23. $x^2 + y^2 < 16$ **24.** $x + y \leq 2$
       $y > x^2 - 4$ $-x + y \leq 2$
                                   $y \geq -2$

Solve each system of equations. [10.3]

25. $x^2 + y^2 = 16$ **26.** $x^2 + y^2 = 4$
      $2x + y = 4$ $y = x^2 - 2$

27. $9x^2 - 4y^2 = 36$
      $9x^2 + 4y^2 = 36$

Simplify.

1. $\dfrac{-9 - 6}{4 - 7}$

2. $\dfrac{5(-6) + 3(-2)}{4(-3) + 3}$

3. $\dfrac{18a^7 b^{-4}}{36a^2 b^{-8}}$

4. $\dfrac{x^{3/4} y^2}{x^{1/2} y^{1/4}}$

5. $\dfrac{12x^2 y^3 - 16x^2 y + 8xy^3}{4xy}$

6. $\dfrac{y^2 - y - 6}{y^2 - 4}$

7. $\log_4 16$

8. $\log_3 27$

Factor completely.

9. $ab^3 + b^3 + 6a + 6$

10. $8x^2 - 5x - 3$

Solve.

11. $6 - 2(5x - 1) + 4x = 20$

12. $3 - 2(3x - 4) = -1$

13. $(x + 1)(x + 2) = 12$

14. $1 - \dfrac{2}{x} = \dfrac{8}{x^2}$

15. $t - 6 = \sqrt{t - 4}$

16. $\sqrt{7x - 4} = -2$

17. $(4x - 3)^2 = -50$

18. $(x - 3)^2 = -3$

Solve and graph the solution on the number line.

19. $|4x + 3| - 6 > 5$

20. $|2x + 5| - 2 < 9$

Graph on a rectangular coordinate system.

21. $y < \dfrac{1}{2}x + 3$

22. $3x - y < -2$

23. $y = (x - 2)^2 - 3$

24. $y = x^2 - 2x - 3$

Multiply.

25. $\dfrac{3y^2 - 3y}{3y - 12} \cdot \dfrac{y^2 - 2y - 8}{y^2 + 3y + 2}$

26. $\dfrac{x^2 - 16}{x^2 + 5x + 6} \cdot \dfrac{x^2 + 6x + 9}{x^3 + 4x^2}$

27. $(2 + 3i)(1 - 4i)$

28. $(2 + 3i)(1 + 4i)$

Rationalize the denominator.

29. $\dfrac{3}{\sqrt{7} - \sqrt{3}}$

30. $\dfrac{5\sqrt{6}}{2\sqrt{6} + 7}$

Find the inverse, $f^{-1}(x)$.

31. $f(x) = 4x + 1$

32. $f(x) = \dfrac{1}{2}x + 3$

33. Find x if $\log x = 3.9786$ to the nearest hundreth.

34. Find $\log_6 14$ to the nearest hundredth.

35. Solve $mx + 2 = nx - 3$ for x.

36. Solve $S = 2x^2 + 4xy$ for y.

37. Find the slope of the line through $(2, -3)$ and $(-1, -3)$.

38. Find the slope and y-intercept for $2x - 3y = 12$.

39. If $f(x) = -\dfrac{3}{2}x + 1$, find $f(4)$.

40. If $C(t) = 80\left(\dfrac{1}{2}\right)^{t/5}$, find $C(5)$ and $C(10)$.

41. Find an equation that has solutions $x = 1$ and $x = \dfrac{2}{3}$.

42. Find an equation with solutions $t = -\dfrac{3}{4}$ and $t = \dfrac{1}{5}$.

43. If $f(x) = 3x - 7$ and $g(x) = 4 - x$, find $(f \circ g)(x)$.

44. If $f(x) = 4 - x^2$ and $g(x) = 5x - 1$, find $(g \circ f)(x)$.

45. Add -8 to the difference of -7 and 4.

46. Subtract 9 from the sum of -2 and 5.

47. Variation y varies inversely with the square of x. If y is 3 when x is 3, find y when x is 6.

48. Mixture How many gallons of 20% alcohol solution and 60% alcohol solution must be mixed to get 16 gallons of 30% alcohol solution?

Use the illustration to answer problems 49 and 50. Round your answers to the nearest minute.

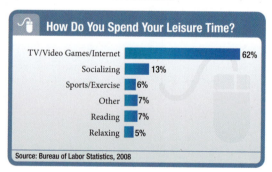

How Do You Spend Your Leisure Time?

TV/Video Games/Internet 62%
Socializing 13%
Sports/Exercise 6%
Other 7%
Reading 7%
Relaxing 5%

Source: Bureau of Labor Statistics, 2008

49. If 150 people were surveyed, how many people spend their leisure time playing video games, watching TV or surfing the internet?

50. If 150 people were surveyed, how many people spend their leisure time reading or relaxing?

Find the distance between the points. [10.1]

1. $(-9, 4)$ and $(3, -1)$ **2.** $(4, -6)$ and $(0, -4)$

3. Find x so that $(x, 2)$ is $2\sqrt{5}$ units from $(-1, 4)$. [10.1]

4. Give the equation of the circle with center at $(-2, 4)$ and radius 3. [10.1]

5. Give the equation of the circle with center at the origin that contains the point $(-3, -4)$. [10.1]

Find the center and radius of the circle. [10.1]

6. $x^2 + (y + 2)^2 = 64$

7. $x^2 + y^2 - 10x + 6y = 5$

Graph each of the following. [10.1, 10.2, 10.3]

8. $(x - 1)^2 + (y + 2)^2 = 9$ **9.** $\dfrac{x^2}{25} + \dfrac{y^2}{4} = 1$

10. $4x^2 - y^2 = 16$

11. $25x^2 + 150x + 4y^2 + 125 = 0$

12. $(x - 2)^2 + (y + 1)^2 \le 9$ **13.** $y > x^2 - 4$

Solve the following systems. [10.3]

14. $x^2 + y^2 = 25$ **15.** $x^2 + y^2 = 16$
      $2x + y = 5$ $y = x^2 - 4$

16. Graph the solution set to the system. [10.3]
$$x^2 + y^2 \ge 1$$
$$y < -x^2 + 1$$

Match each equation with its graph. [10.1, 10.2]

17. $x^2 + 4y^2 = 16$ **18.** $x^2 + y^2 = 16$

19. $x^2 - 4y^2 = 16$ **20.** $4x^2 + y^2 = 16$

A
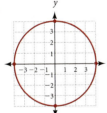

B
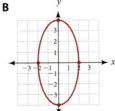

C
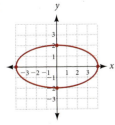

D
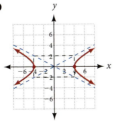

Find an equation for each graph. [10.1, 10.2]

21.
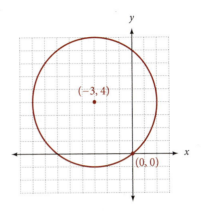

22.
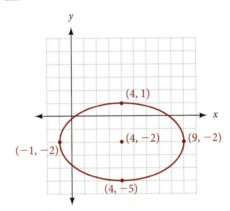

23.
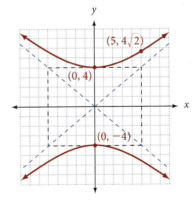

CONIC SECTIONS

Group Project

Constructing Ellipses

Number of People	4
Time Needed	20 minutes
Equipment	Graph paper, pencils, string, and thumbtacks
Background	The geometric definition for an ellipse is the set of points the sum of whose distances from two fixed points (called foci) is a constant. We can use this definition to draw an ellipse using thumbtacks, string, and a pencil.
Procedure	

1. Start with a piece of string 7 inches long. Place thumbtacks through the string $\frac{1}{2}$ inch from each end, then tack the string to a pad of graph paper so that the tacks are 4 inches apart. Pull the string tight with the tip of a pencil, then trace all the way around the two tacks. (See Figure 1.) The resulting diagram will be an ellipse.

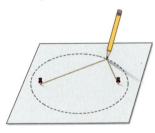

FIGURE 1

2. The line segment that passes through the tacks (these are the foci) and connects the opposite ends of the ellipse is called the major axis. The line segment perpendicular to the major axis that passes through the center of the ellipse and connects the opposite ends of the ellipse is called the minor axis. (See Figure 2.) Measure the length of the major axis and the length of the minor axis. Record your results in Table 1.

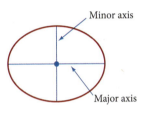

Minor axis

Major axis

FIGURE 2

3. Explain how drawing the ellipse as you have in step 2 shows that the geometric definition of an ellipse given at the beginning of this project is, in fact, correct.

4. Next, move the tacks so that they are 3 inches apart. Trace out that ellipse. Measure the length of the major axis and the length of the minor axis, and record your results in Table 1.

5. Repeat step 4 with the tacks 2 inches apart.

6. If the length of the string between the tacks stays at 6 inches, and the tacks were placed 6 inches apart, then the resulting ellipse would be a _____. If the tacks were placed 0 inches apart, then the resulting ellipse would be a _____.

TABLE 1			
ELLIPSES (ALL LENGTHS ARE INCHES)			
Length of String	Distance Between Foci	Length of Major Axis	Length of Minor Axis
6	4		
6	3		
6	2		

Hypatia of Alexandria

The first woman mentioned in the history of mathematics is Hypatia of Alexandria. Research the life of Hypatia, and then write an essay that begins with a description of the time and place in which she lived. Continue with an indication of the type of person she was, her accomplishments in areas other than mathematics, and how she was viewed by her contemporaries.

Bettmann/Corbis

PRESIDENT'S PARK: THE ELLIPSE

Gray Buildings © District of Columbia DC GIS) & CyberCity
Gray Buildings © 2008 Sanborn
President's Park, Washington D.C.

President's Park in Washington, DC is a public park, located just south of the White House. The park is often called The Ellipse because of its shape. An ellipse is created when a plane intersects a cone creating a closed curve. A circle is special case of an ellipse. Now let's apply our knowledge of conic sections to President's Park.

STEP 1 Use Google Earth to find President's Park in Washington, DC.

STEP 2 Use the **Ruler** tool set to **Feet** to measure the major and minor axes of the park. In this case, the major axis is the maximum horizontal line with a midpoint at the origin of the ellipse. The minor axis is the maximum vertical line with a midpoint at the origin of the ellipse. (Answers will vary slightly.)

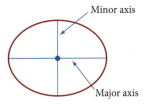

STEP 3 Using your measurements, find an equation for the elliptical path around the park. (Again, answers will vary slightly.)

Introduction to Determinants

Determinants were originally known as resultants, a name given to them by French mathematician Pierre Simon Laplace (1749-1827); however, the work of German mathematician Gottfried Wilhelm Leibniz contains the beginnings of the original idea of resultants, or determinants.

In this section, we will expand and evaluate *determinants*. The purpose of this section is simply to be able to find the value of a given determinant. As we will see in Appendix B, determinants are very useful in solving systems of linear equations. Before we apply determinants to systems of linear equations, however, we must practice calculating the value of some determinants.

A 2×2 Determinants

> **Definition**
>
> The value of the **2 × 2** (read as "2 by 2") **determinant**
>
> $$\begin{vmatrix} a & c \\ b & d \end{vmatrix}$$
>
> is given by
>
> $$\begin{vmatrix} a & c \\ b & d \end{vmatrix} = ad - bc$$

From the preceding definition we see that a determinant is simply a square array of numbers with two vertical lines enclosing it. The value of a 2×2 determinant is found by cross-multiplying on the diagonals and then subtracting, a diagram that looks like

$$\begin{vmatrix} a & c \\ b & d \end{vmatrix} = ad - bc$$

EXAMPLE 1 Find the value of the following 2×2 determinants.

a. $\begin{vmatrix} 1 & 2 \\ 3 & 4 \end{vmatrix} = 1(4) - 3(2) = 4 - 6 = -2$

b. $\begin{vmatrix} 3 & -2 \\ 5 & 7 \end{vmatrix} = 3(7) - (-2)5 = 21 + 10 = 31$ ∎

EXAMPLE 2 Solve for x if

$$\begin{vmatrix} x & 2 \\ x & 4 \end{vmatrix} = 8$$

SOLUTION We expand the determinant on the left side to get

$$x(4) - x(2) = 8$$

$$4x - 2x = 8$$

$$2x = 8$$

$$x = 4$$ ∎

PRACTICE PROBLEMS

1. Find the value of each determinant.

a. $\begin{vmatrix} 2 & 1 \\ 4 & 3 \end{vmatrix}$

b. $\begin{vmatrix} 4 & -2 \\ 0 & 3 \end{vmatrix}$

2. Solve for x if

$$\begin{vmatrix} -3 & x \\ 2 & x \end{vmatrix} = 20$$

Answers
1. a. 2 **b.** 12
2. −4

B 3 × 3 Determinants

We now turn our attention to 3×3 determinants. A 3×3 determinant is also a square array of numbers enclosed by a vertical line, the value of which is given by the following definition.

> **Definition**
>
> The value of the **3 × 3 determinant**
> $$\begin{vmatrix} a_1 & b_1 & c_1 \\ a_2 & b_2 & c_2 \\ a_3 & b_3 & c_3 \end{vmatrix}$$
> is given by
> $$\begin{vmatrix} a_1 & b_1 & c_1 \\ a_2 & b_2 & c_2 \\ a_3 & b_3 & c_3 \end{vmatrix} = a_1b_2c_3 + a_3b_1c_2 + a_2b_3c_1 - a_3b_2c_1 - a_1b_3c_2 - a_2b_1c_3$$

At first glance, the expansion of a 3×3 determinant looks a little complicated. There are actually two different methods used to find the six products in the preceding definition, which simplifies matters somewhat.

METHOD I We begin by writing the determinant with the first two columns repeated on the right.

$$\begin{vmatrix} a_1 & b_1 & c_1 \\ a_2 & b_2 & c_2 \\ a_3 & b_3 & c_3 \end{vmatrix}\begin{matrix} a_1 & b_1 \\ a_2 & b_2 \\ a_3 & b_3 \end{matrix}$$

The positive products in the definition come from multiplying down the three full diagonals:

$$\begin{vmatrix} a_1 & b_1 & c_1 \\ a_2 & b_2 & c_2 \\ a_3 & b_3 & c_3 \end{vmatrix}\begin{matrix} a_1 & b_1 \\ a_2 & b_2 \\ a_3 & b_3 \end{matrix}$$

The negative products come from multiplying up the three full diagonals.

$$\begin{vmatrix} a_1 & b_1 & c_1 \\ a_2 & b_2 & c_2 \\ a_3 & b_3 & c_3 \end{vmatrix}\begin{matrix} a_1 & b_1 \\ a_2 & b_2 \\ a_3 & b_3 \end{matrix}$$

NOTE

Check the products found by multiplying up and down the diagonals given here with the products given in the definition of a 3×3 determinant to see that they match.

3. Find the value of
$$\begin{vmatrix} 2 & 0 & -1 \\ 3 & 1 & 2 \\ 5 & -2 & 1 \end{vmatrix}$$

EXAMPLE 3   Find the value of
$$\begin{vmatrix} 1 & 3 & -2 \\ 2 & 0 & 1 \\ 4 & -1 & 1 \end{vmatrix}$$

Answer

3. 21

SOLUTION   Repeating the first two columns and then finding the products down the diagonals and the products up the diagonals as given in Method 1, we have

$$
\begin{vmatrix} 1 & 3 & -2 \\ 2 & 0 & 1 \\ 4 & -1 & 1 \end{vmatrix} \begin{matrix} 1 & 3 \\ 2 & 0 \\ 4 & -1 \end{matrix}
$$

$$
= 1(0)(1) + 3(1)(4) + (-2)(2)(-1) - 4(0)(-2) - (-1)(1)(1) - 1(2)(3)
$$

$$
= 0 + 12 + 4 - 0 + 1 - 6
$$

$$
= 11 \qquad\blacksquare
$$

METHOD 2   The second method of evaluating a 3×3 determinant is called *expansion by minors*.

Definition

The **minor** for an element in a 3×3 determinant is the determinant consisting of the elements remaining when the row and column to which the element belongs are deleted. For example, in the determinant

$$
\begin{vmatrix} a_1 & b_1 & c_1 \\ a_2 & b_2 & c_2 \\ a_3 & b_3 & c_3 \end{vmatrix}
$$

$$
\text{Minor for element } a_1 = \begin{vmatrix} b_2 & c_2 \\ b_3 & c_3 \end{vmatrix}
$$

$$
\text{Minor for element } b_2 = \begin{vmatrix} a_1 & c_1 \\ a_3 & c_3 \end{vmatrix}
$$

$$
\text{Minor for element } c_3 = \begin{vmatrix} a_1 & b_1 \\ a_2 & b_2 \end{vmatrix}
$$

Before we can evaluate a 3×3 determinant by Method 2, we must first define what is known as the *sign array* for a 3×3 determinant.

Definition

The **sign array** for a 3×3 determinant is a 3×3 array of signs in the following pattern:

$$
\begin{vmatrix} + & - & + \\ - & + & - \\ + & - & + \end{vmatrix}
$$

The sign array begins with a plus sign in the upper left-hand corner. The signs then alternate between plus and minus across every row and down every column.

NOTE
If you have read this far and are confused, hang on. After you have done a couple of examples you will find expansion by minors to be a fairly simple process. It just takes a lot of writing to explain it.

To Evaluate a 3 X 3 Determinant by Expansion of Minors

We can evaluate a 3×3 determinant by expanding across any row or down any column as follows:

Step 1 Choose a row or column to expand.

Step 2 Write the product of each element in the row or column chosen in step 1 with its minor.

Step 3 Connect the three products in step 2 with the signs in the corresponding row or column in the sign array.

To illustrate the procedure, we will use the same determinant we used in Example 3.

4. Expand across the first row.

$$\begin{vmatrix} 2 & 0 & -1 \\ 3 & 1 & 2 \\ 5 & -2 & 1 \end{vmatrix}$$

EXAMPLE 4   Expand across the first row.

$$\begin{vmatrix} 1 & 3 & -2 \\ 2 & 0 & 1 \\ 4 & -1 & 1 \end{vmatrix}$$

SOLUTION   The products of the three elements in row 1 with their minors are

$$1 \begin{vmatrix} 0 & 1 \\ -1 & 1 \end{vmatrix} \qquad 3 \begin{vmatrix} 2 & 1 \\ 4 & 1 \end{vmatrix} \qquad (-2) \begin{vmatrix} 2 & 0 \\ 4 & -1 \end{vmatrix}$$

Connecting these three products with the signs from the first row of the sign array, we have

$$+1 \begin{vmatrix} 0 & 1 \\ -1 & 1 \end{vmatrix} \qquad -3 \begin{vmatrix} 2 & 1 \\ 4 & 1 \end{vmatrix} \qquad +(-2) \begin{vmatrix} 2 & 0 \\ 4 & -1 \end{vmatrix}$$

We complete the problem by evaluating each of the three 2×2 determinants and then simplifying the resulting expression.

$$+1[0 - (-1)] - 3(2 - 4) + (-2)(-2 - 0)$$
$$= 1(1) - 3(-2) + (-2)(-2)$$
$$= 1 + 6 + 4$$
$$= 11 \qquad \blacksquare$$

NOTE
This method of evaluating a determinant is actually more valuable than our first method because it works with any size determinant from 3×3 to 4×4 to any higher order determinant. Method 1 works only on 3×3 determinants. It cannot be used on a 4×4 determinant.

The results of Examples 3 and 4 match. It makes no difference which method we use—the value of a 3×3 determinant is unique.

5. Expand down column 2.

$$\begin{vmatrix} 0 & 4 & -2 \\ 3 & 1 & 1 \\ 1 & -2 & 0 \end{vmatrix}$$

EXAMPLE 5   Expand down column 2.

$$\begin{vmatrix} 2 & 3 & -2 \\ 1 & 4 & 1 \\ 1 & 5 & -1 \end{vmatrix}$$

SOLUTION   We connect the products of elements in column 2 and their minors with the signs from the second column in the sign array.

$$\begin{vmatrix} 2 & 3 & -2 \\ 1 & 4 & 1 \\ 1 & 5 & -1 \end{vmatrix} = -3 \begin{vmatrix} 1 & 1 \\ 1 & -1 \end{vmatrix} + 4 \begin{vmatrix} 2 & -2 \\ 1 & -1 \end{vmatrix} - 5 \begin{vmatrix} 2 & -2 \\ 1 & 1 \end{vmatrix}$$
$$= -3(-1 - 1) + 4[-2 - (-2)] - 5[2 - (-2)]$$
$$= -3(-2) + 4(0) - 5(4)$$
$$= 6 + 0 - 20$$
$$= -14 \qquad \blacksquare$$

Answers
4. 21
5. 18

Problem Set A

A Find the value of the following 2×2 determinants. [Example 1]

1. $\begin{vmatrix} 1 & 0 \\ 2 & 3 \end{vmatrix}$

2. $\begin{vmatrix} 5 & 3 \\ 3 & 2 \end{vmatrix}$

3. $\begin{vmatrix} 2 & 1 \\ 3 & 4 \end{vmatrix}$

4. $\begin{vmatrix} 4 & 1 \\ 5 & 2 \end{vmatrix}$

5. $\begin{vmatrix} 0 & 1 \\ 1 & 0 \end{vmatrix}$

6. $\begin{vmatrix} 1 & 0 \\ 0 & 1 \end{vmatrix}$

7. $\begin{vmatrix} -3 & 2 \\ 6 & -4 \end{vmatrix}$

8. $\begin{vmatrix} 8 & -3 \\ -2 & 5 \end{vmatrix}$

A Solve each of the following for x. [Example 2]

9. $\begin{vmatrix} 2x & 1 \\ x & 3 \end{vmatrix} = 10$

10. $\begin{vmatrix} 3x & -2 \\ 2x & 3 \end{vmatrix} = 26$

11. $\begin{vmatrix} 1 & 2x \\ 2 & -3x \end{vmatrix} = 21$

12. $\begin{vmatrix} -5 & 4x \\ 1 & -x \end{vmatrix} = 27$

13. $\begin{vmatrix} 2x & -4 \\ x & 2 \end{vmatrix} = -16$

14. $\begin{vmatrix} 3x & -2 \\ x & 4 \end{vmatrix} = -28$

15. $\begin{vmatrix} 11x & -7x \\ 3 & -2 \end{vmatrix} = 3$

16. $\begin{vmatrix} -3x & -5x \\ 4 & 6 \end{vmatrix} = -14$

B Find the value of each of the following 3×3 determinants by using Method 1 of this section. [Example 3]

17. $\begin{vmatrix} 1 & 2 & 0 \\ 0 & 2 & 1 \\ 1 & 1 & 1 \end{vmatrix}$

18. $\begin{vmatrix} -1 & 0 & 2 \\ 3 & 0 & 1 \\ 0 & 1 & 3 \end{vmatrix}$

19. $\begin{vmatrix} 1 & 2 & 3 \\ 3 & 2 & 1 \\ 1 & 1 & 1 \end{vmatrix}$

20. $\begin{vmatrix} -1 & 2 & 0 \\ 3 & -2 & 1 \\ 0 & 5 & 4 \end{vmatrix}$

B Find the value of each determinant by using Method 2 and expanding across the first row. [Example 4]

21. $\begin{vmatrix} 0 & 1 & 2 \\ 1 & 0 & 1 \\ -1 & 2 & 0 \end{vmatrix}$

22. $\begin{vmatrix} 3 & -2 & 1 \\ 0 & -1 & 0 \\ 2 & 0 & 1 \end{vmatrix}$

23. $\begin{vmatrix} 3 & 0 & 2 \\ 0 & -1 & -1 \\ 4 & 0 & 0 \end{vmatrix}$

24. $\begin{vmatrix} 1 & 1 & 1 \\ 1 & -1 & 1 \\ 1 & 1 & -1 \end{vmatrix}$

B Find the value of each of the following determinants. [Examples 3–5]

25. $\begin{vmatrix} 2 & -1 & 0 \\ 1 & 0 & -2 \\ 0 & 1 & 2 \end{vmatrix}$

26. $\begin{vmatrix} 5 & 0 & -4 \\ 0 & 1 & 3 \\ -1 & 2 & -1 \end{vmatrix}$

27. $\begin{vmatrix} 1 & 3 & 7 \\ -2 & 6 & 4 \\ 3 & 7 & -1 \end{vmatrix}$

28. $\begin{vmatrix} 2 & 1 & 5 \\ 6 & -3 & 4 \\ 8 & 9 & -2 \end{vmatrix}$

Cramer's Rule

OBJECTIVES

A Solve a system of linear equations in two variables using Cramer's rule.

B Solve a system of linear equations in three variables using Cramer's rule.

Cramer's rule is named after the Swiss mathematician Gabriel Cramer (1704–1752). Cramer's rule appeared in the appendix of an algebraic work of his classifying curves, but the basic idea behind his now-famous rule was formulated earlier by Gottfried Leibniz and Chinese mathematicians. It was actually Cramer's superior notation that helped to popularize the technique.

Cramer had very broad interests. He wrote on philosophy, law, and government, as well as mathematics; served in public office; and was an expert on cathedrals, often instructing workers about their repair and coordinating excavations to recover cathedral archives. Cramer never married, and a fall from a carriage eventually led to his death.

We begin this section with a look at how determinants can be used to solve a system of linear equations in two variables. We will use Cramer's rule to do so, but first we state it here as a theorem without proof.

A Solving Two-Variable Systems with Cramer's Rule

> **Cramer's Rule I**
>
> The solution to the system
>
> $$a_1x + b_1y = c_1$$
> $$a_2x + b_2y = c_2$$
>
> is given by
>
> $$x = \frac{D_x}{D}, \ y = \frac{D_y}{D}$$
>
> where
>
> $$D = \begin{vmatrix} a_1 & b_1 \\ a_2 & b_2 \end{vmatrix} \qquad D_x = \begin{vmatrix} c_1 & b_1 \\ c_2 & b_2 \end{vmatrix} \qquad D_y = \begin{vmatrix} a_1 & c_1 \\ a_2 & c_2 \end{vmatrix} \qquad (D \neq 0)$$

The determinant D is made up of the coefficients of x and y in the original system. The determinants D_x and D_y are found by replacing the coefficients of x or y by the constant terms in the original system. Notice also that Cramer's rule does not apply if $D = 0$. In this case the equations are dependent, or the system is inconsistent.

EXAMPLE 1 Use Cramer's rule to solve.

$$2x - 3y = 4$$
$$4x + 5y = 3$$

SOLUTION We begin by calculating the determinants D, D_x, and D_y.

$$D = \begin{vmatrix} 2 & -3 \\ 4 & 5 \end{vmatrix} = 2(5) - 4(-3) = 22$$

$$D_x = \begin{vmatrix} 4 & -3 \\ 3 & 5 \end{vmatrix} = 4(5) - 3(-3) = 29$$

$$D_y = \begin{vmatrix} 2 & 4 \\ 4 & 3 \end{vmatrix} = 2(3) - 4(4) = -10$$

PRACTICE PROBLEMS

1. Use Cramer's rule to solve:
$$3x - 5y = 2$$
$$2x + 4y = 1$$

Answer

1. $\left(\frac{13}{22}, -\frac{1}{22}\right)$

$$x = \frac{D_x}{D} = \frac{29}{22} \quad \text{and} \quad y = \frac{D_y}{D} = \frac{-10}{22} = -\frac{5}{11}$$

The solution set for the system is $\left\{\left(\frac{29}{22}, -\frac{5}{11}\right)\right\}$. ∎

B Solving Three-Variable Systems with Cramer's Rule

Cramer's rule can also be applied to systems of linear equations in three variables.

Cramer's Rule II

The solution set to the system

$$a_1x + b_1y + c_1z = d_1$$

$$a_2x + b_2y + c_2z = d_2$$

$$a_3x + b_3y + c_3z = d_3$$

is given by

$$x = \frac{D_x}{D}, \quad y = \frac{D_y}{D}, \quad \text{and} \quad z = \frac{D_z}{D}$$

where

$$D = \begin{vmatrix} a_1 & b_1 & c_1 \\ a_2 & b_2 & c_2 \\ a_3 & b_3 & c_3 \end{vmatrix} \quad D_x = \begin{vmatrix} d_1 & b_1 & c_1 \\ d_2 & b_2 & c_2 \\ d_3 & b_3 & c_3 \end{vmatrix} \quad (D \neq 0)$$

$$D_y = \begin{vmatrix} a_1 & d_1 & c_1 \\ a_2 & d_2 & c_2 \\ a_3 & d_3 & c_3 \end{vmatrix} \quad D_z = \begin{vmatrix} a_1 & b_1 & d_1 \\ a_2 & b_2 & d_2 \\ a_3 & b_3 & d_3 \end{vmatrix}$$

Again, the determinant D consists of the coefficients of x, y, and z in the original system. The determinants D_x, D_y, and D_z are found by replacing the coefficients of x, y, and z, respectively, with the constant terms from the original system. If $D = 0$, there is no unique solution to the system.

EXAMPLE 2 Use Cramer's rule to solve.

$$x + y + z = 6$$
$$2x - y + z = 3$$
$$x + 2y - 3z = -4$$

SOLUTION This is the same system used in Example 1 in the second section of Chapter 4, so we can compare Cramer's rule with our previous methods of solving a system in three variables. We begin by setting up and evaluating D, D_x, D_y, and D_z. (Recall that there are a number of ways to evaluate a 3×3 determinant. Since we have four of these determinants, we can use both Methods 1 and 2 from the previous section.) We evaluate D using Method 1 from the previous section.

$$D = \begin{vmatrix} 1 & 1 & 1 \\ 2 & -1 & 1 \\ 1 & 2 & -3 \end{vmatrix} \begin{matrix} 1 & 1 \\ 2 & -1 \\ 1 & 2 \end{matrix}$$

$$= 3 + 1 + 4 - (-1) - (2) - (-6)$$
$$= 13$$

---

2. Use Cramer's rule to solve.

$$x + 2y + z = -2$$
$$x + 2y - z = -6$$
$$x - 2y + z = -4$$

NOTE
When we are solving a system of linear equations by Cramer's rule, it is best to find the determinant D first. If $D = 0$, then there is no unique solution to the system and we may not want to go further.

Answer
2. $(-5, \frac{1}{2}, 2)$

We evaluate D_x using Method 2 from the previous section and expanding across row 1.

$$D_x = \begin{vmatrix} 6 & 1 & 1 \\ 3 & -1 & 1 \\ -4 & 2 & -3 \end{vmatrix} = 6\begin{vmatrix} -1 & 1 \\ 2 & -3 \end{vmatrix} - 1\begin{vmatrix} 3 & 1 \\ -4 & -3 \end{vmatrix} + 1\begin{vmatrix} 3 & -1 \\ -4 & 2 \end{vmatrix}$$

$$= 6(1) - 1(-5) + 1(2)$$

$$= 13$$

Find D_y by expanding across row 2.

$$D_y = \begin{vmatrix} 1 & 6 & 1 \\ 2 & 3 & 1 \\ 1 & -4 & -3 \end{vmatrix} = -2\begin{vmatrix} 6 & 1 \\ -4 & -3 \end{vmatrix} + 3\begin{vmatrix} 1 & 1 \\ 1 & -3 \end{vmatrix} - 1\begin{vmatrix} 1 & 6 \\ 1 & -4 \end{vmatrix}$$

$$= -2(-14) + 3(-4) - 1(-10)$$

$$= 26$$

Find D_z by expanding down column 1.

$$D_z = \begin{vmatrix} 1 & 1 & 6 \\ 2 & -1 & 3 \\ 1 & 2 & -4 \end{vmatrix} = 1\begin{vmatrix} -1 & 3 \\ 2 & -4 \end{vmatrix} - 2\begin{vmatrix} 1 & 6 \\ 2 & -4 \end{vmatrix} + 1\begin{vmatrix} 1 & 6 \\ -1 & 3 \end{vmatrix}$$

$$= 1(-2) - 2(-16) + 1(9)$$

$$= 39$$

Now find $x, y,$ and z.

$$x = \frac{D_x}{D} = \frac{13}{13} = 1 \qquad y = \frac{D_y}{D} = \frac{26}{13} = 2 \qquad z = \frac{D_z}{D} = \frac{39}{13} = 3$$

The solution set is $\{(1, 2, 3)\}$. ∎

NOTE
We are finding each of these determinants by expanding about different rows or columns just to show the different ways these determinants can be evaluated.

EXAMPLE 3   Use Cramer's rule to solve.

$$x + y = -1$$
$$2x - z = 3$$
$$y + 2z = -1$$

SOLUTION   It is helpful to rewrite the system using zeros for the coefficients of those variables not shown.

$$x + y + 0z = -1$$
$$2x + 0y - z = 3$$
$$0x + y + 2z = -1$$

The four determinants used in Cramer's rule are

$$D = \begin{vmatrix} 1 & 1 & 0 \\ 2 & 0 & -1 \\ 0 & 1 & 2 \end{vmatrix} = -3$$

$$D_x = \begin{vmatrix} -1 & 1 & 0 \\ 3 & 0 & -1 \\ -1 & 1 & 2 \end{vmatrix} = -6$$

$$D_y = \begin{vmatrix} 1 & -1 & 0 \\ 2 & 3 & -1 \\ 0 & -1 & 2 \end{vmatrix} = 9$$

$$D_z = \begin{vmatrix} 1 & 1 & -1 \\ 2 & 0 & 3 \\ 0 & 1 & -1 \end{vmatrix} = -3$$

3. Use Cramer's rule to solve.
$$x + y = 3$$
$$2x - z = 3$$
$$y + 2z = 9$$

Answer
3. $(4, -1, 5)$

$$x = \frac{D_x}{D} = \frac{-6}{-3} = 2 \qquad y = \frac{D_y}{D} = \frac{9}{-3} = -3 \qquad z = \frac{D_z}{D} = \frac{-3}{-3} = 1$$

The solution set is $\{(2, -3, 1)\}$. ∎

Finally, we should mention the possible situations that can occur when the determinant D is 0 and we are using Cramer's rule. If $D = 0$ and at least one of the other determinants, D_x or D_y (or D_z), is not 0, then the system is inconsistent. In this case, there is no solution to the system.

However, if $D = 0$ and both D_x and D_y (and D_z in a system of three equations in three variables) are 0, then the equations are dependent.

Problem Set B

A Solve each of the following systems using Cramer's rule. [Example 1]

1. $2x - 3y = 3$
$4x - 2y = 10$

2. $3x + y = -2$
$-3x + 2y = -4$

3. $5x - 2y = 4$
$-10x + 4y = 1$

4. $-4x + 3y = -11$
$5x + 4y = 6$

5. $4x - 7y = 3$
$5x + 2y = -3$

6. $3x - 4y = 7$
$6x - 2y = 5$

7. $9x - 8y = 4$
$2x + 3y = 6$

8. $4x - 7y = 10$
$-3x + 2y = -9$

B [Examples 2, 3]

9. $x + y + z = 4$
$x - y - z = 2$
$2x + 2y - z = 2$

10. $-x + y + 3z = 6$
$x + y + 2z = 7$
$2x + 3y + z = 4$

11. $x + y - z = 2$
$-x + y + z = 3$
$x + y + z = 4$

12. $-x - y + z = 1$
$x - y + z = 3$
$x + y - z = 4$

13. $3x - y + 2z = 4$
$6x - 2y + 4z = 8$
$x - 5y + 2z = 1$

14. $2x - 3y + z = 1$
$3x - y - z = 4$
$4x - 6y + 2z = 3$

15. $2x - y + 3z = 4$
$x - 5y - 2z = 1$
$-4x - 2y + z = 3$

16. $4x - y + 5z = 1$
$2x + 3y + 4z = 5$
$x + y + 3z = 2$

17. $-x - 7y = 1$
$x + 3z = 11$
$2y + z = 0$

18. $x + y = 2$
$-x + 3z = 0$
$2y + z = 3$

19. $x - y = 2$
$3x + z = 11$
$y - 2z = -3$

20. $4x + 5y = -1$
$2y + 3z = -5$
$x + 2z = -1$

Synthetic Division

Synthetic division is a short form of long division with polynomials. We will consider synthetic division only for those cases in which the divisor is of the form $x + k$, where k is a constant.

A Using Synthetic Division

Let's begin by looking over an example of long division with polynomials.

$$
\begin{array}{r}
3x^2 - 2x + 4 \\
x + 3\overline{)\,3x^3 + 7x^2 - 2x - 4} \\
\underline{3x^3 + 9x^2} \\
-2x^2 - 2x \\
\underline{-2x^2 - 6x} \\
4x - 4 \\
\underline{4x + 12} \\
-16
\end{array}
$$

We can rewrite the problem without showing the variable since the variable is written in descending powers and similar terms are in alignment. It looks like this:

$$
\begin{array}{r}
3 \quad -2 \quad +4 \\
1 + 3\overline{)\,3 \quad\; 7 \quad -2 \quad -4} \\
\underline{(3) + 9} \\
-2 \;\; (-2) \\
\underline{(-2) \; -6} \\
4 \;\; (-4) \\
\underline{(4) \;\; 12} \\
-16
\end{array}
$$

We have used parentheses to enclose the numbers that are repetitions of the numbers above them. We can compress the problem by eliminating all repetitions except the first one:

$$
\begin{array}{r}
3 \quad -2 \quad\; 4 \\
1 + 3\overline{)\,3 \quad\; 7 \quad -2 \quad -4} \\
\underline{9 \quad -6 \quad\; 12} \\
3 \quad -2 \quad\; 4 \;\; -16
\end{array}
$$

The top line is the same as the first three terms of the bottom line, so we eliminate the top line. Also, the 1 that was the coefficient of x in the original problem can be eliminated since we will consider only division problems where the divisor is of the form $x + k$. The following is the most compact form of the original division problem.

$$
\begin{array}{r}
+3\overline{)\,3 \quad\; 7 \quad -2 \quad -4} \\
\underline{9 \quad -6 \quad\; 12} \\
3 \quad -2 \quad\; 4 \;\; -16
\end{array}
$$

If we check over the problem, we find that the first term in the bottom row is exactly the same as the first term in the top row—and it always will be in problems of this type. Also, the last three terms in the bottom row come from multiplication by 3 and then subtraction. We can get an equivalent result by multiplying by -3 and adding. The problem would then look like this:

$$
\begin{array}{r|rrrr}
-3 & 3 & 7 & -2 & -4 \\
 & \downarrow & -9 & 6 & -12 \\
\hline
 & 3 & -2 & 4 & \boxed{-16}
\end{array}
$$

We have used the brackets ⌐⌐ to separate the divisor and the remainder. This last expression is synthetic division. It is an easy process to remember. Simply change the sign of the constant term in the divisor, then bring down the first term of the dividend. The process is then just a series of multiplications and additions, as indicated in the following diagram by the arrows:

$$
\begin{array}{r|rrrr}
-3 & 3 & 7 & -2 & -4 \\
 & \downarrow & -9 & 6 & -12 \\
\hline
 & 3 & -2 & 4 & \boxed{-16}
\end{array}
$$

The last term of the bottom row is always the remainder.

Here are some additional examples of synthetic division with polynomials.

EXAMPLE 1   Divide $x^4 - 2x^3 + 4x^2 - 6x + 2$ by $x - 2$.

SOLUTION   We change the sign of the constant term in the divisor to get 2 and then complete the procedure.

$$
\begin{array}{r|rrrrr}
+2 & 1 & -2 & 4 & -6 & 2 \\
 & \downarrow & 2 & 0 & 8 & 4 \\
\hline
 & 1 & 0 & 4 & 2 & \boxed{6}
\end{array}
$$

From the last line we have the answer:

$$
1x^3 + 0x^2 + 4x + 2 + \frac{6}{x - 2}
$$

∎

EXAMPLE 2   Divide $\dfrac{3x^3 - 4x + 5}{x + 4}$.

SOLUTION   Since we cannot skip any powers of the variable in the polynomial $3x^3 - 4x + 5$, we rewrite it as $3x^3 + 0x^2 - 4x + 5$ and proceed as we did in Example 1.

$$
\begin{array}{r|rrrr}
-4 & 3 & 0 & -4 & 5 \\
 & \downarrow & -12 & 48 & -176 \\
\hline
 & 3 & -12 & 44 & \boxed{-171}
\end{array}
$$

From the synthetic division, we have

$$
\frac{3x^3 - 4x + 5}{x + 4} = 3x^2 - 12x + 44 - \frac{171}{x + 4}
$$

∎

EXAMPLE 3   Divide $\dfrac{x^3 - 1}{x - 1}$.

SOLUTION   Writing the numerator as $x^3 + 0x^2 + 0x - 1$ and using synthetic division, we have

$$
\begin{array}{r|rrrr}
+1 & 1 & 0 & 0 & -1 \\
 & \downarrow & 1 & 1 & 1 \\
\hline
 & 1 & 1 & 1 & \boxed{0}
\end{array}
$$

which shows

$$
\frac{x^3 - 1}{x - 1} = x^2 + x + 1
$$

∎

PRACTICE PROBLEMS

1. Divide $x^3 + 2x^2 - 8x + 1$ by $x + 2$.

2. Divide $\dfrac{2x^3 - 5x^2 + 3}{x - 3}$.

3. Divide $\dfrac{x^3 + 8}{x + 2}$.

Answers

1. $x^2 - 8 + \dfrac{17}{x + 2}$

2. $2x^2 + x + 3 + \dfrac{12}{x - 3}$

3. $x^2 - 2x + 4$

Problem Set C

A Use synthetic division to find the following quotients. [Examples 1–3]

1. $\dfrac{x^2 - 5x + 6}{x + 2}$

2. $\dfrac{x^2 + 8x - 12}{x - 3}$

3. $\dfrac{3x^2 - 4x + 1}{x - 1}$

4. $\dfrac{4x^2 - 2x - 6}{x + 1}$

5. $\dfrac{x^3 + 2x^2 + 3x + 4}{x - 2}$

6. $\dfrac{x^3 - 2x^2 - 3x - 4}{x - 2}$

7. $\dfrac{3x^3 - x^2 + 2x + 5}{x - 3}$

8. $\dfrac{2x^3 - 5x^2 + x + 2}{x - 2}$

9. $\dfrac{2x^3 + x - 3}{x - 1}$

10. $\dfrac{3x^3 - 2x + 1}{x - 5}$

11. $\dfrac{x^4 + 2x^2 + 1}{x + 4}$

12. $\dfrac{x^4 - 3x^2 + 1}{x - 4}$

13. $\dfrac{x^5 - 2x^4 + x^3 - 3x^2 - x + 1}{x - 2}$

14. $\dfrac{2x^5 - 3x^4 + x^3 - x^2 + 2x + 1}{x + 2}$

15. $\dfrac{x^2 + x + 1}{x - 1}$

16. $\dfrac{x^2 + x + 1}{x + 1}$

17. $\dfrac{x^4 - 1}{x + 1}$

18. $\dfrac{x^4 + 1}{x - 1}$

19. $\dfrac{x^3 - 1}{x - 1}$

20. $\dfrac{x^3 - 1}{x + 1}$

Introduction to Conditional Statements

A Conditional Statements

Consider the two statements below:

Statement 1: If I study, then I will get good grades.

Statement 2: If x is a negative number, then $-x$ is a positive number.

In Statement 1, if we let A represent the phrase "I study" and B represent the phrase "I will get good grades," then Statement 1 has the form

If A, then B

Likewise, in Statement 2, if A is the phrase "x is a negative number" and B is "$-x$ is a positive number," then Statement 2 has the form

If A, then B

Each statement has the same form: If A, then B. We call this the "if/then" form, and any statement that has this form is called a *conditional statement.* For every conditional statement, the first phrase, A, is called the *hypothesis,* and the second phrase, B, is called the *conclusion.* All conditional statements can be written in the form

If *hypothesis*, then *conclusion*

> **Notation**
>
> A shorthand way to write an "if/then" statement is with the implies symbol
>
> $$A \Rightarrow B$$
>
> This statement is read "A implies B." It is equivalent to saying "If A, then B."

EXAMPLE 1 Identify the hypothesis and conclusion in each statement.

  a. If a and b are positive numbers, then $-a(-b) = ab$.

  b. If it is raining, then the streets are wet.

  c. $C = 90° \Rightarrow c^2 = a^2 + b^2$.

SOLUTION   **a.** *Hypothesis:* a and b are positive numbers

      Conclusion: $-a(-b) = ab$

  b. *Hypothesis:* It is raining

      Conclusion: The streets are wet

  c. *Hypothesis:* $C = 90°$

      Conclusion: $c^2 = a^2 + b^2$ ∎

> **Definition**
>
> For every conditional statement $A \Rightarrow B$, there exist the following associated statements:
>
> | | | |
> |---|---|---|
> | The converse: | $B \Rightarrow A$ | If B, then A |
> | The inverse: | Not $A \Rightarrow$ not B | If not A, then not B |
> | The contrapositive: | Not $B \Rightarrow$ not A | If not B, then not A |

PRACTICE PROBLEMS

1. Identify the hypothesis and conclusion in each conditional statement.

  a. If a and b are negative numbers, then $a + b < 0$.

  b. If it is a bird, then it can fly.

  c. $a = -5 \Rightarrow a^2 = 25$.

Answers

1. **a.** Hypothesis: a and b are negative numbers.
   Conclusion: $a + b < 0$

  b. Hypothesis: It is a bird.
   Conclusion: It can fly.

  c. Hypothesis: $a = -5$
   Conclusion: $a^2 = 25$

For each conditional statement $A \Rightarrow B$, we can find three related statements that may or may not be true depending on whether or not the original conditional statement is true.

2. For the statement below, write the converse, the inverse, and the contrapositive.

If it is a bird, then it can fly.

EXAMPLE 2   For the statement below, write the converse, the inverse, and the contrapositive.

If it is raining, then the streets are wet.

SOLUTION   It is sometimes easier to work a problem like this if we write out the phrases, A, B, not A, and not B.

Let A = it is raining; then not A = it is not raining.

Let B = the streets are wet; then not B = the streets are not wet.

Here are the three associated statements:

The converse: (If B, then A.) If the streets are wet, then it is raining.

The inverse: (If not A, then not B.) If it is not raining, then the streets are not wet.

The contrapositive: (If not B, then not A.) If the streets are not wet, then it is not raining. ∎

B True or False Statements

Next, we want to answer this question: "If a conditional statement is true, which, if any, of the associated statements are true also?" Consider the statement below.

If it is a square, then it has four sides.

We know from our experience with squares that this is a true statement. Now, is the converse necessarily true? Here is the converse:

If it has four sides, then it is a square.

Obviously the converse is *not* true because there are many four-sided figures that are not squares—rectangles, parallelograms, and trapezoids, to mention a few. So, the converse of a true conditional statement is not necessarily true itself.

Next, we consider the inverse of our original statement.

If it is not a square, then it does not have four sides.

Again, the inverse is not true since there are many nonsquare figures that do have four sides. For example, a 3-inch by 5-inch rectangle fits that description.

Finally, we consider the contrapositive:

If it does not have four sides, then it is not a square.

As you can see, the contrapositive is true; that is, if something doesn't have four sides, it can't possibly be a square.

The preceding discussion leads us to the following theorem.

Theorem

If a conditional statement is true, then so is its contrapositive; that is, the two statements

If A, then B and   If not B, then not A

are equivalent. One can't be true without the other being true also; that is, they are either both true or both false.

Answer

2. See Solutions Section.

The theorem doesn't mention the inverse and the converse because they are true or false independent of the original statement; that is, knowing that a conditional statement is true tells us that the contrapositive is also true—but the truth of the inverse and the converse does not follow from the truth of the original statement.

The next two examples are intended to clarify the preceding discussion and our theorem. As you read through them, be careful not to let your intuition, experience, or opinion get in the way.

EXAMPLE 3   If the statement "If you are guilty, then you will be convicted" is true, give another statement that also must be true.

SOLUTION   From our theorem and the discussion preceding it, we know that the contrapositive of a true conditional statement is also true. Here is the contrapositive of our original statement:

> If you are not convicted, then you are not guilty.

Remember, we are not asking for your opinion; we are simply asking for another conditional statement that must be true if the original statement is true. The answer is *always* the contrapositive. Now, you may be wondering about the converse:

> If you are convicted, then you must be guilty.

It may be that the converse is actually true. But if it is, it is not because of the original conditional statement; that is, the truth of the converse *does not follow* from the truth of the original statement.   ■

EXAMPLE 4   If the following statement is true, what other conditional statement also must be true?

> If $a = b$, then $a^2 = b^2$.

SOLUTION   Again, every true conditional statement has a true contrapositive. Therefore, the statement below is also true:

> If $a^2 \neq b^2$, then $a \neq b$.

In this case, we know from experience that the original statement is true; that is, if two numbers are equal, then so are their squares. We also know from experience that the contrapositive is true; if the squares of two numbers are not equal, then the numbers themselves can't be equal. Do you think the inverse and converse are true also? Here is the converse:

> If $a^2 = b^2$, then $a = b$.

The converse is not true. If a is -3 and b is 3, then a^2 and b^2 are equal, but a and b are not. This same kind of reasoning will show that the inverse is not necessarily true. This example then gives further evidence that our theorem is true: A true conditional statement has a true contrapositive. No conclusion can be drawn about the inverse or the converse.   ■

3. If the statement "If you are asleep, then your eyes are closed" is true, give another statement that also must be true.

4. If the statement below is true, what other conditional statement also must be true?

> If $a = 3$, then $a^2 = 9$

Answers
3–4. See Solutions Section.

Everyday Language

In everyday life, we don't always use the "if/then" form exactly as we have illustrated it here. Many times we use shortened, reversed, or otherwise altered forms of "if/then" statements. For instance, each of the following statements is a variation of the "if/then" form, and each carries the same meaning.

If it is raining, then the streets are wet.

If it is raining, the streets are wet.

When it rains, the streets get wet.

The streets are wet if it is raining.

The streets are wet because it is raining.

Rain will make the streets wet.

5. Write the following statement in "if/then" form.

History repeats itself.

EXAMPLE 5   Write the following statement in "if/then" form.

Romeo loves Juliet.

SOLUTION   We must be careful that we do not change the meaning of the statement when we write it in "if/then" form. Here is an "if/then" form that has the same meaning as the original statement:

If he is Romeo, then he loves Juliet.

We can see that it would be incorrect to rewrite the original statement as

If she is Juliet, then she loves Romeo.

This is because the original statement is Romeo loves Juliet, not Juliet loves Romeo. It would also be incorrect to rewrite our statement as either

If he is not Romeo, then he does not love Juliet.

or

If he loves Juliet, then he is Romeo.

This is because people other than Romeo may also love Juliet. (The preceding statements are actually the inverse and converse, respectively, of the original statement.) Finally, another statement that has the same meaning as our original statement is

If he does not love Juliet, then he is not Romeo.

This is because this is the contrapositive of our original statement, and we know that the contrapositive is always true when the original statement is true.   ■

Answers

5. See Solutions Section.

Problem Set D

A For each conditional statement below, state the hypothesis and the conclusion. [Example 1]

1. If you argue for your limitations, then they are yours.

2. If you think you can, then you can.

3. If x is an even number, then x is divisible by 2.

4. If x is an odd number, then x is not divisible by 2.

5. If a triangle is equilateral, then all of its angles are equal.

6. If a triangle is isosceles, then two of its angles are equal.

7. If $x + 5 = -2$, then $x = -7$.

8. If $x - 5 = -2$, then $x = 3$.

For each of the following conditional statements, give the converse, the inverse, and the contrapositive. [Example 2]

9. If $a = 8$, then $a^2 = 64$.

10. If $x = y$, then $x^2 = y^2$.

11. If $\dfrac{a}{b} = 1$, then $a = b$.

12. If $a + b = 0$, then $a = -b$.

13. If it is a square, then it is a rectangle.

14. If you live in a glass house, then you shouldn't throw stones.

15. If better is possible, then good is not enough.

16. If a and b are positive, then ab is positive.

For each statement below, write an equivalent statement in "if/then" form.

17. $E \Rightarrow F$

18. $a^3 = b^3 \Rightarrow a = b$

19. Misery loves company.

20. Rollerblading is not a crime.

21. The squeaky wheel gets the grease.

22. The girl who can't dance says the band can't play.

B In each of the following problems a conditional statement is given. If the conditional statement is true, which of the three statements that follow it also must be true? [Examples 3–5]

23. If you heard it first, then you heard it on Eyewitness News.
 a. If you heard it on Eyewitness News, then you heard it first.
 b. If you didn't hear it first, then you didn't hear it on Eyewitness News.
 c. If you didn't hear it on Eyewitness News, then you didn't hear it first.

24. If it is raining, then the streets are wet.
 a. If the streets are wet, then it is raining.
 b. If the streets are not wet, then it is not raining.
 c. If it is not raining, then the streets are not wet.

25. If $C = 90°$, then $c^2 = a^2 + b^2$.
 a. If $c^2 \neq a^2 + b^2$, then $C \neq 90°$.
 b. If $c^2 = a^2 + b^2$, then $C = 90°$.
 c. If $C \neq 90°$, then $c^2 \neq a^2 + b^2$.

26. If you graduate from college, then you will get a good job.
 a. If you get a good job, then you graduated from college.
 b. If you do not get a good job, then you did not graduate from college.
 c. If you do not graduate from college, then you will not get a good job.

27. If you get a B average, then your car insurance will cost less.
 a. If your car insurance costs less, then you got a B average.
 b. If you do not get a B average, then your car insurance will not cost less.
 c. If your car insurance does not cost less, then you did not get a B average.

28. If you go out without a sweater, then you will get sick.
 a. If you do not get sick, then you went out with a sweater.
 b. If you go out with a sweater, then you will not get sick.
 c. If you get sick, then you went out without a sweater.

29. If a and b are negative, then $a + b$ is negative.
 a. If a and b are not both negative, then $a + b$ is not negative.
 b. If $a + b$ is not negative, then a and b are not both negative.
 c. If $a + b$ is negative, then a and b are negative.

30. If it is a square, then all its sides have the same length.
 a. If all its sides are not the same length, then it is not a square.
 b. If all its sides have the same length, then it is a square.
 c. If it is not a square, then its sides are not all the same length.

Solutions to Selected Practice Problems

Solutions to most of the practice problems are shown here. Before you look here to see where you have made a mistake, you should try the problem you are working on twice. If you do not get the correct answer the second time you work the problem, then the solution here should show you where you went wrong.

Chapter 1

Section 1.1

2. $4^2 = 4 \cdot 4 = 16$

3. $2^4 = 2 \cdot 2 \cdot 2 \cdot 2 = 16$

4. $3^3 = 3 \cdot 3 \cdot 3 = 27$

5. $6 + 2(3 + 4) = 6 + 2(7)$
$= 6 + 14$
$= 20$

6. $5 \cdot 3^2 - 2 \cdot 4^2 = 5 \cdot 9 - 2 \cdot 16$
$= 45 - 32$
$= 13$

7. $30 - (2 \cdot 3^2 - 8) = 30 - (2 \cdot 9 - 8)$
$= 30 - (18 - 8)$
$= 30 - 10$
$= 20$

8. $60 + 20 \div 2 - 40 = 60 + 10 - 40$
$= 70 - 40$
$= 30$

9. $3 + 5[2 + (7 \cdot 2 - 10)]$
$= 3 + 5[2 + (14 - 10)]$
$= 3 + 5(2 + 4)$
$= 3 + 5(6)$
$= 3 + 30$
$= 33$

10. When $a = 0$, $(0 + 4)^2 = 4^2 = 16$
When $a = 1$, $(1 + 4)^2 = 5^2 = 25$
When $a = 2$, $(2 + 4)^2 = 6^2 = 36$
When $a = 3$, $(3 + 4)^2 = 7^2 = 49$
When $a = 4$, $(4 + 4)^2 = 8^2 = 64$

11. When $x = 4$ and $y = 1$, then
$2x + 7y - 5 = 2 \cdot 4 + 7 \cdot 1 - 5$
$= 8 + 7 - 5$
$= 15 - 5$
$= 10$

Section 1.2

1. True. The dots mean the set continues indefinitely in the same manner.

2. True, since every member of the finite set $\{2, 4, 8\}$ is a member of the infinite set $\{2, 4, 6, \ldots\}$.

3. The union of A and B is the set of all elements that are either in A or in B.
Therefore, $A \cup B = \{1, 2, 3\} \cup \{0, 3, 7\} = \{0, 1, 2, 3, 7\}$.

4. The intersection of A and B is the set of elements in both A and B. Therefore, $A \cap B = \{1, 2, 3\} \cap \{0, 3, 7\} = \{3\}$.

5. The intersection of A and C is the set of elements in both A and C.
Therefore, $A \cap C = \{1, 2, 3\} \cap \{0, 1, 2, 3, \ldots\} = \{1, 2, 3\}$.

6. The intersection of B and C is the set of elements in both B and C.
Therefore, $B \cap C = \{0, 3, 7\} \cap \{0, 1, 2, \ldots\} = \{0, 3, 7\}$.

7. With $A = \{2, 4, 6, 8, 10\}$ $B = \{x | x \in A \text{ and } x < 8\} = $ All elements in A less than 8. So, $B = \{2, 4, 6\}$

8.

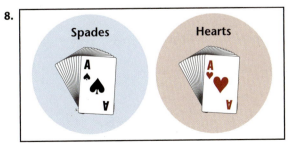

9.

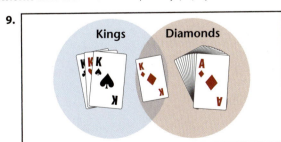

10. *A*:   *B* ∩ *C*:   *A* ∪ (*B* ∩ *C*):

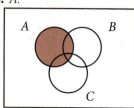   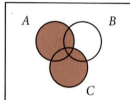   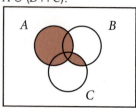

Comparing *A* ∪ (*B* ∩ *C*) and (*A* ∪ *B*) ∩ (*A* ∪ *C*), we see that they are the same.

A ∪ *B*:   *A* ∪ *C*:   (*A* ∪ *B*) ∩ (*A* ∪ *C*):

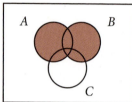   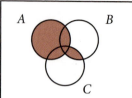   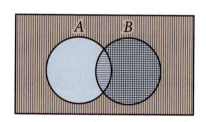

11. Using vertical lines to indicate all elements not in *A* and horizontal lines to show everything in *B* we have:

Since the connecting word is "or" we want the regions with either horizontal, vertical, or both lines.

Section 1.3

1.

3. $420 = 42 \cdot 10$
$= 6 \cdot 7 \cdot 2 \cdot 5$
$= 2 \cdot 3 \cdot 7 \cdot 2 \cdot 5$
$= 2^2 \cdot 3 \cdot 5 \cdot 7$

4. $\dfrac{231}{770} = \dfrac{3 \cdot 7 \cdot 11}{2 \cdot 5 \cdot 7 \cdot 11}$
$= \dfrac{3}{2 \cdot 5}$
$= \dfrac{3}{10}$

Section 1.4

1. **2.** **3.**

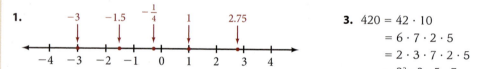

4. **5.** **6.**

7. If *t* is more than 10 but less than 20, then $t > 10$ and $t < 20$, or equivalently, $10 < t < 20$.

8. If *W* ranges from a low of 125 to a high of 131, then *W* is greater than or equal to 125 and less than or equal to 131, that is, $W \geq 125$ and $W \leq 131$, or equivalently, $125 \leq W \leq 131$.

Section 1.5

2. The reciprocal of 5 is $\frac{1}{5}$ because $5 \cdot \frac{1}{5} = \frac{5}{1} \cdot \frac{1}{5} = \frac{5}{5} = 1$.   **3.** The reciprocal of $\frac{1}{4}$ is 4 because $\frac{1}{4} \cdot 4 = \frac{1}{4} \cdot \frac{4}{1} = \frac{4}{4} = 1$.

4. The reciprocal of $\frac{3}{4}$ is $\frac{4}{3}$ because $\frac{3}{4} \cdot \frac{4}{3} = \frac{12}{12} = 1$.   **5.** The reciprocal of *x* is $\frac{1}{x}$ because $x \cdot \frac{1}{x} = \frac{x}{1} \cdot \frac{1}{x} = \frac{x}{x} = 1$.

13. $7 - 4 = 7 + (-4) = 3$ **14.** $-6 - 3 = -6 + (-3) = -9$ **15.** $8 - (-2) = 8 + 2 = 10$

16. $-4 - (-6) = -4 + 6 = 2$ **17.** $-3 - (-5) = -3 + 5 = 2$ **18.** $(-7 - 4) + (-8) = -11 + (-8) = -19$

21. $\dfrac{3}{7} \cdot \dfrac{2}{5} = \dfrac{3 \cdot 2}{7 \cdot 5} = \dfrac{6}{35}$

22. $\dfrac{3}{5} \div \dfrac{6}{7} = \dfrac{3}{5} \cdot \dfrac{7}{6}$

$= \dfrac{21}{30}$

$= \dfrac{7}{10}$

23. $\dfrac{3}{14} + \dfrac{7}{30} = \dfrac{3}{2 \cdot 7} + \dfrac{7}{2 \cdot 3 \cdot 5}$

$= \dfrac{\mathbf{3 \cdot 5}}{\mathbf{3 \cdot 5}} \cdot \dfrac{3}{2 \cdot 7} + \dfrac{7}{2 \cdot 3 \cdot 5} \cdot \dfrac{\mathbf{7}}{\mathbf{7}}$

$= \dfrac{45}{210} + \dfrac{49}{210}$

$= \dfrac{94}{210}$

$= \dfrac{47}{105}$

Section 1.6

3. $5 + (7 + y) = (5 + 7) + y$
$= 12 + y$

4. $3(2x) = (3 \cdot 2)x$
$= 6x$

5. $\dfrac{1}{3}(3a) = \left(\dfrac{1}{3} \cdot 3\right)a$
$= a$

6. $5\left(\dfrac{1}{5}x\right) = \left(5 \cdot \dfrac{1}{5}\right)x$
$= 1x$
$= x$

7. $9\left(\dfrac{2}{3}x\right) = \left(9 \cdot \dfrac{2}{3}\right)x$
$= 6x$

8. $7(6x + 8) = 7(6x) + 7(8)$
$= 42x + 56$

9. $9(7x + 11y) = 9(7x) + 9(11y)$
$= 63x + 99y$

10. $\dfrac{1}{3}(3x + 6) = \dfrac{1}{3}(3x) + \dfrac{1}{3}(6)$
$= x + 2$

11. $4(5y + 2) + 8 = 4(5y) + 4(2) + 8$
$= 20y + 8 + 8$
$= 20y + 16$

12. $a\left(2 - \dfrac{1}{a}\right) = a \cdot 2 - a \cdot \dfrac{1}{a} = 2a - 1$

13. $5\left(\dfrac{1}{5}x + 8\right) = 5 \cdot \dfrac{1}{5}x + 5 \cdot 8 = x + 40$

14. $12\left(\dfrac{2}{3}x + \dfrac{3}{4}y\right) = 12 \cdot \dfrac{2}{3}x + 12 \cdot \dfrac{3}{4}y$
$= 8x + 9y$

15. $4x + 9x = (4 + 9)x$
$= 13x$

16. $y + 5y = (1 + 5)y$
$= 6y$

18. $(-6 - 4)(8 - 12) = (-10)(-4) = 40$

19. $3 - 7(9 - 5) - 2 = 3 - 7(4) - 2 = 3 - 28 - 2 = -25 - 2 = -27$

20. $3(5 - 8)^3 - 2(-1 - 1)^2 = 3(-3)^3 - 2(-2)^2 = 3(-27) - 2(4) = -89$

21. $10x + 3 + 7x + 12 = (10x + 7x) + (3 + 12)$
$= (10 + 7)x + (3 + 12)$
$= 17x + 15$

22. $9 + 5(4y + 8) + 10y = 9 + 20y + 40 + 10y$
$= (20y + 10y) + (9 + 40)$
$= 30y + 49$

23. $2\left(\dfrac{x}{2}\right) = 2\left(\dfrac{1}{2}x\right)$
$= \left(2 \cdot \dfrac{1}{2}\right)x$
$= x$

24. $2\left(\dfrac{x}{2} - 3\right) = 2 \cdot \dfrac{x}{2} - 2 \cdot 3$
$= x - 6$

25. $\dfrac{-6 - 6}{-5 - 3} = \dfrac{-12}{-8}$
$= \dfrac{3}{2}$

26. $\dfrac{5(-6) + 3(-2)}{4(-3) + 3} = \dfrac{-30 + (-6)}{-12 + 3}$
$= \dfrac{-36}{-9}$
$= 4$

27. $\dfrac{3^3 - 4^3}{3^2 + 4^2} = \dfrac{27 - 64}{9 + 16}$
$= \dfrac{-37}{25}$
$= -\dfrac{37}{25}$

28. $2(5y - 1) - y = 10y - 2 - y$
$= 9y - 2$

29. $6 - 2(5x + 1) + 4x = 6 - 10x - 2 + 4x = -6x + 4$

30. $4(3a + 1) - (7a - 6) = 12a + 4 - 7a + 6 = 5a + 10$

31. a. $7[4(-1) - 6] = 7(-4 - 6)$
$= 7(-10)$
$= -70$

b. $28(-1) - 42 = -28 - 42$
$= -70$

c. $28(-1) - 6 = -28 - 6$
$= -34$

Section 1.7

1. a. If the sequence is formed by multiplying by 3 each time, the next term will be $3(135) = 405$.

   b. Turning the triangle a quarter turn to the right each time, the next triangle will point to the right: $\triangleright$

   c. The sequence of cubes. The next term will be $5^3 = 125$.

2. a. Add 10 each time; result is 36, 46.

   b. Add $\frac{1}{2}$ each time; result is 1, $\frac{3}{2}$.

   c. Add 5 each time; result is 10, 15.

3. a. Multiply by 2 each time; result is 80.

   b. Multiply by -3 each time; result is -135

   c. Multiply by $\frac{1}{2}$ each time; result is $\frac{1}{4}$

Chapter 2

Section 2.1

1.
$$4(3) - 2 \overset{?}{=} 10$$
$$12 - 2 = 10$$
$$10 = 10$$

2.
$$3(5) + 1 \overset{?}{=} 16$$
$$15 + 1 = 16$$
$$16 = 16$$
$$4(5) - 6 \overset{?}{=} 14$$
$$20 - 6 = 14$$
$$14 = 14$$

3.
$$\frac{2}{3}x + 4 = -8$$
$$\frac{2}{3}x + 4 - \mathbf{4} = -8 - \mathbf{4}$$
$$\frac{2}{3}x = -12$$
$$\frac{\mathbf{3}}{\mathbf{2}} \cdot \frac{2}{3}x = \frac{\mathbf{3}}{\mathbf{2}}(-12)$$
$$x = -18$$

Or eliminating the fractions in the beginning we can solve it this way:
$$30 \cdot \frac{3}{5}x + 30 \cdot \frac{1}{3} = 30\left(-\frac{5}{6}\right)$$
$$18x + 10 = -25$$
$$18x = -35$$
$$x = -\frac{35}{18}$$

4.
$$3a - 3 = -5a + 9$$
$$3a + \mathbf{5a} - 3 = -5a + \mathbf{5a} + 9$$
$$8a - 3 = 9$$
$$8a - 3 + \mathbf{3} = 9 + \mathbf{3}$$
$$8a = 12$$
$$\frac{\mathbf{1}}{\mathbf{8}} \cdot 8a = \frac{\mathbf{1}}{\mathbf{8}} \cdot 12$$
$$a = \frac{12}{8} = \frac{3}{2}$$

5. a.
$$-x = \frac{2}{3}$$
$$-1(-x) = -1\left(\frac{2}{3}\right)$$
$$x = -\frac{2}{3}$$

   b.
$$-y = -4$$
$$-1(-y) = -1(-4)$$
$$y = 4$$

6.
$$\frac{3}{5}x + \frac{1}{3} = -\frac{5}{6}$$
$$\frac{3}{5}x + \frac{1}{3} + \left(-\frac{\mathbf{1}}{\mathbf{3}}\right) = -\frac{5}{6} + \left(-\frac{\mathbf{1}}{\mathbf{3}}\right)$$
$$\frac{3}{5}x = -\frac{7}{6}$$
$$\frac{\mathbf{5}}{\mathbf{3}} \cdot \frac{3}{5}x = \frac{\mathbf{5}}{\mathbf{3}}\left(-\frac{7}{6}\right)$$
$$x = -\frac{35}{18}$$

7. Method 1

Working with the decimals.
$$0.08x + 0.10(8{,}000 - x) = 680$$
$$0.08x + 800 - 0.10x = 680$$
$$-0.02x + 800 = 680$$
$$-0.02x + 800 + (-\mathbf{800}) = 680 + (-\mathbf{800})$$
$$-0.02x = -120$$
$$x = \frac{-120}{-0.02} = 6{,}000$$

Method 2

Eliminating the decimals by multiplying each side by 100.
$$100(0.08x) + 100(0.10)(8{,}000 - x) = 100(680)$$
$$8x + 10(8{,}000 - x) = 68{,}000$$
$$8x + 80{,}000 - 10x = 68{,}000$$
$$-2x + 80{,}000 = 68{,}000$$
$$-2x = -12{,}000$$
$$x = \frac{-12{,}000}{-2}$$
$$= 6{,}000$$

8.
$$6 - 2(5x - 1) + 4x = 20$$
$$6 - 10x + 2 + 4x = 20$$
$$-6x + 8 = 20$$
$$-6x + 8 + (-\mathbf{8}) = 20 + (-\mathbf{8})$$
$$-6x = 12$$
$$-\frac{\mathbf{1}}{\mathbf{6}}(-6x) = -\frac{\mathbf{1}}{\mathbf{6}}(12)$$
$$x = -2$$

9.
$$3(5x + 1) = 10 + 15x$$
$$15x + 3 = 10 + 15x$$
$$15x + 3 + (-\mathbf{3}) = 10 + 15x + (-\mathbf{3})$$
$$15x = 7 + 15x$$
$$15x + (-\mathbf{15x}) = 7 + 15x + (-\mathbf{15x})$$
$$0 = 7$$

No solution

10.
$$-4 + 8x = 2(4x - 2)$$
$$-4 + 8x = 8x - 4$$
$$-4 + 8x + \mathbf{4} = 8x - 4 + \mathbf{4}$$
$$8x = 8x$$
$$8x + (-\mathbf{8x}) = 8x + (-\mathbf{8x})$$
$$0 = 0$$

All real numbers

Section 2.2

1. When $x = -3$, we have
$$2(-3) - 3y = 6$$
$$-6 - 3y = 6$$
$$-3y = 12$$
$$y = -4$$

2. When $x = 375$, we have
$$375 = 900 - 300p$$
$$-525 = -300p$$
$$p = \frac{-525}{-300} = \$1.75$$

3. $60 = (18 - c) \cdot 5$
$$60 = 90 - 5c$$
$$30 = -5c$$
$$6 = c$$

4.
$$P = 2w + 2l$$
$$P - 2w = 2l$$
$$\frac{P - 2w}{2} = l \quad \text{or} \quad \frac{P}{2} - w = l$$

5. $C = 200\pi \approx 200(3.14) = 628$ ft
$$r = \frac{d}{t} = \frac{628 \text{ ft}}{18 \text{ min}} = 34.9 \text{ ft/min}$$

6. $ax + 5 = cx + 3$
$$ax - cx = 3 - 5$$
$$(a - c)x = -2$$
$$x = \frac{-2}{a - c}$$

7. $\dfrac{y - 2}{x - 0} = \dfrac{4}{3}$
$$\frac{y - 2}{x} = \frac{4}{3}$$
$$x \cdot \frac{y - 2}{x} = \frac{4}{3} \cdot x$$
$$y - 2 = \frac{4}{3}x$$
$$y = \frac{4}{3}x + 2$$

8. $\dfrac{y + 8}{x - 7} = 6$
$$(x - 7) \cdot \frac{y + 8}{x - 7} = 6 \cdot (x - 7)$$
$$y + 8 = 6x - 42$$
$$y = 6x - 50$$

9. $x = 0.25(74)$
$$x = 18.5$$

10. $x \cdot 84 = 21$
$$x = \frac{21}{84}$$
$$x = 0.25$$
$$x = 25\%$$

11. $35 = 0.40x$
$$\frac{35}{0.40} = x$$
$$87.5 = x$$

Section 2.3

1. $2x + 2(4x - 10) = 12.5$
$$2x + 8x - 20 = 12.5$$
$$10x - 20 = 12.5$$
$$10x = 32.5$$
$$x = 3.25$$
The length is $x = 3.25$ feet; the width is $4x - 10 = 4(3.25) - 10 = 3$ feet.

2. $x + 0.0725x = 23{,}466.30$
$$1.0725x = 23{,}466.30$$
$$x = \frac{23{,}466.30}{1.0725}$$
$$= 21{,}880$$
The price of the car is \$21,880.00.

3. $x + 5x = 180$
$$6x = 180$$
$$x = 30$$
The angles are 30° and 150°.

4.

	Dollars at 8%	Dollars at 10%	Total
Number	x	$8{,}000 - x$	8,000
Interest	$0.08x$	$0.10(8{,}000 - x)$	680

$$0.08x + 0.10(8{,}000 - x) = 680$$
$$8x + 10(8{,}000 - x) = 68{,}000$$
$$8x + 80{,}000 - 10x = 68{,}000$$
$$-2x + 80{,}000 = 68{,}000$$
$$-2x = -12{,}000$$
$$x = 6{,}000$$
\$6,000 at 8% and \$2,000 at 10%

5.

Width	Length	Area (in²)
1	$l = 7 - 1 = 6$	6
2	$l = 7 - 2 = 5$	10
3	$l = 7 - 3 = 4$	12
4	$l = 7 - 4 = 3$	12
5	$l = 7 - 5 = 2$	10
6	$l = 7 - 6 = 1$	6

Section 2.4

1.
$$4x - 2 > 3x + 4$$
$$4x + (-3x) - 2 > 3x + (-3x) + 4$$
$$x - 2 > 4$$
$$x - 2 + 2 > 4 + 2$$
$$x > 6$$

Interval notation $(6, \infty)$

2.
$$-3y - 2 < 7$$
$$-3y < 9$$
$$-\frac{1}{3}(-3y) > -\frac{1}{3}(9)$$
$$y > -3$$

Interval notation $(-3, \infty)$

3.
$$3(2x + 5) \le -3x$$
$$6x + 15 \le -3x$$
$$9x + 15 \le 0$$
$$9x \le -15$$
$$\frac{1}{9}(9x) \le \frac{1}{9}(-15)$$
$$x \le -\frac{5}{3}$$

Interval notation $\left(-\infty, -\frac{5}{3}\right]$

4.
$$-7 \le 2x + 1 \le 7$$
$$-8 \le 2x \le 6$$
$$-4 \le x \le 3$$

Interval notation $[-4, 3]$

5.
$$3t - 6 \le -3 \quad \text{or} \quad 3t - 6 \ge 3$$
$$3t \le 3 \quad \text{or} \quad 3t \ge 9$$
$$t \le 1 \quad \text{or} \quad t \ge 3$$

Interval notation $(-\infty, 1] \cup [3, \infty)$

6.
$$1{,}300 - 100p \ge 800$$
$$-100p \ge -500$$
$$p \le 5$$

The price is $5 or less.

7.
$$-18 \le \quad F \quad \le 22$$
$$-18 \le \frac{9}{5}C + 32 \le 22$$
$$-50 \le \frac{9}{5}C \le -10$$
$$\frac{5}{9}(-50) \le \frac{5}{9}\left(\frac{9}{5}C\right) \le \frac{5}{9}(-10)$$
$$-27.8° \le \quad C \quad \le -5.6° \text{ to the nearest tenth}$$

Section 2.5

1. If $|x| = 3$
then $x = 3$ or $x = -3$

2. If $|3x - 6| = 9$
then $3x - 6 = 9$ or $3x - 6 = -9$
$$3x = 15 \text{ or} \quad 3x = -3$$
$$x = 5 \quad \text{or} \quad x = -1$$

3. If $|4x - 3| + 2 = 3$
then $|4x - 3| = 1$
$$4x - 3 = 1 \quad \text{or} \quad 4x - 3 = -1$$
$$4x = 4 \quad \text{or} \quad 4x = 2$$
$$x = 1 \quad \text{or} \quad x = \frac{1}{2}$$

4. $|7a - 1| = -2$ has no solution
since the left side is positive
or zero and the right side is
negative.

5. $|x + 3| = |x + 8|$
$$x + 3 = x + 8 \quad \text{or } x + 3 = -(x + 8)$$
$$3 = 8 \qquad\qquad x + 3 = -x - 8$$
No solution $\qquad\quad 2x + 3 = -8$
$$2x = -11$$
$$x = -\frac{11}{2}$$

Section 2.6

1. If $|2x + 1| \leq 7$
then $-7 \leq 2x + 1 \leq 7$
$-8 \leq 2x \leq 6$
$-4 \leq x \leq 3$

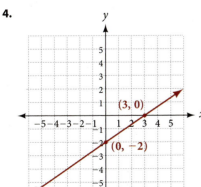

2. If $|4a - 3| < 5$
then $-5 < 4a - 3 < 5$
$-2 < 4a < 8$
$-\dfrac{1}{2} < a < 2$

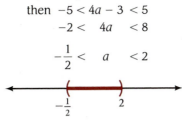

3. If $|x + 2| > 7$
then $x + 2 < -7$ or $x + 2 > 7$
$x < -9$ or $x > 5$

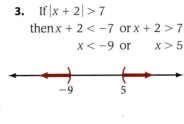

4. If $|3t - 6| \geq 3$
then
$3t - 6 \leq -3$ or $3t - 6 \geq 3$
$3t \leq 3$ or $3t \geq 9$
$t \leq 1$ or $t \geq 3$

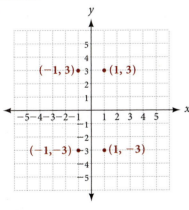

5. If $|2x + 5| - 2 < 9$
then $|2x + 5| < 11$
$-11 < 2x + 5 < 11$
$-16 < 2x < 6$
$-8 < x < 3$

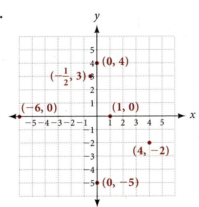

6. If $|5 - 2t| > 3$
then $5 - 2t < -3$ or $5 - 2t > 3$
$-2t < -8$ or $-2t > -2$

$-\dfrac{1}{2}(-2t) > -\dfrac{1}{2}(-8)$ or $-\dfrac{1}{2}(-2t) < -\dfrac{1}{2}(-2)$

$t > 4$ or $t < 1$

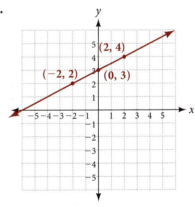

7. $|8y + 3| \leq -3$ has no solution since the left side cannot be negative and any number less than -3 must be negative.

8. $|2x - 9| \geq -7$ All real numbers satisfy this inequality since the left side is positive or zero, either of which is always greater than -7.

Chapter 3

Section 3.1

1.

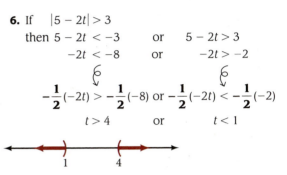

2.

3.

4.

5.

Section 3.2

1. When $x = 0$, $y = -5$, so the point $(0, -5)$ is on the graph. When $x = 2$, $y = 1$, so the point $(2, 1)$ is a second point on the graph. The ratio of rise to run going from $(0, -5)$ to $(2, 1)$ is $\frac{6}{2} = 3$. The slope of $y = 3x - 5$ is 3.

2. $m = \dfrac{4 - (-2)}{3 - 1} = \dfrac{6}{2} = 3$

3. $m = \dfrac{-3 - (-3)}{2 - (-1)} = \dfrac{0}{3} = 0$

4. $m = \dfrac{2.93 - 1.25}{2005 - 1985} = \dfrac{1.68}{20}$

$= 0.084$ dollars/year

$= 8.4$ cents/year

5. Using the second and third parts; $(3, 210)$ and $(6, 420)$.

We see:

$m = \dfrac{\text{rise}}{\text{run}} = \dfrac{420 - 210}{6 - 3}$

$= \dfrac{210}{3}$

$= 70$ mph

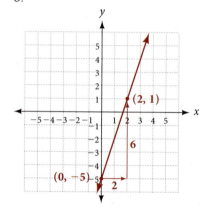

Section 3.3

1.

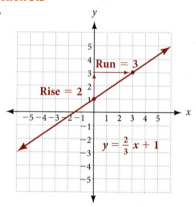

$y = \frac{2}{3}x + 1$

2. $4x - 5y = 7$

$-5y = -4x + 7$

$y = \dfrac{4}{5}x - \dfrac{7}{5}$

Slope $= \dfrac{4}{5}$, y-intercept $= -\dfrac{7}{5}$

3. $-3x + 2y = -6$

$2y = 3x - 6$

$y = \dfrac{3}{2}x - 3$

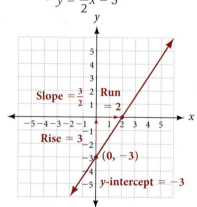

4. $m = 3$, $(x_1, y_1) = (-1, 2)$

$y - y_1 = m(x - x_1)$

$y - 2 = 3(x + 1)$

$y - 2 = 3x + 3$

$y = 3x + 5$

5. $m = \dfrac{5 - (-3)}{2 - 6} = \dfrac{8}{-4} = -2$

let $(x_1, y_1) = (2, 5)$

then $y - 5 = -2(x - 2)$

$y - 5 = -2x + 4$

$y = -2x + 9$

6. $3x - y = 2$

$-y = -3x + 2$

$y = 3x - 2$

The slope of a line perpendicular to this line is $-\dfrac{1}{3}$. Using $m = -\dfrac{1}{3}$ and $(x_1, y_1) = (3, 2)$ we have

$y - 2 = -\dfrac{1}{3}(x - 3)$

$y - 2 = -\dfrac{1}{3}x + 1$

$y = -\dfrac{1}{3}x + 3$

$3y = -x + 9$

$x + 3y = 9$

Section 3.4

1.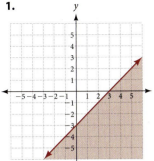

2.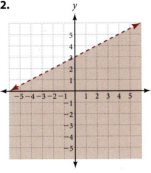

3.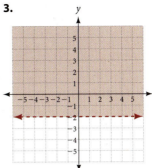

4.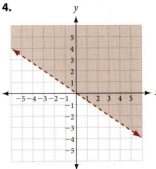

Section 3.5

1.

x	y
0	0
10	105
20	210
30	315
40	420

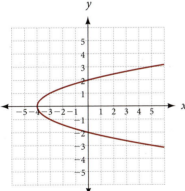

4.

t	h
0	0
$\frac{1}{2}$	20
1	32
$\frac{3}{2}$	36
2	32
$\frac{5}{2}$	20
3	0

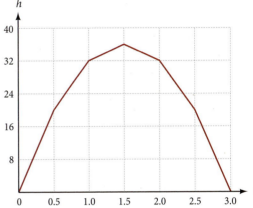

5.

x	y
5	−3
0	−2
−3	−1
−4	0
−3	1
0	2
5	3

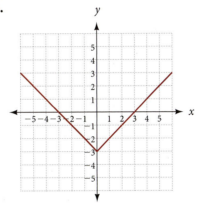

6.

Section 3.6

1. a. $f(0) = 8(0)$
$= 0$

b. $f(5) = 8(5)$
$= 40$

c. $f(10.5) = 8(10.5)$
$= 84$

2. a. $s(8) = \dfrac{88}{8}$
$= 11$

b. $s(11) = \dfrac{88}{11}$
$= 8$

3. a. $C(5) = 80\left(\dfrac{1}{2}\right)^{5/5}$
$= 80\left(\dfrac{1}{2}\right)^{1}$
$= 40$

b. $C(10) = 80\left(\dfrac{1}{2}\right)^{10/5}$
$= 80\left(\dfrac{1}{2}\right)^{2}$
$= 80\left(\dfrac{1}{4}\right)$
$= 20$

4. a. $C(5) = 2\pi(5)$
$= 10\pi$
≈ 31.4 inches

b. $A(5) = \pi(5)^2$
$= 25\pi$
≈ 78.5 inches²

5. a. $f(0) = 4(0)^2 - 3$
$= -3$
b. $f(3) = 4(3)^2 - 3$
$= 4(9) - 3$
$= 36 - 3$
$= 33$
c. $f(-2) = 4(-2)^2 - 3$
$= 4(4) - 3$
$= 16 - 3$
$= 13$

6. a. $f(5) = 2(5) + 1$
$= 10 + 1$
$= 11$
b. $g(5) = (5)^2 - 3$
$= 25 - 3$
$= 22$
c. $f(-2) = 2(-2) + 1$
$= -4 + 1$
$= -3$

d. $g(-2) = (-2)^2 - 3$
$= 4 - 3$
$= 1$
e. $f(a) = 2(a) + 1$
$= 2a + 1$
f. $g(a) = (a)^2 - 3$
$= a^2 - 3$

Section 3.7

1. a. $(f + g)(x) = f(x) + g(x)$
$= (x^2 - 4) + (x + 2)$
$= x^2 + x - 2$
b. $(f - g)(x) = f(x) - g(x)$
$= (x^2 - 4) - (x + 2)$
$= x^2 - x - 6$

2. a. $(f + g)(x) = f(x) + g(x)$
$= 3x + 2 + 3x^2 - 10x - 8$
$= 3x^2 - 7x - 6$
b. $(fg)(x) = f(x)g(x)$
$= (3x + 2)(3x^2 - 10x - 8)$
$= 9x^3 - 24x^2 - 44x - 16$

c. $\left(\dfrac{g}{f}\right)(x) = \dfrac{g(x)}{f(x)}$
$= \dfrac{3x^2 - 10x - 8}{3x + 2}$
$= \dfrac{(3x + 2)(x - 4)}{3x + 2}$
$= x - 4$
$= h(x)$

3. a. $(f \circ g)(x) = f(g(x))$
$= f(x^2 + 3x)$
$= (x^2 + 3x) - 4$
$= x^2 + 3x - 4$
b. $(g \circ f)(x) = g(f(x))$
$= g(x - 4)$
$= (x - 4)^2 + 3(x - 4)$
$= x^2 - 8x + 16 + 3x - 12$
$= x^2 - 5x + 4$

Section 3.8

1. $y = Kx$
when $y = 24$
and $x = 8$
the equation $y = Kx$
becomes $24 = K \cdot 8$
$K = 3$

$y = 3x$
letting $x = 2$
$y = 3 \cdot 2$
$y = 6$

2. $d(t) = Kx^2$
since $d(2) = 64$
$64 = k(2)^2$
$K = 16$

$d(t) = 16t^2$
$d(2.5) = 16(2.5)^2$
$d(2.5) = 100$ ft

3. $V = \dfrac{K}{P}$
$50 = \dfrac{K}{48}$
$50 \cdot 48 = K$
$2{,}400 = K$
$V = \dfrac{2{,}400}{P}$
$150 = \dfrac{2{,}400}{P}$
$150P = 2{,}400$
$P = 16$ lbs/in^2

4. $y = Kxz^2$
substituting
$y = 81, x = 2, z = 9$
$81 = K \cdot 2 \cdot 9^2$
$81 = K \cdot 162$
$\dfrac{1}{2} = K$
$y = \dfrac{1}{2}x \cdot z^2$
$y = \dfrac{1}{2} \cdot 4 \cdot 4^2$
$y = 32$

5. $R = \dfrac{Kl}{d}$
$2 = \dfrac{K(100)}{(.5)^2}$
$K = .0005$
$R = \dfrac{.0005\,l}{d^2}$
$R = \dfrac{.0005(300)}{(.25)^2}$
$R = 2.4$ ohms

Chapter 4

Section 4.1

1. $2x - y = 7$ $\xrightarrow{\ \text{4 times each side}\ }$ $8x - 4y = 28$
$3x + 4y = -6$ $\xrightarrow{\ \text{no change}\ }$ $3x + 4y = -6$
$\quad\overline{11x \quad = 22}$
$\quad x \quad = 2$

Substituting $x = 2$ into $2x - y = 7$ gives us $y = -3$.
The solution is $(2, -3)$.

2. $3x - 2y = -8$ $\xrightarrow{\ \text{2 times each side}\ }$ $6x - 4y = -16$
$-2x + 3y = 7$ $\xrightarrow{\ \text{3 times each side}\ }$ $-6x + 9y = 21$
$\quad\overline{5y = 5}$
$\quad y = 1$

Substituting $y = 1$ into $3x - 2y = -8$ gives us $x = -2$.
The solution is $(-2, 1)$.

3. $3x - 5y = 2$ $\xrightarrow{\text{2 times each side}}$ $6x - 10y = 4$

$2x + 4y = 1$ $\xrightarrow[-3\text{ times each side}]{}$ $-6x - 12y = -3$

$$-22y = 1$$
$$y = -\frac{1}{22}$$

$3x - 5y = 2$ $\xrightarrow{\text{4 times each side}}$ $12x - 20y = 8$

$2x + 4y = 1$ $\xrightarrow[5\text{ times each side}]{}$ $10x + 20y = 5$

$$22x = 13$$
$$x = \frac{13}{22}$$

Solution: $\left(\dfrac{13}{22}, -\dfrac{1}{22}\right)$

4. $2x + 7y = 3$ $\xrightarrow{-2\text{ times each side}}$ $-4x - 14y = -6$

$4x + 14y = 1$ $\xrightarrow[\text{no change}]{}$ $4x + 14y = 1$

$$0 = -5$$

We have eliminated both variables and are left with a false statement, indicating that the lines are parallel. There is no solution to the system.

5. $2x + 7y = 3$ $\xrightarrow{-2\text{ times each side}}$ $-4x - 14y = -6$

$4x + 14y = 6$ $\xrightarrow[\text{no change}]{}$ $4x + 14y = 6$

$$0 = 0$$

We have eliminated both variables and are left with a true statement. The lines coincide. Any ordered pair that satisfies one of the equations will satisfy the other.

6. $\dfrac{1}{3}x + \dfrac{1}{2}y = 4$ $\xrightarrow{\text{times 6}}$ $2x + 3y = 24$

$\dfrac{2}{3}x - \dfrac{1}{4}y = 3$ $\xrightarrow[\text{times 12}]{}$ $8x - 3y = 36$

$$10x = 60$$
$$x = 6$$

Substituting $x = 6$ into any equation with both variables gives $y = 4$. The solution is (6, 4).

7. $4x - 2y = -2$

$y = x + 3$

Substituting $x + 3$ for y in the first equation gives us

$4x - 2(x + 3) = -2$

$4x - 2x - 6 = -2$

$2x - 6 = -2$

$2x = 4$

$x = 2$

When $x = 2$, $y = 2 + 3 = 5$. The solution is (2, 5).

8. $5x - 3y = -4$

$x + 2y = 7$

Solving the second equation for x gives $x = -2y + 7$. Substituting this for x in the first equation we have

$5(-2y + 7) - 3y = -4$

$-10y + 35 - 3y = -4$

$-13y + 35$

Putting $y = 3$ in
the first two equa
$x = 1$. The solution

Section 4.2

1. $x + 2y + z = 2$ (1)

$x + y - z = 6$ (2)

$x - y + 2z = -7$ (3)

Adding equations (1) and (2) yields

$2x + 3y = 8$ (4)

Adding twice (2) to (3) yields

$3x + y = 5$ (5)

We solve the system made up of equations (4) and (5) as follows.

$2x + 3y = 8$ $\xrightarrow{\text{no change}}$ $2x + 3y = 8$

$3x + y = 5$ $\xrightarrow[\text{times} -3]{}$ $-9x - 3y = -15$

$$-7x = -7$$
$$x = 1$$

Substituting $x = 1$ into equation (4) or (5) gives us $y = 2$. Substituting $x = 1$ and $y = 2$ into equation (1), (2), or (3) gives us $z = -3$. The solution is (1, 2, -3).

2. $3x - 2y + z = 2$ (1)

$3x + y + 3z = 7$ (2)

$x + 4y - z = 4$ (3)

Adding (1) and (3) yields

$4x + 2y = 6$ (4)

Adding three times (3) to (2) yields

$6x + 13y = 19$ (5)

We solve the system made up of (4) and (5) as follows:

$4x + 2y = 6$ $\xrightarrow{\text{times 6}}$ $24x + 12y = 36$

$6x + 13y = 19$ $\xrightarrow[\text{times} -4]{}$ $-24x - 52y = -76$

$$-40y = -40$$
$$y = 1$$

Putting $y = 1$ into equation (4) or (5) gives us $x = 1$. When x is 1 and y is 1 in equation (1), (2), or (3), then $z = 1$. The solution is (1, 1, 1).

3. $3x + 5y - 2z = 1$ (1)
  $6x + 10y - 4z = 2$ (2)
  $x - 8y + z = 4$ (3)

Equation (2) is twice equation (1), indicating that the planes they represent coincide with one another. There is either no solution to the system or an infinite number of solutions.

4. $3x - y + 2z = 4$ (1)
  $6x - 2y + 4z = 2$ (2)
  $5x - 3y + 7z = 5$ (3)

If we add -2 times equation (1) to equation (2), we are left with the false statement $0 = -6$. This means that the planes represented by these equations are parallel. There is therefore no solution to the system.

5. $x + 2y = 0$ (1)
  $3y + z = -3$ (2)
  $2x - z = 5$ (3)

Adding equations (2) and (3) gives us equation (4) which is

$2x + 3y = 2$ (4)

Equations (4) and (1) form a system in two variables.

$$\begin{array}{ll} x + 2y = 0 \xrightarrow{\text{times} -2} & -2x - 4y = 0 \\ 2x + 3y = 2 \xrightarrow{\text{no change}} & \underline{2x + 3y = 2} \\ & -y = 2 \\ & y = -2 \end{array}$$

When $y = -2$ in equation (1), x becomes 4. When we substitute -2 for y in equation (2), z is 3. The solution is $(4, -2, 3)$.

Section 4.3

1. Let x and y represent the two numbers. Then,

  rs are 5 and 9.

2. Let x = the number of adult tickets sold and y = the number of children's tickets sold.

  $x + y = 750$
  $2x + 1y = 1{,}090$

The number of adult tickets is 340 and the number of children's tickets is 410.

3. Let x = the amount invested at 6% and y = the amount invested at 7%.

  $x + y = 12{,}000$
  $0.06x + 0.07y = 790$

To solve this system, multiply the top equation by -6 and the bottom equation by 100 to get

  $-6x - 6y = -72{,}000$
  $6x + 7y = 79{,}000$

The amount invested at 6% is $5,000 and the amount invested at 7% is $7,000.

t x = the number of gallons 30% solution and y = the number of gallons of 70% solution.

  $x + y = 16$
  $0.30x + 0.70y = 0.60(16) = 9.6$

Multiply the top equation by -3 and the bottom equation by 10 to get

  $-3x - 3y = -48$
  $3x + 7y = 96$

which yields 4 gallons of 30% solution and 12 gallons of 70% solution.

5. Let x = the speed of the boat in still water and y = the speed of the current.

	d	r	t
Upstream	18	$x - y$	3
Downstream	20	$x + y$	2

$18 = 3(x - y)$
$20 = 2(x + y)$

or $3x - 3y = 18$
    $2x + 2y = 20$

The speed of the boat in still water is 8 miles per hour, whereas the speed of the current is 2 miles per hour.

6. Let x = the number of nickels, y = the number of dimes, and z = the number of quarters.

  $x + y + z = 15$
  $5x + 10y + 25z = 110$
  $x = 4y - 1$

Rewriting the third equation, the system can be written as

  $x + y + z = 15$
  $5x + 10y + 25z = 110$
  $x - 4y = -1$

There are 11 nickels, 3 dimes, and 1 quarter.

7. The new data are summarized in Table 1.

TABLE 1

CORRESPONDING TEMPERATURES	
In Degrees Fahrenheit	**In Degrees Celsius**
95	35
167	75

$$35 = 95m + b \xrightarrow{\text{Times } -1} -35 = -95m - b$$
$$75 = 167m + b \xrightarrow{\text{No change}} 75 = 167m + b$$
$$40 = 72m$$

$$m = \frac{40}{72} = \frac{5}{9}$$

If we assume the relationship is linear, then the formula that relates the two temperature scales can be written in slope-intercept form as

$$C = mF + b$$

Substituting the values from the table into this equation gives us two equations in two unknowns.

To find the value of b, we substitute $m = \frac{5}{9}$ into $35 = 95m + b$ and solve for b.

$$35 = 95\left(\frac{5}{9}\right) + b$$

$$35 = \frac{475}{9} + b$$

$$b = 35 - \frac{475}{9} = \frac{315}{9} - \frac{475}{9} = -\frac{160}{9}$$

The equation that gives C in terms of F is

$$C = \frac{5}{9}F - \frac{160}{9}$$

Section 4.4

1. Rewrite as a matrix:

$$\begin{bmatrix} 1 & -1 & 1 & | & -6 \\ 2 & 1 & -2 & | & 12 \\ 2 & -2 & 1 & | & -9 \end{bmatrix}$$ Multiply row 1 by -2 and add to row 2

$$\begin{bmatrix} 1 & -1 & 1 & | & -6 \\ 0 & 3 & -4 & | & 24 \\ 2 & -2 & 1 & | & -9 \end{bmatrix}$$ Multiply row 1 by -2 and add to row 3

$$\begin{bmatrix} 1 & -1 & 1 & | & -6 \\ 0 & 3 & -4 & | & 24 \\ 0 & 0 & -1 & | & 3 \end{bmatrix}$$

Convert back to a system of equations:
$$x - y + z = -6$$
$$3y - 4z = 24$$
$$z = -3$$

Substituting $z = -3$ into the second equation, $y = 4$.
Substituting $z = -3$ and $y = 4$ into the first equation, $x = 1$.
Solution: $(1, 4, -3)$

Section 4.5

1.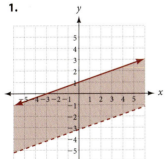

2.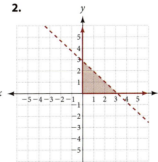

3.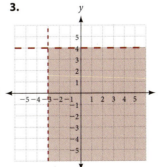

4.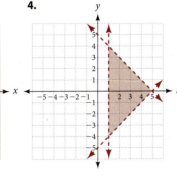

5. The horizontal line would move down to $y = 300$ from $y = 500$.

Chapter 5

Section 5.1

1. $x^5 \cdot x^4 = (x \cdot x \cdot x \cdot x \cdot x)(x \cdot x \cdot x \cdot x)$
$= x^9$

2. $(4^5)^2 = 4^5 \cdot 4^5 = 4^{10}$

3. $2x^3 = (2x)(2x)(2x)$
$= 8x^3$

4. $(-2x^3)(4x^5) = (-2 \cdot 4)(x^3 \cdot x^5)$
$= -8x^8$

5. $(-3y^3)^3(2y^6) = (-27y^9)(2y^6)$
$= -54y^{15}$

6. $(a^2)^3(a^3b^4)^2(b^5)^2 = a^6 \cdot a^6 \cdot b^8 \cdot b^{10}$
$= a^{12}b^{18}$

7. $3^{-2} = \dfrac{1}{3^2} = \dfrac{1}{9}$

8. $(-5)^{-3} = \dfrac{1}{(-5)^3} = -\dfrac{1}{125}$

9. $\left(\dfrac{2}{3}\right)^{-2} = \left(\dfrac{3}{2}\right)^2 = \dfrac{9}{4}$

10. a. $\dfrac{3^7}{3^4} = 3^{7-4} = 3^3 = 27$

   b. $\dfrac{x^3}{x^{10}} = x^{3-10} = x^{-7} = \dfrac{1}{x^7}$

   c. $\dfrac{a^5}{a^{-7}} = a^{5-(-7)} = a^{12}$

   d. $\dfrac{m^{-3}}{m^{-5}} = m^{-3-(-5)} = m^{-3+5} = m^2$

11. a. $(3xy^5)^0 = 1$

   b. $(3xy^5)^1 = 3xy^5$

12. $\dfrac{(x^2)^{-4}(x^3)^6}{(x^{-3})^5} = \dfrac{x^{-8}x^{18}}{x^{-15}}$
$= \dfrac{x^{10}}{x^{-15}}$
$= x^{10-(-15)}$
$= x^{25}$

13. $\dfrac{18a^7b^{-4}}{36a^2b^{-8}} = \dfrac{18}{36} \cdot \dfrac{a^7}{a^2} \cdot \dfrac{b^{-4}}{b^{-8}}$
$= \dfrac{1}{2} \cdot a^5 \cdot b^4 = \dfrac{a^5b^4}{2}$

14. $\dfrac{(3x^{-2}y^7)^2}{(x^5y^{-2})^{-4}} = \dfrac{9x^{-4}y^{14}}{x^{-20}y^8} = 9x^{16}y^6$

15. $27{,}400 = 2.74 \times 10^4$

Move decimal 4 places to the right.   Keep track of the four places we moved the decimal

16. $3.91 \times 10^4 = 3.91 \times 10{,}000$
$= 39{,}100$

17. a. $(3 \times 10^7)(2 \times 10^{-4}) = (3 \cdot 2) \times (10^7)(10^{-4}) = 6 \times 10^3$

   b. $\dfrac{3.9 \times 10^6}{1.3 \times 10^{-4}} = \dfrac{3.9}{1.3} \times \dfrac{10^6}{10^{-4}} = 3 \times 10^{10}$

   c. $\dfrac{(2.4 \times 10^6)(1.8 \times 10^{-4})}{1.2 \times 10^{-3}} = \dfrac{(2.4)(1.8)}{1.2} \times \dfrac{10^6 \cdot 10^{-4}}{10^{-3}} = 3.6 \times 10^5$

Section 5.2

2. $(3x^2 + 2x - 5) + (2x^2 - 7x + 3) = (3x^2 + 2x^2) + (2x - 7x) + (-5 + 3)$
$= (3 + 2)x^2 + (2 - 7)x + (-5 + 3)$
$= 5x^2 - 5x - 2$

3. $x^3 + 7x^2 + 3x + 2$
   $-3x^3 - 2x^2 + 3x - 1$
   $\overline{-2x^3 + 5x^2 + 6x + 1}$

4. $(4x^2 - 2x + 7) - (7x^2 - 3x + 1) = 4x^2 - 2x + 7 - 7x^2 + 3x - 1$
$= -3x^2 + x + 6$

5. $(7x - 4) - (3x + 5) = 7x - 4 - 3x - 5$
$= 4x - 9$

6. $2x - 4[6 - (5x + 3)]$
$= 2x - 4(6 - 5x - 3)$
$= 2x - 4(-5x + 3)$
$= 2x + 20x - 12$
$= 22x - 12$

7. $(9x - 4) - [(2x + 5) - (x + 3)]$
$= 9x - 4 - (2x + 5 - x - 3)$
$= 9x - 4 - (x + 2)$
$= 9x - 4 - x - 2$
$= 8x - 6$

8. $2(-2)^3 - 3(-2)^2 + 4(-2) - 8$
$= 2(-8) - 3(4) + 4(-2) - 8$
$= -16 - 12 - 8 - 8$
$= -44$

Section 5.3

1. $5x^2(3x^2 - 4x + 2) = 5x^2(3x^2) + 5x^2(-4x) + 5x^2(2)$
$$= 15x^4 - 20x^3 + 10x^2$$

2. $(4x - 1)(x + 3) = (4x - 1)x + (4x - 1)3$
$$= 4x^2 - x + 12x - 3$$
$$= 4x^2 + 11x - 3$$

3.
$$
\begin{array}{r}
x^2 - 5x + 6 \\
3x + 2 \\
\hline
3x^3 - 15x^2 + 18x \\
2x^2 - 10x + 12 \\
\hline
3x^3 - 13x^2 + 8x + 12
\end{array}
$$

4. $(2a - 3b)(5a - b) = 10a^2 - 2ab - 15ab + 3b^2$
$$ \text{F} \qquad \text{O} \qquad \text{I} \qquad \text{L}$$
$$= 10a^2 - 17ab + 3b^2$$

5. $(6 - 3t)(2 + 5t) = 12 + 30t - 6t - 15t^2$
$$ \text{F} \qquad \text{O} \qquad \text{I} \qquad \text{L}$$
$$= 12 + 24t - 15t^2$$

6. $\left(3x + \dfrac{1}{4}\right)\left(4x - \dfrac{1}{3}\right) = 12x^2 - x + x - \dfrac{1}{12}$
$$ \text{F} \qquad \text{O} \quad \text{I} \qquad \text{L}$$
$$= 12x^2 - \dfrac{1}{12}$$

7. $(a^4 - 2)(a^4 + 5) = a^8 + 5a^4 - 2a^4 - 10$
$$ \text{F} \qquad \text{O} \qquad \text{I} \qquad \text{L}$$
$$= a^8 + 3a^4 - 10$$

8. $(7x - 2)(3y + 8) = 21xy + 56x - 6y - 16$
$$ \text{F} \qquad \text{O} \qquad \text{I} \qquad \text{L}$$

9. $(3x - 2)^2 = (3x - 2)(3x - 2)$
$$= 9x^2 - 6x - 6x + 4$$
$$= 9x^2 - 12x + 4$$

10. a. $(x - y)^2 = x^2 - (2)xy + y^2$
$$= x^2 - 2xy + y^2$$

b. $(4t + 5)^2 = (4t)^2 + 2(4t)(5) + 5^2$
$$= 16t^2 + 40t + 25$$

c. $(5x - 3y)^2 = (5x)^2 - 2(5x)(3y) + (3y)^2$
$$= 25x^2 - 30xy + 9y^2$$

d. $(6 - a^4)^2 = 6^2 - 2(6)(a^4) + (a^4)^2$
$$= 36 - 12a^4 + a^8$$

11. $(4x - 3)(4x + 3) = 16x^2 + 12x - 12x - 9$
$$= 16x^2 - 9$$

13. $R(p) = xp = (250 - 25p)p$
$$= 250p - 25p^2$$

To find $R(x)$, we solve the equation $x = 250 - 25p$ for p.
$$250 - 25p = x$$
$$-25p = x - 250$$
$$p = \dfrac{x - 250}{-25}$$
$$= 10 - 0.04x$$

We find $R(x)$ by substituting $10 - 0.04x$ for p in the formula
$R(x) = xp$;
$$R(x) = xp = x(10 - 0.04x)$$
$$= 10x - 0.04x^2$$

14. $P(x) = R(x) - C(x)$
$$= 10x - 0.04x^2 - (210 + 1.5x)$$
$$= -0.04x^2 + 8.5x - 210$$
$$P(130) = -0.04(130)^2 + 8.5(130) - 210$$
$$= -0.04(16,900) + 8.5(130) - 210$$
$$= -676 + 1,105 - 210$$
$$= \$219$$

Section 5.4

1. $15a^7 - 25a^5 + 30a^3$
$= 5a^3(3a^4) - 5a^3(5a^2) + 5a^3(6)$
$= 5a^3(3a^4 - 5a^2 + 6)$

2. $12x^4y^5 - 9x^3y^4 - 15x^5y^3$
$= 3x^3y^3(4xy^2) - 3x^3y^3(3y) - 3x^3y^3(5x^2)$
$= 3x^3y^3(4xy^2 - 3y - 5x^2)$

3. $4(a + b)^4 - 6(a + b)^3 + 16(a + b)^2$
$= 2(a + b)^2 \cdot 2(a + b)^2 - 2(a + b)^2 \cdot 3(a + b) + 2(a + b)^2 \cdot 8$
$= 2(a + b)^2[2(a + b)^2 - 3(a + b) + 8]$

4. $R = 80.5x - 0.35x^2$
$= x(80.5 - 0.35x)$
$p = 80.5 - 0.35x$

5. $ab^3 + b^3 + 6a + 6 = b^3(a + 1) + 6(a + 1)$
$= (a + 1)(b^3 + 6)$

6. $15 - 3y^2 - 5x^2 + x^2y^2 = 3(5 - y^2) - x^2(5 - y^2)$
$= (5 - y^2)(3 - x^2)$

7. $x^3 + 5x^2 + 3x + 15 = x^2(x + 5) + 3(x + 5)$
$= (x + 5)(x^2 + 3)$

Section 5.5

1. The leading coefficient is 1. We need two numbers whose sum is -1 and whose product is -12. The numbers are -4 and 3.

$$x^2 - x - 12 = (x - 4)(x + 3)$$

2. We need two numbers whose product is $-15y^2$ and whose sum is $2y$. The numbers are $5y$ and $-3y$.

$$x^2 + 2xy - 15y^2 = (x + 5y)(x - 3y)$$

3. Since there is no pair of integers whose product is 1 and whose sum is 1, the trinomial $x^2 + x + 1$ is not factorable. We say it is a prime polynomial.

4. $5x^2 + 25x + 30 = 5(x^2 + 5x + 6)$
$= 5(x + 3)(x + 2)$

5. We list all possible factors along with their products as follows:

Possible Factors	First Term	Middle Term	Last Term
$(3x + 2)(x - 1)$	$3x^2$	$-x$	-2
$(3x - 2)(x + 1)$	$3x^2$	$+x$	-2
$(3x + 1)(x - 2)$	$3x^2$	$-5x$	-2
$(3x - 1)(x + 2)$	$3x^2$	$+5x$	-2

The first line has the correct middle term: $3x^2 - x - 2 = (3x + 2)(x - 1)$

6. $15x^4 + x^2 - 2 = (5x^2 + 2)(3x^2 - 1)$

7. $3x^2(x - 2) - 7x(x - 2) + 2(x - 2)$
$= (x - 2)(3x^2 - 7x + 2)$
$= (x - 2)(3x - 1)(x - 2)$
$= (x - 2)^2(3x - 1)$

8. $ac = -24$
$b = -5$

Two numbers whose product is -24 and whose sum is -5 are -8 and 3.
$8x^2 - 5x - 3 = 8x^2 - 8x + 3x - 3$
$= 8x(x - 1) + 3(x - 1)$
$= (x - 1)(8x + 3)$

Section 5.6

2. d. $(y + 2)^2 + 8(y + 2) + 16 = [(y + 2) + 4]^2$
$$= (y + 6)^2$$

3. $27x^2 - 36x + 12 = 3(9x^2 - 12x + 4)$
$$= 3(3x - 2)^2$$

4. a. $x^2 - 16 = x^2 - 4^2 = (x + 4)(x - 4)$

b. $64 - t^2 = 8^2 - t^2 = (8 + t)(8 - t)$

c. $25x^2 - 36y^2 = (5x)^2 - (6y)^2 = (5x + 6y)(5x - 6y)$

d. $9x^6 - 1 = (3x^3)^2 - 1^2 = (3x^3 + 1)(3x^3 - 1)$

e. $x^2 - \dfrac{25}{64} = x^2 - \left(\dfrac{5}{8}\right)^2 = \left(x + \dfrac{5}{8}\right)\left(x - \dfrac{5}{8}\right)$

5. $x^4 - 81 = (x^2)^2 - 9^2$
$$= (x^2 - 9)(x^2 + 9)$$
$$= (x - 3)(x + 3)(x^2 + 9)$$

6. $(x - 4)^2 - 9 = [(x - 4) - 3][(x - 4) + 3]$
$$= (x - 7)(x - 1)$$

7. $x^2 - 6x + 9 - y^2 = (x - 3)^2 - y^2$
$$= (x - 3 + y)(x - 3 - y)$$

8. $x^3 + 5x^2 - 4x - 20 = x^2(x + 5) - 4(x + 5)$
$$= (x + 5)(x^2 - 4)$$
$$= (x + 5)(x - 2)(x + 2)$$

9. We use the distributive property to expand the product:
$$(x - 3)(x^2 + 3x + 9) = x^3 + 3x^2 + 9x$$
$$\underline{+ -3x^2 - 9x - 27}$$
$$= x^3 - 27$$

10. The first term is the cube of 3 and the second term is the cube of x. Therefore,
$$27 + x^3 = 3^3 + x^3 = (3 + x)[3^2 - 3(x) + x^2]$$
$$= (3 + x)(9 - 3x + x^2)$$

11. $8x^3 + y^3 = (2x)^3 + y^3$
$$= (2x + y)[(2x)^2 - (2x)(y) + y^2]$$
$$= (2x + y)(4x^2 - 2xy + y^2)$$

12. $a^3 - \dfrac{1}{27} = a^3 - \left(\dfrac{1}{3}\right)^3 = \left(a - \dfrac{1}{3}\right)\left[a^2 + a\left(\dfrac{1}{3}\right) + \left(\dfrac{1}{3}\right)^2\right]$
$$= \left(a - \dfrac{1}{3}\right)\left(a^2 + \dfrac{1}{3}a + \dfrac{1}{9}\right)$$

13. $x^6 - 1 = (x^3)^2 - 1^2 = (x^3 + 1)(x^3 - 1)$
$$= (x + 1)(x^2 - x + 1)(x - 1)(x^2 + x + 1)$$

Section 5.7

1. $3x^8 - 27x^6 = 3x^6(x^2 - 9)$
$$= 3x^6(x + 3)(x - 3)$$

2. $4x^4 + 40x^3 + 100x^2 = 4x^2(x^2 + 10x + 25)$
$$= 4x^2(x + 5)^2$$

3. $y^4 + 36y^2 = y^2(y^2 + 36)$

4. $6x^2 - x - 15 = (3x - 5)(2x + 3)$

5. $3x^5 - 81x^2 = 3x^2(x^3 - 27)$
$$= 3x^2(x - 3)(x^2 + 3x + 9)$$

6. $3a^2b^3 + 6a^2b^2 - 3a^2b =$
$$3a^2b(b^2 + 2b - 1)$$

Section 5.8

1. $x^2 - x - 6 = 0$
$(x - 3)(x + 2) = 0$
$x - 3 = 0$ or $x + 2 = 0$
$x = 3$ or $x = -2$

2. $6 \cdot \dfrac{1}{2}x^3 = 6 \cdot \dfrac{5}{6}x^2 + 6 \cdot \dfrac{1}{3}x$
$3x^3 = 5x^2 + 2x$
$3x^3 - 5x^2 - 2x = 0$
$x(3x^2 - 5x - 2) = 0$
$x(3x + 1)(x - 2) = 0$
$x = 0$ or $3x + 1 = 0$ or $x - 2 = 0$
$x = 0$ or $x = -\dfrac{1}{3}$ or $x = 2$

3. $100x^2 = 500x$
$100x^2 - 500x = 0$
$100x(x - 5) = 0$
$100x = 0$ or $x - 5 = 0$
$x = 0$ or $x = 5$

4. $(x + 1)(x + 2) = 12$
$x^2 + 3x + 2 = 12$
$x^2 + 3x - 10 = 0$
$(x + 5)(x - 2) = 0$
$x + 5 = 0$ or $x - 2 = 0$
$x = -5$ or $x = 2$

5. $x^3 + 5x^2 - 4x - 20 = 0$
$x^2(x + 5) - 4(x + 5) = 0$
$(x + 5)(x^2 - 4) = 0$
$(x + 5)(x + 2)(x - 2) = 0$
$x + 5 = 0$ or $x + 2 = 0$ or $x - 2 = 0$
$x = -5$ or $x = -2$ or $x = 2$

6. Let $x =$ one integer and $x + 3 =$ the other.
$x^2 + (x + 3)^2 = 29$
$x^2 + x^2 + 6x + 9 = 29$
$2x^2 + 6x + 9 = 29$
$2x^2 + 6x - 20 = 0$
$x^2 + 3x - 10 = 0$
$(x + 5)(x - 2) = 0$
$x + 5 = 0$ or $x - 2 = 0$
$x = -5$ or $x = 2$
$x + 3 = -5 + 3$ $x + 3 = 2 + 3$
$= -2$ $= 5$
The two integers are -5 and -2 or 2 and 5.

7.

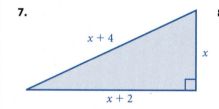

$(x + 4)^2 = (x + 2)^2 + x^2$
$x^2 + 8x + 16 = x^2 + 4x + 4 + x^2$
$x^2 + 8x + 16 = 2x^2 + 4x + 4$
$0 = x^2 - 4x - 12$
$0 = (x - 6)(x + 2)$
$x - 6 = 0$ or $x + 2 = 0$
$x = 6$ or $x = -2$

Since the length of a side cannot be negative, the shortest side is 6. The other two sides are $6 + 2 = 8$ and $6 + 4 = 10$.

8.

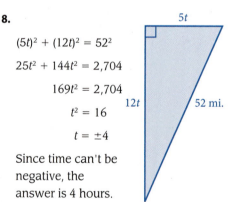

$(5t)^2 + (12t)^2 = 52^2$
$25t^2 + 144t^2 = 2{,}704$
$169t^2 = 2{,}704$
$t^2 = 16$
$t = \pm 4$

Since time can't be negative, the answer is 4 hours.

9. If $v = 64$ and $h = 64$
then $h = vt - 16t^2$
becomes $64 = 64t - 16t^2$
$16t^2 - 64t + 64 = 0$
$t^2 - 4t + 4 = 0$
$(t - 2)^2 = 0$
$t = 2$ seconds

10. $R = (1{,}300 - 100p)p$
If $R = 3{,}600$, then
$3{,}600 = (1{,}300 - 100p)p$
$3{,}600 = 1{,}300p - 100p^2$
$100p^2 - 1{,}300p + 3{,}600 = 0$
$p^2 - 13p + 36 = 0$
$(p - 4)(p - 9) = 0$
$p - 4 = 0$ or $p - 9 = 0$
$p = 4$ or $p = 9$

The price can be set at either $4 or $9.

Chapter 6

Section 6.1

1. $\dfrac{x^2 - 9}{x + 3} = \dfrac{(\cancel{x+3})(x - 3)}{\cancel{x+3}}$

$\qquad = x - 3$

2. $\dfrac{y^2 - y - 6}{y^2 - 4} = \dfrac{(y - 3)(\cancel{y+2})}{(y - 2)(\cancel{y+2})}$

$\qquad = \dfrac{y - 3}{y - 2}$

3. $\dfrac{3a^3 + 3}{6a^2 - 6a + 6} = \dfrac{3(a^3 + 1)}{6(a^2 - a + 1)}$

$\qquad = \dfrac{3(a + 1)(\cancel{a^2 - a + 1})}{6(\cancel{a^2 - a + 1})}$

$\qquad = \dfrac{a + 1}{2}$

4. $\dfrac{x^2 + 4x + ax + 4a}{x^2 + ax + 4x + 4a} = \dfrac{x(x + 4) + a(x + 4)}{x(x + a) + 4(x + a)}$

$\qquad = \dfrac{(\cancel{x+4})(x+a)}{(x+a)(\cancel{x+4})}$

$\qquad = 1$

5. When $a = 7$ and $b = 4$ the expression $\dfrac{a - b}{b - a}$ becomes

$\dfrac{7 - 4}{4 - 7} = \dfrac{3}{-3} = -1$

6. $\dfrac{7 - x}{x^2 - 49} = \dfrac{-1(\cancel{x-7})}{(x + 7)(\cancel{x-7})} = \dfrac{-1}{x + 7}$

7. $C \approx 165(3.14) = 518$ feet to the nearest foot

$r = \dfrac{d}{t} = \dfrac{518}{40} = 13.0$ feet per second

9. a. $f(0) = \dfrac{0 + 6}{0 - 3} = \dfrac{6}{-3} = -2$

b. $f(-6) = \dfrac{-6 + 6}{-6 - 3}$

$\qquad = \dfrac{0}{-9}$

$\qquad = 0$

c. $f(6) = \dfrac{6 + 6}{6 - 3}$

$\qquad = \dfrac{12}{3}$

$\qquad = 4$

d. $f(3) = \dfrac{3 + 6}{3 - 3}$

$\qquad = \dfrac{9}{0}$

$\qquad$ Undefined

10.

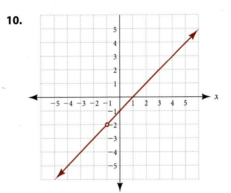

This graph does not contain the point $(-1, -2)$, whereas the graph of $y = x - 1$ does.

Section 6.2

1. $\dfrac{12x^4 - 18x^3 + 24x^2}{6x} = \dfrac{12x^4}{6x} - \dfrac{18x^3}{6x} + \dfrac{24x^2}{6x}$

$\qquad = 2x^3 - 3x^2 + 4x$

2. $\dfrac{27x^4y^7 - 81x^5y^3}{-9x^3y^2} = \dfrac{27x^4y^7}{-9x^3y^2} - \dfrac{81x^5y^3}{-9x^3y^2}$

$\qquad = -3xy^5 + 9x^2y$

3. $\dfrac{12a^5 + 8a^4 + 16a^3 + 4a^2}{8a^4} = \dfrac{12a^5}{8a^4} + \dfrac{8a^4}{8a^4} + \dfrac{16a^3}{8a^4} + \dfrac{4a^2}{8a^4}$

$\qquad = \dfrac{3a}{2} + 1 + \dfrac{2}{a} + \dfrac{1}{2a^2}$

4. $\dfrac{2x^2 - 5xy + 3y^2}{x - y} = \dfrac{(2x - 3y)(\cancel{x-y})}{\cancel{x-y}}$

$\qquad = 2x - 3y$

5. $\dfrac{f(x) - f(a)}{x - a} = \dfrac{(2x + 3) - (2a + 3)}{x - a}$

$\qquad = \dfrac{2x - 2a}{x - a}$

$\qquad = \dfrac{2(x - a)}{x - a}$

$\qquad = 2$

6. $\dfrac{f(x) - f(a)}{x - a} = \dfrac{(x^2 - 1) - (a^2 - 1)}{x - a}$

$\qquad = \dfrac{x^2 - a^2}{x - a}$

$\qquad = \dfrac{(x + a)(x - a)}{x - a}$

$\qquad = x + a$

7. $f(x) = (2x + 3)$

The expression $f(x + h)$ is given by

$f(x + h) = 2(x + h) + 3$

$\qquad = 2x + 2h + 3$

Using this result gives us

$\dfrac{f(x + h) - f(x)}{h} = \dfrac{2x + 2h + 3 - (2x + 3)}{h}$

$\qquad = \dfrac{2x + 2h + 3 - 2x - 3}{h}$

$\qquad = \dfrac{2h}{h}$

$\qquad = 2$

8. $f(x) = x^2 - 1$

The expression $f(x + h)$ is given by

$f(x + h) = (x + h)^2 - 1$

$\qquad = x^2 + 2xh + h^2 - 1$

Using this result gives us

$\dfrac{f(x + h) - f(x)}{h} = \dfrac{x^2 + 2xh + h^2 - 1 - (x^2 - 1)}{h}$

$\qquad = \dfrac{x^2 + 2xh + h^2 - 1 - x^2 + 1}{h}$

$\qquad = \dfrac{2xh + h^2}{h}$

$\qquad = 2x + h$

9.
$$
\begin{array}{r}
215 \\
35\overline{)7{,}546} \\
7\ 0\downarrow \\
\overline{54} \\
35\downarrow \\
\overline{196} \\
175 \\
\overline{21}
\end{array}
$$
Answer: $215 + \dfrac{3}{5}$

10.
$$
\begin{array}{r}
3x + 1 \\
x - 3\overline{)3x^2 - 8x - 1} \\
-\quad\ +\quad\ - \\
\overline{\diagup 3x^2\ \diagup 9x\ -} \\
0\ +\ x - 1 \\
-\quad\ + \\
\overline{\diagup\ x\diagup 3} \\
2
\end{array}
$$
Answer: $3x + 1 + \dfrac{2}{x - 3}$

11.
$$
\begin{array}{r}
3x^2 + 6x + 15 \\
x - 2\overline{)3x^3 + 0x^2 + 3x + 1} \\
-\quad\ + \\
\overline{\diagup 3x^3\ \diagup 6x^2}\ \downarrow \\
+\ 6x^2 +\ 3x \\
-\quad\ + \\
\overline{\diagup 6x^2\ \diagup 12x}\ \downarrow \\
+\ 15x + 1 \\
-\quad\ + \\
\overline{\diagup 15x \diagup 30} \\
31
\end{array}
$$
Answer: $3x^2 + 6x + 15 + \dfrac{31}{x - 2}$

12.
$$
\begin{array}{r}
2x - 3y \\
x - y\overline{)2x^2 - 5xy + 3y^2} \\
-\quad\ + \\
\overline{\diagup 2x^2\ \diagup 2xy}\ \downarrow \\
-3xy + 3y^2 \\
+\quad\ - \\
\overline{\diagup 3xy \diagup 3y^2} \\
0\ +\ 0
\end{array}
$$
Answer: $2x - 3y$

13.
$$
\begin{array}{r}
x^2 + 5x\ + 6 \\
x + 1\overline{)x^3 + 6x^2 + 11x + 6} \\
-\quad\ - \\
\overline{\diagup x^3\ \diagup 1x^2}\ \downarrow \\
5x^2 + 11x \\
-\quad\ - \\
\overline{\diagup 5x^2\ \diagup\ 5x}\ \downarrow \\
6x + 6 \\
-\quad\ - \\
\overline{\diagup 6x \diagup 6} \\
0
\end{array}
$$
Answer: $x^3 + 6x^2 + 11x + 6 = (x + 1)(x^2 + 5x + 6)$

$\qquad\qquad\qquad\qquad\qquad = (x + 1)(x + 2)(x + 3)$

Section 6.3

1. $\dfrac{3}{4} \cdot \dfrac{12}{27} = \dfrac{3 \cdot 12}{4 \cdot 27}$

$\phantom{\dfrac{3}{4} \cdot \dfrac{12}{27}} = \dfrac{3 \cdot 2 \cdot 2 \cdot 3}{2 \cdot 2 \cdot 3 \cdot 3 \cdot 3}$

$\phantom{\dfrac{3}{4} \cdot \dfrac{12}{27}} = \dfrac{1}{3}$

2. $\dfrac{6x^4}{4y^9} \cdot \dfrac{12y^5}{3x^2} = \dfrac{\overset{2}{6} \cdot \overset{3}{12}\,x^4 y^5}{3 \cdot 4x^2 y^9}$

$\phantom{\dfrac{6x^4}{4y^9} \cdot \dfrac{12y^5}{3x^2}} = \dfrac{6x^2}{y^4}$

3. $\dfrac{x+5}{x^2-25} \cdot \dfrac{x-5}{x^2-10x+25} = \dfrac{\cancel{(x+5)}\,\cancel{(x-5)}}{\cancel{(x+5)}\,\cancel{(x-5)}\,(x-5)^2}$

$\phantom{\dfrac{x+5}{x^2-25} \cdot \dfrac{x-5}{x^2-10x+25}} = \dfrac{1}{(x-5)^2}$

4. $\dfrac{3y^2-3y}{3y-12} \cdot \dfrac{y^2-2y-8}{y^2+3y+2} = \dfrac{3y(y-1)\,\cancel{(y-4)}\,\cancel{(y+2)}}{3\cancel{(y-4)}\,(y+1)\,\cancel{(y+2)}}$

$\phantom{\dfrac{3y^2-3y}{3y-12} \cdot \dfrac{y^2-2y-8}{y^2+3y+2}} = \dfrac{y(y-1)}{y+1}$

5. $\dfrac{9x^4}{4y^3} \div \dfrac{3x^2}{8y^5} = \dfrac{9x^4}{4y^3} \cdot \dfrac{8y^5}{3x^2}$

$\phantom{\dfrac{9x^4}{4y^3} \div \dfrac{3x^2}{8y^5}} = \dfrac{\overset{3}{9} \cdot \overset{2}{8}x^4 y^5}{3 \cdot 4x^2 y^3}$

$\phantom{\dfrac{9x^4}{4y^3} \div \dfrac{3x^2}{8y^5}} = 6x^2 y^2$

6. $\dfrac{xy^2-y^3}{x^2-y^2} \div \dfrac{x^3+y^3}{x^2+2xy+y^2} = \dfrac{xy^2-y^3}{x^2-y^2} \cdot \dfrac{x^2+2xy+y^2}{x^3+y^3}$

$\phantom{\dfrac{xy^2-y^3}{x^2-y^2} \div \dfrac{x^3+y^3}{x^2+2xy+y^2}} = \dfrac{y^2\cancel{(x-y)}\,\cancel{(x+y)}\,\cancel{(x+y)}}{\cancel{(x+y)}\,\cancel{(x-y)}\,\cancel{(x+y)}\,(x^2-xy+y^2)}$

$\phantom{\dfrac{xy^2-y^3}{x^2-y^2} \div \dfrac{x^3+y^3}{x^2+2xy+y^2}} = \dfrac{y^2}{x^2-xy+y^2}$

7. $\dfrac{a^2+3a-4}{a-4} \cdot \dfrac{a+3}{a^2-4a+3} \div \dfrac{a+1}{a^2-2a-3}$

$= \dfrac{(a^2+3a-4)(a+3)(a^2-2a-3)}{(a-4)(a^2-4a+3)(a+1)}$

$= \dfrac{(a+4)\cancel{(a-1)}(a+3)\cancel{(a-3)}\cancel{(a+1)}}{(a-4)\cancel{(a-3)}\cancel{(a-1)}\cancel{(a+1)}}$

$= \dfrac{(a+4)(a+3)}{a-4}$

8. $\dfrac{xa+xb-ya-yb}{xa+2x+ya+2y} \cdot \dfrac{xa+2x+ya+2y}{xa+xb+ya+yb}$

$= \dfrac{x(a+b)-y(a+b)}{x(a+2)+y(a+2)} \cdot \dfrac{x(a+2)+y(a+2)}{x(a+b)+y(a+b)}$

$= \dfrac{\cancel{(a+b)}(x-y)\,\cancel{(a+2)}\,\cancel{(x+y)}}{\cancel{(a+2)}(x+y)\,\cancel{(a+b)}\,\cancel{(x+y)}}$

$= \dfrac{x-y}{x+y}$

9. $(5x^2-45) \cdot \dfrac{3}{5x-15} = \dfrac{5x^2-45}{1} \cdot \dfrac{3}{5x-15}$

$\phantom{(5x^2-45) \cdot \dfrac{3}{5x-15}} = \dfrac{5(x+3)\cancel{(x-3)}3}{5\cancel{(x-3)}}$

$\phantom{(5x^2-45) \cdot \dfrac{3}{5x-15}} = 3(x+3)$

Section 6.4

1. $\dfrac{3}{8} + \dfrac{1}{8} = \dfrac{3+1}{8}$

$\phantom{\dfrac{3}{8} + \dfrac{1}{8}} = \dfrac{4}{8}$

$\phantom{\dfrac{3}{8} + \dfrac{1}{8}} = \dfrac{1}{2}$

2. $\dfrac{x}{x^2-9} + \dfrac{3}{x^2-9} = \dfrac{x+3}{x^2-9}$

$\phantom{\dfrac{x}{x^2-9} + \dfrac{3}{x^2-9}} = \dfrac{\cancel{x+3}}{\cancel{(x+3)}(x-3)}$

$\phantom{\dfrac{x}{x^2-9} + \dfrac{3}{x^2-9}} = \dfrac{1}{x-3}$

3. $\dfrac{2x-7}{x-2} - \dfrac{x-5}{x-2} = \dfrac{2x-7-(x-5)}{x-2}$

$\phantom{\dfrac{2x-7}{x-2} - \dfrac{x-5}{x-2}} = \dfrac{2x-7-x+5}{x-2}$

$\phantom{\dfrac{2x-7}{x-2} - \dfrac{x-5}{x-2}} = \dfrac{x-2}{x-2}$

$\phantom{\dfrac{2x-7}{x-2} - \dfrac{x-5}{x-2}} = 1$

4. $\dfrac{3}{10} + \dfrac{11}{42} = \dfrac{3}{2 \cdot 5} + \dfrac{11}{2 \cdot 3 \cdot 7}$

$\qquad = \dfrac{3}{2 \cdot 5} \cdot \dfrac{\mathbf{3 \cdot 7}}{\mathbf{3 \cdot 7}} + \dfrac{11}{2 \cdot 3 \cdot 7} \cdot \dfrac{\mathbf{5}}{\mathbf{5}}$

$\qquad = \dfrac{63}{2 \cdot 3 \cdot 5 \cdot 7} + \dfrac{55}{2 \cdot 3 \cdot 5 \cdot 7}$

$\qquad = \dfrac{118}{2 \cdot 3 \cdot 5 \cdot 7}$

$\qquad = \dfrac{\cancel{2} \cdot 59}{\cancel{2} \cdot 3 \cdot 5 \cdot 7}$

$\qquad = \dfrac{59}{105}$

5. $\dfrac{-3}{x^2 - 2x - 8} + \dfrac{4}{x^2 - 16} = \dfrac{-3}{(x-4)(x+2)} + \dfrac{4}{(x-4)(x+4)}$

$\qquad = \dfrac{-3}{(x-4)(x+2)} \cdot \dfrac{\mathbf{x+4}}{\mathbf{x+4}} + \dfrac{4}{(x-4)(x+4)} \cdot \dfrac{\mathbf{x+2}}{\mathbf{x+2}}$

$\qquad = \dfrac{-3x - 12 + 4x + 8}{(x-4)(x+4)(x+2)}$

$\qquad = \dfrac{\cancel{x-4}}{\cancel{(x-4)}(x+4)(x+2)}$

$\qquad = \dfrac{1}{(x+4)(x+2)}$

6. $\dfrac{x-4}{2x-6} + \dfrac{3}{x^2 - 9} = \dfrac{x-4}{2(x-3)} + \dfrac{3}{(x+3)(x-3)}$

$\qquad = \dfrac{x-4}{2(x-3)} \cdot \dfrac{\mathbf{x+3}}{\mathbf{x+3}} + \dfrac{3}{(x+3)(x-3)} \cdot \dfrac{\mathbf{2}}{\mathbf{2}}$

$\qquad = \dfrac{(x-4)(x+3) + 3 \cdot 2}{2(x+3)(x-3)}$

$\qquad = \dfrac{x^2 - x - 12 + 6}{2(x+3)(x-3)}$

$\qquad = \dfrac{x^2 - x - 6}{2(x+3)(x-3)}$

$\qquad = \dfrac{\cancel{(x-3)}(x+2)}{2(x+3)\cancel{(x-3)}}$

$\qquad = \dfrac{x+2}{2(x+3)}$

7. $\dfrac{2x-4}{x^2 + 5x + 4} - \dfrac{x-4}{x^2 + 6x + 8} = \dfrac{2x-4}{(x+4)(x+1)} - \dfrac{x-4}{(x+4)(x+2)}$

$\qquad = \dfrac{2x-4}{(x+4)(x+1)} \cdot \dfrac{\mathbf{x+2}}{\mathbf{x+2}} - \dfrac{x-4}{(x+4)(x+2)} \cdot \dfrac{\mathbf{x+1}}{\mathbf{x+1}}$

$\qquad = \dfrac{(2x-4)(x+2) - (x-4)(x+1)}{(x+4)(x+1)(x+2)}$

$\qquad = \dfrac{(2x^2 - 8) - (x^2 - 3x - 4)}{(x+4)(x+1)(x+2)}$

$\qquad = \dfrac{x^2 + 3x - 4}{(x+4)(x+1)(x+2)}$

$\qquad = \dfrac{\cancel{(x+4)}(x-1)}{\cancel{(x+4)}(x+1)(x+2)}$

$\qquad = \dfrac{x-1}{(x+1)(x+2)}$

8. $\dfrac{x^2}{x-4} + \dfrac{x+12}{4-x} = \dfrac{x^2}{x-4} + \dfrac{x+12}{4-x} \cdot \dfrac{-1}{-1}$

$\qquad = \dfrac{x^2}{x-4} + \dfrac{-x-12}{x-4}$

$\qquad = \dfrac{x^2 - x - 12}{x-4}$

$\qquad = \dfrac{\cancel{(x-4)}(x+3)}{\cancel{x-4}}$

$\qquad = x + 3$

9. $2 + \dfrac{25}{5x-1} = \dfrac{2}{1} + \dfrac{25}{5x-1}$

$\qquad = \dfrac{2}{1} \cdot \dfrac{\mathbf{5x-1}}{\mathbf{5x-1}} + \dfrac{25}{5x-1}$

$\qquad = \dfrac{10x - 2 + 25}{5x-1}$

$\qquad = \dfrac{10x + 23}{5x-1}$

10. $\dfrac{1}{x} + \dfrac{1}{3x} = \dfrac{3}{3x} + \dfrac{1}{3x} = \dfrac{4}{3x}$

Section 6.5

1. Method 1 $\dfrac{\frac{2}{3}}{\frac{5}{6}} \cdot \dfrac{\mathbf{6}}{\mathbf{6}} = \dfrac{4}{5}$

Method 2 $\dfrac{\frac{2}{3}}{\frac{5}{6}} = \dfrac{2}{3} \cdot \dfrac{6}{5} = \dfrac{12}{15} = \dfrac{4}{5}$

2. $\dfrac{\frac{1}{x} - \frac{1}{3}}{\frac{1}{x} + \frac{1}{3}} = \dfrac{\left(\frac{1}{x} - \frac{1}{3}\right)\mathbf{3x}}{\left(\frac{1}{x} + \frac{1}{3}\right)\mathbf{3x}}$

$\qquad = \dfrac{\frac{1}{x}(3x) - \frac{1}{3}(3x)}{\frac{1}{x}(3x) + \frac{1}{3}(3x)}$

$\qquad = \dfrac{3 - x}{3 + x}$

3. $\dfrac{\frac{x+5}{x^2-16}}{\frac{x^2-25}{x-4}} = \dfrac{x+5}{x^2-16} \cdot \dfrac{x-4}{x^2-25}$

$\qquad = \dfrac{(x+5)\cancel{(x-4)}}{(x+4)\cancel{(x-4)}(x+5)(x-5)}$

$\qquad = \dfrac{1}{(x+4)(x-5)}$

4. $\dfrac{1 - \frac{9}{x^2}}{1 - \frac{1}{x} - \frac{6}{x^2}} = \dfrac{x^2 \cdot \left(1 - \frac{9}{x^2}\right)}{x^2 \cdot \left(1 - \frac{1}{x} - \frac{6}{x^2}\right)}$

$= \dfrac{x^2 \cdot 1 - x^2 \cdot \frac{9}{x^2}}{x^2 \cdot 1 - x^2 \cdot \frac{1}{x} - x^2 \cdot \frac{6}{x^2}}$

$= \dfrac{x^2 - 9}{x^2 - x - 6}$

$= \dfrac{(x + 3)(\cancel{x - 3})}{(\cancel{x - 3})(x + 2)}$

$= \dfrac{x + 3}{x + 2}$

5. $2 + \dfrac{5}{x - \frac{1}{5}} = 2 + \dfrac{5}{x - \frac{1}{5}} \cdot \dfrac{5}{5}$

$= \dfrac{2}{1} + \dfrac{25}{5x - 1}$

$= \dfrac{2}{1} \cdot \dfrac{5x - 1}{5x - 1} + \dfrac{25}{5x - 1}$

$= \dfrac{10x - 2 + 25}{5x - 1}$

$= \dfrac{10x + 23}{5x - 1}$

Section 6.6

1. $\dfrac{x}{3} + 1 = \dfrac{1}{2}$ LCD = 6

$6\left(\dfrac{x}{3} + 1\right) = 6 \cdot \dfrac{1}{2}$

$6 \cdot \dfrac{x}{3} + 6 \cdot 1 = 6 \cdot \dfrac{1}{2}$

$2x + 6 = 3$

$2x = -3$

$x = -\dfrac{3}{2}$

2. $\dfrac{2}{a + 5} = \dfrac{1}{3}$

LCD = $3(a + 5)$

$3(a + 5) \cdot \dfrac{2}{a + 5} = 3(a + 5) \cdot \dfrac{1}{3}$

$6 = a + 5$

$1 = a$

3. $\dfrac{x}{x + 1} - \dfrac{1}{2} = \dfrac{-1}{x + 1}$ LCD = $2(x + 1)$

$2(x + 1)\left(\dfrac{x}{x + 1} - \dfrac{1}{2}\right) = 2(x + 1) \cdot \dfrac{-1}{x + 1}$

$2(x + 1) \cdot \dfrac{x}{x + 1} - 2(x + 1) \cdot \dfrac{1}{2} = 2(x + 1) \cdot \dfrac{-1}{x + 1}$

$2x - (x + 1) = 2(-1)$

$2x - x - 1 = -2$

$x - 1 = -2$

$x = -1$

The only possible solution is $x = -1$, but when $x = -1$, the original equation has two undefined terms. There is no solution to the equation.

4. $\dfrac{x}{x^2 - 9} - \dfrac{1}{x + 3} = \dfrac{1}{4x - 12}$

$\dfrac{x}{(x + 3)(x - 3)} - \dfrac{1}{x + 3} = \dfrac{1}{4(x - 3)}$ LCD = $4(x + 3)(x - 3)$

$4(x + 3)(x - 3) \cdot \dfrac{x}{(x + 3)(x - 3)} - 4(x + 3)(x - 3) \cdot \dfrac{1}{x + 3} = 4(x + 3)(x - 3) \cdot \dfrac{1}{4(x - 3)}$

$4x - 4(x - 3) = x + 3$

$4x - 4x + 12 = x + 3$

$12 = x + 3$

$9 = x$

5. $1 - \dfrac{2}{x} = \dfrac{8}{x^2}$ LCD = x^2

$x^2\left(1 - \dfrac{2}{x}\right) = x^2 \cdot \dfrac{8}{x^2}$

$x^2 \cdot 1 - x^2 \cdot \dfrac{2}{x} = x^2 \cdot \dfrac{8}{x^2}$

$x^2 - 2x = 8$

$x^2 - 2x - 8 = 0$

$(x - 4)(x + 2) = 0$

$x - 4 = 0$ or $x + 2 = 0$

$x = 4$ or $x = -2$

6. $\dfrac{y + 1}{3(y + 4)} = \dfrac{8}{(y + 4)(y - 4)}$

The LCD is $3(y + 4)(y - 4)$.
Multiplying each side by the LCD gives us

$(y + 1)(y - 4) = 8 \cdot 3$

$y^2 - 3y - 4 = 24$

$y^2 - 3y - 28 = 0$

$(y - 7)(y + 4) = 0$

$y = 7$ or $y = -4$

The only solution is 7 because the original equation is undefined when y is -4.

7. $x = \dfrac{y + 2}{y - 1}$

$x(y - 1) = y + 2$

$xy - x = y + 2$

$xy - y = x + 2$

$y(x - 1) = x + 2$

$y = \dfrac{x + 2}{x - 1}$

8. $\dfrac{1}{a} = \dfrac{1}{x} + \dfrac{1}{b}$

$axb \cdot \dfrac{1}{a} = axb \cdot \dfrac{1}{x} + axb \cdot \dfrac{1}{b}$

$xb = ab + ax$

$xb - ax = ab$

$x(b - a) = ab$

$x = \dfrac{ab}{b - a}$

9.

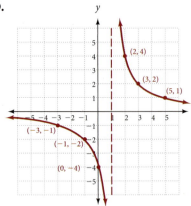

10.

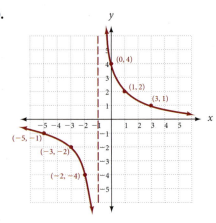

Section 6.7

1. Let x = one of the numbers and $3x$ = the other number.

$\dfrac{1}{x} + \dfrac{1}{3x} = \dfrac{4}{3}$ LCD = $3x$

$\mathbf{3x} \cdot \dfrac{1}{x} + \mathbf{3x} \cdot \dfrac{1}{3x} = \mathbf{3x} \cdot \dfrac{4}{3}$

$3 + 1 = 4x$

$4 = 4x$

$1 = x$

The two numbers are 1 and 3.

2.

	d	r	t
First Family	420	$r + 10$	$\dfrac{420}{r + 10}$
Second Family	420	r	$\dfrac{420}{r}$

$\dfrac{420}{r + 10} + 1 = \dfrac{420}{r}$

$r \cdot (r + 10)\left(\dfrac{420}{r + 10} + 1\right) = \dfrac{420}{r} \cdot r \cdot (r + 10)$

$420r + r^2 + 10r = 420r + 4{,}200$

$r^2 + 10r - 4{,}200 = 0$

$(r + 70)(r - 60) = 0$

$r = -70, 60$

Since r can't be negative, $r = 60$. So, the first family drives $60 + 10 = 70$ mph, and the second family drives 60 mph.

3.

	d	r	t
Upstream	1	$15 - x$	$\dfrac{1}{15 - x}$
Downstream	2	$15 + x$	$\dfrac{2}{15 + x}$

$\dfrac{1}{15 - x} = \dfrac{2}{15 + x}$ LCD = $(15 - x)(15 + x)$

$1(15 + x) = 2(15 - x)$

$15 + x = 30 - 2x$

$3x = 15$

$x = 5$ miles per hour

The speed of the current is 5 miles per hour.

4. $\dfrac{8}{x - 2} + \dfrac{8}{x + 2} = 3$ LCD = $(x + 2)(x - 2)$

$(\mathbf{x + 2})(\mathbf{x - 2}) \cdot \dfrac{8}{x - 2} + (\mathbf{x + 2})(\mathbf{x - 2}) \cdot \dfrac{8}{x + 2} = (\mathbf{x + 2})(\mathbf{x - 2}) \cdot 3$

$(x + 2) \cdot 8 + (x - 2) \cdot 8 = (x + 2)(x - 2) \cdot 3$

$8x + 16 + 8x - 16 = 3x^2 - 12$

$16x = 3x^2 - 12$

$0 = 3x^2 - 16x - 12$

$0 = (3x + 2)(x - 6)$

$3x + 2 = 0$ or $x - 6 = 0$

$x = -\dfrac{2}{3}$ or $x = 6$

	d	r	t
Upstream	8	$x - 2$	$\dfrac{8}{x - 2}$
Downstream	8	$x + 2$	$\dfrac{8}{x + 2}$

The speed of the boat in still water is 6 miles per hour. (The $-\dfrac{2}{3}$ cannot be a solution because it is negative.)

5. Let x = the length of time it takes to fill the sink with both the drain and faucet open.

$$\frac{1}{3} - \frac{1}{4} = \frac{1}{x} \qquad \text{LCD} = 12x$$

$$4x - 3x = 12$$
$$x = 12 \text{ minutes}$$

6. 13.0 feet per second $= \dfrac{13.0 \text{ feet}}{1 \text{ second}} \cdot \dfrac{1 \text{ mile}}{5,280 \text{ feet}} \cdot \dfrac{60 \text{ second}}{1 \text{ minute}} \cdot \dfrac{60 \text{ minutes}}{1 \text{ hour}}$

$$= \frac{13.0 \cdot 60 \cdot 60 \text{ miles}}{5,280 \text{ hours}}$$

$$= 8.9 \text{ miles per hour to the nearest tenth}$$

7. 1,100 feet per minute $= \dfrac{1,100 \text{ feet}}{1 \text{ minute}} \cdot \dfrac{1 \text{ mile}}{5,280 \text{ feet}} \cdot \dfrac{60 \text{ minutes}}{1 \text{ hour}}$

$$= \frac{1,100 \cdot 60 \text{ miles}}{5,280 \text{ hours}}$$

$$= 12.5 \text{ miles per hour}$$

Yes, 12.5 miles per hour is a reasonable speed for a chairlift.

Chapter 7

Section 7.1

1. The positive square root of 36 is 6 because 6 is the positive number with the property $6^2 = 36$. The negative square root of 36 is -6 since -6 is the negative number whose square is 36. The square roots of 36 are 6 and -6.

2. a. $\sqrt[3]{-64} = -4$ because $(-4)^3 = (-4)(-4)(-4) = -64$.
   b. $\sqrt{-25}$ is not a real number since there is no real number whose square is -25.
   c. $-\sqrt{4} = -2$; this is the negative square root of 4.
   d. $\sqrt[5]{-1} = -1$ because $(-1)^5 = (-1)(-1)(-1)(-1)(-1) = -1$.
   e. $\sqrt[4]{-16}$ is not a real number since there is no real number we can raise to the fourth power and obtain -16.

3. a. $\sqrt{81a^4b^8} = 9a^2b^4$ because $(9a^2b^4)^2 = 81a^4b^8$

   b. $\sqrt[3]{8x^3y^9} = 2xy^3$ because $(2xy^3)^3 = 8x^3y^9$

   c. $\sqrt[4]{81a^4b^8} = 3ab^2$ because $(3ab^2)^4 = 81a^4b^8$

4. a. $9^{1/2} = \sqrt{9} = 3$

   b. $27^{1/3} = \sqrt[3]{27} = 3$

   c. $-49^{1/2} = -\sqrt{49} = -7$

   d. $(-49)^{1/2} = \sqrt{-49}$ which is not a real number.

   e. $\left(\dfrac{16}{25}\right)^{1/2} = \sqrt{\dfrac{16}{25}} = \dfrac{4}{5}$

5. $\sqrt[3]{8x^3y^9} = (8x^3y^9)^{1/3}$
$$= 8^{1/3}(x^3)^{1/3}(y^9)^{1/3}$$
$$= 2xy^3$$

6. $\sqrt[4]{81a^4b^8} = (81a^4b^8)^{1/4}$
$$= 81^{1/4}(a^4)^{1/4}(b^8)^{1/4}$$
$$= 3ab^2$$

7. $9^{3/2} = (9^{1/2})^3 = 3^3 = 27$

8. $16^{3/4} = (16^{1/4})^3 = 2^3 = 8$

9. $8^{-2/3} = (8^{1/3})^{-2} = 2^{-2} = \dfrac{1}{4}$

10. $\left(\dfrac{16}{81}\right)^{-3/4} = \left(\dfrac{81}{16}\right)^{3/4} = \left[\left(\dfrac{81}{16}\right)^{1/4}\right]^3 = \left(\dfrac{3}{2}\right)^3 = \dfrac{27}{8}$

11. $x^{1/2} \cdot x^{1/4} = x^{1/2+1/4} = x^{3/4}$

12. $(y^{3/5})^{5/6} = y^{(3/5)(5/6)} = y^{1/2}$

13. $\dfrac{z^{3/4}}{z^{2/3}} = z^{3/4-2/3} = z^{1/12}$

14. $\dfrac{(x^{1/3}y^{-3})^6}{x^4y^{10}} = \dfrac{x^2y^{-18}}{x^4y^{10}}$
$$= x^{-2}y^{-28}$$
$$= \dfrac{1}{x^2y^{28}}$$

Section 7.2

1. $x^{1/3}(x^{2/3} - x^{1/3}) = x^{1/3} \cdot x^{2/3} - x^{1/3} \cdot x^{1/3}$
$$= x^1 - x^{2/3}$$
$$= x - x^{2/3}$$

2. $(x^{3/5} + 2)(x^{3/5} - 7) = x^{3/5}x^{3/5} - 7x^{3/5} + 2x^{3/5} - 14$
$$= x^{6/5} - 5x^{3/5} - 14$$

3. $(5a^{1/2} - 4b^{1/2})(3a^{1/2} - b^{1/2}) = 15a - 5a^{1/2}b^{1/2} - 12a^{1/2}b^{1/2} + 4b$
$$= 15a - 17a^{1/2}b^{1/2} + 4b$$

4. $(t^{1/3} + 3)^2 = (t^{1/3})^2 + 6t^{1/3} + 9$
$$= t^{2/3} + 6t^{1/3} + 9$$

5. $(x^{5/2} - 3^{1/2})(x^{5/2} + 3^{1/2}) = (x^{5/2})^2 - (3^{1/2})^2$
$$= x^5 - 3$$

6.
$$
\begin{array}{r}
a^{2/3} - a^{1/3}b^{1/3} + b^{2/3} \\
\underline{a^{1/3} + b^{1/3}} \\
a \quad - a^{2/3}b^{1/3} + a^{1/3}b^{2/3} \quad\quad \\
\underline{\quad + a^{2/3}b^{1/3} - a^{1/3}b^{2/3} \quad\quad + b} \\
a \quad\quad\quad\quad\quad\quad\quad\quad + b
\end{array}
$$

7. $\dfrac{36x^{3/4}y^{1/4} - 18x^{5/4}y^{1/4}}{6x^{1/4}y^{1/4}} = \dfrac{36x^{3/4}y^{1/4}}{6x^{1/4}y^{1/4}} - \dfrac{18x^{5/4}y^{1/4}}{6x^{1/4}y^{1/4}}$
$$= 6x^{1/2} - 3x$$

8. $8(x - 3)^{3/2} - 6(x - 3)^{1/2} = 2(x - 3)^{1/2}[4(x - 3) - 3]$
$$= 2(x - 3)^{1/2}(4x - 15)$$

9. $x^{2/3} - 4x^{1/3} - 21$
$$= (x^{1/3})^2 - 4x^{1/3} - 21$$
$$= (x^{1/3} - 7)(x^{1/3} + 3)$$

10. $6x^{2/3} + 19x^{1/3} + 10$
$$= 6(x^{1/3})^2 + 19x^{1/3} + 10$$
$$= (3x^{1/3} + 2)(2x^{1/3} + 5)$$

11. $(x^2 - 3)^{1/2} - \dfrac{x^2}{(x^2 - 3)^{1/2}} = \dfrac{(x^2 - 3)^{1/2}}{1} \cdot \dfrac{(x^2 - 3)^{1/2}}{(x^2 - 3)^{1/2}} - \dfrac{x^2}{(x^2 - 3)^{1/2}}$
$$= \dfrac{x^2 - 3 - x^2}{(x^2 - 3)^{1/2}} = \dfrac{-3}{(x^2 - 3)^{1/2}}$$

12. $r = \left(\dfrac{800}{600}\right)^{1/3} - 1$
$$= 1.101 - 1$$
$$= 0.101 \quad \text{or} \quad 10.1\%$$

Section 7.3

1. $\sqrt{18} = \sqrt{9 \cdot 2}$
$$= \sqrt{9}\,\sqrt{2}$$
$$= 3\sqrt{2}$$

2. $\sqrt{50x^2y^3} = \sqrt{25x^2y^2 \cdot 2y}$
$$= \sqrt{25x^2y^2}\,\sqrt{2y}$$
$$= 5xy\,\sqrt{2y}$$

3. $\sqrt[3]{54a^4b^3} = \sqrt[3]{27a^3b^3 \cdot 2a}$
$$= \sqrt[3]{27a^3b^3}\,\sqrt[3]{2a}$$
$$= 3ab\,\sqrt[3]{2a}$$

4. $\sqrt{75x^5y^8} = \sqrt{25x^4y^8 \cdot 3x}$
$$= \sqrt{25x^4y^8}\,\sqrt{3x}$$
$$= 5x^2y^4\,\sqrt{3x}$$

5. $\sqrt[4]{48a^8b^5c^4} = \sqrt[4]{16a^8b^4c^4 \cdot 3b}$
$$= 2a^2bc\sqrt[4]{3b}$$

6. $\sqrt{\dfrac{5}{9}} = \dfrac{\sqrt{5}}{\sqrt{9}}$
$$= \dfrac{\sqrt{5}}{3}$$

7. $\sqrt{\dfrac{2}{3}} = \dfrac{\sqrt{2}}{\sqrt{3}}$
$$= \dfrac{\sqrt{2}}{\sqrt{3}} \cdot \dfrac{\sqrt{3}}{\sqrt{3}}$$
$$= \dfrac{\sqrt{6}}{3}$$

8. $\dfrac{5}{\sqrt{2}} = \dfrac{5}{\sqrt{2}} \cdot \dfrac{\sqrt{2}}{\sqrt{2}}$
$$= \dfrac{5\sqrt{2}}{2}$$

9. $\dfrac{3\sqrt{5x}}{\sqrt{2y}} = \dfrac{3\sqrt{5x}}{\sqrt{2y}} \cdot \dfrac{\sqrt{2y}}{\sqrt{2y}}$

$\qquad = \dfrac{3\sqrt{10xy}}{2y}$

10. $\dfrac{5}{\sqrt[3]{9}} = \dfrac{5}{\sqrt[3]{3^2}}$

$\qquad = \dfrac{5}{\sqrt[3]{3^2}} \cdot \dfrac{\sqrt[3]{3}}{\sqrt[3]{3}}$

$\qquad = \dfrac{5\sqrt[3]{3}}{\sqrt[3]{3^3}}$

$\qquad = \dfrac{5\sqrt[3]{3}}{3}$

11. $\sqrt{\dfrac{48x^3y^4}{7z}} = \dfrac{\sqrt{48x^3y^4}}{\sqrt{7z}}$

$\qquad = \dfrac{\sqrt{16x^2y^4}\,\sqrt{3x}}{\sqrt{7z}}$

$\qquad = \dfrac{4xy^2\,\sqrt{3x}}{\sqrt{7z}}$

$\qquad = \dfrac{4xy^2\,\sqrt{3x}}{\sqrt{7z}} \cdot \dfrac{\sqrt{7z}}{\sqrt{7z}}$

$\qquad = \dfrac{4xy^2\,\sqrt{21xz}}{7z}$

12a. $\sqrt{16x^2} = 4|x|$

b. $\sqrt{25x^3} = 5\,|x|\,\sqrt{x}$

c. $\sqrt{x^2 + 10x + 25}$

$\qquad = \sqrt{(x + 5)^2}$

$\qquad = |x + 5|$

d. $\sqrt{2x^3 + 7x^2}$

$\qquad = \sqrt{x^2(2x + 7)}$

$\qquad = |x|\,\sqrt{2x + 7}$

13a. $\sqrt[3]{(-3)^3}$

$\qquad = \sqrt[3]{-27}$

$\qquad = -3$

b. $\sqrt[3]{(-1)^3}$

$\qquad = \sqrt[3]{-1}$

$\qquad = -1$

Section 7.4

1. $3\sqrt{5} - 2\sqrt{5} + 4\sqrt{5} = (3 - 2 + 4)\sqrt{5}$

$\qquad\qquad\qquad\qquad = 5\sqrt{5}$

2. $4\sqrt{50} + 3\sqrt{8} = 4\sqrt{25 \cdot 2} + 3\sqrt{4 \cdot 2}$

$\qquad\qquad\qquad = 4\sqrt{25}\,\sqrt{2} + 3\sqrt{4}\,\sqrt{2}$

$\qquad\qquad\qquad = 4 \cdot 5\sqrt{2} + 3 \cdot 2\sqrt{2}$

$\qquad\qquad\qquad = 20\sqrt{2} + 6\sqrt{2}$

$\qquad\qquad\qquad = 26\sqrt{2}$

3. $\sqrt{18x^2y} - 3x\sqrt{50y} = \sqrt{9x^2 \cdot 2y} - 3x\sqrt{25 \cdot 2y}$

$\qquad\qquad\qquad\qquad = \sqrt{9x^2}\,\sqrt{2y} - 3x\sqrt{25}\,\sqrt{2y}$

$\qquad\qquad\qquad\qquad = 3x\sqrt{2y} - 3x \cdot 5\sqrt{2y}$

$\qquad\qquad\qquad\qquad = 3x\sqrt{2y} - 15x\sqrt{2y}$

$\qquad\qquad\qquad\qquad = -12x\sqrt{2y}$

4. $2\sqrt[3]{27a^2b^4} + 3b\sqrt[3]{125a^2b} = 2\sqrt[3]{27b^3 \cdot a^2b} + 3b\sqrt[3]{125 \cdot a^2b}$

$\qquad\qquad\qquad\qquad = 2\sqrt[3]{27b^3}\,\sqrt[3]{a^2b} + 3b\sqrt[3]{125}\,\sqrt[3]{a^2b}$

$\qquad\qquad\qquad\qquad = 2 \cdot 3b\sqrt[3]{a^2b} + 3b \cdot 5\sqrt[3]{a^2b}$

$\qquad\qquad\qquad\qquad = 6b\sqrt[3]{a^2b} + 15b\sqrt[3]{a^2b}$

$\qquad\qquad\qquad\qquad = 21b\sqrt[3]{a^2b}$

5. $\dfrac{\sqrt{5}}{3} + \dfrac{1}{\sqrt{5}} = \dfrac{\sqrt{5}}{3} + \dfrac{1}{\sqrt{5}} \cdot \dfrac{\sqrt{5}}{\sqrt{5}}$

$\qquad\qquad = \dfrac{\sqrt{5}}{3} + \dfrac{\sqrt{5}}{5}$

$\qquad\qquad = \left(\dfrac{1}{3} + \dfrac{1}{5}\right)\sqrt{5}$

$\qquad\qquad = \left(\dfrac{5}{15} + \dfrac{3}{15}\right)\sqrt{5}$

$\qquad\qquad = \dfrac{8}{15}\,\sqrt{5}$

$\qquad\qquad = \dfrac{8\sqrt{5}}{15}$

6. First we construct a golden rectangle from a square of side 6.

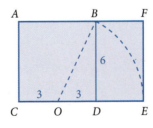

The length of the diagonal OB is found from the Pythagorean theorem.

$OB = \sqrt{3^2 + 6^2}$

$\qquad = \sqrt{9 + 36}$

$\qquad = \sqrt{45}$

$\qquad = 3\sqrt{5}$

The ratio of the length to the width for the rectangle is the golden ratio.

$$\text{Golden ratio} = \dfrac{CE}{EF} = \dfrac{3 + 3\sqrt{5}}{6}$$

$$= \dfrac{3(1 + \sqrt{5})}{3 \cdot 2}$$

$$= \dfrac{1 + \sqrt{5}}{2}$$

Section 7.5

1. We can rearrange the order and grouping of the numbers by applying the commutative and associative properties of multiplication.

$$(7\sqrt{3})(5\sqrt{11}) = (7 \cdot 5)\,\sqrt{3} \cdot \sqrt{11}$$
$$= 35\sqrt{33}$$

2.
$$\sqrt{2}(3\sqrt{5} - 4\sqrt{2}) = \sqrt{2} \cdot 3\sqrt{5} - \sqrt{2} \cdot 4\sqrt{2}$$
$$= 3\sqrt{10} - 4\sqrt{4}$$
$$= 3\sqrt{10} - 4 \cdot 2$$
$$= 3\sqrt{10} - 8$$

3.
$$(\sqrt{2} + \sqrt{7})(\sqrt{2} - 3\sqrt{7})$$
$$= \sqrt{2}\sqrt{2} - \sqrt{2} \cdot 3\sqrt{7} + \sqrt{7}\sqrt{2} - \sqrt{7} \cdot 3\sqrt{7}$$
$$= 2 - 3\sqrt{14} + \sqrt{14} - 21$$
$$= -19 - 2\sqrt{14}$$

4.
$$(\sqrt{x} + 5)^2 = (\sqrt{x} + 5)(\sqrt{x} + 5)$$
$$= \sqrt{x}\sqrt{x} + 5\sqrt{x} + 5\sqrt{x} + 25$$
$$= x + 10\sqrt{x} + 25$$

5.
$$(5\sqrt{a} - 3\sqrt{b})^2 = (5\sqrt{a})^2 - 2 \cdot 5\sqrt{a} \cdot 3\sqrt{b} + (3\sqrt{b})^2$$
$$= 25a - 30\sqrt{ab} + 9b$$

6.
$$(\sqrt{x + 3} - 1)^2 = (\sqrt{x + 3} - 1)(\sqrt{x + 3} - 1)$$
$$= x + 3 - 2\sqrt{x + 3} + 1$$
$$= x + 4 - 2\sqrt{x + 3}$$

7.
$$(\sqrt{5} + \sqrt{3})(\sqrt{5} - \sqrt{3}) = (\sqrt{5})^2 - (\sqrt{3})^2$$
$$= 5 - 3$$
$$= 2$$

8.
$$\frac{3}{\sqrt{7} - \sqrt{3}} = \frac{3}{\sqrt{7} - \sqrt{3}} \cdot \frac{\sqrt{7} + \sqrt{3}}{\sqrt{7} + \sqrt{3}}$$
$$= \frac{3\sqrt{7} + 3\sqrt{3}}{7 - 3}$$
$$= \frac{3\sqrt{7} + 3\sqrt{3}}{4}$$

9.
$$\frac{\sqrt{10} - 3}{\sqrt{10} + 3} = \frac{\sqrt{10} - 3}{\sqrt{10} + 3} \cdot \frac{\sqrt{10} - 3}{\sqrt{10} - 3}$$
$$= \frac{10 - 6\sqrt{10} + 9}{10 - 9}$$
$$= \frac{19 - 6\sqrt{10}}{1}$$
$$= 19 - 6\sqrt{10}$$

Section 7.6

1.
$$\sqrt{2x + 4} = 4$$
$$(\sqrt{2x + 4})^2 = 4^2$$
$$2x + 4 = 16$$
$$2x = 12$$
$$x = 6$$

2. $\sqrt{7x - 3} = -5$ has no solution since the left side is a positive number or 0 and the right side is a negative number.

3.
$$\sqrt{4x + 5} + 2 = 7$$
$$\sqrt{4x + 5} = 5$$
$$(\sqrt{4x + 5})^2 = 5^2$$
$$4x + 5 = 25$$
$$4x = 20$$
$$x = 5$$

4.
$$t - 6 = \sqrt{t - 4}$$
$$(t - 6)^2 = (\sqrt{t - 4})^2$$
$$t^2 - 12t + 36 = t - 4$$
$$t^2 - 13t + 40 = 0$$
$$(t - 5)(t - 8) = 0$$
$$t - 5 = 0 \quad \text{or} \quad t - 8 = 0$$
$$t = 5 \quad \text{or} \quad t = 8$$

Only 8 checks in the original equation.

5.
$$\sqrt{x - 9} = \sqrt{x} - 3$$
$$(\sqrt{x - 9})^2 = (\sqrt{x} - 3)^2$$
$$x - 9 = x - 6\sqrt{x} + 9$$
$$-9 = -6\sqrt{x} + 9$$
$$-18 = -6\sqrt{x}$$
$$3 = \sqrt{x}$$
$$3^2 = (\sqrt{x})^2$$
$$9 = x$$

6.
$$\sqrt{x + 4} = 2 - \sqrt{3x}$$
$$(\sqrt{x + 4})^2 = (2 - \sqrt{3x})^2$$
$$x + 4 = 4 - 4\sqrt{3x} + 3x$$
$$-2x = -4\sqrt{3x}$$
$$x = 2\sqrt{3x}$$
$$x^2 = (2\sqrt{3x})^2$$
$$x^2 = 4 \cdot 3x$$
$$x^2 - 12x = 0$$
$$x(x - 12) = 0$$
$$x = 0 \quad \text{or} \quad x - 12 = 0$$

Checking each solution in the original equation shows that $x = 12$ is extraneous. The only solution is $x = 0$.

7. $\sqrt{x+2} = \sqrt{x+3} - 1$
$(\sqrt{x+2})^2 = (\sqrt{x+3} - 1)^2$
$x + 2 = x + 3 - 2\sqrt{x+3} + 1$
$2 = 3 - 2\sqrt{x+3} + 1$
$2 = 4 - 2\sqrt{x+3}$
$-2 = -2\sqrt{x+3}$
$1 = \sqrt{x+3}$
$1^2 = (\sqrt{x+3})^2$
$1 = x + 3$
$-2 = x$

8. $\sqrt[3]{3x-7} = 2$
$(\sqrt[3]{3x-7})^3 = 2^3$
$3x - 7 = 8$
$3x = 15$
$x = 5$

9. $y = \sqrt{x} + 3$ 　　　　$y = \sqrt{x+3}$

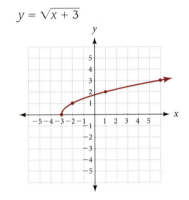

Section 7.7

1a. $\sqrt{-36} = i\sqrt{36} = i \cdot 6 = 6i$

b. $-\sqrt{-64} = -i\sqrt{64} = -i \cdot 8 = -8i$

c. $\sqrt{-18} = i\sqrt{18} = i \cdot 3\sqrt{2} = 3i\sqrt{2}$

d. $-\sqrt{-19} = -i\sqrt{19}$

2a. $i^{20} = (i^2)^{10} = (-1)^{10} = 1$

b. $i^{23} = (i^2)^{11} \cdot i = (-1)^{11} \cdot i = -i$

c. $i^{50} = (i^2)^{25} = (-1)^{25} = -1$

3. $4x + 7i = 8 - 14yi$
$\Rightarrow 4x = 8$　and　$7 = -14y$
$x = 2;$　　　$y = -\dfrac{1}{2}$

4. $(2x - 1) + 9i = 5 + (4y + 1)i$
$\Rightarrow 2x - 1 = 5$　and　$9 = 4y + 1$
$2x = 6$　　　$8 = 4y$
$x = 3$　　　$2 = y$

5a. $(2 + 6i) + (3 - 4i) = (2 + 3) + (6i - 4i)$
$= 5 + 2i$

b. $(6 + 5i) - (4 + 3i) = 6 + 5i - 4 - 3i$
$= 2 + 2i$

c. $(7 - i) - (8 - 2i) = 7 - i - 8 + 2i$
$= -1 + i$

6. $(2 + 3i)(1 - 4i)$
$= 2 \cdot 1 - 2 \cdot 4i + 3i \cdot 1 - 3i \cdot 4i$
$= 2 - 8i + 3i - 12i^2$
$= 2 - 5i - 12(-1)$
$= 2 - 5i + 12$
$= 14 - 5i$

7. $-3i(2 + 3i) = -3i \cdot 2 - 3i \cdot 3i$
$= -6i - 9i^2$
$= -6i - 9(-1)$
$= -6i + 9$
$= 9 - 6i$

8. $(2 + 4i)^2 = 2^2 + 2 \cdot 2 \cdot 4i + (4i)^2$
$= 4 + 16i - 16$
$= -12 + 16i$

9. $(3 - 5i)(3 + 5i) = 3^2 - (5i)^2$
$= 9 - 25i^2$
$= 9 + 25$
$= 34$

10. $\dfrac{3 + 2i}{2 - 5i} = \dfrac{3 + 2i}{2 - 5i} \cdot \dfrac{\mathbf{2 + 5i}}{\mathbf{2 + 5i}}$
$= \dfrac{6 + 15i + 4i + 10i^2}{4 - 25i^2}$
$= \dfrac{6 + 19i - 10}{4 + 25}$
$= \dfrac{-4 + 19i}{29}$
$= -\dfrac{4}{29} + \dfrac{19}{29}i$

11. $\dfrac{3 + 2i}{i} = \dfrac{3 + 2i}{i} \cdot \dfrac{\mathbf{-i}}{\mathbf{-i}}$
$= \dfrac{-3i - 2i^2}{-i^2}$
$= \dfrac{-3i + 2}{1}$
$= 2 - 3i$

Chapter 8

Section 8.1

1. $(3x + 2)^2 = 16$

$3x + 2 = \pm 4$

$3x = -2 \pm 4$

$x = \dfrac{-2 \pm 4}{3}$

$x = \dfrac{-2 + 4}{3}$ or $x = \dfrac{-2 - 4}{3}$

$x = \dfrac{2}{3}$ or $x = -2$

2. $(4x - 3)^2 = -50$

$4x - 3 = \pm\sqrt{-50}$

$4x - 3 = \pm 5i\sqrt{2}$

$4x = 3 \pm 5i\sqrt{2}$

$x = \dfrac{3 \pm 5i\sqrt{2}}{4}$

3. $x^2 + 10x + 25 = 20$

$(x + 5)^2 = 20$

$x + 5 = \pm\sqrt{20} = \pm 2\sqrt{5}$

$x = -5 \pm 2\sqrt{5}$

4. $x^2 + 3x - 4 = 0$

$x^2 + 3x = 4$

$x^2 + 3x + \dfrac{\mathbf{9}}{\mathbf{4}} = 4 + \dfrac{\mathbf{9}}{\mathbf{4}}$

$\left(x + \dfrac{3}{2}\right)^2 = \dfrac{25}{4}$

$x + \dfrac{3}{2} = \pm\dfrac{5}{2}$

$x = -\dfrac{3}{2} \pm \dfrac{5}{2}$

$x = -\dfrac{3}{2} + \dfrac{5}{2}$ or $x = -\dfrac{3}{2} - \dfrac{5}{2}$

$x = 1$ or $x = -4$

5. $5x^2 - 3x + 2 = 0$

$5x^2 - 3x = -2$

$x^2 - \dfrac{3}{5}x = -\dfrac{2}{5}$

$x^2 - \dfrac{3}{5}x + \dfrac{\mathbf{9}}{\mathbf{100}} = -\dfrac{2}{5} + \dfrac{\mathbf{9}}{\mathbf{100}}$

$\left(x - \dfrac{3}{10}\right)^2 = -\dfrac{31}{100}$

$x - \dfrac{3}{10} = \pm\sqrt{-\dfrac{31}{100}}$

$x - \dfrac{3}{10} = \pm\dfrac{i\sqrt{31}}{10}$

$x = \dfrac{3}{10} \pm \dfrac{i\sqrt{31}}{10}$

$x = \dfrac{3 \pm i\sqrt{31}}{10}$

6. Using Figure 3 in the text as a guide, we have $AC = 2$, so $AB = 4$. If $BC = x$, then by the Pythagorean theorem we have

$x^2 + 2^2 = 4^2$

$x^2 + 4 = 16$

$x^2 = 12$

$x = 2\sqrt{3}$

The sides are 2, $2\sqrt{3}$, and 4 inches.

7.

$4{,}330^2 = x^2 + 960^2$

$18{,}748{,}900 = x^2 + 921{,}600$

$x^2 = 18{,}748{,}900 - 921{,}600$

$x^2 = 17{,}827{,}300$

$x = \sqrt{17{,}827{,}300}$

$x = 4{,}222$ feet to the nearest foot

Section 8.2

1. $6x^2 + 7x + 2 = 0$

$a = 6, b = 7, c = 2$

$x = \dfrac{-7 \pm \sqrt{7^2 - 4(6)(2)}}{2(6)}$

$= \dfrac{-7 \pm \sqrt{49 - 48}}{12}$

$= \dfrac{-7 \pm \sqrt{1}}{12}$

$= \dfrac{-7 \pm 1}{12}$

$x = \dfrac{-7 + 1}{12}$ or $x = \dfrac{-7 - 1}{12}$

$x = -\dfrac{1}{2}$ or $x = -\dfrac{2}{3}$

2. $\dfrac{x^2}{2} + x = \dfrac{1}{3}$ $\qquad$ LCD $= 6$

$\mathbf{6} \cdot \dfrac{x^2}{2} + \mathbf{6} \cdot x = \mathbf{6} \cdot \dfrac{1}{3}$

$3x^2 + 6x = 2$

$3x^2 + 6x - 2 = 0$

$a = 3, b = 6, c = -2$

$x = \dfrac{-6 \pm \sqrt{6^2 - 4(3)(-2)}}{2(3)}$

$= \dfrac{-6 \pm \sqrt{36 + 24}}{6}$

$= \dfrac{-6 \pm \sqrt{60}}{6}$

$= \dfrac{-6 \pm 2\sqrt{15}}{6}$

$= \dfrac{2(-3 \pm \sqrt{15})}{2 \cdot 3}$

$= \dfrac{-3 \pm \sqrt{15}}{3}$

3. $\dfrac{1}{x+4} - \dfrac{1}{x} = \dfrac{1}{2}$ LCD $= 2x(x+4)$

$$2x(x+4)\left(\dfrac{1}{x+4} - \dfrac{1}{x}\right) = 2x(x+4) \cdot \dfrac{1}{2}$$

$$2x(x+4) \cdot \dfrac{1}{x+4} - 2x(x+4) \cdot \dfrac{1}{x} = 2x(x+4) \cdot \dfrac{1}{2}$$

$$2x - 2(x+4) = x(x+4)$$
$$2x - 2x - 8 = x^2 + 4x$$
$$-8 = x^2 + 4x$$
$$0 = x^2 + 4x + 8$$

$a = 1, b = 4, c = 8$

$$x = \dfrac{-4 \pm \sqrt{4^2 - 4(1)(8)}}{2(1)}$$

$$= \dfrac{-4 \pm \sqrt{16 - 32}}{2}$$

$$= \dfrac{-4 \pm \sqrt{-16}}{2}$$

$$= \dfrac{-4 \pm 4i}{2}$$

$$= \dfrac{2(-2 \pm 2i)}{2}$$

$$= -2 \pm 2i$$

4.

$$8t^3 - 27 = 0$$
$$(2t - 3)(4t^2 + 6t + 9) = 0$$
$$2t - 3 = 0 \quad \text{or} \quad 4t^2 + 6t + 9 = 0$$
$$2t = 3$$

$$t = \dfrac{3}{2}$$

$$t = \dfrac{-6 \pm \sqrt{36 - 4(4)(9)}}{2(4)}$$

$$= \dfrac{-6 \pm \sqrt{-108}}{8}$$

$$= \dfrac{-6 \pm 6i\sqrt{3}}{8}$$

$$= \dfrac{-3 \pm 3i\sqrt{3}}{4}$$

5.

$$12 = 32t - 16t^2$$
$$16t^2 - 32t + 12 = 0$$
$$4t^2 - 8t + 3 = 0 \qquad \text{Divide by 4}$$

$$t = \dfrac{8 \pm \sqrt{64 - 4(4)(3)}}{2(4)}$$

$$= \dfrac{8 \pm \sqrt{16}}{8}$$

$$= \dfrac{8 \pm 4}{8}$$

$$t = \dfrac{8 + 4}{8} = \dfrac{12}{8} = \dfrac{3}{2} \qquad \text{or} \qquad t = \dfrac{8 - 4}{8} = \dfrac{4}{8} = \dfrac{1}{2}$$

Note: Since the solutions are rational numbers, we could have solved the equation by factoring.

6.

$$P = \$1{,}320$$
$$P = -500 + 27x - 0.1x^2$$
$$1{,}320 = -500 + 27x - 0.1x^2$$
$$0.1x^2 - 27x + 1{,}820 = 0$$

$$x = \dfrac{27 \pm \sqrt{(-27)^2 - 4(0.1)(1{,}820)}}{(2)(0.1)}$$

$$= \dfrac{27 \pm \sqrt{729 - 728}}{0.2}$$

$$= \dfrac{27 \pm \sqrt{1}}{0.2}$$

$$= \dfrac{27 + 1}{0.2} \qquad \text{or} \qquad = \dfrac{27 - 1}{0.2}$$

$$= \dfrac{28}{0.2} \qquad \text{or} \qquad = \dfrac{26}{0.2}$$

$$x = 140 \qquad \text{or} \qquad x = 130$$

Section 8.3

1. $x^2 - 3x - 28 = 0$
$$b^2 - 4ac = (-3)^2 - 4(1)(-28)$$
$$= 9 + 112$$
$$= 121$$

which is a positive number that is a perfect square. Our equation, therefore, has two rational solutions.

2. $x^2 - 6x + 9 = 0$
$$b^2 - 4ac = (-6)^2 - 4(1)(9)$$
$$= 36 - 36$$
$$= 0$$

which means our equation has exactly one rational solution.

3. $3x^2 - 2x + 4 = 0$
$$b^2 - 4ac = (-2)^2 - 4(3)(4)$$
$$= 4 - 48$$
$$= -44$$

which is a negative number, implying that our equation has two complex solutions.

4. $x^2 + 1 = 4x$
$$x^2 - 4x + 1 = 0$$
$$b^2 - 4ac = (-4)^2 - 4(1)(1)$$
$$= 16 - 4$$
$$= 12$$

which is a positive number but not a perfect square. Our equation, therefore, has two irrational solutions.

5. $9x^2 + kx = -4$
$$9x^2 + kx + 4 = 0$$
$$b^2 - 4ac = k^2 - 4(9)(4) = 0$$
$$k^2 - 144 = 0$$
$$k^2 = 144$$
$$k = \pm 12$$

6. $t = -2, t = 2, t = 3$
$$t + 2 = 0, t - 2 = 0, t - 3 = 0$$
$$(t + 2)(t - 2)(t - 3) = 0$$
$$(t^2 - 4)(t - 3) = 0$$
$$t^3 - 3t^2 - 4t + 12 = 0$$

7. $x = -\dfrac{3}{4}$ $x = \dfrac{1}{5}$

$4x = -3$ $5x = 1$

$4x + 3 = 0$ $5x - 1 = 0$

$(4x + 3)(5x - 1) = 0$

$20x^2 + 11x - 3 = 0$

Section 8.4

1. $(x - 2)^2 - 3(x - 2) - 10 = 0$

Let $y = x - 2$.
$$y^2 - 3y - 10 = 0$$
$$(y - 5)(y + 2) = 0$$
$y - 5 = 0$ or $y + 2 = 0$
$y = 5$ or $y = -2$

Now replace y with $x - 2$.
$x - 2 = 5$ or $x - 2 = -2$
$x = 7$ or $x = 0$

2. $6x^4 - 13x^2 = 5$
$$6x^4 - 13x^2 - 5 = 0$$
$$(3x^2 + 1)(2x^2 - 5) = 0$$
$3x^2 + 1 = 0$ or $2x^2 - 5 = 0$
$3x^2 = -1$ or $2x^2 = 5$

$x^2 = -\dfrac{1}{3}$ or $x^2 = \dfrac{5}{2}$

$x = \pm\dfrac{i\sqrt{3}}{3}$ or $x = \pm\dfrac{\sqrt{10}}{2}$

3. $x - \sqrt{x} - 12 = 0$

Let $y = \sqrt{x}$.
$$y^2 - y - 12 = 0$$
$$(y - 4)(y + 3) = 0$$
$y - 4 = 0$ or $y + 3 = 0$
$y = 4$ or $y = -3$

Now replace y with $\sqrt{x}$.
$\sqrt{x} = 4$ or $\sqrt{x} = -3$
$x = 16$ or $x = 9$

Checking each solution in the original equation will show that 9 is extraneous. The only solution is $x = 16$.

4. $16t^2 - 10t - h = 0$

$$t = \dfrac{10 \pm \sqrt{100 - 4(16)(-h)}}{2(16)}$$

$$= \dfrac{10 \pm \sqrt{100 + 64h}}{32}$$

$$= \dfrac{10 \pm 2\sqrt{25 + 16h}}{32}$$

$$= \dfrac{5 \pm \sqrt{25 + 16h}}{16}$$

5. $AB = 2, BC = x$

If $\dfrac{AB}{BC} = \dfrac{BC}{AC}$

then $\dfrac{2}{x} = \dfrac{x}{x + 2}$

$$2(x + 2) = x^2$$
$$2x + 4 = x^2$$
$$0 = x^2 - 2x - 4$$

$$x = \dfrac{2 \pm \sqrt{4 - 4(1)(-4)}}{2}$$

$$= \dfrac{2 \pm \sqrt{20}}{2}$$

$$= \dfrac{2 \pm 2\sqrt{5}}{2}$$

$$= \dfrac{2(1 \pm \sqrt{5})}{2}$$

$$= 1 \pm \sqrt{5}$$

Since x must be positive, we have $x = 1 + \sqrt{5}$, which is twice the golden ratio.

Section 8.5

1. To find the x-intercepts we let $y = 0$

$$x^2 - 2x - 3 = 0$$
$$(x - 3)(x + 1) = 0$$
$$x = 3, \; x = -1$$

The x-coordinate of the vertex is

$$x = \frac{-b}{2a} = \frac{-(-2)}{2(1)} = 1$$

The y-coordinate of the vertex is

$$y = 1^2 - 2(1) - 3 = -4$$

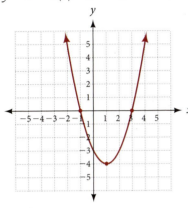

2. x-intercepts:

$$-x^2 + 2x + 8 = 0$$
$$x^2 - 2x - 8 = 0$$
$$(x - 4)(x + 2) = 0$$
$$x = 4, x = -2$$

Vertex:

$$x = \frac{-b}{2a} = \frac{-2}{2(-1)} = 1$$

$$y = -(1)^2 + 2(1) + 8 = 9$$

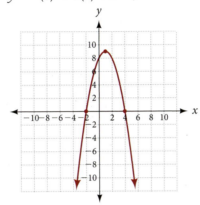

3. The x-intercepts are found by using the quadratic formula on $2x^2 - 4x + 1 = 0$. They are irrational numbers, approximately 0.3 and 1.7. Two other points on the graph are $(0, 1)$ and $(2, 1)$.

Vertex:

$$x = \frac{-b}{2a} = \frac{-(-4)}{2(2)} = 1$$

$$y = 2(1)^2 - 4(1) + 1 = -1$$

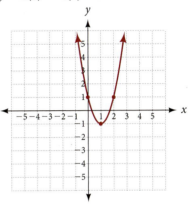

4. The graph does not cross the x-axis because the equation $-x^2 + 4x - 5 = 0$ does not have real solutions. Two points on the graph are $(0, -5)$ and $(4, -5)$.

Vertex:

$$x = \frac{-b}{2a} = \frac{-4}{2(-1)} = 2$$

$$y = -2^2 + 4(2) - 5 = -1$$

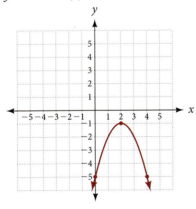

5. $0 = y^2 + 6y + 5$

$$0 = (y + 1)(y + 5)$$
$$y = -1 \quad \text{or} \quad y = -5$$

$$x = (y^2 + 6y + 9) + 5 - 9$$
$$x = (y + 3)^2 - 4$$

so vertex is at $(-4, -3)$

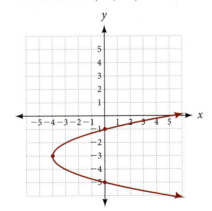

6. x-intercept: When $y = 0$, $x = 3$

y-intercept: When $x = 0$, we have $-y^2 + 2y + 3 = 0$, which yields solutions $y = -1$, and $y = 3$

vertex: $y = \frac{-b}{2a} = \frac{-2}{-2} = 1$

Substituting 1 for y gives us $x = 4$. The vertex is at $(4, 1)$.

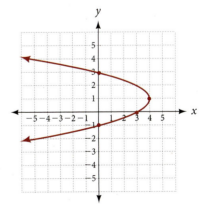

7.

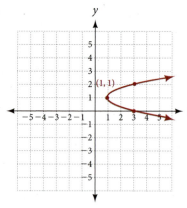

8. $y = -0.01x^2 + 25x - 400$

$$X = \frac{-b}{2a} = \frac{-12}{2(-0.01)} = \frac{-25}{-0.02} = 1250$$

$$
\begin{aligned}
y &= -0.01(1250)^2 + 25(1250) - 400 \\
&= -0.01(1562500) + 31250 - 400 \\
&= -15625 + 31250 - 400 \\
&= 15225
\end{aligned}
$$

9.

From the graph we see that the maximum value of R occurs when $p = \$2.00$. We can calculate the maximum value of R from the equation:

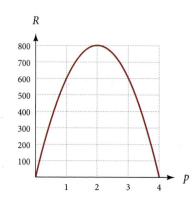

When $p = 2$
the equation $R = (800 - 200p)p$
becomes $R = (800 - 200 \cdot 2)2$
         $= (800 - 400)2$
         $= 400 \cdot 2$
         $= 800$

The maximum revenue is \$800. It is obtained by setting the price of each sketch pad at $p = \$2.00$.

10. The vertex is $(90, 100)$. Thus we have $y = a(x - 90)^2 + 100$.
Substitute $(180, 0)$ for x and y.
$0 = a(180 - 90)^2 + 100$
$0 = a(90)^2 + 100$
$0 = 8100a + 100$

Solve for a to get $a = -\dfrac{1}{81}$.

Equation is $y = -\dfrac{1}{81}(x - 90)^2 + 100$.

Section 8.6

1. $x^2 + 2x - 15 \le 0$
$(x + 5)(x - 3) \le 0$

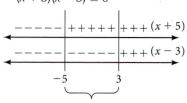

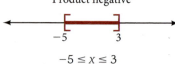

$-5 \le x \le 3$

2. $2x^2 + 3x > 2$
         $2x^2 + 3x - 2 > 0$
         $(2x - 1)(x + 2) > 0$

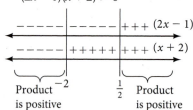

$x < -2$ or $x > -\dfrac{1}{2}$

3. $x^2 + 10x + 25 < 0$
      $(x + 5)^2 < 0$

No value of x will satisfy this inequality because the left side always will be positive or 0, but never less than 0. No solution.

4. $\dfrac{x+3}{x-2} \ge 0$

```
----- +++++ +++ (x + 3)
----- ----- +++ (x − 2)
        −3      2
```

Product is positive   Product is positive

```
        −3        2
```

$x \le -3 \text{ or } x > 2$

5. $\dfrac{1}{x-4} - \dfrac{2}{x-3} > 0$

$\dfrac{1}{x-4} \cdot \dfrac{(x-3)}{(x-3)} - \dfrac{2}{x-3} \cdot \dfrac{(x-4)}{(x-4)} > 0$

$\dfrac{x-3-2x+8}{(x-4)(x-3)} > 0$

$\dfrac{-x+5}{(x-4)(x-3)} > 0$

$\dfrac{-(x-5)}{(x-4)(x-3)} > 0$

```
+++++ ++ ++ -- −(x − 5)
----- -- ++ +++ (x − 4)
----- ++ ++ +++ (x − 3)
            3  4  5
```

Product is positive   Product is positive

```
            3  4  5
```

$x < 3 \text{ or } 4 < x < 5$

Chapter 9

Section 9.1

1. a. $f(0) = 4^0 = 1$
b. $f(1) = 4^1 = 4$
c. $f(2) = 4^2 = 16$
d. $f(3) = 4^3 = 64$
e. $f(-1) = 4^{-1} = \dfrac{1}{4}$
f. $f(-2) = 4^{-2} = \dfrac{1}{4^2} = \dfrac{1}{16}$

2. $A(12) = 1{,}200 \cdot 2^{-12/8}$
$\approx 424.3 \text{ micrograms}$
$A(24) = 1{,}200 \cdot 2^{-24/8}$
$= 150 \text{ micrograms}$

3.

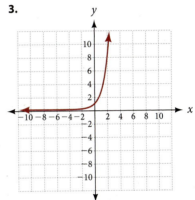

4.

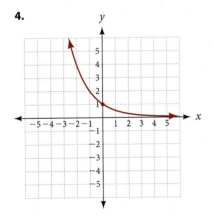

5. a. $A(5) = 600\left(1 + \dfrac{0.06}{12}\right)^{12 \cdot 5}$

$= 600\left(1 + \dfrac{0.06}{12}\right)^{60}$

$\approx \$809.31$

b.

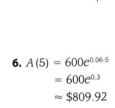

From the graph, $x \approx 8.5$ when $y = 1{,}000$.

6. $A(5) = 600e^{0.06 \cdot 5}$
$= 600e^{0.3}$
$\approx \$809.92$

Section 9.2

1. The inverse of $y = 4x + 1$ is $x = 4y + 1$. Solving for y we have

$$4y + 1 = x$$
$$4y = x - 1$$
$$y = \frac{x - 1}{4}$$

2. The equation of the inverse is $x = y^2 + 1$ or $y = \pm\sqrt{x - 1}$.

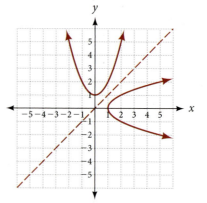

3. $x = \dfrac{y - 3}{y + 1}$ is the inverse of g.

$$x(y + 1) = y - 3$$
$$xy + x = y - 3$$
$$xy - y = -x - 3$$
$$y(x - 1) = -x - 3$$
$$y = \frac{-x - 3}{x - 1}$$

Multiply numerator and denominator by -1 to get $y = \dfrac{3 + x}{1 - x}$.

4.

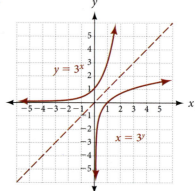

Section 9.3

1. $\log_2 x = 3$ is equivalent to

$$x = 2^3$$
$$x = 8$$

2. $\log_x 5 = 2$ is equivalent to

$$x^2 = 5$$
$$x = \sqrt{5} \text{ only}$$

since the base in a logarithmic statement cannot be negative.

3. $\log_9 27 = x$ is equivalent to

$$9^x = 27$$
$$(3^2)^x = 3^3$$
$$3^{2x} = 3^3$$

So $2x = 3$

$$x = \frac{3}{2}$$

4. First graph $y = 3^x$ and then reflect it about the line $y = x$ to obtain the graph of $y = \log_3 x$.

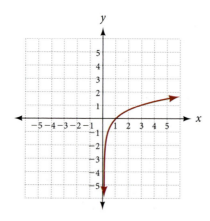

5. $\log_3 27 = \log_3 3^3$
$= 3$

6. $\log_{10} 1{,}000 = \log_{10} 10^3$
$= 3$

7. $\log_6 6 = \log_6 6^1$
$= 1$

8. $\log_3 1 = \log_3 3^0$
$= 0$

9. $\log_2 (\log_8 8) = \log_2 1$
$= 0$

10. $\log_{10} T = 6$
$T = 10^6 = 1{,}000{,}000$

Section 9.4

1. $\log_3 \dfrac{5a}{b} = \log_3 5a - \log_3 b$
$= \log_3 5 + \log_3 a - \log_3 b$

2. $\log_{10} \dfrac{x^2}{\sqrt[3]{y}} = \log_{10} \dfrac{x^2}{y^{1/3}}$
$= \log_{10} x^2 - \log_{10} y^{1/3}$
$= 2 \log_{10} x - \dfrac{1}{3} \log_{10} y$

3. $3 \log_4 x + \log_4 y - 2 \log_4 z$
$= \log_4 x^3 + \log_4 y - \log_4 z^2$
$= \log_4 x^3 y - \log_4 z^2$
$= \log_4 \dfrac{x^3 y}{z^2}$

4. $\log_2 (x + 3) + \log_2 x = 2$
$\log_2 (x + 3)x = 2$
$(x + 3)x = 2^2$
$x^2 + 3x = 4$
$x^2 + 3x - 4 = 0$
$(x - 1)(x + 4) = 0$
$x = 1 \qquad \text{or} \qquad x = -4$

Since $x = -4$ would make both terms of our original equation undefined, the only solution is $x = 1$.

Section 9.5

1. When using a scientific calculator, we simply enter the number 27,600 and press the key labeled $\boxed{\log}$.

$\log 27{,}600 \approx 4.4409$

4. When using a scientific calculator, we enter the number 3.9786 and press the key labeled $\boxed{10^x}$.

$\log x = 3.9786$
$x = 10^{3.9786}$
$\approx 9{,}519$

5. $\log x = -1.5901$
$x = 10^{-1.5901}$
≈ 0.0257

6. $\log T = 5.5$
$T = 10^{5.5} \approx 3.16 \times 10^5$

7. $4.1 = -\log[H^+]$
$\log[H^+] = -4.1$
$[H^+] = 10^{-4.1}$
$\approx 7.9 \times 10^{-5}$

8. $pH = -\log[1.8 \times 10^{-5}]$
$\approx -(-4.7)$
$= 4.7$

9. a. $\ln e^2 = 2 \ln e = 2(1) = 2$
b. $\ln e^4 = 4 \ln e = 4(1) = 4$
c. $\ln e^{-2} = -2 \ln e = -2(1) = -2$
d. $\ln e^x = x \ln e = x(1) = x$

10. $\ln Pe^{rt} = \ln P + \ln e^{rt}$
$= \ln P + rt \ln e$
$= \ln P + rt$

11. a. $\ln 35 = \ln 5 \cdot 7 = \ln 5 + \ln 7$
$= 1.6094 + 1.9459 = 3.5553$

b. $\ln 0.2 = \ln \dfrac{2}{10} = \ln \dfrac{1}{5} = \ln 1 - \ln 5$
$= 0 - 1.6094 = -1.6094$

c. $\ln 49 = \ln 7^2 = 2 \ln 7 = 2(1.9459) = 3.8918$

Section 9.6

1.
$$12^{x+2} = 20$$
$$\log 12^{x+2} = \log 20$$
$$(x + 2) \log 12 = \log 20$$
$$x + 2 = \frac{\log 20}{\log 12}$$
$$x = \frac{\log 20}{\log 12} - 2$$
$$\approx \frac{1.3010}{1.0792} - 2$$
$$\approx -0.7945 \text{ or } -0.7944$$

2.
$$10{,}000 = 5{,}000(1 + 0.11)^t$$
$$10{,}000 = 5{,}000(1.11)^t$$
$$2 = (1.11)^t$$
$$\log 2 = t \log 1.11$$
$$t = \frac{\log 2}{\log 1.11}$$
$$\approx \frac{0.3010}{0.0453}$$
$$\approx 6.64 \text{ years}$$

3.
$$\log_6 14 = \frac{\log 14}{\log 6}$$
$$\approx \frac{1.1461}{0.7782}$$
$$\approx 1.4728 \text{ or } 1.4729$$

depending on whether you round your intermediate answers.

4.
$$75{,}000 = 32{,}000x^{0.05t}$$
$$2.34 \approx e^{0.05t}$$
$$\ln 2.34 \approx 0.05t \ln e$$
$$\ln 2.34 \approx 0.05t$$
$$t \approx \frac{\ln 2.34}{0.05} \approx 17.04$$

Chapter 10

Section 10.1

1. $d = \sqrt{(-4 - 2)^2 + (1 - 5)^2}$
$= \sqrt{36 + 16}$
$= \sqrt{52}$
$= 2\sqrt{13}$

2. $\sqrt{(x - 3)^2 + (2 + 1)^2} = \sqrt{10}$
$(x - 3)^2 + 9 = 10$
$(x - 3)^2 = 1$
$x - 3 = \pm 1$
$x = 3 \pm 1$

$x = 3 + 1$	or	$x = 3 - 1$
$= 4$		$= 2$

Both solutions check.

3. $(x - 4)^2 + [y - (-3)]^2 = 2^2$
$(x - 4)^2 + (y + 3)^2 = 4$

4. Center $= (0, 0)$, radius $= 5$
$(x - 0)^2 + (y - 0)^2 = 5^2$
$x^2 + y^2 = 25$

5. Center $= (3, 4)$; radius $= 3$

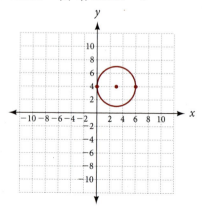

6. $x^2 + y^2 - 6x + 4y - 3 \qquad = 0$
$x^2 - 6x \qquad + y^2 + 4y \qquad = 3$
$x^2 - 6x + \mathbf{9} + y^2 + 4y + \mathbf{4} = 3 + \mathbf{9}$
$\qquad\qquad + \mathbf{4}$
$(x - 3)^2 + (y + 2)^2 = 16$
Center $= (3, -2)$, radius $= 4$

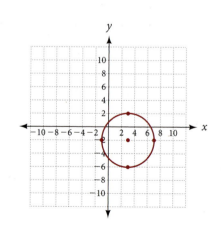

Section 10.2

1.

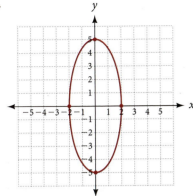

2.

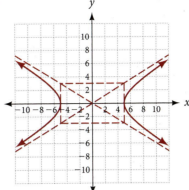

3.

$$4x^2 + 9y^2 - 24x + 36y + 36 = 0$$
$$4x^2 - 24x + 9y^2 + 36y = -36$$
$$4(x^2 - 6x) + 9(y^2 + 4y) = -36$$
$$4(x^2 - 6x + \mathbf{9}) + 9(y^2 + 4y + \mathbf{4}) = -36 + \mathbf{36} + \mathbf{36}$$

$$4(x - 3)^2 + 9(y + 2)^2 = 36$$

$$\frac{(x - 3)^2}{9} + \frac{(y + 2)^2}{4} = 1$$

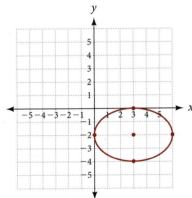

4.

$$9y^2 - x^2 - 36y - 6x + 18 = 0$$
$$9y^2 - 36y - x^2 - 6x = -18$$
$$9(y^2 - 4y) - (x^2 + 6x) = -18$$
$$9(y^2 - 4y + \mathbf{4}) - (x^2 + 6x + \mathbf{9}) = -18 + \mathbf{36} - \mathbf{9}$$

$$9(y - 2)^2 - (x + 3)^2 = 9$$

$$\frac{(y - 2)^2}{1} - \frac{(x + 3)^2}{9} = 1$$

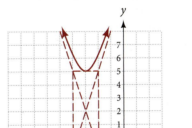

Section 10.3

1.

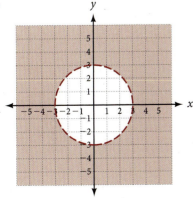

2.

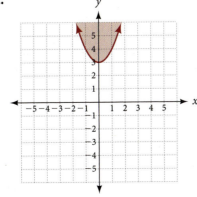

3.

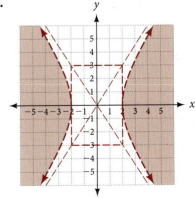

4. $x^2 + y^2 = 9$

$x - y = 3$

Solving the bottom equation for x gives $x = y + 3$. Substituting $y + 3$ for x in the top equation we have

$(y + 3)^2 + y^2 = 9$

$y^2 + 6y + 9 + y^2 = 9$

$2y^2 + 6y = 0$

$2y(y + 3) = 0$

$y = 0$ or $y = -3$

Substituting these values of y into $x = y + 3$ gives $x = 3$ and $x = 0$, respectively. The two solutions are $(3, 0)$ and $(0, -3)$.

5. $16x^2 - 4y^2 = 64 \xrightarrow{\text{no change}} 16x^2 - 4y^2 = 64$

$x^2 + y^2 = 4 \xrightarrow{\text{times 4}} 4x^2 + 4y^2 = 16$

$\underline{\qquad\qquad\qquad\qquad}$

$20x^2 = 80$

$x^2 = 4$

$x = \pm 2$

Using the second equation in the original system, we have

When $x = 2$ $2^2 + y^2 = 4 \Rightarrow y = 0$

When $x = -2$ $(-2)^2 + y^2 = 4 \Rightarrow y = 0$

The two solutions are $(2, 0)$ and $(-2, 0)$.

6. $x^2 + y^2 = 4$

$y = x^2 - 4$

Substituting $x^2 - 4$ from the second equation for y in the first equation yields

$x^2 + (x^2 - 4)^2 = 4$

$x^2 + x^4 - 8x^2 + 16 = 4$

$x^4 - 7x^2 + 16 = 4$

$x^4 - 7x^2 + 12 = 0$

$(x^2 - 4)(x^2 - 3) = 0$

$x^2 = 4$ or $x^2 = 3$

$x = \pm 2$ or $x = \pm\sqrt{3}$

Substituting these four values of x into the second equation in our system gives us four solutions:

$(2, 0), (-2, 0), (\sqrt{3}, -1), (-\sqrt{3}, -1)$.

7. Let x and y represent the two numbers.

$x^2 + y^2 = 58$

$y = x^2 - 2$

Substituting $x^2 - 2$ into the first equation in place of y gives us

$x^2 + (x^2 - 2)^2 = 58$

$x^2 + x^4 - 4x^2 + 4 = 58$

$x^4 - 3x^2 - 54 = 0$

$(x^2 - 9)(x^2 + 6) = 0$

$x^2 = 9$ or $x^2 = -6$

$x = \pm 3$ or $x = \pm i\sqrt{6}$

When $x = \pm 3, y = 7$

When $x = \pm i\sqrt{6}, y = -8$

but since we have no way of comparing the size of a complex number like $i\sqrt{6}$ with a real number like -8, we cannot check the complex solutions. The only solutions are 3 and 7, and -3, and 7.

8. $x^2 + y^2 \geq 9$

$\dfrac{x^2}{4} + \dfrac{y^2}{25} \leq 1$

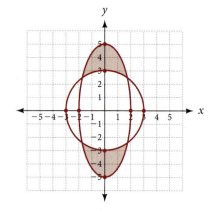

Appendix A

1. a. $\begin{vmatrix} 2 & 1 \\ 4 & 3 \end{vmatrix} = 6 - 4 = 2$

2. $\begin{vmatrix} -3 & x \\ 2 & x \end{vmatrix} = 20$

b. $\begin{vmatrix} 4 & -2 \\ 0 & 3 \end{vmatrix} = 12 - 0 = 12$

$-3x - 2x = 20$

$-5x = 20$

$x = -4$

3. $\begin{vmatrix} 2 & 0 & -1 \\ 3 & 1 & 2 \\ 5 & -2 & 1 \end{vmatrix} = 2 + 0 + 6 - (-5 - 8 + 0) = 8 - (-13)$

$= 21$

4. $\begin{vmatrix} 2 & 0 & -1 \\ 3 & 1 & 2 \\ 5 & -2 & 1 \end{vmatrix} = 2\begin{vmatrix} 1 & 2 \\ -2 & 1 \end{vmatrix} - 0\begin{vmatrix} 3 & 2 \\ 5 & 1 \end{vmatrix} - 1\begin{vmatrix} 3 & 1 \\ 5 & -2 \end{vmatrix}$

$= 2(1 + 4) - 0(3 - 10) - 1(-6 - 5)$

$= 2(5) - 0 - 1(-11)$

$= 10 + 11$

$= 21$

5. $\begin{vmatrix} 0 & 4 & -2 \\ 3 & 1 & 1 \\ 1 & -2 & 0 \end{vmatrix} = -4\begin{vmatrix} 3 & 1 \\ 1 & 0 \end{vmatrix} + 1\begin{vmatrix} 0 & -2 \\ 1 & 0 \end{vmatrix} + 2\begin{vmatrix} 0 & -2 \\ 3 & 1 \end{vmatrix}$

$= -4(0 - 1) + 1(0 + 2) + 2(0 + 6)$

$= -4(-1) + 1(2) + 2(6)$

$= 4 + 2 + 12$

$= 18$

Appendix B

1. $3x - 5y = 2$

$2x + 4y = 1$

$D = \begin{vmatrix} 3 & -5 \\ 2 & 4 \end{vmatrix} = 22$

$D_x = \begin{vmatrix} 2 & -5 \\ 1 & 4 \end{vmatrix} = 13$

$D_y = \begin{vmatrix} 3 & 2 \\ 2 & 1 \end{vmatrix} = -1$

$x = \dfrac{D_x}{D} = \dfrac{13}{22}$

$y = \dfrac{D_y}{D} = -\dfrac{1}{22}$

The solution is $\left(\dfrac{13}{22}, -\dfrac{1}{22}\right)$.

2. $x + 2y + z = -2$

$x + 2y - z = -6$

$x - 2y + z = -4$

$D = \begin{vmatrix} 1 & 2 & 1 \\ 1 & 2 & -1 \\ 1 & -2 & 1 \end{vmatrix} = -8$

$D_x = \begin{vmatrix} -2 & 2 & 1 \\ -6 & 2 & -1 \\ -4 & -2 & 1 \end{vmatrix} = 40$

$D_y = \begin{vmatrix} 1 & -2 & 1 \\ 1 & -6 & -1 \\ 1 & -4 & 1 \end{vmatrix} = -4$

$D_z = \begin{vmatrix} 1 & 2 & -2 \\ 1 & 2 & -6 \\ 1 & -2 & -4 \end{vmatrix} = -16$

$x = \dfrac{D_x}{D} = \dfrac{40}{-8} = -5$

$y = \dfrac{D_y}{D} = \dfrac{-4}{-8} = \dfrac{1}{2}$

$z = \dfrac{D_z}{D} = \dfrac{-16}{-8} = 2$

The solution is $\left(-5, \dfrac{1}{2}, 2\right)$.

3. $x + y = 3$

$2x - z = 3$

$y + 2z = 9$

$D = \begin{vmatrix} 1 & 1 & 0 \\ 2 & 0 & -1 \\ 0 & 1 & 2 \end{vmatrix} = -3$

$D_x = \begin{vmatrix} 3 & 1 & 0 \\ 3 & 0 & -1 \\ 9 & 1 & 2 \end{vmatrix} = -12$

$D_y = \begin{vmatrix} 1 & 3 & 0 \\ 2 & 3 & -1 \\ 0 & 9 & 2 \end{vmatrix} = 3$

$D_z = \begin{vmatrix} 1 & 1 & 3 \\ 2 & 0 & 3 \\ 0 & 1 & 9 \end{vmatrix} = -15$

$x = \dfrac{D_x}{D} = \dfrac{-12}{-3} = 4$

$y = \dfrac{D_y}{D} = \dfrac{3}{-3} = -1$

$z = \dfrac{D_z}{D} = \dfrac{-15}{-3} = 5$

The solution is $(4, -1, 5)$.

Appendix C

1. $-2\rfloor$ $\begin{array}{rrrr} 1 & 2 & -8 & 1 \\ & -2 & 0 & 16 \\ \hline 1 & 0 & -8 & \underline{17} \end{array}$ **2.** $3\rfloor$ $\begin{array}{rrrr} 2 & -5 & 0 & 3 \\ & 6 & 3 & 9 \\ \hline 2 & 1 & 3 & \underline{12} \end{array}$ **3.** $-2\rfloor$ $\begin{array}{rrrr} 1 & 0 & 0 & 8 \\ & -2 & 4 & -8 \\ \hline 1 & -2 & 4 & \underline{0} \end{array}$

Answer: $x^2 - 8 + \dfrac{17}{x + 2}$ Answer: $2x^2 + x + 3 + \dfrac{12}{x - 3}$ Answer: $x^2 - 2x + 4$

Appendix D

2. The converse: (If B, then A.) If it can fly, then it is a bird.
The inverse: (If not A, then not B.) If it is not a bird, then it cannot fly.
The contrapositive: (If not B, then not A.) If it cannot fly, then it is not a bird.
3. The contrapositive also must be true: If your eyes are opened, then you are not asleep.
4. The contrapositive also must be true: If $a^2 \neq 9$, then $a \neq 3$.
5. If it is history, then it repeats itself.

Answers to Odd-Numbered Problems

Chapter 1

Problem Set 1.1

1. $x + 5 = 2$ **3.** $6 - x = y$ **5.** $2t < y$ **7.** $x + y < x - y$ **9.** $3(x - 5) > y$ **11.** 36   **13.** 100   **15.** 8   **17.** 16
19. 10,000   **21.** 121   **23. a.** 19 **b.** 27 **c.** 27   **25. a.** 16 **b.** 12 **c.** 18   **27. a.** 33 **b.** 33   **29. a.** 144 **b.** 74 **c.** 144
31. a. 23 **b.** 41 **c.** 65   **33. a.** 39 **b.** 7 **c.** 5   **35. a.** 48 **b.** 24   **37. a.** 41 **b.** 95   **39.** 5,431   **41.** 138   **43.** 78
45. 152   **47.** 11   **49.** 87   **51.** 4   **53.** 22   **55.** 16   **57.** 625   **59.** 1   **61.** 0.6   **63.** 8,700   **65.** 5.046
67. 5   **69.** 128   **71.** 625   **73.** 31   **75.** 1,000   **77.** 750   **79.** 0.5   **81.** 1.6   **83.** 320   **85.** 1.7917
87. 2.0793   **89.** 785   **91.** 185.12   **93.** 12.3106   **95.** 0.1587   **97.** 196   **99.** 650   **101. a.** 7 **b.** 10 **c.** 25 **d.** 32
103. a. 121 **b.** 121 **c.** 101 **d.** 81   **105. a.** 1 **b.** 2,400 **c.** 0.52   **107. a.** -5 **b.** 23 **c.** 7   **109. a.** 8 **b.** 14 **c.** 14

Problem Set 1.2

1. {0, 1, 2, 3, 4, 5, 6}   **3.** {2, 4}   **5.** {1, 3, 5}   **7.** {0, 1, 2, 3, 4, 5, 6}   **9.** {0, 2}   **11.** {0, 6}   **13.** {0, 1, 2, 3, 4, 5, 6, 7}
15. {1, 2, 4, 5}

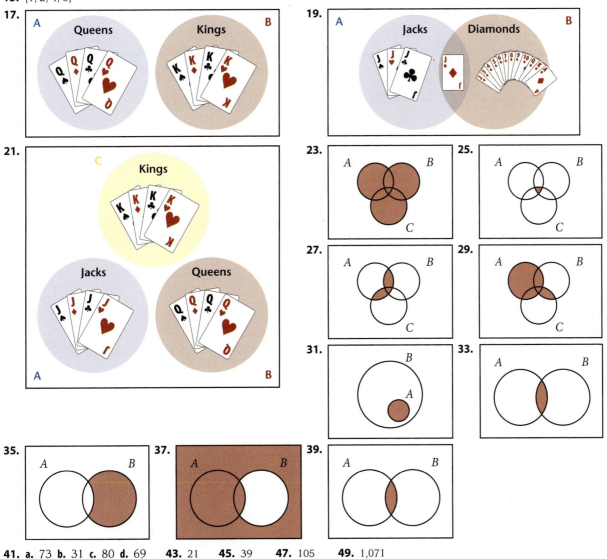

41. a. 73 **b.** 31 **c.** 80 **d.** 69   **43.** 21   **45.** 39   **47.** 105   **49.** 1,071

Problem Set 1.3

1. $1, 2$ **3.** $-6, -5.2, 0, 1, 2, 2.3, \frac{9}{2}$ **5.** $-\sqrt{7}, -\pi, \sqrt{17}$ **7.** $0, 1, 2$ **9.** $2^2 \cdot 3 \cdot 5$ **11.** $2 \cdot 7 \cdot 19$ **13.** $3 \cdot 37$ **15.** $3^2 \cdot 41$

17. $\frac{3}{7}$ **19.** $\frac{11}{21}$ **21.** $\frac{3}{5}$ **23.** $\frac{5}{6}$ **25.** $\frac{5}{9}$ **27.** $\frac{3}{4}$ **29.** 40 **31.** $\frac{5}{11}$

33. a. $10 = 3 + 7$, other answers are possible.   **b.** $16 = 5 + 11$, other answers are possible.
c. $24 = 5 + 19$, other answers are possible.   **d.** $36 = 17 + 19$, other answers are possible.

35. $x = 200, x = 100$ **37.** 36 feet

Problem Set 1.4

1.   **3.**   **5.**

7.   **9.**   **11.**

13.   **15.**   **17.**

19.   **21.**   **23.**

25.   **27.**   **29.**

31.   **33.**   **35.**

37. $x \geq 5$ **39.** $x \leq -3$ **41.** $x \leq 4$ **43.** $-4 < x < 4$ **45.** $-4 \leq x \leq 4$ **47.** -3 and 7 **49.** $3 < x < 13$

51. $x < 3$ or $x > 13$ **53.** $50 \leq F \leq 270$ **55.**

Problem Set 1.5

1. $4, -4, \frac{1}{4}$ **3.** $-\frac{1}{2}, \frac{1}{2}, -2$ **5.** $5, -5, \frac{1}{5}$ **7.** $\frac{3}{8}, -\frac{3}{8}, \frac{8}{3}$ **9.** $-\frac{1}{6}, \frac{1}{6}, -6$ **11.** $3, -3, \frac{1}{3}$ **13.** $-1, 1$ **15.** 0

17. 2 **19.** $\frac{3}{4}$ **21.** π **23.** -4 **25.** -2 **27.** $-\frac{3}{4}$ **29.** 4 **31.** -4 **33.** 12 **35.** -13 **37.** -10

39. -4 **41.** $\frac{19}{12}$ **43.** $-\frac{32}{105}$ **45.** -8 **47.** -12 **49.** $-7x$ **51.** 13 **53.** -14 **55.** $6a$

57.

a	b	Sum $a + b$	Difference $a - b$	Product ab	Quotient $\frac{a}{b}$
3	12	15	-9	36	$\frac{1}{4}$
-3	12	9	-15	-36	$-\frac{1}{4}$
3	-12	-9	15	-36	$-\frac{1}{4}$
-3	-12	-15	9	36	$\frac{1}{4}$

59. -15 **61.** 15 **63.** -24 **65.** $-10x$ **67.** x **69.** y **71.** $\frac{21}{40}$ **73.** 2 **75.** $\frac{8}{27}$ **77.** $\frac{1}{10,000}$ **79.** $\frac{72}{385}$

81. 1 **83.** -2 **85.** Undefined   **87.** 0 **89.** $-\frac{2}{3}$ **91.** 32 **93.** 64 **95.** $-\frac{1}{18}$ **97.** $\frac{7}{15}$ **99.** $\frac{29}{35}$ **101.** $\frac{35}{144}$

103. $\frac{949}{1260}$ **105.** $\frac{101}{504}$ **107.** $\frac{41}{195}$ **109.** $\frac{133}{540}$ **111.** $\frac{47}{105}$ **113.** 15 **115.** 26 **117.** 16 **119.** -14

121. -1 **123.** 31 **125.** -2 **127.** 4 **129.** 0 **131.** Undefined   **133.** $-\frac{2}{3}$

Problem Set 1.6

1. $6 + x$ **3.** $a + 8$ **5.** $15y$ **7.** x **9.** a **11.** x **13.** $3x + 18$ **15.** $12x + 8$ **17.** $15a + 10b$ **19.** $\frac{4}{3}x + 2$

21. $2 + y$ **23.** $40t + 8$ **25.** $9x + 3y - 6z$ **27.** $3x + 7y$ **29.** $6x + 7y$ **31.** $3x + 1$ **33.** $2x - 1$ **35.** $x + 2$

37. $a - 3$ **39.** $x + 24$ **41.** $3x - 2y$ **43.** $3x + 8y$ **45.** $8x + 5y$ **47.** $15x + 10$ **49.** $8y + 32$ **51.** $15t + 9$

53. $28x + 11$ **55.** $14a + 7$ **57.** $6y + 6$ **59.** $12x + 2$ **61.** $8y + 11$ **63.** $24a + 15$ **65.** $11x + 20$

67. Commutative property of addition   **69.** Commutative property of multiplication   **71.** Additive Inverse

73. Commutative property of addition   **75.** Associative and commutative properties of multiplication

77. Commutative and associative properties of addition   **79.** Distributive property   **81.** -14 **83.** 18

85. 16   **87.** -19 **89.** 26   **91.** 20   **93.** -2 **95.** 1   **97.** -30 **99.** 18   **101.** 277   **103.** -73 **105.** 39

107. 11   **109.** $b - a$ **111.** $y - 6$ **113.** $-x + 2z$ **115.** $2x + 6$ **117.** $5a + 30$ **119.** $5x - 90$ **121.** $8y$

123. $27x + 12$ **125.** $5m - 22$ **127.** $-6x + 23$ **129.** $18y + 18$ **131.** $-4x + 3$ **133.** $-8x$

135. -14 **137.** -59 **139.** 15   **141.**

x	$(x + 1)^2$	$x^2 + 1$	$x^2 + 2x + 1$
-2	1	5	1
-1	0	2	0
0	1	1	1
1	4	2	4
2	9	5	9

143. a. -44 **b.** 121   **c.** 0   **d.** 49   **145.** 1.5439   **147.** -0.3974 **149.** 0.0909   **151.** 3.6   **153.** 3.84

155. 1,200   **157.** 219   **159.** $9x + 27$ **161.** $3a + ab$

Problem Set 1.7

1. 5   **3.** 10   **5.** 25   **7.** 29   **9.** 23   **11.** △   **13.** ⊙   **15.** 17, 21   **17.** $-2, -3$ **19.** $-4, -7$ **21.** $-\frac{1}{2}, -\frac{3}{4}$

23. $\frac{5}{2}, 3$ **25.** 27   **27.** -270 **29.** $\frac{1}{8}$ **31.** $\frac{5}{2}$ **33.** -625 **35.** $-\frac{1}{125}$ **37. a.** 12   **b.** 16   **39.** 144

41. 2, 3, 5, among others   **43.** 2, 8, 34   **45.**

Two numbers a and b	Their product ab	Their sum $a + b$
1, -24	-24	-23
-1, 24	-24	23
2, -12	-24	-10
-2, 12	-24	10
3, -8	-24	-5
-3, 8	-24	5
4, -6	-24	-2
-4, 6	-24	2

47. 41, 37.5, 34, 30.5, 27, 23.5; yes   **49.** 41, 45.5, 50, 54.5, 59, 63.5; yes

51. The patient on Antidepressant 1 misses his morning dose; less than half of the antidepressant will remain in the body. The patient on Antidepressant 2 still has most of the medication in his body even after missing a dose because of the relatively long (5-day) half-life.

53.

Hours Since Discontinuing	Concentration (ng/mL)
0	60
4	30
8	15
12	7.5
16	3.75

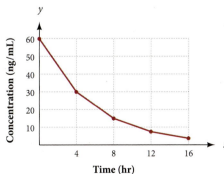

55.

Elevation (ft)	Boiling Point (°F)
$-2,000$	215.6
$-1,000$	213.8
0	212
1,000	210.2
2,000	208.4
3,000	206.6

Chapter 1 Review

1. $x + 2$ **2.** $2x + y$ **3.** 27   **4.** 125   **5.** 32   **6.** 81   **7.** 17   **8.** 4   **9.** 30   **10.** 43   **11.** $\{1, 2, 3, 4, 5, 6\}$

12. $\{1, 3\}$ **13.** $\{5\}$ **14.** $\{6\}$ **15.**

16. $-7, 0, 5$ **17.** $-7, -4.2, 0, \frac{3}{4}, 5$ **18.** $-\sqrt{3}, \pi$ **19.** $2^2 \cdot 3^2 \cdot 11^2$ **20.** $\frac{11}{13}$

21.

22.

23.

24. $0 < x < 8$ **25.** $0 \le x \le 8$ **26.** 1   **27.** $\frac{27}{64}$ **28.** $-2, \frac{1}{2}$

29. $\frac{2}{5}, -\frac{5}{2}$ **30.** 4   **31.** 6   **32.** -42 **33.** 30   **34.** $21x$ **35.** $-6x$ **36.** $2y$ **37.** $3x$ **38.** a

39. c   **40.** a   **41.** b, d   **42.** a, c   **43.** f   **44.** e   **45.** g   **46.** 2   **47.** -2 **48.** 3   **49.** $-x + 3$

50. $-15x + 3$ **51.** $-\frac{5}{6}$ **52.** $-\frac{2}{7}$ **53.** -13 **54.** -17 **55.** 2   **56.** 39   **57.** -1, arithmetic   **58.** 8

Chapter 1 Test

1. $2(3x + 4y)$ **2.** $2a - 3b < 2a + 3b$ **3.** 57   **4.** 10   **5.** 16   **6.** 0   **7.** $\{2, 4\}$ **8.** $\varnothing$ **9.** $-5, 0, 1, 4$

10. $-5, -4.1, -\frac{5}{6}, 0, 1, 1.8, 4$ **11.** $-\sqrt{2}, \sqrt{3}$ **12.** $0, 1, 4$ **13.** $3^2 \cdot 5 \cdot 13$ **14.** $2^2 \cdot 5 \cdot 31$ **15.** $\frac{117}{124}$

16. $\{x \mid x > 3\}$ **17.** $\{x \mid x \le -5\}$ **18.** $\{x \mid -5 < x < 3\}$ **19.** $\{x \mid x \le -5 \text{ or } x \ge 3\}$

20.

21.

22. a. 3 **b.** -2 **23.** Commutative property of addition

24. Multiplicative identity property   **25.** Associative property of multiplication   **26.** Associative property of addition

27. 2   **28.** 0   **29.** $\frac{59}{72}$ **30.** $-4x$ **31.** $-5x - 8$ **32.** $4y - 10$ **33.** $3x - 17$ **34.** $11a - 10$ **35.** $-\frac{7}{3}$ **36.** $-\frac{5}{2}$

37. -320, geometric   **38.** -2, arithmetic   **39.** 17   **40.** $\frac{1}{125}$, geometric   **41.** 74   **42.** 46

Chapter 2

Problem Set 2.1

1. 8   **3.** 5   **5.** 2   **7.** -7 **9.** $-\frac{9}{2}$ **11.** -4 **13.** $-\frac{4}{3}$ **15.** -4 **17.** -2 **19.** $\frac{3}{4}$ **21.** 12   **23.** -10

25. 7   **27.** -7 **29.** 3   **31.** $\frac{4}{5}$ **33.** 1   **35.** 4   **37.** 17   **39.** $2 -$ **41.** 6   **43.** $-\frac{4}{3}$ **45.** $-\frac{3}{2}$ **47.** $\frac{5}{3}$

49. $\frac{3}{2}$ **51.** 1   **53.** No solution   **55.** All real numbers are solutions.   **57.** No solution   **59.** $-\frac{2}{3}$ **61.** $-\frac{7}{640}$

63. 2   **65.** 20   **67.** 0.7   **69.** 2,400   **71.** 24   **73.** 6   **75.** 3   **77.** 8.5   **79.** 2   **81.** 5   **83.** 18   **85.** 36

87. 4   **89.** 30   **91.** 6,000   **93.** 5,000   **95.** $\frac{8}{15}$ **97.** $-\frac{5}{11}$ **99. a.** $\frac{5}{8}$ **b.** 0 **c.** $10x - 10$ **d.** 0 **e.** $8x - 40$ **f.** 5

101. a. $\$25.45 = \$1.80n + \$3.85$ **b.** 12 miles   **103. a.** $3,937,000 = 1,125A$ **b.** 3,500 square miles

105. 50 miles per hour   **107.** Commutative   **109.** Associative   **111.** Commutative and associative

113. Multiplicative identity   **115.** Commutative   **117.** Additive identity   **119.** $7x - 3y - 2$ **121.** $y - 5$

123. $-4x + 24$ **125.** 0.5   **127.** 62.5   **129.** 0   **131.** 1.25   **133.** 13   **135. a.** 3 **b.** 5 **c.** 9   **137.** 2

139. $-\frac{3}{2}$ **141.** $-\frac{21}{2}$ **143.** No Solution

Problem Set 2.2

1. -3 **3.** 0 **5.** $\frac{3}{2}$ **7.** 4 **9.** $\frac{8}{5}$ **11.** 2 **13.** $-\frac{7}{640}$ **15.** 675 **17. a.** 0 **b.** 0 **19. a.** 23 **b.** 23 **21.** $-\frac{1}{2}$

23. 3 **25.** $2{,}400$ **27.** $\ell = \frac{A}{w}$ **29.** $t = \frac{I}{pr}$ **31.** $T = \frac{PV}{nR}$ **33.** $x = \frac{y-b}{m}$ **35.** $F = \frac{9}{5}C + 32$ **37.** $v = \frac{h - 16t^2}{t}$

39. $d = \frac{A-a}{n-1}$ **41.** $y = -\frac{2}{3}x + 2$ **43.** $y = \frac{3}{5}x + 3$ **45.** $y = \frac{1}{3}x + 2$ **47.** $x = \frac{5}{a-b}$ **49.** $P = \frac{A}{1+rt}$

51. $y = -\frac{1}{4}x + 2$ **53.** $y = \frac{3}{5}x - 3$ **55.** $y = \frac{1}{2}x + \frac{3}{2}$ **57.** $y = -2x - 5$ **59.** $y = -\frac{2}{3}x + 1$ **61.** $y = -\frac{1}{2}x + \frac{7}{2}$

63. a. $y = 4x - 1$ **b.** $y = -\frac{1}{2}x$ **c.** $y = -3$ **65.** 20.52 **67.** 25% **69.** 925

71. a. $-\frac{15}{4}$ **b.** -7 **c.** $y = \frac{4}{5}x + 4$ **d.** $x = \frac{5}{4}y - 5$ **73.** 16% **75.** $\$3.25$ **77.** 40% **79.** 20.6 square miles

81. 7.5 hours   **83.** $\$30.00$ **85.** $\$34.00$ **87.** 6 mph   **89.** 42 mph   **91.** 55 mph   **93.** 84 kmph

95. 6.8 feet/second   **97.** 20 **99.** 31 **101.** 8 **103.** 1 **105.** -3 **107.** 2 **109.** 13 **111.** $2x - 3$

113. $x + y = 180$ **115.** 30 **117.** 8.5 **119.** $6{,}000$ **121.** $x = -\frac{a}{b}y + a$ **123.** $a = \frac{bc}{b-c}$ **125.** $a = \frac{bc}{b-c}$

127. $a = \frac{bc}{b - 2c}$

Problem Set 2.3

1. 10 feet by 20 feet   **3.** 7 feet   **5.** 5 inches   **7.** 4 meters   **9.** $\$92.00$ **11.** $\$200.00$ **13.** $\$56.39$ per book

15. $\$25.65$ **17.** $20°, 160°$ **19. a.** $20.4°, 69.6°$ **b.** $38.4°, 141.6°$ **21.** $27°, 72°, 81°$ **23.** $102°, 44°, 34°$

25. $43°, 43°, 94°$ **27.** $\$6{,}000$ at 8%, $\$3{,}000$ at 9% **29.** $\$5{,}000$ at 12%, $\$10{,}000$ at 10% **31.** $\$4{,}000$ at 8%, $\$2{,}000$ at 9%

33. 16 adults, 22 children   **35.** $\$54$

37.

t	0	$\frac{1}{4}$	1	$\frac{7}{4}$	2
h	0	7	16	7	0

39.

HORSE RACING	
Year	Bets (millions of dollars)
1985	20.2
1990	34.4
1995	44.8
2000	65.4
2005	104
2010	163

41.

HOT COFFEE SALES	
Year	Sales (billions of dollars)
2005	7
2006	7.5
2007	8
2008	8.6
2009	9.2

43.

DISTANCE	
Speed (mph)	Distance (miles)
20	10
30	15
40	20
50	25
60	30
70	35

45.

Time (hours)	Distance Upstream (miles)	Distance Downstream (miles)
1	6	14
2	12	28
3	18	42
4	24	56
5	30	70
6	36	84

47. 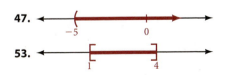   **49.**    **51.**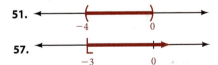

53.   **55.**   **57.**

59. -5 **61.** 6

Problem Set 2.4

1. $x \le \frac{3}{2}$

3. $x > 4$

5. $x \ge -5$

7. $x < 4$

9. $x \ge -6$

11. $x \ge 4$

13. $x < -3$

15. $m \ge -1$

17. $x \ge -3$

19. $y \le \frac{7}{2}$

21. $x < 6$

23. $y \ge -52$

25. $(-\infty, -2]$ **27.** $[1, \infty)$

29. $(-\infty, 3)$ **31.** $(-\infty, -1]$ **33.** $[-17, \infty)$ **35.** $x > 435$ **37.** $[3, 7]$

39. $(-4, 2)$

41. $[4, 6]$

43. $(-4, 2)$

45. $(-3, 3)$

47. $(-\infty, -7] \cup [-3, \infty)$

49. $(-\infty, -1] \cup [\frac{3}{5}, \infty)$

51. $(-\infty, -10) \cup (6, \infty)$

53. $-2 < x \le 4$ **55.** $x < -4$ or $x \ge 1$

57. a. $x > 0$ **b.** $x \ge 0$ **c.** $x \ge 0$ **59.** $\varnothing$ **61.** All real numbers   **63.** $\varnothing$ **65. a.** 1   **b.** 16   **c.** no   **d.** $x > 16$

67. $31 \le d \le 49$ **69.** 3   **71.** -3 **73.** The distance between x and 0 on the number line   **75.** -3

77. 15   **79.** No solution   **81.** -1 **83.** $x < \frac{c-b}{a}$ **85.** $\frac{-c-b}{a} < x < \frac{c-b}{a}$

Problem Set 2.5

1. $-4, 4$ **3.** $-2, 2$ **5.** $\varnothing$ **7.** $-1, 1$ **9.** $\varnothing$ **11.** $-6, 6$ **13.** $-3, 7$ **15.** $\frac{17}{3}, \frac{7}{3}$ **17.** $2, 4$ **19.** $-\frac{5}{2}, \frac{5}{6}$

21. $-1, 5$ **23.** $\varnothing$ **25.** $-4, 20$ **27.** $-4, 8$ **29.** $-\frac{10}{3}, \frac{2}{3}$ **31.** $\varnothing$ **33.** $-1, \frac{3}{2}$ **35.** $5, 25$ **37.** $-30, 26$

39. $-12, 28$ **41.** $-5, \frac{3}{5}$ **43.** $1, \frac{1}{9}$ **45.** $-\frac{1}{2}$ **47.** 0 **49.** $-\frac{1}{2}$ **51.** $-\frac{1}{6}, -\frac{7}{4}$ **53.** All real numbers

55. All real numbers   **57. a.** $\frac{5}{4} = 1.25$ **b.** $\frac{5}{4} = 1.25$ **c.** 2   **d.** $\frac{1}{2}, 2$ **e.** $\frac{1}{3}, 4$ **59.** $|3{,}500 - 4{,}200| = x$

61.

63.

65. $\frac{19}{15}$ **67.** 40   **69.** -2 **71.** 0

73. $x < 4$ **75.** $-\frac{11}{3} \le a$ **77.** $t \le -\frac{3}{2}$ **79.** $x = a \pm b$ **81.** $x = \frac{-b \pm c}{a}$ **83.** $x = -\frac{a}{b}y \pm a$

Problem Set 2.6

1. $-3 < x < 3$

3. $x \le -2$ or $x \ge 2$

5. $-3 < x < 3$

7. $t < -7$ or $t > 7$

9. $\varnothing$

11. All real numbers

13. $-4 < x < 10$

15. $a \le -9$ or $a \ge -1$

17.

$\varnothing$

19. $-1 < x < 5$

21. $y \le -5$ or $y \ge -1$ 　　**23.** $k \le -5$ or $k \ge 2$

25. $-1 < x < 7$ 　　**27.** $a \le -2$ or $a \ge 1$

29. $-6 < x < \frac{8}{3}$ 　　**31.** $x < 2$ or $x > 8$

33. $x \le -3$ or $x \ge 12$ 　　**35.** $x < 2$ or $x > 6$

37. $0.99 < x < 1.01$　**39.** $x \le -\frac{3}{5}$ or $x \ge -\frac{2}{5}$　**41.** $-\frac{1}{6} \le x \le \frac{3}{2}$　**43.** $-0.05 < x < 0.25$　**45.** $|x - 5| \le 1$

47. no　**49.** $x < -2$ or $x > \frac{4}{5}$　**51.** -6　**53.** 66　**55.** -13　**57.** -4　**59.** 13　**61.** 16　**63.** -1

65. $a - b < x < a + b$　**67.** $x < \frac{b - c}{a}$ or $x > \frac{b + c}{a}$　**69.** $\frac{-b - c}{a} \le x \le \frac{c - b}{a}$

Chapter 2 Review

1. 10　**2.** 2　**3.** 2　**4.** -3　**5.** -3　**6.** 0　**7.** $\frac{2}{3}$　**8.** $\frac{10}{13}$　**9.** $\frac{5}{2}$　**10.** 1　**11.** $-\frac{5}{11}$　**12.** $-\frac{2}{9}$

13. $h = 17$　**14.** $t = 20$　**15.** $p = \frac{I}{rt}$　**16.** $x = \frac{y - b}{m}$　**17.** $y = \frac{4}{3}x - 4$　**18.** $v = \frac{d - 16t^2}{t}$　**19.** $y = -\frac{5}{3}x + 2$

20. $y = -\frac{2}{3}x + 2$　**21.** $y = -2x - 1$　**22.** $y = -3x + 1$　**23.** 4 feet by 12 feet　**24.** 3 meters, 4 meters, 5 meters

25. \$24,875.24　**26. a.** 3,780 bricks　**b.** 625 feet　**27.** $(-\infty, \frac{1}{2})$　**28.** $(-\infty, 8]$　**29.** $(-\infty, 12]$　**30.** $(-1, \infty)$

31. $[2, 6]$　**32.** $[0, 1]$　**33.** $(-\infty, -\frac{3}{2}] \cup [3, \infty)$　**34.** $(-\infty, 1) \cup [4, \infty)$　**35.** $-2, 2$　**36.** $-4, 4$　**37.** $2, 4$

38. $-1, 4$　**39.** $-\frac{3}{2}, 3$　**40.** $5, 9$　**41.** $-1, 1$　**42.** 0　**43.** $-5 < x < 5$

44. $a \ge 500$ or $a \le -500$ 　　**45.** $\varnothing$　**46.** $-3 < t < 2$

47. $\varnothing$　**48.** $\varnothing$　**49.** $\varnothing$　**50.** All real numbers except 0　**51.** $\varnothing$　**52.** All real numbers

Chapter 2 Cumulative Review

1. 49　**2.** 3　**3.** -6　**4.** -18　**5.** 13　**6.** 16　**7.** 25　**8.** 36　**9.** 3　**10.** 123　**11.** 2　**12.** 3　**13.** $\frac{6}{7}$　**14.** $\frac{3}{4}$

15. $\frac{71}{105}$　**16.** $\frac{23}{56}$　**17.** -9　**18.** -34　**19.** $\{2, 3, 4, 6, 8, 9\}$　**20.** $\{6\}$　**21.** $\{4, 5\}$　**22.** $\{8\}$　**23.** $-13, -6.7, 0, \frac{1}{2}, 2, \frac{5}{2}$

24. $-13, 0, 2$　**25.** $-\sqrt{5}, \pi, \sqrt{13}$　**26.** $0, 2$　**27.** -54　**28.** $\frac{1}{16}$　**29.** $2x - 12$　**30.** $10x + 26$　**31.** $x - 4$

32. $7x + 6y$　**33.** $-\frac{7}{3}$　**34.** 5　**35.** 10　**36.** $0, 5$　**37.** $-5, 5$　**38.** $-\frac{5}{3}, 3$　**39.** 10　**40.** 5　**41.** $x = \frac{3}{4}y + 3$

42. $C = \frac{5}{9}(F - 32)$　**43.** $h = \frac{2A}{b}$　**44.** $x = \frac{2}{b - a}$　**45.** $25°, 65°$　**46.** $l = 26$ ft; $w = 13$ ft　**47.** $(-1, \infty)$　**48.** $[-3, 2]$

49. $x < -\frac{2}{5}$ or $x > \frac{4}{5}$ 　　**50.** $-\frac{7}{2} < t < \frac{5}{2}$

51. Gulf Coast < Midwest, or $\$3.18 < \$4.01$　**52.** $A = \{$Rocky Mountains, East Coast$\}$

Chapter 2 Test

1. 12　**2.** $-\frac{4}{3}$　**3.** $-\frac{7}{4}$　**4.** 2　**5.** -2　**6.** 0　**7.** 4　**8.** 8　**9.** $w = \frac{P - 2l}{2}$　**10.** $B = \frac{2A - hb}{h}$　**11.** $y = \frac{3}{2}x + 2$

12. $y = -\frac{5}{3}x + 2$　**13.** $y = \frac{2}{3}x + 7$　**14.** $y = 3x - 6$　**15.** 60%　**16.** $44\frac{4}{9}\%$　**17.** 55　**18.** 26　**19.** 6 cm by 24 cm

20. 6 meters　**21.** $55°, 125°$　**22.** $30°, 60°$　**23.** \$56.25　**24.** \$2,000 at 11%, \$6,000 at 12%

25. $[-6, \infty)$ 　　**26.** $(-\infty, 4)$ 　　**27.** $(-\infty, 6)$

28. $[-52, \infty)$ 　　**29.** $2, 6$　**30.** $-15, 3$　**31.** $\varnothing$　**32.** $-5, 1$

33. $x < -1$ or $x > \frac{4}{3}$ 　　**34.** $-\frac{2}{3} \le x \le 4$

35. All real numbers　**36.** $\varnothing$

Chapter 3

Problem Set 3.1

3. A. (4, 1)   B. (−4, 3)   C. (−2, −5)   D. (2, −2)   E. (0, 5)   F. (−4, 0)   G. (1, 0)

1.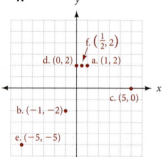

5.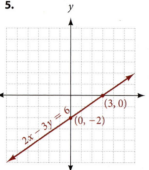

7.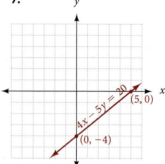

9.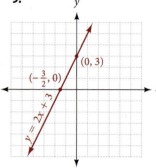

11.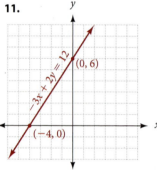

13.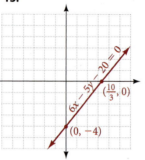

15.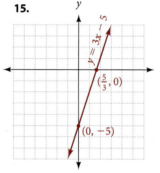

17.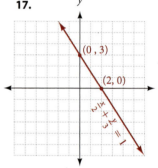

19. b

21.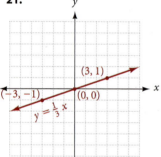

23.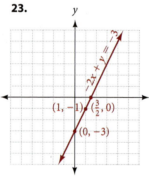

25.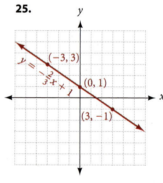

27.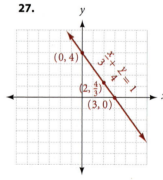

29. b   **31. a.** −7   **b.** −4   **c.** $-\frac{4}{3}$ **d.** See graph below.   **e.** $y = -\frac{1}{3}x - \frac{4}{3}$

31. d.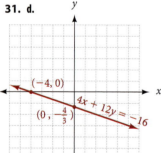

33. a.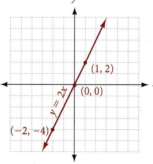

b.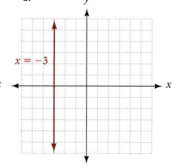

c.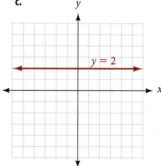

35. a.

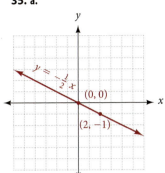

b.

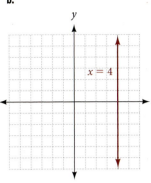

c.

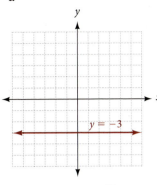

37.

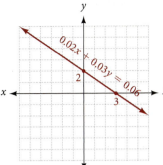

39.

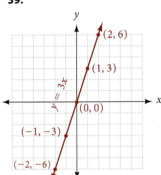

41. a. Answers will vary
   b. $440
   c. 25 hours
   d. No, if she works 35 hours she should be paid $385

43.

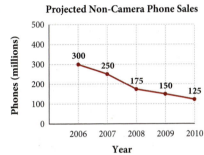

45. (2006, 30), (2007, 33), (2008, 37), (2009, 45), (2010, 48)

47.

x (year)	2006	2007	2008	2009
y (age)	370	450	580	820

49. $A = (1, 2)$, $B = (6, 7)$ **51.** $A = (3, 3)$, $B = (3, 6)$, $C = (8, 6)$ **53. a.** Yes   **b.** No   **c.** No   **55.** 2   **57.** -2

59. 1,200   **61.** 2   **63.** $\frac{8}{15}$ **65.** $-\frac{6}{100}$ **67.** 1   **69.** $-\frac{4}{3}$ **71.** Undefined   **73. a.** $\frac{3}{2}$ **b.** $-\frac{3}{2}$

75. x-intercept $= \frac{c}{a}$, y-intercept $= \frac{c}{b}$,   **77.** x-intercept $= a$, y-intercept $= b$,

Problem Set 3.2

1. $\frac{3}{2}$ **3.** No slope   **5.** $\frac{2}{3}$

7.

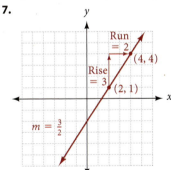

9.

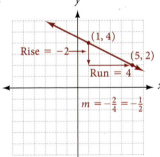

11.

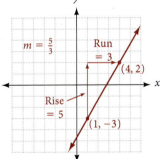

13.

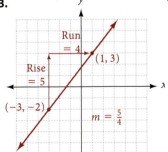

15.

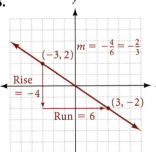

17.

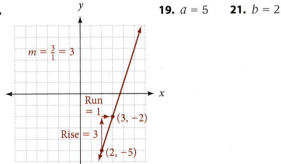

19. $a = 5$ **21.** $b = 2$

23.

x	y
0	2
3	0

Slope $= -\frac{2}{3}$

25.

x	y
0	-5
3	-3

Slope $= \frac{2}{3}$ **27.** $-\frac{1}{2}$ **29.** $-\frac{8}{3}$ **31.** -2 **33.** $\frac{2}{5}$ **35.** $\frac{4}{5}$

37. a. yes   **b.** no   **39.** 8   **41. a.** 17.5 mph   **b.** 40 km/h   **c.** 120 ft/sec   **d.** 28 m/min

43. Slopes: A: -50; B: -75, C: -25 **45. a.** 15,800 ft.   **b.** $-\frac{7}{100}$

47. 818.18 solar thermal collector shipments/year. Between 1997 and 2008 the number of solar thermal collector shipments increased an average of 818.18 shipments per year.   **49.** $-\frac{3}{20}$

51. a. \$2.84 million/year. From the years 1985 to 1990, the amount of money bet on horse races increased at an average of \$2.84 million a year.

b. \$7.72 million/year. From the years 2000 to 2005, the amount of money bet on horse races increased at an average of \$7.72 million a year.

53. 0   **55.** $y = -\frac{3}{2}x + 6$ **57.** $t = \frac{A - P}{Pr}$ **59.** -1 **61.** -2 **63.** $y = mx + b$ **65.** $y = -2x - 5$ **67.** 5

Problem Set 3.3

1. $y = 2x + 3$ **3.** $y = x - 5$ **5.** $y = \frac{1}{2}x + \frac{3}{2}$ **7.** $y = 4$

9. Slope $= 3$,
y-intercept $= -2$,
perpendicular slope $= -\frac{1}{3}$

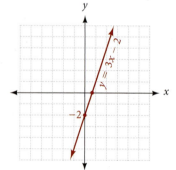

11. Slope $= \frac{2}{3}$,
y-intercept $= -4$,
perpendicular slope $= -\frac{3}{2}$

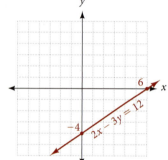

13. Slope $= -\frac{4}{5}$,
y-intercept $= 4$,
perpendicular slope $= \frac{5}{4}$

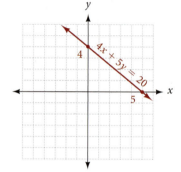

15. Slope $= \frac{1}{2}$, y-intercept $= -4$, $y = \frac{1}{2}x - 4$ **17.** Slope $= -\frac{2}{3}$, y-intercept $= 3$, $y = -\frac{2}{3}x + 3$ **19.** $y = 2x - 1$

21. $y = -\frac{1}{2}x - 1$ **23.** $y = -3x + 1$ **25.** $y = \frac{2}{3}x + \frac{2}{3}$ **27.** $y = -\frac{1}{4}x + \frac{3}{4}$ **29.** $x - y = 2$ **31.** $2x - y = 3$

33. $6x - 5y = 3$ **35. a.** 3   **b.** $-\frac{1}{3}$ **37. a.** -3 **b.** $\frac{1}{3}$ **39. a.** $-\frac{2}{5}$ **b.** $\frac{5}{2}$

41. a. x: $\frac{10}{3}$ y: -5 **b.** $(4, 1)$; answers will vary   **c.** $y = \frac{3}{2}x - 5$ **d.** No

43. a. x: $\frac{4}{3}$ y: -2 **b.** $(2, 1)$; answers will vary   **c.** $y = \frac{3}{2}x - 2$ **d.** No

45. a. 2   **b.** $\frac{3}{2}$ **c.** -3 **d.** (graph below)   **e.** $y = 2x - 3$ **47.** $(0, -4), (2, 0)$; $y = 2x - 4$ **49.** $(-2, 0)$ $(0, 4)$; $y = 2x + 4$

51. Slope $= 0$, y-intercept $= -2$ (graph below)   **53.** $y = 3x + 7$ **55.** $y = -\frac{5}{2}x - 13$

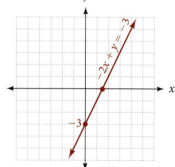

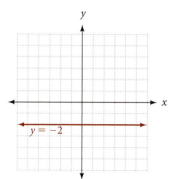

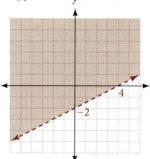

57. $y = \frac{1}{4}x + \frac{1}{4}$ **59.** $y = -\frac{2}{3}x + 2$ **61.** $y = 4x - 7{,}997$ **63. b.** $86°$ F

65.a. \$1,500   **b.** \$2,000   **c.** $m = 25$ **d.** \$25   **e.** $y = 25x + 1000$

67. a. \$5,000   **b.** $y = -3000x + 20000$ **c.** $-\$10{,}000$. When we try to estimate the cars worth, we get a negative number, which can't happen.

69. 23 inches, 5 inches   **71.** \$46.50   **73.** $(0, 0), (4, 0)$ **75.** $(0, 0), (2, 0)$ **77.** $y = -\frac{3}{2}x + 12$ **79.** $y = -\frac{4}{3}x - \frac{3}{2}$

Problem Set 3.4

1. 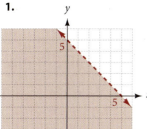 **3.** 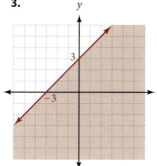 **5.** 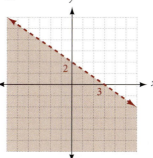 **7.**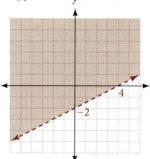

9. 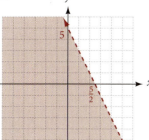 **11.** 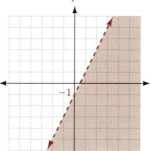 **13.** 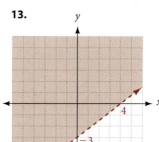 **15.**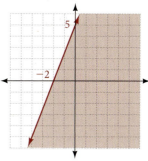

17. 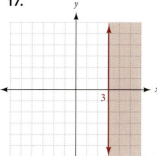 **19.** 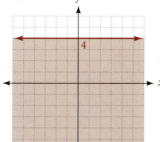 **21.** 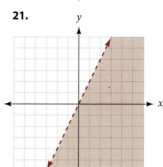 **23.**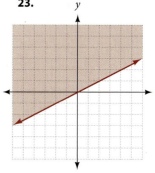

25. 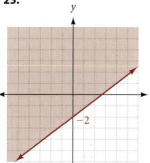 **27.** 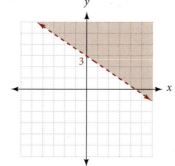 **29.** 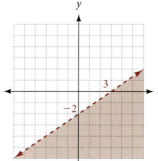 **31.**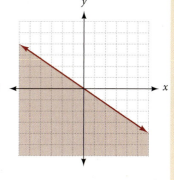

33.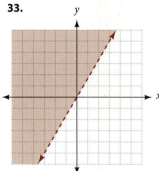

35.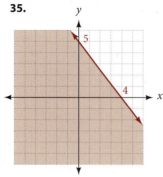

37. $y > -x + 4$ **39.** $y \le \frac{1}{2}x + 2$ **41.** $y \ge \frac{2}{3}x + 2$

43. **a.** $y < \frac{4}{3}$ **b.** $y > -\frac{4}{3}$ **c.** $y = -\frac{2}{3}x + 2$ **d.**

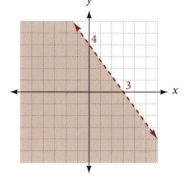

45.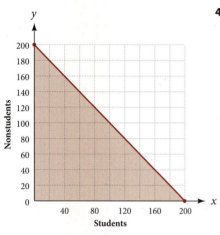

47.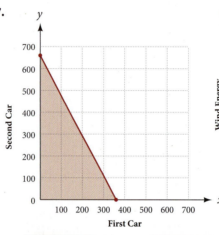

49.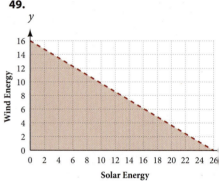

51. $y \le 7$ **53.** $t > -2$ **55.** $-1 < t < 2$

57.

x	y
0	0
10	75
20	150

59.

x	y
0	0
1	1
1	-1

61.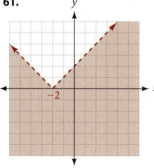

63.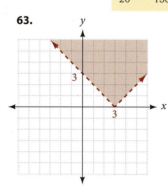

65.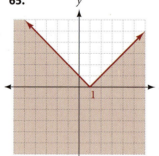

67.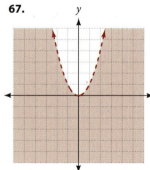

69.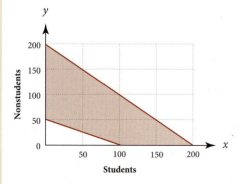

Problem Set 3.5

1. Domain = {1, 2, 4}; Range = {3, 5, 1}; a function   **3.** Domain = {−1, 1, 2}; Range = {3, −5}; a function

5. Domain = {7, 3}; Range = {−1, 4}; not a function   **7.** Domain = {a, b, c, d}; Range = {3, 4, 5}; a function

9. Domain = {a}; Range = {1, 2, 3, 4}; not a function   **11.** Domain = {$x \mid -5 \leq x \leq 5$}; Range = {$y \mid 0 \leq y \leq 5$}

13. Domain = {$x \mid -5 \leq x \leq 3$}; Range = {$y \mid y = 4$}

15. Domain = All real numbers;
Range = {$y \mid y \geq -1$};
a function

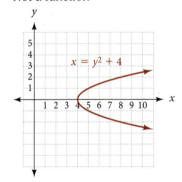

17. Domain = All real numbers;
Range = {$y \mid y \geq 4$};
a function

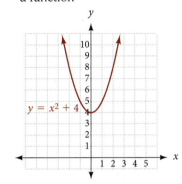

19. Domain = {$x \mid x \geq -1$};
Range = All real numbers;
not a function

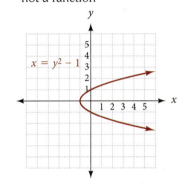

21. Domain = {$x \mid x \geq 4$};
Range = All real numbers;
Not a function

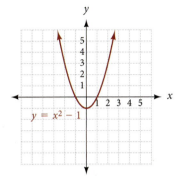

23. Domain = All real numbers;
Range = {$y \mid y \geq 0$};
A function

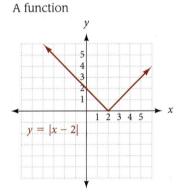

25. Domain = All real numbers;
Range = {$y \mid y \geq -2$};
A function

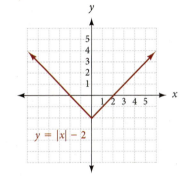

27. Yes   **29.** No   **31.** No   **33.** Yes   **35.** Yes

37. Domain = {$x \mid 2002 \leq x \leq 2010$}; Range = {$y \mid 31{,}000 \leq y \leq 204{,}000$};

39. a. $y = 8.5x$ for $10 \leq x \leq 40$ **b.** (table below)   **c.**

Hours Worked	Function Rule	Gross Pay ($)
x	$y = 8.5x$	y
10	$y = 8.5(10) = 85$	85
20	$y = 8.5(20) = 170$	170
30	$y = 8.5(30) = 255$	255
40	$y = 8.5(40) = 340$	340

d. Domain = {$x \mid 10 \leq x \leq 40$}; Range = {$y \mid 85 \leq y \leq 340$}

e. Minimum = $85; Maximum = $340

41. a.

Time (sec)	Function Rule	Height (ft)
t	$h = 16t - 16t^2$	h
0	$h = 16(0) - 16(0)^2$	0
0.1	$h = 16(0.1) - 16(0.1)^2$	1.44
0.2	$h = 16(0.2) - 16(0.2)^2$	2.56
0.3	$h = 16(0.3) - 16(0.3)^2$	3.36
0.4	$h = 16(0.4) - 16(0.4)^2$	3.84
0.5	$h = 16(0.5) - 16(0.5)^2$	4
0.6	$h = 16(0.6) - 16(0.6)^2$	3.84
0.7	$h = 16(0.7) - 16(0.7)^2$	3.36
0.8	$h = 16(0.8) - 16(0.8)^2$	2.56
0.9	$h = 16(0.9) - 16(0.9)^2$	1.44
1	$h = 16(1) - 16(1)^2$	0

b. Domain = $\{t \mid 0 \le t \le 1\}$; Range = $\{h \mid 0 \le h \le 4\}$

c.

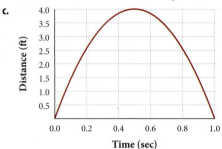

43. a.

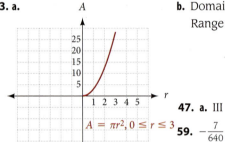

$A = \pi r^2, 0 \le r \le 3$

b. Domain = $\{r \mid 0 \le r \le 3\}$
Range = $\{A \mid 0 \le A \le 9\pi\}$

45. a. Yes
b. Domain = $\{t \mid 0 \le t \le 6\}$; Range = $\{h \mid 0 \le h \le 60\}$
c. $t = 3$
d. $h = 60$
e. $t = 6$

47. a. III **b.** I **c.** II **d.** IV   **49.** 10   **51.** -14 **53.** 1   **55.** -3 **57.** $-\frac{6}{5}$

59. $-\frac{7}{640}$ **61.** 150   **63.** 113   **65.** -9 **67. a.** 6 **b.** 7.5   **69. a.** 27 **b.** 6

71. Domain = All real numbers
Range = $\{y \mid y \le 5\}$
A Function

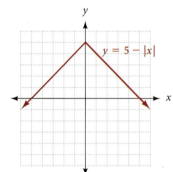

$y = 5 - |x|$

73. Domain = $\{x \mid x \ge 3\}$
Range = All real numbers
Not a Function

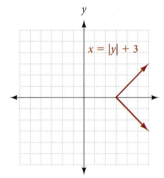

$x = |y| + 3$

75. Domain = $\{x \mid -4 \le x \le 4\}$
Range = $\{y \mid -4 \le y \le 4\}$
Not a Function

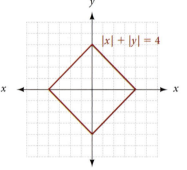

$|x| + |y| = 4$

Problem Set 3.6

1. -1 **3.** -11 **5.** 2   **7.** 4   **9.** 35   **11.** -13 **13.** 1   **15.** -9 **17.** 8   **19.** 19   **21.** 16   **23.** 0

25. $3a^2 - 4a + 1$ **27.** -8 **29.** -1 **31.** $2a^2 - 8$ **33.** $2b^2 - 8$

35.

$f(x) = \frac{1}{2}x + 2$

$f(4)$

37. $x = 4$ **39.**

41. 4   **43.** 0   **45.** 2 **47.** $V(3) = 300$, the painting is worth $300 in 3 years; $V(6) = 600$, the painting is worth $600 in 6 years.

49. a. True **b.** False **c.** True   **51. a.** True **b.** True **c.** False **d.** True

53. a. $5,625   **b.** $1,500   **c.** $\{t \mid 0 \le t \le 5\}$ **d.**

e. $\{V(t) \mid 1,500 \le V(t) \le 18,000\}$

f. About 2.42 years

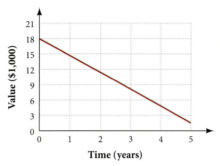

55. $-0.1x^2 + 27x - 500$ **57.** $6x^2 - 2x - 4$ **59.** $2x^2 + 8x + 8$ **61.** $0.6M - 42$ **63.** $4x^2 - 7x + 3$ **65.** $-\frac{2}{3}, 4$

67. $-2, 1$ **69.** $\varnothing$ **71. a.** 2 **b.** 0 **c.** 1 **d.** 4

73. a.

Weight (ounces)	0.6	1.0	1.1	2.5	3.0	4.8	5.0	5.3
Cost (cents)	88	88	105	122	122	156	156	173

b. More than 2 ounces, but not more than 3 ounces; $2 < x \le 3$

c. Domain: $\{x \mid 0 < x \le 6\}$

d. Range: $\{C \mid C = 88, 105, 122, 156, 173\}$

Problem Set 3.7

1. $6x + 2$ **3.** $-2x + 8$ **5.** $4x - 7$ **7.** $3x^2 + x - 2$ **9.** $-2x + 3$ **11.** $9x^3 - 15x^2$ **13.** $\frac{3x^2}{3x - 5}$ **15.** $\frac{3x - 5}{3x^2}$

17. $3x^2 + 4x - 7$ **19.** 15   **21.** 98   **23.** $\frac{3}{2}$ **25.** 1   **27.** 40   **29.** 147   **31. a.** 81 **b.** 29 **c.** $(x + 4)^2$ **d.** $x^2 + 4$

33. a. -2 **b.** -1 **c.** $4x^2 + 12x - 1$ **35.** $(f \circ g)(x) = 5\left(\frac{x + 4}{5}\right) - 4 = x + 4 - 4 = x; \ (g \circ f)(x) = \frac{(5x - 4) + 4}{5} = \frac{5x}{5} = x$

37. 6   **39.** 2   **41.** 3   **43.** -8

45. a. $R(x) = 11.5x - 0.05x^2$ **b.** $C(x) = 2x + 200$ **c.** $P(x) = -0.05x^2 + 9.5x - 200$ **d.** $\overline{C}(x) = 2 + \frac{200}{x}$

47. $T(M) = 24.8 + 0.6M$ **a.** $M(x) = 220 - x$ **b.** $M(24) = 196$ **c.** 142 beats per minute

d. 135 beats per minute **e.** 128 beats per minute

49. $2 < x < 4$ **51.** $x < 4$ or $x > 8$ **53.** $-\frac{5}{7} \le x \le 1$ **55.** 196   **57.** 4   **59.** 1.6   **61.** 3   **63.** 2,400

Problem Set 3.8

1. Direct   **3.** Direct   **5.** Direct   **7.** Direct   **9.** Direct   **11.** Inverse   **13.** 30   **15.** 5   **17.** 10

19. 16   **21.** 40   **23.** 225   **25.** -6 **27.** $\frac{1}{2}$ **29.** $\frac{81}{5}$ **31.** $\frac{81}{2}$ **33.** 64   **35.** 8   **37.** $\frac{50}{3}$ pounds

39. $371 \frac{1}{15} \approx 371.1$ feet/second

41. a. $T = 4P$ **c.** 70 pounds per square inch   **43.** 12 pounds per square inch   **45. a.** $f = \frac{80}{d}$ **c.** An f-stop of 8

b.

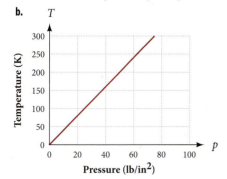

Temperature (K) vs Pressure (lb/in²)

b.

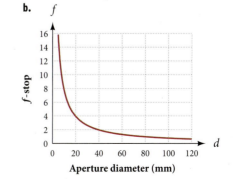

f-stop vs Aperture diameter (mm)

47. $\frac{1,504}{15}$ square inches   **49.** 1.5 ohms   **51.** 1.28 meters   **53.** $F = G\,\frac{m_1 m_2}{d^2}$ **55.** 12   **57.** 28   **59.** $-\frac{7}{4}$

61. $w = \frac{P - 21}{2}$ **63.** $t \geq -6$ **65.** $x < 6$ **67.** 6, 2   **69.** $\varnothing$

71. $x \leq -9$ or $x \geq -1$

73. All real numbers   **75. a.** Square of speed   **b.** $d = 0.0675s^2$ **c.** About 204 feet from the cannon
     d. About 7.5 feet farther away

Chapter 3 Review

1.

2.

3.

4. -2 **5.** 0   **6.** 3   **7.** 5   **8.** -5 **9.** 6   **10.** $y = 3x + 5$ **11.** $y = -2x$ **12.** $m = 3, b = -6$

13. $m = \frac{2}{3}, b = -3$ **14.** $y = 2x$ **15.** $y = -\frac{1}{3}x$ **16.** $y = 2x + 1$ **17.** $y = 7$ **18.** $y = -\frac{3}{2}x - \frac{17}{2}$ **19.** $y = 2x - 7$

20. $y = \frac{1}{3}x - \frac{2}{3}$

21.

22.

23. Domain = {2, 3, 4}; Range = {4, 3, 2}; a function
24. Domain = {6, -4, -2}; Range = {3, 0}; a function
25. 0   **26.** 1   **27.** 1   **28.** $3a + 2$ **29.** 1   **30.** 31
31. 24   **32.** 6   **33.** 4   **34.** 25   **35.** 84 pounds
36. 16 foot-candles

Chapter 3 Cumulative Review

1. -25 **2.** -6 **3.** 74   **4.** 144   **5.** 36   **6.** 25   **7.** 27   **8.** 27   **9.** 1   **10.** -9 **11.** $\frac{7}{9}$ **12.** $\frac{7}{9}$ **13.** $\frac{1}{10}$

14. $\frac{2}{21}$ **15.** -2 **16.** $-\frac{8}{5}$ **17.** $6x - 4y$ **18.** $10x + 9y$ **19.** 15   **20.** $\frac{7}{12}$ **21.** $\frac{5}{9}$ **22.** -2 **23.** $\frac{11}{2}, -\frac{5}{2}$ **24.** $\varnothing$

25. $y = -3x - 1$ **26.** $y = \frac{3}{2}x - 5$

27. $-\frac{5}{3} \leq x \leq 3$ **28.** $x < -\frac{14}{3}$ or $x > 2$

29.   **30.**   **31.**

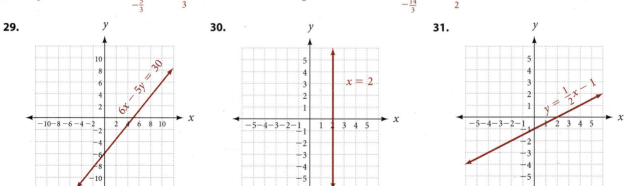

32.

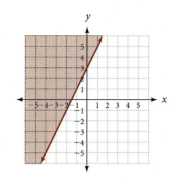

33. $-\frac{5}{2}$ **34.** 6 **35.** 0 **36.** $y = -\frac{3}{4}x + 2$ **37.** $m = \frac{4}{5}$; $b = -4$ **38.** $y = -\frac{1}{2}x - 1$

39. $y = -\frac{5}{2}x - 12$ **40.** $y = \frac{7}{3}x - 6$ **41.** -1 **42.** 7 **43.** 9 **44.** $2x$

45. Domain = $\{-1, 2, 3\}$; Range = $\{-1, 3\}$; a function

46. $\frac{16}{9}$ **47.** 20% **48.** $9.43 **49.** 25% more **50.** 8% less

Chapter 3 Test

1. x-intercept = 3;
y-intercept = 6; slope = -2

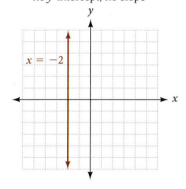

2. x-intercept = $-\frac{3}{2}$;
y-intercept = -3; slope = -2

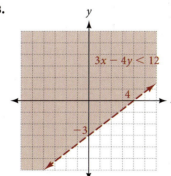

3. x-intercept = $-\frac{8}{3}$;
y-intercept = 4; slope = $\frac{3}{2}$

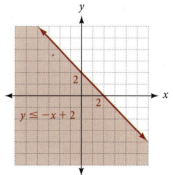

4. x-intercept = -2;
no y-intercept; no slope

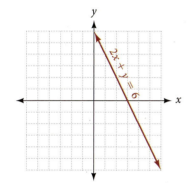

5. $y = 3x + 3$ **6.** $y = -\frac{5}{3}x + 5$ **7.** $y = 2x + 5$ **8.** $y = -\frac{3}{7}x + \frac{5}{7}$

9. $y = \frac{2}{5}x - 5$ **10.** $y = -\frac{1}{3}x - \frac{7}{3}$ **11.** $y \geq -x + 3$ **12.** $y < -x + 3$

13.

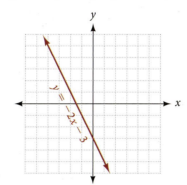

14.

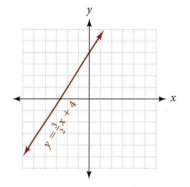

15. Domain = $\{-2, -3\}$; Range = $\{0, 1\}$; not a function **16.** Domain = All real numbers; Range = $\{y \mid y \geq -9\}$; a function

17. Domain = $\{0, 1, 2\}$; Range = $\{0, 3, 5\}$; a function **18.** Domain = All real numbers; Range = All real numbers; a function

19. Domain = $\{x \mid -4 \leq x \leq 4\}$; Range = $\{y \mid -1 \leq y \leq 3\}$ **20.** Domain = $\{x \mid -3 \leq x \leq 3\}$; Range = $\{y \mid -4 \leq y \leq 4\}$

21. 11 **22.** -4 **23.** 8 **24.** 4 **25.** 5 **26.** 0 **27.** 3 **28.** 0 **29.** 18 **30.** $\frac{81}{4}$

Chapter 4

Problem Set 4.1

1. (4, 3)

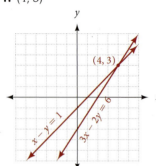

3. (−5, −6)

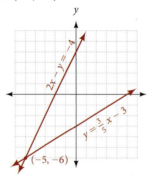

5. (4, 2)

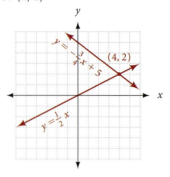

7. Lines are parallel; no solution.

9. $\{(x, y) \mid 2x − y = 5\}$

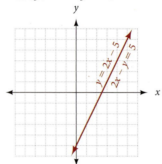

11. (2, 3)   **13.** (1, 1)   **15.** Lines coincide: $\{(x, y) \mid 3x − 2y = 6\}$ **17.** $\left(1, −\frac{1}{2}\right)$ **19.** $\left(\frac{1}{2}, −3\right)$ **21.** $\left(−\frac{8}{3}, 5\right)$

23. (2, 2)   **25.** Parallel lines; ∅   **27.** (12, 30)   **29.** (10, 24)   **31.** (3, −3)   **33.** $\left(\frac{4}{3}, −2\right)$ **35.** $y = 5, z = 2$

37. (2, 4)   **39.** Lines coincide: $\{(x, y) \mid 2x − y = 5\}$ **41.** Lines coincide: $\left\{(x, y) \mid x = \frac{3}{2}y\right\}$ **43.** $\left(−\frac{15}{43}, −\frac{27}{43}\right)$

45. $\left(\frac{60}{43}, \frac{46}{43}\right)$ **47.** $\left(\frac{9}{41}, −\frac{11}{41}\right)$ **49.** $\left(−\frac{11}{7}, −\frac{20}{7}\right)$ **51.** $\left(2, \frac{4}{3}\right)$ **53.** (−12, −12)   **55.** Lines are parallel; ∅

57. $y = 5, z = 2$ **59.** $\left(\frac{3}{2}, \frac{3}{8}\right)$ **61.** $\left(\frac{12}{5}, \frac{8}{15}\right)$

63. $\left(\frac{24}{11}, \frac{12}{11}\right)$ **65.** $\left(\frac{43}{19}, \frac{25}{19}\right)$ **67. a.** $−y$ **b.** $−2$ **c.** $−2$ **d.** graph below **e.** (0, −2)

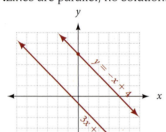

69. (6,000, 4,000)   **71.** 2   **73.** (4, 0)   **75.** Burrito \$3; Taco \$2   **77.** 3

79. $m = \frac{2}{3}, b = −2$ **81.** $y = \frac{2}{3}x + 6$ **83.** $y = \frac{2}{3}x − 2$ **85.** $y = \frac{4}{3}x + \frac{17}{3}$

87. −10   **89.** $3y + 2z$ **91.** 1   **93.** 3   **95.** $10x − 2z$ **97.** $9x + 3y − 6z$

99. $a = 3, b = −2$ **101.** $m = \frac{2}{3}, b = 4$

Problem Set 4.2

1. (1, 2, 1)   **3.** (2, 1, 3)   **5.** (2, 0, 1)   **7.** $\left(\frac{1}{2}, \frac{2}{3}, −\frac{1}{2}\right)$ **9.** No solution, inconsistent system   **11.** (4, −3, −5)

13. No unique solution   **15.** (4, −5, −3)   **17.** No unique solution   **19.** $\left(\frac{1}{2}, 1, 2\right)$ **21.** $\left(\frac{1}{2}, \frac{1}{3}, \frac{1}{4}\right)$

23. $\left(\frac{10}{3}, −\frac{5}{3}, −\frac{1}{3}\right)$ **25.** $\left(\frac{1}{4}, −\frac{1}{3}, \frac{1}{8}\right)$ **27.** (6, 8, 12)   **29.** (3, 6, 4)   **31.** (1, 3, 1)   **33.** (−1, 2, −2)

35. 4 amp, 3 amp, 1 amp   **37.** 30 sodas, 30 hot dogs, 25 bags of chips   **39.** −2   **41.** 11   **43.** −14   **45.** −4

47. 147   **49.** 50   **51.** 100   **53.** (1, 2, 3, 4)

Problem Set 4.3

1. 5, 13 **3.** 10, 16 **5.** 1, 3, 4 **7.** 225 adult and 700 children's tickets **9.** $12,000 at 6% and $8,000 at 7%.

11. $4,000 at 6%, $8,000 at 7.5% **13.** $200 at 6%, $1,400 at 8%, $600 at 9% **15.** $300 at 6%, $600 at 8%, $400 at 12%

17. 6 gallons of 20%, 3 gallons of 50% **19.** 5 gallons of 20%, 10 gallons of 14% **21.** 12.5 lbs of 40%, 37.5 lbs of 60%

23. boat; 9 mph; current: 3 mph **25.** airplane: 270 mph; wind: 30 mph **27.** 12 nickels, 8 dimes

29. 3 nickels, 3 dimes, 3 quarters **31.** 110 nickels **33.** 5 nickels, 10 dimes, 16 quarters

35. $x = -200p + 700$; 100 items **37.** $h = -16t^2 + 64t + 80$ **39.** Highway: 5 gallons; City: 20 gallons **41.** No

43. (4, 0) **45.** $x > 435$

47.

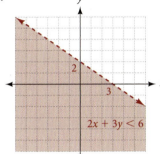

49.

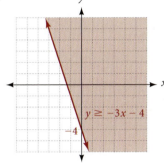

51.

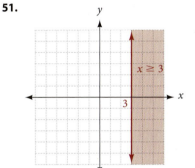

53.

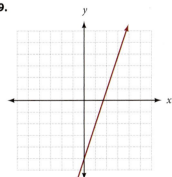

55. a. $M = \frac{1}{5}x - 412$ **b.** $M = -\frac{2}{5}x + 810$ **c.** 2036 or 2037

Problem Set 4.4

1. (2, 3) **3.** (−1, −2) **5.** (7, 1) **7.** (−3, 4) **9.** (0, −9) **11.** (−4, 3) **13.** (−5, 7) **15.** (0, −8)

17. (8, 4) **19.** (1, 2, 1) **21.** (2, 0, 1) **23.** (0, 1, −2) **25.** (0, −2, 4) **27.** (2, 1, 1) **29.** (1, 1, 2) **31.** (4, 1, 5)

33. (5, 4, −2) **35.** (−4, 3, 5) **37.** (0, 3, −1) **39.** (1, 3, −4) **41.** $(4, \frac{10}{3})$ **43.** (6, 4)

49.

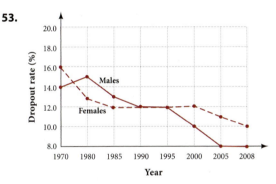

51.

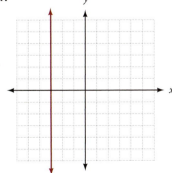

Problem Set 4.5

1.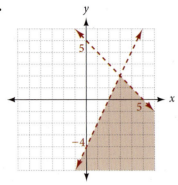

3.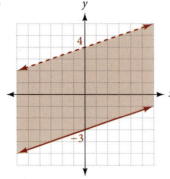

5.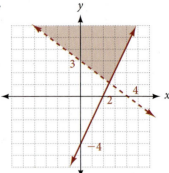

7.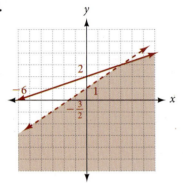

9.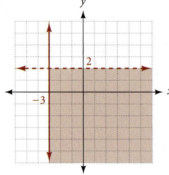

11.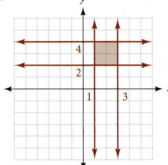

13.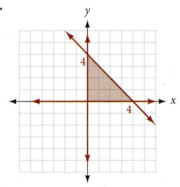

15.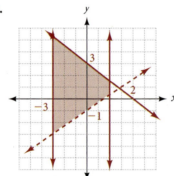

17.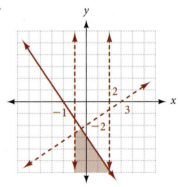

19.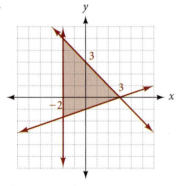

21.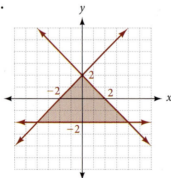

23.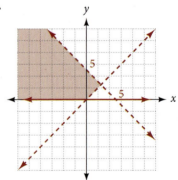

25.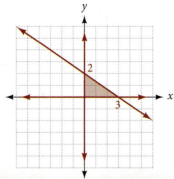

27. $x + y \le 4$
    $-x + y < 4$

29. $x + y \ge 4$
    $-x + y < 4$

31. a. $0.55x + 0.65y \leq 40$
$x \geq 2y$
$x > 15$
$y \geq 0$

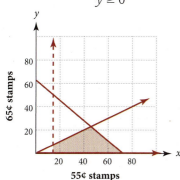

55¢ stamps

b. Ten 65-cent stamps

33. a. $x + y < 25$
$20x + 14.5y \leq (x + y)17$
$x \geq 0$
$y \geq 0$

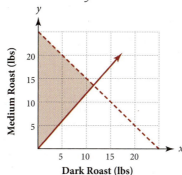

Dark Roast (lbs)

b. 17 pounds

35. $x + y \geq 35$
$7.75x + 13y < 10(x + y)$
$x \geq y + 6$
$x \geq 0$
$y \geq 0$

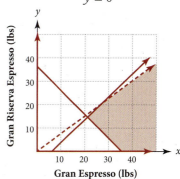

Gran Espresso (lbs)

37. x-intercept $= 3$, y-intercept $= 6$, slope $= -2$

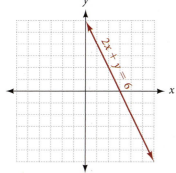

39. x-intercept $= -2$, no y-intercept, no slope

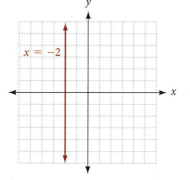

41. $y = -\frac{3}{7}x + \frac{5}{7}$ **43.** $x = 4$ **45.** Domain = All real numbers; Range = $\{y \mid y \geq -9\}$; a function   **47.** -4 **49.** 4

51. $\frac{81}{4}$

Chapter 4 Review

1. $(6, -2)$ **2.** $(2, -4)$ **3.** No solution (parallel lines)   **4.** $(1, -1)$ **5.** $(0, 1)$

6. No unique solution (dependent system)   **7.** $(3, 1)$ **8.** $\left(\frac{3}{2}, 0\right)$ **9.** $(3, 5)$ **10.** $(4, 8)$ **11.** $(3, 12)$ **12.** $(3, -2)$

13. $\left(\frac{3}{2}, \frac{1}{2}\right)$ **14.** $(4, 1)$ **15.** $(6, -2)$ **16.** $(7, -4)$ **17.** $(-5, 3)$ **18.** No solution (parallel lines)

19. $(3, -1, 4)$ **20.** $\left(2, \frac{1}{2}, -3\right)$ **21.** $\left(-1, \frac{1}{2}, \frac{3}{2}\right)$ **22.** $\left(2, \frac{1}{3}, \frac{2}{3}\right)$ **23.** No unique solution (dependent system)

24. No unique solution (dependent system)   **25.** $(2, -1, 4)$ **26.** No unique solution (dependent system)

27. 47 adults, 80 children   **28.** 12 dimes, 8 quarters   **29.** $5,000 at 12%, $7,000 at 15%

30. Boat: 12 mph; current: 2 mph   **31.** $(2, -1)$ **32.** $(-3, 4)$ **33.** $(2, -2, 2)$ **34.** $(-3, -2, 1)$

35.

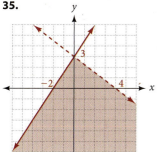

36.

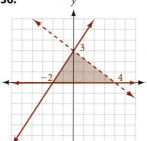

37.

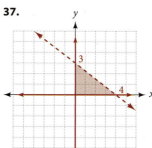

38.

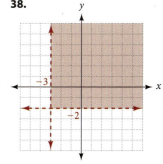

Chapter 4 Cumulative Review

1. -48 **2.** 16 **3.** 6 **4.** -93 **5.** -42 **6.** -12 **7.** -63 **8.** -27 **9.** $18x + 7$ **10.** $6x - 19$ **11.** $-3x + 6y$

12. $-10x - 3y$ **13.** -1 **14.** 0 **15.** $\varnothing$ **16.** $\varnothing$ **17.** $y = -5x + 13$ **18.** $y = \frac{2}{3}x - 1$ **19.** $x = \frac{2}{a - b}$

20. $x = \frac{9}{c - a}$ **21.** $(-\infty, -4]$ **22.** $(-\infty, -2] \cup [3, \infty)$ **23.** $\{0, 2, 3, 4, 6, 9\}$ **24.** $\{6\}$ **25.** $0, -1, 2.35, 4$ **26.** $-1, 0, 4$

27. $(4,3)$ **28.** Lines are parallel; $\varnothing$ **29.** $\{(x, y) \mid 2x + y = 3\}$ **30.** $\left(-\frac{45}{31}, -\frac{57}{31}, \frac{51}{62}\right)$

31. 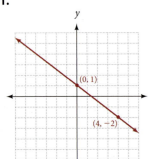   **32.** 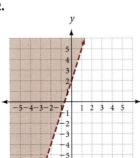   **33.**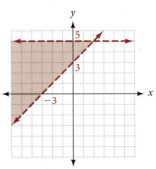

34. 2 **35.** $m = \frac{3}{5}, b = -3$ **36.** $y = -\frac{2}{3}x - 3$ **37.** $y = -\frac{5}{2}x + 1$ **38.** $y = 3x + 1$ **39.** -1 **40.** 5 **41.** $-x^2 + 2x + 8$

42. $4x^2 + 12x + 4$ **43.** 3 **44.** 2 **45.** -1 **46.** -7

Chapter 4 Test

1. $(1, 2)$ **2.** $(15, 12)$ **3.** $\left(-\frac{54}{13}, -\frac{58}{13}\right)$ **4.** $(1, 2)$ **5.** $(3, -2, 1)$ **6.** $\left(\frac{20}{7}, \frac{16}{7}, \frac{4}{7}\right)$ **7.** $5, 9$

8. \$4,000 at 5%, \$8,000 at 6%   **9.** 340 adult tickets, 410 children's tickets   **10.** Boat: 8 mph; Current: 2 mph

11. 1 quarter, 3 dimes, 11 nickels   **12.** $(-5, 3)$ **13.** $(3, 0, 4)$

14.   **15.**   **16.** $y \geq \frac{4}{3}x - 4, y > -\frac{4}{3}x - 4$ **17.** $y \leq \frac{4}{3}x - 4, y > -\frac{4}{3}x - 4$

Chapter 5

Problem Set 5.1

1. 16 **3.** -16 **5.** -0.027 **7.** 32 **9.** $\frac{1}{8}$ **11.** $-\frac{25}{36}$ **13.** x^9 **15.** 64 **17.** $-\frac{8}{27}x^6$ **19.** $-6a^6$ **21.** $\frac{1}{9}$

23. $-\frac{1}{32}$ **25.** $\frac{16}{9}$ **27.** 17 **29.** x^3 **31.** $\frac{a^6}{b^{15}}$ **33.** $\frac{8}{125y^{18}}$ **35.** $\frac{1}{5}$ **37.** $\frac{24a^{12}c^6}{b^3}$ **39.** $\frac{8x^{22}}{81y^{23}}$ **41.** $\frac{1}{x^{10}}$ **43.** a^{10}

45. $\frac{1}{t^6}$ **47.** x^{12} **49.** x^{18} **51.** $\frac{1}{x^{22}}$ **53.** $\frac{a^3b^7}{4}$ **55.** $\frac{y^{38}}{x^{16}}$ **57.** $\frac{16y^{16}}{x^8}$ **59.** x^4y^6 **61.** x^2y **63.** $3ab^2$ **65.** $2a$

67. $4xy^4$ **69. a.** 32 **b.** 64 **c.** 32 **d.** 64 **71. a.** 8 **b.** 8 **c.** $\frac{1}{16}$ **d.** $\frac{1}{16}$ **73. a.** $\frac{1}{7}$ **b.** $\frac{1}{11}$ **c.** $\frac{1}{2x}$ **d.** $\frac{1}{8x^2}$ **75.** 2^n

77. r^3 **79.** 3.78×10^5 **81.** 4.9×10^3 **83.** 3.7×10^{-4} **85.** 4.95×10^{-3} **87.** $5,340$ **89.** $7,800,000$

91. 0.00344 **93.** 0.49 **95.** 8×10^4 **97.** 2×10^9 **99.** 2.5×10^{-6} **101.** 275,625 square feet

103. 1.2×10^7 **105.** Two hundred billion; 2.0×10^{11} **107.** 1.003×10^{19} miles   **109.** $5x$ **111.** $8x^2$

113. $2x^3$ **115.** -5 **117.** $-2x + 3$ **119.** $9x + 6$ **121.** $1,200$ **123.** $(1, 2)$ **125.** $\left(\frac{5}{2}, -1\right)$ **127.** $(0, 3)$

129. $(3, 4)$ **131.** $(2, -1)$ **133.** No unique solution (dependent system)   **135.** $\frac{1}{x^3}$ **137.** y^3 **139.** x^5

Problem Set 5.2

1. Trinomial, 2, 5   **3.** Binomial, 1, 3   **5.** Trinomial, 2, 8   **7.** Polynomial, 3, 4   **9.** Monomial, 0, $-\frac{3}{4}$

11. Trinomial, 3, 6   **13.** $7x + 1$ **15.** $2x^2 + 7x - 15$ **17.** $12a^2 - 7ab - 10b^2$ **19.** $x^2 - 13x + 3$ **21.** $\frac{1}{4}x^2 - \frac{7}{12}x - \frac{1}{4}$

23. $-y^3 - y^2 - 4y + 7$ **25.** $2x^3 + x^2 - 3x - 17$ **27.** $\frac{1}{14}x^2 + \frac{1}{7}xy + \frac{5}{7}y^2$ **29.** $-3a^3 + 6a^2b - 5ab^2$ **31.** $-3x$

33. $3x^2 - 12xy$ **35.** $17x^5 - 12$ **37.** $14a^2 - 2ab + 8b^2$ **39.** $2 - x$ **41.** $10x - 5$ **43.** $9x - 35$ **45.** $9y - 4x$

47. $9a + 2$ **49.** -2 **51. a.** 208 **b.** 103 **53. a.** 51 **b.** -15 **55. a.** 110 **b.** -120 **57. a.** $-5,000$ **b.** $3,000$

59. 1st year $51,568x + 67,073y$; 2nd year $51,568(2x) + 67,073y$; total $154,704x + 134,146y$ **61.** 240 feet for both

63. $P(x) = -300 + 40x - 0.5x^2$; $P(60) = \$300$ **65.** $P(x) = -800 + 3.5x - 0.002x^2$; $P(1,000) = \$700$ **67.** $2x^2 + 7x - 15$

69. $6x^3 - 11x^2y + 11xy^2 - 12y^3$ **71.** $-12x^4$ **73.** $20x^5$ **75.** a^6 **77.** 650 **79.** $x - 5$ **81.** $x - 7$ **83.** $5x - 1$

85. $x - 1$ **87.** $3x + 8y$ **89. a.** 5 **b.** -4 **c.** -3 **d.** 3 **e.** 4 **f.** 0 **g.** -4 **h.** 4

Problem Set 5.3

1. $12x^3 - 10x^2 + 8x$ **3.** $-3a^5 + 18a^4 - 21a^2$ **5.** $2a^5b - 2a^3b^2 + 2a^2b^4$ **7.** $x^2 - 2x - 15$ **9.** $6x^4 - 19x^2 + 15$

11. $x^3 + 9x^2 + 23x + 15$ **13.** $a^3 - b^3$ **15.** $8x^3 + y^3$ **17.** $2a^3 - a^2b - ab^2 - 3b^3$ **19.** $x^2 + x - 6$

21. $6a^2 + 13a + 6$ **23.** $20 - 2t - 6t^2$ **25.** $x^6 - 2x^3 - 15$ **27.** $20x^2 - 9xy - 18y^2$ **29.** $18t^2 - \frac{2}{9}$

31. $25x^2 + 20xy + 4y^2$ **33.** $25 - 30t^3 + 9t^6$ **35.** $4a^2 - 9b^2$ **37.** $9r^4 - 49s^2$ **39.** $y^2 + 3y + \frac{9}{4}$ **41.** $a^2 - a + \frac{1}{4}$

43. $x^2 + \frac{1}{2}x + \frac{1}{16}$ **45.** $t^2 + \frac{2}{3}t + \frac{1}{9}$ **47.** $\frac{1}{9}x^2 - \frac{4}{25}$ **49.** $x^3 - 6x^2 + 12x - 8$ **51.** $x^3 - \frac{3}{2}x^2 + \frac{3}{4}x - \frac{1}{8}$

53. $3x^3 - 18x^2 + 33x - 18$ **55.** $a^2b^2 + b^2 + 8a^2 + 8$ **57.** $3x^2 + 12x + 14$ **59.** $24x$

61. $x(6,200 - 25x) = 6,200x - 25x^2$; $\$18,375$ **63.** $R(p) = 900p - 300p^2$; $R(x) = 3x - \frac{x^2}{300}$; $\$672$

65. $R(p) = 350p - 10p^2$; $R(x) = 35x - \frac{x^2}{10}$; $\$1,852.50$ **67.** $P(x) = -\frac{x^2}{10} + 30x - 500$; $P(60) = \$940$

69. $A = 100 + 400r + 600r^2 + 400r^3 + 100r^4$ **71.** $8a^2$ **73.** -48 **75.** $2a^3b$ **77.** $-3b^2$ **79.** $-y^4$ **81.** $(1, 2, 3)$

83. $(1, 3, 1)$ **85.** $(x + y)^2 + (x + y) - 20 = x^2 + 2xy + y^2 + x + y - 20$ **87.** $x^{2n} - 5x^n + 6$ **89.** $10^{2n} + 13x^n - 3$

91. $x^{2n} + 10x^n + 25$ **93.** $x^{3n} + 1$

Problem Set 5.4

1. $5x^2(2x - 3)$ **3.** $9y^3(y^3 + 2)$ **5.** $3ab(3a - 2b)$ **7.** $7xy^2(3y^2 + x)$ **9.** $3(a^2 - 7a + 10)$ **11.** $4x(x^2 - 4x - 5)$

13. $10x^2y^2(x^2 + 2xy - 3y^2)$ **15.** $xy(-x + y - xy)$ **17.** $2xy^2z(2x^2 - 4xz + 3z^2)$ **19.** $5abc(4abc - 6b + 5ac)$

21. $(a - 2b)(5x - 3y)$ **23.** $3(x + y)^2(x^2 - 2y^2)$ **25.** $(x + 5)(2x^2 + 7x + 6)$ **27.** $(x + 1)(3y + 2a)$

29. $(xy + 1)(x + 3)$ **31.** $(x - 2)(3y^2 + 4)$ **33.** $(x - a)(x - b)$ **35.** $(b + 5)(a - 1)$ **37.** $(b^2 + 1)(a^4 - 5)$

39. $(x + 3)(x^2 - 4)$ **41.** $(x + 2)(x^2 - 25)$ **43.** $(2x + 3)(x^2 - 4)$ **45.** $(x + 3)(4x^2 - 9)$ **47.** 6

49. $P(1 + r) + P(1 + r)r = (1 + r)(P + Pr) = (1 + r)P(1 + r) = P(1 + r)^2$ **51.** $p = 11.5 - 0.05x$; $\$5.25$

53. $\$28.50$ **55.** $3x^2(x^2 - 3xy - 6y^2)$ **57.** $(x - 3)(2x^2 - 4x - 3)$ **59.** $3x^2 + 5x - 2$ **61.** $3x^2 - 5x + 2$

63. $x^2 + 5x + 6$ **65.** $6y^2 + y - 35$ **67.** $20 - 19a + 3a^2$ **69.** $(-3, 2)$ **71.** $(9, -5)$

Problem Set 5.5

1. $(x + 3)(x + 4)$ **3.** $(x + 3)(x - 4)$ **5.** $(y + 3)(y - 2)$ **7.** $(2 - x)(8 + x)$ **9.** $(2 + x)(6 + x)$ **11.** $(x + 2y)(x + y)$

13. $(a + 6b)(a - 3b)$ **15.** $(x - 8a)(x + 6a)$ **17.** $(x - 6b)^2$ **19.** $3(a - 2)(a - 5)$ **21.** $4x(x - 5)(x + 1)$

23. $3(x - 3y)(x + y)$ **25.** $2a^3(a^2 + 2ab + 2b^2)$ **27.** $10x^2y^2(x + 3y)(x - y)$ **29.** $(2x - 3)(x + 5)$ **31.** $(2x - 5)(x + 3)$

33. $(2x - 3)(x - 5)$ **35.** $(2x - 5)(x - 3)$ **37.** Prime   **39.** $(2 + 3a)(1 + 2a)$ **41.** $15(4y + 3)(y - 1)$

43. $x^2(3x - 2)(2x + 1)$ **45.** $10r(2r - 3)^2$ **47.** $(4x + y)(x - 3y)$ **49.** $(2x - 3a)(5x + 6a)$ **51.** $(3a + 4b)(6a - 7b)$

53. $200(1 + 2t)(3 - 2t)$ **55.** $y^2(3y - 2)(3y + 5)$ **57.** $2a^2(3 + 2a)(4 - 3a)$ **59.** $2x^2y^2(4x + 3y)(x - y)$

61. $100(3x^2 + 1)(x^2 + 3)$ **63.** $(5a^2 + 3)(4a^2 + 5)$ **65.** $3(3 + 4r^2)(1 - r^2)$ **67.** $(x + 5)(2x + 3)(x + 2)$

69. $(2x + 3)(x + 5)(x + 2)$ **71.** $(3x - 5)(x + 4)(x - 3)$ **73.** $(2x - 3)(3x - 4)(x - 2)$ **75.** $(3x - 5)(4x + 9)(x + 3)$

77. $(3x + 2)(2x - 5)(5x - 2)$ **79.** $(5x + 3)(4x + 7)(2x + 3)$ **81.** $9x^2 - 25y^2$ **83.** $a + 250$ **85.** $12x - 35$

87. $9x + 8$ **89.** $7x + 8$ **91.** $y = 2(2x - 1)(x + 5)$; $y = 0$ when $x = \frac{1}{2}$ or $x = -5$, $y = 42$ when $x = 2$ **93.** $4x^2 - 9$

95. $4x^2 - 12x + 9$ **97.** $8x^3 - 27$ **99.** $\frac{5}{8}$ **101.** x^3 **103.** $4x^2$ **105.** $\frac{1}{2}$ **107.** x^2 **109.** $3x$ **111.** $2y$

113. $(2x^3 + 5y^2)(4x^3 + 3y^2)$ **115.** $(3x - 5)(x + 100)$ **117.** $\left(\frac{1}{4}x + 1\right)\left(\frac{1}{2}x + 2\right)$ **119.** $(2x + 0.5)(x + 0.5)$

Problem Set 5.6

1. $(x - 3)^2$ **3.** $(a - 6)^2$ **5.** $(5 - t)^2$ **7.** $\left(\frac{1}{3}x + 3\right)^2$ **9.** $(2y^2 - 3)^2$ **11.** $(4a + 5b)^2$ **13.** $\left(\frac{1}{5} + \frac{1}{4}t^2\right)^2$

15. $\left(y + \frac{3}{2}\right)^2$ **17.** $\left(a - \frac{1}{2}\right)^2$ **19.** $\left(x - \frac{1}{4}\right)^2$ **21.** $\left(t + \frac{1}{3}\right)^2$ **23.** $4(2x - 3)^2$ **25.** $3a(5a + 1)^2$

27. $(x + 2 + 3)^2 = (x + 5)^2$ **29.** $(x + 3)(x - 3)$ **31.** $(7x + 8y)(7x - 8y)$ **33.** $\left(2a + \frac{1}{2}\right)\left(2a - \frac{1}{2}\right)$ **35.** $\left(x + \frac{3}{5}\right)\left(x - \frac{3}{5}\right)$

37. $(3x + 4y)(3x - 4y)$ **39.** $10(5 + t)(5 - t)$ **41.** $(x^2 + 9)(x + 3)(x - 3)$ **43.** $(3x^3 + 1)(3x^3 - 1)$

45. $(4a^2 + 9)(2a + 3)(2a - 3)$ **47.** $\left(\frac{1}{9} + \frac{y^2}{4}\right)\left(\frac{1}{3} + \frac{y}{2}\right)\left(\frac{1}{3} - \frac{y}{2}\right)$ **49.** $(x - y)(x + y)(x^2 + xy + y^2)(x^2 - xy + y^2)$

51. $2a(a - 2)(a + 2)(a^2 + 2a + 4)(a^2 - 2a + 4)$ **53.** $(x + 1)(x - 5)$ **55.** $y(y + 8)$ **57.** $(x - 5 + y)(x - 5 - y)$

59. $(a + 4 + b)(a + 4 - b)$ **61.** $(x + y + a)(x + y - a)$ **63.** $(x + 3)(x + 2)(x - 2)$ **65.** $(x + 2)(x + 5)(x - 5)$

67. $(2x + 3)(x + 2)(x - 2)$ **69.** $(x + 3)(2x + 3)(2x - 3)$ **71.** $(2x - 15)(2x + 5)$ **73.** $(a - 3 - 4b)(a - 3 + 4b)$

75. $(a - 3 - 4b)(a - 3 + 4b)$ **77.** $(x + 4)(x - 3)^2$ **79.** $(x - y)(x^2 + xy + y^2)$ **81.** $(a + 2)(a^2 - 2a + 4)$

83. $(3 + x)(9 - 3x + x^2)$ **85.** $(y - 1)(y^2 + y + 1)$ **87.** $10(r - 5)(r^2 + 5r + 25)$ **89.** $(4 + 3a)(16 - 12a + 9a^2)$

91. $(2x - 3y)(4x^2 + 6xy + 9y^2)$ **93.** $\left(t + \frac{1}{3}\right)\left(t^2 - \frac{1}{3}t + \frac{1}{9}\right)$ **95.** $\left(3x - \frac{1}{3}\right)\left(9x^2 + x + \frac{1}{9}\right)$

97. $(4a + 5b)(16a^2 - 20ab + 25b^2)$ **99.** 30 and -30 **101.** $y(y^2 + 25)$ **103.** $2ab^3(b^2 + 4b + 1)$

105. $(2x - 3)(2x + a)$ **107.** $(x + 2)(x - 2)$ **109.** $(x - 3)^2$ **111.** $(3a - 4)(2a - 1)$ **113.** $(x + 2)(x^2 - 2x + 4)$

115. $(5, 3)$ **117.** $(1, 3, 1)$ **119.** $(a - b + 3)(a + b - 3)$ **121.** $(x - y - 8)(x + y + 2)$ **123.** $k = 144$

125. $k = \pm126$

Problem Set 5.7

1. $(x + 9)(x - 9)$ **3.** $(x - 3)(x + 5)$ **5.** $(x + 2)(x + 3)^2$ **7.** $(x^2 + 2)(y^2 + 1)$ **9.** $2ab(a^2 + 3a + 1)$

11. Does not factor: prime   **13.** $3(2a + 5)(2a - 5)$ **15.** $(3x - 2y)^2$ **17.** $(5 - t)^2$ **19.** $4x(x^2 + 4y^2)$ **21.** $2y(y + 5)^2$

23. $a^4(a + 2b)(a^2 - 2ab + 4b^2)$ **25.** $(t + 3 + x)(t + 3 - x)$ **27.** $(x + 5)(x + 3)(x - 3)$ **29.** $5(a + b)^2$

31. Does not factor: prime   **33.** $3(x + 2y)(x + 3y)$ **35.** $\left(3a + \frac{1}{3}\right)^2$ **37.** $(x - 3)(x - 7)^2$ **39.** $(x + 8)(x - 8)$

41. $(2 - 5x)(4 + 3x)$ **43.** $a^5(7a + 3)(7a - 3)$ **45.** $\left(r + \frac{1}{5}\right)\left(r - \frac{1}{5}\right)$ **47.** Does not factor: prime

49. $100(x - 3)(x + 2)$ **51.** $a(5a + 3)(5a + 1)$ **53.** $(3x^2 + 1)(x^2 - 5)$ **55.** $3a^2b(2a - 1)(4a^2 + 2a + 1)$

57. $(4 - r)(16 + 4r + r^2)$ **59.** $5x^2(2x + 3)(2x - 3)$ **61.** $100(2t + 3)(2t - 3)$ **63.** $2x^3(4x - 5)(2x - 3)$

65. $(y + 1)(y - 1)(y^2 - y + 1)(y^2 + y + 1)$ **67.** $2(5 + a)(5 - a)$ **69.** $3x^2y^2(2x + 3y)^2$ **71.** $(x - 2 + y)(x - 2 - y)$

73. $\left(a - \frac{2}{3}b\right)^2$ **75.** $\left(x - \frac{2}{5}y\right)^2$ **77.** $\left(a - \frac{5}{6}b\right)^2$ **79.** $\left(x - \frac{4}{5}y\right)^2$ **81.** $(2x - 3)(x - 5)(x + 2)$ **83.** $(x - 4)^3(x - 3)$

85. $2(y - 3)(y^2 + 3y + 9)$ **87.** $2(a - 4b)(a^2 + 4ab + 16b^2)$ **89.** $2(x + 6y)(x^2 - 6xy + 36y^2)$ **91.** 60 geese; 48 ducks

93. 150 oranges; 144 apples   **95.** $2x^2 + 2x + 1$ **97.** $t^2 - 4t + 3$ **99.** $(x - 6)(x + 4)$ **101.** $x(2x + 1)(x - 3)$

103. $(x + 2)(x - 3)(x + 3)$ **105.** 6   **107.** $-\frac{1}{2}$

Problem Set 5.8

1. $6, -1$ **3.** $0, 2, 3$ **5.** $\frac{1}{3}, -4$ **7.** $\frac{2}{3}, \frac{3}{2}$ **9.** $5, -5$ **11.** $0, -3, 7$ **13.** $-4, \frac{5}{2}$ **15.** $0, \frac{4}{3}$ **17.** $-\frac{1}{5}, \frac{1}{3}$

19. $0, -\frac{4}{3}, \frac{4}{3}$ **21.** $-10, 0$ **23.** $-5, 1$ **25.** $1, 2$ **27.** $-2, 3$ **29.** $-2, \frac{1}{4}$ **31.** $-3, -2, 2$ **33.** $-2, -5, 5$

35. $-\frac{3}{2}, -2, 2$ **37.** $-3, -\frac{3}{2}, \frac{3}{2}$ **39.** $-2, \frac{5}{3}$ **41.** $-1, 9$ **43.** $0, -3$ **45.** $-4, -2$ **47.** $-4, 2$ **49.** $-\frac{3}{2}$

51. $-2, 8$ **53.** -3 **55.** $-7, 1$ **57.** $0, 5$ **59.** $-1, 8$ **61. a.** $\frac{25}{9}$ **b.** $-\frac{5}{3}, \frac{5}{3}$ **c.** $-3, 3$ **d.** $\frac{5}{3}$ **63.** 3 hours

65. $-5, -4$ or $4, 5$ **67.** 24 feet   **69.** $6, 8, 10$ **71.** 2 feet, 8 feet   **73.** 18 inches, 4 inches

75. 0 and 2 seconds   **77.** 1 and 2 seconds   **79.** 0 and $\frac{3}{2}$ seconds   **81.** 2 and 3 seconds   **83.** $4 or $8

85. $7 or $10   **87.** $(1, 2)$ **89.** $(15, 12)$ **91.** $(3, -2, 1)$ **93.** $4,000 at 5%, $8,000 at 6%

95.

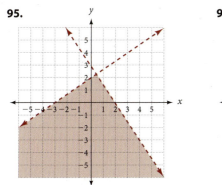

97.

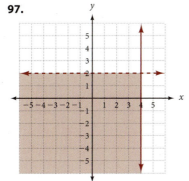

Chapter 5 Review

1. x^{10} **2.** $25x^6$ **3.** $-32x^{18}y^8$ **4.** $\frac{1}{8}$ **5.** $\frac{9}{4}$ **6.** $\frac{1}{2}$ **7.** 3.45×10^7 **8.** 3.57×10^{-3} **9.** 44,500

10. 0.000445   **11.** $\frac{1}{a^9}$ **12.** $\frac{x^{12}}{4}$ **13.** x^2 **14.** 8×10^{-2} **15.** 4×10^{-10} **16.** $2x^2 - 5x + 7$ **17.** $2x^3 - 2x^2 - 2x - 4$

18. $x^2 - 2x - 3$ **19.** $30x + 12$ **20.** 15   **21.** $12x^3 - 6x^2 + 3x$ **22.** $2a^4b^3 + 4a^3b^4 + 2a^2b^5$ **23.** $18 - 9y + y^2$

24. $6x^4 + 5x^2 - 4$ **25.** $8x^3 - 27$ **26.** $a^4 - 4a^2 + 4$ **27.** $x^2 - \frac{1}{9}$ **28.** $4a^2 - b^2$ **29.** $3xy(2x^3 - 3y^3 + 6x^2y^2)$

30. $4(x + y)^2(x^2 - 2y^2)$ **31.** $(4x^2 + 5)(2 - y)$ **32.** $(1 - y)(1 + y)(x^3 + 8b^2)$ **33.** $(x - 2)(x - 3)$ **34.** $2x(x + 5)(x - 3)$

35. $(5a - 4b)(4a - 5b)$ **36.** $x^2(3x + 2)(2x - 5)$ **37.** $3y(4x + 5)(2x - 3)$ **38.** $(x^2 + 4)(x + 2)(x - 2)$ **39.** $3(a^2 + 3)^2$

40. $(a - 2)(a^2 + 2a + 4)$ **41.** $5x(x + 3y)^2$ **42.** $(6 - 5a)(6 + 5a)$ **43.** $(x + 3)(x - 3)(x + 4)$ **44.** $-3, -2$

45. $-\frac{1}{2}, \frac{4}{5}$ **46.** $-\frac{5}{3}, \frac{5}{3}$ **47.** $0, -2$ **48.** $-3, 6$ **49.** $-4, 3, -3$ **50.** $-10, -8$ or $8, 10$ **51.** $-5, -4$ or $4, 5$

52. $3, 4, 5$ **53.** $6, 8, 10$

Chapter 5 Cumulaive Review

1. -125 **2.** $-\frac{1}{125}$ **3.** 60   **4.** 48   **5.** 48   **6.** 121   **7.** -1 **8.** 2   **9.** 11.49   **10.** -11.51 **11.** 22   **12.** 16

13. $6x - 1$ **14.** $20x - 9y$ **15.** $-2x - 4$ **16.** $5a - 2$ **17.** $8x^3 - 7x^2 - 14$ **18.** $\frac{1}{3}x^2 - \frac{17}{6}x$ **19.** $81x^8y^4$ **20.** $-15a^5b^5$

21. $-2xy^4$ **22.** $\frac{81a^2}{b^4}$ **23.** $15x^2 - 4x - 4$ **24.** $15x^3 + 10x^2 + 35x$ **25.** $x^2 - \frac{4}{3}x + \frac{4}{9}$ **26.** $t^3 + 2t^2 + \frac{4}{3}t + \frac{8}{27}$

27. 2.8×10^{10} **28.** 4×10^2 **29.** $(x - a)(x - y)$ **30.** $(3a - 7)(2a + 5)$ **31.** $(x - 2)(x + 8)$ **32.** $\left(\frac{1}{2} + t\right)\left(\frac{1}{4} - \frac{1}{2}t + t^2\right)$

33. 2   **34.** $-1, 5$ **35.** -36 **36.** $-\frac{3}{4}$ **37.** $0, -\frac{6}{5}, \frac{6}{5}$ **38.** -4 **39.** $(-2, 2)$ **40.** $(12, 5)$ **41.** $-\frac{5}{7}$ **42.** $\frac{1}{3}$

43. -13 **44.** -7 **45.** 2   **46.** 3   **47.** $\frac{16}{9}$ **48.** -108 **49.** $b = 10$ ft, $h = 15$ ft   **50.** $20°, 40°, 120°$ **51.** 74   **52.** 6

Chapter 5 Test

1. $\frac{16}{9}$ **2.** $\frac{1}{32}$ **3.** x^8 **4.** a^2 **5.** $32x^{12}y^{11}$ **6.** $\frac{a^{14}}{4b^{18}}$ **7.** 6.53×10^6 **8.** 8.7×10^{-4} **9.** 8.7×10^7 **10.** 3×10^8

11. 149 **12.** -151 **13.** $\frac{3}{4}x^3 - \frac{5}{4}x^2 - 2x - 1$ **14.** $4x + 75$ **15.** $25x - 0.2x^2$ **16.** $P(x) = -0.2x^2 + 23x - 100$

17. $\$500$ **18.** $\$300$ **19.** $\$200$ **20.** $6y^2 + y - 35$ **21.** $2x^3 + 3x^2 - 26x + 15$ **22.** $64 - 48t^3 + 9t^6$

23. $1 - 36y^2$ **24.** $4x^3 - 2x^2 - 30x$ **25.** $10t^4 - \frac{1}{10}$ **26.** $(x + 4)(x - 3)$ **27.** $2(3x^2 - 1)(2x^2 + 5)$

28. $(2a - 3y)(2a + 3y)(4a^2 + 9y^2)$ **29.** $(7a - b^2)(x^2 - 2y)$ **30.** $\left(t + \frac{1}{2}\right)\left(t^2 - \frac{1}{2}t + \frac{1}{4}\right)$ **31.** $4a^3b(a - 8b)(a + 2b)$

32. $(x - 5 - b)(x - 5 + b)$ **33.** $(3 - x)(3 + x)(9 + x^2)$ **34.** $-\frac{1}{3}, 2$ **35.** $0, 5$ **36.** $-5, 2$ **37.** $-4, -2, 4$ **38.** $2, 3$

39. $2, 4$ **40.** 9 **41.** 5 **42.** 6

Chapter 6

Problem Set 6.1

1. **a.** $\frac{1}{5}$ **b.** $\frac{1}{x - 3}; x \neq \pm 3$ **c.** $\frac{x}{x + 3}; x \neq \pm 3$ **d.** $\frac{x^2 + 3x + 9}{x + 3}; x \neq \pm 3$

3. $h(0) = -3, h(-3) = 3, h(3) = 0, h(-1)$ is undefined, $h(1) = -1$ **5.** $\frac{x - 4}{6}$ **7.** $\frac{4x - 3y}{x(x + y)}$ **9.** $(a^2 + 9)(a + 3)$

11. $\frac{a - 6}{a + 6}$ **13.** $\frac{2y + 3}{y + 1}$ **15.** $\frac{a^2 - ab + b^2}{a - b}$ **17.** $\frac{2(x - 1)}{x}$ **19.** $\frac{2x + 3y}{2x + y}$ **21.** $\frac{x + 3}{y - 4}$ **23.** $\frac{x + b}{x - 2b}$ **25.** $x + 2$

27. $\frac{2x^2 - 5}{3x - 2}$ **29.** $\frac{x^2 + 2x + 4}{x + 2}$ **31.** $4 + t$ **33.** $\frac{4x^2 + 6x + 9}{2x + 3}$ **35.** -1 **37.** $-(y + 6)$ **39.** $-\frac{3a + 1}{3a - 1}$

41. 3 **43.** $x + a$ **45.** $\{x \mid x \neq 1\}$ **47.** $\{t \mid t \neq 4, t \neq -4\}$ **49.** $\{x \mid x \neq 5\}$ **51.** All real numbers

53. $\{x \mid x \neq 0\}$ **55.** $\{x \mid x \neq -4, x \neq 5\}$ **57.** $2; 2$ **59.** Undefined; 4   **61.** $1; 1$ **63.** $3; 3$

65.

Weeks	Weight (pounds)
x	$W(x)$
0	200
1	194
4	184
12	173
24	168

67. 6.8 feet per second   **69.** **a.** Domain $= \{t \mid 20 \leq t \leq 50\}$

b.

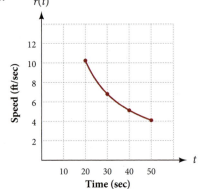

71. **a.** Domain $= \{d \mid 1 \leq d \leq 6\}$

b.

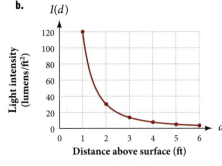

73. 3,768 inches per minute; 2,826 inches per minute   **75.** $3x^2 - 7x + 4$ **77.** 9 **79.** $8x^2$ **81.** $(-3, 6)$

83. $\left(16, -\frac{13}{2}\right)$ **85.** $(-4, -2)$ **87.** $(0, 2, -4)$ **89.** $2x^3$ **91.** $-2x^2y^2$ **93.** 185.12 **95.** $4x^3 - 8x^2$

97. $4x^3 - 6x - 20$

Problem Set 6.2

1. $2x^2 - 4x + 3$ **3.** $-2x^2 - 3x + 4$ **5.** $2y^2 + \frac{5}{2} - \frac{3}{2y^2}$ **7.** $-\frac{5}{2}x + 4 + \frac{3}{x}$ **9.** $4ab^3 + 6a^2b$ **11.** $-xy + 2y^2 + 3xy^2$

13. $x + 2$ **15.** $a - 3$ **17.** $5x + 6y$ **19.** $x^2 + xy + y^2$ **21.** $(y^2 + 4)(y + 2)$ **23.** $(x + 2)(x + 5)$ **25. a.** 4 **b.** 4

27. a. 5 **b.** 5 **29. a.** $2x + h$ **b.** $x + a$ **31. a.** $2x + h$ **b.** $x + a$ **33. a.** $2x + h - 3$ **b.** $x + a - 3$

35. a. $4x + 2h + 3$ **b.** $2x + 2a + 3$ **37. a.** $-6x - 3h + 2$ **b.** $-3x - 3a + 2$ **39.** $x - 7 + \frac{7}{x + 2}$

41. $2x^2 - 5x + 1 + \frac{4}{x + 1}$ **43.** $y^2 - 3y - 13$ **45.** $3y^2 + 6y + 8 + \frac{37}{2y - 4}$ **47.** $a^3 + 2a^2 + 4a + 6 + \frac{17}{a - 2}$

49. $y^3 + 2y^2 + 4y + 8$ **51.** $h(x) = \frac{x + 6}{4}; \{x \mid x \neq 6\}$ **53.** $h(x) = \frac{x - 8}{x + 4}; \{x \mid x \neq -4, x \neq 8\}$

55. $h(x) = x^2 + 3x + 9; \{x \mid x \neq 3\}$ **59. a.** $(x + 4)(x + 5)(x + 1)$ **b.** $(x + 5)(x + 1)$

61. a. $(x + 2)(x + 4)(x - 3)$ **b.** $(x + 4)(x - 3)$ **63.** 13 is the same as the remainder

65. a.

x	1	5	10	15	20
$C(x)$	2.15	2.75	3.50	4.25	5.00

b. $\overline{C}(x) = \frac{2}{x} + 0.15$

c.

x	1	5	10	15	20
$\overline{C}(x)$	2.15	0.55	0.35	0.28	0.25

d. It decreases.

67. $\frac{21}{10}$ **69.** $\frac{11}{8}$ **71.** $\frac{1}{18}$ **73.** 32 **75.** 17 **77.** $\frac{x^{16}}{y^{22}}$ **79.** $\frac{2}{3}$ **81.** $20x^2y^2$ **83.** $72x^4y^5$

85. $(x + 2)(x - 2)$ **87.** $x^2(x - y)$ **89.** $2(y - 1)(y + 1)$ **91.** $4x^3 - x^2 + 3$ **93.** $0.5x^2 - 0.4x + 0.3$

95. $\frac{3}{2}x - \frac{5}{2} + \frac{1}{2x + 4}$ **97.** $\frac{2}{3}x + \frac{1}{3} + \frac{2}{3x - 1}$

Problem Set 6.3

1. $\frac{1}{6}$ **3.** $\frac{9}{4}$ **5.** $\frac{1}{2}$ **7.** $\frac{15y}{x^2}$ **9.** $\frac{b}{a}$ **11.** $\frac{2y^5}{z^3}$ **13.** $\frac{x + 3}{x + 2}$ **15.** $y + 1$ **17.** $\frac{3(x + 4)}{x - 2}$ **19.** 1 **21.** $\frac{(a - 2)(a + 2)}{a - 5}$

23. $\frac{9t^2 - 6t + 4}{4t^2 - 2t + 1}$ **25.** $\frac{x + 3}{x + 4}$ **27.** $\frac{5a - b}{9a^2 + 15ab + 25b^2}$ **29.** 2 **31.** $\frac{x(x - 1)}{x^2 + 1}$ **33.** $\frac{(a + 4b)(a - 3b)}{(a - 4b)(a + 5b)}$ **35.** $\frac{2y - 1}{2y - 3}$

37. $\frac{(y - 2)(y + 1)}{(y + 2)(y - 1)}$ **39.** $\frac{x - 1}{x + 1}$ **41.** $\frac{x - 2}{x + 3}$ **43. a.** $\frac{5}{21}$ **b.** $\frac{5x + 3}{25x^2 + 15x + 9}$ **c.** $\frac{5x - 3}{25x^2 + 15x + 9}$ **d.** $\frac{5x + 3}{5x - 3}$

45. a. $\frac{(x + 2)^2}{(x - 1)^2}$ **b.** $(x - 3)^2$ **47.** $\frac{(x + 2)(x - 4)}{4(x + 9)}$ **49.** $(x + 3)(x + 4)$ **51.** $3x$ **53.** $2(x + 5)$ **55.** $x - 2$ **57.** $-(y - 4)$

59. $(a - 5)(a + 1)$ **61.** $\frac{(x - 4)^2}{x - 3}$ **63.** $-\frac{x - 1}{x + 3}$ **65.** $(y - 2)(x - 7)$ **67.** $\frac{(x + 3)^2}{x - 4}$ **69.** $\frac{(2x - 3)(x - 1)}{x - 2}$

71.

Number of Copies	Price per Copy ($)
1	20.33
10	9.33
20	6.40
50	4.00
100	3.05

73. $\$304.76$ **75.**

Number of Copies	Price per Copy ($)
1	20.50
10	10.00
20	7.20
50	4.91
100	4.00

77. $\$400$ **79.** $x^2 - 6x + 9$

81. $10x^5 + 8x^3 - 6x^2$ **83.** $12a^2 + 11a - 5$ **85.** $12xy - 6x + 28y - 14$ **87.** $9 - 6t^2 + t^4$ **89.** $3x^3 + 18x^2 + 33x + 18$

91. $\frac{2}{3}$ **93.** $\frac{47}{105}$ **95.** $x - 7$ **97.** $(x + 1)(x - 1)$ **99.** $2(x + 5)$ **101.** $(a - b)(a^2 + ab + b^2)$

103. $2(2x + 3y)(4x^2 - 6xy + 9y^2)$ **105.** $\frac{x^4 - x^2y^2 + y^4}{x^2 + y^2}$ **107.** $\frac{(a + 5)(a - 1)}{3a^2 - 2a + 1}$ **109.** $\frac{a(c - 1)}{a - b}$ **111.** x

Problem Set 6.4

1. 1   **3.** -1 **5.** $\dfrac{1}{x+y}$ **7.** 1   **9.** $\dfrac{a^2+2a-3}{a^3}$ **11.** 1   **13. a.** $\dfrac{1}{16}$ **b.** $\dfrac{9}{4}$ **c.** $\dfrac{13}{24}$ **d.** $\dfrac{5x+15}{(x-3)^2}$ **e.** $\dfrac{x+3}{5}$ **f.** $\dfrac{x-2}{x-3}$

15. $\dfrac{5}{4}$ **17.** $\dfrac{1}{3}$ **19.** $\dfrac{41}{24}$ **21.** $\dfrac{19}{144}$ **23.** $\dfrac{31}{24}$ **25.** $\dfrac{4-3t}{2t^2}$ **27.** $\dfrac{1}{2}$ **29.** $\dfrac{x+3}{2(x+1)}$ **31.** $\dfrac{a-b}{a^2+ab+b^2}$

33. $\dfrac{2y-3}{4y^2+6y+9}$ **35.** $\dfrac{2(2x-3)}{(x-3)(x-2)}$ **37.** $\dfrac{1}{2t-7}$ **39.** $\dfrac{4}{(a-3)(a+1)}$ **41.** $\dfrac{-4x^2}{(2x+1)(2x-1)(4x^2+2x+1)}$ **43.** $\dfrac{2}{(2x+3)(4x+3)}$

45. $\dfrac{a}{(a+4)(a+5)}$ **47.** $\dfrac{x+1}{(x-2)(x+3)}$ **49.** $\dfrac{x-1}{(x+1)(x+2)}$ **51.** $\dfrac{1}{(x+2)(x+1)}$ **53.** $\dfrac{1}{(x+2)(x+3)}$ **55.** $\dfrac{4x+5}{2x+1}$ **57.** $\dfrac{22-5t}{4-t}$

59. $\dfrac{2x^2+3x-4}{2x+3}$ **61.** $\dfrac{2x-3}{2x}$ **63.** $\dfrac{1}{2}$ **65.** $-\dfrac{(2x-11)}{4(x-4)}$ **67.** $\dfrac{3}{x+4}$ **69.** $\dfrac{(2x+1)(x+5)}{(x-2)(x+1)(x+3)}$ **71.** $\dfrac{(x+2)(x-1)}{(3x-2)(3x+2)(x-2)}$

73. $\dfrac{3(x-2)(x+1)}{(2x+1)(x-1)(x-3)}$ **75.** $\dfrac{2x}{(x-9)(x-7)}$ **77.** $x+\dfrac{4}{x}=\dfrac{x^2+4}{x}$ **79.** $\dfrac{51}{10}=5.1$

81. a. $T=120$ months **b.** The two objects will never meet.   **83.** 5.4×10^4 **85.** 3.4×10^{-4} **87.** 6,440

89. 0.00644   **91.** 1.2×10^4 **93.** $8x-5$ **95.** $2x-1$ **97.** 2   **99.** $\dfrac{6}{5}$ **101.** $\dfrac{3}{4}$ **103.** $x+2$ **105.** $3-x$

107. $(x+2)(x-2)$

Problem Set 6.5

1. $\dfrac{9}{8}$ **3.** $\dfrac{2}{15}$ **5.** $\dfrac{119}{20}$ **7.** $\dfrac{1}{x+1}$ **9.** $\dfrac{a+1}{a-1}$ **11.** $\dfrac{y-x}{y+x}$ **13.** $\dfrac{1}{(x+5)(x-2)}$ **15.** $\dfrac{1}{a^2-a+1}$ **17.** $\dfrac{x+3}{x+2}$ **19.** $\dfrac{a+3}{a-2}$

21. $\dfrac{9x^2+6x+4}{x(x+1)}$ **23.** $\dfrac{x}{x-2}$ **25.** $\dfrac{x-3}{x}$ **27.** $\dfrac{x+4}{x+2}$ **29.** $\dfrac{a-1}{a+1}$ **31.** $-\dfrac{x}{3}$ **33.** $\dfrac{y^2+1}{2y}$ **35.** $\dfrac{-x^2+x-1}{x-1}$ **37.** $\dfrac{5}{3}$

39. $\dfrac{2x-1}{2x+3}$ **41.** $-\dfrac{1}{x(x+h)}$ **43.** $\dfrac{3c+4a-2b}{5}$ **45.** $\dfrac{(t-4)(t+1)}{(t+6)(t-3)}$ **47.** $\dfrac{(5b-1)(b+5)}{2(2b-11)}$ **49.** $-\dfrac{3}{2x+14}$ **51.** $2m-9$

53. a. $-\dfrac{4}{ax}$ **b.** $-\dfrac{1}{(x+1)(a+1)}$ **c.** $-\dfrac{a+x}{a^2x^2}$ **55.** $(a^{-1}+b^{-1})^{-1}=\left(\dfrac{1}{a}+\dfrac{1}{b}\right)^{-1}=\left(\dfrac{a+b}{ab}\right)^{-1}=\dfrac{ab}{a+b}$

57. a. As v approaches 0, the denominator approaches 1. **b.** $v=\dfrac{s(f-h)}{h}$ **59.** -15 **61.** 5   **63.** 1   **65.** $-3,4$

67. 2, 3   **69.** $xy-2x$ **71.** $3x-18$ **73.** ab **75.** $(y+5)(y-5)$ **77.** $x(a+b)$ **79.** 2   **81.** $\dfrac{1}{3}$ **83.** $\dfrac{40}{243}$

85. $\dfrac{a-2b}{a+2b}$ **87.** $\dfrac{a+b}{2}$ **89.** $-\dfrac{qt}{q+t}$

Problem Set 6.6

1. $-\dfrac{35}{3}$ **3.** $-\dfrac{18}{5}$ **5.** $\dfrac{36}{11}$ **7.** 2   **9.** 5   **11.** 2   **13.** $-3,4$ **15.** $1,-\dfrac{4}{3}$ **17.** $-\dfrac{9}{5},5$ **19.** $\dfrac{9}{2}$ **21.** $\varnothing$

23. $3,-\dfrac{4}{3}$ **25.** $-\dfrac{4}{3}$ **27.** $\varnothing$ **29. a.** $\dfrac{1}{3}$ **b.** 3 **c.** 9 **d.** 4 **e.** $\dfrac{1}{3},3$

31. a. $\dfrac{6}{(x-4)(x+3)}$ **b.** $\dfrac{x-3}{x-4}$ **c.** $5;-2$ does not check   **33.** $\varnothing;-1$ does not check   **35.** 5   **37.** $-\dfrac{1}{2},\dfrac{5}{3}$ **39.** $\dfrac{2}{3}$

41. 18   **43.** $\varnothing;4$ does not check   **45.** $-4;3$ does not check   **47.** -6 **49.** -5 **51.** $\dfrac{53}{17}$

53. 2; 1 does not check   **55.** $\varnothing;3$ does not check   **57.** $\dfrac{22}{3}$ **59.** 2   **61.** 1, 5   **63.** $x=\dfrac{ab}{a-b}$ **65.** $R=\dfrac{R_1R_2}{R_1+R_2}$

67. $y=\dfrac{x-3}{x-1}$ **69.** $y=\dfrac{1-x}{3x-2}$

71.   **73.**   **75.**   **77.**

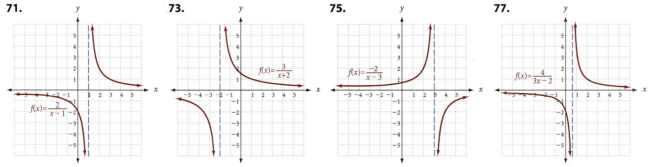

79.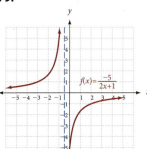

81.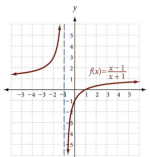

83.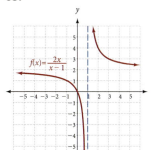

85.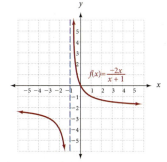

87.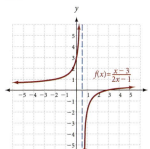

89.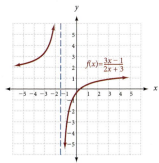

93.

Time t(sec)	Speed of Kayak in Still Water v(m/sec)	Current of the River c(m/sec)
240	4	1
300	4	2
514.29	4	3
337.5	3	1
540	3	2
N/A	3	3

95. 5   **97.** 8 meters, 13 meters   **99.** $-6, -5$ or $5, 6$ **101.** 3, 4, 5   **103.** 2,358   **105.** 12.3   **107.** 3   **109.** 9, -1

111. 60   **113.** $-\frac{2}{3}, -\frac{5}{2}, \frac{5}{2}$ **115.** $-2, 2, 3$ **117.** $-\frac{a}{5}$ **119.** $v = \frac{16t^2 + s}{t}$ **121.** $f = \frac{pg}{g - p}$

Problem Set 6.7

As you can see, in addition to the answers to the problems we have included some of the equations used to solve the problems. Remember, you should attempt the problems on your own before looking here to check your answers or equations.

1. $\frac{1}{x} + \frac{1}{3x} = \frac{20}{3}; \frac{1}{5}$ and $\frac{3}{5}$ **3.** $x + \frac{1}{x} = \frac{10}{3}$; 3 or $\frac{1}{3}$ **5.** $\frac{1}{x} + \frac{1}{x + 1} = \frac{7}{12}$; 3, 4   **7.** $\frac{7 + x}{9 + x} = \frac{5}{6}$; 3   **9.** $\frac{9 + x}{11 + x} = \frac{6}{7}$; 3

11. Let x = speed of current; $\frac{1.5}{5 - x} = \frac{3}{5 + x}; \frac{5}{3}$ miles per hour   **13.** $\frac{8}{x + 2} + \frac{8}{x - 2} = 3$; 6 miles per hour

15. Train A: 75 miles per hour, Train B: 60 miles per hour; let x = speed of B; $\frac{150}{x + 15} = \frac{120}{x}$

17. 540 miles per hour; let x = speed of 747; $\frac{810}{270} - 1\frac{1}{2} = \frac{810}{x}$ **19.** 54 miles per hour; let x = usual speed; $\frac{270}{x + 6} + \frac{1}{2} = \frac{270}{x}$

21. 75 mph: 12 minutes faster; 80 mph: 22 minutes faster

23. Let x = time to fill the tank with both open; $\frac{1}{8} - \frac{1}{16} = \frac{1}{x}$; 16 hours   **25.** $\frac{1}{3} - \frac{1}{5} = \frac{1}{x}; \frac{15}{2}$ minutes

27. Let x = time to fill with hot water; $\frac{1}{x} + \frac{1}{3.5} = \frac{1}{2.1}; 5\frac{1}{4}$ minutes

29. Let x = number of hours to fill using pipe B; $\frac{1}{x} + \frac{1}{3x} = \frac{1}{4}; \frac{16}{3}$ hours   **31.** $\frac{36}{7}$ or 5.14 hours   **33.** 62.0 acres

35. 5.9 miles per hour   **37.** 20.7 miles per hour   **39.** 4.6 miles per hour   **41.** 3.6 miles per hour   **43.** 8,278.2 mph

45. $\frac{1}{3}\left[\left(x + \frac{2}{3}x\right) + \frac{1}{3}\left(x + \frac{2}{3}x\right)\right] = 10, x = \frac{27}{2}$ **47.** $\frac{2}{3a}$ **49.** $(x - 3)(x + 2)$ **51.** 1   **53.** $\frac{3 - x}{3 + x}$

55. $\varnothing$; 3 does not check

Chapter 6 Review

1. $\frac{25x^2}{7y^3}$ **2.** $\frac{a(a - b)}{4}$ **3.** $\frac{x - 5}{x + 5}$ **4.** $\frac{a + 1}{a - 1}$ **5.** $3x + 2 + \frac{4}{x}$ **6.** $-9b + 5a - 7a^2b^2$ **7.** $x^{3n} - x^{2n}$ **8.** $x + 2$ **9.** $5x + 6y$

10. $y^3 + 2y^2 + 4y + 8$ **11.** $4x + 1 - \frac{2}{2x - 7}$ **12.** $y^2 - 3y - 13$ **13.** $\frac{9}{5}$ **14.** $\frac{3x}{4y^2}$ **15.** $\frac{x - 1}{x^2 + 1}$ **16.** 1   **17.** $\frac{x + 2}{x - 2}$

18. $(2x - 3)(x + 3)$ **19.** $\frac{31}{30}$ **20.** -1 **21.** $\frac{x^2 + x + 1}{x^3}$ **22.** $\frac{1}{(y + 4)(y + 3)}$ **23.** $\frac{x - 1}{2(x + 1)(x + 2)}$ **24.** $\frac{15x - 2}{5x - 2}$ **25.** 5

26. $\frac{1}{a^2 - a + 1}$ **27.** $\frac{x^2 + x + 1}{x^2 + 1}$ **28.** $\frac{x + 3}{x + 2}$ **29.** 6   **30.** 1   **31.** -6 **32.** $\varnothing$; -3 does not check   **33.** $\frac{22}{3}$

34. 4; -5 does not check   **35.** Car: 30 miles per hour; truck: 20 miles per hour   **36.** 7.5 miles per hour   **37.** 742 miles per hour

Chapter 6 Cumulative Review

1. $-\frac{27}{8}$ **2.** $\frac{36}{25}$ **3.** 47   **4.** -55 **5.** 16   **6.** 25   **7.** $\frac{3}{5}$ **8.** 2   **9.** -2 **10.** $5a - 7b > 5a + 7b$ **11.** 159

12. -161 **13.** 90   **14.** 80   **15.** $-3x - 12$ **16.** $12x$ **17.** $\frac{1}{(y+3)(y+2)}$ **18.** $-\frac{1}{x+y}$ **19.** $12t^4 - \frac{1}{12}$ **20.** $x + 2$

21. $2x^{2n} - 3x^{3n}$ **22.** $a^3 - a^2 + 2a - 4 + \frac{7}{a+2}$ **23.** x^3 **24.** $\frac{x^3}{y^7}$ **25.** $x + 3$ **26.** $\frac{1}{a^2 + a + 1}$ **27.** $y = -\frac{4}{3}x - 9$

28. $x = -2$ **29.** $2^3 \cdot 3 \cdot 7$ **30.** $(x + 7)(x - 10)$ **31.** $(x + 5 - y)(x + 5 + y)$ **32.** $\left(y + \frac{2}{3}x\right)\left(y^2 - \frac{2}{3}xy + \frac{4}{9}x^2\right)$ **33.** -20

34. 3   **35.** 1   **36.** $\frac{5}{6}$ **37.** 5   **38.** $\frac{3}{2}, 4$ **39.** $(-1, -2)$ **40.** $\left(-\frac{8}{17}, \frac{7}{34}\right)$ **41.** $\left(-\frac{33}{5}, -\frac{52}{5}\right)$ **42.** $(-2, 1, 3)$

43. $x \geq -1$    **44.** $-4 < x < 1$

45. 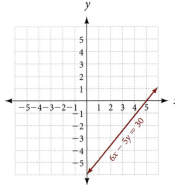   **46.**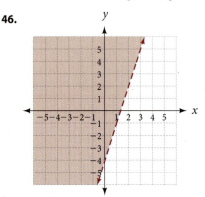

47. -6 **48.** 6   **49.** 4   **50.** 8

Chapter 6 Test

1. $x + y$ **2.** $\frac{x-1}{x+1}$ **3.** 5.9 feet/second   **4.** $6x^2 + 3xy - 4y^2$ **5.** $x^2 - 4x - 2 + \frac{8}{2x-1}$ **6.** $2(a + 4)$ **7.** $4(a + 3)$ **8.** $x + 3$ **9.** $\frac{38}{105}$

10. $\frac{7}{8}$ **11.** $\frac{1}{a-3}$ **12.** $\frac{3(x-1)}{x(x-3)}$ **13.** $\frac{x}{(x+4)(x+5)}$ **14.** $\frac{x+4}{(x+1)(x+2)}$ **15.** $\frac{3a+8}{3a+10}$ **16.** $\frac{x-3}{x-2}$ **17. a.** 2 **b.** 2

18. a. $h + 2x + 5$ **b.** $x + a + 5$ **19.** $-\frac{3}{5}$ **20.** $\varnothing$ **21.** $\frac{3}{13}$ **22.** $-2, 3$ **23.** -7 **24.** 6 mph   **25.** 15 hrs

26. 2.7 miles   **27.** K2, 5.3 miles; Mount Everest, 5.5 miles; Kangchenjunga, 5.3 miles   **28.** 1,012 mph

Chapter 7

Problem Set 7.1

1. 12   **3.** Not a real number   **5.** -7 **7.** -3 **9.** 2   **11.** Not a real number   **13.** 0.2   **15.** 0.2   **17.** 5

19. -6 **21.** $\frac{1}{6}$ **23.** $\frac{2}{5}$ **25.** $6a^4$ **27.** $3a^4$ **29.** $2x^2y$ **31.** $2a^3b^5$ **33.** 6   **35.** -3 **37.** 2   **39.** -2

41. 2   **43.** $\frac{9}{5}$ **45.** 9   **47.** 125   **49.** $\frac{1}{3}$ **51.** $\frac{1}{27}$ **53.** $\frac{6}{5}$ **55.** $\frac{8}{27}$ **57.** 7   **59.** $\frac{3}{4}$ **61.** $x^{4/5}$ **63.** a

65. $\frac{1}{x^{2/5}}$ **67.** $x^{1/6}$ **69.** $x^{9/25}y^{1/2}z^{1/5}$ **71.** $\frac{b^{7/4}}{a^{1/8}}$ **73.** $y^{3/10}$ **75.** $\frac{1}{a^2b^4}$ **77.** $\frac{s^{1/2}}{r^{20}}$ **79.** $10b^3$ **81.** 25 mph

83. a. 2.25 in. **b.** 3.2 in. **c.** 8.13×10^{-5} km   **85. a.** B **b.** A **c.** C **d.** $(0, 0)$ and $(1, 1)$ **87.** $x^6 - x^3$

89. $x^2 + 2x - 15$ **91.** $x^4 - 10x^2 + 25$ **93.** $x^3 - 27$ **95.** $x^6 - x^5$ **97.** $12a^2 - 11ab + 2b^2$ **99.** $x^6 - 4$

101. $3x - 4x^3y$ **103.** $(x - 5)(x + 2)$ **105.** $(3x - 2)(2x + 5)$ **107.** x^2 **109.** t **111.** $x^{1/3}$

113. $(9^{1/2} + 4^{1/2})^2 = (3 + 2)^2 = 5^2 = 25 \neq 9 + 4$ **115.** $\sqrt{\sqrt{a}} = (a^{1/2})^{1/2} = a^{1/4} = \sqrt[4]{a}$

Problem Set 7.2

1. $x + x^2$ **3.** $a^2 - a$ **5.** $6x^3 - 8x^2 + 10x$ **7.** $12x^2 - 36y^2$ **9.** $x^{4/3} - 2x^{2/3} - 8$ **11.** $a - 10a^{1/2} + 21$

13. $20y^{2/3} - 7y^{1/3} - 6$ **15.** $10x^{4/3} + 21x^{2/3}y^{1/2} + 9y$ **17.** $t + 10t^{1/2} + 25$ **19.** $x^3 + 8x^{3/2} + 16$ **21.** $a - 2a^{1/2}b^{1/2} + b$

23. $4x - 12x^{1/2}y^{1/2} + 9y$ **25.** $a - 3$ **27.** $x^3 - y^3$ **29.** $t - 8$ **31.** $4x^3 - 3$ **33.** $x + y$ **35.** $a - 8$ **37.** $8x + 1$

39. $t - 1$ **41.** $2x^{1/2} + 3$ **43.** $3x^{1/3} - 4y^{1/3}$ **45.** $3a - 2b$ **47.** $3(x - 2)^{1/2}(4x - 11)$ **49.** $5(x - 3)^{7/5}(x - 6)$

51. $3(x + 1)^{1/2}(3x^2 + 3x + 2)$ **53.** $(x^{1/3} - 2)(x^{1/3} - 3)$ **55.** $(a^{1/5} - 4)(a^{1/5} + 2)$ **57.** $(2y^{1/3} + 1)(y^{1/3} - 3)$

59. $(3t^{1/5} + 5)(3t^{1/5} - 5)$ **61.** $(2x^{1/7} + 5)^2$ **63.** -8 **65.** 90 **67.** -3 **69.** 0 **71.** $\dfrac{3 + x}{x^{1/2}}$ **73.** $\dfrac{x + 5}{x^{1/3}}$

75. $\dfrac{x^3 + 3x^2 + 1}{(x^3 + 1)^{1/2}}$ **77.** $\dfrac{-4}{(x^2 + 4)^{1/2}}$ **79.** 15.8% **81.** 14% **83.** $\dfrac{1}{x^2 + 9}$ **85.** $3x - 4x^3y$ **87.** $5x - 4$ **89.** $x^2 + 5x + 25$

91. 5 **93.** 6 **95.** $4x^2y$ **97.** $5y$ **99.** 3 **101.** 2 **103.** $2ab$ **105.** 25 **107.** $48x^4y^2$ **109.** $4x^6y^6$

Problem Set 7.3

1. $2\sqrt{2}$ **3.** $7\sqrt{2}$ **5.** $12\sqrt{2}$ **7.** $4\sqrt{5}$ **9.** $4\sqrt{3}$ **11.** $15\sqrt{3}$ **13.** $3\sqrt[3]{2}$ **15.** $4\sqrt[3]{2}$ **17.** $6\sqrt[3]{2}$ **19.** $2\sqrt[5]{2}$

21. $3x\sqrt{2x}$ **23.** $2y\sqrt[4]{2y^3}$ **25.** $2xy^2\sqrt[3]{5xy}$ **27.** $4abc^2\sqrt{3b}$ **29.** $2bc\sqrt[3]{6a^2c}$ **31.** $2xy^2\sqrt[5]{2x^3y^2}$ **33.** $3xy^2z\sqrt[5]{x^2}$

35. $2\sqrt{3}$ **37.** $\sqrt{-20}$, which is not a real number   **39.** $\dfrac{\sqrt{11}}{2}$ **41.** $\dfrac{2\sqrt{3}}{3}$ **43.** $\dfrac{5\sqrt{6}}{6}$ **45.** $\dfrac{\sqrt{2}}{2}$ **47.** $\dfrac{\sqrt{5}}{5}$

49. $2\sqrt[3]{4}$ **51.** $\dfrac{2\sqrt[3]{3}}{3}$ **53.** $\dfrac{\sqrt[4]{24x^2}}{2x}$ **55.** $\dfrac{\sqrt[4]{8y^3}}{y}$ **57.** $\dfrac{\sqrt[3]{36xy^2}}{3y}$ **59.** $\dfrac{\sqrt[3]{6xy^2}}{3y}$ **61.** $\dfrac{\sqrt[4]{2x}}{2x}$ **63.** $\dfrac{3x\sqrt{15xy}}{5y}$

65. $\dfrac{5xy\sqrt{6xz}}{2z}$ **67.** $\dfrac{2ab\sqrt[3]{6ac^2}}{3c}$ **69.** $\dfrac{2xy^2\sqrt[3]{3z^2}}{3z}$ **71.** $\sqrt{x}$ **73.** $\sqrt[3]{xy}$ **75.** $\sqrt[12]{a}$ **77.** $x\sqrt[9]{6x}$ **79.** $ab^2c\sqrt[6]{c}$

81. $abc^2\sqrt[15]{3a^2b}$ **83.** $2b^2\sqrt[3]{a}$ **85.** $5|x|$ **87.** $3|xy|\sqrt{3x}$ **89.** $|x - 5|$ **91.** $|2x + 3|$ **93.** $2|a(a + 2)|$

95. $2|x|\sqrt{x - 2}$ **97.** $\sqrt{9 + 16} = \sqrt{25} = 5$; $\sqrt{9} + \sqrt{16} = 3 + 4 = 7$ **99.** $5\sqrt{13}$ feet   **101.** $r = \dfrac{7\sqrt{11}}{44}$ meters

103. a. 13 feet **b.** $2\sqrt{14} \approx 7.5$ feet   **107.** $7x$ **109.** $27xy^2$ **111.** $\dfrac{5}{6}x$ **113.** $3\sqrt{2}$ **115.** $5y\sqrt{3xy}$ **117.** $2a\sqrt[3]{ab^2}$

119. $\dfrac{y^3}{x^2}$ **121.** 1 **123.** $\dfrac{4x^2 - 6x + 9}{9x^2 - 3x + 1}$ **125.** $12\sqrt[3]{5}$ **127.** $6\sqrt[3]{49}$ **129.** $\dfrac{\sqrt[10]{a^7}}{a}$ **131.** $\dfrac{\sqrt[20]{a^9}}{a}$

133.

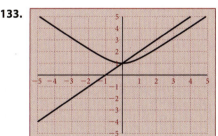

135. About $\dfrac{3}{4}$ of a unit   **137.** $x = 0$

Problem Set 7.4

1. $7\sqrt{5}$ **3.** $-x\sqrt{7}$ **5.** $\sqrt[3]{10}$ **7.** $9\sqrt[5]{6}$ **9.** 0 **11.** $\sqrt{5}$ **13.** $-32\sqrt{2}$ **15.** $-3x\sqrt{2}$ **17.** $-2\sqrt[3]{2}$

19. $8x\sqrt[3]{xy^2}$ **21.** $3a^2b\sqrt{3ab}$ **23.** $11ab\sqrt[3]{3a^2b}$ **25.** $10xy\sqrt[4]{3y}$ **27.** $\sqrt{2}$ **29.** $\dfrac{2\sqrt{3}}{3}$ **31.** $\dfrac{(x - 1)\sqrt{x}}{x}$

33. $\dfrac{3\sqrt{2}}{2}$ **35.** $\dfrac{5\sqrt{6}}{6}$ **37.** $\dfrac{8\sqrt[3]{25}}{5}$ **39. a.** $8\sqrt{2x}$ **b.** $-4\sqrt{2x}$ **41. a.** $4\sqrt{2x}$ **b.** $2\sqrt{2x}$ **43. a.** $3x\sqrt{2}$ **b.** $-x\sqrt{2}$

45. a. $3\sqrt{2x} + 3$ **b.** $-\sqrt{2x} - 7$ **47.** $\sqrt{12} \approx 3.464$; $2\sqrt{3} \approx 2(1.732) = 3.464$

49. $\sqrt{8} + \sqrt{18} \approx 2.828 + 4.243 = 7.071$; $\sqrt{50} \approx 7.071$; $\sqrt{26} \approx 5.099$ equal to $\sqrt{50}$ **55.** $\dfrac{\sqrt{3}}{2}$ **57.** $\dfrac{\sqrt{6}}{3}$ **59.** 1

61. $\dfrac{13 - 3t}{3 - t}$ **63.** $\dfrac{6}{4x + 3}$ **65.** $\dfrac{x - y}{x^2 + xy + y^2}$ **67.** 6 **69.** $4x^2 + 3xy - y^2$ **71.** $x^2 + 6x + 9$ **73.** $x^2 - 4$ **75.** $6\sqrt{2}$

77. 6 **79.** $9x$ **81.** $\dfrac{\sqrt{6}}{2}$ **83.** $-xy$ **85.** $-xy^2z^3$ **87.** $5b^2c\sqrt[4]{2a^3b}$ **89.** $-3ab\sqrt[3]{2a}$

Problem Set 7.5

1. $3\sqrt{2}$ **3.** $10\sqrt{21}$ **5.** 720   **7.** 54   **9.** $\sqrt{6}-9$ **11.** $24+6\sqrt[3]{4}$ **13.** $2+2\sqrt[3]{3}$ **15.** $xy\sqrt[3]{y}+x^2\sqrt[3]{y}$

17. $2x^2\sqrt[4]{x}+2x^3\sqrt[4]{2}$ **19.** $7+2\sqrt{6}$ **21.** $x+2\sqrt{x}-15$ **23.** $34+20\sqrt{3}$ **25.** $19+8\sqrt{3}$ **27.** $x-6\sqrt{x}+9$

29. $4a-12\sqrt{ab}+9b$ **31.** $x+4\sqrt{x-4}$ **33.** $x+4-6\sqrt{x-5}$ **35.** 1   **37.** $a-49$ **39.** $25-x$

41. $x-8$ **43.** $10+6\sqrt{3}$ **45.** $5+\sqrt[3]{12}+\sqrt[3]{18}$ **47.** $x^2+x\sqrt[3]{x^2y^2}+\sqrt[3]{xy}+y$ **49.** $\frac{\sqrt{2}}{2}$ **51.** $\frac{\sqrt{x}}{x}$

53. $\frac{4\sqrt{3}}{3}$ **55.** $\frac{x\sqrt{6}}{3}$ **57.** $\frac{2\sqrt{10x}}{5x}$ **59.** $\frac{\sqrt{2}}{2}$ **61.** $\frac{2x\sqrt{2y}}{y}$ **63.** $\frac{a\sqrt{6c}}{2bc}$ **65.** $\frac{2ac\sqrt[3]{b^2}}{b}$ **67.** $\frac{1+\sqrt{3}}{2}$ **69.** $\frac{5-\sqrt{5}}{4}$

71. $\frac{x+3\sqrt{x}}{x-9}$ **73.** $\frac{10+3\sqrt{5}}{11}$ **75.** $\frac{3\sqrt{x}+3\sqrt{y}}{x-y}$ **77.** $2+\sqrt{3}$ **79.** $\frac{11-4\sqrt{7}}{3}$ **81.** $\frac{a+2\sqrt{ab}+b}{a-b}$ **83.** $\frac{x+4\sqrt{x}+4}{x-4}$

85. $\frac{5-\sqrt{21}}{4}$ **87.** $\frac{\sqrt{x}-3x+2}{1-x}$ **89.** $\frac{8\sqrt{3}}{3}$ **91.** $\frac{11\sqrt{5}}{5}$ **93.** $-4\sqrt{3}$ **95.** $5\sqrt{3}$ **99.** $10\sqrt{3}$ **101.** $x+6\sqrt{x}+9$

103. 75   **105.** $\frac{5\sqrt{2}}{4}$ seconds; $\frac{5}{2}$ seconds   **111.** $-\frac{1}{8}$ **113.** $\frac{y-2}{y+2}$ **115.** $\frac{2x+1}{2x-1}$ **117.** $t^2+10t+25$ **119.** x

121. 7   **123.** $-4,-3$ **125.** $-6,-3$ **127.** $-5,-2$ **129.** $\frac{x(\sqrt{x-2}-4)}{x-18}$ **131.** $\frac{x(\sqrt{x+5}+5)}{x-20}$ **133.** $\frac{3(\sqrt{5x}-x)}{5-x}$

Problem Set 7.6

1. 4   **3.** $\varnothing$ **5.** 5   **7.** $\varnothing$ **9.** $\frac{39}{2}$ **11.** $\varnothing$ **13.** 5   **15.** 3   **17.** $-\frac{32}{3}$ **19.** 3, 4   **21.** $-1,-2$

23. $11, \frac{1}{2}$ does not check   **25.** 3, 7   **27.** $-\frac{1}{4}, 14$ **29.** -1 **31.** $\varnothing$ **33.** 7   **35.** 0, 3   **37.** -4 **39.** 8

41. 0   **43.** 9   **45.** 0   **47.** 8   **49.** $\varnothing$; 9 does not check   **51.** 0; 32 does not check   **53.** 6; -2 does not check

55. $\frac{1}{2}$ **57.** $\frac{1}{2}, 1$ **59.** 5, 13   **61.** $-\frac{3}{2}$ **63.** $\varnothing$; 1 does not check   **65.** $\frac{1}{2}$ **67.** $\varnothing$; 2 does not check   **69.** 1   **71.** -11

73.
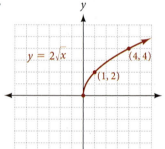

75.
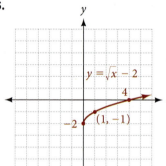

77.
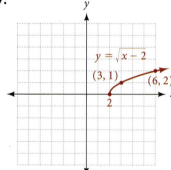

79.
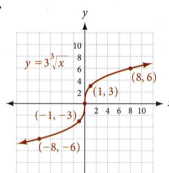

81.
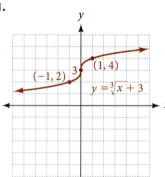

83.
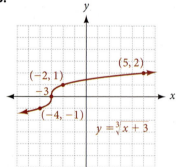

85. $h=100-16t^2$ **87.** $\frac{392}{121}\approx 3.24$ feet   **89.** 5 meters   **91.** 2,500 meters   **93.** $\sqrt{6}-2$ **95.** $x+10\sqrt{x}+25$

97. $\frac{x-3\sqrt{x}}{x-9}$ **99.** 5   **101.** $2\sqrt{3}$ **103.** -1 **105.** 1   **107.** 4   **109.** 2   **111.** $10-2x$ **113.** $2-3x$

115. $6+7x-20x^2$ **117.** $8x-12x^2$ **119.** $4+12x+9x^2$ **121.** $4-9x^2$

Problem Set 7.7

1. $6i$ **3.** $-5i$ **5.** $6i\sqrt{2}$ **7.** $-2i\sqrt{3}$ **9.** 1 **11.** -1 **13.** $-i$ **15.** $x = 3, y = -1$ **17.** $x = -2, y = -\dfrac{1}{2}$

19. $x = -8, y = -5$ **21.** $x = 7, y = \dfrac{1}{2}$ **23.** $x = \dfrac{3}{7}, y = \dfrac{2}{5}$ **25.** $5 + 9i$ **27.** $5 - i$ **29.** $2 - 4i$ **31.** $1 - 6i$

33. $2 + 2i$ **35.** $-1 - 7i$ **37.** $6 + 8i$ **39.** $2 - 24i$ **41.** $-15 + 12i$ **43.** $18 + 24i$ **45.** $10 + 11i$ **47.** $21 + 23i$

49. $-2 + 2i$ **51.** $2 - 11i$ **53.** $-21 + 20i$ **55.** $-2i$ **57.** $-7 - 24i$ **59.** 5 **61.** 40 **63.** 13 **65.** 164

67. $-3 - 2i$ **69.** $-2 + 5i$ **71.** $\dfrac{8}{13} + \dfrac{12}{13}i$ **73.** $-\dfrac{18}{13} - \dfrac{12}{13}i$ **75.** $-\dfrac{5}{13} + \dfrac{12}{13}i$ **77.** $\dfrac{13}{15} - \dfrac{2}{5}i$ **79.** $R = -11 - 7i$ ohms

81. $-\dfrac{3}{2}$ **83.** $-3, \dfrac{1}{2}$ **85.** $\dfrac{5}{4}$ or $\dfrac{4}{5}$ **87.** $\dfrac{1}{i} \cdot \dfrac{i}{i} = \dfrac{i}{i^2} = \dfrac{i}{-1} = -i$

89.
$$(1 + i)^2 - 2(1 + i) + 2 \overset{?}{=} 0$$
$$1 + 2i + i^2 - 2 - 2i + 2 \overset{?}{=} 0$$
$$1 - 1 - 2 + 2 + 2i - 2i \overset{?}{=} 0$$
$$0 = 0$$

91.
$$(2 + i)^3 - 11(2 + i) + 20 \overset{?}{=} 0$$
$$8 + 12i + 6i^2 + i^3 - 22 - 11i + 20 \overset{?}{=} 0$$
$$8 - 6 - 22 + 20 + 12i - i - 11i \overset{?}{=} 0$$
$$0 = 0$$

Chapter 7 Review

1. 7 **2.** -3 **3.** 2 **4.** 27 **5.** $2x^3y^2$ **6.** $\dfrac{1}{16}$ **7.** x^2 **8.** a^2b^4 **9.** $a^{7/20}$ **10.** $a^{5/12}b^{8/3}$

11. $12x + 11x^{1/2}y^{1/2} - 15y$ **12.** $a^{2/3} - 10a^{1/3} + 25$ **13.** $4x^{1/2} + 2x^{5/6}$ **14.** $2(x - 3)^{1/4}(4x - 13)$ **15.** $\dfrac{x + 5}{x^{1/4}}$

16. $2\sqrt{3}$ **17.** $5\sqrt{2}$ **18.** $2\sqrt[3]{2}$ **19.** $3x\sqrt{2}$ **20.** $4ab^2c\sqrt{5a}$ **21.** $2abc\sqrt[4]{2bc^2}$ **22.** $\dfrac{3\sqrt{2}}{2}$ **23.** $3\sqrt[3]{4}$

24. $\dfrac{4x\sqrt{21xy}}{7y}$ **25.** $\dfrac{2y\sqrt[3]{45x^2z^2}}{3z}$ **26.** $-2x\sqrt{6}$ **27.** $3\sqrt{3}$ **28.** $\dfrac{8\sqrt{5}}{5}$ **29.** $7\sqrt{2}$ **30.** $11a^2b\sqrt{3ab}$

31. $-4xy\sqrt[3]{xz^2}$ **32.** $\sqrt{6} - 4$ **33.** $x - 5\sqrt{x} + 6$ **34.** $3\sqrt{5} + 6$ **35.** $6 + \sqrt{35}$ **36.** $\dfrac{63 + 12\sqrt{7}}{47}$ **37.** 0 **38.** 3

39. 5 **40.** $\varnothing$

41.

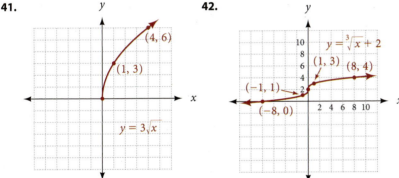

$y = 3\sqrt{x}$, points $(1, 3)$ and $(4, 6)$

42.

$y = \sqrt[3]{x} + 2$, points $(-8, 0)$, $(-1, 1)$, $(1, 3)$, $(8, 4)$

43. 1 **44.** $-i$ **45.** $x = -\dfrac{3}{2}, y = -\dfrac{1}{2}$ **46.** $x = -2, y = -4$ **47.** $9 + 3i$ **48.** $-7 + 4i$ **49.** $-6 + 12i$

50. $5 + 14i$ **51.** $12 + 16i$ **52.** 25 **53.** $1 - 3i$ **54.** $-\dfrac{6}{5} + \dfrac{3}{5}i$

Chapter 7 Cumulative Review

1. -17 **2.** 12 **3.** 5 **4.** $2\sqrt[4]{3}$ **5.** 6 **6.** $\dfrac{1}{4}$ **7.** $-\dfrac{2}{3}$ **8.** 13 **9.** $\dfrac{5}{9}$ **10.** $3x + 4y$ **11.** $\dfrac{x + 2}{x + 1}$ **12.** $\dfrac{a^2}{6b}$

13. $2x - 11\sqrt{x} + 15$ **14.** $22 + 4i$ **15.** $\dfrac{19}{25} - \dfrac{8}{25}i$ **16.** $\dfrac{2\sqrt{3} + 1}{11}$ **17.** $6x^2 - 19xy + 10y^2$ **18.** $3x^2 + 20x + 25$

19. $\dfrac{6}{y^2}$ **20.** $\dfrac{x + 2\sqrt{xy} + y}{x - y}$ **21.** $\dfrac{2}{3}$ **22.** $2, 3$ **23.** $\dfrac{5 \pm 2\sqrt{3}}{2}$ **24.** $-\dfrac{4}{3}, 2$ **25.** 4 **26.** $-4, 0, \dfrac{7}{3}$ **27.** $-\dfrac{5}{4}, \dfrac{1}{2}$

28. $\dfrac{5}{6}$ **29.** $-\dfrac{7}{3}, 3$ **30.** 3 **31.** $8 \le x \le 16$ **32.** $x < -2$ or $x > 5$

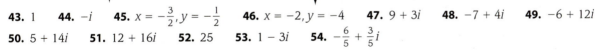

33. $(-2, 4)$ **34.** $(-8, -6)$ **35.** $\left(-\dfrac{36}{35}, \dfrac{116}{35}, -\dfrac{26}{5}\right)$ **36.** $\left(\dfrac{1}{2}, -\dfrac{3}{2}\right)$

37.

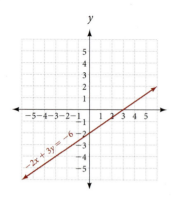

38.

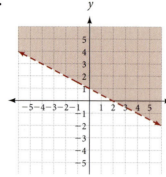

39. -7 **40.** -10 **41.** $3\sqrt[4]{2}$ **42.** $\dfrac{\sqrt[3]{2x^2y}}{x}$

43. $(5a - 2b)(5a + 2b)(25a^2 + 4b^2)$

44. $2(3a^2 - 1)(4a^2 + 3)$ **45.** $y = 1$ **46.** $z = -6$

47. $b^2 - 3b$ **48.** 2 **49.** -1 **50.** 28

51. 22,000,000 arrivals   **52.** 5,000,000 people

Chapter 7 Test

1. $\dfrac{1}{9}$ **2.** $\dfrac{7}{5}$ **3.** $a^{5/12}$ **4.** $\dfrac{x^{13/12}}{y}$ **5.** $7x^4y^5$ **6.** $2x^2y^4$ **7.** $2a$ **8.** $x^{n^2-n}y^{1-n^3}$ **9.** $6a^2 - 10a$ **10.** $16a^3 - 40a^{3/2} + 25$

11. $(3x^{1/3} - 1)(x^{1/3} + 2)$ **12.** $(3x^{1/3} - 7)(3x^{1/3} + 7)$ **13.** $5xy^2\sqrt{5xy}$ **14.** $2x^2y^2\sqrt[3]{5xy^2}$ **15.** $\dfrac{\sqrt{6}}{3}$ **16.** $\dfrac{2a^2b\sqrt{15bc}}{5c}$

17. $\dfrac{4 + x}{x^{1/2}}$ **18.** $\dfrac{3}{(x^2 - 3)^{1/2}}$ **19.** $-6\sqrt{3}$ **20.** $-ab\sqrt[3]{3}$ **21.** $x + 3\sqrt{x} - 28$ **22.** $21 - 6\sqrt{6}$ **23.** $\dfrac{5\sqrt{3} + 5}{2}$

24. $\dfrac{x - 2\sqrt{2x} + 2}{x - 2}$ **25.** 8; 1 does not check   **26.** -4 **27.** -3

28.

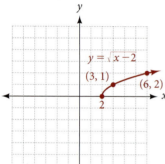

29.

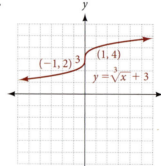

30. $x = \dfrac{1}{2}; y = 7$ **31.** $6i$ **32.** $17 - 6i$

33. $9 - 40i$ **34.** $-\dfrac{5}{13} - \dfrac{12}{13}i$

35. $i^{38} = (i^2)^{19} = (-1)^{19} = -1$

36. E   **37.** A   **38.** F   **39.** B   **40.** D   **41.** C

Chapter 8

Problem Set 8.1

1. ± 5 **3.** $\pm\dfrac{\sqrt{3}}{2}$ **5.** $\pm 2i\sqrt{3}$ **7.** $\pm\dfrac{3\sqrt{5}}{2}$ **9.** $-2, 3$ **11.** $\dfrac{-3 \pm 3i}{2}$ **13.** $-4 \pm 3i\sqrt{3}$ **15.** $\dfrac{3 \pm 2i}{2}$

17. $x^2 + 12x + 36 = (x + 6)^2$ **19.** $x^2 - 4x + 4 = (x - 2)^2$ **21.** $a^2 - 10a + 25 = (a - 5)^2$ **23.** $x^2 + 5x + \dfrac{25}{4} = \left(x + \dfrac{5}{2}\right)^2$

25. $y^2 - 7y + \dfrac{49}{4} = \left(y - \dfrac{7}{2}\right)^2$ **27.** $x^2 + \dfrac{1}{2}x + \dfrac{1}{16} = \left(x + \dfrac{1}{4}\right)^2$ **29.** $x^2 + \dfrac{2}{3}x + \dfrac{1}{9} = \left(x + \dfrac{1}{3}\right)^2$ **31.** $-3, -9$

33. $1 \pm 2i$ **35.** $4 \pm \sqrt{15}$ **37.** $\dfrac{5 \pm \sqrt{37}}{2}$ **39.** $1 \pm \sqrt{5}$ **41.** $\dfrac{4 \pm \sqrt{13}}{3}$ **43.** $\dfrac{3 \pm i\sqrt{71}}{8}$ **45. a.** No   **b.** $\pm 3i$

47. a. $0, 6$ **b.** $0, 6$ **49. a.** $-7, 5$ **b.** $-7, 5$ **51.** No   **53. a.** $\dfrac{7}{5}$ **b.** 3 **c.** $\dfrac{7 \pm 2\sqrt{2}}{5}$ **d.** $\dfrac{7}{5}$ **e.** 3 **55.** ± 1

57. ± 1 **59.** ± 1

61.

x	$f(x)$	$g(x)$	$h(x)$
-2	49	49	25
-1	25	25	13
0	9	9	9
1	1	1	13
2	1	1	25

63. 3   **65. a.** $-1, 6$ **b.** $-1, 6$ **c.** -6 **d.** -10 **67.** $\dfrac{\sqrt{3}}{2}$ inch, 1 inch

69. $\sqrt{2}$ inches   **71.** 781 feet   **73.** 7.3% to the nearest tenth

75. $20\sqrt{2} \approx 28$ feet   **77.** $3\sqrt{5}$ **79.** $3y^2\sqrt{3y}$ **81.** $3x^2y\sqrt[3]{2y^2}$ **83.** $\dfrac{3\sqrt{2}}{2}$

85. $\sqrt[3]{2}$ **87.** 13   **89.** 7   **91.** $\dfrac{1}{2}$ **93.** 13   **95.** $(3t - 2)(9t^2 + 6t + 4)$

97. $x = \pm 2a$ **99.** $x = \dfrac{-p \pm \sqrt{p^2 - 4q}}{2}$ **101.** $x = \dfrac{-p \pm \sqrt{p^2 - 12q}}{6}$

103. $(x - 5)^2 + (y - 3)^2 = 2^2$

Problem Set 8.2

1. a. $\dfrac{-2 \pm \sqrt{10}}{3}$ **b.** $\dfrac{2 \pm \sqrt{10}}{3}$ **c.** $\dfrac{-2 \pm i\sqrt{2}}{3}$ **d.** $\dfrac{-2 \pm \sqrt{10}}{2}$ **e.** $\dfrac{2 \pm i\sqrt{2}}{2}$ **3. a.** $1 \pm i$ **b.** $1 \pm 2i$ **c.** $-1 \pm i$ **5.** $1, 2$ **7.** $\dfrac{2 \pm i\sqrt{14}}{3}$

9. $0, 5$ **11.** $0, -\dfrac{4}{3}$ **13.** $\dfrac{3 \pm \sqrt{5}}{4}$ **15.** $-3 \pm \sqrt{17}$ **17.** $\dfrac{-1 \pm i\sqrt{5}}{2}$ **19.** 1 **21.** $\dfrac{1 \pm i\sqrt{47}}{6}$ **23.** $4 \pm \sqrt{2}$

25. $\dfrac{1}{2}, 1$ **27.** $-\dfrac{1}{2}, 3$ **29.** $\dfrac{-1 \pm i\sqrt{7}}{2}$ **31.** $1 \pm \sqrt{2}$ **33.** $\dfrac{-3 \pm \sqrt{5}}{2}$ **35.** $3, -5$ **37.** $2, -1 \pm i\sqrt{3}$

39. $-\dfrac{3}{2}, \dfrac{3 \pm 3i\sqrt{3}}{4}$ **41.** $\dfrac{1}{5}, \dfrac{-1 \pm i\sqrt{3}}{10}$ **43.** $0, \dfrac{-1 \pm i\sqrt{5}}{2}$ **45.** $0, 1 \pm i$ **47.** $0, \dfrac{-1 \pm i\sqrt{2}}{3}$ **49.** a and b

51. a. $\dfrac{5}{3}, 0$ **b.** $\dfrac{5}{3}, 0$ **53.** No, $2 \pm i\sqrt{3}$ **55.** Yes   **57. a.** $-1, 3$ **b.** $1 \pm i\sqrt{7}$ **c.** ± 2 **d.** $2 \pm 2\sqrt{2}$

59. a. $-2, \dfrac{5}{3}$ **b.** $\dfrac{3}{2}, \dfrac{-3 \pm 3i\sqrt{3}}{4}$ **c.** $\varnothing$ **d.** ± 1 **61.** $\dfrac{-5 + \sqrt{505}}{16} \approx 1.09$ seconds   **63.** 20 or 60 items

65. 0.49 centimeter (8.86 cm is impossible)   **67. a.** $\ell + w = 10$, $\ell w = 15$ **b.** 8.16 yards, 1.84 yards

   c. Two answers are possible because either dimension (long or short) may be considered the length.

69. $4y + 1 - \dfrac{2}{2y - 7}$ **71.** $x^2 + 7x + 12$ **73.** 5 **75.** $\dfrac{27}{125}$ **77.** $\dfrac{1}{4}$ **79.** $21x^3y$ **81.** 169 **83.** 0 **85.** ± 12

87. $x^2 - x - 6$ **89.** $x^3 - 4x^2 - 3x + 18$ **91.** $-2\sqrt{3}, \sqrt{3}$ **93.** $\dfrac{-\sqrt{2} \pm \sqrt{6}}{2}$ **95.** $-2i, i$

Problem Set 8.3

1. $D = 16$, two rational   **3.** $D = 0$, one rational   **5.** $D = 5$, two irrational   **7.** $D = 17$, two irrational

9. $D = 36$, two rational   **11.** $D = 116$, two irrational   **13.** ± 10 **15.** ± 12 **17.** 9 **19.** -16 **21.** $\pm 2\sqrt{6}$

23. $x^2 - 7x + 10 = 0$ **25.** $t^2 - 3t - 18 = 0$ **27.** $y^3 - 4y^2 - 4y + 16 = 0$ **29.** $2x^2 - 7x + 3 = 0$

31. $4t^2 - 9t - 9 = 0$ **33.** $6x^3 - 5x^2 - 54x + 45 = 0$ **35.** $10a^2 - a - 3 = 0$ **37.** $9x^3 - 9x^2 - 4x + 4 = 0$

39. $x^4 - 13x^2 + 36 = 0$ **41.** $f(x) = x^2 + x - 2$ **43.** $f(x) = x^2 - 6x + 7$ **45.** $f(x) = x^2 - 4x + 5$

47. $f(x) = 2x^2 - 10x + 9$ **49.** $f(x) = 2x^2 - 6x + 7$ **51.** $f(x) = x^3 - 4x^2 + 7x$ **53. a.** $g(x)$ **b.** $h(x)$ **c.** $f(x)$

55. $a^{11/2} - a^{9/2}$ **57.** $x^3 - 6x^{3/2} + 9$ **59.** $6x^{1/2} - 5x$ **61.** $5(x - 3)^{1/2}(2x - 9)$ **63.** $(2x^{1/3} - 3)(x^{1/3} - 4)$

65. $x^2 + 4x - 5$ **67.** $4a^2 - 30a + 56$ **69.** $32a^2 + 20a - 18$ **71.** $\pm \dfrac{1}{2}$ **73.** No solution   **75.** 1 **77.** $-2, 4$

79. $-2, \dfrac{1}{4}$ **81.** $-1, 5$ **83.** $1 + \sqrt{2} \approx 2.41; 1 - \sqrt{2} \approx -0.41$ **85.** $-1, \dfrac{1}{2}, 1$ **87.** $\dfrac{1}{2}, \sqrt{2} \approx 1.41, -\sqrt{2} \approx -1.41$

89. $D = -316$, no it is not possible

Problem Set 8.4

1. $1, 2$ **3.** $-8, -\dfrac{5}{2}$ **5.** $\pm 3, \pm 1$ **7.** $\pm 2, \pm\sqrt{3}$ **9.** $\pm 3, \pm i\sqrt{3}$ **11.** $\pm 2i, \pm i\sqrt{5}$ **13.** $\dfrac{7}{2}, 4$ **15.** $-\dfrac{9}{8}, \dfrac{1}{2}$

17. $\pm\dfrac{\sqrt{30}}{6}, \pm i$ **19.** $\pm\dfrac{\sqrt{21}}{3}, \pm\dfrac{i\sqrt{21}}{3}$ **21.** $4, 25$ **23.** 25; 9 does not check   **25.** $\dfrac{25}{9}; \dfrac{49}{4}$ does not check

27. 16; 4 does not check   **29.** 9; 36 does not check   **31.** $\dfrac{1}{4}$; 25 does not check   **33.** $27, 38$ **35.** $4, 12$

37. $\pm 2, \pm 2i$ **39.** $\pm 2, \pm 2i\sqrt{2}$ **41.** $\dfrac{1}{2}, 1$ **43.** $\pm\dfrac{\sqrt{5}}{5}$ **45.** $-216, 8$ **47.** $\pm 3i\sqrt{3}$ **49.** $t = \dfrac{v \pm \sqrt{v^2 + 64h}}{32}$

51. $x = \dfrac{-4 \pm 2\sqrt{4 - k}}{k}$ **53.** $x = -y$ **55.** $t = \dfrac{1 \pm \sqrt{1 + h}}{4}$

57. a.

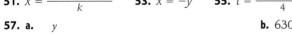

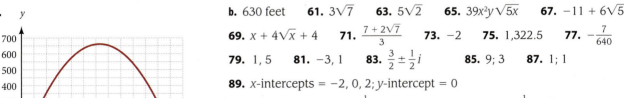

b. 630 feet   **61.** $3\sqrt{7}$ **63.** $5\sqrt{2}$ **65.** $39x^2y\sqrt{5x}$ **67.** $-11 + 6\sqrt{5}$

69. $x + 4\sqrt{x} + 4$ **71.** $\dfrac{7 + 2\sqrt{7}}{3}$ **73.** -2 **75.** $1{,}322.5$ **77.** $-\dfrac{7}{640}$

79. $1, 5$ **81.** $-3, 1$ **83.** $\dfrac{3}{2} \pm \dfrac{1}{2}i$ **85.** 9; 3   **87.** 1; 1

89. x-intercepts $= -2, 0, 2$; y-intercept $= 0$

91. x-intercepts $= -3, -\dfrac{1}{3}, 3$; y-intercept $= -9$ **93.** $\dfrac{1}{2}$ and -1

Problem Set 8.5

1. x-intercepts $= -3, 1$;
Vertex $= (-1, -4)$

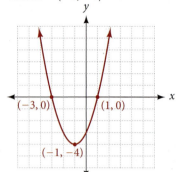

3. x-intercepts $= -5, 1$;
Vertex $= (-2, 9)$

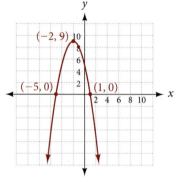

5. x-intercepts $= -1, 1$;
Vertex $= (0, -1)$

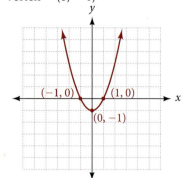

7. x-intercepts $= 3, -3$;
Vertex $= (0, 9)$

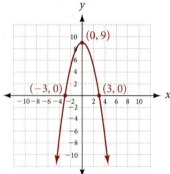

9. x-intercepts $= -1, 3$;
Vertex $= (1, -8)$

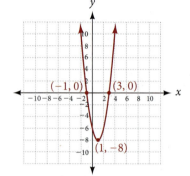

11. x-intercepts $= 1 - \sqrt{5}, 1 + \sqrt{5}$;
Vertex $= (1, -5)$

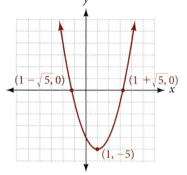

13. Vertex $= (2, -8)$

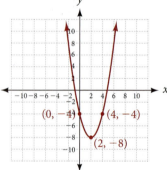

15. Vertex $= (1, -4)$

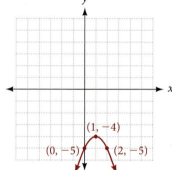

17. Vertex $= (0, 1)$

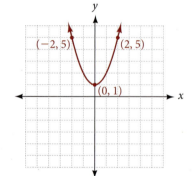

19. Vertex $= (0, -3)$

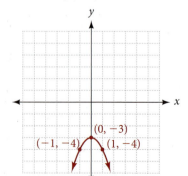

21. $(3, -4)$ lowest   **23.** $(1, 9)$ highest   **25.** $(2, 16)$ highest   **27.** $(-4, 16)$ highest

29.

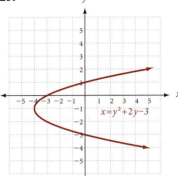

31.

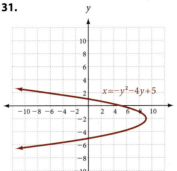

33.

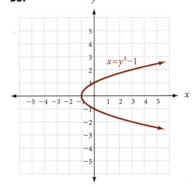

35.

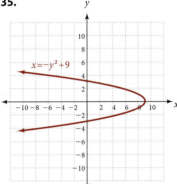

37.

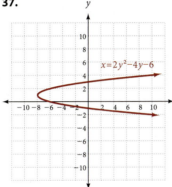

39.

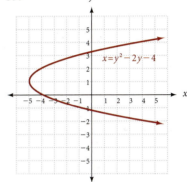

41. 875 patterns; maximum profit $731.25

43. The ball is in her hand when $h(t) = 0$, which means $t = 0$ or $t = 2$ seconds. Maximum height is $h(1) = 16$ feet.

45. Maximum $R = \$3{,}600$ when $p = \$6.00$ **47.** Maximum $R = \$7{,}225$ when $p = \$8.50$

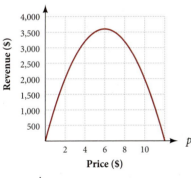

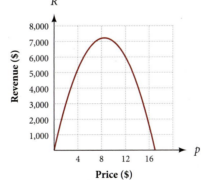

49. $y = -\dfrac{1}{135}(x - 90)^2 + 60$

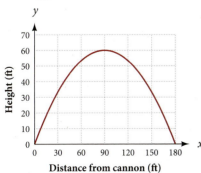

51. $-2, 4$ **53.** $-\dfrac{1}{2}, \dfrac{2}{3}$ **55.** 3   **57.** $1 - i$ **59.** $27 + 5i$ **61.** $\dfrac{1}{10} + \dfrac{3}{10}i$ **63.** $y = (x - 2)^2 - 4$

65. a. $20°$ or $70°$ **b.** At approximately 170 feet   **c.** Approximately 42 or 43 feet   **d.** Approximately $10°$ **e.** Quadratic

Problem Set 8.6

1.  -3 2

3.  -3 4

5.  -3 -2

7. $\frac{1}{3}$ $\frac{1}{2}$

9.  -3 3

11.  $-\frac{3}{2}$ $\frac{3}{2}$

13. -1 $\frac{3}{2}$

15. All real numbers   **17.** No solution; $\varnothing$ **19.** 2 3 4

21. -3 -2 -1

23. $\varnothing$ **25.** $x = 1$ **27.** All real numbers   **29.** $x \neq 1$ **31.** $\varnothing$

33. $x \neq 3$ **35.** -4 1 **37.** -6 $\frac{8}{3}$

39. 2 6 **41.** -3 2 4 **43.** 2 3 4

45. a. $x - 1 > 0$ **b.** $x - 1 \geq 0$ **c.** $x - 1 \geq 0$ **47. a.** $-2 < x < 2$ **b.** $x < -2$ or $x > 2$ **c.** $x = -2$ or $x = 2$

49. a. $-2 < x < 5$ **b.** $x < -2$ or $x > 5$ **c.** $x = -2$ or $x = 5$

51. a. $x < -1$ or $1 < x < 3$ **b.** $-1 < x < 1$ or $x > 3$ **c.** $x = -1$ or $x = 1$ or $x = 3$

53. $x \geq 4$; the width is at least 4 inches   **55.** $5 \leq p \leq 8$; charge at least \$5 but no more than \$8 for each set

57. 1.5625   **59.** 0.6549   **61.** $\frac{5}{3}$ **63.** 6; 1 does not check   **65.**

67. $1 - \sqrt{2}$ $1 + \sqrt{2}$

69. $4 - \sqrt{3}$ $4 + \sqrt{3}$

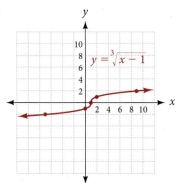

$y = \sqrt[3]{x - 1}$

Chapter 8 Review

1. $0, 5$ **2.** $0, \frac{4}{3}$ **3.** $\frac{4 \pm 7i}{3}$ **4.** $-3 \pm \sqrt{3}$ **5.** $-5, 2$ **6.** $-3, -2$ **7.** 3 **8.** 2 **9.** $\frac{-3 \pm \sqrt{3}}{2}$ **10.** $\frac{3 \pm \sqrt{5}}{2}$

11. $-5, 2$ **12.** $0, \frac{9}{4}$ **13.** $\frac{2 \pm i\sqrt{15}}{2}$ **14.** $1 \pm \sqrt{2}$ **15.** $-\frac{4}{3}, \frac{1}{2}$ **16.** 3 **17.** $5 \pm \sqrt{7}$ **18.** $0, \frac{5 \pm \sqrt{21}}{2}$

19. $-1, \frac{3}{5}$ **20.** $3, \frac{-3 \pm 3i\sqrt{3}}{2}$ **21.** $\frac{1 \pm i\sqrt{2}}{3}$ **22.** $\frac{3 \pm \sqrt{29}}{2}$ **23.** 100 or 170 items   **24.** 20 or 60 items

25. $D = 0$; 1 rational   **26.** $D = 0$; 1 rational   **27.** $D = 25$; 2 rational   **28.** $D = 361$; 2 rational

29. $D = 5$; 2 irrational   **30.** $D = 5$; 2 irrational   **31.** $D = -23$; 2 complex   **32.** $D = -87$; 2 complex

33. ± 20 **34.** ± 20 **35.** 4   **36.** 4   **37.** 25   **38.** 49   **39.** $x^2 - 8x + 15 = 0$ **40.** $x^2 - 2x - 8 = 0$

41. $2y^2 + 7y - 4 = 0$ **42.** $t^3 - 5t^2 - 9t + 45 = 0$ **43.** $-4, 12$ **44.** $-\frac{3}{4}, -\frac{1}{6}$ **45.** $\pm 2, \pm i\sqrt{3}$

46. 4; 1 does not check   **47.** $\frac{9}{4}, 16$ **48.** 4   **49.** 4   **50.** 7   **51.** $t = \frac{5 \pm \sqrt{25 + 16h}}{16}$ **52.** $t = \frac{v \pm \sqrt{v^2 + 640}}{32}$

53.

$y = x^2 - 6x + 8$

$(3, -1)$

54.

$y = x^2 - 4$

55. -1 2

56. $\frac{2}{3}$ 4

57. -4 $\frac{3}{2}$

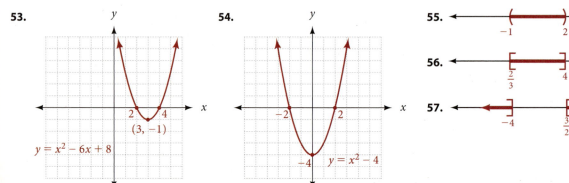

Chapter 8 Cumulative Review

1. 0   **2.** 88   **3.** $-\frac{8}{27}$ **4.** $\frac{25}{16}$ **5.** $-7x + 10$ **6.** $10x - 37$ **7.** $\frac{x^3}{y^7}$ **8.** $\frac{a^8}{b^2}$ **9.** $2\sqrt[3]{4}$ **10.** $5x\sqrt{2x}$ **11.** $\frac{9}{20}$

12. $\frac{9}{4}$ **13.** $\frac{1}{7}$ **14.** $\frac{33}{5}$ **15.** $x - 6y$ **16.** $x - 3$ **17.** $1 - 3i$ **18.** $\frac{23}{13} + \frac{11i}{13}$ **19.** 15   **20.** $\frac{35}{6}$ **21.** $-9, 9$

22. No solution   **23.** $-5, -3$ **24** $a = 2$ **25.** $\frac{4 \pm 3\sqrt{2}}{3}$ **26.** $0, \frac{5 \pm \sqrt{37}}{6}$ **27.** $-\frac{3}{2}, -1$ **28.** $\frac{-1 \pm i\sqrt{47}}{12}$ **29.** $-2, -3$

30. $\frac{9}{4}$ **31.** 81   **32.** $-8, 27$ **33.** $8 \le x \le 20$    **34.** $x \le -\frac{1}{2}, x \ge 2$

35. $-4 < x < 2$ **36.** $x < -3$ or $x \ge 1$ **37.** No solution   **38.** No solution

39. 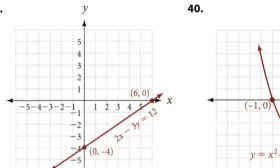   **40.**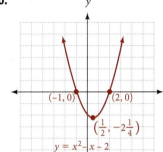

41. $y = \frac{4}{3}x - \frac{2}{3}$ **42.** $y = -2x + 7$

43. $(x + 4 - y)(x + 4 + y)$ **44.** $(x - 5)(x - 2)$

45. $\frac{7\sqrt[3]{3}}{3}$ **46.** $\frac{3 - \sqrt{3}}{2}$ **47.** $\{2, 4\}$ **48.** $\{0, 6\}$

49. 25°, 55°, 100°   **50.** $6,000 at 5%; $2,000 at 4%

51. $\frac{25}{16}$ **52.** 6   **53.** $3,012.50   **54.** $3,594

Chapter 8 Test

1. $-\frac{9}{2}, \frac{1}{2}$ **2.** $3 \pm i\sqrt{2}$ **3.** $5 \pm 2i$ **4.** $1 \pm i\sqrt{2}$ **5.** $\frac{5}{2}, \frac{-5 \pm 5i\sqrt{3}}{4}$ **6.** $-1 \pm i\sqrt{5}$ **7.** $r = -1 \pm \frac{\sqrt{A}}{8}$ **8.** $2 \pm \sqrt{2}$

9. 2 in., $2\sqrt{3}$ in.   **10.** $\frac{1}{2}$ second, $\frac{3}{2}$ seconds   **11.** 15 or 100 cups   **12.** 9   **13.** Two rational solutions

14. $3x^2 - 13x - 10 = 0$ **15.** $x^3 - 7x^2 - 4x + 28 = 0$ **16** $\pm\frac{1}{2}i, \pm\sqrt{2}$ **17.** $\frac{1}{2}, 1$ **18.** $\frac{1}{4}, 9$ **19.** $t = \frac{7 \pm \sqrt{16h + 49}}{16}$

20. 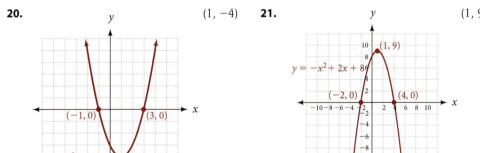 $(1, -4)$ **21.**  $(1, 9)$

22. $y = -(x + 1)^2 + 4$ **23.** $900   **24.** $-2 \le x \le 3$

25. $x < -3$ or $x > \frac{1}{2}$

26. a. $x < -2$ or $-1 < x < 1$

  b. $-2 < x < -1$ or $x > 1$

  c. $x = -2$ or $x = -1$ or $x = 1$

Chapter 9

Problem Set 9.1

1. 1   **3.** 2   **5.** $\frac{1}{27}$ **7.** 13

9.

11.

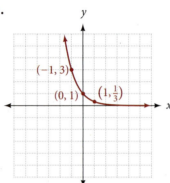

13.

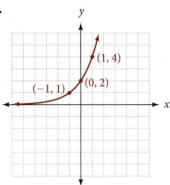

15.

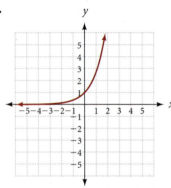

17.

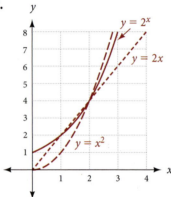

19.

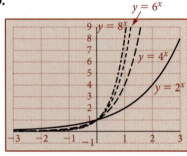

21. $h = A\left(\frac{1}{2}\right)^n$; $h(8) = 0.039$ feet   **23.** After 8 days, 700 micrograms; After 11 days, $1{,}400 \cdot 2^{-11/8} \approx 539.8$ micrograms

25. After 4 hours, 150 mg; after 8 hours, 37.5 mg

27. a. $A(t) = 1{,}200\left(1 + \frac{.06}{4}\right)^{4t}$ **b.** \$1,932.39 **c.** About 11.64 years **d.** \$1,939.29

29. a. The function underestimated the expenditures by \$69 billion.   **b.** \$4,123 billion; \$4,577 billion; \$5,080 billion

31. $V(t)$

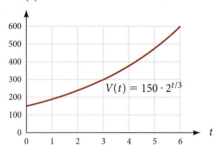

33. After 6 years   **35. a.** \$129,138.48   **b.** $\{t \mid 0 \le t \le 6\}$

c.

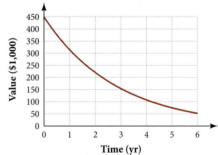

d. $\{V(t) \mid 52{,}942.05 \le V(t) \le 450{,}000\}$

e. After approximately 4 years 8 months

37. $D = \{1, 3, 4\}$, $R = \{2, 4, 1\}$, a function   **39.** $\left\{x \mid x \ge -\frac{1}{3}\right\}$ **41.** -18 **43.** 2   **45.** $(2x - 3)(x + 4)$

47. $x(5x - 2)(x - 4)$ **49.** $x(x - 4)(x^2 + 4x + 16)$ **51.** $y = \frac{x + 3}{2}$ **53.** $y = \pm\sqrt{x + 2}$ **55.** $y = \frac{2x - 4}{x - 1}$

57. $y = x^2 + 3$

Problem Set 9.2

1. $f^{-1}(x) = \frac{x+1}{3}$ **3.** $f^{-1}(x) = \sqrt[3]{x}$ **5.** $f^{-1}(x) = \frac{x-3}{x-1}$ **7.** $f^{-1}(x) = 4x + 3$ **9.** $f^{-1}(x) = 2(x + 3) = 2x + 6$

11. $f^{-1}(x) = \frac{1-x}{3x-2}$

13.
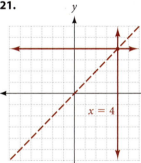

15.
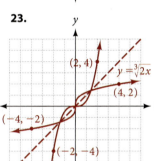

17.
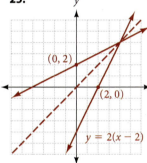

19.
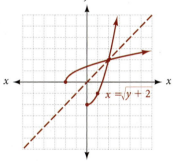

21.
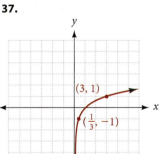

23.
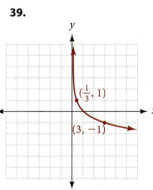

25.
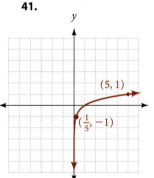

27.
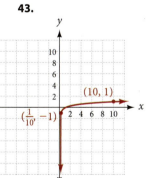

29. a. Yes **b.** No **c.** Yes   **31. a.** 4 **b.** $\frac{4}{3}$ **c.** 2 **d.** 2   **33.** $f^{-1}(x) = \frac{1}{x}$ **35.** $-2, 3$ **37.** 25   **39.** $5 \pm \sqrt{17}$

41. $1 \pm i$ **43.** $\frac{1}{9}$ **45.** $\frac{2}{3}$ **47.** $\sqrt[3]{4}$ **49.** 3   **51.** 4   **53.** 4   **55.** 1   **57.** $f^{-1}(x) = \frac{x-5}{3}$

59. $f^{-1}(x) = \sqrt[3]{x-1}$ **61.** $f^{-1}(x) = \frac{2x-4}{x-1}$

Problem Set 9.3

1. $\log_2 16 = 4$ **3.** $\log_5 125 = 3$ **5.** $\log_{10} 0.01 = -2$ **7.** $\log_2 \frac{1}{32} = -5$ **9.** $\log_{1/2} 8 = -3$ **11.** $\log_3 27 = 3$

13. $10^2 = 100$ **15.** $2^6 = 64$ **17.** $8^0 = 1$ **19.** $10^{-3} = 0.001$ **21.** $6^2 = 36$ **23.** $5^{-2} = \frac{1}{25}$ **25.** 9   **27.** $\frac{1}{125}$

29. 4   **31.** $\frac{1}{3}$ **33.** 2   **35.** $\sqrt[3]{5}$

37.

39.

41.

43.

45. 4   **47.** $\frac{3}{2}$ **49.** 3   **51.** 1   **53.** 0   **55.** 0   **57.** $\frac{1}{2}$ **59.** 7   **61.** 10^{-6} **63.** 2   **65.** 10^8 times as large

67. $25; 5$ **69.** $1; 1$ **71.** $\pm 3i$ **73.** $\pm 2i$ **75.** $\dfrac{-2 \pm \sqrt{10}}{2}$ **77.** $\dfrac{2 \pm i\sqrt{3}}{2}$ **79.** $5, \dfrac{-5 \pm 5i\sqrt{3}}{2}$ **81.** $0, 5, -1$

83. $\dfrac{3 \pm \sqrt{29}}{2}$ **85.** 4 **87.** $-4, 2$ **89.** $-\dfrac{11}{8}$ **91.** $2^3 = (x + 2)(x)$ **93.** $3^4 = \dfrac{x - 2}{x + 1}$

95. a.

x	$f(x)$
-1	$\frac{1}{8}$
0	1
1	8
2	64

b.

x	$f^{-1}(x)$
$\frac{1}{8}$	-1
1	0
8	1
64	2

c. $f(x) = 8^x$ **d.** $f^{-1}(x) = \log_8 x$

Problem Set 9.4

1. $\log_3 4 + \log_3 x$ **3.** $\log_6 5 - \log_6 x$ **5.** $5 \log_2 y$ **7.** $\dfrac{1}{3} \log_9 z$ **9.** $2 \log_6 x + 4 \log_6 y$ **11.** $\dfrac{1}{2} \log_5 x + 4 \log_5 y$

13. $\log_b x + \log_b y - \log_b z$ **15.** $\log_{10} 4 - \log_{10} x - \log_{10} y$ **17.** $2 \log_{10} x + \log_{10} y - \dfrac{1}{2} \log_{10} z$

19. $3 \log_{10} x + \dfrac{1}{2} \log_{10} y - 4 \log_{10} z$ **21.** $\dfrac{2}{3} \log_b x + \dfrac{1}{3} \log_b y - \dfrac{4}{3} \log_b z$ **23.** $\log_b xz$ **25.** $\log_3 \dfrac{x^2}{y^3}$ **27.** $\log_{10} \sqrt{x} \sqrt[3]{y}$

29. $\log_2 \dfrac{x^3\sqrt{y}}{z}$ **31.** $\log_2 \dfrac{\sqrt{x}}{y^3 z^4}$ **33.** $\log_{10} \dfrac{x^{3/2}}{y^{3/4} z^{4/5}}$ **35.** $\dfrac{2}{3}$ **37.** 18 **39.** $3; -1$ does not check   **41.** 3

43. $4; -2$ does not check   **45.** $4; -1$ does not check   **47.** $\dfrac{5}{3}; -\dfrac{5}{2}$ does not check

51. $N = 2$ for both   **53.** $\dfrac{x^2}{3} + \dfrac{y^2}{36}$ **55.** $\dfrac{x^2}{4} + \dfrac{y^2}{25}$ **57.** $D = -7;$ two complex   **59.** $x^2 - 2x - 15 = 0$

61. $3y^2 - 11y + 6 = 0$ **63.** 1 **65.** 1 **67.** 4 **69.** 2.5×10^{-6} **71.** 51

Problem Set 9.5

1. 2.5775 **3.** 1.5775 **5.** 3.5775 **7.** -1.4225 **9.** 4.5775 **11.** 2.7782 **13.** 3.3032 **15.** -2.0128

17. -1.5031 **19.** -0.3990 **21.** 759 **23.** 0.00759 **25.** $1,430$ **27.** 0.00000447 **29.** 0.0000000918

31. 10^{10} **33.** 10^{-10} **35.** 10^{20} **37.** $\dfrac{1}{100}$ **39.** $1,000$ **41.** 1 **43.** 5 **45.** x **47.** $\ln 10 + 3t$ **49.** $\ln A - 2t$

51. 2.7080 **53.** -1.0986 **55.** 2.1972 **57.** 2.7724 **59. a.** 1.1886 **b.** 0.2218 **61. a.** 0.6667 **b.** -0.3010

63. 3.19 **65.** 1.78×10^{-5} **67.** 3.16×10^5 **69.** 2.00×10^8 **71.** 10 times larger

73.

Location	Date	Magnitude, M	Shock Wave, T
Moresby Island	January 23	4.0	1.00×10^4
Vancouver Island	April 30	5.3	1.99×10^5
Quebec City	June 29	3.2	1.58×10^3
Mould Bay	November 13	5.2	1.58×10^5
St. Lawrence	December 14	3.7	5.01×10^3

75. 12.9% **77.** 5.3% **79**

x	$(1 + x)^{1/x}$
1	2
0.5	2.25
0.1	2.5937
0.01	2.7048
0.001	2.7169
0.0001	2.7181
0.00001	2.7183

81. $0, -3$ **83.** $-4, -2$ **85.** $\pm 2, \pm i\sqrt{2}$ **87.** $1, \dfrac{9}{4}$ **89.** 0.7 **91.** 3.125 **93.** 1.2575

95. $t \cdot \log(1.05)$ **97.** $0.05t$

Problem Set 9.6

1. 1.4650 **3.** 0.6826 **5.** -1.5440 **7.** -0.6477 **9.** -0.3333 **11.** 2 **13.** -0.1845 **15.** 0.1845

17. 1.6168 **19.** 2.1131 **21.** 1.3333 **23.** 0.75 **25.** 1.3917 **27.** 0.7186 **29.** 2.6356 **31.** 4.1632

33. 5.8435 **35.** -1.0642 **37.** 2.3026 **39.** 10.7144 **41.** 11.7 years   **43.** 9.25 years   **45.** 8.75 years

47. 18.58 years   **49.** 11.55 years   **51.** 18.3 years   **53.** $(-2, -23);$ lowest point   **55.** $\left(\dfrac{3}{2}, 9\right);$ highest point

57. 2 seconds, 64 feet   **59.** 13.9 years later or toward the end of 2023   **61.** 27.5 years   **63.** $t = \dfrac{1}{r} \ln \dfrac{A}{P}$

65. $t = \dfrac{1}{k} \dfrac{\log P - \log A}{\log 2}$ **67.** $t = \dfrac{\log A - \log P}{\log (1 - r)}$ **69. a.** 1 **b.** 2 **c.** 5 **d.** 0 **e.** 1 **f.** 2 **g.** 2 **h.** 5

Chapter 9 Review

1. 16   **2.** $\frac{1}{2}$ **3.** $\frac{1}{9}$ **4.** -5 **5.** $\frac{5}{6}$ **6.** 7

7.
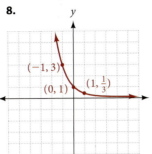

8.
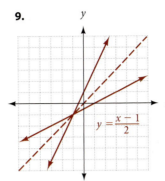

9.
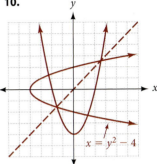

10.

11. $f^{-1}(x) = \dfrac{x-3}{2}$ **12.** $y = \pm\sqrt{x+1}$ **13.** $f^{-1}(x) = 2x - 4$ **14.** $y = \pm\sqrt{\dfrac{4-x}{2}}$ **15.** $\log_3 81 = 4$ **16.** $\log_7 49 = 2$

17. $\log_{10} 0.01 = -2$ **18.** $\log_2 \dfrac{1}{8} = -3$ **19.** $2^3 = 8$ **20.** $3^2 = 9$ **21.** $4^{1/2} = 2$ **22.** $4^1 = 4$ **23.** 25

24. $\frac{3}{4}$ **25.** 10

26.
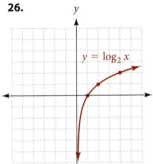

27.
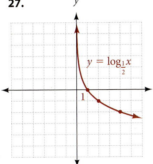

28. 2   **29.** $\frac{2}{3}$ **30.** 0   **31.** $\log_2 5 + \log_2 x$

32. $\log_{10} 2 + \log_{10} x - \log_{10} y$

33. $\frac{1}{2}\log_a x + 3\log_a y - \log_a z$

34. $2\log_{10} x - 3\log_{10} y - 4\log_{10} z$ **35.** $\log_2 xy$

36. $\log_3 \dfrac{x}{4}$ **37.** $\log_a \dfrac{25}{3}$ **38.** $\log_2 \dfrac{x^3 y^2}{z^4}$ **39.** 2

40. 6   **41.** 3; -1 does not check   **42.** 3

43. 3; -2 does not check   **44.** 4; -1 does not check

45. 2.5391   **46.** -0.1469 **47.** 9,230   **48.** 0.0251

49. 1   **50.** 0   **51.** 2   **52.** -4 **53.** 2.1   **54.** 5.1   **55.** 2.0×10^{-3} **56.** 3.2×10^{-8} **57.** $\frac{3}{2}$

58. $x = \dfrac{1}{3}\left(\dfrac{\log 5}{\log 4} - 2\right) \approx -0.28$ **59.** 0.75   **60.** 2.43   **61.** About 4.67 years

Chapter 9 Cumulative Review

1. 32   **2.** $24x - 26$ **3.** 9   **4.** $\frac{17}{8}$ **5.** $3\sqrt{3}$ **6.** $18 + 14\sqrt{3}$ **7.** $-2 + 7i$ **8.** $2i$ **9.** 0   **10.** 0   **11.** $\dfrac{x-4}{x-2}$

12. $\dfrac{x^2 + x + 1}{x+1}$ **13.** $\frac{4}{7}$ **14.** $\frac{4}{5}$ **15.** $12t^4 - \dfrac{1}{12}$ **16.** $3x^3 - 11x^2 + 4$ **17.** $y + 5 - \dfrac{3}{y+2}$ **18.** $3x + 7 + \dfrac{10}{3x-4}$

19. $\dfrac{1}{(x+4)(x+2)}$ **20.** $\dfrac{24}{(4x+3)(4x-3)}$ **21.** $\frac{8}{3}$ **22.** 4   **23.** $-\frac{1}{7}, \frac{1}{4}$ **24.** 1   **25.** $\dfrac{1 \pm \sqrt{29}}{2}$ **26.** $-\dfrac{12}{5}$ **27.** 1, 4

28. $-\frac{1}{4}, \frac{2}{3}$ **29.** 27   **30.** 10   **31.** $\varnothing$ **32.** 7   **33.** $\left(\frac{8}{3}, \frac{23}{3}\right)$ **34.** $(8, -5)$ **35.** $\left(8, -2, -\frac{7}{2}\right)$ **36.** $\left(3, 0, \frac{1}{2}\right)$

37.
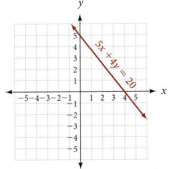

38.
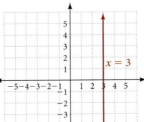

39. $7 - y = -x$ **40.** $9a - 4b < 9a + 4b$ **41.** 9.72×10^{-5}

42. 4.7×10^7 **43.** 3.64×10^{13} **44.** 8.6×10^{-15}

45. $(2a - 5)(4a^2 + 10a + 25)$ **46.** $2(5a^2 + 2)(5a^2 - 1)$

47. Domain = $\{2, -3\}$; Range = $\{-3, -1, 3\}$; not a function

48. Domain = $\{3, 4, 6\}$; Range = $\{4, 5\}$; a function

49. 300 million   **50.** 6,550 million

Chapter 9 Test

1.

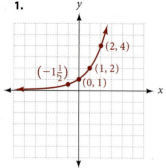

2.

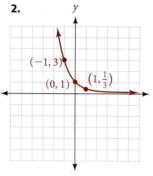

3.

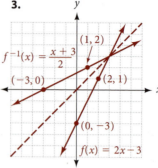

4.

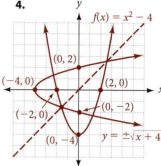

5. 64   **6.** $\sqrt{5}$

7.

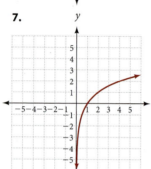

8.

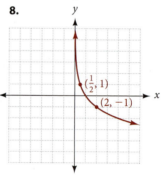

9. 0.67   **10.** 1.56   **11.** 4.37   **12.** -1.91 **13.** 3.83   **14.** -3.07 **15.** $\log_2 6 + 2\log_2 x - \log_2 y$

16. $\frac{1}{2}\log x - 4\log y - \frac{1}{5}\log z$ **17.** $\log_3 \frac{x^2}{\sqrt{y}}$ **18.** $\log \frac{\sqrt[3]{x}}{yz^2}$ **19.** 70,404.432   **20.** 0.002   **21.** 1.465   **22.** 1.25   **23.** 15

24. 8   **25.** 6.18   **26.** \$651.56   **27.** 13.87 years   **28.** yes   **29.** no   **30.** yes   **31.** yes   **32.** no   **33.** no

Chapter 10

Problem Set 10.1

1. 5   **3.** $\sqrt{106}$ **5.** $\sqrt{61}$ **7.** $\sqrt{130}$ **9.** 3 or -1 **11.** 3   **13.** $(x - 2)^2 + (y - 3)^2 = 16$

15. $(x - 3)^2 + (y + 2)^2 = 9$ **17.** $(x + 5)^2 + (y + 1)^2 = 5$ **19.** $x^2 + (y + 5)^2 = 1$ **21.** $x^2 + y^2 = 4$

23. Center $= (0, 0)$,
Radius $= 2$

25. Center $= (1, 3)$,
Radius $= 5$

27. Center $= (-2, 4)$,
Radius $= 2\sqrt{2}$

29. Center $= (-1, -1)$
Radius $= 1$

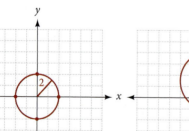

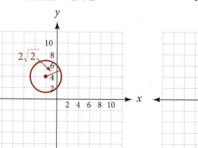

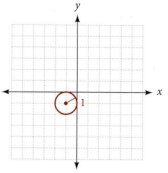

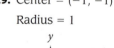

31. Center = $(0, 3)$,
Radius = 4

33. Center = $(2, 3)$,
Radius = 3

35. Center = $\left(-1, -\frac{1}{2}\right)$,
Radius = 2

37. $(x - 3)^2 + (y - 4)^2 = 25$

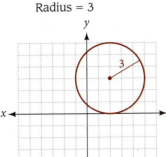

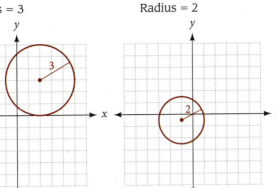

39. A: $\left(x - \frac{1}{2}\right)^2 + (y - 1)^2 = \frac{1}{4}$
B: $(x - 1)^2 + (y - 1)^2 = 1$
C: $(x - 2)^2 + (y - 1)^2 = 4$

41. A: $(x + 8)^2 + y^2 = 64$
B: $x^2 + y^2 = 64$
C: $(x - 8)^2 + y^2 = 64$

43. A: $(x + 4)^2 + (y - 4)^2 = 16$
B: $x^2 + y^2 = 16$
C: $(x - 4)^2 + (y + 4)^2 = 16$

45. $(x - 500)^2 + (y - 132)^2 = 120^2$ **47.** $f^{-1}(x) = \log_3 x$ **49.** $f^{-1}(x) = \frac{x - 3}{2}$ **51.** $f^{-1}(x) = 5x - 3$ **53.** $y = \pm 3$

55. $y = \pm 2i$ **57.** $x = \pm 3i$ **59.** $\frac{x^2}{9} + \frac{y^2}{4}$ **61.** x-intercept 4, y-intercept -3 **63.** ± 2.4 **65.** $(x - 2)^2 + (y - 3)^2 = 4$

67. $(x - 2)^2 + (y - 3)^2 = 4$ **69.** 5   **71.** 5

73. $y = \sqrt{9 - x^2}$ corresponds to the top half; $y = -\sqrt{9 - x^2}$ to the bottom half.

Problem Set 10.2

1.

3.

5.

7.

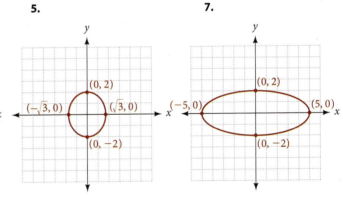

9.

11.

13.

15.

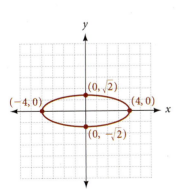

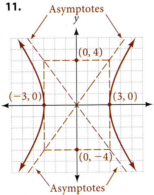

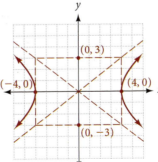

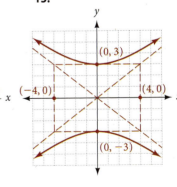

17.

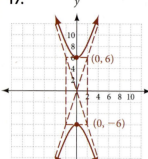

19.

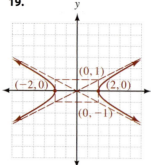

21.

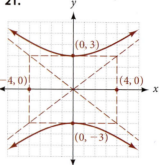

23. x-intercepts $= \pm 3$, y-intercepts $= \pm 2$ **25.** x-intercepts $= \pm 0.2$, no y-intercepts

27. x-intercepts $= \pm\frac{3}{5}$, y-intercepts $= \pm\frac{2}{5}$

29.

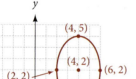

31.

33.

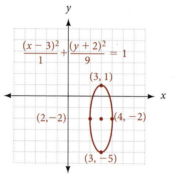

35.

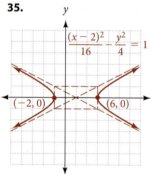

37.

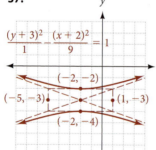

39.

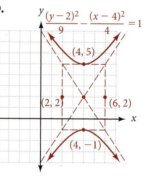

41. $\pm\frac{9}{5}$ **43.** ± 3.2 **45.** $y = \frac{3}{4}x,\ y = -\frac{3}{4}x$ **47.** $\frac{x^2}{16} + \frac{y^2}{4} = 1$

49. The equation is $\frac{x^2}{20^2} + \frac{y^2}{10^2} = 1$. A 6-foot man could walk upright under the arch anywhere between 16 feet to the left and 16 feet to the right of the center.

51. About 5.3 feet wide   **53.** 1   **55.** -2 **57.** $\frac{4x}{(x-2)(x+2)}$ **59.** $(0, 0)$ **61.** $4y^2 + 16y + 16$

63. $x = 2y + 4$ **65.** $-x^2 + 6$ **67.** $(5y + 6)(y + 2)$ **69.** ± 2 **71.** ± 2

Problem Set 10.3

1.
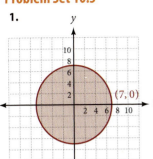

3.
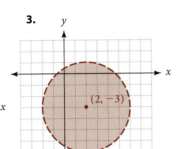

5.
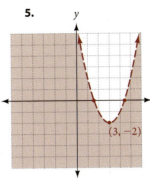

7.
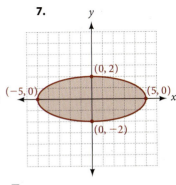

9. $(0, 3), \left(\frac{12}{5}, -\frac{9}{5}\right)$ **11.** $(0, 4), \left(\frac{16}{5}, \frac{12}{5}\right)$ **13.** $(5, 0), (-5, 0)$ **15.** $(0, -3), (\sqrt{5}, 2), (-\sqrt{5}, 2)$

17. $(0, -4), (\sqrt{7}, 3), (-\sqrt{7}, 3)$ **19.** $(-4, 11), \left(\frac{5}{2}, \frac{5}{4}\right)$ **21.** $(-4, 5), (1, 0)$ **23.** $(2, -3), (5, 0)$ **25.** $(3, 0), (-3, 0)$

27. $(4, 0), (0, -4)$ **29. a.** $(-4, 4\sqrt{3})$ and $(-4, -4\sqrt{3})$ **b.** $(4, 4\sqrt{3})$ and $(4, -4\sqrt{3})$

31.

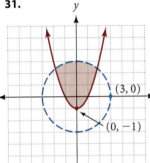

33.

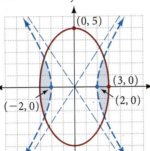

35. No intersection
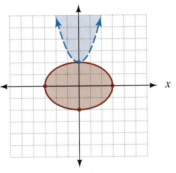

37. $x^2 + y^2 < 16, y > 4 - \frac{1}{4}x^2$ **39.** 8, 5 or $-8, -5$ or $8, -5$ or $-8, 5$ **41.** 6, 3 or 13, -4 **43.** $-\frac{2}{5}$ **45.** $-5, 1$

47. $-2, -1$

Chapter 10 Review

1. $\sqrt{10}$ **2.** $\sqrt{13}$ **3.** 5   **4.** 9   **5.** $-2, 6$ **6.** $-12, 4$ **7.** $x^2 + y^2 = 25$ **8.** $x^2 + y^2 = 9$

9. $(x + 2)^2 + (y - 3)^2 = 25$ **10.** $(x + 6)^2 + (y - 8)^2 = 100$

11. $(0, 0); r = 2$
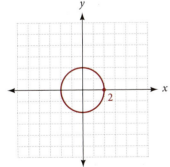

12. $(3, -1); r = 4$
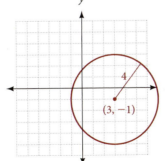

13. $(3, -2); r = 3$
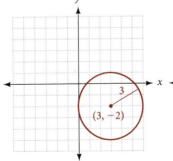

14. $(-2, 1); r = 3$
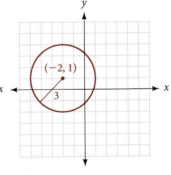

15.
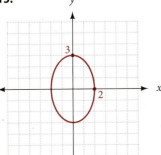

16.
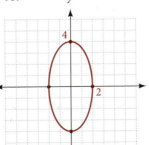

17.
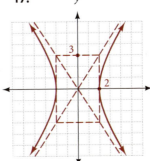

18.
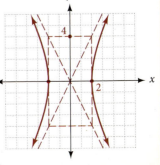

19.
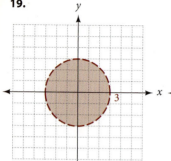

20.
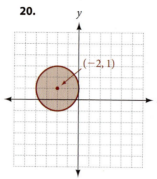

21.
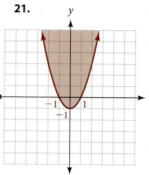

22.
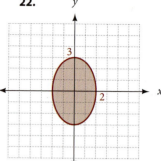

23.
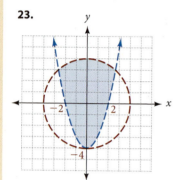

24.
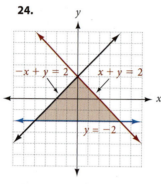

25. $(0, 4), \left(\frac{16}{5}, -\frac{12}{5} \right)$ **26.** $(0, -2), (\sqrt{3}, 1), (-\sqrt{3}, 1)$

27. $(-2, 0), (2, 0)$

Chapter 10 Cumulative Review

1. 5   **2.** 4   **3.** $\frac{a^5 b^4}{2}$ **4.** $x^{1/4} y^{7/4}$ **5.** $3xy^2 - 4x + 2y^2$ **6.** $\frac{y-3}{y-2}$ **7.** 2   **8.** 3   **9.** $(b^3 + 6)(a + 1)$ **10.** $(8x + 3)(x - 1)$

11. -2 **12.** 2   **13.** $-5, 2$ **14.** $-2, 4$ **15.** 8; 5 does not check   **16.** No solution   **17.** $\frac{3 \pm 5i\sqrt{2}}{4}$ **18.** $3 \pm i\sqrt{3}$

19. $x < -\frac{7}{2}$ or $x > 2$ **20.** $-8 < x < 3$

21.
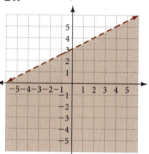

22.
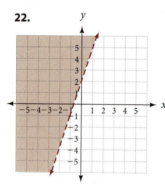

23.
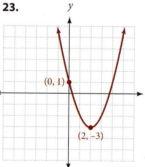

24.

25. $\frac{y(y-1)}{y+1}$ **26.** $\frac{(x-4)(x+3)}{x^2(x+2)}$ **27.** $14 - 5i$ **28.** $-10 + 11i$ **29.** $\frac{3\sqrt{7} + 3\sqrt{3}}{4}$ **30.** $\frac{-12 + 7\sqrt{6}}{5}$ **31.** $f^{-1}(x) = \frac{1}{4}x - \frac{1}{4}$

32. $f^{-1}(x) = 2x - 6$ **33.** 9,519.19   **34.** 1.47   **35.** $x = -\frac{5}{m-n}$ **36.** $y = \frac{S - 2x^2}{4x}$ **37.** 0

38. slope $= \frac{2}{3}$; y-intercept $= -4$ **39.** -5 **40.** $C(5) = 40; C(10) = 20$ **41.** $3x^2 - 5x + 2 = 0$ **42.** $20t^2 + 11t - 3 = 0$

43. $-3x + 5$ **44.** $-5x^2 + 19$ **45.** -19 **46.** -6 **47.** $\frac{3}{4}$ **48.** 12 gal of 20%; 4 gal of 60%   **49.** 93 people

50. 18 people

Chapter 10 Test

1. 13   **2.** $2\sqrt{5}$ **3.** $-5, 3$ **4.** $(x + 2)^2 + (y - 4)^2 = 9$ **5.** $x^2 + y^2 = 25$ **6.** Center = $(0, -2)$; Radius = 8

7. Center = $(5, -3)$; Radius = $\sqrt{39}$

8.
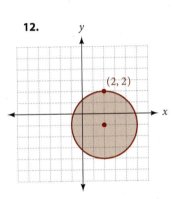

9.
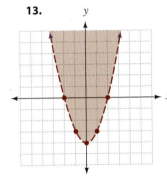

10.
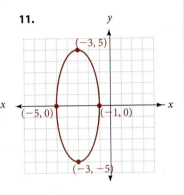

11.
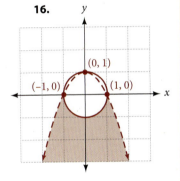

12.

13.

14. $(0, 5)$, $(4, -3)$

15. $(\pm\sqrt{7}, 3)$, $(0, -4)$

16.

17. C   **18.** A   **19.** D   **20.** B   **21.** $(x + 3)^2 + (y - 4)^2 = 25$ **22.** $\dfrac{(x - 4)^2}{25} + \dfrac{(y + 2)^2}{9} = 1$ **23.** $\dfrac{y^2}{16} - \dfrac{x^2}{25} = 1$

Appendix A

1. 3   **3.** 5   **5.** -1 **7.** 0   **9.** 2   **11.** -3 **13.** -2 **15.** -3 **17.** 3   **19.** 0   **21.** 3   **23.** 8

25. 6   **27.** -228

Appendix B

1. $(3, 1)$ **3.** Inconsistent System $\varnothing$ **5.** $\left(-\dfrac{15}{43}, -\dfrac{27}{43}\right)$ **7.** $\left(\dfrac{60}{43}, \dfrac{46}{43}\right)$ **9.** $(3, -1, 2)$ **11.** $\left(\dfrac{1}{2}, \dfrac{5}{2}, 1\right)$

13. No unique solution   **15.** $\left(-\dfrac{10}{91}, -\dfrac{9}{13}, \dfrac{107}{91}\right)$ **17.** $\left(\dfrac{71}{13}, -\dfrac{12}{13}, \dfrac{24}{13}\right)$ **19.** $(3, 1, 2)$

Appendix C

1. $x - 7 + \dfrac{20}{x + 2}$ **3.** $3x - 1$ **5.** $x^2 + 4x + 11 + \dfrac{26}{x - 2}$ **7.** $3x^2 + 8x + 26 + \dfrac{83}{x - 3}$ **9.** $2x^2 + 2x + 3$

11. $x^3 - 4x^2 + 18x - 72 + \dfrac{289}{x + 4}$ **13.** $x^4 + x^2 - x - 3 - \dfrac{5}{x - 2}$ **15.** $x + 2 + \dfrac{3}{x - 1}$ **17.** $x^3 - x^2 + x - 1$ **19.** $x^2 + x + 1$

Appendix D

1. Hypothesis: You argue for your limitations.   Conclusion: They are yours.

3. Hypothesis: x is an even number.   Conclusion: x is divisible by 2.

5. Hypothesis: A triangle is equilateral.   Conclusion: All of its angles are equal.

7. Hypothesis: $x + 5 = -2$ Conclusion: $x = -7$

9. Converse: If $a^2 = 64$, then $a = 8$ Inverse: If $a \neq 8$, then $a^2 \neq 64$ Contrapositive: If $a^2 \neq 64$, then $a \neq 8$

11. Converse: If $a = b$, then $\dfrac{a}{b} = 1$ Inverse: If $\dfrac{a}{b} \neq 1$, then $a \neq b$ Contrapositive: If $a \neq b$, then $\dfrac{a}{b} \neq 1$

13. Converse: If it is a rectangle, then it is a square.   Inverse: If it is not a square, then it is not a rectangle.
Contrapositive: If it is not a rectangle, then it is not a square.

15. Converse: If good is not enough, then better is possible.   Inverse: If better is not possible, then good is enough.
Contrapositive: If good is enough, then better is not possible.   **17.** If E, then F

19. If it is misery, then it loves company.   **21.** If it is the squeaky wheel, then it gets the grease.

23. c   **25.** a   **27.** c   **29.** b

Index